T0180314

Lecture Notes in Computer Science 13110

More information about this subseries at https://link.springer.com/bookseries/7407

Teddy Mantoro · Minho Lee ·
Media Anugerah Ayu · Kok Wai Wong ·
Achmad Nizar Hidayanto (Eds.)

Neural Information Processing

28th International Conference, ICONIP 2021
Sanur, Bali, Indonesia, December 8–12, 2021
Proceedings, Part III

Springer

Editors
Teddy Mantoro 🆔
Sampoerna University
Jakarta, Indonesia

Media Anugerah Ayu 🆔
Sampoerna University
Jakarta, Indonesia

Achmad Nizar Hidayanto 🆔
Universitas Indonesia
Depok, Indonesia

Minho Lee 🆔
Kyungpook National University
Daegu, Korea (Republic of)

Kok Wai Wong 🆔
Murdoch University
Murdoch, WA, Australia

ISSN 0302-9743   ISSN 1611-3349 (electronic)
Lecture Notes in Computer Science
ISBN 978-3-030-92237-5   ISBN 978-3-030-92238-2 (eBook)
https://doi.org/10.1007/978-3-030-92238-2

LNCS Sublibrary: SL1 – Theoretical Computer Science and General Issues

This Springer imprint is published by the registered company Springer Nature Switzerland AG
The registered company address is: Gewerbestrasse 11, 6330 Cham, Switzerland

Preface

Welcome to the proceedings of the 28th International Conference on Neural Information Processing (ICONIP 2021) of the Asia-Pacific Neural Network Society (APNNS), held virtually from Indonesia during December 8–12, 2021.

The mission of the Asia-Pacific Neural Network Society is to promote active interactions among researchers, scientists, and industry professionals who are working in neural networks and related fields in the Asia-Pacific region. APNNS has Governing Board Members from 13 countries/regions – Australia, China, Hong Kong, India, Japan, Malaysia, New Zealand, Singapore, South Korea, Qatar, Taiwan, Thailand, and Turkey. The society's flagship annual conference is the International Conference of Neural Information Processing (ICONIP).

The ICONIP conference aims to provide a leading international forum for researchers, scientists, and industry professionals who are working in neuroscience, neural networks, deep learning, and related fields to share their new ideas, progress, and achievements. Due to the current COVID-19 pandemic, ICONIP 2021, which was planned to be held in Bali, Indonesia, was organized as a fully virtual conference.

The proceedings of ICONIP 2021 consists of a four-volume set, LNCS 13108–13111, which includes 226 papers selected from 1093 submissions, representing an acceptance rate of 20.86% and reflecting the increasingly high quality of research in neural networks and related areas in the Asia-Pacific. The conference had four main themes, i.e., "Theory and Algorithms," "Cognitive Neurosciences," "Human Centred Computing," and "Applications."

The four volumes are organized in topical sections which comprise the four main themes mentioned previously and the topics covered in three special sessions. Another topic is from a workshop on Artificial Intelligence and Cyber Security which was held in conjunction with ICONIP 2021. Thus, in total, eight different topics were accommodated at the conference. The topics were also the names of the 20-minute presentation sessions at ICONIP 2021. The eight topics in the conference were: Theory and Algorithms; Cognitive Neurosciences; Human Centred Computing; Applications; Artificial Intelligence and Cybersecurity; Advances in Deep and Shallow Machine Learning Algorithms for Biomedical Data and Imaging; Reliable, Robust, and Secure Machine Learning Algorithms; and Theory and Applications of Natural Computing Paradigms.

Our great appreciation goes to the Program Committee members and the reviewers who devoted their time and effort to our rigorous peer-review process. Their insightful reviews and timely feedback ensured the high quality of the papers accepted for publication. Finally, thank you to all the authors of papers, presenters, and participants at the conference. Your support and engagement made it all worthwhile.

December 2021

Teddy Mantoro
Minho Lee
Media A. Ayu
Kok Wai Wong
Achmad Nizar Hidayanto

Organization

Honorary Chairs

Jonathan Chan — King Mongkut's University of Technology Thonburi, Thailand

Lance Fung — Murdoch University, Australia

General Chairs

Teddy Mantoro — Sampoerna University, Indonesia
Minho Lee — Kyungpook National University, South Korea

Program Chairs

Media A. Ayu — Sampoerna University, Indonesia
Kok Wai Wong — Murdoch University, Australia
Achmad Nizar — Universitas Indonesia, Indonesia

Local Arrangements Chairs

Linawati — Universitas Udayana, Indonesia
W. G. Ariastina — Universitas Udayana, Indonesia

Finance Chairs

Kurnianingsih — Politeknik Negeri Semarang, Indonesia
Kazushi Ikeda — Nara Institute of Science and Technology, Japan

Special Sessions Chairs

Sunu Wibirama — Universitas Gadjah Mada, Indonesia
Paul Pang — Federation University Australia, Australia
Noor Akhmad Setiawan — Universitas Gadjah Mada, Indonesia

Tutorial Chairs

Suryono — Universitas Diponegoro, Indonesia
Muhammad Agni Catur Bhakti — Sampoerna University, Indonesia

Proceedings Chairs

Adi Wibowo   Universitas Diponegoro, Indonesia
Sung Bae Cho   Yonsei University, South Korea

Publicity Chairs

Dwiza Riana   Universitas Nusa Mandiri, Indonesia
M. Tanveer   Indian Institute of Technology, Indore, India

Program Committee

Abdulrazak Alhababi   Universiti Malaysia Sarawak, Malaysia
Abhijit Adhikary   Australian National University, Australia
Achmad Nizar Hidayanto   University of Indonesia, Indonesia
Adamu Abubakar Ibrahim   International Islamic University Malaysia, Malaysia
Adi Wibowo   Diponegoro University, Indonesia
Adnan Mahmood   Macquarie University, Australia
Afiyati Amaluddin   Mercu Buana University, Indonesia
Ahmed Alharbi   RMIT University, Australia
Akeem Olowolayemo   International Islamic University Malaysia, Malaysia
Akira Hirose   University of Tokyo, Japan
Aleksandra Nowak   Jagiellonian University, Poland
Ali Haidar   University of New South Wales, Australia
Ali Mehrabi   Western Sydney University, Australia
Al-Jadir   Murdoch University, Australia
Ana Flavia Reis   Federal Technological University of Paraná, Brazil
Anaissi Ali   University of Sydney, Australia
Andrew Beng Jin Teoh   Yonsei University, South Korea
Andrew Chiou   Central Queensland University, Australia
Aneesh Chivukula   University of Technology Sydney, Australia
Aneesh Krishna   Curtin University, Australia
Anna Zhu   Wuhan University of Technology, China
Anto Satriyo Nugroho   Agency for Assessment and Application of
                              Technology, Indonesia
Anupiya Nugaliyadde   Sri Lanka Institute of Information Technology,
                              Sri Lanka
Anwesha Law   Indian Statistical Institute, India
Aprinaldi Mantau   Kyushu Institute of Technology, Japan
Ari Wibisono   Universitas Indonesia, Indonesia
Arief Ramadhan   Bina Nusantara University, Indonesia
Arit Thammano   King Mongkut's Institute of Technology Ladkrabang,
                              Thailand
Arpit Garg   University of Adelaide, Australia
Aryal Sunil   Deakin University, Australia
Ashkan Farhangi   University of Central Florida, USA

Atul Negi	University of Hyderabad, India
Barawi Mohamad Hardyman	Universiti Malaysia Sarawak, Malaysia
Bayu Distiawan	Universitas Indonesia, Indonesia
Bharat Richhariya	IISc Bangalore, India
Bin Pan	Nankai University, China
Bingshu Wang	Northwestern Polytechnical University, Taicang, China
Bonaventure C. Molokwu	University of Windsor, Canada
Bo-Qun Ma	Ant Financial
Bunthit Watanapa	King Mongkut's University of Technology Thonburi, Thailand
Chang-Dong Wang	Sun Yat-sen University, China
Chattrakul Sombattheera	Mahasarakham University, Thailand
Chee Siong Teh	Universiti Malaysia Sarawak, Malaysia
Chen Wei Chén	Chongqing Jiaotong University, China
Chengwei Wu	Harbin Institute of Technology, China
Chern Hong Lim	Monash University, Australia
Chih-Chieh Hung	National Chung Hsing University, Taiwan
Chiranjibi Sitaula	Deakin University, Australia
Chi-Sing Leung	City University of Hong Kong, Hong Kong
Choo Jun Tan	Wawasan Open University, Malaysia
Christoph Bergmeir	Monash University, Australia
Christophe Guyeux	University of Franche-Comté, France
Chuan Chen	Sun Yat-sen University, China
Chuanqi Tan	BIT, China
Chu-Kiong Loo	University of Malaya, Malaysia
Chun Che Fung	Murdoch University, Australia
Colin Samplawski	University of Massachusetts Amherst, USA
Congbo Ma	University of Adelaide, Australia
Cuiyun Gao	Chinese University of Hong Kong, Hong Kong
Cutifa Safitri	Universiti Teknologi Malaysia, Malaysia
Daisuke Miyamoto	University of Tokyo, Japan
Dan Popescu	Politehnica University of Bucharest
David Bong	Universiti Malaysia Sarawak, Malaysia
David Iclanzan	Sapientia Hungarian Science University of Transylvania, Romania
Debasmit Das	IIT Roorkee, India
Dengya Zhu	Curtin University, Australia
Derwin Suhartono	Bina Nusantara University, Indonesia
Devi Fitrianah	Universitas Mercu Buana, Indonesia
Deyu Zhou	Southeast University, China
Dhimas Arief Dharmawan	Universitas Indonesia, Indonesia
Dianhui Wang	La Trobe University, Australia
Dini Handayani	Taylors University, Malaysia
Dipanjyoti Paul	Indian Institute of Technology, Patna, India
Dong Chen	Wuhan University, China

Donglin Bai	Shanghai Jiao Tong University, China
Dongrui Wu	Huazhong University of Science & Technology, China
Dugang Liu	Shenzhen University, China
Dwina Kuswardani	Institut Teknologi PLN, Indonesia
Dwiza Riana	Universitas Nusa Mandiri, Indonesia
Edmund Lai	Auckland University of Technology, New Zealand
Eiji Uchino	Yamaguchi University, Japan
Emanuele Principi	Università Politecnica delle Marche, Italy
Enmei Tu	Shanghai Jiao Tong University, China
Enna Hirata	Kobe University, Japan
Eri Sato-Shimokawara	Tokyo Metropolitan University, Japan
Fajri Koto	University of Melbourne, Australia
Fan Wu	Australian National University, Australia
Farhad Ahamed	Western Sydney University, Australia
Fei Jiang	Shanghai Jiao Tong University, China
Feidiao Yang	Microsoft, USA
Feng Wan	University of Macau, Macau
Fenty Eka Muzayyana Agustin	UIN Syarif Hidayatullah Jakarta, Indonesia
Ferda Ernawan	Universiti Malaysia Pahang, Malaysia
Ferdous Sohel	Murdoch University, Australia
Francisco J. Moreno-Barea	Universidad de Málaga, Spain
Fuad Jamour	University of California, Riverside, USA
Fuchun Sun	Tsinghua University, China
Fumiaki Saitoh	Chiba Institute of Technology, Japan
Gang Chen	Victoria University of Wellington, New Zealand
Gang Li	Deakin University, Australia
Gang Yang	Renmin University of China
Gao Junbin	Huazhong University of Science and Technology, China
George Cabral	Universidade Federal Rural de Pernambuco, Brazil
Gerald Schaefer	Loughborough University, UK
Gouhei Tanaka	University of Tokyo, Japan
Guanghui Wen	RMIT University, Australia
Guanjin Wang	Murdoch University, Australia
Guoqiang Zhong	Ocean University of China, China
Guoqing Chao	East China Normal University, China
Sangchul Hahn	Handong Global University, South Korea
Haiqin Yang	International Digital Economy Academy, China
Hakaru Tamukoh	Kyushu Institute of Technology, Japan
Hamid Karimi	Utah State University, USA
Hangyu Deng	Waseda University, Japan
Hao Liao	Shenzhen University, China
Haris Al Qodri Maarif	International Islamic University Malaysia, Malaysia
Haruhiko Nishimura	University of Hyogo, Japan
Hayaru Shouno	University of Electro-Communications, Japan

He Chen   Nankai University, China
He Huang   Soochow University, China
Hea Choon Ngo   Universiti Teknikal Malaysia Melaka, Malaysia
Heba El-Fiqi   UNSW Canberra, Australia
Heru Praptono   Bank Indonesia/Universitas Indonesia, Indonesia
Hideitsu Hino   Institute of Statistical Mathematics, Japan
Hidemasa Takao   University of Tokyo, Japan
Hiroaki Inoue   Kobe University, Japan
Hiroaki Kudo   Nagoya University, Japan
Hiromu Monai   Ochanomizu University, Japan
Hiroshi Sakamoto   Kyushu Institute of Technology, Japan
Hisashi Koga   University of Electro-Communications, Japan
Hiu-Hin Tam   City University of Hong Kong, Hong Kong
Hongbing Xia   Beijing Normal University, China
Hongtao Liu   Tianjin University, China
Hongtao Lu   Shanghai Jiao Tong University, China
Hua Zuo   University of Technology Sydney, Australia
Hualou Liang   Drexel University, USA
Huang Chaoran   University of New South Wales, Australia
Huang Shudong   Sichuan University, China
Huawen Liu   University of Texas at San Antonio, USA
Hui Xue   Southeast University, China
Hui Yan   Shanghai Jiao Tong University, China
Hyeyoung Park   Kyungpook National University, South Korea
Hyun-Chul Kim   Kyungpook National University, South Korea
Iksoo Shin   University of Science and Technology, South Korea
Indrabayu Indrabayu   Universitas Hasanuddin, Indonesia
Iqbal Gondal   RMIT University, Australia
Iuliana Georgescu   University of Bucharest, Romania
Iwan Syarif   PENS, Indonesia
J. Kokila   Indian Institute of Information Technology, Allahabad,
                            India
J. Manuel Moreno   Universitat Politècnica de Catalunya, Spain
Jagdish C. Patra   Swinburne University of Technology, Australia
Jean-Francois Couchot   University of Franche-Comté, France
Jelita Asian   STKIP Surya, Indonesia
Jennifer C. Dela Cruz   Mapua University, Philippines
Jérémie Sublime   ISEP, France
Jiahuan Lei   Meituan, China
Jialiang Zhang   Alibaba, China
Jiaming Xu   Institute of Automation, Chinese Academy of Sciences
Jianbo Ning   University of Science and Technology Beijing, China
Jianyi Yang   Nankai University, China
Jiasen Wang   City University of Hong Kong, Hong Kong
Jiawei Fan   Australian National University, Australia
Jiawei Li   Tsinghua University, China

Jiaxin Li	Guangdong University of Technology, China
Jiaxuan Xie	Shanghai Jiao Tong University, China
Jichuan Zeng	Bytedance, China
Jie Shao	University of Science and Technology of China, China
Jie Zhang	Newcastle University, UK
Jiecong Lin	City University of Hong Kong, Hong Kong
Jin Hu	Chongqing Jiaotong University, China
Jin Kyu Kim	Facebook, USA
Jin Ren	Beijing University of Technology, China
Jin Shi	Nanjing University, China
Jinfu Yang	Beijing University of Technology, China
Jing Peng	South China Normal University, China
Jinghui Zhong	South China University of Technology, China
Jin-Tsong Jeng	National Formosa University, Taiwan
Jiri Sima	Institute of Computer Science, Czech Academy of Sciences, Czech Republic
Jo Plested	Australian National University, Australia
Joel Dabrowski	CSIRO, Australia
John Sum	National Chung Hsing University, China
Jolfaei Alireza	Federation University Australia, Australia
Jonathan Chan	King Mongkut's University of Technology Thonburi, Thailand
Jonathan Mojoo	Hiroshima University, Japan
Jose Alfredo Ferreira Costa	Federal University of Rio Grande do Norte, Brazil
Ju Lu	Shandong University, China
Jumana Abu-Khalaf	Edith Cowan University, Australia
Jun Li	Nanjing Normal University, China
Jun Shi	Guangzhou University, China
Junae Kim	DST Group, Australia
Junbin Gao	University of Sydney, Australia
Junjie Chen	Inner Mongolia Agricultural University, China
Junya Chen	Fudan University, China
Junyi Chen	City University of Hong Kong, Hong Kong
Junying Chen	South China University of Technology, China
Junyu Xuan	University of Technology, Sydney
Kah Ong Michael Goh	Multimedia University, Malaysia
Kaizhu Huang	Xi'an Jiaotong-Liverpool University, China
Kam Meng Goh	Tunku Abdul Rahman University College, Malaysia
Katsuhiro Honda	Osaka Prefecture University, Japan
Katsuyuki Hagiwara	Mie University, Japan
Kazushi Ikeda	Nara Institute of Science and Technology, Japan
Kazuteru Miyazaki	National Institution for Academic Degrees and Quality Enhancement of Higher Education, Japan
Kenji Doya	OIST, Japan
Kenji Watanabe	National Institute of Advanced Industrial Science and Technology, Japan

Kok Wai Wong	Murdoch University, Australia
Kitsuchart Pasupa	King Mongkut's Institute of Technology Ladkrabang, Thailand
Kittichai Lavangnananda	King Mongkut's University of Technology Thonburi, Thailand
Koutsakis Polychronis	Murdoch University, Australia
Kui Ding	Nanjing Normal University, China
Kun Zhang	Carnegie Mellon University, USA
Kuntpong Woraratpanya	King Mongkut's Institute of Technology Ladkrabang, Thailand
Kurnianingsih Kurnianingsih	Politeknik Negeri Semarang, Indonesia
Kusrini	Universitas AMIKOM Yogyakarta, Indonesia
Kyle Harrison	UNSW Canberra, Australia
Laga Hamid	Murdoch University, Australia
Lei Wang	Beihang University, China
Leonardo Franco	Universidad de Málaga, Spain
Li Guo	University of Macau, China
Li Yun	Nanjing University of Posts and Telecommunications, China
Libo Wang	Xiamen University of Technology, China
Lie Meng Pang	Southern University of Science and Technology, China
Liew Alan Wee-Chung	Griffith University, Australia
Lingzhi Hu	Beijing University of Technology, China
Linjing Liu	City University of Hong Kong, Hong Kong
Lisi Chen	Hong Kong Baptist University, Hong Kong
Long Cheng	Institute of Automation, Chinese Academy of Sciences, China
Lukman Hakim	Hiroshima University, Japan
M. Tanveer	Indian Institute of Technology, Indore, India
Ma Wanli	University of Canberra, Australia
Man Fai Leung	Hong Kong Metropolitan University, Hong Kong
Maram Mahmoud A. Monshi	Beijing Institute of Technology, China
Marcin Wozniak	Silesian University of Technology, Poland
Marco Anisetti	Università degli Studi di Milano, Italy
Maria Susan Anggreainy	Bina Nusantara University, Indonesia
Mark Abernethy	Murdoch University, Australia
Mark Elshaw	Coventry University, UK
Maruno Yuki	Kyoto Women's University, Japan
Masafumi Hagiwara	Keio University, Japan
Masataka Kawai	NRI SecureTechnologies, Ltd., Japan
Media Ayu	Sampoerna University, Indonesia
Mehdi Neshat	University of Adelaide, Australia
Meng Wang	Southeast University, China
Mengmeng Li	Zhengzhou University, China

Miaohua Zhang	Griffith University, Australia
Mingbo Zhao	Donghua University, China
Mingcong Deng	Tokyo University of Agriculture and Technology, Japan
Minghao Yang	Institute of Automation, Chinese Academy of Sciences, China
Minho Lee	Kyungpook National University, South Korea
Mofei Song	Southeast University, China
Mohammad Faizal Ahmad Fauzi	Multimedia University, Malaysia
Mohsen Marjani	Taylor's University, Malaysia
Mubasher Baig	National University of Computer and Emerging Sciences, Lahore, Pakistan
Muhammad Anwar Ma'Sum	Universitas Indonesia, Indonesia
Muhammad Asim Ali	Shaheed Zulfikar Ali Bhutto Institute of Science and Technology, Pakistan
Muhammad Fawad Akbar Khan	University of Engineering and Technology Peshawar, Pakistan
Muhammad Febrian Rachmadi	Universitas Indonesia, Indonesia
Muhammad Haris	Universitas Nusa Mandiri, Indonesia
Muhammad Haroon Shakeel	Lahore University of Management Sciences, Pakistan
Muhammad Hilman	Universitas Indonesia, Indonesia
Muhammad Ramzan	Saudi Electronic University, Saudi Arabia
Muideen Adegoke	City University of Hong Kong, Hong Kong
Mulin Chen	Northwestern Polytechnical University, China
Murtaza Taj	Lahore University of Management Sciences, Pakistan
Mutsumi Kimura	Ryukoku University, Japan
Naoki Masuyama	Osaka Prefecture University, Japan
Naoyuki Sato	Future University Hakodate, Japan
Nat Dilokthanakul	Vidyasirimedhi Institute of Science and Technology, Thailand
Nguyen Dang	University of Canberra, Australia
Nhi N. Y. Vo	University of Technology Sydney, Australia
Nick Nikzad	Griffith University, Australia
Ning Boda	Swinburne University of Technology, Australia
Nobuhiko Wagatsuma	Tokyo Denki University, Japan
Nobuhiko Yamaguchi	Saga University, Japan
Noor Akhmad Setiawan	Universitas Gadjah Mada, Indonesia
Norbert Jankowski	Nicolaus Copernicus University, Poland
Norikazu Takahashi	Okayama University, Japan
Noriyasu Homma	Tohoku University, Japan
Normaziah A. Aziz	International Islamic University Malaysia, Malaysia
Olarik Surinta	Mahasarakham University, Thailand

Olutomilayo Olayemi Petinrin	Kings University, Nigeria
Ooi Shih Yin	Multimedia University, Malaysia
Osamu Araki	Tokyo University of Science, Japan
Ozlem Faydasicok	Istanbul University, Turkey
Parisa Rastin	University of Lorraine, France
Paul S. Pang	Federation University Australia, Australia
Pedro Antonio Gutierrez	Universidad de Cordoba, Spain
Pengyu Sun	Microsoft
Piotr Duda	Institute of Computational Intelligence/Czestochowa University of Technology, Poland
Prabath Abeysekara	RMIT University, Australia
Pui Huang Leong	Tunku Abdul Rahman University College, Malaysia
Qian Li	Chinese Academy of Sciences, China
Qiang Xiao	Huazhong University of Science and Technology, China
Qiangfu Zhao	University of Aizu, Japan
Qianli Ma	South China University of Technology, China
Qing Xu	Tianjin University, China
Qing Zhang	Meituan, China
Qinglai Wei	Institute of Automation, Chinese Academy of Sciences, China
Qingrong Cheng	Fudan University, China
Qiufeng Wang	Xi'an Jiaotong-Liverpool University, China
Qiulei Dong	Institute of Automation, Chinese Academy of Sciences, China
Qiuye Wu	Guangdong University of Technology, China
Rafal Scherer	Częstochowa University of Technology, Poland
Rahmadya Handayanto	Universitas Islam 45 Bekasi, Indonesia
Rahmat Budiarto	Albaha University, Saudi Arabia
Raja Kumar	Taylor's University, Malaysia
Rammohan Mallipeddi	Kyungpook National University, South Korea
Rana Md Mashud	CSIRO, Australia
Rapeeporn Chamchong	Mahasarakham University, Thailand
Raphael Couturier	Université Bourgogne Franche-Comté, France
Ratchakoon Pruengkarn	Dhurakij Pundit University, Thailand
Reem Mohamed	Mansoura University, Egypt
Rhee Man Kil	Sungkyunkwan University, South Korea
Rim Haidar	University of Sydney, Australia
Rizal Fathoni Aji	Universitas Indonesia, Indonesia
Rukshima Dabare	Murdoch University, Australia
Ruting Cheng	University of Science and Technology Beijing, China
Ruxandra Liana Costea	Polytechnic University of Bucharest, Romania
Saaveethya Sivakumar	Curtin University Malaysia, Malaysia
Sabrina Fariza	Central Queensland University, Australia
Sahand Vahidnia	University of New South Wales, Australia

Saifur Rahaman	City University of Hong Kong, Hong Kong
Sajib Mistry	Curtin University, Australia
Sajib Saha	CSIRO, Australia
Sajid Anwar	Institute of Management Sciences Peshawar, Pakistan
Sakchai Muangsrinoon	Walailak University, Thailand
Salomon Michel	Université Bourgogne Franche-Comté, France
Sandeep Parameswaran	Myntra Designs Pvt. Ltd., India
Sangtae Ahn	Kyungpook National University, South Korea
Sang-Woo Ban	Dongguk University, South Korea
Sangwook Kim	Kobe University, Japan
Sanparith Marukatat	NECTEC, Thailand
Saptakatha Adak	Indian Institute of Technology, Madras, India
Seiichi Ozawa	Kobe University, Japan
Selvarajah Thuseethan	Sabaragamuwa University of Sri Lanka, Sri Lanka
Seong-Bae Park	Kyung Hee University, South Korea
Shan Zhong	Changshu Institute of Technology, China
Shankai Yan	National Institutes of Health, USA
Sheeraz Akram	University of Pittsburgh, USA
Shenglan Liu	Dalian University of Technology, China
Shenglin Zhao	Zhejiang University, China
Shing Chiang Tan	Multimedia University, Malaysia
Shixiong Zhang	Xidian University, China
Shreya Chawla	Australian National University, Australia
Shri Rai	Murdoch University, Australia
Shuchao Pang	Jilin University, China/Macquarie University, Australia
Shuichi Kurogi	Kyushu Institute of Technology, Japan
Siddharth Sachan	Australian National University, Australia
Sirui Li	Murdoch University, Australia
Sonali Agarwal	Indian Institute of Information Technology, Allahabad, India
Sonya Coleman	University of Ulster, UK
Stavros Ntalampiras	University of Milan, Italy
Su Lei	University of Science and Technology Beijing, China
Sung-Bae Cho	Yonsei University, South Korea
Sunu Wibirama	Universitas Gadjah Mada, Indonesia
Susumu Kuroyanagi	Nagoya Institute of Technology, Japan
Sutharshan Rajasegarar	Deakin University, Australia
Takako Hashimoto	Chiba University of Commerce, Japan
Takashi Omori	Tamagawa University, Japan
Tao Ban	National Institute of Information and Communications Technology, Japan
Tao Li	Peking University, China
Tao Xiang	Chongqing University, China
Teddy Mantoro	Sampoerna University, Indonesia
Tedjo Darmanto	STMIK AMIK Bandung, Indonesia
Teijiro Isokawa	University of Hyogo, Japan

Thanh Tam Nguyen	Leibniz University Hannover, Germany
Thanh Tung Khuat	University of Technology Sydney, Australia
Thaweesak Khongtuk	Rajamangala University of Technology Suvarnabhumi, Thailand
Tianlin Zhang	University of Chinese Academy of Sciences, China
Timothy McIntosh	Massey University, New Zealand
Toan Nguyen Thanh	Ho Chi Minh City University of Technology, Vietnam
Todsanai Chumwatana	Murdoch University, Australia
Tom Gedeon	Australian National University, Australia
Tomas Maul	University of Nottingham, Malaysia
Tomohiro Shibata	Kyushu Institute of Technology, Japan
Tomoyuki Kaneko	University of Tokyo, Japan
Toshiaki Omori	Kobe University, Japan
Toshiyuki Yamane	IBM, Japan
Uday Kiran	University of Tokyo, Japan
Udom Silparcha	King Mongkut's University of Technology Thonburi, Thailand
Umar Aditiawarman	Universitas Nusa Putra, Indonesia
Upeka Somaratne	Murdoch University, Australia
Usman Naseem	University of Sydney, Australia
Ven Jyn Kok	National University of Malaysia, Malaysia
Wachira Yangyuen	Rajamangala University of Technology Srivijaya, Thailand
Wai-Keung Fung	Robert Gordon University, UK
Wang Yaqing	Baidu Research, Hong Kong
Wang Yu-Kai	University of Technology Sydney, Australia
Wei Jin	Michigan State University, USA
Wei Yanling	TU Berlin, Germany
Weibin Wu	City University of Hong Kong, Hong Kong
Weifeng Liu	China University of Petroleum, China
Weijie Xiang	University of Science and Technology Beijing, China
Wei-Long Zheng	Massachusetts General Hospital, Harvard Medical School, USA
Weiqun Wang	Institute of Automation, Chinese Academy of Sciences, China
Wen Luo	Nanjing Normal University, China
Wen Yu	Cinvestav, Mexico
Weng Kin Lai	Tunku Abdul Rahman University College, Malaysia
Wenqiang Liu	Southwest Jiaotong University, China
Wentao Wang	Michigan State University, USA
Wenwei Gu	Chinese University of Hong Kong, Hong Kong
Wenxin Yu	Southwest University of Science and Technology, China
Widodo Budiharto	Bina Nusantara University, Indonesia
Wisnu Ananta Kusuma	Institut Pertanian Bogor, Indonesia
Worapat Paireekreng	Dhurakij Pundit University, Thailand

Xiang Chen	George Mason University, USA
Xiao Jian Tan	Tunku Abdul Rahman University College, Malaysia
Xiao Liang	Nankai University, China
Xiaocong Chen	University of New South Wales, Australia
Xiaodong Yue	Shanghai University, China
Xiaoqing Lyu	Peking University, China
Xiaoyang Liu	Huazhong University of Science and Technology, China
Xiaoyang Tan	Nanjing University of Aeronautics and Astronautics, China
Xiao-Yu Tang	Zhejiang University, China
Xin Liu	Huaqiao University, China
Xin Wang	Southwest University, China
Xin Xu	Beijing University of Technology, China
Xingjian Chen	City University of Hong Kong, Hong Kong
Xinyi Le	Shanghai Jiao Tong University, China
Xinyu Shi	University of Science and Technology Beijing, China
Xiwen Bao	Chongqing Jiaotong University, China
Xu Bin	Northwestern Polytechnical University, China
Xu Chen	Shanghai Jiao Tong University, China
Xuan-Son Vu	Umeå University, Sweden
Xuanying Zhu	Australian National University, Australia
Yanling Zhang	University of Science and Technology Beijing, China
Yang Li	East China Normal University, China
Yantao Li	Chongqing University, China
Yanyan Hu	University of Science and Technology Beijing, China
Yao Lu	Beijing Institute of Technology, China
Yasuharu Koike	Tokyo Institute of Technology, Japan
Ya-Wen Teng	Academia Sinica, Taiwan
Yaxin Li	Michigan State University, USA
Yifan Xu	Huazhong University of Science and Technology, China
Yihsin Ho	Takushoku University, Japan
Yilun Jin	Hong Kong University of Science and Technology, Hong Kong
Yiming Li	Tsinghua University, China
Ying Xiao	University of Birmingham, UK
Yingjiang Zhou	Nanjing University of Posts and Telecommunications, China
Yong Peng	Hangzhou Dianzi University, China
Yonghao Ma	University of Science and Technology Beijing, China
Yoshikazu Washizawa	University of Electro-Communications, Japan
Yoshimitsu Kuroki	Kurume National College of Technology, Japan
Young Ju Rho	Korea Polytechnic University, South Korea
Youngjoo Seo	Ecole Polytechnique Fédérale de Lausanne, Switzerland

Yu Sang	PetroChina, China
Yu Xiaohan	Griffith University, Australia
Yu Zhou	Chongqing University, China
Yuan Ye	Xi'an Jiaotong University, China
Yuangang Pan	University of Technology Sydney, Australia
Yuchun Fang	Shanghai University, China
Yuhua Song	University of Science and Technology Beijing
Yunjun Gao	Zhejiang University, China
Zeyuan Wang	University of Sydney, Australia
Zhen Wang	University of Sydney, Australia
Zhengyang Feng	Shanghai Jiao Tong University, China
Zhenhua Wang	Zhejiang University of Technology, China
Zhenqian Wu	University of Electronic Science and Technology of China, China
Zhenyu Cui	University of Chinese Academy of Sciences, China
Zhenyue Qin	Australian National University, Australia
Zheyang Shen	Aalto University, Finland
Zhihong Cui	Shandong University, China
Zhijie Fang	Chinese Academy of Sciences, China
Zhipeng Li	Tsinghua University, China
Zhiri Tang	City University of Hong Kong, Hong Kong
Zhuangbin Chen	Chinese University of Hong Kong, Hong Kong
Zongying Liu	University of Malaya, Malaysia

Contents – Part III

Cognitive Neurosciences

Reliable, Robust, and Secure Machine Learning Algorithms

Theory and Applications of Natural Computing Paradigms

Advances in Deep and Shallow Machine Learning Algorithms for Biomedical Data and Imaging

Applications

Cognitive Neurosciences

A Novel Binary BCI Systems Based on Non-oddball Auditory and Visual Paradigms

Madina Saparbayeva(ID), Adai Shomanov(ID), and Min-Ho Lee(✉)(ID)

School of Engineering and Digital Sciences, Nazarbayev University,
Kabanbay Batyr Ave. 53, Nur-Sultan 010000, Kazakhstan
{madina.saparbayeva,adai.shomanov,minho.lee}@nu.edu.kz

Abstract. Event-Related Potentials (ERPs) based binary BCI systems help enable users to control external devices through brain signals responding to stimulus. However, the external properties of the auditory or visual stimuli in the typical oddball-paradigm are loud and large for a user, which often brings psychological discomfort. In this study, we proposed novel non-oddball BCI paradigms where the intensity of external properties is greatly minimized while maintaining the system performance. To compensate for the loss of accuracy from the diminutive stimulus, users were instructed to generate discriminant ERP responses by performing a voluntary mental task. As the result, task-relevant endogenous components were investigated by the certain mental task and greatly enhanced system performance. The decoding accuracies of proposed CNN with data augmentation technique were 77.8% and 76.7% for the non-oddball visual and auditory paradigms, respectively, which significantly outperformed the linear classifier model. These results open up novel avenues for practical ERP systems, which could increase the usability of current brain-computer interfaces remarkably.

Keywords: Brain-computer interface (BCI) · Event-related potential (ERP) · Active mental task · Non-oddball paradigm · Convolutional neural networks (CNN)

1 Introduction

A brain-computer interface (BCI) [1] allows the user to control an external device for the users with diagnoses such as locked-in-syndrome, paralysis, or spinal cord injury. Electroencephalography (EEG) is widely used for BCI-purpose because of its non-invasive, low-risk, and easy-to-use method [2].

The oddball paradigm is the foundation for most ERP-based BCI applications. The oddball paradigm was first applied during an experiment where an "odd" event in a stream of typical events would elicit a distinct scalp-recorded potential pattern while the subject was concentrating on the stream of external auditory or visual stimuli [3].

© Springer Nature Switzerland AG 2021
T. Mantoro et al. (Eds.): ICONIP 2021, LNCS 13110, pp. 3–14, 2021.
https://doi.org/10.1007/978-3-030-92238-2_1

A representative BCI application based on the ERP paradigm is a binary decision system, which is simple, but extremely useful for patients with a later stage of a severe locked-in syndrome (LIS). This system has primarily been used for binary communication, emergency calls, or prior steps for multi-class BCI systems. Moreover, many studies have demonstrated that feature discrimination of ERP components in the oddball paradigm highly depends on the external properties of target stimulus [4–6]. Therefore, a common approach to enhance the performance of ERP-based BCI systems is primarily a development of novel manipulations of the stimuli parameters; volume, spatial arrangement, modulated frequency in the auditory BCIs, and shape, color, intensity, or highly recognizable images [4,6,7] in the visual BCIs. Commonly, those studies have been demanded the current auditory or visual stimuli to be louder or larger to have a greater impact on the user.

Therefore, current ERP-based BCI systems often create stress, annoyance, and fatigue by exposing the user to such intensively repetitive stimuli. These systems also require the users to distinguish the differences between individual stimuli to select a specific target class which may be problematic for some patient populations [8].

The goal of this study is to propose a novel binary BCI system that extremely minimizes the impact of external stimulus (e.g., size or volume) where the user is not uncomfortable to the given stimulus but maintaining comparable performance to the conventional approaches. Proposed binary BCI system is not relying on its performance from the characteristics of oddball paradigm where the ERP responses are passively evoked by the deviant stream of stimuli, but the user actively performs the particular mental task according to the continuously presented cue-signs. We hypothesized that the user's voluntary mental task could generate strong endogenous potentials which can be alternatively utilized with the typical ERP components which are passively evoked by the conventional oddball paradigm.

Therefore, proposed non-oddball cues were rather designed to eliminate the passively evoked ERP components from an oddball paradigm and thereby to maximize the feature discriminant between the exogenous (passive) and endogenous (active) brain components. The intensity (i.e., size and volume) of proposed visual and auditory cues were extremely reduced as its role is priorly letting the user know the timing of performing a mental task.

Proposed non-oddball paradigm consisted of visual and auditory binary systems. In both experiments, users were instructed to perform certain tasks following the fully predictable and non-impactful visual or auditory stimuli. To investigate the neurophysiological differences on ERP responses by user's mental states, three particular tasks have defined: non-intention (NI), passive concentration (PC), and active concentration (AC).

Previous ERP studies have investigated the evidence that diminishing stimuli effect can lead to a decrease in the ERP amplitude and therefore to degrade system performance [4]. In this study, Convolutional Neural Network (CNN) [9, 10] with self-average statistical generative model [11] have applied to overcome the limited performance of previous linear classification [4, 6, 12]. We assumed that the discriminative features can be induced not only in the time-series ERP responses but also in the spatial and spectral domain as the user performed a particular mental task [13]. Data-driven optimistic filter [14] sets were validated from the training dataset and Gaussian Autoregressive models were constructed in each spectrally-filtered EEG signal. A sufficient number of EEG trials have augmented based on the Autoregressive models and the generated EEG trials were then concatenated upon each other. Finally, temporal-spectral features were represented by CNN.

Our results investigated a late positive potential (LPP) by the active mental task and sufficiently decodes the user's intention in the non-oddball paradigms. The maximum decoding accuracies of the non-oddball visual and auditory systems were 77.8% and 76.7%, respectively, that significantly outperformed the linear classifier model.

2   Materials and Methods

2.1   Participants and Data Acquisition

Fourteen healthy subjects participated in this experiment: 25–33 years old, 4 women. No psychiatric or neurological disorders were mentioned by subjects. Five subjects were newbies in the BCI field and the other nine knew some aspect of BCI tasks. The environment of the experiment was equipped with a chair and monitor. The monitor was 19 in. LCD monitor (60 Hz refresh rate, 1280×1024 resolution). The aim of experiment and the process of gathering data were explained to the subjects.

Thirty-two channels (Fp1-2, F3-4, Fz, F7-8, FC5-6, FC1-2, T7-8, C3-4, Cz, CP1-2, CP5-6, TP9-10, P3-4, P7-8, Pz, PO9-10, O1-2, and Oz) of EEG acquired data with ActiCap EEG amplifier (Brain Products, Munich, Germany). EEG amplifier referenced on the nose with a forehead ground and used Ag/AgCl electrodes, which fit in the international 10–20 system. An impedance was 10k ohm or less. The sampling rate 1000 Hz. The DC artifact was removed 60 Hz notch filter. A 0.5 30 Hz band-pass filter using a 5th order Butterworth filter was applied to EEG signal.

2.2   Experimental Paradigms and Task Definitions

Two experiments were carried out: the binary systems in non-oddball visual-cue and auditory-cue conditions. The experiments were mainly designed to eliminate the odd-ball characteristics in order to fully derive the endogenous potentials. Therefore, the visual and auditory stimuli were presented fully predictable to the user with an identical interval.

The experiments consisted of two phases: a training and a test phase. All trials in the training phase were used for estimating the classifier parameters, and the test dataset was used to validate the decoding accuracy over the individual sequences. These two experiments were conducted on the same day as these two experiments had technically identical procedures.

Three mental tasks were defined: 1) non-intension (NI), 2) passive concentration (PC), and 3) active concentration (AC). The number of trials in the three defined tasks was balanced. Commonly in the three tasks, an identical stimulus was presented to the subjects, and they were instructed to gaze (or listen) the stimulus in different mental states. In the NI condition, subjects focused on the stimuli without any intention, i.e., resting state, for the given binary selection. On the contrary, the subjects were instructed to attend to the target selection by passively or actively focusing on the stimulus in the PC and the AC conditions. The PC condition required the subject to simply concentrate on a given stimulus which has been the normal instruction in previous ERP-based BCI studies. In the AC condition, subjects were instructed to perform a sound imagery task. To do this, beep sounds at a frequency of $(8000\,Hz)$ were presented to the participants for approximately 1 min before the experiment. Subject remembered this certain tone of sound and repetitively imagined it along with the sequential cues.

All experimental paradigms were developed with the Psychophysics Toolbox (http://psychtoolbox.com) and OpenBMI [15] in Matlab (MathWorks; MA, USA). This study was reviewed and approved by the Institutional Review Board at Korea University [1040548-KUIRB-16-159-A-2], and written informed consent was obtained from all participants before the experiments.

2.3 Experiment I: Non-oddball Visual Cue (NV)

We designed a bar-stimulus that continuously moves along its x-axis in one direction (right to left). A fixed cross-symbol was located in the center. In the training phase, individual bars had different colors according to their experimental condition: gray, blue, and red colors referred to NI, PC, and AC, respectively. Participants were instructed to fix their eyes on the cross-symbol and perform the designated tasks when the bar-stimulus exactly overlapped the fixed cross-symbol (see Fig. 1a). The individual bars were equally spaced with an ISI of 1 s, i.e., the subject performed a certain task every second. Sufficient resting periods were given to the subjects after every 50 trials. At the end of the training phase, a total of 1080 trials comprised of 360 trials for each NI, PC and AC were collected. Two binary classifiers were then constructed: NI vs. PC and IG vs. AC.

In the test phase, ten bars with a gray-color were consecutively presented in a single attempt. Subjects were instructed to perform a specific task ten times designated by the given voice cue 5s before the visual stimuli presentation (see Fig. 1b). For instance, subjects performed sound-imagery task ten times every second by following the visual stimuli when the sound-imagery cue was given. Contrarily, subjects unintentionally gazed at the visual stimulus when given the ignoring-cue. EEG data were acquired in real-time and fed to the classifiers. The classification result was given to the subject as feedback, therefore subjects see what specific action was selected by subject, since in the training phase system

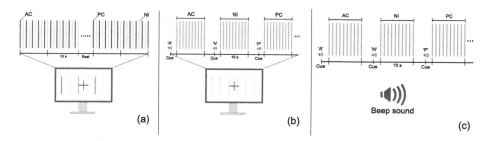

Fig. 1. Illustration of three experiments implemented in our study. (a) Training phase of visual cue experiment. (b) Testing phase of visual cue experiment. (c) Training and test phases of non-oddball auditory cue experiment. (Color figure online)

evaluates brain signals without the feedback. Subjects performed 10 attempts in each class, so 300 trials (10 attempts × 10 sequences × 3 classes) were collected in the test phase.

2.4 Experiment II: Non-oddball Auditory Cue (NA)

The auditory experiment was designed to explore the ERP responses in visually blinded conditions. The participant sat in a comfortable chair with armrests and closed their eyes. Before the experiment, a beep-type auditory stimulus at a frequency 8000 Hz was presented to the participant, and the volume of the auditory stimulus was adjusted as low as possible, to the point where the subject was only able to recognize the moment of the given auditory stimulus (i.e., timing). In the training phase, this auditory stimulus was presented ten times with an ISI of 1 second. The participant was instructed to repetitively perform the task by the following stimuli as notified by the auditory voice 5 s before the stimulus (see Fig. 1c). Similar to the design of the NV experiment above, 360 trials for each task were collected.

In the test phase, the subject closed their eyes and performed a certain task ten times as instructed by the given voice cue 5s before the auditory stimuli. Again, like in the NV experiment, the subjects performed 10 attempts in each class, and a total of 300 trials were collected in the test phase.

3 Data Analysis and Performance Evaluations

The acquired EEG data were down-sampled 100 Hz. From −200 to 1000 ms from stimulus onset segmented EEG data into individual trials. Subtracting the mean amplitudes in the −100 to 0 ms pre-stimulus interval makes baseline correction on EEG data.

To visually investigate the ERP responses for *NI*, *PC*, and *AC* tasks, all trials in the training and test phases were concatenated along with each condition.

Grand averaged ERP patterns and signed r-squared value [16] were applied to evaluate the temporal and statistical differences in the ERP patterns across all the subjects.

For the performance evaluation, the conventional linear model [4, 12] and the proposed CNN approach were constructed based on the training data. In the test phase, certain target stimuli were presented 10 times (i.e., 10 sequences), and this is the typical method in ERP studies to achieve a reliable result by accumulative averaging the current trials with the previous trials [12, 17, 18]. Therefore, decoding accuracy was calculated in individual sequences from one to a maximum of ten. For instance, the decoding accuracy at the 10th sequence was estimated by the averaging of epochs accumulatively through all ten sequences.

Let us denote $X = \{x_n\}_{n=1}^{N}, x \in \mathbb{R}^{T \times D}$ as a set of single-trial EEG, where N is the total number of trial, T and D are time-series data samples and number of channels, respectively. $\Omega \in \{1, 2\}$ and ω_n denote a set of class labels corresponding to target (AC or PC) and non-target (NI) class, and a particular class label for individual trials, respectively.

3.1   Linear Classifier Model

We defined k time intervals from the stimulus onset to 1000 ms with a length of 100 ms and a step size of 50 ms (i.e., $\{[0 - 100], [50 - 150], ..., [900 - 1000]\}$). From the EEG trials X, mean amplitude features in the specific time intervals were calculated across all channels, and then concatenated. The feature vector set $V = \{v\}_{n=1}^{N}$ were therefore formed as $\mathbb{R}^{(D \times k) \times N}$. From the extracted feature set V, a regularized linear discriminant analysis (RLDA) [19] classifier was generated. The decision function $f(\mathbf{v}) = \mathbf{w}^{\mathrm{T}} \cdot v + b$ is defined, where $\mathbf{w}$ is the hyper-plane for separation of binary classes and b is a bias term.

3.2   ERP Data Augmentation

The size of acquired EEG data are generally small due to the expensive cost for human behavior tasks, and therefore, we used the data augmentation method to increase the amount of training dataset to efficiently train the CNN model parameters. Diverse approaches for augmentation techniques have been proposed in machine learning studies: geometric transformations, adding noise to existing data, or newly created synthetic data. ERP potential is continuous signals that include context meaning over time (e.g., N200, P300), and therefore, the geometric transformations (e.g., shift, scale, or rotation) are not a suitable approach as they could distort the characteristics of ERP components. Based on these considerations, our approach was to add Gaussian random noise to the task-relevant ERP signals.

The ERP signal x_n can be assumed as $x_n = p(t) + r_n(t)$, where p is the task-relevant component and r_n is the residual/noise signals with $\mathcal{N}(0, \sigma^2)$. EEG signal has strong randomness and non-stationary, therefore, it is almost difficult to divide the given EEG signals into task-relevant components and noise, especially on a single-trial basis. Therefore, the previous study has been used a heuristic or empirical parameter for estimating the noise signal [20]. In principle, if we approximate the signal parameter of r_n, then an unlimited number of synthetic ERP trials can be created by adding Gaussian random noise to the task-relevant component p. In this study, we approximately estimate the r_n signal parameter based on the characteristics of the ERP paradigm.

Note that the amplitude of noise r_n is decreased (canceled) by the averaging procedure across the trials N, i.e., $r(t) := 1/N \sum_{n=1}^{N} r_n(t) \sim \mathcal{N}(0, \sigma^2/N)$. From the given assumption, σ^2 of the noise signal was acquired as follows:

- We estimated the task-relevant signal $\tilde{p}(t)$ by averaging all training trials, i.e., $\tilde{p}(t) = \bar{X} = 1/N \sum_{n=1}^{N} x_n$, where we assumed that the $r(t)$ is approximated to zero by sufficient averaging procedure.
- The certain number of trials was randomly selected from the X and then averaged (denoted by $\bar{Z} = 1/K \sum_{k=1}^{K} z_k$).
- We calculated $\bar{Z} - \bar{X} = \tilde{p}(t) - p(t) + r_k(t)$. The output is then $r_k(t) \sim \mathcal{N}(0, \sigma^2/K)$ by our previous assumption.
- The noise parameter (i.e., σ^2) was then estimated by $K \cdot var(r(t))$.

The random variable v were defined with the probability density function P as follows:

$$P(v) = \frac{1}{\sigma\sqrt{2\pi}} \cdot e^{(v-\mu)^2/2\sigma^2}, \tag{1}$$

where the μ is zero and the $\sigma^2 = K \cdot var(r(t))$.

Finally, the new data samples were created by adding the Gaussian noise to $\bar{X}_t$ and $\bar{Z}_t$, i.e., $\tilde{p}(t) + v(t)$, $p(t) + r_k(t) - r_k(t) + v(t)$, respectively. This augmentation algorithm iteratively performed until it confirmed the sufficient training data for the CNN model.

3.3   Classification with CNN

A set of $\mathbb{X} = \{X\}_{f=1}^{fn}$ was constructed by concatenating the spectrally filtered EEG trials into third dimension. As a result, the input image is formed as $\mathbb{X} \in \mathbb{R}^{T \times D \times fn}$ corresponding to temporal (T), spatial (D), and frequency domains (fn). The CNN have three convolutional layers with (80, 5), (40, 3), (40, 5) dimensions with 16, 32, 64 depth respectively to extract the spatial and temporal features from the EEG signal. The output size is calculated by the formula:

$$(n + 2p - f)/s + 1, \tag{2}$$

where n represents the number of filters, p is padding size, f is filter size, and s is the amount of stride. CNN uses four ReLU layers function as a threshold operation to accelerate the training process and two dropout layer to prevent the network from overfitting. Max pooling is used to down-sampling a feature map which reduced the computational cost.

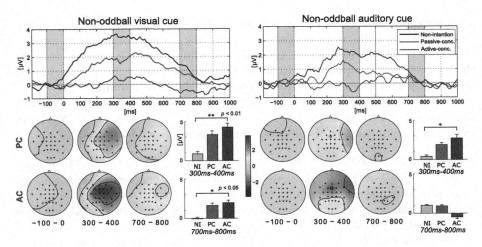

Fig. 2. Averaged ERP responses at Cz electrode for three sessions, i.e., non-oddball visual/auditory cue. The scalp plots indicate the distribution of signal response for the three different conditions, i.e., *NI*, *PC* and *AC* in the non-oddball paradigms. The corresponding responses for the auditory cue experiment are less pronounced compared to the visual cue experiment.

The output layer is constructed from the FC layer with softmax activation layer. The classification output layer is the last layer of the network which computes the cross-entropy loss. The network is trained with a 0.001 learning rate and Root mean square propagation (RMSProp) optimization with 40 maximum epochs.

4   Results

4.1   ERP Responses in Non-oddball Visual/Auditory Paradigms

Figure 2 indicates the grand averaged ERP responses of the individual tasks (i.e., *NI*, *PC*, and *AC*).

In both non-oddball visual and auditory conditions, highly discriminative ERP responses were investigated in the three tasks, and these patterns were in discord with previous knowledge of the typical ERP components. First of all, the ERPs gradually increased from the stimulus onset and peaked in the 300–400 ms interval. These ERPs then gradually decreased until 800 ms after stimulus onset. The sound imagery task induced the largest amplitude, and passively concentrating as well as the ignoring task barely induced ERP responses by the non-oddball paradigms. These ERP components were induced at the central cortex (Cz), however, the ERP responses were not evoked in the occipital or temporal lobes.

The means of peak amplitude in the interval of 300–400 ms for NV were 0.558 (±2.083 uV), 2.277 (±3.461) uV, and 3.547 (±3.637) uV, and NA were

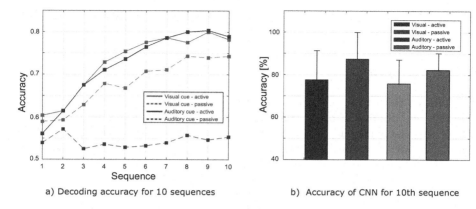

a) Decoding accuracy for 10 sequences   b) Accuracy of CNN for 10th sequence

Fig. 3. Decoding accuracy of non-oddball visual and auditory paradigms. a) For the given number of sequences, the graph indicates the decoding accuracy of active and passive tasks for both non-oddball visual and non-oddball auditory paradigms. Much higher accuracy is achieved for active tasks compared to passive tasks. b) The graph indicates the 10th sequence CNN accuracy for active and passive tasks.

0.204 ($\pm$1.079) uV, 1.210 ($\pm$1.683) uV, and 2.257 ($\pm$3.090) uV for NI, PC, and AC conditions, respectively.

We performed one-way, Bonferroni corrected ANOVA for the three conditions with the null hypothesis of equal means to the maximum amplitude within the two specific intervals. Before the statistical test, we validated the data distributions in all conditions with the Jarque-Bera test for normality. In both non-oddball visual and auditory paradigms, the active task showed a significant difference with the NI condition in the 300–400 interval ($p < 0.05$).

4.2   Decoding Accuracy of Non-oddball Visual/Auditory Paradigm

Figure 3a indicates the LDA decoding accuracy of the proposed non-oddball visual and auditory paradigms. The decoding accuracy (y-axis) is calculated with respect to the number of sequences (x-axis).

To reach an accuracy level of 70% which is known as an efficient communication rate in BCI studies [21], the active tasks in both paradigms required 4 sequences. The passive task in the visual paradigm required 6 sequences while the passive task in the auditory paradigm never reached an accuracy level of 60%.

Proposed CNN with augmentation indicated 77.8%, 76.7% accuracy at the 10th sequence (see Fig. 3b) for the active visual and auditory paradigm respectively. The 10th sequence was the mean value of the previous sequences, which was the accurate value of paradigms and was used for the CNN training process. The accuracies of passive tasks were 87.5% and 85.7% for visual and auditory paradigm respectively. The improved accuracy of passive tasks slightly differentiates from active tasks. The paired t-test of both active and passive tasks in the CNN model indicated that there were not any significant differences for 10th sequence accuracy ($p > 0.5$).

5   Discussion

The aim of this study was to investigate real-world BCI applications and to this end, we proposed a binary visual/auditory system with minuscule stimulus effects. Robust endogenous potentials were validated where the users elicited significant ERP components by themselves. Consequently, we pursue a stimulus-free ERP paradigm to overcome the current limitation where system performance is excessively dependent on external factors.

In our experiment, we present unobtrusive continuous visual and auditory stimuli where the magnitude of the auditory and visual stimuli is reduced to the point that they are barely noticeable. This approach greatly reduces the intensity of external stimulation, which is unlike previous approaches that relied primarily on high impact stimulation to generate a robust ERP response. The continuous low magnitude stimulation is not the source of the ERP itself; instead, it is merely a timing mechanism that allows the user to create their own willful, endogenous potential for the system to detect. Therefore, these proposed systems are highly appropriate for the severely debilitated or patients at the later stages of a disease who might be sensitive to intense external stimulation. The proposed non-oddball paradigms will not generate any command unless she/he performs a particular mental task, and the stimulus' intensity is at an appropriate level for use in everyday life.

The grand averaged ERP responses of AC and PC for the visual cue were greater than those resulting from the auditory cue. This is due to the different experimental protocols of each paradigm. Specifically, subjects were instructed to perform the mental task 10 times, once every second in the auditory paradigm, while they alternatively performed the mental tasks and ignoring state in the visual paradigm (see Fig. 1). A sufficient inter-stimulus interval between the active mental tasks not only benefits mental task preparation but also leads to less jitter of the brain response and thus higher amplitudes in the average. The most remarkable result is that CNN model showed more than 75% of accuracy for all cases. Although our CNN results differ slightly from Linear classifier model results. With a higher than average decoding accuracy, our system still retains a very reasonable performance rate for a BCI communication system [21].

6   Conclusion

Contrary to conventional oddball-based ERP approaches, our approach doesn't rely on robust stimulation for increased system performance. Our design sought to increase the user's endogenous potential with less intense visual or auditory stimuli. To achieve that goal we focused on the LPP response which is a result of the neural activity from an active cognitive process. Our results are important for future practical BCI applications because our system: 1.) can be utilized in a more convenient smartphone environment, 2.) reduces interference from peripheral stimulation, 3.) minimizes the size of the stimuli, and 4.) reduces unintentional command signaling.

Recent studies have investigated novel approaches to boost the performance of conventional ERP-based BCI systems such as a calibration-free classifier model [22], tactile paradigm [23], among others [24, 25]. Importantly, the investigation of our study could have an important role in future applications as this approach can be easily applied to many existing ERP systems for additional performance improvement. We hope that our study will enable other researchers to create novel and reliable BCI spelling systems, and that their application in a real-world environment becomes more viable.

Acknowledgments. This work was supported by Faculty Development Competitive Research Grant Program (No. 080420FD1909) at Nazarbayev University and by Institute for Information & communications Technology Planning & Evaluation (IITP) grant funded by the Korea government (MSIT) (No. 2017-0-00451).

References

1. Nicolas-Alonso, L.F., Gomez-Gil, J.: Brain-computer interfaces, a review. Sensors **12**(2), 1211–1279 (2012)
2. Pfurtscheller, G., Neuper, C.: Motor imagery and direct brain-computer communication. Proc. IEEE **89**(7), 1123–1134 (2001)
3. Squires, N.K., Squires, K.C., Hillyard, S.A.: Two varieties of long-latency positive waves evoked by unpredictable auditory stimuli in man. Electroencephalogr. Clin. Neurophysiol. **38**(4), 387–401 (1975)
4. Yeom, S.K., Fazli, S., Müller, K.R., Lee, S.W.: An efficient ERP-based brain-computer interface using random set presentation and face familiarity. PloS one **9**(11), e111157 (2014)
5. Li, Q., Liu, S., Li, J., Bai, O.: Use of a green familiar faces paradigm improves P300-speller brain-computer interface performance. PloS one **10**(6), e0130325 (2015)
6. Li, Q., Lu, Z., Gao, N., Yang, J.: Optimizing the performance of the visual P300-speller through active mental tasks based on color distinction and modulation of task difficulty. Front. Human Neurosci. **13**, 130 (2019)
7. Lee, M.H., Williamson, J., Kee, Y.J., Fazli, S., Lee, S.W.: Robust detection of event-related potentials in a user-voluntary short-term imagery task. PloS one **14**(12), e0226236 (2019)
8. Lee, M.H., Williamson, J., Lee, Y.E., Lee, S.W.: Mental fatigue in central-field and peripheral-field steady-state visually evoked potential and its effects on event-related potential responses. NeuroReport **29**(15), 1301 (2018)
9. Hara, K., Kataoka, H., Satoh, Y.: Learning spatio-temporal features with 3D residual networks for action recognition. In: Proceedings of the IEEE International Conference on Computer Vision Workshops, pp. 3154–3160 (2017)
10. Kwon, O.Y., Lee, M.H., Guan, C., Lee, S.W.: Subject-independent brain-computer interfaces based on deep convolutional neural networks. IEEE Trans. Neural Netw. Learn. Syst. **31**(10), 3839–3852 (2019)
11. Kang, Y., Hyndman, R.J., Li, F.: Gratis: Generating time series with diverse and controllable characteristics. Stat. Anal. Data Mining ASA Data Sci. J. **13**(4), 354–376 (2020)

12. Lee, M.H., et al.: EEG dataset and OpenBMI toolbox for three BCI paradigms: an investigation into BCI illiteracy. GigaScience **8**(5), giz002 (2019)
13. Bang, J.S., Lee, M.H., Fazli, S., Guan, C., Lee, S.W.,: Spatio-spectral feature representation for motor imagery classification using convolutional neural networks. IEEE Trans. Neural Netw. Learn. Syst. (2021)
14. Suk, H.I., Lee, S.W.: A novel bayesian framework for discriminative feature extraction in brain-computer interfaces. IEEE Trans. Pattern Anal. Mach. Intell. **35**(2), 286–299 (2012)
15. Lee, M.H., et al.: OpenBMI: a real-time data analysis toolbox for brain-machine interfaces. In: 2016 IEEE International Conference on Systems, Man, and Cybernetics (SMC), pp. 001884–001887. IEEE (2016)
16. Blankertz, B., Lemm, S., Treder, M., Haufe, S., Muller, K.R.: Single-trial analysis and classification of ERP components:a tutorial. NeuroImage **56**(2), 814–825 (2011)
17. Wenzel, M.A., Almeida, I., Blankertz, B.: Is neural activity detected by ERP-based brain-computer interfaces task specific? PloS One **11**(10), e0165556 (2016)
18. Lee, M.H., Williamson, J., Won, D.O., Fazli, S., Lee, S.W.: A high performance spelling system based on EEG-EOG signals with visual feedback. IEEE Trans. Neural Syst. Rehabil. Eng. **26**(7), 1443–1459 (2018)
19. Friedman, J.H.: Regularized discriminant analysis. J. Am. Stat. Assoc. **84**(405), 165–175 (1989)
20. Wang, F., Zhong, S., Peng, J., Jiang, J., Liu, Y.: Data augmentation for EEG-based emotion recognition with deep convolutional neural networks. In: Schoeffmann, K., et al. (eds.) MMM 2018. LNCS, vol. 10705, pp. 82–93. Springer, Cham (2018). https://doi.org/10.1007/978-3-319-73600-6_8
21. Kübler, A., Birbaumer, N.: Brain-computer interfaces and communication in paralysis: extinction of goal directed thinking in completely paralysed patients. Clin. Neurophysiol. **119**(11), 2658–2666 (2008)
22. Jin, J., et al.: The study of generic model set for reducing calibration time in P300-based brain-computer interface. IEEE Trans. Neural Syst. Rehabil. Eng. **28**(1), 3–12 (2019)
23. Jin, J., Chen, Z., Xu, R., Miao, Y., Wang, X., Jung, T.P.: Developing a novel tactile P300 brain-computer interface with a cheeks-stim paradigm. IEEE Trans. Biomed. Eng. **67**(9), 2585–2593 (2020)
24. Li, A., Alimanov, K., Fazli, S., Lee, M.H.: Towards paradigm-independent brain computer interfaces. In: 2020 8th International Winter Conference on Brain-Computer Interface (BCI), pp. 1–6. IEEE (2020)
25. Lee, M.H., Fazli, S., Mehnert, J., Lee, S.W.: Subject-dependent classification for robust idle state detection using multi-modal neuroimaging and data-fusion techniques in BCI. Pattern Recogn. **48**(8), 2725–2737 (2015)

A Just-In-Time Compilation Approach for Neural Dynamics Simulation

Chaoming Wang[1,2], Yingqian Jiang[1], Xinyu Liu[1], Xiaohan Lin[1],
Xiaolong Zou[1,3], Zilong Ji[1], and Si Wu[1,3(✉)]

[1] School of Psychology and Cognitive Sciences, IDG/McGovern Institute for Brain
Research, Peking-Tsinghua Center for Life Sciences, Academy for Advanced
Interdisciplinary Studies, School of Electronics Engineering and Computer Science,
Peking University, Beijing 100871, China
`siwu@pku.edu.cn`
[2] Chinese Institute for Brain Research, Beijing 102206, China
[3] Beijing Academy of Artificial Intelligence, Beijing 100083, China

Abstract. As the bridge between brain science and brain-inspired computation, computational neuroscience has been attracting more and more attention from researchers in different disciplines. However, the current neural simulators based on low-level language programming or pseudo-programming using high-level descriptive language can not full fill users' basic requirements, including easy-to-learn-and-use, high flexibility, good transparency, and high-speed performance. Here, we introduce a Just-In-Time (JIT) compilation approach for neural dynamics simulation. The core idea behind the JIT approach is that any dynamical model coded with a high-level language can be just-in-time compiled into efficient machine codes running on a device of CPU or GPU. Based on the JIT approach, we develop a neural dynamics simulator in the Python framework called BrainPy, which is available publicly at https://github.com/PKU-NIP-Lab/BrainPy. BrainPy provides a friendly and highly flexible interface for users to define an arbitrary dynamical system, and the JIT compilation enables the defined model to run efficiently. We hope that BrainPy can serve as a general software for both research and education in computational neuroscience.

Keywords: Neural dynamics · Spiking neural networks · Neural simulator · Computational neuroscience · Just-In-Time compilation

1 Introduction

The prosperity of deep learning research in this round can be partly attributed to the popularity of machine learning software packages, such as NumPy [11], TensorFlow [1], and PyTorch [15]. These softwares based on high-level programming languages (especially Python) lower the entry barrier of normal users and

---

Y. Jiang and X. Liu—Equal contribution.

© Springer Nature Switzerland AG 2021
T. Mantoro et al. (Eds.): ICONIP 2021, LNCS 13110, pp. 15–26, 2021.
https://doi.org/10.1007/978-3-030-92238-2_2

greatly save their programming efforts. As the bridge between brain science and brain-inspired computation, computational neuroscience aims to use mathematical models to elucidate the principles of brain functions, and thus inspire the development of artificial intelligence. Nowadays, computational neuroscience is becoming more and more important, whose development needs more researchers coming from different disciplines, such as biology, mathematics, physics etc. However, compared to the deep learning community, a general programming framework based on a high-level language for neural dynamics simulation is still lacking.

Currently, most tools for simulating neural dynamics provide two types of methods: low-level programming or descriptive language. The representative frameworks in the first category include NEURON [6], NEST [9], GENESIS [3], and CARLsim [7]. These simulators offer a library of standard models and the corresponding convenient interface based on Python. Users can create networks in Python by grouping the models in the library. However, once a new model is needed, the user must learn to program using a low-level language, such as SLI in NEST, C++ in CARLsim, and NMODL in NEURON, which increases the learning cost of the user dramatically and restricts the flexibility of the simulator to define new models [21]. Another approach for model definition is code generation based on a descriptive language [2]. The representative frameworks employing this approach include Brian2 [18,19], ANNarchy [22], Brain Modeling ToolKit [8], GeNN [23], NESTML [16], NeuroML [5,10], and NineML [17]. These simulators allow users to create new models based on descriptions, such as text, JSON or XML files, in high-level programming languages (like Python). Afterwards, high-level model descriptions are translated into a low-level code to speed up the running. The code generation approach has made great success as it enables rapid prototyping for conventional models. However, its coding flexibility is limited by the anticipation of possible models. Once the variety of models goes beyond the anticipation, extension to accommodate new components must be made in both high-level and low-level languages. This is hard or nearly impossible for normal users. Moreover, the generation procedure and the generated code are usually hidden, making the logic controlling, model debugging, and model error correction not transparent to users. Generally speaking, the code generation approach has intrinsic limitations on flexibility and transparency. This is because the coding based on descriptive languages is not real programming where users can freely define variables and express their thoughts.

Motivated by solving the foregoing shortcomings, we introduce a Just-In-Time (JIT) compilation approach for neural dynamics simulation and develop a simulator based on Python called **BrainPy**. *Easy-to-learn-and-use* is important for most users starting to program models in computational neuroscience. Rooted in Python, BrainPy provides a user-friendly interface to program dynamical models. Compared to the frameworks based on low-level programming languages, a BrainPy user only needs to learn the NumPy array programming [11] and the basic Python syntax, such as the if-else condition and the for-loop control. *Flexibility* in model building is increasingly important for a researcher in computational neuroscience, as new models and concepts emerge constantly in the field. Different from simulators based on descriptive languages, BrainPy

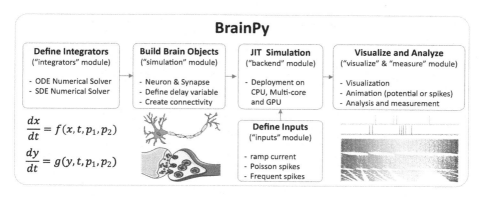

Fig. 1. The overall workflow of the simulator `BrainPy`.

endows users with the full data/logic flow control. No assumption is imposed on the model form. Any dynamical model can be coded in a Python class with inheritance from the base `DynamicSystem`. In this way, the *performance* can also be guaranteed. Because the child class of `DynamicSystem` in BrainPy can be just-in-time compiled into efficient machine codes running on a device of CPU or GPU.

2   Results

2.1   The Overview of the Framework

The overall workflow of BrainPy consists of four stages: (1) defining numerical integrators for differential equations, (2) building brain objects for simulation, (3) running models with the JIT acceleration and (4) analyzing and visualizing the simulation results (Fig. 1). In BrainPy, the fist stage involves defining the numerical integrators for all differential equations. This step is important but very easy, since BrainPy provides a simple and intuitive way to help the user. The next stage involves defining all the wanted brain models with `NeuGroup`, `TwoEndConn` and `Network`. These three objects are children classes of `DynamicSystem`. In this step, BrainPy provides a highly flexible way which allows users to define any variables and update logic. The third stage is accomplished with a single function named `run(duration, inputs)`, where `duration` refers to the start and end time points of each simulation, and `inputs` specifies the input data to model variables. At this stage, BrainPy gathers all defined neurons and connections across neuron groups, and just-in-time compiles them into machine codes, and then deploys the codes to the target hardware specified by the user. At the final state, the user can perform various analyses and visualizations of the simulation output.

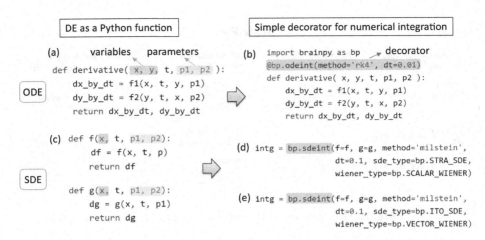

Fig. 2. Numerical solvers provided in BrainPy for ODEs and SDEs.

2.2 Stage 1: Defining Numerical Solvers

The essence of neural dynamics simulation is to solve differential equations. BrainPy provides various numerical solvers for ordinary differential equations (ODEs) and stochastic differential equations (SDEs). Specifically, BrainPy supports explicit Runge-Kutta numerical methods and the Exponential Euler method for ODE and SDE numerical integration. It also supports a variety of adaptive Runge-Kutta methods for ODEs.

In BrainPy, any differential equation can be defined as a Python function. For example, for a two-dimensional ODE system,

$$\frac{dx}{dt} = f_1(x, t, y, p_1), \tag{1a}$$

$$\frac{dy}{dt} = f_2(y, t, x, p_2), \tag{1b}$$

we can define it as a function illustrated in Fig. 2(a), where t denotes the time variable, p1 and p2 after t denote parameters needed in this system, and x and y before t denote dynamical variables. In the function body, the derivative for each variable can be customized by the user's need (f1 and f2). After defining the derivative function, numerical integration can be implemented with a simple decorator @odeint(method, dt) (Fig. 2(b)), where method denotes the numerical method used to integrate the ODE function, and dt controls the numerical integration precision.

Same as the ODE system, any SDE system expressed as

$$dx = f(x, t, p_1)dt + g(x, t, p_2)dW, \tag{2}$$

can be defined as two Python functions as shown in Fig. 2(c). Specifically, the drift and diffusion coefficients in the SDE system are coded as f and g functions.

For the SDE function with a scalar noise, the size of the return data df and dg should be the same, e.g., $df \in R^d, dg \in R^d$. However, for a more general SDE system, it usually has multi-dimensional driving Wiener processes:

$$dx = f(x, t, p_1)dt + \sum_{\alpha=1}^{m} g_\alpha(x, t, p_2)dW^\alpha. \tag{3}$$

For such a m-dimensional noise system, the coding schema is the same as the scalar one, except that the return size of dg is multi-dimensional, e.g., $df \in R^d, dg \in R^{d \times m}$. SDEs have two ways of integral: Itô and Stratonovich stochastic integrals [14], and BrainPy supports both of them. Specifically, the numerical integration of SDEs are performed by `sdeint(f, g, method, dt, wiener_type, sde_type)` (Fig. 2(d) and Fig. 2(e)), where `wiener_type` denotes the type of Wiener process (`SCALAR_WIENER` or `VECTOR_WIENER`) and `sde_type` the integral type (`ITO_SDE` or `STRA_SDE`).

2.3   Stage 2: Building Brain Objects

BrainPy is designed to facilitate flexible model customization. A set of pre-defined models about neurons, synapses, connectivities and networks can not allow us to capture the full range of possible model logic. Alternatively, we should provide a concise but productive and expressive interface to help users code new models with the logic even never seen before. In BrainPy, this step can be implemented by object-oriented programming with the inheritance from the base class `DynamicSystem`.

Generally speaking, any dynamical system can evolve continuously or discontinuously. Discontinuous evolution may be triggered by events, such as the reset of membrane potential, the pre-synaptic spike, the intracellular calcium concentration, or the firing rate intensity. Moreover, it is common in a neural system that a dynamical system has different states, such as the excitable or refractory state in a leaky integrate-and-fire (LIF) model. In this section, we will illustrate how to use two children classes of `DynamicSystem` (specifically, `NeuGroup` and `TwoEndConn`) to capture these complexity in dynamics modeling.

LIF model has the lowest computation cost but is difficult to implement, since it has a continuous potential evolution and a discontinuous reset operation, along with a refractory period. Specifically, it is given by,

$$\tau \frac{dV(t)}{dt} = -(V(t) - V_r) + RI(t), \tag{4a}$$

$$\text{if } V(t) \geq V_{th}, V(t) \leftarrow V_r, \text{ and last } \tau_{ref} \text{ ms}, \tag{4b}$$

where V is the membrane potential, τ the membrane time constant, and R the membrane resistance. Whenever V reaches a fixed threshold V_{th} (i.e., $V \geq V_{th}$), the neuron generates a spike and its potential is reset to the rest value V_r, followed by the refractory period τ_{ref}. In BrainPy, the built-in class `NeuGroup` can be used to describe the kinetics of a population of neurons.

```
1  import brainpy as bp
2  import numpy as np
3
4  class LIF(bp.NeuGroup):
5    @staticmethod
6    # integrate V with Exponential Euler method
7    @bp.odeint(method='exponential_euler')
8    def int_V(V, t, Iext, Vr, R, tau):
9      dV = (- (V - Vr) + R * Iext) / tau
10     return dV
11
12   def __init__(self, num, **kwargs):
13     super(LIF, self).__init__(size=num, **kwargs)
14     # parameters
15     self.Vr, self.Vth = -60, -50
16     self.tau, self.R, self.t_ref = 20., 1., 5.
17     # variables
18     self.V = np.ones(num) * self.Vr # potential variable
19     self.t_last_spike = np.ones(num) * -1e7 # last spike time
20     self.spikes = np.zeros(num, dtype=bool) # spike state
21     self.ref = np.zeros(num, dtype=bool) # refractory state
22     self.I = np.zeros(num) # external inputs
23
24   def update(self, _t, _i):
25     for i in range(self.num):
26       # reset the spike STATE
27       self.spikes[i] = False
28       # reset the refractory STATE
29       self.ref[i] = True
30       # if neuron i is not in refractory STATE
31       if _t - self.t_last_spike[i] > self.t_ref:
32         V = self.int_V(self.V[i], _t, self.I[i],
33                        self.Vr, self.R, self.tau)
34         if V >= self.Vth:  # if generate a spike EVENT
35           self.V[i] = self.Vr
36           self.spikes[i] = True
37           self.t_last_spike[i] = _t
38         else:
39           self.V[i] = V
40           self.ref[i] = False
41       self.I[i] = 0. # reset the input variable
```

Listing 1.1. Simulation code for a population of LIF neurons.

The above code snippet (Listing 1.1) demonstrates the creation of a LIF neuron group by inheriting from NeuGroup. The model parameters and variables are initialized in __init__ function, the continuous potential evolution is defined in int_V function with the help of odeint decorator, and the state switch and event processing in update function are intuitively coded with the basic Python if-else condition and for-loop syntax.

Different from the neuron modeling, synaptic models usually need to (1) create connectivity between neuronal populations and (2) process the transmission delay from the source to the target neurons. BrainPy provides TwoEndConn to capture these features. We illustrate this with the exponential synapse, whose kinetics is given by:

$$\frac{ds(t)}{dt} = -\frac{s(t)}{\tau} + \sum_k \delta(t - t^k), \tag{5a}$$

$$I(t) = g_{max}s(t - D)(V(t) - E), \tag{5b}$$

with s the synaptic state, τ the decay time, t^k the pre-synaptic spike time, D the synaptic delay, V the post-synaptic potential, g_{max} the synaptic conductance, and E the reversal potential.

```
1   class ExpSyn(bp.TwoEndConn):
2     @staticmethod
3     # integrate s with Exponential Euler method
4     @bp.odeint(method='exponential_euler')
5     def int_s(s, t, tau):
6       ds = - s / tau
7       return ds
8
9     def __init__(self, pre, post, conn, tau,
10                g_max, E, delay=0., **kwargs):
11      # parameters
12      self.tau, self.g_max, self.E = tau, g_max, E
13      # connections
14      self.conn = conn(pre.size, post.size)
15      self.pre_ids, self.post_ids = \
16        conn.requires('pre_ids', 'post_ids')
17      self.num = len(self.pre_ids)
18      # variables
19      self.s = bp.ops.zeros(self.num) # synaptic state
20      # the delayed synaptic state
21      self.g=self.register_constant_delay('g', self.num, delay)
22      # initialize the base class
23      super(ExpSyn,self).__init__(pre=pre, post=post, **kwargs)
24
25    def update(self, _t):
26      # continuous synaptic state evolution
27      self.s = self.int_s(self.s, _t, self.tau)
28      # traverse each connected neuron pair
29      for i in range(self.num):
30        pre_i, post_i = self.pre_ids[i], self.post_ids[i]
31        # update synaptic state by the pre-synaptic spike
32        if self.pre.spikes[pre_i]: # if has spike EVENT
33          self.s[i] += 1.
34        # push synaptic state to delay
35        self.g.push(i, self.s[i])
36        # output currents to the post-synaptic neuron
37        g = self.g.pull(i) * self.g_max # get delayed state
38        self.post.I[post_i] += g*(self.E-self.post.V[post_i])
```

Listing 1.2. Simulation code for exponential synapses between two populations.

Listing 1.2 shows the implementation of the exponential synapse model. Compared with the neuron model coding in Listing 1.1, there are two major differences. First, the synaptic connections `self.pre_ids` and `self.post_ids` are constructed by the instance of **Connector** (`conn`, see Listing 1.3). Second, synaptic delays are registered with the function call by `self.register_constant_delay`. Such registered delay variables will be automatically updated in simulation.

To simulate a neural system, we usually need to group a large number of neuron populations as well as synaptic connections between them to form a network. **Network** in BrainPy thus is provided to manage this burdensome task. Users only need instantiate the customized models and then pack them into the **Network** container. The syntax parsing, code compilation, and device deployment etc. will then be automatically processed. For example, the following Python code implements a conductance-based E/I balanced network (COBA-LIF) [4].

```
1  # excitatory neuron group
2  E = LIF(3200, monitors=['spike'])
3  E.V = np.random.randn(E.num) * 5 + E.Vr
4  # inhibitory neuron group
5  I = LIF(800, monitors=['spike'])
6  I.V = np.random.randn(I.num) * 5 + I.Vr
7  # excitatory synapses from E group to E group
8  E2E = ExpSyn(pre=E, post=E, conn=bp.connect.FixedProb(0.02),
9               tau=5., g_max=0.6, E=0.)
10 # excitatory synapses from E group to I group
11 E2I = ExpSyn(pre=E, post=I, conn=bp.connect.FixedProb(0.02),
12              tau=5., g_max=0.6, E=0.)
13 # inhibitory synapses from I group to E group
14 I2E = ExpSyn(pre=I, post=E, conn=bp.connect.FixedProb(0.02),
15              tau=10., g_max=6.7, E=-80.)
16 # inhibitory synapses from I group to I group
17 I2I = ExpSyn(pre=I, post=I, conn=bp.connect.FixedProb(0.02),
18              tau=10., g_max=6.7, E=-80.)
19 # pack neurons and synapses into a network
20 net = bp.Network(E, I, E2E, E2I, I2E, I2I)
```

Listing 1.3. Simulation code for a conductance-based E/I balanced network model.

2.4   Stage 3: Simulation with JIT Acceleration

The magic of **Network** is not limited to the convenient brain object packing, but it also provides the JIT compilation for the defined model.

Python is known highly convenient but simultaneously very slow. This is because all variables in Python are dynamically typed. In order to accelerate the running speed, many JIT compilers, such as PyPy [20], Pyston[13] and Numba [12], have been introduced in Python. Especially, in the Python scientific computing ecosystem, Numba has been a standard for JIT compilation, in which Python functions can be just-in-time compiled to either CPU machine codes using industry-standard LLVM compiler library or GPU codes using NVIDIA PTX compiler. Thus, it may be very suitable for JIT compilation in neural dynamics simulation, and there is no need to create another JIT compiler by ourself. Unfortunately, Numba's JIT compilation works best for functions, not Python classes. In order to make BrainPy's object-oriented programming be compatible with Numba's functional programming, we build a simple Abstract Syntax Trees (AST) based compiler to automatically translate users' classes into functions. Simply state, all the data and functions accessed by `self.` will be interpreted as the function arguments in our AST compiler. In such a way, coding in BrainPy not only keeps the advantages of object-oriented programming, but also obtains the running efficiency of JIT compilation. In practice, users can firstly configure the simulation backend by calling:

```
1  # set simulation backend to single-threaded CPU
2  bp.backend.set(jit=True, device='cpu')
3
4  # set simulation backend to paralleled CPU
5  bp.backend.set(jit=True, device='multi-cpu')
6
7  # set simulation backend to paralleled GPU
8  bp.backend.set(jit=True, device='cuda')
```

Then, the simulation is started by a single functional call:

```
1  # start a simulation
2  net.run(duration=100., # simulation time length
3          inputs=[(E, 'I', 20.), # inputs to E.I variable
4                  (I, 'I', 20.)], # inputs to I.I variable
5          report=True) # turn on the progress report
```

in which **inputs** can be used to mimic the patch clamp or transcranial magnetic stimulation in a physiological experiment.

2.5   Stage 4: Analysis and Visualization of Simulation Output

To gain insight from neural dynamics simulation, it is necessary to further analyze and visualize the simulation output. BrainPy provides useful analyzing functions and visualization tools in **measure** and **visualize** modules, respectively. **measure** module implements a wide range of analyzing functions commonly used in neuroscience (like **cross_correlation()** and **firing_rate()**). Moreover, generic plotting functions in **visualize** module, e.g., **line_plot()**, **animate_1D()** and **animate_2D()**, are able to statically or interactively display spikes, membrane potentials, etc. For example, a single line of code can effectively present the spiking results after simulation in the above stages:

```
1  bp.visualize.raster_plot(E.mon.ts, E.mon.spikes, show=True)
```

The visualization output of the above function call is shown in Fig. 3.

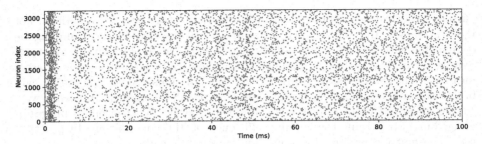

Fig. 3. The raster plot of the COBA-LIF network model in a 100-ms run.

3   Comparison of Running Efficiency

A key factor in neural dynamics simulation is the running efficiency. To compare this with currently popular neural simulators, we adopt two benchmark network models [4]: a COBA-LIF network and a Hodgkin–Huxley model based network (COBA-HH). We implemented two benchmarks on Brian2, NEST and NEURON simulators. All experiments had the duration of several seconds. Dynamical equations were solved using the Exponential Euler method with a step size of 0.1 ms. All simulations were run on Python 3.8 installed on Ubuntu 20.04.2 LTS, with Intel(R) Core(TM) i7-6700K CPU @ 4.00 GHz.

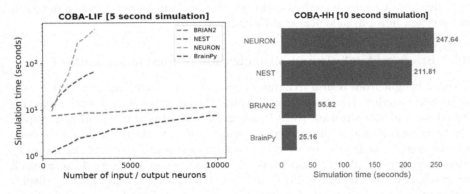

Fig. 4. Speed comparison between NEURON, NEST, Brian2 and BrainPy. (a) The network-size scaling test based on the COBA-LIF network. The x-axis denotes the number of neurons and the y-axis the (logarithmic) simulation time. (b) The speed comparison on the COBA-HH network.

As shown in Fig. 4, due to the JIT compilation, BrainPy has the best performance on both benchmarks. Notably, NEST and NEURON are significantly slower than other simulators, but this may only demonstrate that they are not suitable to simulate point neuron models, as NEURON is designed to simulate biological-detailed multi-compartment models, while NEST has focused on simulating large-scale network models on distributed computing systems. Nevertheless, it demonstrates that BrainPy is efficient for simulating medial-size networks with simplified neurons.

4   Conclusion

We have developed a neural simulator **BrainPy**, which exploits the JIT compilation approach for neural dynamics simulation. We demonstrated that BrainPy can effectively satisfy the needs of users in computational neuroscience. First, by providing a concise and intuitive interface for neural modeling, BrainPy is *easy to learn and use*. Second, with the minimal assumption and restriction in the design, BrainPy endows the user with the full *flexibility and transparency* to define arbitrary dynamical systems. Third, by incorporating the JIT compilation, BrainPy achieves highly *efficient* dynamics simulation. Overall, BrainPy exhibits significant advantages on neural dynamics simulation, as the incorporated JIT approach preserves the intrinsic productive and expressive power of Python programming, and meanwhile realizes efficient running with the speed approaching to that of C or FORTRAN. We hope that BrainPy can serve as a general platform to support users programming efficiently in computational neuroscience. BrainPy is open source and can be available at https://github.com/PKU-NIP-Lab/BrainPy.

References

1. Abadi, M., et al.: Tensorflow: large-scale machine learning on heterogeneous distributed systems (2016). arXiv preprint arXiv:1603.04467
2. Blundell, I., et al.: Code generation in computational neuroscience: a review of tools and techniques. Front. Neuroinf. **12**, 68 (2018)
3. Bower, J.M., Beeman, D.: The Book of GENESIS: Exploring Realistic Neural Models with the GEneral NEural SImulation System. Springer, Heidelberg (2012). https://doi.org/10.1007/978-1-4612-1634-6
4. Brette, R., et al.: Simulation of networks of spiking neurons: a review of tools and strategies. J. Comput. Neurosci. **23**(3), 349–398 (2007)
5. Cannon, R.C., et al.: Lems: a language for expressing complex biological models in concise and hierarchical form and its use in underpinning neuroml 2. Front. Neuroinf. **8**, 79 (2014)
6. Carnevale, N.T., Hines, M.L.: The NEURON Book. Cambridge University Press, Cambridge (2006)
7. Chou, T.S., et al.: Carlsim 4: an open source library for large scale, biologically detailed spiking neural network simulation using heterogeneous clusters. In: 2018 International Joint Conference on Neural Networks (IJCNN), pp. 1–8. IEEE (2018)
8. Dai, K., et al.: Brain modeling toolkit: an open source software suite for multiscale modeling of brain circuits. PLOS Comput. Biol. **16**(11), e1008386 (2020)
9. Gewaltig, M.O., Diesmann, M.: Nest (neural simulation tool). Scholarpedia **2**(4), 1430 (2007)

10. Gleeson, P., et al.: Neuroml: a language for describing data driven models of neurons and networks with a high degree of biological detail. PLoS Comput. Biol. **6**(6), e1000815 (2010)
11. Harris, C.R., et al.: Array programming with numpy. Nature **585**(7825), 357–362 (2020)
12. Lam, S.K., Pitrou, A., Seibert, S.: Numba: a llvm-based python jit compiler. In: Proceedings of the Second Workshop on the LLVM Compiler Infrastructure in HPC, pp. 1–6 (2015)
13. Modzelewski, K., Wachtler, M., Galindo, P.: Pyston (2021). https://github.com/pyston/pyston
14. Øksendal, B.: Stochastic Differential Equations: An Introduction with Applications. Springer, Heidelberg (2003). https://doi.org/10.1007/978-3-642-14394-6
15. Paszke, A., et al.: Pytorch: an imperative style, high-performance deep learning library. In: Wallach, H., Larochelle, H., Beygelzimer, A., d'Alché-Buc, F., Fox, E., Garnett, R. (eds.) Advances in Neural Information Processing Systems 32, pp. 8024–8035. Curran Associates, Inc. (2019). https://www.pytorch.org/
16. Plotnikov, D., Rumpe, B., Blundell, I., Ippen, T., Eppler, J.M., Morrison, A.: Nestml: a modeling language for spiking neurons (2016). arXiv preprint arXiv:1606.02882
17. Raikov, I., et al.: Nineml: the network interchange for ne uroscience modeling language. BMC Neurosci. **12**(1), 1–2 (2011)
18. Stimberg, M., Brette, R., Goodman, D.F.: Brian 2, an intuitive and efficient neural simulator. Elife **8**, e47314 (2019)
19. Stimberg, M., Goodman, D.F., Benichoux, V., Brette, R.: Equation-oriented specification of neural models for simulations. Front. Neuroinf **8**, 6 (2014)
20. Team, T.P.: (2019). https://www.pypy.org/
21. Tikidji-Hamburyan, R.A., Narayana, V., Bozkus, Z., El-Ghazawi, T.A.: Software for brain network simulations: a comparative study. Front. Neuroinf. **11**, 46 (2017)
22. Vitay, J., Dinkelbach, H.Ü., Hamker, F.H.: Annarchy: a code generation approach to neural simulations on parallel hardware. Front. Neuroinf. **9**, 19 (2015)
23. Yavuz, E., Turner, J., Nowotny, T.: Genn: a code generation framework for accelerated brain simulations. Sci. Rep. **6**(1), 1–14 (2016)

STCN-GR: Spatial-Temporal Convolutional Networks for Surface-Electromyography-Based Gesture Recognition

Zhiping Lai[1,2,3,4,5], Xiaoyang Kang[1,2,3,4,5](✉), Hongbo Wang[1,6](✉),
Weiqi Zhang[1], Xueze Zhang[1], Peixian Gong[1], Lan Niu[3], and Huijie Huang[5]

[1] Engineering Research Center of AI and Robotics, Ministry of Education,
Shanghai Engineering Research Center of AI and Robotics, MOE Frontiers Center
for Brain Science, Laboratory for Neural Interface and Brain Computer Interface,
Institute of AI and Robotics, Academy for Engineering and Technology,
Fudan University, Shanghai, China
{zplai19,xiaoyang_kang,Wanghongbo}@fudan.edu.cn
[2] Yiwu Research Institute of Fudan University, Yiwu City, China
[3] Ji Hua Laboratory, Foshan, China
[4] Research Center for Intelligent Sensing, Zhejiang Lab, Hangzhou, China
[5] Shanghai Robot Industrial Technology Research Institute, Shanghai, China
[6] Shanghai Clinical Research Center for Aging and Medicine, Shanghai, China

Abstract. Gesture recognition using surface electromyography (sEMG) is the technical core of muscle-computer interface (MCI) in human-computer interaction (HCI), which aims to classify gestures according to signals obtained from human hands. Since sEMG signals are characterized by spatial relevancy and temporal nonstationarity, sEMG-based gesture recognition is a challenging task. Previous works attempt to model this structured information and extract spatial and temporal features, but the results are not satisfactory. To tackle this problem, we proposed *spatial-temporal convolutional networks for sEMG-based gesture recognition (STCN-GR)*. In this paper, the concept of the sEMG graph is first proposed by us to represent sEMG data instead of image and vector sequence adopted by previous works, which provides a new perspective for the research of sEMG-based tasks, not just gesture recognition. Graph convolutional networks (GCNs) and temporal convolutional networks (TCNs) are used in STCN-GR to capture spatial-temporal information. Additionally, the connectivity of the graph can be adjusted adaptively in different layers of networks, which increases the flexibility of networks compared with the fixed graph structure used by original GCNs. On two high-density sEMG (HD-sEMG) datasets and a sparse armband dataset, STCN-GR outperforms previous works and achieves the state-of-the-art, which shows superior performance and powerful generalization ability.

Keywords: Gesture recognition · Surface electromyography · Human-computer interaction · sEMG graph · Spatial-temporal convolutional networks

© Springer Nature Switzerland AG 2021
T. Mantoro et al. (Eds.): ICONIP 2021, LNCS 13110, pp. 27–39, 2021.
https://doi.org/10.1007/978-3-030-92238-2_3

1   Introduction

The technology of human-computer interaction (HCI) allows human to interact with computers via speech, touch, or gesture [14], which promotes the prosperity of rehabilitation robots [3] and virtual reality [10]. With the development of HCI, a new technology called muscle-computer interface (MCI), which used surface electromyography (sEMG) to recognize gestures and realized natural interaction with human, has emerged and used in many applications, especially rehabilitation robots. sEMG is a bio-signal derived from the muscle fibers' action potential [11], which is recorded by electrodes placed on the skin. According to the number of electrodes, sEMG can be categorized into sparse sEMG and high-density sEMG (HD-sEMG), both of them record spatial and temporal changes of muscle activities when gestures are performed. Since sEMG signals provide sufficient information to decode muscle activities and hand movements, gesture recognition based on surface electromyography (sEMG) forms the technical core of non-intrusive MCIs [1].

Gesture recognition based on sEMG can be divided into two categories: conventional machine learning (ML) approaches and novel deep learning (DL) approaches. ML approaches (e.g., SVM) depend heavily on hand-crafted features (e.g., root mean square), which limits their wider application. As revolutionary ML approaches, DL approaches have achieved great success on the sEMG-based gesture recognition tasks. In existing DL-based recognition approaches, sEMG data are represented as images ([4,6,17]) or sequences ([13]), and are fed into convolutional neural networks (CNNs) or recurrent neural networks (RNNs) to extract high-level features for gesture classification tasks. Since the superposition effect of muscle fibers' action potential, neither sEMG images nor sequences can reveal this characteristic. Moreover, most of the previous works only focus on spatial information or temporal information and have not considered them together [4,7,13,17]. The potential spatial-temporal information is not fully utilized, which further limits the performance of these approaches.

To bridge the gaps mentioned above, we proposed *spatial-temporal convolutional networks for sEMG-based gesture recognition* called *STCN-GR*, in which spatial information and temporal information are taken into consideration together by using graph convolutional networks (GCNs) and temporal convolutional networks (TCNs). Instead of the representation of images or sequences for sEMG data, we propose the concept of sEMG graph and use graph neural networks for sEMG-based gesture recognition, in which the topology of the graph can be learned on different layers of networks. To our knowledge, it is the first time that graph neural networks have been applied in the sEMG-based gesture recognition tasks. Our work makes it possible for graph neural networks to be used in sEMG-based gesture recognition tasks and provides a new perspective for the research of sEMG-based tasks.

The main contributions of our work can be summarized as:

- We propose the concept of sEMG graph and use graph neural networks to solve the task of sEMG-based gesture recognition for the first time, in which the connectivity of graph can be learned automatically to suit the hierarchical structure of networks.

- We propose spatial-temporal convolutional networks *STCN-GR* which uses spatial graph convolutions and temporal convolutions to capture spatial-temporal structured information for gesture recognition.
- On three public sEMG datasets for gesture recognition, the proposed model exceeds all previous approaches and achieves the state-of-the-art, which verifies the superiority of STCN-GR.

The remainder of this paper is organized as follows. Section 2 provides an overview of the DL approaches for gesture recognition and the neural networks on graph. Section 3 describes the proposed model. Section 4 is the experimental details followed by the conclusion in Sect. 5.

2   Related Work

DL-Based Gesture Recognition. sEMG signals are time-series data with high correlation in spatial and temporal dimensions, which reflect the activities of gesture-related muscles. Given by a sequence (i.e., window) of sEMG data, the object of the gesture recognition task is to determine the gesture corresponding to these data. As a leading technology to solve gesture recognition tasks, the deep learning approaches are categorized into CNN approaches, RNN approaches, and hybrid approaches. CNN approaches describe each frame of sEMG data as an image to extract spatial features and turn the gesture classification task into an image classification task. The recognition result obtains by performing a simple majority vote over all frames of a window [2,4,17]. There also exist works that use CNNs directly on the whole window data of sEMG [11]. RNN approaches treat sEMG data as vector sequences and directly feed them into RNN to obtain the recognition results [7,8], in which the temporal information is mainly utilized. Hybrid approaches use CNN, RNN, or other experiential knowledge, simultaneously. Hybrid CNN-RNN architecture [6] has been used and achieves 99.7% recognition accuracy. By integrating experience knowledge into deep models [16,20], good outcomes also are achieved.

However, all of these works are failed to capture structured spatial-temporal information, especially spatial information. Since the superposition effect of muscle fibers' action potential, correlations exist between majority channels. Graph-based approaches may be more appropriate for sEMG-based gesture recognition.

Graph Convolutional Network. Graph neural network (GNN) is a kind of network used to solve the tasks based on graph structure, such as text classification [19], recommender system [9], point cloud generation [15], and action recognition [12,18]. As a typical GNN, graph convolutional network (GCN) is the most widely used one and follows two streams: the spectral perspective and the spatial perspective. The spectral perspective approaches consider graph convolution operations in the form of spectral analysis in the frequency domain. The spatial perspective approaches define graph nodes and their neighbors, where convolutional operations are performed directly using defined rules. Our work follows the second stream. Graph convolutions are performed on constructed sEMG graph followed by temporal convolutions. More details will be introduced in Sect. 3.

3   Spatial-Temporal Convolutional Networks

When performing gestures, human muscles (e.g., extensor) in the arm are involved at the same time. Motor units (MU) in muscles "discharge" or "fire" and generate "motor unit action potential" (MUAP) [4]. The superposition of MUAPs forms sEMG signals. Usually, a gesture is relevant to a sEMG window that contains a sequence of frames, i.e., the sEMG signal shows temporal and spatial correlation. In tasks such as skeleton-based action recognition, this spatial-temporal feature can be extracted using graph convolutional networks (GCNs) and temporal convolutional networks (TCNs) jointly [12,18]. Motivated by them, we introduce GCNs and TCNs into sEMG-based gesture recognition and propose our STCN-GR model. The pipeline of gesture recognition using STCN-GR is presented in Fig. 1. Given a sEMG window, spatial graph convolution and temporal convolution will be performed several times alternately after graph construction to obtain high-level features. Then the corresponding gesture category will be obtained by the softmax classifier.

Graph Construction. Surface electromyography is usually acquired as muti-channel temporal signals, To model this complex spatial-temporal structured information appropriately, for the first time, we propose the concept of *sEMG graph* and create a sEMG graph $G = (V, E)$ for sEMG signals.

In constructed sEMG graph, the states of sEMG channels are represented as the vertex set $V = \{v_i | i = 1, 2, ..., N\}$, N is the number of sEMG channels , and each frame of gesture windows shares this graph. Particularly, the states of sEMG channels will be referred to as vertices for distinguishing them from the channels of the feature map below. Given a vertex v_i and its neighbor v_j, the connectivity between vertex v_i and vertex v_j can be denoted as a spatial edge e_{v_i, v_j}, and all the edges form the spatial edge set $E = \{e_{v_i, v_j} | v_i, v_j \in V\}$. It is worth noting that every spatial edge $e \in E$ (dark blue solid line in Fig. 1) can be learned and updated using a learnable offset δ with the parameters of networks dynamically. For each graph convolutional network layer, a unique topology is learned to suit hierarchical structure base on the original graph.

Spatial Graph Convolution. In graph convolutional networks (GCNs), vertices (dark blue circles in Fig. 1) are updated by aggregating neighbor vertices' information along the spatial edges. Each vertex in the sEMG graph will go through multiple layers and be updated several times. In the $(m + 1)^{th}$ layer, the process of vertex feature aggregation can be formulated as [12]:

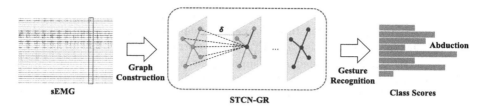

Fig. 1. Pipeline of gesture recognition. Data of a sEMG window (red rectangle) are input to STCN-GR after sEMG graph construction. Then gesture category will be classified by the softmax classifier. The edges of the sEMG graph (dark blue solid lines) are updated using a learnable offset δ with parameters of networks and the vertices (dark blue circles) are also updated according to their neighbors and themselves. (Color figure online)

$$h_i^{m+1} = \sum_{j \in \mathcal{B}_i} \frac{1}{c_{ij}} h_j^m w(l_i(j)) \tag{1}$$

where h_i^{m+1} is the feature representation of vertex i through the aggregation of the $(m+1)^{th}$ layer, $i = 1, 2, ..., N$, $m = 0, 1, 2, ..., M-1$), N and M are the numbers of vertices and the total number of graph convolution layers. h_i^0 denotes the initial state of vertex i. c_{ij} is a normalization factor. $w(\cdot)$ is the weighting function, which is similar to the original convolution. l_i is a mapping function to map vertex j with a unique weight vector [12]. $\mathcal{B}_i$ is the neighbors of vertex i. It can be considered that neighbors are connected to each other. For a standard 3×3 convolution operation, the number of neighbors $|\mathcal{B}|$ can be considered as 9. More generally, the adjacency relationship of vertices can be denoted as an *adjacency matrix*. The spatial graph convolution can be rewritten as [12] in a matrix form:

$$H_{m+1} = \sum_{k=1}^{K} W_{m+1,k} (H_m \tilde{A}_{m+1,k}) \tag{2}$$

where $H_m \in \mathbb{R}^{C_m \times T \times N}$ is the input feature map, $H_{m+1} \in \mathbb{R}^{C_{m+1} \times T \times N}$ is the output feature map after aggregation. C, T and N are the channels of the feature map, length of the window and the number of vertices, respectively. K is the spatial kernel size of the graph convolution, in our work, it is set to 1. $W_{m+1,k} \in \mathbb{R}^{C_{m+1} \times C_m}$ is a weight matrix that can realize a mapping: $\mathbb{R}^{C_m} \to \mathbb{R}^{C_{m+1}}$. $A_{m+1,k} \in \mathbb{R}^{N \times N}$ is the adjacency matrix, to note that, $\tilde{A}_{m,k}$ denotes the "soft" connectivity of sEMG graph learned by the networks, which is a significant improvement compared with original graph convolutional network that uses "hard" fixed topology. The connectivity of sEMG graph is parameterized and can be optimized together with the other parameters of networks, which increases the flexibility of the networks. $\tilde{A}_{m,k}$ is calculated by:

$$\tilde{A}_{mk} = \bar{A}_{mk} + \Delta A_{mk} \tag{3}$$

where $\bar{A}_{mk} = \Lambda_{mk}^{-\frac{1}{2}} A_{mk} \Lambda_{mk}^{-\frac{1}{2}}$, A_{mk} is the connection relationship between vertices (includes self loops), and the fully connected relationship is used as the basic topology in our work. $\Lambda_{mk}^{i,i} = \sum_j A_{mk}^{i,j}$ is the normalized diagonal matrix. As for ΔA_{mk}, it can be regarded as a supplement of $\bar{A}_{mk}$, each element $\delta_{i,j}$ of ΔA_{mk} is a learnable parameter that learns an offset for each spatial edge $e_{i,j}$ and captures import information for gesture recognition (illustrated in Fig. 1). After M updates, vertices in the spatial graph include task-related information. Combined with temporal features, networks can obtain high-level features, which is beneficial for gesture classification.

We can find that graph convolution is similar to traditional convolutional operation, but graph convolution is more flexible, its neighbors can be determined according to actual situation or tasks (traditional convolution has only local grid neighbors), that's why it can achieve good performance on the gesture recognition tasks.

Temporal Convolution. For a T-frames data of a sEMG window, a spatial feature map $S \in \mathbb{R}^{C \times T \times N}$ is obtained after graph convolution is finished, and it is input to a temporal convolution network (TCN) to extract the temporal features. At this stage, temporal feature extraction which uses temporal convolution operation is performed on every vertex (i.e., state sequence $s \in \mathbb{R}^{C \times T \times 1}$). In practice, $K \times 1$ convolution kernel is used to perform temporal convolution, K and 1 are the kernel size along the temporal axis and spatial axis, respectively. By changing the kernel size, the receptive field in the temporal dimension can be adjusted, which means that it can process sequences of arbitrary lengths. In our work, $K = 9$, the stride of 1, and zero paddings are utilized. Using dilated convolutions and stacking TCN layers, history information can be seen in the current time step. However, unlike [13] performs standard temporal convolution operations on the overall sequence, for convenience, our temporal convolution operations use simple convolution operations along the time dimension without dilated convolutions or causal convolutions. In this way, temporal convolutions can be simple enough to be embedded anywhere.

Spatial-Temporal Convolutional Networks. Our spatial-temporal convolutional networks for gesture recognition (STCN-GR) follow similar architectures as [12,18]. As shown in Fig. 2, a basic spatial-temporal convolution block (STCB, box with blue dashed line) includes one GCN block and one TCN block to capture spatial and temporal information together. Besides, batch normalization (BN) layers and ReLU layers are followed to speed up convergence and improve the expression ability of networks. Residual blocks (RBs) are used to stabilize the training, which uses 1×1 kernels to match input channels and out channels (if need).

The overall architecture of networks is shown in Fig. 3. The STCN-GR is stacked by M STCBs, in our work, $M = 4$. c_1, c_2, c_3, c_4 denote the number of output channels of STCBs, which are set to $4, 8, 8, G$ (the number of gestures), respectively. A global average pooling (GAP) layer is added after the last STCB

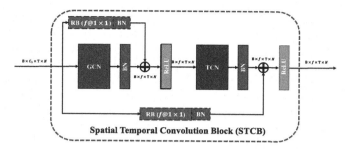

Fig. 2. Illustration of spatial-temporal convolution block (STCB). GCN and TCN are the graph convolutional network and temporal convolutional network, respectively, and both of which are followed by batch normalization (BN) and ReLU. RB stands for residual connection block. (Color figure online)

to improve generalization ability and get the final features, which replaces the full connection layer and reduces the number of parameters. Then, an optional dropout operation is performed. Through a softmax layer, class-conditional probability $\log p(y_j|x_i, \theta)$ can be get to predict gestures. The loss of the i_{th} sample is defined as:

$$\mathcal{L}_i = -\sum_{j=1}^{G} \mathbb{1}_i(y_j) \log p(y_j|x_i, \theta) \tag{4}$$

Where $\mathbb{1}$ is the indicator function, G is the number of gestures, y_j is the j_{th} labels. x_i and θ are the input sEMG signals and parameters of networks, respectively.

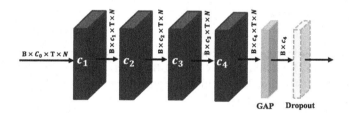

Fig. 3. The overall architecture of STCN-GR. STCN-GR comprises $M(M = 4)$ STCBs, a global average pooling (GAP) layer, and an optional dropout layer. In this architecture, c_1, c_2, c_3, c_4 are 4, 8, 8, G, respectively.

4   Experiments

4.1   Datasets and Settings

To evaluate the performance of STCN-GR, experiments are conducted on three sEMG datasets for gesture recognition: CapgMyo DB-a, CapgMyo DB-b and BandMyo. Different experiments are performed on these datasets to illustrate the superior performance of the proposed STCN-GR.

CapgMyo. CapgMyo [1] is a high-density surface electromyography (HD-sEMG) database for gesture recognition, which is recorded by two-dimensional arrays (8 × 16, total 128) of closely spaced electrodes. This dataset sampled 1000 Hz has three sub-datasets: DB-a, DB-b, and DB-c. 18, 10 and 10 subjects are recruited for DB-a, DB-b and DB-c, respectively. DB-a is designed for evaluating the intra-session performance and fine-tuning hyper-parameters of models, DB-b and DB-c are used for inter-session and inter-subject evaluation [1]. In this work, to compare performance with most existing works, DB-a and DB-b are used. DB-a and DB-b both contain 8 isometric and isotonic hand gestures, each gesture in them is held for 3–10 s and 10 trials are performed for each gesture. We followed the pre-processing procedure like [1,7,17] and used the preprocessed data which use a 45–55 Hz second-order Butterworth band-stop filter to remove the power-line interference and only include the middle one-second data, 1000 frames of data for each trial.

BandMyo. BandMyo dataset [20] is a sparse armband dataset collected by a Myo armband wore on the forearm. This dataset is comprised of finger movement and wrist movement, other movements, a total of 15 gestures. Significantly, the Myo armband just has 8 channels, which means much fewer channels compared with HD-sEMG. 6 subjects are recruited to perform all of 15 gestures by following video guidance, and all of 15 gestures are performed one by one in a trial. When a trial is finished, the armband is taken off and participants will be given a short rest. Briefly, participants wear the armband again and the acquisition process is repeated 8 times (i.e., 8 trials). From the acquisition process, we can know that domain shift [7] exists in one subject's data for the slight change of armband position. No preprocessing is used on this dataset, which is different from CapgMyo datasets.

Experimental Settings. All experiments were conducted on a Linux server (16 Intel(R) Xeon(R) Gold 5222 CPU @ 3.80 GHz) with a NVIDIA GeForce RTX 3090 GPU. All the details in this paper are implemented by using the PyTorch deep learning framework.

For all the experiments, STCN-GR was trained using Adam optimizer, and a weight decay of 0.0001. The base learning rate was set to 0.01 and was divided by 10 after the 5th, 10th, and 25th epochs on three datasets. For CapgMyo DB-a and DB-b, the number of epochs and the batch size were 30 and 16, respectively. 30 epochs and a batch size of 32 were utilized on BandMyo. To get enough samples, the sliding window strategy is used like most works [6,7,11,20]. The window size and window step are 150 ms and 70 ms [7] on CapgMyo DB-a and DB-b, respectively. Since the detailed parameters and training details are not clear [20], 150 and 10 are set as window size and window step on BandMyo. Following the same evaluation method [1,2,6,16,17] on CapgMyo, the model is trained on the odd trials and tested on the even trials. The same evaluation method is used on BandMyo like [20]. Before training, all sEMG signals were normalized in the temporal dimension. No pre-training process has been adopted in our experiments, which is different from other works [2,4,6,17].

4.2   Comparison Results

To evaluate the overall performance of the proposed model in this paper, we compare it with existing approaches on all three datasets. The best results reported literatures are summarized in Table 1, which are all from the latest approaches or the existing state-of-the-art approaches that can be found. Comparison results show that our STCN-GR achieves state-of-the-art performance on all three sEMG datasets, which verifies the superiority of the proposed model.

Table 1. Comparison results on three datasets. The results of the other approaches are the best results reported in literatures. The results in bold show that STCN-GR achieves the best performances on all three datasets.

	Accuracy (%)		
	CapgMyo DB-a	CapgMyo DB-b	BandMyo
GengNet [4]	99.5	98.6	57.8
DMA [1]	99.5	98.6	–
SSL-GR [2]	99.6	98.7	–
CNN-RNN [6]	99.7	–	–
MS-CNN [17]	99.7	–	–
2sRNN [7]	97.1	97.1	–
SA-CNN [5]	96.1	–	–
SVM [20]	71.0	70.8	59.4
RF [20]	83.2	76.2	68.1
STF-GR [20]	91.7	90.3	71.7
STCN-GR (ours)	**99.8**	**99.4**	**75.8**

The experimental results in Table 1 show that, on three public sEMG datasets for gesture recognition (CapgMyo DB-a, CapgMyo DB-b and BandMyo), the results of STCN-GR are superior to that of previous best approaches by **0.1%**, **0.7%** and **4.1%**, respectively. Since accuracy on CapgMyo DB-a is almost saturated [5], it is a significant improvement on this dataset, though the improvement is only 0.1%. The same reason can be found on CapgMyo DB-b. Since the results of the other approaches shown in Table 1 are the best ones in their reports, which means the window size may be 200 ms, 300 ms, even the entire trail. However, just 150 ms is used as window size on CapgMyo in our STCN-GR, and the best performance is achieved. In other words, better performance is achieved using less data, which shows advantages both in accuracy and speed. Due to inter-session domain shift [7] which is a very common phenomenon in practical applications, gesture recognition can be more difficult on the BandMyo dataset compared with the CapgMyo dataset, and our STCN-GR also achieves the state-of-the-art, which shows the powerful generalization ability.

4.3   Ablation Study

We examine the effectiveness of the proposed components by conducting experiments on the first subject of three datasets. As is shown in Fig. 4. "stcn-gr" is the proposed complete model STCN-GR, "tcn" and "gcn" are the models that remove GCN and TCN components, respectively. Particularly, 1×1 convolutions are applied to replace the GCNs to match the output channels. As seen from Fig. 4, the "stcn-gr" always outperforms the other two models. As the number of STCBs (Fig. 2) increases, the performance of "gcn" gradually approaches "stcn-gr", which indicates the core role of GCNs based on the learnable graph. What's more, the performance of "stcn-gr" decreases on some datasets (e.g., CapgMyo DB-b) with the number of layers deepens, and the reason for it may be over-fitting. From this ablation experiment, it can be seen that STCN-GR with 4 STCBs performs well on three datasets, which achieves superior performance while maintains the uncomplicated structure of the networks.

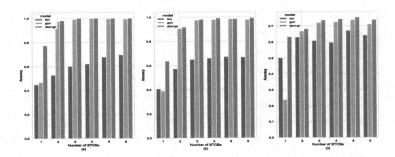

Fig. 4. Ablation study on three datasets. $(a) \sim (c)$ are the results on CapgMyo DB-a, Capgmyo DB-b and BandMyo, respectively. The "stcn-gr" is the proposed complete model, the "tcn" and the "gcn" are the models that remove GCNs and TCNs, respectively.

4.4   Visualization of the Learned Graphs

Figure 5 gives an illustration of learned adjacency matrices by our model based on the first subject of BandMyo. The far left is the original graph, which are followed by learned graphs of 4 STCBs. The darker color represents the stronger connectivity. The visualization of learned graphs indicates that the connectivity with significant vertices (i.e., sEMG channels) will be strengthened, e.g., the vertex 0 and vertex 6, while the connectivity with insignificant vertices will be weakened with the deepening of network layers. Hence, important information will be gathered on a small number of vertices, which is significant for the following classification task.

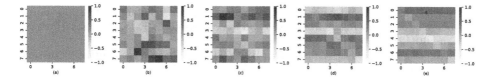

Fig. 5. Visualization of the learned graph. (*a*) is the original adjacency matrix, (*b*) ∼ (*e*) are adjacency matrices learned by 4 STCB layers of STCN-GR. The darker color indicates stronger connectivity. (Color figure online)

5   Conclusion

In this paper, we propose the spatial-temporal convolutional networks for sEMG-based gesture recognition. The concept of sEMG graph is first proposed by us to describe structured sEMG information, which provides a new perspective for the research of sEMG-based tasks. The novel learnable topology of graph can adjust the strength of connectivity between sEMG channels and gathers important information on a small number of vertices. Spatial graph convolutional are performed on the constructed sEMG graph followed by temporal convolution. The proposed networks can fully utilize spatial-temporal information and extract task-related features. The experimental results show that our model outperforms the other approaches and achieves the state-of-the-art on all three datasets. In our feature work, we will concentrate on solving the domain adaptation problem [1,7], includes the adaptation of inter-session domain shift and inter-subject domain shift. Based on this work, the self-supervised and semi-supervised learning framework will be also taken into our consideration.

Acknowledgments. This research is supported by the National Natural Science Foundation of China, grant no. 61904038 and no. U1913216; National Key R&D Program of China, grant no. 2021YFC0122702 and no. 2018YFC1705800; Shanghai Sailing Program, grant no. 19YF1403600; Shanghai Municipal Science and Technology Commission, grant no. 19441907600, no.19441908200, and no. 19511132000; Opening Project of Zhejiang Lab, grant no. 2021MC0AB01; Fudan University-CIOMP Joint Fund, grant no.FC2019-002; Opening Project of Shanghai Robot R&D and Transformation Functional Platform, grant no. KEH2310024; Ji Hua Laboratory, grant no. X190021TB190; Shanghai Municipal Science and Technology Major Project, grant no. 2021SHZDZX0103 and no. 2018SHZDZX01; ZJ Lab, and Shanghai Center for Brain Science and Brain-Inspired Technology.

References

1. Du, Y., Jin, W., Wei, W., Hu, Y., Geng, W.: Surface EMG-based inter-session gesture recognition enhanced by deep domain adaptation. Sensors **17**(3), 458 (2017)
2. Du, Y., et al.: Semi-supervised learning for surface EMG-based gesture recognition. In: IJCAI, pp. 1624–1630 (2017)

3. Fan, Y., Yin, Y.: Active and progressive exoskeleton rehabilitation using multi-source information fusion from EMG and force-position EPP. IEEE Trans. Biomed. Eng. **60**(12), 3314–3321 (2013)
4. Geng, W., Du, Y., Jin, W., Wei, W., Hu, Y., Li, J.: Gesture recognition by instantaneous surface EMG images. Sci. Rep. **6**(1), 1–8 (2016)
5. Hao, S., Wang, R., Wang, Y., Li, Y.: A spatial attention based convolutional neural network for gesture recognition with HD-sEMG signals. In: 2020 IEEE International Conference on E-health Networking, Application & Services (HEALTHCOM), pp. 1–6. IEEE (2021)
6. Hu, Y., Wong, Y., Wei, W., Du, Y., Kankanhalli, M., Geng, W.: A novel attention-based hybrid CNN-RNN architecture for sEMG-based gesture recognition. PloS one **13**(10), e0206049 (2018)
7. Ketykó, I., Kovács, F., Varga, K.Z.: Domain adaptation for sEMG-based gesture recognition with recurrent neural networks. In: 2019 International Joint Conference on Neural Networks (IJCNN), pp. 1–7. IEEE (2019)
8. Koch, P., Brügge, N., Phan, H., Maass, M., Mertins, A.: Forked recurrent neural network for hand gesture classification using inertial measurement data. In: ICASSP 2019–2019 IEEE International Conference on Acoustics, Speech and Signal Processing (ICASSP), pp. 2877–2881. IEEE (2019)
9. Monti, F., Bronstein, M.M., Bresson, X.: Deep geometric matrix completion: a new way for recommender systems. In: 2018 IEEE International Conference on Acoustics, Speech and Signal Processing (ICASSP), pp. 6852–6856. IEEE (2018)
10. Muri, F., Carbajal, C., Echenique, A.M., Fernández, H., López, N.M.: Virtual reality upper limb model controlled by EMG signals. In: Journal of Physics: Conference Series, vol. 477, p. 012041. IOP Publishing (2013)
11. Rahimian, E., Zabihi, S., Atashzar, S.F., Asif, A., Mohammadi, A.: Xceptiontime: independent time-window xceptiontime architecture for hand gesture classification. In: ICASSP 2020–2020 IEEE International Conference on Acoustics, Speech and Signal Processing (ICASSP), pp. 1304–1308. IEEE (2020)
12. Shi, L., Zhang, Y., Cheng, J., Lu, H.: Two-stream adaptive graph convolutional networks for skeleton-based action recognition. In: Proceedings of the IEEE/CVF Conference on Computer Vision and Pattern Recognition, pp. 12026–12035 (2019)
13. Tsinganos, P., Cornelis, B., Cornelis, J., Jansen, B., Skodras, A.: Improved gesture recognition based on sEMG signals and TCN. In: ICASSP 2019–2019 IEEE International Conference on Acoustics, Speech and Signal Processing (ICASSP), pp. 1169–1173. IEEE (2019)
14. Turk, M.: Perceptual user interfaces. In: Earnshaw, R.A., Guedj, R.A., Dam, A., Vince, J.A. (eds.) Frontiers of Human-Centered Computing, Online Communities and Virtual Environments, pp. 39–51. Springer, Heidelberg (2001). https://doi.org/10.1007/978-1-4471-0259-5_4
15. Valsesia, D., Fracastoro, G., Magli, E.: Learning localized generative models for 3D point clouds via graph convolution. In: International Conference on Learning Representations (2018)
16. Wei, W., Dai, Q., Wong, Y., Hu, Y., Kankanhalli, M., Geng, W.: Surface-electromyography-based gesture recognition by multi-view deep learning. IEEE Trans. Biomed. Eng. **66**(10), 2964–2973 (2019)
17. Wei, W., Wong, Y., Du, Y., Hu, Y., Kankanhalli, M., Geng, W.: A multi-stream convolutional neural network for sEMG-based gesture recognition in muscle-computer interface. Pattern Recogn. Lett. **119**, 131–138 (2019)

18. Yan, S., Xiong, Y., Lin, D.: Spatial temporal graph convolutional networks for skeleton-based action recognition. In: Proceedings of the AAAI Conference on Artificial Intelligence, vol. 32 (2018)
19. Yao, L., Mao, C., Luo, Y.: Graph convolutional networks for text classification. In: Proceedings of the AAAI Conference on Artificial Intelligence, vol. 33, pp. 7370–7377 (2019)
20. Zhang, Y., Chen, Y., Yu, H., Yang, X., Lu, W.: Learning effective spatial-temporal features for sEMG armband-based gesture recognition. IEEE Internet Things J. **7**(8), 6979–6992 (2020)

Gradient Descent Learning Algorithm Based on Spike Selection Mechanism for Multilayer Spiking Neural Networks

Xianghong Lin[✉], Tiandou Hu, Xiangwen Wang, and Han Lu

College of Computer Science and Engineering, Northwest Normal University,
Lanzhou 730070, China
linxh@nwnu.edu.cn

Abstract. Gradient descent is one of the significant research contents in supervised learning of spiking neural networks (SNNs). In order to improve the performance of gradient descent learning algorithms for multilayer SNNs, this paper proposes a spike selection mechanism to select optimal presynaptic spikes to participate in computing the change amount of synaptic weights during the process of weight adjustment. The proposed spike selection mechanism comprehensively considers the desired and actual output spikes of the network. The presynaptic spikes involved in the calculation are determined within a certain time interval, so that the network output spikes matches the desired output spikes perfectly as far as possible. The proposed spike selection mechanism is used for the gradient descent learning algorithm for multilayer SNNs. The experimental results show that our proposed mechanism can make the gradient descent learning algorithm for multilayer SNNs have higher learning accuracy, fewer learning epochs and shorten the running time. It indicates that the spike selection mechanism is very effective for improving gradient descent learning performance.

Keywords: Supervised learning · Gradient descent · Spike selection mechanism · Spiking neural networks

1 Introduction

Spiking neural network (SNN) is a new biologically plausible connectionism model in brain-inspired artificial intelligence, which can simulate human's intelligent behaviors. SNN belongs to the third generation artificial neural network (ANN) model [1]. Due to its rationality in biology and its powerful ability in computation and information representation, it has received extensive attention in the field of artificial intelligence [2]. The SNNs usually use temporal coding, and directly use the spike firing time of neurons as the input and output of the network model [3]. Compared with traditional ANNs, SNNs more truly describes the structure and running mode of the biological nervous system and simulates

© Springer Nature Switzerland AG 2021
T. Mantoro et al. (Eds.): ICONIP 2021, LNCS 13110, pp. 40–51, 2021.
https://doi.org/10.1007/978-3-030-92238-2_4

the intelligent activities of the biological brain, which can realize efficient information processing [4].

Supervised learning is one of the significant research content in the field of SNNs. In recent years, according to different SNNs structures and ways of thinking, various supervised learning algorithms have been proposed [5–7]. Supervised learning involves a mechanism of finding an appropriate set of parameters to achieve a perfect match between the output spikes of the SNN and the desired spikes [8,9]. For the SNNs, it's more hard to construct effective and widely applicable supervised learning algorithms.

The gradient descent method is one of the core algorithms in supervised learning for SNNs. Supervised learning algorithm based on gradient descent combines the error backpropagation (BP) mechanism and gradient calculation to update synaptic weights to minimize the value of the error function. Bohte et al. [10] first proposed the SpikeProp algorithm. However, the SpikeProp algorithm limits neurons to fire only one spike. Based on SpikeProp, various kinds of algorithms have been proposed to analyze and improve its learning performance different aspects [11,12]. However, SpikeProp and its simple extensions still limit neurons to fire only one spike. Moreover, the algorithm proposed in [13] enables the neurons in the input and hidden layers to fire multiple spikes, but the output neurons are still limited to fire only one spike. More importantly, Xu et al. [14] proposed a multi-spike learning algorithm, where neurons in all layers can emit multiple spikes, realizing the spike train spatio-temporal pattern learning.

Presynaptic spikes before the current actual output spike moment will participate in the weight adjustment calculation in the conventional gradient descent learning process. However, not all presynaptic spikes before the current actual output spike moment need to participate in the weight adjustment calculation. It has been proved that choosing appropriate presynaptic spikes to participate in the weight adjustment calculation is beneficial for improving the learning efficiency of the gradient descent learning algorithms [15]. But the study on the presynaptic spike selection mechanism is very rare in current research of gradient descent learning, and there is no research on the spike selection mechanism of gradient descent learning algorithms for multilayer SNNs.

This paper proposes a novel spike selection mechanism of gradient descent learning algorithm for multilayer SNNs to enhance their learning performance. The proposed spike selection mechanism comprehensively considers the desired and actual output spikes, and can automatically determine the optimal number of presynaptic spikes participated in the weight adjustment calculation within a certain time interval. Then the proposed spike selection mechanism is applied on ST-SpikeProp [14] to improve its learning performance. The main contributions of this paper are summarized as follows:

- A spike selection mechanism for gradient descent learning algorithm is proposed to determine the optimal presynaptic spikes the participated in computing the change amount of synaptic weights during the process of synaptic weight adjustment in multilayer SNNs.

- The gradient descent learning algorithm, ST-SpikeProp, is improved using the spike selection mechanism to enhance its learning performance. The improved algorithm can achieve higher learning accuracy, fewer learning epochs and shorter running time compared with the ST-SpikeProp algorithm in various parameter settings.

2   SNN Architecture and spiking Neuron Model

2.1   SNN Architecture

Similar to ANN, multilayer feedforward SNN is one of the most commonly used network architectures. As shown in Fig. 1, the fully connected feedforward SNN with multiple synaptic connections between neurons in the previous and posterior layers is used in this paper. The network architecture contains three layers, and each layer plays a different functional role in the learning system. The input layer receives the encoded samples and generates a set of specific spike patterns that can represent external stimuli with various attributes. The hidden layer modulates the synaptic weights between neurons to ensure that the spike patterns are mapped correctly. The output layer is used to extract information from the given input spatio-temporal patterns. The difference in the feedforward SNN used in this paper is that there is more than one synaptic connection between neurons in the previous and posterior layers. Every synaptic connection is equipped with a modifiable synaptic weight and a modifiable synaptic delay. This kind of multiple synaptic connections make the spike trains of presynaptic neurons affect the spike trains of postsynaptic neurons in a longer range of time.

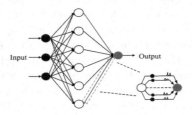

Fig. 1. The diagram of a fully connected feedforward SNN architecture with multiple synaptic connections between neurons in the previous and posterior layers.

2.2   Spiking Neuron Model

In this paper, we employ the spike response model (SRM) to construct SNNs because the SRM's internal state can be expressed intuitively [16]. Suppose that the postsynaptic neuron j connected with N_I presynaptic neurons and there are K synapses between the presynaptic neuron i ($i \in 1, 2, \ldots, N_I$) and the postsynaptic neuron j. The spike train transmitted over the kth ($k \in 1, 2, \ldots, K$)

synapse between the neurons i and j is $F_i = \{t_i^1, t_i^2, \ldots, t_i^{F_i}\}$. The fth spike ($f \in [1, F_i]$) is fired at time t_i^f. The weight and the delay of the kth synapse between neurons i and j are w_{ij}^k and d^k, respectively. The membrane potential $u(t)$ at time t can be expressed as

$$u(t) = \sum_{i=1}^{N_I} \sum_{k=1}^{K} \sum_{f=1}^{F_i} w_{ij}^k \varepsilon \left(t - t_i^f - d^k\right) + \eta \left(t - t_j^{f-1}\right) \tag{1}$$

where t_j^{f-1} is the most recent postsynaptic spike before time t. The function $\varepsilon(s)$ in (1) is called the spike response function, which expresses the influence of the presynaptic spike on u. It can be defined as follows with a time decay constant τ:

$$\varepsilon(s) = \begin{cases} \frac{s}{\tau} \exp\left(1 - \frac{s}{\tau}\right), & s > 0 \\ 0, & s \leq 0 \end{cases} \tag{2}$$

The refractoriness function $\eta(s)$ is defined as follows with a time decay constant τ_R:

$$\eta(s) = \begin{cases} -\theta \exp\left(-\frac{s}{\tau_R}\right), & s > 0 \\ 0, & s \leq 0 \end{cases} \tag{3}$$

where θ is the spike firing threshold.

3 Spike Selection Mechanism and learning Rules of the improved Algorithm

3.1 Spike Selection Mechanism

Gradient descent algorithms realize spike train learning by minimizing the error function between desired and actual output spikes. Due to the existence of the absolute refractory period, presynaptic spikes involved in the weight adjustment calculation are chose within the time interval between the current and previous actual output spikes. The time interval contains the presynaptic spikes involved in the weight adjustment of a postsynaptic spike is named the spike selection time interval (SSTI) in this paper.

For the ST-SpikeProp learning algorithm, the presynaptic spike that participated in the weight adjustment calculation is selected in the time interval between two adjacent actual output spikes. As shown in Fig. 2, S_i, S_h, S_d, S_o represent the input spikes, hidden spikes, desired output spikes and actual output spikes of SNNs respectively. The spikes in S_i are marked as t_i^1, t_i^2, ... , t_i^6 in the order of firing. The spikes in S_h are marked as t_h^1, t_h^2, ... , t_h^6 in the order of firing. The three spikes t_o^1, t_o^2 and t_o^3 in S_o correspond to the three spikes t_d^1, t_d^2 and t_d^3 in S_d one-to-one to construct an error function. In the process of adjusting the weight of the output layer, when the t_o^1 is emitted, the SSTI of t_o^1 is between t_o^0 and t_o^1, which is marked as H_1. Selecting that the hidden spike t_h^1 and t_h^2 belonging to H_1 to adjust the weight w_h of S_h. For the second actual output spike t_o^2, SSTI is H_2 and t_h^3 is selected to adjust the weight w_h. Next, for

t_o^3, select the hidden spike t_h^4, t_h^5 and t_h^6 belonging to H_3 adjust w_h. Similarly, in the process of adjusting the weight of the hidden layer, when the t_o^1 is emitted, select the input spike t_i^1 and t_i^2 belonging to T_1 to adjust the weight w_i of S_i. For the second actual output spike t_o^2, SSTI is T_2 and t_i^3 is selected to adjust the weight w_i. Next, for t_o^3, select the input spike t_i^4, t_i^5 and t_i^6 belonging to T_3 adjust w_i.

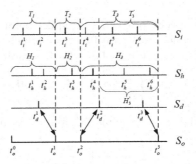

Fig. 2. Input spikes, hidden spikes, desired and actual output spikes during SNNs running process. ST-SpikeProp selects the input spikes and hidden spikes in the time interval between two adjacent actual output spikes.

For multiple synapses, the increase and decrease of synaptic weights may affect each other, resulting in a decrease in the performance of the ST-SpikeProp algorithm. To improve the learning performance of the ST-SpikeProp, we change the spike selection mechanism in the process of synaptic weight adjustment, which comprehensively considers the actual and the desired spikes. For the convenience of description, the new algorithm obtained by applying the new mechanism to ST-SpikeProp is called ST-SpikeProp_SSTI in this paper. In adjusting the output layer weight of ST-SpikeProp_SSTI algorithm, when the actual output spike t_o^f is emitted, the SSTI of t_o^f is H_f. In adjusting the weight of hidden layer, the SSTI of t_o^f is T_f. The end point of T_f and H_f is t_o^f. The starting point $\bar{t}$ of T_f and H_f has the following three situations, as shown in Fig. 3: (a) If t_d^{f-1} is earlier than t_o^{f-1}, $\bar{t}$ is t_o^{f-1}, where t_o^{f-1} is the $f-1$th actual output spike and t_d^{f-1} is the $f-1$th desired output spike; (b) If t_d^{f-1} is between t_o^{f-1} and t_o^f, $\bar{t}$ is t_d^{f-1}; (c) If t_d^{f-1} is later than t_o^f, $\bar{t}$ is t_o^{f-1}.

Let us continue to take Fig. 2 as an example. In the process of adjusting the synaptic weights of the ST-SpikeProp_SSTI algorithm, at t_o^1 and t_o^2, the number of presynaptic spikes participate in the weight adjustment calculation is the same as that selected by the ST-SpikeProp algorithm. At the time t_o^3 of the third actual output spike, t_d^2 is between t_o^2 and t_o^3, so $\bar{t}$ is t_d^2. The spikes for adjusting the weights of the hidden layer and the output layer are marked as T_3' and H_3' respectively in the Fig. 2, where t_i^4 and t_h^4 are ignored.

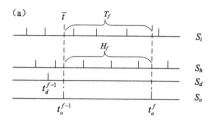

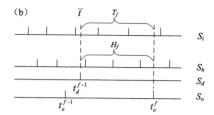

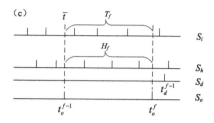

Fig. 3. ST-SpikeProp_SSTI algorithm comprehensively considers actual and desired output spikes. (a) t_d^{f-1} is earlier than t_o^{f-1}. (b) t_d^{f-1} is between t_o^{f-1} and t_o^f. (c) t_d^{f-1} is later than t_o^f.

3.2 Learning Rules of the improved Algorithm

Assuming that the output neuron o in the output layer corresponding to an input is a spike train composed of F_o spikes, the network error function is defined as

$$E = \frac{1}{2} \sum_{o=1}^{N_O} \sum_{f=1}^{F_o} \left(t_o^f - t_d^f \right)^2 \tag{4}$$

where N_O represents the number of output neurons, t_o^f and t_d^f represent actual and desired output spike respectively.

According to the gradient descent mechanism, the change of the kth synaptic weight Δw_{ij}^k between the presynaptic neuron i and the postsynaptic neuron j is calculated as follows:

$$\Delta w_{ij}^k = -\alpha \nabla E_{ij}^k = -\alpha \frac{\partial E}{\partial w_{ij}^k} \tag{5}$$

where α represents the learning rate.

In this paper, the core research content is spike selection mechanism, and then this new spike selection mechanism is applied to the ST-SpikeProp algorithm. Because the weight adjustment rule of the ST-SpikeProp algorithm has been given in [14], so we will not deduce the ST-SpikeProp_SSTI algorithm in detail, only give the formula which can reflect the spike selection mechanism.

In the ST-SpikeProp_SSTI algorithm, the adjustment formula of synaptic weight from hidden layer neuron h to output layer neuron o is

$$\Delta w_{ho}^k = -\alpha \sum_{f=1}^{F_o} \frac{\partial E}{\partial t_o^f} \frac{\partial t_o^f}{\partial w_{ho}^k}$$

$$= \alpha \sum_{f=1}^{F_o} \frac{\left(t_o^f - t_d^f\right) \left[\displaystyle\sum_{t_h^g \in \left(\bar{t}, t_o^f\right)} \varepsilon\left(t_o^f - t_h^g - d^k\right) + \frac{1}{\tau_R}\eta\left(t_o^f - t_o^{f-1}\right) \frac{\partial t_o^{f-1}}{\partial w_{ho}} \right]}{\displaystyle\sum_{h=1}^{N_H} \sum_{k=1}^{K} \sum_{t_h^g \in \left(\bar{t}, t_o^f\right)} w_{ho}^k \varepsilon\left(t_o^f - t_h^g - d^k\right) \left[\frac{1}{\left(t_o^f - t_h^g - d^k\right)} - \frac{1}{\tau} \right] - \frac{1}{\tau_R}\eta\left(t_o^f - t_o^{f-1}\right)}$$

(6)

where N_H represents the number of hidden neurons. t_h^g is the gth spike of hidden layer neuron h.

The adjustment formula of synaptic weight from input layer neuron i to hidden layer neuron h is

$$\Delta w_{ih}^k = -\alpha \nabla E_{ih}^k = -\alpha \frac{\partial E}{\partial w_{ih}^k} = -\alpha \sum_{g=1}^{N_h} \frac{\partial E}{\partial t_h^g} \frac{\partial t_h^g}{\partial w_{ih}^k}$$

(7)

where

$$\frac{\partial E}{\partial t_h^g} = \sum_{o=1}^{N_O} \sum_{f=1}^{F_o} \frac{\partial E}{\partial t_o^f} \frac{\partial t_o^f}{\partial u_o\left(t_o^f\right)} \frac{\partial u_o\left(t_o^f\right)}{\partial t_h^g}$$

(8)

$$\frac{\partial t_h^g}{\partial w_{ih}^k} = \frac{-\displaystyle\sum_{t_i^f \in \left(\bar{t}, t_o^f\right)} \varepsilon\left(t_h^g - t_i^f - d^k\right) - \frac{1}{\tau_R}\eta\left(t_h^g - t_h^{g-1}\right) \frac{\partial t_h^{g-1}}{\partial w_{ih}}}{\displaystyle\sum_{i=1}^{N_I} \sum_{k=1}^{K} \sum_{t_i^f \in \left(\bar{t}, t_o^f\right)} w_{ih}^k \varepsilon\left(t_h^g - t_i^f - d^k\right) \left[\frac{1}{\left(t_h^g - t_i^f - d^k\right)} - \frac{1}{\tau} \right] - \frac{1}{\tau_R}\eta\left(t_h^g - t_h^{g-1}\right)}$$

(9)

where N_I is the number of input neurons. t_i^f represents the fth spike of input layer neuron i.

4 Simulation Results

4.1 Parameter Settings

All simulation experiments use Windows 10 environment with a six-core CPU and 16-GB RAM, running on the Eclipse software platform with java 1.8. The clock-driven simulation strategy is used to run the SNN, where the time-step is set as 0.1 ms. The parameters of the fully connected feedforward SNN are: $N_I = 50$, $N_H = 30$ and $N_O = 1$. There are $K = 5$ synapses between each presynaptic neuron and each postsynaptic neuron. In SRM, the time constants $\tau = 5$ ms and $\tau_R = 50$ ms, the length of the absolute refractory period is 1 ms, and the spike firing threshold $\theta = 1$. The Poisson process is used for generating input spikes 40 Hz and desired output spikes 100 Hz. The spike train length is 300 ms. The synaptic weight is generated randomly in the interval $[0, 0.5]$. The learning rate of the ST-SpikeProp_SSTI and ST-SpikeProp algorithm is 0.0005. Unless otherwise specified, our simulation results are the average of 30 trials. The learning epochs of the algorithm are 500. The values of all the above parameters are empirical values and base settings. The correlation-based metric C [5] is used for measuring the spike train learning accuracy.

4.2 Spike Train Learning

Learning rate is an important parameter for learning algorithms. We determine the optimal learning rate of the ST-SpikeProp_SSTI and ST-SpikeProp algorithms through a spike train learning task under benchmark parameters. The learning rate takes 0.00001, 0.00005, 0.0001, 0.0005, 0.005 in total of five values. Figure 4 shows the learning results. Figure 4(a) shows the learning accuracy C. When the learning rate is 0.0005, the learning accuracy of two algorithms reaches the maximum $C = 0.9583$ and $C = 0.9039$ respectively. Therefore, we choose 0.0005 as the optimal learning rate for the ST-SpikeProp_SSTI and ST-SpikeProp algorithms. Figure 4(b) shows the learning epochs during the learning process.

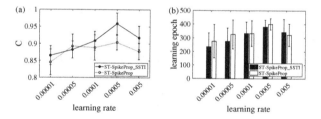

Fig. 4. Learning results for the different learning rates under benchmark parameters. (a)Learning accuracy C when the learning rate is varied. (b) The learning epochs when the training algorithms achieve the highest learning accuracy.

Next, we use four spike train learning tasks to verify the effectiveness of our proposed spike selection mechanism. The first spike train learning task tests the influence of the spike train lengths. The learning results of the ST-SpikeProp_SSTI algorithm and ST-SpikeProp algorithm with different spike train lengths are shown in Fig. 5. As shown in Fig. 5(a), the learning accuracy C shows a decreasing trend for both ST-SpikeProp_SSTI and ST-SpikeProp. It can be found that C of ST-SpikeProp_SSTI is higher than that of ST-SpikeProp. This is because the length of the spike train is a factor that affects how many spikes are fired in the network simulation for the given input spike rate. A longer spike train means the SNN may emit more spikes, where the weight may be adjusted more difficult, so the learning accuracy of algorithms will decrease. The learning epochs during the learning process are represented in Fig. 5(b). It shows that the learning epoch of the ST-SpikeProp_SSTI and ST-SpikeProp algorithms increases firstly and then decreases, and the learning epoch of the ST-SpikeProp_SSTI algorithm is less than that of the ST-SpikeProp algorithm. The average running time of the two algorithms are shown Fig. 5(c). The average running time of the ST-SpikeProp_SSTI is shorter than ST-SpikeProp algorithm for the different spike train lengths.

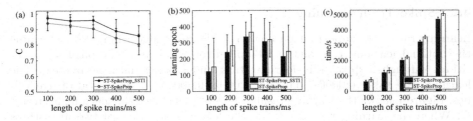

Fig. 5. The learning results of ST-SpikeProp_SSTI and the ST-SpikeProp with different spike train lengths. (a) The learning accuracy C when the simulation length is varied. (b) The learning epochs when the training algorithms achieve the highest learning accuracy. (c) The average running time of the algorithm.

The second spike train learning task tests the influence of the firing rates of spike trains. For the different spike train firing rates, the results of ST-SpikeProp_SSTI and ST-SpikeProp algorithms are shown in Fig. 6. The learning accuracy C is represented Fig. 6(a). It shows that both ST-SpikeProp_SSTI and ST-SpikeProp algorithms can achieve high learning accuracy. Learning accuracy of the two algorithms first increases and then decreases with the increase of the spike firing rates, but the ST-SpikeProp_SSTI algorithm achieves higher accuracy than the ST-SpikeProp algorithm. Figure 6(b) represents the learning epoch during the learning process for the different spike train firing rates. It shows that the ST-SpikeProp_SSTI algorithm requires fewer learning epochs than the ST-SpikeProp algorithm. As shown in Fig. 6(c), we can find that the running time of ST-SpikeProp_SSTI is shorter than ST-SpikeProp for the different spike train firing rates.

The third spike train learning task tests the influence of the number of synapses. The learning results of the ST-SpikeProp_SSTI algorithm and ST-SpikeProp algorithm with the different numbers of synapses are shown Fig. 7. Figure 7(a) represents the learning accuracy C. When the number of synapses is increasing, the learning accuracy C of the ST-SpikeProp_SSTI algorithm and ST-SpikeProp algorithm first increases and then decreases. However, the learning accuracy of ST-SpikeProp_SSTI is higher. This is because more synapses make it more difficult to adjust the weight of the algorithm and the results of weight adjustment will affect each other, so the learning accuracy will decrease. As shown in Fig. 7(b), ST-SpikeProp_SSTI can reach the highest accuracy with

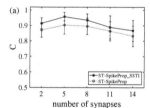 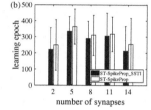 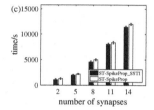

Fig. 6. The learning results of ST-SpikeProp_SSTI algorithm and ST-SpikeProp algorithm with different spike train firing rates. (a) The learning accuracy C. (b) The learning epochs when the training algorithms achieve the highest learning accuracy. (c) The average running time of the algorithm.

fewer learning epochs than ST-SpikeProp. Figure 7(c) shows the average running time of two algorithms for the different number of synapses. We find that the ST-SpikeProp_SSTI is faster than the ST-SpikeProp.

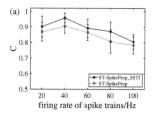 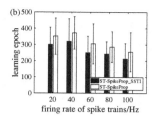 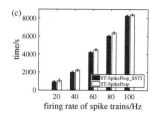

Fig. 7. The learning results of ST-SpikeProp_SSTI algorithm and ST-SpikeProp algorithm with different numbers of synapses. (a) The Learning accuracy C when the number of synapses gradually increases. (b) The learning epochs when the training algorithms achieve the highest learning accuracy. (c) The average running time of the algorithm.

The fourth spike train learning task tests the influence of the number of input neurons. The learning results of the ST-SpikeProp_SSTI and ST-SpikeProp with the different numbers of input neurons are shown in Fig. 8. As shown in Fig. 8(a), when the number of input neurons is increasing, the learning accuracy C of these two algorithms is increasing too. Meanwhile, the accuracy C of ST-SpikeProp_SSTI is higher than that of ST-SpikeProp. This is because more input neurons will make the SNN find more potential information from the input spike patterns. Figure 8(b) shows the learning epochs for these two algorithms when the accuracy reaches the highest. Clearly, the ST-SpikeProp_SSTI algorithms need fewer epochs than the ST-SpikeProp algorithms. The average running time shown in Fig. 8(c) indicates that the ST-SpikeProp_SSTI algorithm is faster than the ST-SpikeProp algorithm.

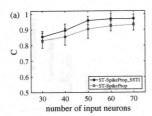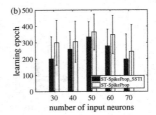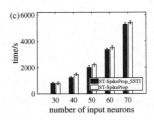

Fig. 8. The learning results of ST-SpikeProp_SSTI algorithm and ST-SpikeProp algorithm with different numbers of input neurons. (a) The Learning accuracy C when the number of input neurons gradually increases. (b) The learning epochs when the training algorithms achieve the highest learning accuracy. (c) The average running time of the algorithm.

5   Conclusion

We propose a novel spike selection mechanism to select the optimal presynaptic spikes participated in computing the synaptic weight adjustment in this paper, which is used to improve the performance and efficiency of the gradient descent learning algorithm for the multilayer SNNs. The spike selection time interval containing appropriate presynaptic spikes. Then the proposed spike selection mechanism is applied on the ST-SpikeProp algorithm, and the novel ST-SpikeProp_SSTI algorithm is obtained. The experimental results show that the ST-SpikeProp_SSTI can achieve higher learning accuracy, fewer learning epochs and shorter running time than the ST-SpikeProp algorithm in various parameter settings. It indicates that the spike selection mechanism can enhance the learning performance and efficiency of gradient descent learning for multilayer SNNs.

Acknowledgments. This work was supported by the National Natural Science Foundation of China under grant no. 61762080, the Key Research and Development Project of Gansu Province under grant no. 20YF8GA049, the Youth Science and Technology Fund Project of Gansu Province under grant no. 20JR10RA097, the Lanzhou Municipal Science and Technology Project under grant no. 2019-1-34.

References

1. Ghosh-Dastidar, S., Adeli, H.: Spiking neural networks. Int. J. Neural Syst. **19**(04), 295–308 (2009)
2. Taherkhani, A., Belatreche, A., Li, Y., et al.: A review of learning in biologically plausible spiking neural networks. Neural Netw. **122**, 253–272 (2020)
3. Skatchkovsky, N., Jang, H., Simeone, O.: Spiking neural networks-part II: detecting spatio-temporal patterns. IEEE Commun. Lett. **25**(6), 1741–1745 (2021)
4. Kulkarni, S.R., Rajendran, B.: Spiking neural networks for handwritten digit recognition-supervised learning and network optimization. Neural Netw. **103**, 118–127 (2018)

5. Wang, X., Lin, X., Dang, X.: Supervised learning in spiking neural networks: a review of algorithms and evaluations. Neural Netw. **125**, 258–280 (2020)
6. Lin, X., Wang, X., Zhang, N., et al.: Supervised learning algorithms for spiking neural networks: a review. Acta Electron. Sin. **43**(3), 577–586 (2015)
7. Lin, X., Wang, X.: Spiking Neural Networks: Principles and Applications. Science Press, China (2018)
8. Comsa, I., Potempa, K., Versari, L., et al.: Temporal coding in spiking neural networks with alpha synaptic function: learning with backpropagation. IEEE Trans. Neural Netw. Learn. Syst., 1–14 (2021)
9. Kheradpisheh, S., Masquelier, T.: Temporal backpropagation for spiking neural networks with one spike per neuron. Int. J. Neural Syst. **30**(06), 2050027 (2020)
10. Bohte, S.M., Kok, J.N., Poutré, H.: Error-backpropagation in temporally encoded networks of spiking neurons. Neurocomputing **48**(1), 17–37 (2002)
11. Zhao, J., Zurada, J.M., Yang, J., Wu, W.: The convergence analysis of SpikeProp algorithm with smoothing $L_{1/2}$ regularization. Neural Netw. **103**, 19–28 (2018)
12. Shrestha, S.B., Song, Q.: Robustness to training disturbances in SpikeProp learning. IEEE Trans. Neural Netw. Learn. Syst. **29**(7), 3126–3139 (2018)
13. Booij, O., tat Nguyen, H.: A gradient descent rule for spiking neurons emitting multiple spikes. Inf. Process. Lett. **95**(6), 552–558 (2005)
14. Xu, Y., Zeng, X., Han, L., Yang, J.: A supervised multi-spike learning algorithm based on gradient descent for spiking neural networks. Neural Netw. **43**, 99–113 (2013)
15. Xu, Y., Yang, J., Zhong, S.: An online supervised learning method based on gradient descent for spiking neurons. Neural Netw. **93**, 7–20 (2017)
16. Gerstner, W., Kistler, W.M.: Spiking Neuron Models: Single Neurons, Populations, Plasticity. Cambridge University Press, Cambridgeshire (2002)

Learning to Coordinate via Multiple Graph Neural Networks

Zhiwei Xu, Bin Zhang, Yunpeng Bai, Dapeng Li, and Guoliang Fan[✉]

Institute of Automation, Chinese Academy of Sciences, School of Artificial Intelligence, University of Chinese Academy of Sciences, Beijing, China
{xuzhiwei2019,zhangbin2020,baiyunpeng2020,lidapeng2020, guoliang.fan}@ia.ac.cn

Abstract. The collaboration between agents has gradually become an important topic in multi-agent systems. The key is how to efficiently solve the credit assignment problems. This paper introduces MGAN for collaborative multi-agent reinforcement learning, a new algorithm that combines graph convolutional networks and value-decomposition methods. MGAN learns the representation of agents from different perspectives through multiple graph networks, and realizes the proper allocation of attention between all agents. We show the amazing ability of the graph network in representation learning by visualizing the output of the graph network, and therefore improve interpretability for the actions of each agent in the multi-agent system.

Keywords: Decision making and control · Multi-agent reinforcement learning · Graph neural network

1 Introduction

In the past decade, multi-agent systems (MAS) have received considerable attention from researchers due to their extensive application scenarios. The change of the environment is no longer determined by a single agent but is the result of the joint actions of all agents in MAS, which results in the traditional single-agent reinforcement learning algorithm cannot be directly applied to the case of Multi-Agent. In the field of cooperative multi-agent reinforcement learning, since the dimensionality of the joint action space of multi-agents will increase exponentially as the number of agents increases, the centralized method of combining multiple agents as a single agent for training cannot achieve desired results. In addition, there is a decentralized approach represented by Independent Q-Learning (IQL) [20], in which each agent learns independently, using other agents as part of the environment, but this method is unstable and easy to overfit. At present, centralized training and distributed execution (CTDE) [9] are the most popular learning paradigms, in which we can use and share some global information during training to make the distributed execution more effective, so as to improve learning efficiency.

© Springer Nature Switzerland AG 2021
T. Mantoro et al. (Eds.): ICONIP 2021, LNCS 13110, pp. 52–63, 2021.
https://doi.org/10.1007/978-3-030-92238-2_5

On the one hand, it's better to learn a centralized action-value function to capture the effects of all agents' actions. On the other hand, such a function is difficult to learn. Even if it can be learned, there is no obvious way for us to extract decentralized policy. Facing this challenge, the COMA [4] algorithm learns a fully centralized Q-value function and uses it to guide the training of decentralized policies in an actor-critic framework. Different from this method, researchers have proposed another value-based algorithm. The main idea is to learn a centralized but decomposable value function. Both Value-Decomposition Network (VDN) [18] and QMIX [14] adopt this idea. VDN approximates joint action-value function as the linear summation of the individual value functions obtained through local observations and actions, but in fact, the relationship between joint action-value and individual action-value is much more complicated than this, besides, VDN ignores any additional state information available during learning. The QMIX algorithm relaxes the restriction on the relationship between the whole and the individual. It approximates joint Q-value function through a neural network and decomposes it into a monotonically increasing function of all individual values. In addition, there are many excellent works in the field of value function decomposition, such as QTRAN [16] that directly learn the joint action value function and then fit residuals with another network.

The above-mentioned value-decomposition methods have achieved good results in the SMAC [15] testbed. But it's worth noting that the aforementioned algorithms mainly focus on the value decomposition for credit assignment, but the underlying topology between agents in the MAS is not paid attention to or utilized. When we take this structure into account, a natural idea is to use graph structure for modeling. For data in an irregular or non-Euclidean domain, graph convolutional networks (GCNs) [3,6,13,21,23–25] can replace traditional convolution operations and perform graph convolutions by taking the weighted average of a node's neighborhood information, so as to use the geometric structure of the graph to learn the embedding feature of each node or the whole graph. Recently, many graph convolutional networks based on different types of aggregators have been proposed, and significant results have been obtained on many tasks such as node classification or graph classification. Since the agents in the MAS can communicate and influence each other, similar to social networks, some works that combines graph networks and multi-agent reinforcement learning have appeared. Most of them can be seen as variants that increase the communication between agents. For example, CommNet [17], BiCNet [12], and DGN [1] all use different convolutional kernels to process the information transmitted by neighbor agents.

In this paper, we propose a multi-agent reinforcement learning algorithm based on the CTDE structure that combines graph convolutional neural networks with value-decomposition method, namely Multi-Graph Attention Network (MGAN). We establish an undirected graph, and each agent acts as a node in the graph. Based on this graph, we build multiple graph convolutional neural networks and the attention mechanism [22] is used in the aggregators. The input of the network is the individual value function obtained by a single agent, and the output of the network is the global value function. At the same time, in order to ensure that the local optimal action is the same as the global optimal action, the MGAN algorithm also satisfies the monotonicity assumption. Graph convolutional network effectively learns the vector representation of the agents in MAS, making the efficiency and accuracy of centralized training higher than other algorithms. Our experiments also show that the MGAN algorithm is superior in performance to the baseline algorithms, especially in the scenarios of a large number of agents.

Contribution

- We propose MGAN, a multi-agent reinforcement learning algorithm that combines graph convolutional networks and value-decomposition methods. The graph network is used to make full use of the topological structure between agents, thereby increasing the speed of training.
- The graph networks can learn the vector representation of each agent in the embedding space. By visualizing these vectors, we can intuitively understand that all agents are divided into several groups at each step, thereby improving interpretability for the agents' behaviors.
- We demonstrate through experiments that the proposed algorithm is comparable to the baseline algorithms in the SMAC environment. In some scenarios with a large number of agents, MGAN significantly outperforms previous state-of-the-art methods.

2   Background

2.1   Dec-POMDP

A fully cooperative multi-agent task can be modeled as a decentralized partially observable Markov decision process (Dec-POMDP) [11] in which each agent only takes a local observation of the environment. A typical Dec-POMDP can be defined by a tuple $G = \langle \mathcal{S}, \mathcal{U}, \mathcal{P}, \mathcal{Z}, r, \mathcal{O}, n, \gamma \rangle$. $s \in \mathcal{S}$ is the global state of the environment. At each timestep, every agent $a \in \mathcal{A} := \{1, ..., n\}$ will choose an individual action $u_a \in \mathcal{U}$. The joint action takes the form of $\boldsymbol{u} \in \boldsymbol{\mathcal{U}} \equiv \mathcal{U}^n$. $\mathcal{P}$ denotes the state transition function. All the agents in Dec-POMDP share the same global reward function $r(s, u) : \mathcal{S} \times \boldsymbol{\mathcal{U}} \to \mathbb{R}$. According to the observation function $\mathcal{O}(s, a) : \mathcal{S} \times \mathcal{A} \to \mathcal{Z}$, each agent a gets local individual partial observation $z \in \mathcal{Z}$. $\gamma \in [0, 1)$ is the discount factor.

In Dec-POMDP, each agent a has its own action-observation history $\tau_a \in T \equiv (\mathcal{Z} \times \mathcal{U})$. The policy of each agent a can be written as $\pi_a(u_a | \tau_a) : T \times \mathcal{U} \to [0, 1]$. Our aim is to maximize the discounted return $R^t = \sum_{l=0}^{\infty} \gamma^l r_{t+l}$. The joint action-value function can be computed by the following equation: $Q^\pi(s_t, \boldsymbol{u}_t) = \mathbb{E}_{s_{t+1:\infty}, \boldsymbol{u}_{t+1:\infty}} [R_t | s_t, \boldsymbol{u}_t]$, where $\boldsymbol{\pi}$ is the joint policy of all agents.

2.2   Value-Decomposition Multi-agent RL

In the cooperative multi-agent reinforcement learning problem, one of the most basic solutions is to learn action-value function of each agent independently. It's more related to the individual agent's observations. However, previous studies indicate that this method is often very unstable and it is very difficult to design an efficient reward function. By contrast, learning the overall joint reward function is the other extreme. A key limitation of this method is that the problem of "lazy agents" often occurs, i.e., only one agent active and the other being "lazy".

To solve this issue, many researchers have proposed various methods lying between the extremes of independent Q-learning and centralized Q-learning, such as VDN, QMIX and QTRAN, which try to achieve automated learning decomposition of joint value function by the CTDE method. These value-decomposition methods are based on the Individual-Global-Max (IGM) [16] assumption that the optimality of each agent is consistent with the optimality of all agents. The equation that describes IGM is as follows:

$$\arg \max_{\boldsymbol{u}} Q_{\text{tot}}(\boldsymbol{\tau}, \boldsymbol{u}) = \begin{pmatrix} \arg \max_{u_1} Q_1(\tau_1, u_1) \\ \vdots \\ \arg \max_{u_n} Q_n(\tau_n, u_n) \end{pmatrix},$$

where Q_{tot} is global action-value function and Q_a is the individual ones.

VDN assumes that the joint value function is linearly decomposable. Each agent learns the additive value function independently. VDN aims to learn the optimal linear value decomposition from the joint action-value function to reflect the value function of each agent. The sum Q_{tot} of all individual value functions is given by

$$Q_{tot}(s, u_a) = \sum_{a=1}^{n} Q_a(s, u_a).$$

By this method, spurious rewards can be avoided and training is easier for each agent. However, because the additivity assumption used by VDN is too simple and there are only few applicable scenarios, a nonlinear global value function is proposed in QMIX. QMIX introduces a new type of value function module named mixing network. In order to satisfy the IGM assumption, it is assumed that the joint action-value function Q_{tot} is monotonic to the individual action-value function Q_a:

$$\frac{\partial Q_{tot}(\boldsymbol{\tau}, \boldsymbol{u})}{\partial Q_a(\tau_a, u_a)} \geq 0, \quad \forall a \in \{1, \dots, n\}.$$

Furthermore, QTRAN uses a new approach that can relax the assumption. However, several studies have indicated that the actual performance of the QTRAN is not very good because of its relaxation.

2.3 Graph Convolutional Networks

Convolutional graph neural network, as a kind of graph neural network, is often used to process data of molecules, social, biological, and financial networks. Convolutional graph neural networks fall into two categories, spectral-based

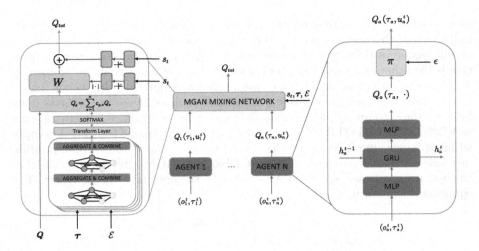

Fig. 1. The overall architecture of MGAN.

and spatial-based. Spectral-based methods analyze data from the perspective of graph signal processing. The spatial-based convolutional graph neural network processes the data of graph by means of information propagation. The emergence of graph convolutional network has well unified these two methods.

Let $G = (V, E)$ be a graph. Each node $v \in V$ in the graph has its own feature, which is denoted as $h_v^{(0)}$. Assuming that a graph convolutional network has a K-layers structure, then the hidden output of the k-th layer of the node v is updated as follows:

$$a_v^{(k)} = AGGREGATE^{(k)}(\{h_u^{(k-1)} | u \in \mathcal{N}(v)\}),$$
$$h_v^{(k)} = COMBINE^{(k)}(a_v^{(k)}, h_v^{(k-1)}), \tag{1}$$

where $COMBINE$ is often a 1-layer MLP, and $\mathcal{N}$ is the neighborhood function to get immediate neighbor nodes. Each node $v \in V$ aggregates the representations of the nodes in its immediate neighborhood to get a new vector representation. With the introduction of different $AGGREGATE$ functions, various variants of the graph convolutional network have obtained desired results on some datasets. For example, in addition to the most common mean aggregators, Graph Attention Network (GAT) [23] uses attention aggregators and Graph Isomorphism Network (GIN) [24] uses sum aggregators, both of which have achieved better results.

3   MGAN

In this section, we will propose a new method called MGAN. By constructing multiple graph convolutional networks at the same time, each graph convolutional network has its own unique insights into the graphs composed of agents.

This algorithm can not only make full use of the information of each agent and the connections between agents, but also improve the robustness of the performance.

3.1 Embedding Generation via Graph Networks

First, we need to construct all agents as a graph $G = (V, E)$, where each agent a can be seen as a node in the graph $v \in V$, i.e., agent a and node v has a one-to-one correspondence. We define the neighborhood function $\mathcal{N}$ to get the immediate neighbor nodes of the specified node. The edge e_{uv} between any two nodes in the graph is defined as:

$$e_{uv} = \begin{cases} 1, & \text{if } u \in \mathcal{N}(v) \text{ or } v \in \mathcal{N}(u) \\ 0, & \text{otherwise} \end{cases} \tag{2}$$

and according to this definition, we get the adjacency matrix $\mathcal{E} \in \mathbb{R}^{n \times n}$. In the reinforcement learning tasks, the adjacency matrix often indicates whether the agents are visible or whether they can communicate with each other. Each node v has its own feature h_v.

Then we build a two-layer graph convolutional network to learn the embedding vector of each agent. To build a graph convolutional network, we need to define the $AGGREGATE$ and $COMBINE$ functions mentioned by Eq. (1). Considering the actual situation, agents often need to pay special attention to a few of all other agents in the real tasks. So mean aggregators are often not qualified for this task. We adopted a simplified dot-product attention mechanism to solve this problem. The vector a_v obtained by the node v through the attention aggregate function can be expressed as:

$$a_v = AGGREGATE(\{h_u | u \in \mathcal{N}(v)\})$$
$$= \sum_{u \in \mathcal{N}(v)} \frac{\exp((h_v)^T \cdot (h_u))}{\sum_u \exp((h_v)^T \cdot (h_u))} \cdot h_u.$$

Then a_v needs to be entered into the $COMBINE$ function. It can be clearly seen that the embedding vectors obtained after the $AGGREGATE$ function processing loses the original characteristics of the node itself, i.e., the feature of the node is over smooth, and the characteristic information of the node itself is lacking. Therefore, we define the next layer's representation h'_v of the node v i.e. output by the $COMBINE$ function as:

$$h'_v = COMBINE(a_v) = ReLU\left(MLP\left(CONCAT(a_v, h_v)\right)\right)$$

This step completes the nonlinear transformation of the features obtained after the node v aggregates its neighbor nodes. Note that the MLP in the $COMBINE$ function of each layer is shared for each node. Similar to the simplified JK-Net [25], the original feature h_v is concatenated with the aggregate feature to ensure that the original node information will not be lost. From another perspective, this is very similar to ResNet [8].

3.2 MGAN Mixing Network

Each agent corresponds to a DRQN [7] network to learn individual action-value Q_a, where $a \in \{1, \ldots, n\}$. We have defined the graph convolutional network used to obtain the embedding vector of the agent, and then we will explain how to construct the network fitting joint action value function Q_{tot}. The embedding vector obtained through graph convolutional network is input into a fully connected layer neural network, which we call a transform layer, so that the embedding vector of each node v is transformed into a scalar c_v through affine transformation. The joint action-value function obtained by this graph convolutional network can be obtained by the following equation:

$$\sum_{a=1}^{n} \left(Q_a \cdot \frac{\exp(c_a)}{\sum_{v \in V} \exp(c_v)} \right),$$

which connects the vectors output by the graph networks with the individual action-values through dot multiplication.

Inspired by the multi-head attention mechanism, we propose to use multiple graph convolutional networks to jointly learn the embedding representation of nodes. Multiple graphs allow the model to jointly attend to information from different embedding spaces. Multiple graph convolutional networks share a transform layer. We set the number of graph convolutional networks to G. Thus, the following equation of the value function corresponding to each graph convolutional network is obtained:

$$Q_g = \sum_{a=1}^{n} \left(Q_a \cdot \frac{\exp(c_{g,a})}{\sum_{v \in V} \exp(c_{g,v})} \right), \quad \forall g \in \{1, \ldots, G\}.$$

where $c_{g,v}$ is the scalar output by the v-th node in the g-th graph convolutional network after the transform layer.

VDN obtains the global action-value by simply summing the individual action-values of all agents. And QMIX uses multiple hypernetworks [5], inputs state s, and outputs network weight parameters to construct a Mixing Network. It should be noted that in order to satisfy the monotonicity assumption proposed by QMIX, the network weight parameters output by hypernetworks are all positive. Our weighted linear factorization lies between the two and has a stronger representational capability for the joint value function than VDN while keeping a linear decomposition structure. This is because we only use hypernetworks to generate a layer of mixing network to linearly combine multiple Q_g. The entire network framework of the MGAN algorithm is shown in the Fig. 1.

3.3 Loss Function

MGAN is the same as other recently proposed MARL algorithms in that they are all trained end-to-end. The loss function is set to TD-error, which is the same as the traditional value-based reinforcement learning algorithm [19]. We denote

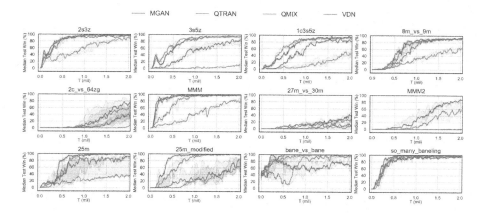

Fig. 2. Overall results in different scenarios.

the parameters of all neural networks as θ and MGAN is trained by minimizing the following loss function:

$$\mathcal{L}(\theta) = (y_{tot} - Q_{tot}(\boldsymbol{\tau}, \boldsymbol{u}|\theta))^2,$$

where y_{tot} is the target joint action-value function and $y_{tot} = r + \gamma \max_{\boldsymbol{u}'} Q_{tot} (\boldsymbol{\tau}', \boldsymbol{u}'|\theta^-)$. θ^- are the parameters of the target network.

4   Experiments

In this section we will evaluate MGAN and other baselines in the Starcraft II decentralized micromanagement tasks. In addition, to illustrate the representation learning capacity of the graph networks, the visualization of the output of the graph network was performed. We can intuitively understand the motivation of the agents' decision from the output of the graph neural network.

4.1   Settings

We use SMAC as the testbed because SMAC is a real-time simulation experiment environment based on Starcraft II. It contains a wealth of micromanagement tasks with varying levels of difficulty. Recently, it has gradually become an important platform for evaluating the coordination capabilities of agents. The scenarios in SMAC include challenges such as asymmetric, heterogeneous, and a large number of agents. We selected more representative scenarios such as *1c3s5z, 3s5z, 2c_vs_64zg, MMM2, bane_vs_bane* and so on. Besides, in order to be able to more conveniently show MGAN's understanding of the agent in decision-making, we have also introduced a new scenario *25m_modified*, which is modified on the basis of the *25m* scenario. The distribution of agents in the *25m_modified* scenario is more dispersed, which makes collaboration more difficult than the original *25m* scenario.

Our experiment is based on Pymarl [15]. We set the hyperparameters of QMIX and VDN to the default in Pymarl. The version of the Starcraft II is 4.6.2(B69232) in our experiments. The feature of each node in the graph network is initialized as its local observation in our proposed MGAN. And according to Eq. (2), the adjacency matrix $\mathcal{E}$ is given by:

$$e_{uv} = \begin{cases} 1, & \text{if } u \text{ is alive and } v \text{ is alive} \\ 0, & \text{otherwise} \end{cases} \qquad \forall e_{uv} \in \mathcal{E}.$$

The number of graph networks G is set to 4, and the other settings are the same as those of other baselines. We run each experiment 5 times independently to alleviate the effects of accidents and outliers. Depending on the complexity of the experimental scenario, the duration of each experiment ranges from 5 to 14 h. Experiments are carried out on Nvidia GeForce RTX 3090 graphics cards and Intel(R) Xeon(R) Platinum 8280 CPU. The model is evaluated every 10,000 steps in the experiment, i.e., 32 episodes are run and the win rate is recorded. The agents follow a completely greedy strategy during evaluation.

4.2 Validation

Figure 2 shows the performance results of MGAN and other baselines in different scenarios. The solid line represents the median win ratio of the five experiments. The 25–75% percentiles of the win ratios are shaded. It can be observed that in some scenarios with a large number of agents, MGAN far exceeds other algorithms in performance. Especially in *bane_vs_bane*, MGAN quickly reached convergence. In other scenarios, MGAN is still comparable to other popular algorithms.

As follows from Fig. 2 shown above, it can be seen intuitively that MGAN performs well in hard and super hard scenarios such as *MMM2*, *bane_vs_bane* and *27m_vs_30m*.

4.3 Graph Embedding and Weight Analysis

In order to understand the working principle of MGAN and explore the reasons for its effect improvement, we visualized the embedding vectors output by the graph network and the scalar weights output by the transform layer. We think these two provide an explanatory basis for the agents' actions.

We choose the *25m* and its variant *25m_modified* scenario with a large number of agents, and show the positions of the agents at each step in the task as a scatter diagram. Meanwhile, t-SNE [10] and MeanShift [2] clustering methods are performed on the graph embedding vector corresponding to each agent in each step, and the corresponding relationship between the position of the agent and the clustering result can be clearly found. This is illustrated in Fig. 3.

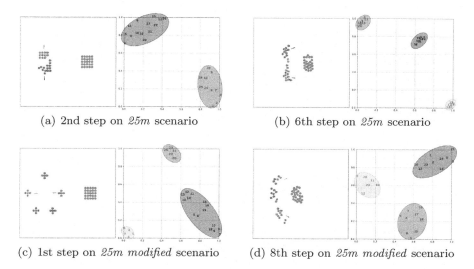

(a) 2nd step on *25m* scenario   (b) 6th step on *25m* scenario

(c) 1st step on *25m modified* scenario   (d) 8th step on *25m modified* scenario

Fig. 3. The agents location map at specific step (left) and the corresponding 2D t-SNE embedding of agents' internal states output by one of graph convolutional networks (right). Gray dots in location map represent the enemy agents and color dots denote the agents controlled by MGAN. Each number in 2D t-SNE embedding corresponds to each color dot in the location map one by one. (Color figure online)

In the *25m* scenario, the key to victory is that our agents can form an arc that surrounds the enemy agents. At the beginning of the episode, all agents gathered together. From the results of dimensionality reduction and clustering of embedding vectors, it can be found that the agents are divided into two groups, one group moves upward and the other moves downward. In the middle of the episode, in order to form a relatively concentrated line of fire, the agents was divided into three parts and moved in three directions respectively. In the *25m_modified* scenario, the agents also need to form the same arc, so the leftmost group of agents needs to move to the right, and the rightmost group of agents needs to move to the left to rendezvous with other agents. And in the middle of the episode, it will still be divided into three parts similar to the *25m* scenario. The finding was quite surprising and suggests that agents in the same subgroup can act together.

For the visualization of the weights, we still use the *25m* scenario for verification. The figure shows the change in the health values of the agents in an episode and the change in the weights of each agent corresponding to the four graph networks. As can be seen from Fig. 4, although the values of the weights given by each graph network is not the same, they all have a relationship with the health values of the agents. For example, Graph network 1 believes that agents with drastic changes in health values are the most important ones, while Graph network 2 believes that agents with more health values are the most important. On the contrary, Graph network 3 and Graph network 4 pay more attention to

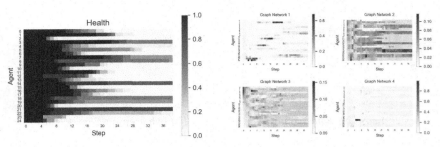

(a) The health values in one episode   (b) The weight values in one episode

Fig. 4. The health values and the weight values on *25m* scenario.

agents whose health values are zero. We guess that this is because these agents cause harm to the enemy and therefore pay more attention.

Through the analysis, we have concluded that the graph network can learn the characteristics of each agent well, and this provides basis for our understanding of the actions of the agents, which improves the interpretability of the motivation of the agents.

5   Conclusion

In this paper, we propose a MARL algorithm called MGAN that combines graph network and value-decomposition. From the outcome of our experiments it is possible to conclude that MGAN is comparable to the common baseline, especially in scenarios with a large number of agents. The figures obtained by visualization indicate that the performance improvement is brought about by the graph networks. The findings suggest that this method could also be useful for the works to understand how agents make decisions and what roles they play.

Since MGAN still needs to satisfy the IGM assumption, in our future research we intend to concentrate on how to relax the restrictions of the mixing networks. On the basis of the promising findings presented in this paper, work on the remaining issues is continuing and will be presented in future papers.

References

1. Böhmer, W., Kurin, V., Whiteson, S.: Deep coordination graphs. arXiv arXiv:1910.00091 (2020)
2. Comaniciu, D., Meer, P.: Mean shift: a robust approach toward feature space analysis. IEEE Trans. Pattern Anal. Mach. Intell. **24**, 603–619 (2002)
3. Defferrard, M., Bresson, X., Vandergheynst, P.: Convolutional neural networks on graphs with fast localized spectral filtering. In: NIPS (2016)
4. Foerster, J.N., Farquhar, G., Afouras, T., Nardelli, N., Whiteson, S.: Counterfactual multi-agent policy gradients. In: AAAI (2018)
5. Ha, D., Dai, A.M., Le, Q.V.: Hypernetworks. arXiv arXiv:1609.09106 (2017)

6. Hamilton, W.L., Ying, Z., Leskovec, J.: Inductive representation learning on large graphs. In: NIPS (2017)
7. Hausknecht, M.J., Stone, P.: Deep recurrent Q-learning for partially observable MDPs. In: AAAI Fall Symposia (2015)
8. He, K., Zhang, X., Ren, S., Sun, J.: Deep residual learning for image recognition. In: 2016 IEEE Conference on Computer Vision and Pattern Recognition (CVPR), pp. 770–778 (2016)
9. Lowe, R., Wu, Y., Tamar, A., Harb, J., Abbeel, P., Mordatch, I.: Multi-agent actor-critic for mixed cooperative-competitive environments. In: NIPS (2017)
10. Maaten, L.V.D., Hinton, G.E.: Visualizing data using t-SNE. J. Mach. Learn. Res. **9**, 2579–2605 (2008)
11. Oliehoek, F.A., Amato, C.: A Concise Introduction to Decentralized POMDPs. SpringerBriefs in Intelligent Systems. Springer, Cham (2016). https://doi.org/10.1007/978-3-319-28929-8
12. Peng, P., et al.: Multiagent Bidirectionally-Coordinated Nets: emergence of human-level coordination in learning to play StarCraft combat games. arXiv: Artificial Intelligence (2017)
13. Perozzi, B., Al-Rfou, R., Skiena, S.: DeepWalk: online learning of social representations. In: Proceedings of the 20th ACM SIGKDD International Conference on Knowledge Discovery and Data Mining (2014)
14. Rashid, T., Samvelyan, M., Witt, C.S., Farquhar, G., Foerster, J.N., Whiteson, S.: QMIX: monotonic value function factorisation for deep multi-agent reinforcement learning. arXiv arXiv:1803.11485 (2018)
15. Samvelyan, M., et al.: The StarCraft multi-agent challenge. arXiv arXiv:1902.04043 (2019)
16. Son, K., Kim, D., Kang, W., Hostallero, D., Yi, Y.: QTRAN: learning to factorize with transformation for cooperative multi-agent reinforcement learning. arXiv arXiv:1905.05408 (2019)
17. Sukhbaatar, S., Szlam, A., Fergus, R.: Learning multiagent communication with backpropagation. In: NIPS (2016)
18. Sunehag, P., et al.: Value-decomposition networks for cooperative multi-agent learning. arXiv arXiv:1706.05296 (2018)
19. Sutton, R., Barto, A.: Reinforcement learning: an introduction. IEEE Trans. Neural Netw. **16**, 285–286 (2005)
20. Tampuu, A., et al.: Multiagent cooperation and competition with deep reinforcement learning. PLOS ONE **12**, e0172395 (2017)
21. Thekumparampil, K.K., Wang, C., Oh, S., Li, L.: Attention-based graph neural network for semi-supervised learning. arXiv arXiv:1803.03735 (2018)
22. Vaswani, A., et al.: Attention is all you need. arXiv arXiv:1706.03762 (2017)
23. Velickovic, P., Cucurull, G., Casanova, A., Romero, A., Liò, P., Bengio, Y.: Graph attention networks. arXiv arXiv:1710.10903 (2018)
24. Xu, K., Hu, W., Leskovec, J., Jegelka, S.: How powerful are graph neural networks? arXiv arXiv:1810.00826 (2019)
25. Xu, K., Li, C., Tian, Y., Sonobe, T., Kawarabayashi, K., Jegelka, S.: Representation learning on graphs with jumping knowledge networks. In: ICML (2018)

A Reinforcement Learning Approach for Abductive Natural Language Generation

Hongru Huang[✉]

Shanghai Jiao Tong University, Shanghai, China
onedesire@sjtu.edu.cn

Abstract. Teaching deep learning models commonsense knowledge is a crucial yet challenging step towards building human-level artificial intelligence. Abductive Commonsense Reasoning ($\mathcal{ART}$) is a benchmark that investigates model's ability on inferencing the most plausible explanation within the given context, which requires model using commonsense knowledge about the world. $\mathcal{ART}$ consists of two datasets, αNLG and αNLI, that challenge models from *generative* and *discriminative* settings respectively. Despite the fact that both of the datasets investigate the same ability, existing work solves them independently. In this work, we address αNLG in a teacher-student setting by getting help from another model with adequate commonsense knowledge fully-trained on αNLI. We fulfill this intuition by representing the desired optimal generation model as an Energy-Based Model and training it using a reinforcement learning algorithm. Experiment results showed that our model achieve state-of-the-art results on both automatic and human evaluation metrics, which have demonstrated the effectiveness and feasibility of our model (Code available in https://github.com/Huanghongru/commonsense-generation).

Keywords: Natural language generation · Reinforcement learning · Commonsense knowledge

1 Introduction

Abductive reasoning is inference to the most plausible explanation for incomplete observations, which is a critical ability that a true artificial intelligence should have. To access such an ability of an artificial intelligence model, two benchmarks are proposed [3] One is αNLG (short for Abductive Natural Language Generation), a conditional text generation task that assesses the generation ability of models on abductively inferencing the most plausible explanation given the context. An illustration example is shown in Fig. 1. Given the observation O_1 at time t_1 and observation O_2 at time $t_2 > t_1$, the model needs to generate a plausible hypothesis H that can consistently explain both of the observations. The other one is αNLI (short for Abductive Natural Language Inference), which

© Springer Nature Switzerland AG 2021
T. Mantoro et al. (Eds.): ICONIP 2021, LNCS 13110, pp. 64–75, 2021.
https://doi.org/10.1007/978-3-030-92238-2_6

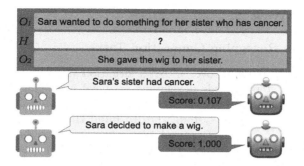

Fig. 1. An illustration example of our intuition in αNLG dataset. Given the two observations O_1 and O_2, the generation model (on the left) need to compose a hypothesis that can consistently connect and explain both of the two observations. The teacher model (on the right) evaluates and gives a score to the generated hypothesis about how coherent and consistent it is, thereby providing a explicit supervised signal for further training of the generative model.

poses the same challenge as αNLG to a model, except that the task is framed in a *discriminative* setting: Given two observations O_1 and O_2 and two possible explanation hypotheses H_1 and H_2, a model is expected to choose the most plausible one explanation. In this work, we focus on αNLG task because we believe it is more challenging in the sense that the model need to generate hypothesis from scratch, while in αNLI the model just need to choose the feasible one from those written by human.

To address αNLG, Existing works [8,21,24,25] resorted to external knowledge source or proposed novel decoding algorithm. However, all of them solve αNLG without explicitly evaluating how well the generated hypothesis fit into the given context, thus solving the task in deductive way instead of abductive way. What's more, they ignored the strong underlying relatedness between αNLG and αNLI and solve them independently. Given the fact that both αNLG and αNLI investigate the same inference ability of models, we propose to boost the NLG performance by taking advantage of NLI models. With such a well-trained NLI model, the underlying objective of this task, i.e., the consistency and coherency of the generated hypothesis, can be using as a training target directly and thus solving the task in truly abductive way.

To fulfill this intuition, we design our model under a teacher-student setting with reinforcement learning algorithm. We firstly train a simple but good enough αNLI model as the teacher model by fine-tuning a commonly-used pre-trained language model, such as BERT or RoBERTa [5,15]. Then we represent the desired optimal αNLG model as an Energy-Based Model [4,7,12] with the teacher model and a base generation model [13,22], which can be regarded as the reward model for reinforcement learning. Finally, we train to reach the optimal αNLG model by the KL-Adaptive Distributional Gradient Policy algorithm. The best of our model outperforms the strong baseline models in varied evaluation metrics including both automatic and human evaluation, which demonstrated the effectiveness and feasibility of our proposed method.

2   Related Work

2.1   Abductive Reasoning Generation Tasks

Teaching a model reasoning over a given situation described in natural language is a critical step towards developing artificial intelligence. [27] It would be beneficial to many downstream natural language generation tasks including summarization and dialogue system. [17] To fulfill this intuition, Bhagavatula et al. proposed αNLG, a new natural language generation benchmark where models need to write down feasible hypotheses that can consistently explain the two given observations [3]. Recent years have witnessed many works focusing on addressing αNLG. Ji et al. [8] resorted to external knowledge graphs like conceptnet [25] and performed reasoning over the extracted subgraph according to the given context using graph neural network (GNN). They finally combined the hidden state of GNN and language model for output generation. Qin et al. [21] addressed this task in unsupervised way by dynamically updating the hidden state of language model. However, all of these works didn't explicitly evaluate how well the generated hypothesis fit into the given context, and thus solving the task in deductive way instead of abductive way.

2.2   Reinforcement Learning in Natural Language Generation

Generally, a supervised training for natural language generation tasks is to learn a language model under the distribution of reference texts. In this procedure, the underlying target of the text generation task (e.g.: the consistency and correctness in αNLG) cannot be explicitly learned. [16] To directly grant models such kind of ability, many works try to model the underlying target as reward using reinforcement learning algorithms. Razoto et al. [23] used automatic metrics like BLEU or ROUGE-2 as training signal by applying REINFORCE [29] for sequence level training. Tambwekar et al. [26] presented a reward-shaping technique that analyzed a story corpus and produced intermediate reward for story plot generation task. Bahdanau et al. [1] proposed a heuristic reward for abstractive summarization. Khalifa et al. [9] combined the language model and reward model together as Energy-Based Model (EBM) [7,12,23], which provides more flexibility to modeling complicated underlying target. Inspired by these work, we explicitly evaluate the consistency of the generated hypothesis and use it as a training signal using reinforcement learning algorithm.

3   Methodology

In this section, we firstly present the problem formulation of abductive natural language generation tasks. Then we split the entire training procedure into three stages (i.e., initial LM training, reward model training and desired LM training) and introduce them one by one. An overview of the training stages is shown in Fig. 2.

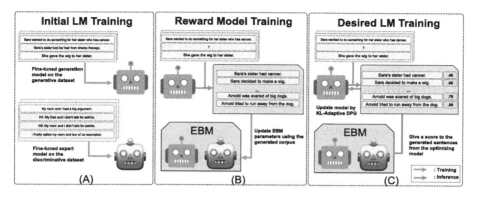

Fig. 2. The training stages overview of our methods. (**A**): We firstly train the initial LM model $a(\boldsymbol{x})$ on αNLG and a scorer model $\phi(\boldsymbol{x})$ on αNLI. (**B**): We sample sentences from a as a corpus to train the EBM. (**C**): The trained EBM scores each sentence generated by the policy model q and instructs it for better generation. We implement this idea by formalizing the task in reinforcement learning setting and using the KL-Adaptive DPG algorithm.

3.1   Problem Formulation

Given two observations O_1 and O_2 happening at t_1 and t_2, where $t_2 > t_1$, the task is to generate a valid hypothesis $\boldsymbol{x} = (x_1, x_2, ..., x_M)$ that can explain these two observations. In that case, our ultimate objective is to find a language model P (i.e., a distribution) over the benchmark corpus $\mathbb{X}$ s.t.:

$$\hat{\boldsymbol{x}} = \arg\max_{\boldsymbol{x}} P(\boldsymbol{x}|O_1, O_2) \tag{1}$$

3.2   Initial LM Training

At first, we need a language model that can write coherent and grammarly correct sentences conditioned on some given context. For this purpose, we fine-tune a language model pretrained on large corpus [13,22] on αNLG as the initial LM model $a(\boldsymbol{x}|O_1, O_2)$[1]. Formally, we concatenate the observations and golden hypothesis by the $[SEP]$ token as $(o_1^1, ..., o_m^1, o_1^2, ..., o_n^2, [SEP], x_1, ..., x_M)$. The training object is the commonly-used cross entropy loss. The initial LM would be used for the following reinforcement learning algorithm and would be abled to mitigate the difficulty of convergence in reinforcement learning [30].

To explicitly evaluate how well the generated hypothesis fit into the given context, we train a scorer model $\phi(\boldsymbol{x}|O_1, O_2)$[2] on αNLI dataset as a binary classification task. Concretely, we concatenate the generated hypotheses $\boldsymbol{x}$ and the given observations O_1 and O_2 as a whole story $([CLS], o_1^1, ..., o_m^1, x_1, ..., x_M, o_1^2, ..., o_n^2)$ following the chronological timeline. Then we feed the story into a fine-tuning

---

[1] We use $a(\boldsymbol{x})$ for short in the rest of the paper.
[2] We use $\phi(\boldsymbol{x})$ for short in the rest of the paper.

large pretrained models [5,15], the output logits of the $[CLS]$ would be projected into a real value between $[0,1]$ as the score for hypothesis $\boldsymbol{x}$. The higher $\phi(x)$ is, the better $\boldsymbol{x}$ fits into the given context as shown in Fig. 1.

3.3 Reward Model Training

Having the initial LM $a(x)$ and the scorer model $\phi(x)$, we define the reward model for reinforcement learning as:

$$\hat{P}(\boldsymbol{x}) = a(\boldsymbol{x})e^{\lambda \cdot \phi(\boldsymbol{x})} \tag{2}$$

where $\boldsymbol{\lambda}$ is a parameter s.t. the expectation of score of generated hypothesis can reach a desired threshold:

$$\mathbb{E}_{\boldsymbol{x} \sim p}\phi(\boldsymbol{x}) \simeq \bar{\mu} \tag{3}$$

The reward model in Eq. 2 is called Energy-Based Model (EBM). It was an abstract model proposed by LeCun et al. [12] and Hinton et al. [7] borrowing from the concepts of thermodynamics. In recent years, EBM attracted more and more interest in reinforcement learning area [4,9,19]. Basically, the merits of using EBM as reward model are two folds. (1) It can take the quality of language model and consistency of generated hypothesis into consideration at the same time. (2) It brings flexibility to use advanced discriminative model(i.e., $\phi(x)$ in our case) compared with traditional reward model using BLEU as heuristic reward.

To train the reward model, we need to find the parameter $\boldsymbol{\lambda}$. We cannot directly sample $\boldsymbol{x}$ from $P(\boldsymbol{x})$ because $P(\boldsymbol{x})$ is not an auto-regressive language model. Therefore we follow [9] and use Self Normalized Importance Sampling(SNIS) [10,18,20] to estimate $\bar{\mu}$ by sampling a large number $\mathcal{N}$ of sequences $\boldsymbol{x}_1, ..., \boldsymbol{x}_i, ..., \boldsymbol{x}_{\mathcal{N}}$ from the initial model a:

$$\hat{\boldsymbol{\mu}}(\boldsymbol{\lambda}) = \frac{\sum_{i=1}^{\mathcal{N}} w_i(\boldsymbol{\lambda})\phi(x_i)}{\sum_{i=1}^{\mathcal{N}} w_i(\boldsymbol{\lambda})} \tag{4}$$

where $w_i(\boldsymbol{\lambda}) = \frac{P(\boldsymbol{x}_i)}{a(\boldsymbol{x}_i)} = e^{\lambda \cdot \phi(\boldsymbol{x}_i)}$ is the *importance weights*. Finally we solve in $\boldsymbol{\lambda}$ by minimizing the L2-norm $||\bar{\mu} - \hat{\mu}(\boldsymbol{\lambda})||_2^2$ using Adam [11] optimizer until the L2-norm less than a tolerant value τ.

3.4 Desired LM Training

At the last stage, we apply the KL-Adaptive Distributional Policy Gradient (DPG) algorithm [9,20] to reach the optimal model P. The whole procedure is summarized in Algorithm 1. As we mentioned before, we can represent the optimal model as an EBM P. However, an EBM is not auto-regressive so that it cannot generate sentences. The KL-Adaptive DPG is a reinforcement learning algorithm whose objective is to obtain an auto-regressive policy π_θ

Algorithm 1. KL-Adaptive DPG

Input: EBM P, initial policy q, learning rate $\alpha^{(\theta)}$

1:   $\pi_\theta \leftarrow q$
2:   $Z_{ma} \leftarrow 0$ // Moving average estimate of Z
3:   **for** each iteration i **do**
4:      **for** each step $k \in [1, K]$ **do**
5:         sample x_k from $q(\cdot)$
6:         $\theta \leftarrow \theta + \alpha^{(\theta)} \frac{P(x_k)}{q(x_k)} \nabla_\theta \log \pi_\theta(x_k)$
7:      **end for**
8:      $\hat{Z}_i \leftarrow \frac{1}{K} \sum_{i=1}^{K} \frac{P(x_k)}{q(x_k)}$
9:      $Z_{ma} \leftarrow \frac{i*Z_{MA}+\hat{Z}_i}{i+1}$
10:     compute $D_{KL}(P||\pi_\theta)$ according to Eq. 5
11:     compute $D_{KL}(P||q)$ according to Eq. 5
12:     **if** $D_{KL}(p||\pi_\theta) < D_{KL}(p||q)$ **then**
13:        $q \leftarrow \pi_\theta$
14:     **end if**
15: **end for**

Output: the approximation π_θ

---

that approximates P by minimizing the cross entropy between P and π_θ:
$CE(p, \pi_\theta) = - \sum_x p(\boldsymbol{x}) \log \pi_\theta$

Basically, we start from the EBM $\hat{P}$ and a proxy language model q to P. We initialize q by the fine-tuned model a, which we can directly sample from. For each input data, we use q to generate up to k_{bs} samples using beam search. Intuitively, if a sample $\boldsymbol{x}_i$ fits its context very well, the expert model will give it a high score $\phi(\boldsymbol{x}_i)$. Hence the EBM $\hat{P}(\boldsymbol{x})$ is also high, the generation model will pay more attention to those well-generated samples.

After each episode, we update q by the optimized policy π_θ if the KL divergence $D_{KL}(P||\pi_\theta)$ is smaller than $D_{KL}(P||q)$. The KL divergence constraint is to avoid π_θ drifting too far from the initial model a [9,32]. We approximate the KL divergence and partition function Z also by importance sampling:

$$D_{KL}(p||q) = -\log Z + \frac{1}{Z} \mathbb{E}_{\boldsymbol{x} \sim q} \frac{P(\boldsymbol{x})}{q(\boldsymbol{x})} \log \frac{P(\boldsymbol{x})}{q(\boldsymbol{x})} \tag{5}$$

4   Experiment and Discussion

We firstly brievely introduce the implementation details of our model. For base generation model $a(\boldsymbol{x})$, we fine-tuned GPT2 [22] and BART-base [13] on αNLG. For teacher model $\phi(\boldsymbol{x})$ in EBM, we fine-tuned BERT-base [5] and RoBERTa-base [15] on αNLI. The tolerant value τ for EBM training is set to $1e-3$. In the KL-Adaptive DPG algorithm, we set $K = 20480$ for the estimation of $\hat{Z}_i$ and D_{KL}. The maximum training step is set to 512k. The beam search size k_{bs} is 4. We use common random seed 42, 52 and 62 for all of the experiments and take average for all of the metrics.

4.1 Automatic Evaluation

We use BLEU [19], ROUGE-L(R-L) [14], METEOR(MT) [2] and CIDEr (CD) [28] for automatic evaluation. Besides of the above metrics, we also evaluate the $\mathcal{ART}$ score of the generation texts, which should be a natural and coherent evaluation metric but is ignored in previous work. We use top-tier models on the αNLI leaderboard[3] to score the generated texts. Since we have used RoBERTa as teacher model in the EBM, we use two more strong αNLI model as scorer for fair comparison. One is DeBERTa [6], a recently published large pre-trained language model reaching to top-1 on varied of benchmarks. Another one is $L2R^2$ [31], which solved αNLI by leveraging ranking loss.

Table 1. †: Baseline Models and results from [3]. ◇: We feed reference hypothesis to $\phi(x)$ to obtain human performance as upper bound. ‡: The base generation model we fine-tuned on αNLG. **Bold face**: highest score. <u>Underline</u>: second highest score.

	BLEU-4	R-L	MT	CD	$\mathcal{ART}$ score		
					RoBERTa	$L2R^2$	DeBERTa
GPT2-fixed†	2.23	22.83	16.71	33.54	42.32	40.19	41.38
GPT2+COMeT txt†	2.29	22.51	16.73	31.99	48.22	42.56	48.53
GPT2+COMeT emb†	3.03	22.93	17.66	32.00	49.34	42.64	49.64
GRF [8]	<u>11.62</u>	34.62	27.76	63.76	66.89	55.76	65.79
GPT2-FT‡	8.49	31.09	22.87	54.10	49.32	53.34	49.54
EBM (GPT2+BERT)	8.59	31.27	23.43	54.80	50.46	53.45	51.70
EBM (GPT2+RoBERTa)	8.86	31.63	23.95	55.25	55.39	54.06	54.77
BART-FT‡	**11.79**	35.57	**28.09**	**67.33**	66.12	<u>57.53</u>	64.98
EBM (BART+BERT)	11.60	**35.62**	27.93	<u>66.80</u>	<u>66.39</u>	57.50	<u>65.84</u>
EBM (BART+RoBERTa)	11.54	<u>35.58</u>	<u>28.00</u>	66.69	**70.01**	**57.77**	**67.08**
Human◇	–	–	–	–	77.31	64.04	79.80

We show the experiment results on the αNLG test set in Table 1. We observe that compared with the initial generation model GPT2 and BART-base, our proposed EBM model has improved the $\mathcal{ART}$ score across all of the scorers, demonstrating the effectiveness of our methods. In particular, the improvement of using RoBERTa as teacher model in EBM more significant than using BERT. Compared with GRF [8], EBM (BART+RoBERTa) obtains a comparable score on the traditional automatic evaluation metrics, but outperforms it on $\mathcal{ART}$ score by 3.28.

4.2 Learning Curves

We plot the learning curve in Fig. 3. We evaluate π_θ on the validation set and compute the corresponding $\mathcal{ART}$ score $\phi(x)$. We find that the $\mathcal{ART}$

---

[3] https://leaderboard.allenai.org/anli/submissions/public.

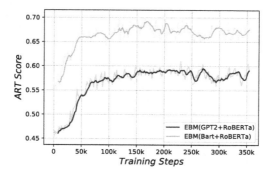

Fig. 3. Learning curve for the EBM on α*NLG*.

score of both GPT2-based and BART-based EBM increase stably during train-ing, demonstrating the effectiveness and feasibility of posing the consistency and coherency of generated hypothesis to training objective. With the same RoBERTa as the $\phi(x)$ scorer model, we find that BART-based generation model significantly outperformed GPT2-based. One possible reason is that BART has a BERT-based encoder, which can capture and encoder the underlying context bidirectionally for abudtive reasoning while GPT2 cannot.

4.3 Effect of the Teacher Model

In our methods, we expect the teacher model can give a reasonable and precise score to evaluate how well the generated text fit into the context. Hence the performance of the teacher model should be a critical point in our methods. From Table 1 we can find that a RoBERTa EBM is better than a BERT-based EBM. We fine-tuned RoBERTa on less training data of α*NLI* and report the improvement of $\mathcal{ART}$ *score* against the base generation model (‡) in Table 1. From Table 2 we can see that the improvements of $\mathcal{ART}$ *score* and the accuracy on α*NLI* dev set are positively related, which demonstrates that a better-trained α*NLI* model can bring more improvement.

4.4 Human Evaluation

We randomly sample 100 inputs from test set to investigate the reliability of the $\mathcal{ART}$ *scores*. Specifically, we ask 3 annotators to make a preference among *win*, *lose* and *tie* given the input observations and two outputs by our model and a baseline respectively, according to the criteria that whether the output sentence explains well the observations. Results are shown in right of Table 2. We can see that the EBM (GPT2+RoBERTa) obtains more *win* than GPT2-FT but less than GRF, while EBM (BART+RoBERTa) obtains more *win* than both baselines, which is consistent with the results in Table 1 and thus supports the reliability of $\mathcal{ART}$ *score*.

Table 2. Left: Improvement of $\mathcal{ART}$ *score* w.r.t. RoBERTa trained on different ratio of αNLI as teacher model. **Right**: Percentage of win (**W**), tie (**T**) and lose (**L**). †: Fleiss' Kappa for annotator agreement.

% of data	GPT2	BART	αNLI dev acc		EBM (GPT2 +RoBERTa)		EBM (BART +RoBERTa)	
					vs GPT2-FT	vs GRF	vs BART-FT	vs GRF
20%	+1.76	+1.12	68.05	W	0.163	0.236	0.203	0.310
40%	+2.20	+1.43	70.28	T	0.753	0.490	0.665	0.460
60%	+2.77	+2.02	72.43	L	0.083	0.277	0.132	0.230
80%	+4.55	+2.57	73.56	$\kappa^\dagger$	0.634	0.602	0.628	0.594
100%	+6.07	+3.89	74.98					

4.5 Showcase Analysis

O_1:	Tim wanted to learn astronomy	O_1:	Dominick used to hate school
O_2:	Tim work hard in school to become one	O_2:	He realized school was good for his brain
BART:	Tim work hard in school	BART:	Dominick went to the school
GRF:	Tim went to Kenya	GRF:	Dominick decided to be better
Ours:	Tim then dreamed of becoming an astronaut	**Ours:**	He went to the school to get better.
O_1:	Brad and Allison love Texas country music	O_1:	George took his niece to the bookstore
O_2:	They both had a good time at the concert	O_2:	George and his niece had a fun outgoing
BART:	They both went to the concert	BART:	They decided to go on outing
GRF:	Brad and Allison decided to go to a concert	GRF:	George and his niece went to the bookstore
Ours:	They both went to a country concert	**Ours:**	They found a funny book in the bookstore

Fig. 4. Randomly sampled examples from GRF [8] and our model (BART+RoBERTa).

We randomly sampled some concrete results in Fig. 4. We can see that the vanilla Bart model can only cover one of the observations and tend to repeat some of the content in the given context. This issue also happened to GRF [8] in some cases. By introducing teacher model $\phi(\boldsymbol{x})$ and training with reinforcement learning algorithm, our model can better take both observations into consideration.

5 Conclusion

We represent a new model with reinforcement learning approach for abductive natural language generation task αNLG. By this algorithm, we can explicitly evaluate how well the generated hypothesis fit into the given context in a teacher-student setting and thus truly solve this problem abductively. Both of automatic metric and human evaluation demonstrated the effectiveness and feasibility of our proposed methods. We also present detailed analysis and some concrete examples to show how our model outperforms state-of-the-art models. In the future, we will focus on extending our work to other language generation tasks that requires external commonsense knowledge.

References

1. Bahdanau, D., et al.: An actor-critic algorithm for sequence prediction. In: 5th International Conference on Learning Representations (Conference Track Proceedings), ICLR 2017, Toulon, France, 24–26 April 2017. OpenReview.net (2017). https://openreview.net/forum?id=SJDaqqveg
2. Banerjee, S., Lavie, A.: METEOR: an automatic metric for MT evaluation with improved correlation with human judgments. In: Proceedings of the ACL Workshop on Intrinsic and Extrinsic Evaluation Measures for Machine Translation and/or Summarization, Ann Arbor, Michigan, pp. 65–72. Association for Computational Linguistics (June 2005). https://www.aclweb.org/anthology/W05-0909
3. Bhagavatula, C., et al.: Abductive commonsense reasoning. OpenReview.net (2020). https://openreview.net/forum?id=Byg1v1HKDB
4. Deng, Y., Bakhtin, A., Ott, M., Szlam, A., Ranzato, M.: Residual energy-based models for text generation. In: 8th International Conference on Learning Representations, ICLR 2020, Addis Ababa, Ethiopia, 26–30 April 2020. OpenReview.net (2020). https://openreview.net/forum?id=B1l4SgHKDH
5. Devlin, J., Chang, M.W., Lee, K., Toutanova, K.: BERT: pre-training of deep bidirectional transformers for language understanding. In: Proceedings of the 2019 Conference of the North American Chapter of the Association for Computational Linguistics: Human Language Technologies, Volume 1 (Long and Short Papers), Minneapolis, Minnesota, pp. 4171–4186. Association for Computational Linguistics (June 2019). https://doi.org/10.18653/v1/N19-1423. https://www.aclweb.org/anthology/N19-1423
6. He, P., Liu, X., Gao, J., Chen, W.: DeBERTa: decoding-enhanced BERT with disentangled attention. CoRR abs/2006.03654 (2020). arXiv arXiv:2006.03654
7. Hinton, G.E.: Training products of experts by minimizing contrastive divergence. Neural Comput. 14(8), 1771–1800 (2002). https://doi.org/10.1162/089976602760128018
8. Ji, H., Ke, P., Huang, S., Wei, F., Zhu, X., Huang, M.: Language generation with multi-hop reasoning on commonsense knowledge graph. In: Proceedings of the 2020 Conference on Empirical Methods in Natural Language Processing (EMNLP), pp. 725–736. Association for Computational Linguistics (November 2020). https://doi.org/10.18653/v1/2020.emnlp-main.54. https://www.aclweb.org/anthology/2020.emnlp-main.54
9. Khalifa, M., Elsahar, H., Dymetman, M.: A distributional approach to controlled text generation. CoRR abs/2012.11635 (2020). arXiv arXiv:2012.11635
10. Kim, T., Bengio, Y.: Deep directed generative models with energy-based probability estimation. CoRR abs/1606.03439 (2016). arXiv arXiv:1606.03439
11. Kingma, D.P., Ba, J.: Adam: a method for stochastic optimization. In: Bengio, Y., LeCun, Y. (eds.) 3rd International Conference on Learning Representations (Conference Track Proceedings), ICLR 2015, San Diego, CA, USA, 7–9 May 2015 (2015). arXiv arXiv:1412.6980
12. LeCun, Y., Chopra, S., Hadsell, R., Huang, F.J., et al.: A tutorial on energy-based learning. In: Predicting Structured Data. MIT Press (2006)
13. Lewis, M., et al.: BART: denoising sequence-to-sequence pre-training for natural language generation, translation, and comprehension. In: Proceedings of the 58th Annual Meeting of the Association for Computational Linguistics, pp. 7871–7880. Association for Computational Linguistics (July 2020). https://doi.org/10.18653/v1/2020.acl-main.703. https://www.aclweb.org/anthology/2020.acl-main.703

14. Lin, C.Y.: ROUGE: a package for automatic evaluation of summaries. In: Text Summarization Branches Out, Barcelona, Spain, pp. 74–81. Association for Computational Linguistics (July 2004). https://www.aclweb.org/anthology/W04-1013

15. Liu, Y., et al: RoBERTa: a robustly optimized BERT pretraining approach. CoRR abs/1907.11692 (2019). arXiv arXiv:1907.11692

16. Luketina, J., et al.: A survey of reinforcement learning informed by natural language. In: Kraus, S. (ed.) Proceedings of the 28th International Joint Conference on Artificial Intelligence, IJCAI 2019, Macao, China, 10–16 August 2019, pp. 6309–6317. ijcai.org (2019). https://doi.org/10.24963/ijcai.2019/880

17. Moore, C.: The development of commonsense psychology (2006). https://doi.org/10.4324/9780203843246

18. Owen, A.B.: Monte Carlo theory, methods and examples (2013). https://statweb.stanford.edu/~owen/mc/Ch-var-is.pdf

19. Papineni, K., Roukos, S., Ward, T., Zhu, W.J.: BLEU: a method for automatic evaluation of machine translation. In: Proceedings of the 40th Annual Meeting of the Association for Computational Linguistics, Philadelphia, Pennsylvania, USA, pp. 311–318. Association for Computational Linguistics (July 2002). https://doi.org/10.3115/1073083.1073135. https://www.aclweb.org/anthology/P02-1040

20. Parshakova, T., Andreoli, J.M., Dymetman, M.: Global autoregressive models for data-efficient sequence learning. In: Proceedings of the 23rd Conference on Computational Natural Language Learning (CoNLL), Hong Kong, China, pp. 900–909. Association for Computational Linguistics (November 2019). https://doi.org/10.18653/v1/K19-1084. https://www.aclweb.org/anthology/K19-1084

21. Qin, L., et al.: Back to the future: unsupervised backprop-based decoding for counterfactual and abductive commonsense reasoning. In: Proceedings of the 2020 Conference on Empirical Methods in Natural Language Processing (EMNLP), pp. 794–805. Association for Computational Linguistics (November 2020). https://doi.org/10.18653/v1/2020.emnlp-main.58. https://www.aclweb.org/anthology/2020.emnlp-main.58

22. Radford, A., Wu, J., Child, R., Luan, D., Amodei, D., Sutskever, I.: Language models are unsupervised multitask learners (2018). https://d4mucfpksywv.cloudfront.net/better-language-models/language-models.pdf

23. Ranzato, M., Chopra, S., Auli, M., Zaremba, W.: Sequence level training with recurrent neural networks. In: Bengio, Y., LeCun, Y. (eds.) 4th International Conference on Learning Representations (Conference Track Proceedings), ICLR 2016, San Juan, Puerto Rico, 2–4 May 2016 (2016). arXiv arXiv:1511.06732

24. Sap, M., et al.: ATOMIC: an atlas of machine commonsense for if-then reasoning. In: The 33rd AAAI Conference on Artificial Intelligence, AAAI 2019, The 31st Innovative Applications of Artificial Intelligence Conference, IAAI 2019, The 9th AAAI Symposium on Educational Advances in Artificial Intelligence, EAAI 2019, Honolulu, Hawaii, USA, 27 January–1 February 2019, pp. 3027–3035. AAAI Press (2019). https://doi.org/10.1609/aaai.v33i01.33013027

25. Speer, R., Chin, J., Havasi, C.: ConceptNet 5.5: an open multilingual graph of general knowledge, pp. 4444–4451. AAAI Press (2017). http://aaai.org/ocs/index.php/AAAI/AAAI17/paper/view/14972

26. Tambwekar, P., Dhuliawala, M., Martin, L.J., Mehta, A., Harrison, B., Riedl, M.O.: Controllable neural story plot generation via reinforcement learning (2019)

27. Tincoff, R., Jusczyk, P.W.: Some beginnings of word comprehension in 6-month-olds. Psycholol. Sci. **10**(2), 172–175 (1999). https://doi.org/10.1111/1467-9280.00127

28. Vedantam, R., Zitnick, C.L., Parikh, D.: CIDEr: consensus-based image description evaluation. In: IEEE Conference on Computer Vision and Pattern Recognition, CVPR 2015, Boston, MA, USA, 7–12 June 2015, pp. 4566–4575. IEEE Computer Society (2015). https://doi.org/10.1109/CVPR.2015.7299087
29. Williams, R.J.: Simple statistical gradient-following algorithms for connectionist reinforcement learning. Mach. Learn. **8**, 229–256 (1992). https://doi.org/10.1007/BF00992696
30. Yang, Z., Xie, Y., Wang, Z.: A theoretical analysis of deep q-learning. CoRR abs/1901.00137 (2019). arXiv arXiv:1901.00137
31. Zhu, Y., Pang, L., Lan, Y., Cheng, X.: L2r^2: leveraging ranking for abductive reasoning, pp. 1961–1964. ACM (2020). https://doi.org/10.1145/3397271.3401332
32. Ziegler, D.M., et al.: Fine-tuning language models from human preferences. CoRR abs/1909.08593 (2019). arXiv arXiv:1909.08593

DFFCN: Dual Flow Fusion Convolutional Network for Micro Expression Recognition

Jinjie Chen[✉], Yuzhuo Fu, YiBo Jin, and Ting Liu

Shanghai Jiao Tong University, Shanghai, China
{cjj82173690,louisa_liu}@sjtu.edu.cn

Abstract. Recently, micro-expression recognition (MER) has attracted much attention due to its wide application in various fields such as crime trials and psychotherapy. However, the short duration and subtle movement of facial muscles make it difficult to extract micro-expression features. In this article, we propose a Dual Flow Fusion Convolutional Network (DFFCN) that combines the learning flow and optical flow to capture spatiotemporal features. Specifically, we adopt a trainable Learning Flow Module to extract the frame-level motion characteristics, fused with the mask generated from hand-crafted optical flow, and finally predict the micro-expression. Additionally, to overcome the shortcomings of limited and imbalanced training samples, we propose a data augmentation strategy based on Generative Adversarial Network (GAN). Comprehensive experiments are conducted on three public micro-expression datasets: CASME II, SAMM and SMIC with Leave-One-Subject-Out (LOSO) cross-validation. The results demonstrated that our method achieves competitive performance when compared with the existing approaches, with the best UF1 (0.8452) and UAR (0.8465).

Keywords: Micro-expression recognition · Micro-expression synthesis · Convolutional neural network

1 Introduction

Facial micro-expressions (ME) are brief and involuntary rapid facial emotions that are elicited to hide a certain true emotion [1]. Haggard and Isaacs [2] first discovered this kind of subtle emotional expression in 1966. Compared to the long-duration and obvious changes of common macro-expressions, the facial muscle movements of ME are subtle and rapid, which makes them difficult to detect and usually imperceptible to the human eyes.

Facial expression is one of the most important non-verbal channels for the expression of human inner emotions. Research on micro-expression recognition

---

This research was supported by the National Natural Science Foundation of China under Project (Grant No. 61977045).

T. Mantoro et al. (Eds.): ICONIP 2021, LNCS 13110, pp. 76–87, 2021.
https://doi.org/10.1007/978-3-030-92238-2_7

(MER) enables people to have a more sensitive understanding of subtle facial movements. It has attracted more and more researchers into the studies due to its potential applications in the field of police inquiry, clinical diagnosis, business negotiation and other fields.

The current MER research is based on public ME datasets. These datasets are generated from subjects in a strictly controlled laboratory environment, and then labeled by coders trained with professional micro-expression training tools (METT). When building a ME dataset, it is both challenging to trigger spontaneous micro-expressions and label the data. Thus the number of samples are always limited. Most of the early methods relied on hand-crafted features such as LBP [9]. In recent years, deep learning has become the most effective learning technology in the fields of detection and pattern recognition. However, the small-scale of the micro-expression dataset and imbalanced categories have brought huge challenges to the training of deep networks.

To address the problem mentioned above, in this paper, we propose our Dual Flow Fusion Convolutional Network (DFFCN) for facial micro-expression recognition in video clips. Inspired from [5], a trainable Learning Flow (LF) Module is inserted between frames with different time intervals to enrich the subtle frame-level representations. In addition, we extract a mask from the hand-crafted optical flow and feed it into the LF Module as attention information. In order to overcome the problem of small dataset scale and imbalanced categories, we adopt a Generative Adversarial Network (GAN) called Ganimation [4] to generate synthetic micro-expression image sequences. Our method makes full use of the frame-level information within a ME video clip, and greatly enrich the ME dataset for training and conquer the problem of imbalanced label classes. The main contributions are summarized as follows:

- We propose a DFFCN model that adopts a trainable LF Module to incorporate both the hand-crafted optical flow features and learning flow (CNN) features for motion extraction.
- We propose a data augmentation strategy based on GAN to greatly enrich the ME dataset for training, and overcome the shortcomings of limited and imbalanced training samples.
- Comprehensive experiments are conducted on three public ME datasets: CASME II, SAMM and SMIC with Leave-One-Subject-Out (LOSO) cross-validation and our method can achieve competitive performance when comparing with the existing approaches, with the best UF1 (0.8452) and UAR (0.8465).

The rest of this paper is organized as follows: We briefly reviews the existing state-of-art approaches for MER in Sect. 2. The details of our proposed method is present in Sect. 3. Further, experimental results and comparative analysis is discussed in Sect. 4. Finally, we conclude our work in Sect. 5.

2   Related Works

The framework of MER systems consists of three parts: data preprocessing, feature extraction, and classification. Among them, feature extraction plays the most important role. According to the method of extraction, ME feature can be divided into two main categories: handcrafted feature, and learning-based feature.

Hand-Crafted Feature. In the early years, most MER methods chose handcrafted features as their feature descriptor. For example, local binary pattern with three orthogonal planes (LBP-TOP) has been applied in the three proposed databases: CASME II [6], SAMM [7], SMIC [8] as their primary baseline feature extractor. LBP-TOP is a spatio-temporal extension of local binary pattern (LBP) [9], which helps in distinguishing local texture feature information by encoding a vector of binary code into histograms. These histograms are performed on each plane (XY, XT, YT) and finally concatenated into a single histogram feature. Several optical flow-based approaches are also proposed by researchers. Liong et al. [10] used bi-weighted oriented optical flow (Bi-WOOF) to enhance the apex frame features to recognize the micro expression with only apex frame.

Learning-Based Feature. Recently, numerous MER works have shifted towards deep convolutional neural networks (CNNs) due to its great successes in computer vision and pattern recognition tasks [11]. Quang et al. [21] applied Capsule Networks (CapsuleNet) for micro-expression recognition using only apex frames. Domain adaptation has also been applied to MER [13], achieving the 1st place in the 2nd Micro-Expression Grand Challenge (MEGC2019). Peng et al. [14] proposed a two-stream 3D CNN model called Dual Temporal Scale Convolutional Neural Network (DTSCNN), which is designed to accommodate different frame-rates of facial micro-expression videos. ICE-GAN [15] was proposed for Identity-aware MEs and Capsule-Enhanced MER, which outperforms the winner of the MEGC2019 Challenge benchmark by a large margin.

However, these deep learning-based methods will inevitably face the problems of limited micro-expression datasets and imbalanced label distribution, and are likely to be over-fitting. In addition, some methods [13] only use the onset frame and the apex frame to extract the motion features, which ignore the information contained in the intermediate frame since not all AUs occur and end simultaneously.

3   Proposed Method

In this section, we propose our DFFCN with LF Module for facial MER. The video sequences of CASME II, SAMM and SMIC dataset are firstly processed with Eulerian Motion Magnification (EVM) [12]. For each sequence, we sample

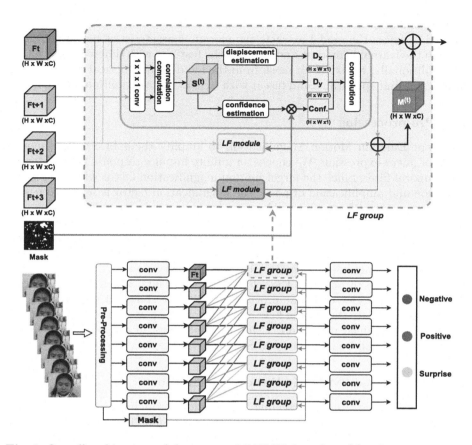

Fig. 1. Overall architecture of the proposed DFFCN, based on [5]. 8 frames are input to the CNN backbone. For frame t, the size of appearance feature map $\mathbf{F}_t$ after convolution is $H \times W \times C$, motion features $\mathbf{M}_{\mathbf{k}}^{(t)}$ are learned between $\mathbf{F}_t$ and $\mathbf{F}_{t+\mathbf{k}}$, where k=1,2,3, through a LF group containing 3 LF modules. Inside the LF module, a correlation tensor $\mathbf{S}^{(t)}$ is obtained by computing correlations, and then a displacement tensor $\mathbf{D}_{(t)}$ and confidence tenor is estimated. Element-wise multiplication ($\otimes$) is performed between the confidence map and the mask generated from the hand-crafted optical flow. After the transformation process of convolution layers, the motion feature $\mathbf{M}_{\mathbf{k}}^{(t)}$ is obtained. Then the final $\mathbf{M}^{(t)}$ is added back to $\mathbf{F}_t$ through element-wise addition ($\oplus$).

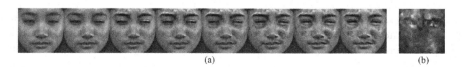

(a)   (b)

Fig. 2. A video clip from onset to apex is sampled into 8 frames (a) after EVM and face-crop. Then the optical flow (b) is calculated based on onset and apex frame.

a fixed number of frames and feed them into our network after face cropping. In the middle layer, the model extracts the motion characteristics between frames through the learnable module and enriches the features with different time interval. Additionally, we extract a mask from the hand-crafted optical flow between onset frame and apex frame and fuse it with LF Module as attention information.

3.1   Pre-processing

We adopt Eulerian Motion Magnification to amplify the subtle muscle movements of micro-expression. We choose an infinite impulse response (IIR) filter as our temporal filter. Since the larger motion magnification factor will bring more distortion and amplification of noise, the magnification factor is set to 10 based on [16] to avoid artifacts in the video.

Face cropping and alignment is an essential step to avoid the interference of irrelevant information such as the background and non-face areas. We detect the coordinates of 68 localized facial landmarks of the onset frame. Then we calculate the transformation matrix between five key points (2 for eyes, 1 for nose, 2 for mouth corner) and the reference facial points, which is used to crop all frames in the video clip. The processed results are shown in the Fig. 2.

We also calculate the optical flow between onset and apex frame. The optical flow will be combined with the LF Module to serve as an attention mechanism and the details will be introduced in Sect. 3.3.

3.2   GAN-Based Synthetic Data Augmentation

Horizontal flipping, rotation, scaling and random cropping are common data enhancement methods for conventional computer vision classification tasks. To a certain extent, they can increase the robustness of the model and reduce the possibility of over-fitting. In addition, considering that GANs have shown their significant generative capabilities in synthetic data generation, we adopt the pre-trained Ganimation model [4] to generate synthetic ME sequences for data augmentation. Ganimation is an anatomically coherent facial expression synthesis method based on Facial Action Coding System (FACS) [3], which is to encode facial muscle activity to emotion states. There are 46 Action Units (AUs) in the FACS system. Given an input image and the target AU activation vector defined by 14 AUs (1, 2, 4, 5, 6, 7, 9, 10, 12, 14, 15, 17, 20, 23, 25, 26, 45), Ganimation can render a novel expression in a continuum by controlling the magnitude of activation of each AU.

Specifically, we count the AU groups and emotion labels of all samples in the CASME II (we do not use SAMM because its annotation is inconsistent with our strategy), and then construct a series of representative mappings between AU group and emotion label. We eliminate the controversial mappings and finally 34 AU-label pair are obtained as shown in the Table 1. Considering that the region where micro-expressions occur is not always symmetrical on the face, the AU occurrence position is retained to mark which half of the face is the AU on. For example, AU L1 means AU 1 (Inner brow raiser)only appears on

Table 1. There are 34 AU group-label pairs obtained from original datasets. Emotion labels are converted to *Negative, Positive* and *Surprise*. The prefix letter *L* or *R* indicates the area of the face where AU takes place. For example, $1 + R2 + 25$ means AU 1 and AU 25 appear on both left and right face while AU 2 only appears on the right face.

AU Group	Label	Count	AU Group	Label	Count	AU Group	Label	Count
4+L10	Negative	2	4	Negative	11	1+4	Negative	1
4+9	Negative	3	6+12	Positive	6	20	Negative	1
4+7	Negative	20	4+5+7+9	Negative	1	17	Negative	13
4+7+9	Negative	3	1+R2+25	Surprise	1	L7+L10	Negative	1
4+L9+14	Negative	1	4+R12	Positive	1	12+15	Positive	1
4+6+7	Negative	1	2	Surprise	1	4+7+10	Negative	2
1+2	Surprise	5	4+6+7+9	Negative	1	L12	Positive	6
14	Positive	3	4+9+17	Negative	1	4+15+17	Negative	1
L2+12	Surprise	1	12+L14	Positive	1	15+17	Negative	4
R4+15	Negative	1	6+7+12	Positive	2	6+12+17	Positive	1
25	Surprise	1	L2+14	Surprise	1			
L10	Negative	1	4+9+10	Negative	2			

the left face. Note that the feature distribution of the synthetic dataset should be close to the original dataset. Besides, we wish that the synthetic dataset can have applicability to all existing micro-expression datasets rather than a specific dataset, so we did not use the human faces from the original dataset. Instead, we choose the images from the synthetic human face data generated by the Stylegan2 model [17] as the input image of Ganimation, which preserve the fidelity of human faces. Then for each AU group-label pair, we encode AU groups into two vectors with shape of 1×14. The element of each vector with range 0 to 1 represents the activation magnitude of specific AU on the left and right face. By multiplying the vectors by different intensity coefficients α, we are able to control the value of the input AU vector to generate a series of images with continuous intensity and fuse left and right part of the face, which simulates the dynamic changes of micro expressions. Figure 3 shows some example of the results we obtain.

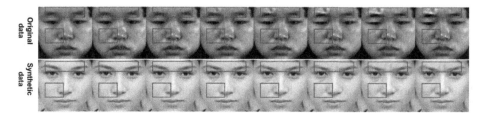

Fig. 3. Visualization of original ME data (top) and synthetic data (bottom) with same AU and label annotations. For GAN model, given one input image and AU-label $(4 + R10, Negative)$, we encode AU groups into two vectors αV_L and αV_R. We set activation factor α in range $(0, 0.3)$ to simulate the intensity of micro-expressions.

3.3   Model Architecture

The overall model architecture modified from [5] is illustrated in Fig. 1. We choose the ImageNet-pretrained ResNet18 [18] as the CNN backbone since deeper model has larger possibility of over-fitting on small scale datasets. The input of the model is the stack of 8 facial images sampled from a video clip and 1 optical flow image, with a shape of $[9 \times 224 \times 224 \times 3]$. Inspired by [5], we insert our LF Module after the first stage of ResNet18 to learn frame-wise motion features. We do not choose deeper position because subtle motion information may disappear as the resolution of feature map decreases after deeper convolution.

Correlation Computation. Suppose there are two input frames at time t and $t + k$, whose feature maps after convolution are denoted by $\mathbf{F}_t$ and $\mathbf{F}_{t+k}$. A correlation score of position $\mathbf{x}$ with respect to displacement $\mathbf{p}$ is defined as

$$s(\mathbf{x}, \mathbf{p}, t) = \mathbf{F}_{\mathbf{x},t} \cdot \mathbf{F}_{\mathbf{x}+\mathbf{p},t+k} \tag{1}$$

where $\cdot$ denotes dot product. For the feature map with shape of $H \times W \times C$, the size of resultant correlation tensor $\mathbf{S}^{(t)}$ is $H \times W \times P^2$. In the experiment, for a feature map with spatial resolution 56×56, we set $\mathbf{p} = 15$ for subtle motion in MER tasks because short-duration micro-expressions do not involve wide-range facial changes.

Displacement Estimation. From the correlation tensor $\mathbf{S}^{(t)}$, we estimate a displacement field for motion information with the concept kernel-soft-argmax [19] as implemented in [5].

$$d(\mathbf{x}, t) = \sum_{\mathbf{p}} \frac{\exp(g(\mathbf{x}, \mathbf{p}, t) s(\mathbf{x}, \mathbf{p}, t)/\tau)}{\sum_{\mathbf{p}'} \exp(g(\mathbf{x}, \mathbf{p}', t) s(\mathbf{x}, \mathbf{p}', t)/\tau)} \mathbf{p} \tag{2}$$

where

$$g(\mathbf{x}, \mathbf{p}, t) = \frac{1}{\sqrt{2\pi}\sigma} \exp(\frac{\mathbf{p} - \arg\max_{\mathbf{p}} s(\mathbf{x}, \mathbf{p}, t)}{\sigma^2}) \tag{3}$$

is the Gaussian kernel. The result $\mathbf{D}_x$ and $\mathbf{D}_y$ with size $H \times W \times 1$ denotes the displacement on x and y direction. The standard deviation σ and temperature factor τ are empirically set to 5 and 0.01 respectively.

The confidence map of correlation tensor can be regarded as auxiliary motion information, which is defined as

$$s^*(\mathbf{x}, t) = \max_{\mathbf{p}} s(\mathbf{x}, \mathbf{p}, t) \tag{4}$$

Mask Generation. Since the facial muscle movement in the micro-expression video is continuous, subtle and only occurs in a specific area, Here we perform a fusion on it with the mask generated from the hand-crafted optical flow to shield redundant information such as the rigid movement of the head, so that the model will pay more attention to the area where the ME-related AU occurs. Specifically, we resize the magnitude map to the same size (56×56) as confidence map and calculate the threshold value of top 30% by sorting all the values. Then the top 30% values are set to 1 and others to 0. After that we perform a element-wise multiplication as shown in Fig. 4.

Feature Transformation and Fusion. In the next step, the displacement map and the confidence map are concatenated into a displacement tensor $\mathbf{D}^{(t)}$ of size $H \times W \times 3$, followed by a feature transformation process with convolution layers, and results in a $H \times W \times C$ output.

$$\mathbf{M}^{(t)} = \sum_{k=1}^{3} \mathbf{M_k}^{(t)} \tag{5}$$

For frame t, with appearance feature map $\mathbf{F}_t$ after convolution, motion features $\mathbf{M_k}^{(t)}$ are learned between $\mathbf{F}_t$ and $\mathbf{F}_{t+k}$, where $k = 1, 2, 3$, through a LF group containing 3 LF modules. This operation retains subtle frame-level motion information, which strengthens the learnability of the module for minute details and maintain the essence of the feature maps. Then the final $\mathbf{M}^{(t)}$, defined as the fusion of multiple time spans motion features, is added back to appearance features $\mathbf{F}_t$ through element-wise addition.

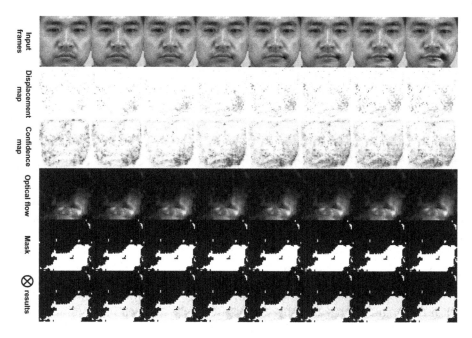

Fig. 4. The input frames, displacement map, confidence map (normalized for visualization), optical flow map, mask, and element-wise multiplication result are shown from the top row.

4   Experiments

In this section, we briefly introduce the datasets and evaluation metrics, and then present the implementation details and experimental results compared to previous state-of-the-art MER methods.

4.1   Datasets and Evaluation Metrics

We validate the performance of the proposed method on three spontaneous facial micro-expression datasets: CASME II [6], SAMM [7] and SMIC [8]. In our experiment, 439 samples (143 from CASME II, 132 from SAMM and 164 form SMIC) from 67 subjects (23 from CASME II, 28 from SAMM and 16 from SMIC) are combined in the final composite dataset, whose labels are mapped into a reduced set of 3 ME classes: Positive (107), Negative (252) and Surprise (80). For our synthetic dataset, theoretically an infinite number of samples can be generated. In the experiment, we randomly select 1000 samples with relatively balanced number for the above three classes and add them to training dataset.

We adopt Leave-One-Subject-Out (LOSO) cross-validation for subject independent evaluation under the composite database evaluation (CDE) protocol as in MEGC2019 challenge. For 67 subjects in total, the training and evaluating experiment is repeated for 67 times, in each of which 1 subject is split alone for testing and the remaining 66 subjects are for training. To avoid the impact of imbalanced class distribution for the composite dataset, the recognition performance is evaluated with two balanced metrics: Unweighted F1-score (UF1) and Unweighted Average Recall (UAR), which are defined as:

$$UF1 = \frac{1}{C} \sum_i^C \frac{2TP_i}{2TP_i + FP_i + FN_i} \tag{6}$$

$$UAR = \frac{1}{C} \sum_i^C \frac{TP_i}{N_i} \tag{7}$$

where C is the number of classes and TP_i, FP_i, TN_i, and N_i denote the number of true positives, false positives, false negatives, and total samples with ground-truth class i, respectively.

4.2   Implementation Details

We calculate the mean and standard deviation of the composite dataset to normalize the input. Each video clip will be uniformly sampled into 8 frames from onset frame to apex frame with the resolution of 224 × 224. Note that SMIC dataset has no apex frame annotations, so we set the middle position of the video clips as the apex frame. We use Adam optimizer with an initial learning rate 0.001 and decaying learning rate weight 0.9 for optimization. The batch size is set to 32. The data augmentation includes horizontal flipping, rotation and color jittering.

4.3 Results and Discussion

Table 2 illustrates the proposed DFFCN and other existing methods in terms of UF1 and UAR. Experiments conducted on benchmarks demonstrate that our DFFCN with synthetic data outperforms (or is on par with) SOTA approaches on the full composite database, achieving the best UF1 (**0.8452**) and UAR (**0.8465**), improved by **25.7%** and **26.8%** compared to the baseline LBP-TOP. More importantly, our UF1 and UAR scores are obviously superior to the existing approaches with a substantial margin of **6.8%** and **5.7%** on CASME II. Additionally, DFFCN also performs best on SMIC. Training with synthetic data, both UF1 and UAR can be improved about **2%** on the composite dataset, **4%** on SMIC and **3%** on CASME II. However, it does not show the same effectiveness on the SAMM dataset. The reason of the inconsistency may be that SAMM is a gray dataset, whose feature distribution is quite different from that of synthetic dataset. Figure 5 shows the confusion matrix of experiment results on CDE protocol.

Table 2. Comparative results for different methods from the corresponding original paper, based on CDE protocol with LOSO cross-evaluation method. Bracketed bold numbers indicate the best score; bold numbers indicate the second best. SD means synthetic data.

Method	Composite		SMIC		CASME II		SAMM	
	UF1	UAR	UF1	UAR	UF1	UAR	UF1	UAR
LBP-TOP [20]	0.5882	0.5785	0.2000	0.5280	0.7026	0.7429	0.3954	0.4102
Bi-WOOF [20]	0.6296	0.6227	0.5727	0.5829	0.7805	0.8026	0.5211	0.5139
CapsuleNet [21]	0.6520	0.6506	0.5820	0.5877	0.7068	0.7018	0.6209	0.5989
OFF-ApexNet [22]	0.7196	0.7096	0.6817	0.6695	0.8764	0.8681	0.5409	0.5392
Dual-Inception [23]	0.7322	0.7278	0.6645	0.6726	0.8621	0.8560	0.5868	0.5663
STSTNet [24]	0.7353	0.7605	0.6801	0.7013	0.8382	0.8686	0.6588	0.6810
EMR [13]	0.7885	0.7824	0.7461	0.7530	0.8293	0.8209	0.7754	0.7152
RCN [25]	0.7052	0.7164	0.5980	0.5991	0.8087	0.8563	0.6771	0.6976
ICE-GAN [15]	**0.8450**	**0.8410**	**0.7900**	**0.7910**	0.8760	0.8680	[**0.8550**]	[**0.8230**]
DFFCN(ours) w/o mask	0.8230	0.8178	0.7563	0.7684	0.9065	0.8681	0.7740	0.7444
DFFCN(ours)	0.8262	0.8211	0.7686	0.7750	**0.9100**	**0.8880**	**0.7820**	**0.7573**
DFFCN(ours)+SD	[**0.8452**]	[**0.8465**]	[**0.8016**]	[**0.8125**]	[**0.9441**]	[**0.9256**]	0.7586	0.7385

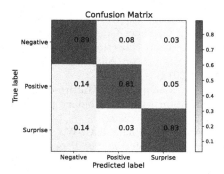

Fig. 5. Confusion matrix of DFFCN on CDE protocol with LOSO cross-evaluation method.

5   Conclusion

In this article, we propose our DFFCN that adopts a trainable LF Module to incorporate both the handcrafted optical flow features and learning flow (CNN) features for motion extraction. We propose a data augmentation strategy based on GAN to greatly enrich the micro-expression dataset for training, and overcome the shortcomings of limited and imbalanced training samples. Comprehensive experiments are conducted on three public micro-expression datasets: CASME II, SAMM and SMIC with LOSO cross-validation and our method with synthetic data outperforms (or is on par with) SOTA approaches on the full composite database, achieving the best UF1 (**0.8452**) and UAR (**0.8465**).

References

1. Ekman, P., Friesen, W.V.: Nonverbal leakage and clues to deception. Psychiatry **32**(1), 88–106 (1969)
2. Haggard E.A., Isaacs K.S.: Micromomentary facial expressions as indicators of ego mechanisms in psychotherapy. In: Methods of Research in Psychotherapy. The Century Psychology Series, pp. 154–165. Springer, Boston (1966). https://doi.org/10.1007/978-1-4684-6045-2_14
3. Ekman, R.: What the Face Reveals: Basic and Applied Studies of Spontaneous Expression Using the Facial Action Coding System (FACS). Oxford University Press, USA (1997)
4. Pumarola, A., Agudo, A., Martinez, A.M., Sanfeliu, A., Moreno-Noguer, F.: GAN-imation: anatomically-aware facial animation from a single image. In: Ferrari, V., Hebert, M., Sminchisescu, C., Weiss, Y. (eds.) ECCV 2018. LNCS, vol. 11214, pp. 835–851. Springer, Cham (2018). https://doi.org/10.1007/978-3-030-01249-6_50
5. Kwon, H., Kim, M., Kwak, S., Cho, M.: MotionSqueeze: neural motion feature learning for video understanding. In: Vedaldi, A., Bischof, H., Brox, T., Frahm, J.-M. (eds.) ECCV 2020. LNCS, vol. 12361, pp. 345–362. Springer, Cham (2020). https://doi.org/10.1007/978-3-030-58517-4_21
6. Yan, W.-J., et al.: CASME II: an improved spontaneous micro-expression database and the baseline evaluation. PLOS ONE **9**(1), e86041 (2014)
7. Davison, A.K., Lansley, C., Costen, N., Tan, K., Yap, M.H.: SAMM: a spontaneous micro-facial movement dataset. IEEE Trans. Affect. Comput. **9**(1), 116–129 (2016)
8. Li, X., Pfister, T., Huang, X., Zhao, G., Pietikäinen, M.: A spontaneous micro-expression database: inducement, collection and baseline. In: 10th IEEE International Conference and Workshops on Automatic Face and Gesture Recognition, FG 2013, pp. 1–6. IEEE (2013)
9. Ojala, T., Pietikainen, M., Maenpaa, T.: Multiresolution gray-scale and rotation invariant texture classification with local binary patterns. IEEE Trans. Pattern Anal. Mach. Intell. **24**(7), 971–987 (2002)
10. Liong, S.-T., See, J., Wong, K., Phan, R.C.-W.: Less is more: micro-expression recognition from video using apex frame. Sig. Process. Image Commun. **62**, 82–92 (2018)
11. Khor, H.-Q., See, J., Phan, R.C.W., Lin, W.: Enriched long-term recurrent convolutional network for facial micro-expression recognition. In: FG 2018, pp. 667–674. IEEE (2018)

12. Wu, H.-Y., Rubinstein, M., Shih, E., Guttag, J., Durand, F., Freeman, W.: Eulerian video magnification for revealing subtle changes in the world. ACM Trans. Graph. (TOG) **31**(4), 1–8 (2012)

13. Liu, Y., Du, H., Zheng, L., Gedeon, T.: A neural micro-expression recognizer. In: FG 2019, pp. 1–4. IEEE (2019)

14. Peng, M., Wang, C., Chen, T., Liu, G., Fu, X.: Dual temporal scale convolutional neural network for micro-expression recognition. Front. Psychol. **8**, 1745 (2017)

15. Yu, J., Zhang, C., Song, Y., Cai, W.: ICE-GAN: identity-aware and capsule-enhanced GAN for micro-expression recognition and synthesis. arXiv preprint arXiv:2005.04370 (2020)

16. Le Ngo, A.C., Johnston, A., Phan, R.C.-W., See, J.: Micro-expression motion magnification: global Lagrangian vs. local Eulerian approaches. In: FG 2018, pp. 650–656. IEEE (2018)

17. Karras, T., Laine, S., Aittala, M., Hellsten, J., Lehtinen, J., Aila, T.: Analyzing and improving the image quality of StyleGAN. In: Proceedings of the IEEE/CVF Conference on Computer Vision and Pattern Recognition, pp. 8110–8119 (2020)

18. He, K., Zhang, X., Ren, S., Sun, J.: Deep residual learning for image recognition. In: CVPR 2016, pp. 770–778 (2016)

19. Lee, J., Kim, D., Ponce, J., Ham, B.: SFNet: learning object-aware semantic correspondence. In: CVPR 2019, pp. 2278–2287 (2019)

20. See, J., Yap, M.H., Li, J., Hong, X., Wang, S.-J.: MEGC 2019 - the second facial micro-expressions grand challenge. In: FG 2019, pp. 1–5. IEEE (2019)

21. Van Quang, N., Chun, J., Tokuyama, T.: CapsuleNet for micro-expression recognition. In: FG 2019, pp. 1–7. IEEE (2019)

22. Gan, Y.S., Liong, S.-T., Yau, W.-C., Huang, Y.-C., Tan, L.-K.: OFF-ApexNet on micro-expression recognition system. Sig. Process. Image Commun. **74**, 129–139 (2019)

23. Zhou, L., Mao, Q., Xue, L.: Dual-inception network for cross-database micro-expression recognition. In: FG 2019, pp. 1–5. IEEE (2019)

24. Liong, S.-T., Gan, Y.S., See, J., Khor, H.-Q., Huang, Y.-C.: Shallow triple stream three-dimensional CNN (STSTNet) for micro-expression recognition. In: FG 2019, pp. 1–5. IEEE (2019)

25. Xia, Z., Peng, W., Khor, H.-Q., Feng, X., Zhao, G.: Revealing the invisible with model and data shrinking for composite-database micro-expression recognition. IEEE Trans. Image Process. **29**, 8590–8605 (2020)

AUPro: Multi-label Facial Action Unit Proposal Generation for Sequence-Level Analysis

Yingjie Chen[1], Jiarui Zhang[1], Diqi Chen[2], Tao Wang[1]([✉]), Yizhou Wang[1], and Yun Liang[1]

[1] Department of Computer Science and Technology, Peking University, Beijing, China
{chenyingjie,zjr954,wangtao,yizhou.wang,ericlyun}@pku.edu.cn
[2] Advanced Institute of Information Technology, Peking University, Hangzhou, China
dqchen@aiit.org.cn

Abstract. Facial action unit (AU) plays an essential role in human facial behavior analysis. Despite the progress made in frame-level AU analysis, the discrete classification results provided by previous work are not explicit enough for the analysis required by many real-world applications, and as AU is a dynamic process, sequence-level analysis maintaining a global view has yet been gravely ignored in the literature. To fill in the blank, we propose a multi-label AU proposal generation task for sequence-level facial action analysis. To tackle the task, we design AUPro, which takes a video clip as input and directly generates proposals for each AU category. Extensive experiments conducted on two commonly used AU benchmark datasets, BP4D and DISFA, show the superiority of our proposed method.

Keywords: Affective computing · Facial action unit · Sequence-level AU analysis

1 Introduction

With the proliferation of user sentiment analysis for social media applications, facial behavior-related research has become increasingly important. Facial Action Coding System (FACS), proposed by Ekman *et al.* [4] in the 1970s, describes facial behavior as a set of facial action units (AUs). AUs are defined according to basic facial muscle movements, and the combination of AUs can form or simulate almost all facial behaviors. AU analysis has seen rapid growth in recent years, and findings of this task are crucial to various real-world scenarios such as online education, remote interview, and auxiliary medical treatment. Hitherto, AU analysis tasks can be roughly summarized into AU recognition [2,9] and AU intensity estimation [21,24]. AU recognition aims to predict frame-level

---

Y. Chen and J. Zhang—Equal Contribution.

© Springer Nature Switzerland AG 2021
T. Mantoro et al. (Eds.): ICONIP 2021, LNCS 13110, pp. 88–99, 2021.
https://doi.org/10.1007/978-3-030-92238-2_8

AU occurrence probabilities, while AU intensity estimation aims to estimate the frame-level AU intensity represented by a discrete intensity label.

Despite their success, these two lines of work do not suffice the ever-increasing need for a lot of real-world applications, due to the lack of a global view of the entire video to ensure high-quality global information required by AU-related applications, such as the occurrence frequency, duration, and the chronological order of each AU's appearance. Taking personal credit assessment as an example, sequence-level AU analysis is required to assist the financial institutions to pay special attention to the fleeting panic under the appearance of calm and warn about the possibility of a defrauding loaner. Since frame-level analysis lacks a global view and highly relies on features of a single frame, it tends to only attach importance to the major facial features in one given frame and take the fleeting anomalies as jittering, and the obtained discrete results are implicit for researchers to draw certain conclusions. On the contrary, sequence-level analysis is capable of treating the sequence as a whole and directly produces target results that provide global information suitable for AU-related applications [3,17].

To fill in the blank, we propose a **multi-label AU proposal generation** task for sequence-level AU analysis, which takes a video clip as input and outputs a set of AU proposals which are supposed to precisely and exhaustively cover all AU instances, which refers to a video segment whose boundaries are determined by the starting and ending time of one AU occurrence. It is worth noticing that although our task may seem to share some similarities with temporal action detection, the advances in the temporal action detection task cannot be directly borrowed to tackle ours. First, compared to coarse-grained body actions, such as running and jumping, AU instances are much harder to capture since facial muscle movements are much subtler and usually of short duration. Second, unlike temporal action detection task which assumes there is at most one action happening at any time point, AU instances of different categories may overlap with each other, due to AU co-occurrences.

There are two schemes for solving this task: one is based on the aforementioned frame-level AU recognition methods by collecting the frame-level AU occurrence probabilities and selecting a set of hyper-parameters to obtain sequence-level AU proposals through a post-processing method; the other is direct AU proposal generation by taking the whole video clip as input and analyzing the clip as a whole. There are two main drawbacks lying in the first scheme: (1) the quality of AU proposals generated in this way suffers from the jittering issue due to the lack of global context from the whole video sequence, and (2) manually selected parameters for post-processing greatly affect the performance. Therefore, we follow the second scheme to tackle the task.

To this end, we propose **AUPro** to utilize the benefit of global context for multi-label AU proposal generation. AUPro takes a video clip of a person's frontal face as input, then analyzes the video clip as a whole, and finally generates AU proposals for each AU category. Specifically, first, a backbone network is used for AU features extraction. Then, for each AU category, AUPro estimates the probabilities of each time point being an AU instance's starting or ending time point and generates a completeness map showing the probability of each temporal range containing one complete AU instance. Based on these, AU proposals for each AU category are generated. After that, a post-processing module is used for filtering redundant or less-confident proposals.

The main contributions of our work can be concluded as:

- We propose a multi-label AU proposal generation task for sequence-level AU analysis, which is more practical for various real-world application scenarios.
- We propose AUPro for multi-label AU proposal generation, which takes a video clip as input, analyzes the clip as a whole, and then outputs AU proposals for each category.
- Extensive experiments are conducted on commonly used AU benchmark datasets, BP4D and DISFA, and the experimental results show the superiority of the proposed method.

2   Related Work

Facial Action Unit Recognition. Research topics about AU have become more and more popular due to the increasing need for user sentimental analysis. Traditional methods mainly use hand-crafted features such as Gabor filters [20] and SIFT features [27] and employ SVM [18] to predict AU occurrence probabilities. Deep learning methods [16,22,25] have proved their efficiency in AU recognition tasks by largely improving performance through end-to-end learning. Considering the fact that AUs are active on sparse facial regions, Zhao *et al.* [25] proposed a method combining region learning and multi-label learning. Since AU is a dynamic process, methods such as [7,10] use CNN for feature extraction and use LSTM to capture image-level temporal information.

Temporal Action Detection. Temporal action detection aims to detect action instances in untrimmed videos, which can be regarded as a combination of class-agnostic proposal generation and proposal classification. Methods for action proposal generation task can be briefly divided into anchor-based methods [5,19], and boundary-based methods [12,13,26]. As an anchor-based method, Shou *et al.* [19] generates anchors with varied lengths via sliding windows. Gao *et al.* [5] proposed Temporal Unit Regression Network, which divides an untrimmed video into fixed-length video units. Unlike anchor-based methods, Zhao *et al.* [26] proposed a method to identify the starting, course, and ending stages of an action, and estimate proposal completeness via structured temporal pyramid pooling. Lin *et al.* [11] proposed a fast and unified dense boundary generator (DBG) to evaluate dense boundary confidence maps for all proposals.

3   Method

In this section, we clarify the task setting of the proposed multi-label AU proposal generation task and describe our AUPro in details.

3.1   Task Setting

A video clip recording facial actions with T frames is denoted as $\mathcal{I} = \{I_t\}_{t=1}^{T}$, where I_t is the t-th frame in $\mathcal{I}$. The annotation of each frame I_t is the

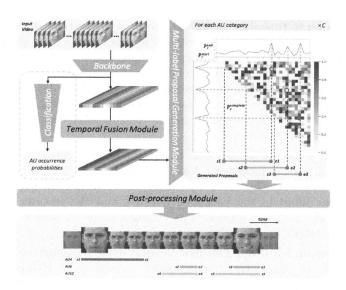

Fig. 1. Overview of the proposed method. Given a video clip, *Backbone* is employed for feature extraction, and *Temporal Fusion Module* enables information exchange through time dimension. The fused features are then fed into one *Multi-label Proposal Generation Module* to estimate the starting and ending probabilities, $\hat{P}^{\mathrm{start}}_{c,t}$ and $\hat{P}^{\mathrm{end}}_{c,t}$, for each time point t and generate a completeness map $\hat{P}^{\mathrm{complete}}_c$ that consists of the completeness score for any proposal (s,e) satisfying $0 \leq s < e \leq T$, for AU category c ($c \in \{1,2,\ldots,\mathcal{C}\}$). The final confidence score for a proposal (s,e) with AU category label c is calculated as $\hat{P}^{\mathrm{start}}_{c,s} \times \hat{P}^{\mathrm{end}}_{c,e} \times \hat{P}^{\mathrm{complete}}_{c,(s,e)}$. After that, *Post-processing Module* is used to filter out redundant proposals. *Classification Branch* in the dashed box is enabled only for the end-to-end training setting to predict AU occurrence probabilities.

binary occurrence labels of $\mathcal{C}$ AU categories, which can be further processed into a set of ground-truth AU instances, $\Phi^{\mathrm{g}} = \{\phi^{\mathrm{g}}_i = (s^{\mathrm{g}}_i, e^{\mathrm{g}}_i, c^{\mathrm{g}}_i)\}^{N^{\mathrm{g}}}_{i=1}$, where $0 \leq s^{\mathrm{g}}_i < e^{\mathrm{g}}_i \leq T$, $c^{\mathrm{g}}_i \in \{1,2,\ldots,\mathcal{C}\}$, N^{g} is the number of ground-truth AU instances, and s^{g}_i, e^{g}_i, c^{g}_i denotes the starting time, ending time and AU category label of ground-truth AU instance ϕ^{g}_i, respectively. Multi-label AU proposal generation for sequence-level AU analysis requires generating a set of AU instances $\Phi^{\mathrm{p}} = \{\phi^{\mathrm{p}}_i = (s^{\mathrm{p}}_i, e^{\mathrm{p}}_i, c^{\mathrm{p}}_i)\}^{N^{\mathrm{p}}}_{i=1}$, which should match Φ^{g} precisely and exhaustively. In the training stage, the set of ground-truth AU instances Φ^{g} is used as supervision for generated proposals Φ^{p}. In the inference stage, the generated proposals Φ^{p} can be further filtered to obtain key information required by AU-related applications, such as the occurrence frequency, duration, and the order of each AU's appearance.

3.2   AUPro Architecture

As shown in Fig. 1, the overall AUPro architecture consists of several parts: (1) *Backbone* is used as the feature extractor. (2) *Temporal Fusion Module* is used for feature fusion in time dimension. (3) *Proposal Generation Module* applies a proposal feature generation layer to generate features with a fixed shape for each duration (s, e), followed by a completeness branch and a boundary branch. The completeness branch estimates the probability of a duration (s, e) containing a complete AU instance, which is called completeness score. And the boundary branch estimates the probabilities of each time point as starting or ending of an AU instance. (4) *Post-processing Module* is used to filter redundant proposals according to confidence scores.

Backbone. The goal of *Backbone* is to extract representative features out of input images for further processing. ResNet [8] without the last two linear layers is used as *Backbone* for AUPro. Specifically, a whole video sequence $\mathcal{I} = \{I_t\}_{t=1}^T$ is taken as the input of the backbone network and the features denoted by $\mathcal{F}^B = \{f_t^B \in \mathbb{R}^{d^B}\}_{t=1}^T$ are generated as output, where d^B denotes the feature dimension corresponding to each frame I_t.

Temporal Fusion Module. This module aims to enable information exchange among features of consecutive frames, which is implemented through two 1D convolution layers with kernel size set to 3 to enlarge the receptive field. The output of the backbone network $\mathcal{F}^B$ is fed into the module to obtain fused features $\mathcal{F}^{TFM} = \{f_t^{TFM} \in \mathbb{R}^{d^{TFM}}\}_{t=1}^T$, where $d^{TFM} = 128$ is the output feature dimension corresponding to each f^B.

Multi-label Proposal Generation Module. For estimating the completeness of each duration (s, e), a proposal feature generation layer [11] is employed to generate $\mathcal{F}^{PFG} = \{f_{(s,e)}^{PFG} \in \mathbb{R}^{N_{loc} \times d^{TFM}} | 0 \leq s, e \leq T\}$, where $f_{(s,e)}^{PFG}$ is the feature for a duration (s, e), $d^{TFM} = 128$, and N_{loc} is the number of sampled temporal locations. In our AUPro, we sample $N_s = 8$ locations around the starting time point s, $N_e = 8$ locations around the ending time point e, $N_d = 16$ locations between s and e, and in this way, $N_{loc} = N_s + N_e + N_d = 32$. Linear interpolation is used to make sure that there are enough features for sampling. And we simply set all elements of $f_{(s,e)}^{PFG}(s \geq e)$ to zero. Then a 3D convolution layer is applied to $\mathcal{F}^{PFG}$ and the proposal features $\mathcal{F}^P \in \mathbb{R}^{d^P \times T \times T}$ is obtained, where $d^P = 512$.

$\mathcal{F}^P$ is further fed into *Boundary Branch* and *Completeness Branch*. *Boundary Branch* aims to estimate the starting and ending probabilities of each time point, resulting in two probability vectors $\hat{P}^{start} \in \mathbb{R}^T$ and $\hat{P}^{end} \in \mathbb{R}^T$. *Completeness Branch* aims to generate a completeness score map $\hat{P}^{complete} \in \mathbb{R}^{T \times T}$, and the value in each location (s, e) represents the probability of the duration (s, e) containing a complete AU instance, *i.e.* the completeness score of (s, e).

In this way, we obtain $\hat{P}_c^{start}$, $\hat{P}_c^{end}$, and $\hat{P}_c^{complete}$ for each category c, and the confidence score map $\hat{P}_c^{conf} \in \mathbb{R}^{T \times T}$ can be calculated as Eq. 1:

$$\hat{P}_{c,(s,e)}^{conf} = \hat{P}_{c,s}^{start} \times \hat{P}_{c,e}^{end} \times \hat{P}_{c,(s,e)}^{complete}. \tag{1}$$

Post-processing Module. This module aims to filter redundant proposals according to the confidence score map $\hat{P}^{conf}$ for each AU category. We employ Non-Maximum Suppression (NMS) algorithm to process the proposals generated for each AU category independently.

3.3   Training of AUPro

AUPro can be trained following a two-stage setting or an end-to-end setting. For the two-stage setting, the backbone network with an extra linear layer is first trained using binary AU occurrence labels for AU recognition in the first stage, and the weights of the backbone network are fixed while training the rest of parts in the second stage. For the end-to-end setting, the output of the backbone network $\mathcal{F}^{B}$ is not only fed into the following *Temporal Fusion Module* but also *Classification Branch* for AU recognition, and an extra classification loss $\mathcal{L}_{clf}$ is enabled for back-propagation.

The loss function of AUPro consists of $\mathcal{L}_{start}$, $\mathcal{L}_{complete}$, $\mathcal{L}_{end}$, and an optional $\mathcal{L}_{clf}$. $\mathcal{L}_{start}$, $\mathcal{L}_{end}$, and $\mathcal{L}_{complete}$ are designed to supervise the estimated starting probabilities, ending probabilities, and completeness scores, respectively. For a ground-truth AU instance $\phi^{g} = (s^{g}, e^{g}, c^{g})$, we define the starting region as $d_{s} = [s^{g} - \Delta/2, s^{g} + \Delta/2]$, the ending region as $d_{e} = [e^{g} - \Delta/2, e^{g} + \Delta/2]$, where Δ denotes the length of a time range for boundary smoothing. For each time point t, we expand the time point to a region $d_{t} = [t - \Delta/4, t + \Delta/4]$. The starting probability labels are calculated by computing the ratio $\gamma_{s} = (d_{t} \cap d_{s})/d_{s}$ between each time point's region and each starting region, and if γ_{s} is larger than a given threshold $\tau = 0.5$, the starting probability label of the time point is set to 1, otherwise 0. The ending probability labels are set in the same way by computing γ_{e}. And for completeness labels, γ_{t} denotes the tIoU of each time point's region d_{t} and each (s, e), and the completeness label of the time point is assigned to 1 if $\gamma_{t} > \tau$, 0 otherwise. In this way, the ground-truth annotations P_{c}^{start}, P_{c}^{end}, and $P_{c}^{complete}$ are obtained for each AU category c.

$\mathcal{L}_{start}$ and $\mathcal{L}_{end}$ are computed as the cross-entropy between ground-truth annotations and the estimated results, as follows:

$$\mathcal{L}_{(\cdot)} = -\frac{1}{T} \sum_{c=1}^{C} \sum_{t=1}^{T} \left[P_{c,t}^{(\cdot)} \log \hat{P}_{c,t}^{(\cdot)} + \left(1 - P_{c,t}^{(\cdot)}\right) \log \left(1 - \hat{P}_{c,t}^{(\cdot)}\right) \right], \qquad (2)$$

where $(\cdot) \in \{start, end\}$ and $\hat{P}_{c}^{(\cdot)}$ denotes the estimated starting or ending probabilities for AU category c, and $\mathcal{L}_{complete}$ is computed as follows:

$$\mathcal{L}_{complete} = -\frac{1}{T^{2}} \sum_{c=1}^{C} \sum_{t_{j}=1}^{T} \sum_{t_{i}=1}^{T} \|P_{c,(t_{i},t_{j})}^{complete} - \hat{P}_{c,(t_{i},t_{j})}^{complete}\|_{1}. \qquad (3)$$

where $\hat{P}_{c}^{complete}$ denotes the estimated completeness score for AU category c.

For the end-to-end training setting, the extra classification loss $\mathcal{L}_{clf}$ is designed as the cross-entropy between frame-level binary AU occurrence annotations and the predicted AU occurrence probabilities, as shown in Eq. 4:

$$\mathcal{L}_{clf} = -\frac{1}{T} \sum_{c=1}^{C} \sum_{t=1}^{T} \left[P_{c,t}^{occur} \log \hat{P}_{c,t}^{occur} + \left(1 - P_{c,t}^{occur}\right) \log \left(1 - \hat{P}_{c,t}^{occur}\right) \right], \qquad (4)$$

where P_c^{occur} is the ground-truth binary occurrence label for the AU category c, and $\hat{P}_c^{\text{occur}}$ denotes the predicted occurrence probability of the AU category c.

The total loss function is as follows:

$$\mathcal{L}_{\text{total}} = \mathcal{L}_{\text{start}} + \mathcal{L}_{\text{end}} + \mathcal{L}_{\text{complete}} + \mathbb{1}_{\{e2e=True\}}\mathcal{L}_{\text{clf}}. \tag{5}$$

4   Evaluation

4.1   Experimental Settings

Datasets. We conduct extensive experiments on two commonly used AU benchmark datasets, BP4D [23] and DISFA [15]. **BP4D** involves 23 female and 18 male young adults. 328 videos are captured, resulting in around 140,000 images. Frame-level AU occurrence labels are given by two FACS coders independently. AU1, 2, 4, 6, 7, 10, 12, 14, 15, 17, 23 and 24 are considered on BP4D. **DISFA** involves 26 adult subjects. Each frame is coded manually by a FACS coder with AU intensity labels within a scale of 0 to 5. We select the frames with AU intensity labels greater than 1 as positive samples. AU1, 2, 4, 6, 9, 12, 25, and 26 are considered on DISFA.

For training data construction, we use observation windows with length T to truncate a whole video sequence into video clips with an overlap of $\frac{T}{2}$. Ground-truth AU instances Φ^g as annotations for each video clip are generated from the frame-level binary AU occurrence labels. We conduct a subject-exclusive 3-fold cross-validation for both datasets.

Evaluation Metrics. For AU proposal generation task, not only high precision but also high recall of proposals are pursued. We draw on the experience of action detection and select a set of metrics suitable for our task. Mean average precision (mAP) and average recall (AR) with average number of proposals (AN) at different tIoU thresholds α are reported. We consider a set of thresholds α_1 of $[0.3 : 0.1 : 0.7]$ for mAP and a set of thresholds α_2 of $[0.5 : 0.05 : 0.95]$ for AR@AN. Area under the AR-AN curve (AUC) is also used for evaluation. Values of metrics are calculated for each AU category respectively, and an overall average of all AU categories is given.

Implementation Details. We choose ResNet18 and ResNet34 [8] without the last two linear layers as our backbone networks. We follow two training settings for comparison. For the two-stage training setting, our backbone networks are pre-trained in the first stage by adding an extra linear layer to predict AU occurrence probabilities using frame-level AU occurrence labels, and the weights of the backbone networks are fixed in the second stage. For the end-to-end training setting, a classification branch is added to the backbone networks, as shown in the dashed part of Fig. 1, for predicting frame-level occurrence probabilities, and the classification loss $\mathcal{L}_{\text{clf}}$ is added to the total loss $\mathcal{L}_{\text{total}}$ for back-propagation.

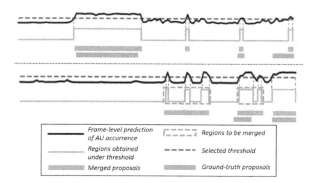

Fig. 2. An example of TAG-based method. First: It is hard to select an appropriate water level for TAG to obtain a true positive for the second ground-truth shown in green line while correctly predicting the boundary of the last ground-truth. Second: Jitters shown in red line are merged into proposals shown in gray bounding boxes through TAG, resulting in one false positive and one true positive of poor quality (Color figure online).

Table 1. Comparison between AUPro and TAG-based method in terms of mAP@tIoU.

Dataset	Backbone	Method	e2e	mAP@0.3	mAP@0.4	mAP@0.5	mAP@0.6	mAP@0.7
BP4D	ResNet18	TAG	✗	10.03	8.97	8.07	7.17	6.22
		AUPro	✗	22.09	19.52	16.71	14.53	12.54
		AUPro	✓	**26.77**	**24.33**	**21.64**	**19.21**	**17.01**
	ResNet34	TAG	✗	10.53	9.04	7.84	6.73	5.69
		AUPro	✗	24.04	21.18	18.15	15.84	13.77
		AUPro	✓	**26.90**	**24.53**	**21.91**	**19.59**	**17.27**
DISFA	ResNet18	TAG	✗	11.06	9.43	7.71	6.21	5.08
		AUPro	✗	14.39	12.98	11.05	9.06	6.15
		AUPro	✓	**15.95**	**13.75**	**11.10**	**9.53**	**8.22**
	ResNet34	TAG	✗	11.20	10.11	8.90	7.70	6.01
		AUPro	✗	16.28	13.98	11.62	8.57	6.59
		AUPro	✓	**16.61**	**14.86**	**13.64**	**12.05**	**10.61**

We train AUPro with AdamW [14] optimizer setting learning rate to 10^{-3} and weight decay to 10^{-4}. Weights are initialized with Xavier init [6]. Batch size is set to 4 and the number of training epochs is set to 15 for each fold. All models are trained on one NVIDIA Tesla V100 GPU.

4.2 AUPro vs. TAG-Based Method

To show the necessity of the proposed task, we implement a frame-level AU recognition algorithm and apply a post-processing method to transfer the predicted frame-level AU occurrence probabilities to proposals, which we refer to as TAG-based method.

Table 2. Comparison between AUPro and TAG-based method in terms of AR@AN and AUC.

Dataset	Backbone	Method	e2e	AR@10	AR@50	AR@100	AUC
BP4D	ResNet18	TAG	✗	3.83	24.15	60.15	27.34
		AUPro	✗	48.22	**73.83**	74.51	68.17
		AUPro	✓	**53.38**	73.52	**79.11**	**69.21**
	ResNet34	TAG	✗	4.80	29.42	64.43	30.52
		AUPro	✗	49.09	73.79	75.85	68.49
		AUPro	✓	**54.62**	**75.75**	**81.25**	**71.18**
DISFA	ResNet18	TAG	✗	10.61	33.42	59.29	33.02
		AUPro	✗	42.40	67.62	74.51	62.35
		AUPro	✓	**47.65**	**72.37**	**76.18**	**67.15**
	ResNet34	TAG	✗	12.14	35.28	59.92	34.64
		AUPro	✗	46.78	68.57	73.61	63.42
		AUPro	✓	**47.76**	**69.06**	**75.62**	**64.47**

TAG-Based Method. In our TAG-based method, a backbone network that is the same as that in AUPro acts as a feature extractor, followed by a classifier for prediction. ResNet18 and ResNet34 are used as backbone networks, and two linear layers are added as a classifier. Temporal Actionness Grouping (TAG) [26] is used for post-processing. TAG applies a classic watershed algorithm to a 1D signal formed by a sequence of AU occurrence probabilities. Two hyper-parameters, *water level* γ and *union threshold* τ, are required for TAG. By horizontally mirroring the signal sequence, a complement signal sequence which looks like terrain with many basins is obtained. After flooding it with different γ, some basins are filled with *water* while some higher terrain not. Adjacent basins are merged until the fraction of the tIoU of these basins drops below τ, resulting in a set of merged basins regarded as proposals. An example is given in Fig. 2.

To obtain sufficient candidate proposals, we sample γ and τ within $(0, 1)$ with a step of 0.05, and transfer frame-level probabilities into proposals under each combination of these two hyper-parameters. The confidence score of each candidate proposal (s, e, c) is set to the average probability for category c from frame s to frame e. Soft-NMS [1] is used to select a final set of proposals.

Comparison. For a fair comparison, we select the top-100 proposals of each category for TAG-based method. As shown in Table 1 and Table 2, AUPro outperforms TAG-based method by a large margin under mAP and AUC given any tIoU threshold, regardless of which backbone network is used. Especially, end-to-end AUPro achieves a performance gain of 16.37% in mAP@0.3 and 40.66% in AUC on BP4D compared with TAG-based method using ResNet34 for feature extraction, which shows the superiority of AUPro. TAG-based method obtains poor results due to the lack of a global view, and the impact of prediction jitters

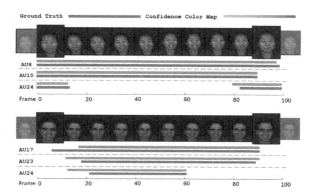

Fig. 3. Visualization of some generated proposals (Color figure online).

on the quality of proposals can not be neglected. These limitations of the first scheme reflect the necessity of sequence-level AU analysis.

Noted that the performance of AUPro following the end-to-end training setting is better than that of the two-stage training setting, because losses for proposal generation can also influence *Backbone*'s weights updating through backpropagation, resulting in a stronger *Backbone* for sequence-level AU analysis.

4.3 Qualitative Results

We visualize several ground-truth AU instances (green bars) and matched proposals (warm color bars) as shown in Fig. 3. The color for each proposal is selected according to its confidence score. From Fig. 3 we can see that for the top video clip, AUPro successfully generates proposals covering the ground-truth AU instances precisely with high confidence scores. Instances of AU6 (Cheek Raiser) and AU10 (Upper Lip Raiser) last nearly the whole time and the proposals generated for these two AUs almost coincide with the corresponding ground-truths. There are two instances for AU24 (Lip Pressor) near both ends of the video clip, and the two matched proposals cover each of them precisely. The bottom part of Fig. 3 shows matched pairs of ground-truths and proposals for AU17 (Chin Raiser), AU23 (Lip Tightener), and AU24. All three AUs cause appearance changes around the mouth and are much harder to distinguish. Our AUPro is still managed to generate precise proposals for AU17 and AU23, while the estimated boundaries of the proposal for AU24 are slightly inaccurate.

5   Conclusion

This paper aims to fill in the blank of sequence-level AU analysis, which are more practical for many real-world application scenarios. We proposed a multi-label AU proposal generation task and provide two schemes to tackle the task. Since there are several drawbacks of the first scheme based on frame-level AU recognition, we designed AUPro following the second scheme to generate AU proposals directly from a whole video clip. Experimental results show that AUPro outperforms TAG-based method following the first scheme by a large margin, which demonstrates the superiority of our method.

Acknowledgements. This work is in part supported by the PKU-NTU Joint Research Institute (JRI) sponsored by a donation from the Ng Teng Fong Charitable Foundation.

References

1. Bodla, N., Singh, B., Chellappa, R., Davis, L.S.: Soft-NMS - improving object detection with one line of code. In: Proceedings of the IEEE International Conference on Computer Vision (ICCV), October 2017 (2017)
2. Chen, Y., Wu, H., Wang, T., Wang, Y., Liang, Y.: Cross-modal representation learning for lightweight and accurate facial action unit detection. IEEE Robot. Autom. Lett. **6**(4), 7619–7626 (2021)
3. Cohn, J.F., Schmidt, K.: The timing of facial motion in posed and spontaneous smiles. In: Active Media Technology, pp. 57–69. World Scientific (2003)
4. Ekman, P., Friesen, W.: Facial action coding system: a technique for the measurement of facial movement (1978)
5. Gao, J., Yang, Z., Chen, K., Sun, C., Nevatia, R.: Turn TAP: temporal unit regression network for temporal action proposals. In: Proceedings of the IEEE International Conference on Computer Vision (ICCV), October 2017 (2017)
6. Glorot, X., Bengio, Y.: Understanding the difficulty of training deep feedforward neural networks. In: Teh, Y.W., Titterington, M. (eds.) Proceedings of the 13th International Conference on Artificial Intelligence and Statistics. Proceedings of Machine Learning Research, Chia Laguna Resort, Sardinia, Italy, 13–15 May 2010, vol. 9, pp. 249–256. PMLR (2010)
7. He, J., Li, D., Yang, B., Cao, S., Sun, B., Yu, L.: Multi view facial action unit detection based on CNN and BLSTM-RNN. In: International Conference on Automatic Face and Gesture Recognition (2017)
8. He, K., Zhang, X., Ren, S., Sun, J.: Deep residual learning for image recognition. In: Proceedings of the IEEE Conference on Computer Vision and Pattern Recognition (2016)
9. Li, G., Zhu, X., Zeng, Y., Wang, Q., Lin, L.: Semantic relationships guided representation learning for facial action unit recognition. In: Proceedings of the AAAI Conference on Artificial Intelligence, vol. 33, pp. 8594–8601 (2019)
10. Li, W., Abtahi, F., Zhu, Z.: Action unit detection with region adaptation, multi-labeling learning and optimal temporal fusing. In: Proceedings of the IEEE Conference on Computer Vision and Pattern Recognition (2017)
11. Lin, C., et al.: Fast learning of temporal action proposal via dense boundary generator. In: Proceedings of the AAAI Conference on Artificial Intelligence, vol. 34, pp. 11499–11506 (2020)
12. Lin, T., Liu, X., Li, X., Ding, E., Wen, S.: BMN: boundary-matching network for temporal action proposal generation (2019)

13. Lin, T., Zhao, X., Su, H., Wang, C., Yang, M.: BSN: boundary sensitive network for temporal action proposal generation. In: Ferrari, V., Hebert, M., Sminchisescu, C., Weiss, Y. (eds.) ECCV 2018. LNCS, vol. 11208, pp. 3–21. Springer, Cham (2018). https://doi.org/10.1007/978-3-030-01225-0_1

14. Loshchilov, I., Hutter, F.: Decoupled weight decay regularization (2019)

15. Mavadati, S.M., Mahoor, M.H., Bartlett, K., Trinh, P., Cohn, J.F.: DISFA: a spontaneous facial action intensity database. IEEE Trans. Affect. Comput. **4**, 151–160 (2013)

16. Niu, X., Han, H., Shan, S., Chen, X.: Multi-label co-regularization for semi-supervised facial action unit recognition. arXiv preprint arXiv:1910.11012 (2019)

17. Schmidt, K.L., Ambadar, Z., Cohn, J.F., Reed, L.I.: Movement differences between deliberate and spontaneous facial expressions: Zygomaticus major action in smiling. J. Nonverbal Behav. **30**(1), 37–52 (2006)

18. Senechal, T., Rapp, V., Salam, H., Seguier, R., Bailly, K., Prevost, L.: Combining AAM coefficients with LGBP histograms in the multi-kernel SVM framework to detect facial action units. In 2011 IEEE International Conference on Automatic Face Gesture Recognition (FG), pp. 860–865 (2011)

19. Shou, Z., Wang, D., Chang, S.-F.: Temporal action localization in untrimmed videos via multi-stage CNNs. In: Proceedings of the IEEE Conference on Computer Vision and Pattern Recognition (CVPR), June 2016 (2016)

20. Tong, Y., Liao, W., Ji, Q.: Facial action unit recognition by exploiting their dynamic and semantic relationships. IEEE Trans. Pattern Anal. Mach. Intell. **29**(10), 1683–1699 (2007)

21. Walecki, R., Rudovic, O., Pavlovic, V., Pantic, M.: Copula ordinal regression framework for joint estimation of facial action unit intensity. IEEE Trans. Affect. Comput. **10**(3), 297–312 (2017)

22. Wang, C., Wang, S.: Personalized multiple facial action unit recognition through generative adversarial recognition network. In: Proceedings of the 26th ACM international conference on Multimedia, pp. 302–310 (2018)

23. Zhang, X., et al.: BP4D-Spontaneous: a high-resolution spontaneous 3D dynamic facial expression database. Image Vis. Comput. **32**(10), 692–706 (2014)

24. Zhang, Y., Jiang, H., Wu, B., Fan, Y., Ji, Q.: Context-aware feature and label fusion for facial action unit intensity estimation with partially labeled data. In: Proceedings of the IEEE/CVF International Conference on Computer Vision, pp. 733–742 (2019)

25. Zhao, K., Chu, W., Zhang, H.: Deep region and multi-label learning for facial action unit detection. In: Proceedings of the IEEE Conference on Computer Vision and Pattern Recognition (2016)

26. Zhao, Y., Xiong, Y., Wang, L., Wu, Z., Tang, X., Lin, D.: Temporal action detection with structured segment networks. In: Proceedings of the IEEE International Conference on Computer Vision (ICCV), October 2017 (2017)

27. Zhu, Y., De la Torre, F., Cohn, J.F., Zhang, Y.J.: Dynamic cascades with bidirectional bootstrapping for action unit detection in spontaneous facial behavior. IEEE Trans. Affect. Comput. **2**(2), 79–91 (2011)

Deep Kernelized Network for Fine-Grained Recognition

M. Amine Mahmoudi[1(✉)], Aladine Chetouani[2], Fatma Boufera[1], and Hedi Tabia[3]

[1] Mustapha Stambouli University of Mascara, Mascara, Algeria
mohamed.mahmoudi@univ-mascara.dz
[2] Laboratoire PRISME, Université d'Orléans, Orléans, France
[3] IBISC laboratory, University of Paris-Saclay, Paris, France

Abstract. Convolutional Neural Networks (CNNs) are based on linear kernel at different levels of the network. Linear kernels are not efficient, particularly, when the original data is not linearly separable. In this paper, we focus on this issue by investigating the impact of using higher order kernels. For this purpose, we replace convolution layers with Kervolution layers proposed in [28]. Similarly, we replace fully connected layers alternatively with Kernelized Dense Layers (KDL) proposed in [16] and Kernel Support vector Machines (SVM) [1]. These kernel-based methods are more discriminative in the way that they can learn more complex patterns compared to the linear one. Those methods first maps input data to a higher space. After that, they learn a linear classifier in that space which is similar to a powerful non-linear classifier in the first space. We have used Fine-Grained datasets namely FGVC-Aircraft, StanfordCars and CVPRIndoor as well as Facial Expression Recognition (FER) datasets namely, RAF-DB, ExpW and FER2013 to evaluate the performance of these methods. The experimental results demonstrate that these methods outperform the ordinary linear layers when used in a deep network fashion.

Keywords: Facial expression recognition · Fine-grained recognition · Kernel function · Deep learning

1 Introduction

Image classification has always been the core operation of computer vision. Several methods have been proposed in the literature addressing this task, which consists in efficiently assigning the correct label to an image. Recently, with the emergence of Convolutional Neural Network (CNN), the computer vision community has witnessed an era of blossoming result thanks to the use of very large training databases. These databases contain a very large number of different images (i.e. objects, animals...etc.). This advance encouraged the computer vision community to go beyond classical image classification that recognizes basic-level categories. The new challenge consists of discriminating categories that were considered previously as a single category and have only small subtle visual differences. This new sub-topic of image classification, called fine-grained image classification, is receiving a special attention from the computer vision community [2,6,10,11,15,27,29,32]. Such methods aim at discriminating

© Springer Nature Switzerland AG 2021
T. Mantoro et al. (Eds.): ICONIP 2021, LNCS 13110, pp. 100–111, 2021.
https://doi.org/10.1007/978-3-030-92238-2_9

between classes in a sub-category of objects like birds species [30], models of cars [13], and facial expressions, which makes the classification more difficult due to the high intra-class and low inter-class variations. State-of-the-art approaches typically rely on CNN as classification backbone and propose a method to improve its awareness to subtle visual details.

CNNs have been used for a multitude of visual tasks. They showed to perform very competitive results while linear operations are used at different layers of the network. Linear functions are efficient, specially, when data can be separated linearly. In such a case, the decision boundary is likely to be representable as a linear combination of the original features. It is worth noting that not all high dimensional problems are linearly separable [23]. For instance, images may have a high dimensional representation, but individual pixels are not very informative. Moreover, taking in consideration only small regions of the image, dramatically reduces their dimension, which makes linear functions less sensitive to subtle changes in input data. The ability of detecting such differences is crucial essentially for fine-grained recognition. To overcome these limitations, some researches investigated different ways to include non-linear functions in CNNs. Starting from non linear activation functions like ReLU, eLU [4], SeLU [12] and more recently [26]. Moreover, some recent work intended to replace the underlying linear function of a CNN by non linear kernel function without resorting to activation functions.

In this paper, we investigate the usage of different kernel methods in a CNN fashion. The goal here is to use the structure of a CNN, but instead of using simple linear kernel function which are usually in CNN layers, we use Higher degree kernel function. These kernel function are more discriminative than linear function, therefore they can bring more discrimination power to the network. However, the impact of these function can vary from a layer to another. For this purpose, we first replace the convolution operation in CNNs by a non-linear kernel function similarly to Kervolution [28]. We also replace fully connected layer alternatively with Kernelized Dense Layers (KDL) [16] and kernel SVM [1]. The remainder of this paper is organized as follows: Sect. 2 reviews similar work that have been proposed for the improvement of CNN using kernel functions. Section 3 introduces the study design. Section 4 presents our experiments setting, the datasets we used and their related results. Finally, Sect. 5 concludes the paper.

2   Related Work

Kernel-based learning machines, like kernel Support Vector Machines (SVMs) [1], kernel Fisher Discriminant (KFD) [21], and kernel Principal Component Analysis (KPCA) [24], have been widely used in literature for various tasks. Their ability to operate in a high-dimensional feature space, allow them to be more accurate than linear models. These approaches have shown practical relevance for some specific fields like computer vision. Even-though CNNs have boosted dramatically the pattern recognition field, some recent works intended to take advantage of kernel-based learning machines to further boost CNN performance.

To enhance the CNN performances, some researches tried either to increase the network size or employ more complex functions. Chen et al. [3] proposed a new designed dynamic convolution that increases the model complexity without increasing the network depth or width. It uses multiple convolution kernels in parallel instead of using a single one per layer. These convolution kernels are aggregated dynamically via input-dependent attention. However, even though dynamic convolution does not increase the output dimension of each layer, it does increase the model size. Haase et al. [8] introduced Blueprint Separable Convolutions (BSConv) which represents each filter kernel using one 2D blueprint kernel which is distributed along the depth axis using a weight vector.

Some other work intended to replace the underlying linear function of a CNN by non-linear kernel functions. For instance, some of them replaced convolution layers [28], while others replaced the pooling layers ([5, 18]). More recently, Mahmoudi et al. [17] proposed a novel pooling layer that extracts non-linear relations between data while down-sampling input the later. It is based on kernel functions which can be considered as a generalisation of linear pooling to capture higher order data. Mahmoudi et al. [16, 19] also incorporated later more complex kernel functiona in FC layer. These layers uses higher degree function on their input data instead of the usual weighted sum. These layers allow to enhance the discrimination power of the full network. This is ensured by the end-to-end learning due to the fact that these layers are completely differentiable. The higher order kernel functions are, the more susceptible they are to fit slight changes in data. It is worth noting that all of these methods have been proposed in the case of fine-grained recognition. This explains our interest for improving CNNs using higher degree kernel function for facial expression recognition problem.

3   Study Design

In this section, we describe the study design that we have followed in order to investigate the impact of kernel function methods when used in a deep network fashion. Figure 1 shows the different deep kernel networks configuration used for this study. In the first configuration (a), we used a Kervolution based network followed by fully connected layers. This first configuration is used only for training the Kervolution based network that will be used in later configurations. In the second configuration (b), we used the same Kervolution based network followed by Kernelized dense layers (KDL). Finally, for the third configuration, we took also the Kervolution based network and plugged a kernel SVM at its end. The two last configurations represent two fully kernelized deep networks that our study will be based on. In the following, we give a brief presentation of these methods.

3.1   Kervolution

Kervolution has been proposed by Wang et al. [28] to replace convolution layers in a deep CNN. It extends the convolution operation which computes the dot product between an input vector X and a weight vector W, and adds eventually a bias term, according to the following Eq. 1:

$$C_{o,i,j} = (X \times W)_{i,j} = \sum_g \sum_h X_{g+i,h+j} W_{o,g,h} + B_o, \tag{1}$$

where o corresponds to the output size, i and j are specific locations in the input, g and h are respectively the width and height of the convolution filter and B is a bias term.

Convolution is a linear operation that usually requires adding an activation function to introduce non-linearity. Without these activation functions the CNN performance drops dramatically. Kervolution leverages this fact and proposes to replace the convolution operation in CNNs by a non-linear function that performs the same task as convolution without resorting to activation functions. In this paper, we have used two polynomial kernels of second and third degree, according to following equation:

$$K_{o,i,j} = \langle X, W \rangle_{i,j} = \sum_g \sum_h (X_{g+i,h+j} W_{o,g,h} + C)^n + B_o \qquad (2)$$

where C ($C \in \mathbb{R}^+$) is a learnable constant and n ($n \in \mathbb{Z}^+$) is the polynomial order. It extends the feature space to n dimensions. It is worth noting that this layer is fully differentiable and can be used at any level of the network, allowing an end-to-end training.

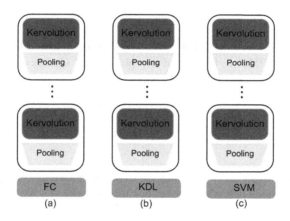

Fig. 1. The deep kernel networks configuration used for this study.

3.2   Kernelized Dense Layer

On the other hand, we replace the fully connected layers by a novel dense layer called Kernelized Dense Layer (KDL). The Kernelized Dense Layer are composed of neuron that uses a kernel function instead of the usual dot product. These Kernelized Dense Layer (KDL), proposed in [16], is similar to a classical neuron layer in the way that it applies a dot product between a vector of weights W and an input vector X, add a bias vector ($B \geq 0$) and eventually applies an activation function σ as follows:

$$Y_i = \sigma(\sum_j X_j W_{i,j} + B_i) \qquad (3)$$

The difference from standard fully-connected layers is that KDL applies higher degree kernel function instead of a simple linear dot product, which allows the model to map the input data to a higher space and thus be more discriminative than a classical linear layer, according to the following equation:

$$Y_i = \sum_j (X_j W_{i,j} + C)^n + B_i \qquad (4)$$

where C $(C \in \mathbb{R}^+)$ is a learnable constant and n $(n \in \mathbb{Z}^+)$ is the polynomial order, it extends the KDL to n dimensions. This layer is also fully differentiable allowing an end-to-end training.

3.3 Support Vector Machine

Finally, we have also replaced the fully connected layers with kernel SVMs. The kernel SVMs return the inner product between two points in a suitable feature space. Thus by defining a notion of similarity, with little computational cost even in very high-dimensional spaces, according to the following equation:

$$K(X_i, X_j) = (X_i X_j + C)^n \qquad (5)$$

where C $(C \in \mathbb{R}^+)$ is constant and n $(n \in \mathbb{Z}^+)$ is the polynomial order. Similarly to Kervolution and KDL, we have used two polynomial kernels of second and third degree. However, unlike Kervolution and KDL, kernel SVM can not be used in an end-to-end training. Therefore, we take the kervolution backbone trained before with fully connected layers and use it as features extractor for the kernel SVM classifier.

4 Experiments

In this section, we explain in more details the experiments we have performed in order to evaluate our methods described above. For this purpose, we have used three well-known fine-grained as well as three well-known FER datasets. Details of these datasets are given below. After that, implementation details are given including the models architectures and the training process. Finally, we discuss the obtained results are discuss the improvement given by each technique.

4.1 Datasets

We evaluated our models on two categories of datasets. First of all, we used three well-known fine-grained datasets, namely, FGVC-Aircraft [20], StanfordCars [13] and CVPRIndoor [22].

- The **FGVC-Aircraft** [20] dataset is composed of 10,200 images of aircraft images. These images are categorized in 102 different aircraft model variants, most of which are airplanes.
- The **StanfordCars** [13] dataset is composed of 16,185 images of 196 categories of cars. These images are divided into 8,144 for training and 8,041 for testing.
- The **CVPRIndoor** [22] dataset is composed of 15620 images categorized in 67 Indoor classes with at least 100 images per category.

In addition, we also tested our approach on three FER datasets, namely, RAF-DB, ExpW and FER2013.

- The **RAF-DB** [14] or Real-world Affective Face DataBase is composed of 29,672 facial in the wild images. This images are categorized in either seven basic classes or eleven compound classes.
- The **ExpW** [31] or EXPression in-the-Wild dataset is composed of 91,793 facial in the wild images. The annotation was done manually on individual images.
- The **FER2013** database was used for the ICML 2013 Challenges in Representation Learning [7]. It is composed of 28,709 training images and 3,589 images for both validation and test.

The intuition behind that was to prove that kernel based models are efficient for recognizing object with subtle visual details. It is worth noting that both categories of the datasets used in this paper meet these criteria.

4.2 Models Architecture and Training Process

To demonstrate the efficiency of the proposed method, we used VGG-16 [25] like models with two fully connected layers of 256 units each and a final softmax layer. In addition, we built a second model from scratch (Fig. 2). This model architecture is composed of five Kervolutional blocks, each followed by a max pooling layer and a dropout layer. A block consists of a Kervolutional layer, batch normalization layers. We will refer to these models as VGG-16-base and Model-1, respectively. For training, we have used Adam optimiser using a learning rate starting from 0.001 to 5e−5. This learning rate is lowered by a factor of 0.5 if the validation accuracy does not improve for five epochs. To avoid overfitting, we first augmented the data using a shear intensity of 0.2, a range degree for random rotations of 20, and randomly flip the inputs horizontally, a range for random zoom of 0.2. We also employed early stopping if the validation accuracy does not increase by a factor of 0.01 over ten epochs. Each layer of our model is initialized with He normal distribution [9] and a weight decay of 0.0001. The only preprocessing we employed on all our experiments is cropping the face region and resizing the resulting images to 100×100 pixels for FER dataset. For both FGVC-Aircraft and StanfordCars, we cropped the object region. We resize the resulting images so that its longer side is 100 while keeping its aspect ratio. We have also zero padded these images to obtain 100×100 pixels.

We first trained our two base kervolutional models, VGG-16-base and Model-1 with fully connected layers. These models use both second and third order polynomial kernels. After that, we took the Kervolution backbones and replaced the fully connected layers with KDL and SVM with the same kernel used in the later (Fig. 1). These resulted in six different configurations for each model.

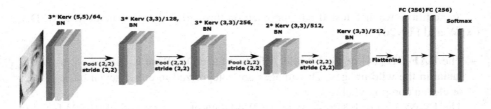

Fig. 2. Model-1 architecture: it is composed of five Kervolutional blocks. Each block consists of a Kervolutional layer, batch normalization layers. Each block is followed by a max pooling layer and a dropout layer. Finally, two fully-connected layers are added on top of these convolution blocks with respectively 256 units and ReLU activation and an output softmax layer.

4.3   Ablation Study

This section explores the impact of kernel-based methods on the overall network accuracy, following the network configurations explained above. The results obtained with these models are reported as VGG-16-base and Model-1 in Table 1 and Table 2 for fine-grained and FER datasets, respectively.

Table 1. Results of the different configurations on fine-grained datasets.

Model	Kernel		FGVC-Aircraft	StanfordCars	IndoorCVPR
VGG-16	FC	2nd order poly	65.42%	62.15%	64.23%
		3rd order poly	65.87%	62.64%	64.92%
	KDL	2nd order poly	66.11%	62.95%	65.06%
		3rd order poly	66.72%	63.36%	65.7%
	SVM	2nd order poly	66.4%	63.31%	65.52%
		3rd order poly	67.08%	63.96%	66.01%
Model-1	FC	2nd order poly	65.88%	62.74%	64.79%
		3rd order poly	66.45%	63.09%	65.32%
	KDL	2nd order poly	65.27%	62.19%	64.08%
		3rd order poly	66.8%	63.68%	65.88%
	SVM	2nd order poly	65.59%	62.79%	64.6%
		3rd order poly	67.03%	63.36%	64.24%

As shown in Table 1, the accuracy rates obtained with VGG-16-base using second order Kervolution and fully connected layers are 65.42%, 62.15%, 64.23 % on FGVC-Aircraft, StanfordCars and IndoorCVPR, respectively. Whereas for third order Kervolution VGG-16-base and fully connected layers, the obtained accuracy rates on FGVC-Aircraft, StanfordCars and IndoorCVPR, respectively, are 65.87%, 62.64%, 64.92%. On the other hand, the accuracy rates obtained with VGG-16 like model are 66.11%

on FGVC-Aircraft, 62.95% on StanfordCars and 65.06% on IndoorCVPR, with second order Kervolution and KDL. This represents an improvement up to 0.8% over the fully connected layers. Whereas, with Kervolution and KDL third degree polynomial, the obtained accuracy rates are 66.72% on FGVC-Aircraft, 63.36% on StanfordCars, 65.7% on IndoorCVPR. This kernel improves further the accuracy of the model up to 0.9% more than its FC counterpart. Lastly, the accuracy rates obtained with VGG-16 like model are 66.4% on FGVC-Aircraft, 63.31% on StanfordCars, 65.52% on Indoor-CVPR, with second order Kervolution and SVM. This represents an improvement up to 1.3% over the fully connected layers. Whereas, with Kervolution and SVM third degree polynomial, the obtained accuracy rates are 67.08% on FGVC-Aircraft, 63.96% on StanfordCars, 66.01% on IndoorCVPR. This kernel improves further the accuracy of the model up to 1.8% more than its FC counterpart.

On the other hand, the accuracy rates obtained with Model-1 using second order Kervolution and fully connected layers are 65.88%, 62.74%, 64.79% on FGVC-Aircraft, StanfordCars and IndoorCVPR, respectively. Whereas for third order Kervolution Model-1 and fully connected layers, the obtained accuracy rates on FGVC-Aircraft, StanfordCars and IndoorCVPR, respectively, are 66.45%, 63.09%, 65.32%. On the other hand, the accuracy rates obtained with Model-1 are 65.27% on FGVC-Aircraft, 62.19% on StanfordCars and 64.08% on IndoorCVPR, with second order Kervolution and KDL. This represents an improvement up to 0.5% over the fully connected layers. Whereas, with Kervolution and KDL third degree polynomial, the obtained accuracy rates are 66.8% on FGVC-Aircraft, 63.68% on StanfordCars, 65.88% on IndoorCVPR. This kernel improves further the accuracy of the model up to 0.6% more than its FC counterpart. Lastly, the accuracy rates obtained with Model-1 are 65.59% on FGVC-Aircraft, 62.79% on StanfordCars, 64.6% on IndoorCVPR, with second order Kervolution and SVM. This represents an improvement up to 0.9% over the fully connected layers. Whereas, with Kervolution and SVM third degree polynomial, the obtained accuracy rates are 67.03% on FGVC-Aircraft, 63.36% on StanfordCars, 64.24% on IndoorCVPR. This kernel improves further the accuracy of the model up to 1.5% more than its FC counterpart.

As shown in Table 2, the accuracy rates obtained with VGG-16-base using second order Kervolution and fully connected layers are 86.42%, 69.57%, 74.13% on RAF-DB, FE2013 and ExpW, respectively. Whereas for third order Kervolution VGG-16-base and fully connected layers, the obtained accuracy rates on RAF-DB, FE2013 and ExpW, respectively, are 87.08%, 69.85%, 74.46%. On the other hand, the accuracy rates obtained with VGG-16 like model are 86.72% for RAF-DB, 69.97% for FER2013, 74.84% for ExpW, with second order Kervolution and KDL. This represents an improvement up to 0.7% over the fully connected layers. Whereas, with Kervolution and KDL third degree polynomial, the obtained accuracy rates are 87.46% for RAF-DB, 70.29% for FER2013, 75.08% for ExpW. This kernel improves further the accuracy of the model up to 0.6% more than its FC counterpart. Lastly, the accuracy rates obtained with VGG-16 like model are 87.11% for RAF-DB, 70.32% for FER2013, 75.27% for ExpW, with second order Kervolution and SVM. This represents an improvement up to 1.2% over the fully connected layers. Whereas, with Kervolution and SVM third degree polynomial, the obtained accuracy rates are 87.82% for RAF-DB, 70.62% for FER2013,

Table 2. Results of the different configurations on FER datasets.

Model	Kernel		RAF - DB	FER2013 - DB	ExpW - DB
VGG-16	FC	2nd order poly	86.42%	69.57%	74.13%
		3rd order poly	87.08%	69.85%	74.46%
	KDL	2nd order poly	86.72%	69.97%	74.84%
		3rd order poly	87.46%	70.29%	75.08%
	SVM	2nd order poly	87.11%	70.32%	75.27%
		3rd order poly	87.82 %	70.62%	76%
Model-1	FC	2nd order poly	87.77%	70.68%	76.25%
		3rd order poly	87.93%	70.95%	76.32%
	KDL	2nd order poly	87.91%	71.01%	76.72%
		3rd order poly	88.14%	71.13%	76.96%
	SVM	2nd order poly	88.52%	71.28%	77.09%
		3rd order poly	88.71 %	71.41%	77.87%

76% for ExpW. This kernel improves further the accuracy of the model up to 1.6% more than its FC counterpart.

The accuracy rates obtained with Model-1 using second order Kervolution and fully connected layers are 87.77%, 70.68%, 76.25% on RAF-DB, FE2013 and ExpW, respectively. Whereas for third order Kervolution Model-1 and fully connected layers, the obtained accuracy rates on RAF-DB, FE2013 and ExpW, respectively, are 87.93% for RAF-DB, 70.95%, for FER 76.32 for ExpW. On the other hand, the accuracy rates obtained with Model-1 are 87.91% for RAF-DB, 71.01% for FER2013, 76.72% for ExpW, with second order Kervolution and KDL. This represents an improvement up to 0.5% over the fully connected layers. Whereas, with Kervolution and KDL third degree polynomial, the obtained accuracy rates are 88.14% for RAF-DB, 71.13% for FER2013, 76.96% for ExpW. This kernel improves further the accuracy of the model up to 0.6% more than its FC counterpart. Lastly, the accuracy rates obtained with Model-1 are 88.52% for RAF-DB, 71.28% for FER2013, 77.09% for ExpW, with second order Kervolution and SVM. This represents an improvement up to 0.8% over the fully connected layers. Whereas, with Kervolution and SVM third degree polynomial, the obtained accuracy rates are 88.71% for RAF-DB, 71.41% for FER2013, 77.87% for ExpW. This kernel improves further the accuracy of the model up to 1.5% more than its FC counterpart.

As discussed before, one can clearly conclude that kernel based methods improve considerably the network performance in terms of accuracy. Indeed, using all configurations and with all the datasets used, the network performance was enhanced. Moreover, the kernel methods used have different impact on the network, as well as, the kernel function itself. For instance, kernel SVM can improve the accuracy from 0.8% to 1.5% using second and third order polynomial kernel respectively. Whereas, KDL can improve the accuracy from 0.5% using second polynomial kernel to 0.7% using third order polynomial kernel respectively.

5   Conclusion

In this paper, we investigated the impact of using higher order kernels at different levels of the network. For this purpose, we replaced convolution layers with Kervolution layers proposed in [28]. Similarly, we replaced fully connected layers alternatively with Kernelized Dense Layers (KDL) proposed in [16] and Kernel Support vector Machines (SVM) [1]. These kernel-based methods are more discriminative in the way that they can learn more complex patterns compared to the linear one. Those methods first maps input data to a higher space. After that, they learn a linear classifier in that space which is similar to a powerful non-linear classifier in the first space. The experimental results performed on challenging Fine-Grained datasets namely, FGVC-Aircraft, StanfordCars and CVPRIndoor as well Facial Expression Recognition (FER) datasets namely, RAF-DB, ExpW and FER2013. demonstrate that these methods outperform the ordinary linear layers when used in a deep network fashion. Finally, all kernel based methods considered in this work allow to improve the network performance. The best result was achieved by kernel SVMs followed by KDL, then Kervolution with FC layers.

As future work, we intend to incorporate more complex kernel functions at different levels of CNNs. Furthermore, We are studying different methods to allow a network to chose automatically the most efficient kernel function to use. This can be performed in a similar manner to the back-propagation algorithm.

References

1. Burges, C.J., Scholkopf, B., Smola, A.J.: Advances in Kernel Methods: Support Vector Learning. MIT Press Cambridge, Cambridge (1999)
2. Chen, S., Zhao, Y., Jin, Q., Wu, Q.: Fine-grained video-text retrieval with hierarchical graph reasoning. In: Proceedings of the IEEE/CVF Conference on Computer Vision and Pattern Recognition, pp. 10638–10647 (2020)
3. Chen, Y., Dai, X., Liu, M., Chen, D., Yuan, L., Liu, Z.: Dynamic convolution: attention over convolution Kernels. In: Proceedings of the IEEE/CVF Conference on Computer Vision and Pattern Recognition (2020)
4. Clevert, D.A., Unterthiner, T., Hochreiter, S.: Fast and accurate deep network learning by exponential linear units (ELUs). arXiv preprint arXiv:1511.07289 (2015)
5. Cui, Y., Zhou, F., Wang, J., Liu, X., Lin, Y., Belongie, S.: Kernel pooling for convolutional neural networks. In: Proceedings of the IEEE Conference on Computer Vision and Pattern Recognition (2017)
6. Gao, Y., Han, X., Wang, X., Huang, W., Scott, M.: Channel interaction networks for fine-grained image categorization. In: AAAI, pp. 10818–10825 (2020)
7. Goodfellow, I.J., et al.: Challenges in representation learning: a report on three machine learning contests. In: Lee, M., Hirose, A., Hou, Z.-G., Kil, R.M. (eds.) ICONIP 2013. LNCS, vol. 8228, pp. 117–124. Springer, Heidelberg (2013). https://doi.org/10.1007/978-3-642-42051-1_16
8. Haase, D., Amthor, M.: Rethinking depthwise separable convolutions: how intra-kernel correlations lead to improved MobileNets. In: Proceedings of the IEEE/CVF Conference on Computer Vision and Pattern Recognition, pp. 14600–14609 (2020)
9. He, K., Zhang, X., Ren, S., Sun, J.: Delving deep into rectifiers: surpassing human-level performance on ImageNet classification. In: Proceedings of the IEEE International Conference on Computer Vision (ICCV), December 2015

10. Huang, Z., Li, Y.: Interpretable and accurate fine-grained recognition via region grouping. In: Proceedings of the IEEE/CVF Conference on Computer Vision and Pattern Recognition, pp. 8662–8672 (2020)
11. Ji, R., et al.: Attention convolutional binary neural tree for fine-grained visual categorization. In: Proceedings of the IEEE/CVF Conference on Computer Vision and Pattern Recognition, pp. 10468–10477 (2020)
12. Klambauer, G., Unterthiner, T., Mayr, A., Hochreiter, S.: Self-normalizing neural networks. In: Advances in Neural Information Processing Systems, pp. 971–980 (2017)
13. Krause, J., Stark, M., Deng, J., Fei-Fei, L.: 3D object representations for fine-grained categorization. In: 4th International IEEE Workshop on 3D Representation and Recognition (3dRR-13), Sydney, Australia (2013)
14. Li, S., Deng, W., Du, J.: Reliable crowdsourcing and deep locality-preserving learning for expression recognition in the wild. In: 2017 IEEE Conference on Computer Vision and Pattern Recognition (Computer Vision and Pattern Recognition), pp. 2584–2593. IEEE (2017)
15. Liu, Z., et al.: 3D part guided image editing for fine-grained object understanding. In: Proceedings of the IEEE/CVF Conference on Computer Vision and Pattern Recognition, pp. 11336–11345 (2020)
16. Mahmoudi, M.A., Chetouani, A., Boufera, F., Tabia, H.: Kernelized dense layers for facial expression recognition. In: 2020 IEEE International Conference on Image Processing (ICIP), pp. 2226–2230 (2020)
17. Mahmoudi, M.A., Chetouani, A., Boufera, F., Tabia, H.: Learnable pooling weights for facial expression recognition. Patt. Recogn. Lett. **138**, 644–650 (2020)
18. Mahmoudi, M.A., Chetouani, A., Boufera, F., Tabia, H.: Improved bilinear model for facial expression recognition. In: Djeddi, C., Kessentini, Y., Siddiqi, I., Jmaiel, M. (eds.) MedPRAI 2020. CCIS, vol. 1322, pp. 47–59. Springer, Cham (2021). https://doi.org/10.1007/978-3-030-71804-6_4
19. Mahmoudi, M.A., Chetouani, A., Boufera, F., Tabia, H.: Taylor series Kernelized layer for fine-grained recognition. In: 2021 IEEE International Conference on Image Processing (ICIP), pp. 1914–1918. IEEE (2021)
20. Maji, S., Rahtu, E., Kannala, J., Blaschko, M., Vedaldi, A.: Fine-grained visual classification of aircraft. arXiv preprint arXiv:1306.5151 (2013)
21. Mika, S., Ratsch, G., Weston, J., Scholkopf, B., Mullers, K.R.: Fisher discriminant analysis with kernels. In: Neural Networks for Signal Processing IX: Proceedings of the 1999 IEEE Signal Processing Society Workshop (cat. no. 98th8468), pp. 41–48. IEEE (1999)
22. Quattoni, A., Torralba, A.: Recognizing indoor scenes. In: 2009 IEEE Conference on Computer Vision and Pattern Recognition, pp. 413–420. IEEE (2009)
23. Robert, C.: Machine learning, a probabilistic perspective (2014)
24. Schölkopf, B., Smola, A., Müller, K.R.: Nonlinear component analysis as a Kernel eigenvalue problem. Neural Comput. **10**(5), 1299–1319 (1998)
25. Simonyan, K., Zisserman, A.: Very deep convolutional networks for large-scale image recognition. arXiv preprint arXiv:1409.1556 (2014)
26. Sitzmann, V., Martel, J.N., Bergman, A.W., Lindell, D.B., Wetzstein, G.: Implicit neural representations with periodic activation functions. arXiv preprint arXiv:2006.09661 (2020)
27. Tang, L., Wertheimer, D., Hariharan, B.: Revisiting pose-normalization for fine-grained few-shot recognition. In: Proceedings of the IEEE/CVF Conference on Computer Vision and Pattern Recognition, pp. 14352–14361 (2020)
28. Wang, C., Yang, J., Xie, L., Yuan, J.: Kervolutional neural networks. In: Proceedings of the IEEE Conference on Computer Vision and Pattern Recognition, pp. 31–40 (2019)

29. Wang, Z., Wang, S., Yang, S., Li, H., Li, J., Li, Z.: Weakly supervised fine-grained image classification via Guassian mixture model oriented discriminative learning. In: Proceedings of the IEEE/CVF Conference on Computer Vision and Pattern Recognition, pp. 9749–9758 (2020)
30. Welinder, P., et al.: Caltech-UCSD Birds 200. Technical report CNS-TR-2010-001, California Institute of Technology (2010)
31. Zhang, Z., Luo, P., Loy, C.C., Tang, X.: From facial expression recognition to interpersonal relation prediction. Int. J. Comput. Vis. **126**(5), 550–569 (2017). https://doi.org/10.1007/s11263-017-1055-1
32. Zhuang, P., Wang, Y., Qiao, Y.: Learning attentive pairwise interaction for fine-grained classification. In: AAAI, pp. 13130–13137 (2020)

Semantic Perception Swarm Policy with Deep Reinforcement Learning

Tianle Zhang[1,2], Zhen Liu[1,2(✉)], Zhiqiang Pu[1,2], and Jianqiang Yi[1,2]

[1] Institute of Automation, Chinese Academy of Sciences, Beijing 100190, China
{liuzhen,zhiqiang.pu,jianqiang.yi}@ia.ac.cn
[2] School of Artificial Intelligence, University of Chinese Academy of Sciences,
Beijing 100049, China

Abstract. Swarm systems with simple, homogeneous and autonomous individuals can efficiently accomplish specified complex tasks. Recent works have shown the power of deep reinforcement learning (DRL) methods to learn cooperative policies for swarm systems. However, most of them show poor adaptability when applied to new environments or tasks. In this paper, we propose a novel semantic perception swarm policy with DRL for distributed swarm systems. This policy implements innovative semantic perception, which enables agents to understand their observation information, yielding semantic information, to promote agents' adaptability. In particular, semantic disentangled representation with posterior distribution and semantic mixture representation with network mapping are realized to represent semantic information of agents' observations. Moreover, in the semantic representation, heterogeneous graph attention network is adopted to effectively model individual-level and group-level relational information. The distributed and transferable swarm policy can perceive the information of uncertain number of agents in swarm environments. Various simulations and real-world experiments on several challenging tasks, i.e., *sheep food collection* and *wolves predation*, demonstrate the superior effectiveness and adaptability performance of our method compared with existing methods.

1 Introduction

In recent years, swarm systems have received increasing attention from researchers due to its unique benefits and great potential applications. For the benefits, they can coordinately complete complex tasks in a low-cost and decentralized form, and realize dynamic adjustment and self-healing combination in complex environments. Their applications can be founded in warehousing logistics [13], satellite cluster, search and rescue scenarios [2], etc. In a swarm system, many identical agents need to collectively accomplish a common goal through interactions among the agents. However, each agent in the swarm system has limited capabilities, e.g., partial perception, local interaction and restricted manipulation. This puts forward high requirements for efficient cooperation among agents. Meanwhile, the number of agents in the swarm system may be uncertain and changing. This requires that the policies of agents have strong adaptability. Therefore, finding a policy that enable the swarm system to reliably and efficiently complete a specified task in uncertain and dynamic environments remains challenging.

© Springer Nature Switzerland AG 2021
T. Mantoro et al. (Eds.): ICONIP 2021, LNCS 13110, pp. 112–124, 2021.
https://doi.org/10.1007/978-3-030-92238-2_10

Recently, deep reinforcement learning (DRL) methods have shown great potential in multi-agent systems. Some multi-agent reinforcement learning methods, that focus on designing global cooperation mechanisms such as centralized critic [14], communication among agents [11], joint value factorization [8], are born. However, most of them cannot be directly applied to swarm systems due to some challenges, including the high dimensionality of observation caused by large-scale agents, and changes in the size of available information sets due to dynamic number of agents. Although there are challenges, some works have made good progress for swarm systems with DRL. A guided approach is proposed to enable a group of agents with limited sensing capabilities to learn cooperative behaviors [10]. But, this approach adopts a histogram over distances with a fixed dimension as observation inputs, which results in additional information not being efficiently encoded due to its poor scalability. Considering this shortcoming, a new state representation for swarm systems with DRL based on mean embeddings of distributions is proposed [9]. This work uses an empirical mean embedding as input for a decentralized policy. However, it ignores different relations among neighboring agents, which may lead to its poor adaptability in new tasks and environments. These works focus on processing observation information of the surrounding environment of agents, where the size of observation information is variable. But, they do not try to consider understanding and interpreting observation information.

In reality, people can perceive the information around them by listening, seeing and touching, thus producing their own understanding to adapt to real life [3]. Naturally, this can be applied to swarm systems: agents understand their perceived observations to generate semantic information to promote their adaptability. Moreover, semantic information is an abstraction obtained by inferring interpretations from agents' observations and updating these interpretations with new information [6].

Motivated by the above discussions, we propose a novel semantic perception swarm policy (SPSP) with DRL to enable distributed swarm systems to reliably and efficiently accomplish specified tasks in this paper. The main innovation of this policy is to adopt semantic perception, which enables agents to understand their observation information, to enhance the agents' adaptability in new tasks or environments. Specifically, semantic disentangled representation with posterior distribution and semantic mixture representation with network mapping are implemented to represent semantic information about the agents' observations. Moreover, in the semantic representation, heterogeneous graph attention network (HGAT) is adopted to effectively model individual and group level relational representation.

2 Background

2.1 Partially Observable Markov Games

In this paper, we consider a partially observable Markov game (POMG) which is an extension of Markov decision processes to a game with multiple agents. A POMG for N agents is defined as follows: $s \in S$ denotes the global state of the game, $o_i \in O_i$ denotes the local observation of agent i, $a_i \in \mathcal{A}_i$ is an action taken by agent i based on the local observation. Agent i obtains the reward by a reward function $r_i : S \times \mathcal{A}_1 \times \ldots \times \mathcal{A}_N \mapsto \mathbb{R}$. The state of the game evolves to next state according to the state transition function $\mathcal{T} : \mathcal{S} \times \mathcal{A}_1 \times \ldots \times \mathcal{A}_N \mapsto \mathcal{S}$. The initial states are determined by a distribution $\rho : \mathcal{S} \mapsto [0, 1]$. Agent i aims to maximize its own expected cumulative return $R_i = \sum_{t=0}^{T} \gamma^t r_i^t$, where $\gamma \in [0, 1]$ is a discount factor and T is the time horizon.

2.2 β-VAE

Variational autoencoder (VAE) can perform efficient approximate inference and learning with directed probabilistic models whose latent variables have intractable posterior distributions [12]. β-VAE is a modification of the VAE framework. It introduces an adjustable hyperparameter β that balances latent channel capacity and independence constraints with reconstruction accuracy. This enables β-VAE to discovery disentangled representation or interpretable factorised latent from raw data in a completely unsupervised manner [7]. The emergence principle of disentangled representation can be explained from the perspective of the information bottleneck. A disentangled representation can be defined as one where single latent units are sensitive to changes in single generative factors, while being relatively invariant to changes in other factors [4]. In this paper, we adopt β-VAE to perform semantic disentangled representation.

2.3 Distributed Swarm Systems

In this paper, we study a distributed swarm system which is a group of self-organizing, homogeneous and autonomous agents that try to collectively achieve a common goal. In the swarm system, agents are modelled as a disc with an actual shape radius. The state of agent i consists of its own position $p_i = [p_i^x, p_i^y]^T$ and velocity $v_i = [v_i^x, v_i^y]^T$, i.e., $s_i = [p_i, v_i]^T$. Meanwhile, the kinematics of agent i is modeled as a double integrator model, and the action of the agent can be denoted as $a_i = [F_i^x, F_i^y]^T$, where F_i^x, F_i^y represent the force applied to the agent in x and y directions. Meanwhile, each agent in the swarm system has partial observability with radius D^O and local interactivity with radius D^I. Furthermore, the swarm system has state semanticity. Specifically, each observation state contains corresponding semantic information. Semantic information can be represented by utilizing different semantic factors. A semantic factor is a latent encoding of an attribute of observation states. For example, in predator-prey games, a predator can obtain semantic information from its observation information. This semantic information consists of semantic factors which are latent encodings for observation attributes, e.g., collaborators' forces level, prey's situation, etc.

3   Approach

In this section, we propose a swarm policy SPSP with good adaptability for the distributed swarm system with n homogeneous agents to reliably and efficiently accomplish specified tasks. Firstly, the overall design of the proposed SPSP is given. Next, semantic perception which is the main innovation of the policy is presented in detail. Finally, the training method of SPSP is shown.

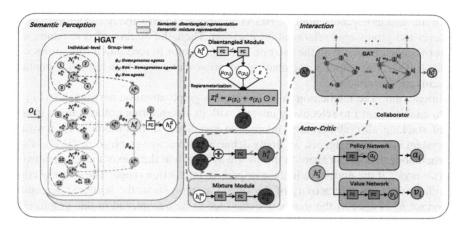

Fig. 1. Semantic perception swarm policy network structure.

3.1 Overall Design of SPSP

As shown in Fig. 1, the network structure of SPSP consists of three modules: a semantic perception module, an interaction module and an actor-critic module. Specifically, the semantic perception module is designed to extract semantic information h_i^o from partial observation o_i for agent i, where $i = 1, ..., n$. Then, the agent regards the semantic information h_i^o as information that can be propagated among agents. The interaction module is adopted to promote cooperation between the agent and neighbor agents through the transmission of semantic information, yielding interaction message h_i^c which implicitly encodes cooperation information. Next, the interaction message h_i^c is used as input to the actor-critic module. In the actor-critic module, the actor generates actions to the swarm environment, where the actions are the results of the SPSP decision. The critic is used to evaluate actions in the training phase. In the following, we present each module in detail.

(1) Semantic Perception: In this module, we hope to enable agents to understand and interpret their observations through semantic perception, yielding semantic information. Therefore, we would like to encode the partial observation o_i of agent i into a disentangled semantic vector Z_i^d and a mixture semantic vector Z_i^m. In the disentangled semantic vector, its semantic factors are sensitive to changes in single observation attribute, such as position or velocity, while being relatively constant to changes in other observation attributes, similar to disentangled representation [7]. Meanwhile, the mixture semantic vector is the mapping result of observation states to semantic information space consisted of semantic factors. These make that the learned knowledge is easier to transfer into new tasks or environments, and new observation states can be interpreted through learned semantic factors. Specifically, semantic disentangled representation with posterior distribution and semantic mixture representation with network mapping are realized to extract the disentangled semantic vector Z_i^d and the mixture semantic vector Z_i^m respectively.

In the semantic representation, HGAT is adopted to effectively model individual-level and group-level relational information. Then, the disentangled semantic vector is concatenated with the mixture semantic vector to feed into a fully-connected layer (FC). The output is a semantic perception vector h_i^o, which implicitly contains semantic information.

(2) Interaction: The interaction module adopts a graph attention network (GAT) [17] to enable agent i to selectively interact with its neighboring collaborators, instead of stacking all collaborators' interaction messages. In this module, for a swarm system including n agents, we firstly define an interactive graph $\mathcal{G} := (\mathcal{V}, \mathcal{E})$, where each node denotes an agent in the swarm system, and there exists an edge between two nodes if the nodes are in their respective interaction range. The existing edges reflect the local interactivity of the swarm system. Next, the semantic perception vector h_i^o is used as the interactive message to be transmitted in this graph. GAT is implemented to selectively attend to extracting the interaction vector h_i^c, which contains effective cooperation information, from the graph by:

$$h_i^c = \sigma\left(\sum_{j \in N_i^I \cup \{i\}} \frac{exp(\sigma_0(a^T[Wh_i^o \| Wh_j^o]))}{\sum_{q \in N_i^I \cup \{i\}} exp(\sigma_0(a^T[Wh_i^o \| Wh_q^o))} * Wh_i^o \right), \tag{1}$$

where N_i^I is some homogeneous neighborhood within the interaction scope D^I of agent i, a is a learnable parameter vector and W is a learnable parameter matrix, σ and σ_0 are ReLU and LeakyReLU activation functions respectively. The fraction in (1) is the attention weight, which acts to attend effective interactions.

(3) Actor-Critic: In actor-critic module shown in Fig. 1, an actor π that generates an action distribution is approximately represented by the policy network F_p, i.e., $\pi(o_i) \approx F_p(h_i^c)$. A critic V that predicts the discounted future returns to evaluate actions is approximately expressed by the value network F_v, i.e., $V(o_i) \approx F_v(h_i^c)$. The actor and critic only use partial observation o_i, which is a completely decentralized form. This helps it apply to large-scale agent scenarios due to avoiding centralized dimensional curses. In addition, in a swarm system, all agents use the same actor and critic networks in the execution and training phase. This idea makes our proposed SPSP be able to satisfy the scalability of swarm systems.

3.2 Semantic Perception

In the semantic perception module, semantic disentangled representation with posterior distribution and semantic mixture representation with network mapping are innovatively designed to represent semantic information. In the following, we present these parts in detail.

Semantics Formulation

Firstly, abstract semantic information needs to be formulated using mathematics. The formulation is based on three innovative assumptions: **Assumption 1.** Semantic information can be represented by a vector $Z = [z_1, ..., z_K]$, which consists of independent semantic factors z_k, $k \in 1, ..., K$. Each semantic factor is a variable, which obeys Gaussian distribution. A semantic factor is a latent encoding of an observation attribute. For simple example, a semantic factor of a three-dimensional object may be a latent encoding for its scale or color. *Assumption 2.* In a task, different observation state o_i implicitly contain finite semantic information Z_i, that contains fixed number of semantic factors. *Assumption 3.* Multiple observation states can be reconstructed from a semantic vector. This is because different observation states may contain similar semantic information. The above assumptions is formally formulated as follows.

Semantic Disentangled Representation

In semantic disentangled representation, we mainly extract a disentangled semantic vector Z_i^d with posterior distribution from observation state o_i. In particular, under the condition of knowing agent's partial observation state o_i, we would like to infer Z_i^d through:

$$p(Z_i^d|o_i) = \frac{p(o_i|Z_i^d)p(Z_i^d)}{p(o_i)} = \frac{p(o_i|Z_i^d)p(Z_i^d)}{\int p(o_i|Z_i^d)p(Z_i^d)dZ_i^d}, \tag{2}$$

where $p(o_i|Z_i^d)$ is a reconstruction process from the disentangled semantic vector. However, the above equation is difficult to calculate directly. Hence, we approximate $p(Z_i^d|o_i)$ by another tractable distribution $q(Z_i^d|o_i)$. The objective is that $q(Z_i^d|o_i)$ needs to be close to $p(Z_i^d|o_i)$. We achieve it through minimizing Kullback-Leibler (KL) divergence [15]: min $KL(q(Z_i^d|o_i)\|p(Z_i^d|o_i))$, which equals to maximize *Evidence Lower BOund* (ELBO) [5]:

$$\max \mathbb{E}_{q(Z_i^d|o_i)}\log p(o_i|Z_i^d) - KL(q(Z_i^d|o_i)\|p(Z_i^d)). \tag{3}$$

In ELBO, the former term denotes reconstruction likelihood, and the latter term guarantees that the posterior distribution $q(Z_i^d|o_i)$ is similar to the true prior distribution $p(Z_i^d)$. This can typically be modeled as a VAE [12]. In this paper, we design a VAE as shown in Fig. 2. The encoder of this VAE learns a posterior distribution mapping $q(Z_i^d|o_i; w)$ from o_i to Z_i^d. Specifically, we firstly adopt HGAT [19] as shown in Fig. 1 to learn state representation with a high dimension. This is because HGAT effectively model individual-level and group-level relational information, and it can process the observation states of different number agents. In HGAT, the first step is to cluster all agents into three groups using prior knowledge, i.e., homogeneous agents ϕ_1, non-homogeneous agents ϕ_2 and non agents ϕ_3. Next, we introduce the modeling process of individual and group levels in HGAT.

Individual Level: Agent i has a partial observation $o_i = [o_i^{\phi_1}, o_i^{\phi_2}, o_i^{\phi_3}]$, where $o_i^{\phi_m} = \{s_{j\phi_m}|j^{\phi_m} \in \mathcal{N}_i^{\phi_m}\}$ $(m = 1, 2, 3)$, $s_{j\phi_m}$ is the local state of agent j belonging to ϕ_m group, and $\mathcal{N}_i^{\phi_m}$ is some neighborhood (include itself) belonging to ϕ_m group within the observation scope D^O of agent i. Now, agent i computes the different individual embedding $h_i^{\phi_m}$ to summarize the individual relations between agent i and other neighboring agents belonging to different groups. The different embedding $h_i^{\phi_m}$ is calculated using different parameter GAT. This calculated process is similar to the Eq. (1).

Group Level: In this step, different individual embedding $h_i^{\phi_m}$ from different groups is aggregated with attention weights as a group embedding h_i^e, i.e., $h_i^e = \sum_{m=1}^{3}\beta_{\phi_m}h_i^{\phi_m}$, where $\beta_{\phi_m} = softmax(f(h_i^{\phi_m}))$ is a attention weight measuring group-level relations, $f(\cdot)$ is a FC layer. Next, the group embedding h_i^e is concatenated with the local state of agent i to be fed into another FC layer. The output is a state representation vector h_i^d.

After obtaining the state representation vector using HGAT, disentangled module is designed to generate the disentangled semantic vector Z_i^d. In particular, we use the reparameterization trick to achieve it, instead of directly outputting Z_i^d. We firstly sample ϵ from a unit Gaussian, and then generate Z_i^d by mean $\mu_{(Z_i)}$ and variance $\sigma_{(Z_i^d)}$ of disentangled module output, i.e., $Z_i^d = \mu_{(Z_i)} + \sigma_{(Z_i^d)} \odot \epsilon$, where $\epsilon \sim N(0, 1)$. The decoder of the VAE learns a distribution mapping $p(\hat{o}_i|Z_i^d; w)$ from Z_i^d back to $\hat{o}_i$ through Bidirectional Long-Short Term Memory (BiLSTM) and FC. The above description is just the construction process of the decoder and encoder in the VAE. Furthermore, we need to train the VAE to generate the disentangled semantic vector. This can be achieved through β-VAE[7], which is a modification of the VAE that introduces an adjustable hyperparameter β to the KL term of original VAE objective to realize disentangled representation. Hence, the loss function for training the VAE is:

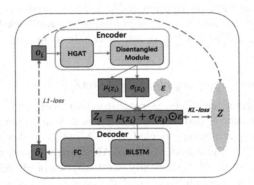

Fig. 2. Variational autoencoder.

$$minL1(o_i, \hat{o}_i; w) + \beta KL(q(Z_i^d|o_i; w)|p(Z_i^d)) \qquad (4)$$

Semantic Mixture Representation

In semantic mixture representation, we need to obtain a mixture semantic vector Z_i^m with network mapping from observation state o_i. Unlike the disentangled representation, each semantic factor of mixture semantic vector is sensitive to changes in multiple observation attributes, but they contain much useful observation information, which can compensate for the information loss in the disentangled semantic vector. Meanwhile, observation states with similar semantics can be mapped into the similar position in semantic information space, which can boost the adaptability of our proposed policy. Therefore, we would like to map observation state space into semantic information space made up of semantic factors by an advanced function Φ. The function is difficult to be constructed directly due to its strong nonlinearity. In this paper, we approximate the function Φ by a novel neural network as shown in Fig. 1. The network consists of HGAT and mixture module. Firstly, HGAT is adopted to realize high dimensional state representation h_i^m that models individual and group relational information. This is similar to the HAGT of semantic disentangled representation, but they have different parameters. Then, the mixture module is used to produce the mixture semantic vector Z_i^m, which has the same number of semantic factors as the disentangled semantic vector.

3.3   Training Method of SPSP

We adopt an improved proximal policy optimization (PPO) algorithm to train our proposed policy SPSP. This algorithm based on an actor-critic architecture is different from traditional PPO algorithm [16]. Specifically, we firstly open multiple threads to simulate the interactions between agents and the swarm environment in parallel. Fusion experiences from multiple parallel environments are collected to train SPSP, which can speed up the convergence. Moreover, SPSP is trained by minimizing an total loss L, which is conducted by the weighted summation of value loss L_V, action loss L_π, action entropy H and disentangled semantic loss L_{ds}, i.e.,

$$L = \beta_1 L_V + \beta_2 L_\pi - \beta_3 H + \beta_4 L_{ds}, \tag{5}$$

where

$$L_V = \mathbb{E}_{(o_i, o_i')}[(r + \gamma V(o_i'; w^-) - V(o_i; w))^2]$$

$$L_\pi = -\mathbb{E}_{(o_i, o_i')}[min(\frac{\pi(o_i; w)}{\pi(o_i; w^-)}, clip(\frac{\pi(o_i; w)}{\pi(o_i; w^-)}, 1 - \epsilon, 1 + \epsilon))$$

$$* (r + \gamma V(o_i'; w^-) - V(o_i; w))] \tag{6}$$

$$H = -\sum \pi(o_i; w) log(\pi(o_i; w))$$

$$L_{ds} = \mathbb{E}_{o_i}[L1(o_i, \hat{o}_i; w) + \beta KL(q(Z_i^d | o_i; w) | p(Z_i^d))]$$

Herein, the action entropy H is specially designed to encourage exploration for agents by penalizing the entropy of actor π. The disentangled semantic loss L_{ds} is implemented to generate the disentangled semantic vector Z_i^d with posterior distribution. This corresponds to Eq. (4).

4   Results

In this section, simulations and real-world experiments, including a fully cooperative *sheep food collection* task and a predatory-prey-style *wolves predation* task, are conducted to verify the effectiveness and adaptability of our proposed SPSP.

4.1   Task Description

Sheep Food Collection: This task has n food locations and n fully cooperative homogeneous agents (sheep). The agents need to collaboratively occupy as many food locations as possible without colliding with each other. **Wolves Predation:** This task is that n slower wolves need to cooperatively hunt m faster sheep with a fixed Voronoi escape strategy [20] in a randomly generated environment with l large static obstacles. Meanwhile, wolves are required to avoid collisions with obstacles and other wolves. Besides, sheep are confined to a closed world. If a sheep is touched by wolves, it will be caught.

4.2   Simulation Results

Simulation Setting

In simulation, all tasks are built on the particle-world environment (MAPE) [14], where agents take actions with discrete timesteps in a continuous 2D world. In each task, the agents have partial observability and local interactivity. We set the observation radius to be 1.5 unit and interaction radius to be 2 unit, i.e., $D^O = 1.5$, $D^I = 2$. For the training phase, each update of SPSP is implemented after accumulating experiences for total 4096 timesteps (128 timesteps on 32 parallel processes). Meanwhile, each episode is terminated by accomplishing the tasks or lasting up to 50 timesteps [1]. In addition, to speed up training process and learn more sophisticated behaviors, we adopted curriculum learning through model reload [18]. Specifically, each task, e.g., *sheep food collection* and *wolves predation*, can be defined as a subtask set containing some subtasks sorted from simple to complex. Due to the transferability of our proposed SPSP, the learned policy in simple subtask can be directly transferred to complex subtask by model reload.

The design of reward is an especially important step for reinforcement learning. In *sheep food collection* task, we set a shape reward with mean distance between sheep and food locations through Hungarian algorithm matching. Meanwhile, sheep get a negative reward when colliding with other sheep. In *wolves predation* task, whenever sheep is eaten (collided) by any wolf, the whole wolves will get a positive reward. Meanwhile, wolves will get a negative reward when colliding with other wolves or static obstacles. In addition, in *wolves predation* task, the state of agent should also include the life state, i.e., $s_i = [p_i, v_i, s_i^l]^T$, where $s_i^l = 1$ represents the state of being alive and $s_i^l = 0$ represents the state of being dead. For implementation details, we set $K = 8$, $\beta = 4$, $\beta_1 = 0.5, \beta_2 = 1, \beta_3 = 0.01, \beta_4 = 0.01$. The dimension of discrete action space is 4.

To fully validate the performance of our method, we use the method proposed in [1] (TRANSFER) as a baseline method. This is because this transferable method also adopts a fully decentralized manner in training and execution. Meanwhile, it uses graph attention networks to perceive agents' observations, without semantic perception.

Effectiveness Analysis

To fully verify the effectiveness of our method SPSP, we set five different scenario simulations for *sheep food collection* and *wolves predation* tasks. Simulation results is shown in Table 1. In the *sheep food collection* task, $n = 3, 6, 12, 18, 24$ sheep scenario is designed respectively. The policy trained by 7000 updates can be transferred to a scenario with more sheep through model reload, that is curriculum learning. The test results show our proposed SPSP has better performance in success rate, reward and episode length metrics than TRANSFER. This is because SPSP can implicitly understand the intention of collaborators by semantic perception to promote cooperation. In the *wolves predation* task, five scenarios, where $n = 3, 6, 9, 12, 15$ wolves hunt $m = 1, 2, 3, 4, 5$ sheep respectively in $l = 2$ static obstacles environments, is designed to realize the task. The policy trained by 5000 updates be transferred to a scenario with more wolves and sheep. As expected, our proposed SPSP obtain higher scores compared with the baseline method. This is due to the semantic perception of SPSP, which can infer the escape strategy of sheep. Therefore, the results fully reflect the good performance and advantages of our method.

Table 1. Simulation test performance of our method and baseline method in sheep food collection and wolves predation tasks. *Success Rate (S%): percentage of the successfully completed-task episodes. Mean Episode Reward (MER): mean of rewards for each episode of the agents. Mean Episode Length (MEL): mean of successful episode length. Score: rewards of each step of agents.*

Sheep food collection	Method	n = 3	n = 6	n = 12
		S%/MER MEL	S%/MER/MEL	S%/MER/MEL
	TRANSFER	97.6/–3.92/14.01	99.4/–2.01/12.69	98.6/–1.24/13.65
	SPSP(ours)	99.2/–3.80/13.62	100.0/–1.91/12.02	100.0/–1.11/12.51
Wolves predation	Method	n = 3, l = 2 (**m = 1**)	n = 6, l = 2 (**m = 2**)	n = 9, l = 2 (**m = 3**)
		Mean score	Mean score	Mean score
		m = 1/m = 2	m = 1/m = 2/m = 3	m = 2/m = 3/m = 4
	TRANSFER	–0.067/–0.092	–0.599/–0.520/–0.479	–0.887/–0.752/–0.692
	SPSP(ours)	0.129/–0.010	–0.287/–0.166/0.018	–0.464/–0.359/–0.252
Sheep food collection	Method	n = 18	n = 24	/
		S%/MER/MEL	S%/MER/MEL	/
	TRANSFER	98.2/–0.92/16.08	97.2/–0.87/18.99	/
	SPSP(ours)	100.0/–0.72/13.34	99.4/–0.64/15.58	/
Wolves predation	Method	n = 12, l = 2 (**m = 4**)	n = 15, l = 2 (**m = 5**)	/
		Mean score	Mean score	/
		m = 3/m = 4/m = 5	m = 4/m = 5/m = 6	/
	TRANSFER	–1.082/–0.967/–0.872	–1.282/–1.158/–1.083	/
	SPSP(ours)	–0.523/–0.466/–0.397	–0.775/–0.710/–0.641	/

Adaptability Analysis

In order to verify the adaptability of the proposed method, we use the learned policy to accomplish the tasks in new scenarios. In *sheep food collection* task, the generalization performance of our method and baseline method is shown in Fig. 3. The results show our method has better adaptability than baseline method, especially in large-scale agent scenarios. This is due to the advantage of our proposed semantic perception. The semantic perception can understand semantic information similar to the old environments or tasks from different observations in new environments or tasks. This strongly improves the adaptability of the policy. In *wolves predation* task, we also conducted generalization tests as shown in Table 1. The learned policy is adopted to enable wolves hunt different numbers of sheep. As expected, our method shows better adaptability

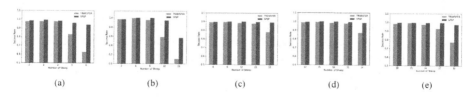

(a)   (b)   (c)   (d)   (e)

Fig. 3. Generalization performance of our method SPSP and baseline method in the *sheep food collection* task, (a) (b) (c) (d) (e) represent generalization results for the policies learned in $n = 3, 9, 12, 18, 24$ sheep scenarios respectively.

than baseline method. This is reflected in our method to getting higher scores in the new environment. This is because different prey has similar escape strategies, which can be inferred by the semantic perception. Hence, though not by a large margin, this adaptability improvement indicates the benefits of the proposed method.

4.3 Real-World Experiment Results

In addition to the simulations, real-world experiments are also carried out to verify the performance of SPSP. Two real-world experiment task systems are conducted, i.e., *sheep food collection* and *wolves predation* task systems. In each physical experiment task system, the kinematic constraints of robots (sheep or wolves) are holonomic constraints, which means that robots can move in any two-dimensional direction. The policies learned by simulation are performed in each robot. The illustration of trajectories is shown in Fig. 4. In *sheep food collection* task, $n = 3$ sheep (green robots) need to occupy three food location (red objects). Each sheep adopts the same policy learned by the simulation of the same scenario. Actual result illustrate that sheep can complete the task of food collection. In *wolves predation* task, $n = 3$ wolves (green robots) need to hunt $m = 1$ sheep (red robot) in environments with $l = 1$ static obstacles (black objects). Experiment result shows wolves with the learned policy can hunt sheep. Hence, although there are sim-to-real problems in the real-world experiment, SPSP can still achieve good results. This is due to SPSP's excellent adaptability brought by semantic perception.

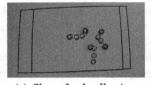   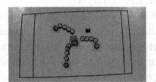

(a) *Sheep food collection*   (b) *Wolves predation*

Fig. 4. Illustration of trajectories in real-world experiments. (Color figure online)

5 Conclusion

In this work, we propose a novel semantic perception swarm policy (SPSP) with DRL for distributed swarm systems to reliably and efficiently accomplish specified tasks. This policy contains innovative semantic perception to promote agents' adaptability by understanding observation information. Specifically, semantic disentangled representation with posterior distribution and semantic mixture representation with network mapping are performed to represent the semantic information of agents' observations. In semantic representation, HGAT is adopted to model individual-level and group-level relational information. Simulation and real-world experiment results verify the effectiveness and adaptability of SPSP in several challenging tasks.

Acknowledgment. This work was supported by the National Key Research and Development Program of China under Grant 2018AAA0102402, the National Natural Science Foundation of China under Grant 62073323, the Strategic Priority Research Program of Chinese Academy of Sciences under Grant No. XDA27030403, and the External Cooperation Key Project of Chinese Academy Sciences No. 173211KYSB20200002.

References

1. Agarwal, A., Kumar, S., Sycara, K.: Learning transferable cooperative behavior in multi-agent teams (2019). arXiv preprint arXiv:1906.01202
2. Arnold, R.D., Yamaguchi, H., Tanaka, T.: Search and rescue with autonomous flying robots through behavior-based cooperative intelligence. J. Int. Humanit. Action **3**(1), 1–18 (2018). https://doi.org/10.1186/s41018-018-0045-4
3. Azzouni, J.: Semantic Perception: How the Illusion of a Common Language Arises and Persists. Oxford University Press, Oxford (2015)
4. Bengio, Y., Courville, A., Vincent, P.: Representation learning: a review and new perspectives. IEEE Trans. Pattern Anal. Mach. Intell. **35**(8), 1798–1828 (2013)
5. Blei, D.M., Kucukelbir, A., McAuliffe, J.D.: Variational inference: a review for statisticians. J. Am. Stat. Assoc **112**(518), 859–877 (2017)
6. Henson, C., Sheth, A., Thirunarayan, K.: Semantic perception: converting sensory observations to abstractions. IEEE Internet Comput. **16**(2), 26–34 (2012)
7. Higgins, I., et al.: beta-vae: learning basic visual concepts with a constrained variational framework (2016)
8. Hostallero, W.J.K.D.E., Son, K., Kim, D., Qtran, Y.Y.: Learning to factorize with transformation for cooperative multi-agent reinforcement learning. In: Proceedings of the 31st International Conference on Machine Learning, Proceedings of Machine Learning Research. PMLR (2019)
9. Hüttenrauch, M., Adrian, S., Neumann, G., et al.: Deep reinforcement learning for swarm systems. J. Mach. Learn. Res. **20**(54), 1–31 (2019)
10. Hüttenrauch, M., Šošić, A., Neumann, G.: Guided deep reinforcement learning for swarm systems (2017). arXiv preprint arXiv:1709.06011
11. Jiang, J., Lu, Z.: Learning attentional communication for multi-agent cooperation. In: Advances in Neural Information Processing Systems, pp. 7254–7264 (2018)
12. Kingma, D.P., Welling, M.: Auto-encoding variational bayes (2013). arXiv preprint arXiv:1312.6114
13. Liu, Y., Wang, L., Huang, H., Liu, M., Xu, C.Z.: A novel swarm robot simulation platform for warehousing logistics. In: 2017 IEEE International Conference on Robotics and Biomimetics (ROBIO), pp. 2669–2674. IEEE (2017)
14. Lowe, R., Wu, Y.I., Tamar, A., Harb, J., Abbeel, O.P., Mordatch, I.: Multi-agent actor-critic for mixed cooperative-competitive environments. In: Advances in Neural Information Processing Systems, pp. 6379–6390 (2017)
15. Mao, H., et al.: Neighborhood cognition consistent multi-agent reinforcement learning (2019). arXiv preprint arXiv:1912.01160
16. Schulman, J., Wolski, F., Dhariwal, P., Radford, A., Klimov, O.: Proximal policy optimization algorithms (2017). arXiv preprint arXiv:1707.06347
17. Veličković, P., Cucurull, G., Casanova, A., Romero, A., Lio, P., Bengio, Y.: Graph attention networks (2017). arXiv preprint arXiv:1710.10903
18. Wang, W., et al.: From few to more: large-scale dynamic multiagent curriculum learning. In: AAAI, pp. 7293–7300 (2020)

19. Wang, X., et al.: Heterogeneous graph attention network. In: The World Wide Web Conference, pp. 2022–2032 (2019)
20. Zhou, Z., Zhang, W., Ding, J., Huang, H., Stipanović, D.M., Tomlin, C.J.: Cooperative pursuit with voronoi partitions. Automatica **72**, 64–72 (2016)

Reliable, Robust, and Secure Machine Learning Algorithms

Open-Set Recognition with Dual Probability Learning

Shanshan Liu and Fenglei Yang$^{(\boxtimes)}$

School of Computer Engineering and Science, Shanghai University,
Shanghai 200444, China
{liuss,flyang}@shu.edu.cn

Abstract. The open-set recognition task is proposed to handle unknown classes that do not belong to any of the classes in training set. The methods should reject unknown samples while maintaining high classification accuracy on the known classes. Previous methods are divided into two stages, including open-set identification and closed-set classification. These methods usually reject unknown samples according to the previous analysis of the known classes. However, this would inevitably cause risks if the discriminative representation from the unknown classes is insufficient. In contrast to the previous methods, we propose a new method which uses a dual probability distribution to represent the unknowns. From the dual distribution, the boundary of known space is naturally derived, thereby helping identify the unknowns without staging or thresholding. Following this formulation, this paper proposed a new method called Dual Probability Learning Model (DPLM). The model built a neural Gaussian Mixed Model for probability estimation. To learn this model, we also added the normalized joint probability of latent representations into the objective function in the training stage. The results showed that the proposed method is highly effective.

Keywords: Open-set recognition · Dual probability · Gaussian mixed model · Deep learning

1 Introduction

Most of the researches on image classification problems are based on the assumption of a closed-set problem [5,6,9,17]. The closed-set means that the classes of training samples in training set contain all of the sample classes that appear in testing set, but this is not applicable to the real world. That is why many methods that perform well in closed-set classification problems show a decrease in accuracy when they are applied in practice. When encountering a sample of an unknown class, the closed-set classification model will incorrectly identify the sample as a known class.

Open-set recognition task aims to identify and reject samples that are not in the known classes, and to accurately classify the rest samples into the known

© Springer Nature Switzerland AG 2021
T. Mantoro et al. (Eds.): ICONIP 2021, LNCS 13110, pp. 127–137, 2021.
https://doi.org/10.1007/978-3-030-92238-2_11

classes. The purpose of closed-set classification is to minimize the empirical risk, and the one of open-set classification is to minimize the unknown risk by minimizing the open space risk. To solve this problem many thresholding methods [14,18] divided training procedure in two sub-tasks. Following the closed-set classification training pipeline, encoder adopting recognition scores learns the first task. Decoder through reconstructing conditioned on class identity learns the second task.

In this article, we apply a new open-set recognition framework, called Dual Probability Learning Model (DPLM). In contrast to thresholding models, our method uses a dual probability distribution to represent the unknowns. With the dual distribution, the boundary of known space is naturally derived, and thereby the training procedure does not need to be divided into stages, nor does it need thresholding. The outlier samples are identified based on the sample distribution peaks learned from the feature space by the model, and the known samples are classified at the same time.

The contributions of this paper are summarized as follows:

- We proposed a new model called DPLM which does not require thresholding. Our method identified the unknowns based on the sample distribution peaks and classified the knowns simultaneously.
- To learn this model, an objective function optimizing the normalized joint probability of latent representations and their corresponding classes are introduced.
- We applied two different evaluations and conducted a series of verification experiments on several standard datasets. In summary, the results showed that the applied method is highly effective.

2   Related Work

Compared with the achievements on the traditional closed-set classification, there is some room for improvement in the open-set recognition. In deep neural network-based methods, Bendale et al. improved the penultimate layer of the network in the deep network and suggested that the OpenMax layer can estimate the probability of unknown samples. Ge et al. [4] proposed the G-OpenMax method that can give explicit probability estimation over the unknown classes, allowing the classifier to find the decision margin. Neal et al. [12] trained and generated counterfactual images that were close to known samples and defined counterfactual images as boundary classes.

Yoshihashi et al. [19] utilized neural network to reconstruct the known samples and proposed the Classification-Reconstruction learning for Open-Set Recognition (CROSR) model. This model also retained the classification accuracy of the original classifier. Oza et al. [14] proposed class conditional auto-encoder (C2AE). The training stage mainly consisted of two steps. To be detailed, the first step was to train the classifier on a closed-set, and the second step was to use extreme value theory to learn sample reconstruction information to generate decisions boundary. Sun et al. [18] learned the idea of variational auto-encoder (VAE) and proposed Conditional Gaussian Distribution Learning (CGDL) method, which classified known classes by forcing potential features to simulate distribution of known sample classes.

Chen *et al.* [3] defined a new open space risk regularization term, which reduced the risk of unknown classes with the help of multi-category interaction. Having combined the generative model with the discriminant model, Zhang *et al.* [21] proposed the OpenHybrid framework. Different from methods mentioned above, we identified the open set into a one-stage task and the outlier samples can be identified based on the distribution peaks learned from the feature space.

3   Proposed Method

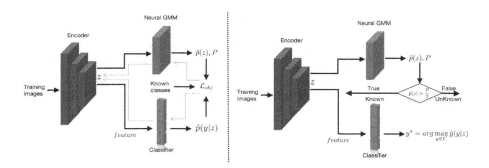

Fig. 1. The block diagram of the proposed method. In the diagram, the encoder is applied the latent features. The Neural GMM takes the latent features as input and estimates marginal distribution $\hat{p}(z)$. Then the model judges whether this sample is unknown by its $\hat{p}(z)$. The classifier will predict its label once the it is a known sample.

In this section, we first give the dual probability formulation of open-set recognition, and then introduce the DPLM in terms of structure, training and inference.

3.1   Dual Probability Formulation

After the joint distribution $p(z, y)$ over y is summed, we define the dual distribution $p(z)$ as the marginal distribution of z and obtain

$$p(z) = \sum_{y \in \mathbf{C}} p(y)p(z|y), \tag{1}$$

where z denotes a latent representation learned from a known sample via an encoder, $y \in \mathbb{N}$ denotes a class, and the set $\mathbf{C} = \{y | 0 \leq y \leq c\}$ denotes all the known classes, and the joint distribution $p(z, y)$ is given by $p(y)p(z|y)$.

Though we have the well-defined joint probability $p(z, y)$ for open-set recognition, the set of all y remains unknown to the algorithm, and thus such an overall joint probability is not available. We use the "non-overall" joint probability, namely the joint probability on the known classes, to define $p(z)$. We call $p(z)$ a dual distribution, since it has a dual property that two opposite probabilities can be derived to respectively represent the likelihoods of z being known and unknown.

We assume that the maximum value P of $p(z)$ indicating the representation $\mathbf{z}$ ($P = p(\mathbf{z})$) is definitely known. By using P as a normalization factor, we can derive the two opposite probabilities $P^{-1}p(z)$ and $[1 - P^{-1}p(z)]$. With the two probabilities, we define open space as

$$\mathcal{O} = S_{\mathcal{O}} - \{z | p(z) > P - p(z)\}, \tag{2}$$

where $\mathcal{O}$ be the open space and $S_{\mathcal{O}}$ be a ball of radius r_o that includes all of the known positive training examples z as well as the open space $\mathcal{O}$ [16].

If $S_{\mathcal{O}}$ is regarded as a complete set, the open space $\mathcal{O}$ is defined as the complement set of the known space occupied by the known representations z with the probabilities $p(z) > (P - p(z))$ or $p(z) > \frac{P}{2}$. Ideally, all the known representations are outside $\mathcal{O}$.

Equipped with the above preliminaries, the probabilistic open space risk $R_{\mathcal{O}}$ for all known classes is defined as

$$R_{\mathcal{O}} = \frac{\int_{\mathcal{O}} \hat{p}_+(z) dz}{\int_{S_{\mathcal{O}}} \hat{p}_+(z) dz}, \tag{3}$$

where $\hat{p}_+(z) = \hat{p}(z)$ if $\hat{p}(z) > (\hat{P} - \hat{p}(z))$, otherwise $\hat{p}_+(z) = 0$. We use $\hat{p}(z)$ and $\hat{P}$ as the approximation of $p(z)$ and P. This probabilistic definition will ensure that open risk computation do not yield large risk values because there is always $\int_{S_{\mathcal{O}}} \hat{p}(z) dz = 1$.

For open-set recognition, we train a model to force $\hat{p}(z)$ to approximate $p(z)$ with the posterior probabilities $\hat{p}(y|z)$ retrieved, so that the model can use the learnt $\hat{p}(z)$ and $\hat{p}(y|z)$ to accurately classify whether a given test sample belongs to a known class or an unknown class.

3.2   Architecture

To approximate the dual distribution $p(z)$, we propose a new method. The overview of the proposed method is in Fig. 1, including the training and testing procedures. There are three modules in this method:

Encoder. Encoder performs information transformation from sample space to latent space. Recently, various types of neural layers of nonlinear information transformation for latent representation acquisition are developed. Among them, ConvNets have become the leading architecture for recognition tasks. A few ConvNets architectures from some excellent classification methods, such as DenseNet are among the choices for latent representation acquisition.

Another benefit of borrowing the CovNets from the existing excellent classification methods is that the knowledge acquired by these methods can be transferred into our model. Specifically, the CovNets trained in these classification methods can be reused in our models. This knowledge transfer may help improve the performance of OSR.

Neural GMM. We use this module to estimate the marginal distribution $\hat{p}(z)$, and the classifier to estimate the posterior probabilities $\hat{p}(y|z)$, and combine the two modules to estimate the joint probability $\hat{p}(zy)(= \hat{p}(z)\hat{p}(y|z))$. In the learning procedure, the joint probability $\hat{p}(zy)$ will be maximized to force $\hat{p}(z)$ to approximate the ideal $p(z)$.

There are multiple options for the probability model to estimate $\hat{p}(z)$. In this study, we use Gaussian Mixture Model(GMM) to approximate $p(z)$, due to the ability of GMM to model arbitrarily complex distributions with multiple modes. The neural GMM [22] is embedded in our neural network for the estimation of $\hat{p}(z)$.

Given a latent representation z produced by the encoder, the neural GMM firstly makes the prediction of mixture membership by $\gamma = softmax\,(MLN(z))$, where MLN is a multi-layer network, and γ is a K-dimensional vector to predict the soft mixture-component membership. As the batch of N features and their membership is available, $\forall 1 \leq k \leq K$, we can further estimate the parameters in GMM as follows

$$\begin{cases} \phi_k = \sum_{i=1}^{N} \dfrac{\gamma_{ik}}{N}, \quad \mu_k = \dfrac{\sum_{i=1}^{N} \gamma_{ik} z_i}{\sum_{i=1}^{N} \gamma_{ik}}, \\[3mm] \Sigma_k = \dfrac{\sum_{i=1}^{N} \gamma_{ik}(z_i - \mu_k)(z_i - \mu_k)^T}{\sum_{i=1}^{N} \gamma_{ik}}, \end{cases} \tag{4}$$

where γ_i is the membership prediction for the feature z_i, and ϕ_k, μ_k, Σ_k are mixture probability, stand for, covariance for component k in GMM, respectively.

With the estimated parameters, the recognition function $\hat{p}(z)$ can be inferred by

$$\hat{p}(z) = \sum_{k=1}^{K} \phi_k \frac{exp\left(-\frac{1}{2}(z - \mu_k)^T \Sigma_k^{-1}(z - \mu_k)\right)}{\sqrt{|2\pi \Sigma_k|}} \tag{5}$$

where $|\cdot|$ indicates the determinant of a matrix.

Classifier. This classifier estimates the probabilities by $\mathbf{p} = softmax\,(LN(z))$, where LN is a linear neural layer, and $\mathbf{p} = [\hat{p}(y|z)]_{y \in \mathbf{C}}$ denotes the vector consisting of posterior probabilities $\hat{p}(y|z)$ of each known class. The posterior probabilities from the trained classifier are kept to classify the known class.

3.3 Training

A sufficient approximation of $p(z)$ requires the coordination of the optimization of relevant components during training, where the loss function plays a vital role.

Our objective function for open classification is theoretically defined as

$$
\begin{aligned}
\mathcal{L}_{obj}(z, \dot{y}) &= -\ln\left(\hat{P}^{-1}\hat{p}(z\dot{y})\right) \\
&= \ln\hat{P} - \ln\hat{p}(z) - \ln\hat{p}(\dot{y}|z),
\end{aligned}
\tag{6}
$$

where $\dot{y} \in \mathcal{C}$ is the true label of feature z from a positive training sample x. $\hat{p}$, on one hand, acts as a normalization factor for $\hat{p}(z)$ and $(\hat{P}-\hat{p}(z))$, and on the other hand, $\hat{p}$ joins in the backward propagation to helps $\hat{p}(z)$ to approximate $p(z)$ by adjusting its maximum value. Here, the joint $\hat{p}(zy)$ is given by $\hat{p}(z)\hat{p}(y|z)$, rather than $\hat{p}(y)\hat{p}(z|y)$. When $\hat{p}(z)$ indicates the likelihood of z being a positive feature, $\hat{p}(y|z)$ explains the likelihood of a feature z belonging to the specific known class y.

Obviously, the optimization of the objective function needs to continuously increase $\hat{p}(z)$ and $\hat{p}(\dot{y}|z)$ and reduce the maximum value of $\hat{p}(z)$, namely $\hat{P}$, in the training procedure.

The goal of increasing $\hat{p}(\dot{y}|z)$ is to increase the discriminability of the classifier so that it can give a higher posterior probability estimation for the true class $\dot{y}$ of a given representation z. The highly discriminative representation z is particularly important for increasing $\hat{p}(\dot{y}|z)$. Certainly, this depends on the optimization of the encoder.

The joint increasing of $\hat{p}(z)$ and $\hat{p}(\dot{y}|z)$ makes $\hat{p}(z)$ change toward a marginal distribution. Reducing P acts as this opposite force in shaping $\hat{p}(z)$. The reduction of $\hat{P}$ will decrease the values of $\hat{p}(z)$ around its peak, and affect the whole distribution of $\hat{p}(z)$.

The training procedure uses opposite sides, namely reduce $\hat{P}$ and increase $\hat{p}(z)$, to shape and force $\hat{p}(z)$ to approximate $p(z)$.

Considering the singularity problem in the neural GMM model for probability estimation, we add a penalty term to the objective function, thus having

$$
\mathcal{L}_{obj}(z, \dot{y}) = \ln\hat{P} - \ln\hat{p}(z) - \ln\hat{p}(\dot{y}|z) + \lambda P(\Sigma),
\tag{7}
$$

where the penalty term $P(\Sigma) = \sum_{k=1}^{K}\sum_{j=1}^{d}\frac{1}{\Sigma_{kjj}}$ is used to penalize small values on the d diagonal entires of the covariance matrixes $\{\Sigma_1, \cdots, \Sigma_K\}$ of K mixture components.

3.4   Testing

Given a test sample x, input it to the trained model to obtain $\hat{p}(z)$ and the posterior probability vector $\mathbf{p}$. This allows us to accept sample as a positive sample and determine its class or to reject it as unknown by the following formula:

$$
\begin{cases}
y^* = \arg\max\limits_{y \in \mathcal{C}} \hat{p}(y|z), & if\ \hat{p}(z) > \hat{P} - \hat{p}(z) \\
unknown, & if\ not
\end{cases}
\tag{8}
$$

4   Experiments and Results

4.1   Experimental Setup

In this part, we use a learning rate of 0.01, the batch size of the single-channel dataset is set to 64, and the rest of the datasets is set to 32. The backbone of the method is improved Densenet, and SGD optimizer. The specific design of the loss function is in Sect. 3.3, where the range of the parameters λ is between 0 and 1, fixed λ as 0.5 in the experiment. The definition of the openness [16] is expressed as follows:

$$O = 1 - \sqrt{\frac{2 \times N_{train}}{N_{test} + N_{target}}} \tag{9}$$

Among them, N_{train} shows how many known classes are put into training N_{target} indicates how many classes in the testing need to be recognized and classified by the model. N_{test} is the classes input model when testing, which includes the total number of known classes and unknown classes, namely $N_{train} = N_{target} \subseteq N_{test}$.

4.2   Ablation Analysis

In this paragraph, we analyze the proposed method, and mainly explore the specific components of the method on CIFAR-10 [7] and CIFAR-100 [8]. The CIFAR-10 dataset has 10 classes as the sample of the known label set. Each class contains 5000 training samples and 1000 test samples. The CIFAR-100 dataset has 100 classes that are only used as the test samples of the unknown label set during testing. Each category has 100 test samples. In the experiment, the 10 classes of CIFAR-10 are regarded as known classes, and the number of unknown classes is randomly selected from CIFAR-100 from 10 to 50, and the resulting openness is from 18% to 46%. A total of 11 classes of macro-average F1-scores with 10 known classes and 1 unknown class are used as the standard to evaluate how the model performs in the experiment.

The contribution of individual components is evaluated by creating multiple baselines of the proposed model. The simplest baseline, namely thresholding SoftMax probabilities of a closed-set method, is used as a start, then each component is added step by step until the final proposed model is built. The relevant upgraded baselines from the simplest baseline are described as follows:

A) Naive: Here, only the encoder and the classifier for open set classification as a simplest baseline. It will recognize the unknown samples when the predicted probability score is under 0.5.

B) Ours method*: The encoder, the classifier and the neural GMM are trained as described in Sect. 3.3 without normalization.

C) Ours method: The method proposed in this work, with normalization described as Sect. 3.3.

The results of the ablation analysis are shown in Fig. 2. From Fig. 2(a), we can see that the model effect is not directly proportional to the component K of the Gaussian mixture model. According to the ablation result, we fixed K at

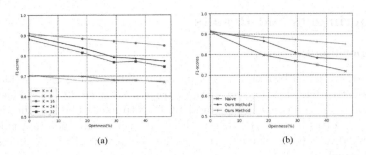

(a)   (b)

Fig. 2. F-measure comparisons for the ablation study. (a) is the impact of hyperparameters and (b) is the impact of component.

16. As seen from Fig. 2(b), the simple Naive thresholding baseline has weakest performance. Since it uses the fixed threshold 0.5, the performance degrades quickly with the increase in openness. However, addition of the neural GMM and the loss function as in *B)* helps to find a threshold value based on raw distribution of latent representations. The performance of our method* remains relatively stable as the openness changes, illustrating the probabilistic modeling of latent space is of great importance. Now, if normalization is added as in *C)*, there is a remarkable improvement over the Naive baseline and our method*. This shows the benefit of the normalization factor and optimization.

4.3 Comparison with State-of-Art Results

Table 1. AUROC for comparisons of our method with recent methods. values other than the proposed methods are taken from [18,21]. The best results are highlighted in bold.

Method	MNIST	SVHN	CIFAR10	CIFAR+10	CIFAR+50	TinyImageNet
SoftMax	0.978	0.886	0.677	0.816	0.805	0.577
OpenMax [2]	0.981	0.894	0.695	0.817	0.796	0.576
G-OpenMax [4]	0.984	0.896	0.675	0.827	0.819	0.580
OSRCI [12]	0.988	0.910	0.699	0.838	0.827	0.586
CROSR [19]	0.991	0.899	0.883	0.912	0.905	0.589
C2AE [14]	0.989	0.922	0.895	0.955	0.937	0.748
CGDL [18]	0.994	0.935	0.903	0.959	0.950	0.762
OpenHybrid [21]	**0.995**	0.947	0.950	0.962	0.955	0.793
Ours:	0.995	**0.950**	**0.963**	**0.984**	**0.972**	**0.801**

Open-Set Detection. We use four standard image datasets for model verification, and the design of experimental follows [12]. The training set of MNIST [11] handwritten digits dataset has 60000 pictures, and the test set has 10000 pictures. When we train the model, 55000 pictures are selected from the training set for training, and the other 5000 pictures are used as the verification model. For single-channel, first convert it to three-channel before experimenting. The MNIST, SVHN [13] and CIFAR-10 datasets all have 10 classes. Each dataset randomly selected 6 classes used as known classes, and the rest as unknown classes. At the same time, for the CIFAR+M experiment, CIFAR-10 and CIFAR-100 are jointly tested. There are 4 classes selected from CIFAR10 as known classes, and 10 classes randomly selected from CIFAR-100 as unknown classes, which are defined as CIFAR+10 experiments. And 50 classes are selected as an unknown classes and defined as a CIFAR+50 experiment. TinyImageNet [10] has 200 classes, 20 of which are randomly selected as known classes, and the remaining 180 classes are used as unknown samples. In the experiment, we use the Area Under the ROC curve (AUROC) to evaluate the model. The specific experimental results are shown in Table 1.

Open-Set Classification. The open-set recognition not only requires the ability to distinguish between the known samples and the unknown samples in the given dataset, but also the ability to correctly classify the known samples. In the open-set classification task, we select the training samples of the known dataset to train the model, and keep part of the validation set. The test dataset contains two parts, one is the test sample in the known classes and the other is the unknown classes. The unknown classes test samples come from other different datasets or synthetic datasets.

Table 2. F1-scores for comparisons of our method with recent methods, values other than the proposed method are taken from [18]. The best results are highlighted in bold.

Method	Omniglot	MNIST-noise	Noise
SoftMax	0.595	0.801	0.829
OpenMax [2]	0.780	0.816	0.826
CROSR [19]	0.793	0.827	0.826
CGDL [18]	0.850	**0.887**	0.859
Ours:	**0.907**	0.887	**0.885**

MNIST. In the open-set classification task, we follow the designed of experimental in [19], using the three datasets of Omniglot [1], Noise, and MNIST-Noise. The Omniglot is a set of handwritten letters in different national languages. Noise is a sample dataset randomly selected from the [0–1] normal distribution. The MNIST-Noise dataset is obtained by adding the MNIST dataset and the Noise dataset. These three datasets are only used as abnormal data in the test set. Each dataset contains the same amount of data as the test sample size of the known category, which ensures that the ratio of the known category to the unknown category is 1:1. The test results on the three datasets are shown in Table 2.

Table 3. F1-scores for comparisons of our method with recent methods, values other than the proposed method are taken from [18]. The best results are highlighted in bold.

Method	ImageNet-crop	ImageNet-resize	LSUN-crop	LSUN-resize
SoftMax	0.639	0.653	0.642	0.647
OpenMax [2]	0.660	0.684	0.657	0.668
CROSR [19]	0.721	0.735	0.720	0.749
C2AE [14]	0.837	0.826	0.783	0.801
CGDL [18]	0.840	0.832	0.806	0.812
Ours:	**0.860**	**0.845**	**0.866**	**0.819**

CIFAR10. The experimental design of CIFAR10 also follows the experimental design in [19]. The samples of known classes are from CIFAR10, the training samples are 50,000, and the test samples are 10,000. Unknown classes samples are selected from the ImageNet [15] and LUSN [20] datasets, and the samples are resized or cropped to ensure that the sample sizes are the same, thereby generating four outline datasets. In order to ensure that the ratio of known classes to unknown classes is 1:1, the number of unknown classes samples in our generated outline datasets is the same as the number of known classes test samples at 10,000. The test results on the four datasets are shown in Table 3. From the experimental results, we can clearly see that the method proposed in this paper is effective.

5   Conclusion

We described a new method DPLM and introduced a novel objective function to guide the model to learn the probability of latent representations. Our extensive experiments showed that the applied method is highly effective. Ablation study suggested that the explicit probability modeling of representations and the optimization of a normalization factor were key to the improved performance of open-set recognition. In our future research, we will consider applying dual probability distribution to various images other than still ones and discovering the new classes among the unknowns.

References

1. Ager, S.: Omniglot writing systems and languages of the world (2008). Accessed 27 Jan 2008
2. Bendale, A., Boult, T.E.: Towards open set deep networks. In: Proceedings of the IEEE Conference on Computer Vision and Pattern Recognition, pp. 1563–1572 (2016)
3. Chen, G., et al.: Learning open set network with discriminative reciprocal points (2020). arXiv preprint arXiv:2011.00178
4. Ge, Z., Demyanov, S., Chen, Z., Garnavi, R.: Generative openmax for multi-class open set classification (2017). arXiv preprint arXiv:1707.07418

5. He, K., Zhang, X., Ren, S., Sun, J.: Deep residual learning for image recognition. In: Proceedings of the IEEE Conference on Computer Vision and Pattern Recognition, pp. 770–778 (2016)
6. Hu, H., Zhang, Z., Xie, Z., Lin, S.: Local relation networks for image recognition. In: Proceedings of the IEEE/CVF International Conference on Computer Vision (ICCV) (2019)
7. Krizhevsky, A., Hinton, G.: Convolutional deep belief networks on cifar-10. Unpublished Manuscript **40**(7), 1–9 (2010)
8. Krizhevsky, A., Hinton, G., et al.: Learning multiple layers of features from tiny images (2009)
9. Krizhevsky, A., Sutskever, I., Hinton, G.E.: Imagenet classification with deep convolutional neural networks. Commun. ACM **60**(6), 84–90 (2017)
10. Le, Y., Yang, X.: Tiny imagenet visual recognition challenge. CS 231N **7**(7), 3 (2015)
11. Lecun, Y., Bottou, L., Bengio, Y., Haffner, P.: The mnist database of handwritten digits. Proc. IEEE **86**(11), 2278–2324 (1998). https://doi.org/10.1109/5.726791
12. Neal, L., Olson, M., Fern, X., Wong, W.K., Li, F.: Open set learning with counterfactual images. In: Proceedings of the European Conference on Computer Vision (ECCV), pp. 613–628 (2018)
13. Netzer, Y., Wang, T., Coates, A., Bissacco, A., Wu, B., Ng, A.Y.: Reading digits in natural images with unsupervised feature learning (2011)
14. Oza, P., Patel, V.M.: C2ae: class conditioned auto-encoder for open-set recognition. In: Proceedings of the IEEE Conference on Computer Vision and Pattern Recognition, pp. 2307–2316 (2019)
15. Russakovsky, O., et al.: Imagenet large scale visual recognition challenge. Int. J. Comput. Vision **115**(3), 211–252 (2015)
16. Scheirer, W.J., de Rezende Rocha, A., Sapkota, A., Boult, T.E.: Toward open set recognition. IEEE Trans. Pattern Anal. Mach. Intell. **35**(7), 1757–1772 (2012)
17. Simonyan, K., Zisserman, A.: Very deep convolutional networks for large-scale image recognition (2014). arXiv preprint arXiv:1409.1556
18. Sun, X., Yang, Z., Zhang, C., Ling, K.V., Peng, G.: Conditional gaussian distribution learning for open set recognition. In: Proceedings of the IEEE/CVF Conference on Computer Vision and Pattern Recognition, pp. 13480–13489 (2020)
19. Yoshihashi, R., Shao, W., Kawakami, R., You, S., Iida, M., Naemura, T.: Classification-reconstruction learning for open-set recognition. In: Proceedings of the IEEE Conference on Computer Vision and Pattern Recognition, pp. 4016–4025 (2019)
20. Yu, F., Seff, A., Zhang, Y., Song, S., Funkhouser, T., Xiao, J.: Lsun: construction of a large-scale image dataset using deep learning with humans in the loop (2015). arXiv preprint arXiv:1506.03365
21. Zhang, H., Li, A., Guo, J., Guo, Y.: Hybrid models for open set recognition (2020). arXiv preprint arXiv:2003.12506
22. Zong, B., et al.: Deep autoencoding gaussian mixture model for unsupervised anomaly detection. In: International Conference on Learning Representations (2018)

How Much Do Synthetic Datasets Matter in Handwritten Text Recognition?

Anna Wróblewska[1]([✉])[iD], Bartłomiej Chechliński[1],
Sylwia Sysko-Romańczuk[2][iD], and Karolina Seweryn[1][iD]

[1] Faculty of Mathematics and Information Science,
Warsaw University of Technology, Warsaw, Poland
`anna.wroblewska1@pw.edu.pl`
[2] Management Faculty, Warsaw University of Technology, Warsaw, Poland

Abstract. This paper explores synthetic image generators in dataset preparation to train models that allow human handwritten character recognition. We examined the most popular deep neural network architectures and presented a method based on autoencoder architecture and a schematic character generator. As a comparative model, we used a classifier trained on the whole NIST set of handwritten letters from the Latin alphabet. Our experiments showed that the 80% synthetic images in the training dataset achieved very high model accuracy, almost the same level as the 100% handwritten images in the training dataset. Our results prove that we can reduce the costs of creating, gathering, and describing human handwritten datasets five times over – with only a 5% loss in accuracy. Our method appears to be beneficial for a part of the training process and avoids unnecessary manual annotation work.

Keywords: Handwritten text · Pattern recognition · Image processing · Data augmentation · Synthetic dataset · Deep learning · Autoencoder

1 Introduction

Identifying and extracting handwritten texts is still challenging in the scanning process [6]. Handwriting styles characterize by high variability across people; moreover, the inferior quality of handwritten text compared to printed text make serious difficulties when transforming it to machine-readable text. Therefore, many industries are concerned with handwritten texts, particularly healthcare and pharmaceutical, insurance, banking, and public services. Over recent years, widespread recognition technologies, like ICR (Intelligent Character Recognition) and IDR (Intelligent Document Recognition), have evolved. Recent advancements in DNN (Deep Neural Networks), such as transformer

Research was funded by the Centre for Priority Research Area Artificial Intelligence and Robotics of Warsaw University of Technology within the Excellence Initiative: Research University (IDUB) programme (grant no 1820/27/Z01/POB2/2021).

T. Mantoro et al. (Eds.): ICONIP 2021, LNCS 13110, pp. 138–149, 2021.
https://doi.org/10.1007/978-3-030-92238-2_12

architectures, have also developed progress in solving handwritten text recognition (HTR) [4]. However, models based on DNNs demand annotated datasets, which takes time and money. The motivation of the paper – to alleviate the high cost associated with annotating data to be used by machine learning algorithms – is aligned with current academic and business concerns.

The paper proposes the unsupervised generation of character image samples to alleviate the cost of manually collecting and annotating images. The proposed method consists of the following steps: (1) textually describing characters employing a set of points (coordinates) and lines linking them (straight or Bezier curves), referred to as "scheme" by the authors; (2) parsing; (3) adding some random shifts to the point coordinates and lines; (4) image rendering; (5) scaling. Then, to improve the variability of the character appearance, we used two approaches. First, the autoencoder is trained with synthetic images (generated with a schema). Then the stable points in its latent space are fixed and used to generate and variate more image samples. Second, the autoencoder is trained with handwritten images, and is then used to process the synthetic images.

In the experimental part, three types of training sets are considered: (H) only handwritten images, (S) only synthetically generated images, and (HS) a mix of handwritten images with synthetically generated images. Using training set (S) yields inferior results compared to using (H), and using training set (HS) yields results closer to those produced using (H). The experiments use a NIST hand-printed letters dataset.

We hypothesize that the model accuracy trained on the partially synthetic dataset is similar to the model trained on a fully real dataset. Nevertheless, preparing the dataset can dramatically save human labour, which significantly impacts on its business applications.

The paper contributions are:

1. synthetic character image generation using our designed schemas,
2. autoencoder-based character generation processes,
3. studying classification results for various proportions of authentic handwritten and generated images in training data.[1]

In the following sections, we review relevant work on state-of-the-art methods and datasets requirements for an HRT task (Sect. 2); describe our solution for data generation (Sect. 3); outline the experimental setup that shows line of the reasoning from the dataset selection and classification training to comparison of solutions (Sect. 4); discuss the results and limitations of this study (Sect. 5); conclude and offer directions for further research (Sect. 6).

2   Related Work

The initial approaches to solving recognition challenges involved machine learning like hidden Markov models (HMM) or support vector machines (SVM) [16].

---

[1] Our source code for the handwritten character generator, schema examples are in: https://github.com/grant-TraDA.

The performance of these techniques is limited due to the manual feature extraction for inputs and their insufficient learning ability. However, RNN/LSTM can deal with sequential data (e.g. handwritten texts) to identify their patterns and generate the data. Even better results have recently been gained with multilayer perceptron neural networks [1]. Thus, sequence-to-sequence (seq2seq) models with encoder-decoder networks and attention mechanisms have gained popularity for solving recognition problems and machine translation, where output generation is essential.

The HRT challenge is to cope with large amounts of labelled data [7,17]. Labels have to be exacted to match each character region with its name. This is done mainly by direct recognition of a word or line. For example, in many datasets, there are annotations of a word and its coordinates [18]. The character level recognition is practically tricky since the first need is to segment a word into characters. Tables 1 and 2 summarize the biggest and most popular datasets of handwritten characters, and then gather, compare, and analyze the requirements for the handwriting recognition datasets. There are synthetic datasets used for handwriting recognition training [2,8,9]. In [9], the authors present a vast dataset, consisting of 9 million images, created with various fonts and augmentation schemes. A similar approach was proposed by Jaderberg et al. [8], who generated the synthetic data engine that assembled data based on a few inputs – randomly chosen parameters – such as colours or fonts. Others [2] proposed a system based on GANs to create synthetic images representing handwritten words. All researchers confirm that enriching a real dataset with artificial observations improves model quality.

Table 1. Handwritten datasets.

Datasets	Description
NIST dataset [12]	Dataset published by The US National Institute of Science. It contains more than 800,000 character images from 3,600 writers. Each person completed a single page questionnaire with a few single words and one paragraph
MNIST database [10]	It is a subset of the NIST Database with only 60,000 images of handwritten digits
Devangri characters [13]	Database of handwritten Devangari characters (Indian alphabet). It contains 1,800 samples from 25 writers
Mathematics expressions [11]	Set of 10,000 expressions with 101 different mathematical symbols
Chinese characters [5]	Nearly 1,000,000 Chinese character images
Arabic printed text [15]	115,000 words images (not single characters) with 10 different Arabic fonts
Chars74K data [3]	74,000 images of digits

3 Our Method – Synthetic Dataset Generation

3.1 Synthetic Schema-Based Character Generator

The synthetic character generator is a fundamental component of our research (see Fig. 1). We provided an initial schema for each character (see the example in Fig. 2). The schema is a text file describing how a character is formed: steps that allowed the natural way of writing characters. For example, to create the character "G", a human has to make an incomplete circle with a lot at the end. The generator loads the schema and randomly modifies it to generate the proper character. All generated characters become an input for the classifier to be trained.

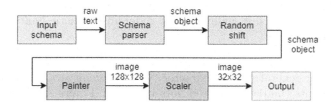

Fig. 1. Diagram of our schema-based generator. The red part was a schema file with a description of a particular character. The blue part was connected to a parser module. It loaded the schema file and translated it to the Painter, which created a character image from the schema. The yellow box represents an output character image. (Color figure online)

Schema Description. The schema is a human-readable description of how a letter should be written, e.g., making a straight line from top to bottom and adding two halves of a circle on the right side of the line. This general description was a source for all letters that followed the same handwriting. The schema contains different keywords understandable by the generator (the Parser and Painter modules) and formed a 'standard' character. The list below indicates primary commands to make up a character:

- *point* $< name >< x >< y >$
  "Point" is used to create a 2D named point with "x" and "y" coordinates.
- *line* $< nameA >< nameB >$
  The generator paints a straight line between two points points "A" and "B".
- *bezier* $< nameA >< nameB >< nameC >< nameD >$.
  "Bezier" creates Bezier's curve between points "A" and "D". Points "B" and "C" are used for deflection, so the curve does not go through them.
- *circle* $< nameA >< nameB >< nameC >$
  Points "A", "B" and "C" are passing by circle (or ellipse). The middle of the centre and radiuses are calculated from the points.

– *connect_left* < *nameA* >
connect_right < *nameA* >

In the commands above, point "A", was used to connect the character to the previous/next character. This keyword was used to merge subsequent letters to generate whole words. This procedure is general, and it allows for the writing of characters in any language or script.

Randomization of Generated Characters. Each schema precisely describes the process of writing only one character, so this deterministic generator created the same image each time. The characters in our datasets should imitate the human way of writing, so the characters should be randomly generated. Firstly, each point from the loaded schema was slightly randomly moved. Secondly, each line was not painted straight – some random movements approximated hand movement inaccuracy. Then, some of the pixels were erased to simulate image scanning errors. As a result, many slightly different synthetic handwritten characters were created.

An Example of a Synthetic Handwritten Character. Figure 2 presents the example of schema for generating the character "G". The generator parses the schema and creates points (marked with red dots in Fig. 2). Points A, B, C, and D are utilised for drawing a Bezier curve [19] (B and C are used to determine the curve, A and D are the ends of the curve). Then the remainder of the straight lines is painted. Figure 2 shows the drawing process of a character with random shifts and examples of generated images.

Fig. 2. Generation of a character "G". On the left side the character with marked points and the schema text. On the right – four different characters "G" generated from the schema. (Color figure online)

A single schema provides an unlimited number of generated characters. All the characters from one schema are similar, but not the same. That allows for the preparation of massive and highly diverse training datasets for neural networks. Figure 3 presents a single sample of each character generated with the above solution.

3.2   Autoencoder-Based Generators

Many different networks can be used for a tandem of encoders and decoders. Usually, the choice is determined by the task of the autoencoder. For a character

ABCDEFSTUVWX
GHIJKLYZMNOP

Fig. 3. Examples of Latin alphabet characters from our generator.

generation task, we used our autoencoder: (1) as a standalone autoencoder – changing the latent representation (the encoder outputs, see Fig. 5), and (2) as a schema-based autoencoder – feeding the autoencoder with synthetic schema-based images (see Fig. 6).

Our autoencoder, convolutional-based (CNN-based) encoder-decoder neural network, was deployed on both encoder and decoder (Fig. 4). Both models operate with two kinds of convolutional filters with sizes 64 and 128. However, the order of filters in the decoder is reversed. The number and size of filters were chosen while fine-tuning. Additional extra layers were added to rescale and reshape data. Filters with smaller sizes could not have learned 25 character classes. On the other hand, increasing the size and number of filters extended computation time without visible impact on the images generated.

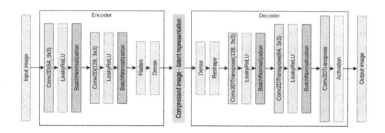

Fig. 4. Autoencoder using CNN.

Standalone Autoencoder. This approach involves only the decoder, using so-called character templates in the latent space to generate character images. The decoder should be trained with the encoder on the handwritten dataset to decode compressed characters, i.e., characters' latent representations in the trained autoencoder. The latent representations are randomly generated with restrictions to use the shared information for each class of letters (so-called character templates, see Fig. 5).

After encoding a few previously prepared samples of each character class, the encoder output comprises a few compressed images of a single letter (their latent representations). By comparing all of the compressed samples, the templates (the latent representation) are prepared and decoded into the image of the chosen character class. Figure 5 illustrates this process: two steps in using autoencoder latent space to generate the character "K". Encoding a few samples of "K" shows that the shared values involve eight different numbers (marked red in

Fig. 5). Other values differ between each sample. This template is the latent representation of "K".

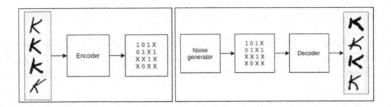

Fig. 5. Sample character generation with the usage of latent space character templates (boxes with red numbers). On the left, the process of generating latent representation templates from the input handwritten characters of the same class (here, "K" character). On the right, the inputs for the decoder during prediction (in generating images) are randomized character templates (boxes with red numbers with injected random numbers instead of crosses). (Color figure online)

The noise generator is fed with latent character representation templates and used for generating compressed data. Decoding such template data with random numbers (in place of the black crosses in Fig. 5) allows for the generation of different versions of each character class.

Schema-Based Autoencoder. This approach involves both encoder and decoder in a single module. The autoencoder trained with a dataset of authentic handwritten images was fed with the synthetic characters, i.e., those produced from our schema-based generator. Hence, the output had features more similar to human handwriting, with more randomness and noise in the output character images (compared with the solely schema-generated images). Figure 6 shows an example of processing generated images with the autoencoder, which is trained with the handwritten dataset and then fed with the schema-based generated character. After processing the image with the trained autoencoder, the character is gaining human handwriting features.

4   Research Design

Our research comprises preparation of partially synthetic handwritten characters datasets with autoencoders and schema-based generator. All the training datasets (also the proportion of real and synthetic data) is tested against the same state-of-the-art classifier architecture.

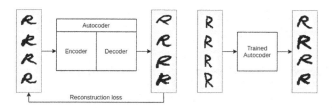

Fig. 6. The autoencoder used to process schema-based generated images. The training is done with a handwritten dataset (on the left). After the training, the schema-based characters become inputs (on the right).

4.1 Experimental Database Preprocessing

We chose NIST [12] dataset because it met all the desirable requirements (see Table 2). Hence, the remaining datasets were rejected. The NIST dataset [12] consisted of handwritten characters (pictures) made by humans (each class consisted of one character).

Table 2. Requirements for handwritten dataset comparison.

Requirement name	Description
English letters	The dataset should have all English letters. In the best case, it should contain both small and capital letters, but only capital letters will be satisfactory. Our approach can be easily extended to other scripts and languages; here, we tested English as a benchmark
Character independence	One input image should consist of one character. Character separation and concatenation are not the main goals of this benchmark research
Unified description	All input images should be prepared in the same way to avoid the unnecessary work of re-description
Number of images	The more, the better
Number of writers	The more, the better
Extras	The dataset can also have digits, non-English letters, words etc

The first layer of the CNN neural network for image recognition was designed with input neurons: one for each pixel of the input image (this is further explained in Sect. 4.3). Images from the NIST dataset, size 128×128, required 16,384 neurons on the first layer. Therefore, memory usage and the large white background were the challenges. Empty pixels in the background were prepossessed in the following steps: (1) cropping an image to a minimal ROI (region of interest); (2) scaling it proportionally to the size of 32 × 32. Image size after scaling: 32 × S or S × 32 (S may vary from 32); (3) increasing the image size to 32 × 32 with empty pixels (white background).

4.2   Training with Synthetic and Handwritten Data

We chose samples from the experimental database to create a training and testing set in the following way. We experimented with different training datasets: (1) a handwritten train dataset, (2) a synthetic (simulated handwritten) train dataset, and (3) a mixed one. These datasets were used to train our classifier before testing on the purely handwritten dataset. The test set was always the same, consisted of the purely handwritten dataset from our database, and used after the classifier was trained to ensure stable and comparable results. The handwritten train dataset was used to compare and measure the classifier's performance in a training environment with all images/characters written by human hands. The third mixed train dataset was changeable and depended on the test scenario: (1) synthetic images from a schema-based generator, (2) synthetic images generated with a standalone autodecoder, and (3) synthetic images generated with a schema-based generator and passed through an autoencoder and autodecoder to randomize characters.

4.3   Base Classifier for Character Recognition

VGG was selected as a classifier to train and test in different combinations for handwritten and synthetic training datasets. Currently, VGG, an object-recognition CNN that supports up to 19 layers, is one of the best models among those from the ImageNet Large Scale Visual Recognition Competition (ILSVRC) [14]. It was designed as an improvement on AlexNet, thanks to the large kernel-size filters being replaced with multiple 3x3 kernel-sized filters [14]. In our study, we utilized VGG16 as a classifier of character images.

5   Experiments and Results

Our challenge was to evaluate how data augmentation with generated synthetic data impacts the handwritten character classification task. Table 3 presents the results.

Handwritten and Synthetic Inputs. The accuracy in the training dataset with only generated inputs was around 40%, whereas solely handwritten inputs allowed for results twice as good (around 80%; see Table 3). It is worth noting that the solely synthetic images were very similar to each other, and the model got trained very fast with the early stopping algorithm (after 1–2 epochs).

Considering that, we trained our classifier on a mixed dataset, which contained both the handwritten and the generated (synthetic) images in various proportions. They were measured between each class of character images: a mixed dataset with 10% handwritten images meant that each character class

Table 3. The classification results for the designed experiments. Note: the results proceed from the test set.

Training dataset	Loss [val]	Accuracy [%]
Solely human handwritten characters	0.71	**79**
Schema-based generator		
Solely schema-based generated images	3.98	38
20% handwritten and 80% generated images	0.89	75
Standalone autoencoder dataset generator		
Only generated images based on autoencoder templates	3.17	39
20% handwritten and 80% generated images	0.91	77
Schema-based generator followed by the autoencoder as a noise generator		
Only generated images	3.11	40
20% handwritten and 80% generated images	0.89	**78**

had 10% handwritten images and 90% generated images in the dataset. The results on that dataset were quite surprising (see Fig. 7). The difference in accuracy between solely handwritten images and the dataset with 80% synthetic images was only four percentage points. This means that we can reduce the cost of creating, gathering, and describing human handwritten datasets five times over with only a 5% accuracy loss. The experiment proved that the assumed hypothesis about generated datasets applied in practice and brought tangible and measurable benefits.

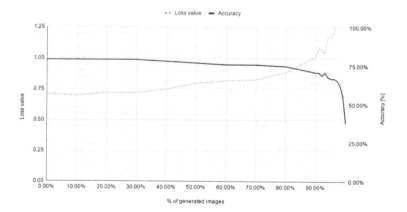

Fig. 7. The classifier results for different proportions of handwritten and synthetic images in the training set. In this test, the synthetic images were produced with our schema-based generator. Note: accuracy – red line; loss value – blue dotted line. (Color figure online)

Autoencoder as Standalone Generator. We also checked how the autoencoder, as a 'standalone' character generator, can be applied in image generation to enrich handwritten images. Hence, letters were generated by the trained decoder with pseudo-random input noise templates achieved from the encoder outputs.

The first step was to train both encoder and decoder with the authentic handwritten dataset. When training is over, the models – the encoder and decoder – were separated. The encoder was used to get template encoded data for each class. So, the average encoded data from a few different images of the same character was calculated. This data was used to prepare inputs for the decoder. The noise generator got the templates for each class, modifying them randomly (each value is multiplied by 0.5–1.5). Such data was utilized as an input to the trained decoder. The decoder reproduced the original image from the input data. Decoded images differed because of random changes in the previous step but allowed us to quickly create a massive dataset of generated character images. In turn, the generated dataset was used to train the classifier in the same way as described in the previous tests with the original NIST dataset and noisy images from the decoder.

It was not surprising that results for the standalone autoencoder generator (see Table 3) were quite similar, but slightly better than the results gathered for schema-based generators. The classifier trained with only generated images achieved nearly 39% classification accuracy, and the mixed dataset got above 77%. The mixed dataset used 20% handwritten images and 80% characters generated with an autoencoder.

Autoencoder as Noise Generator. The subsequent test involved a schema-based generator as a source of character images. Images were generated with our schema-based generator – similar to the first test. Each image was encoded and decoded. Decoded images followed the handwriting of the trained dataset. That allowed the easy creation of a lot of different images. It is worth noting that we did not need the labelled real dataset to train the autoencoder. We needed solely handwritten characters. Finally, the generated dataset was used to train the classifier in the same way as in all our tests. The results were quite similar to the standalone autoencoder generator (see Table 3), even slightly better. It showed that the schema-based generator followed by the autocoder as a noise generator was better, as it needed less effort to create the solution.

6   Conclusions

Image generation can expand the size of existing databases at a low costs (in money and time) and bring satisfying results in sample differentiation. Additionally, image generation can be used to align datasets, where some of the classes have significantly fewer samples than others, e.g., the NIST dataset.

The schema-based solution works for the character recognition task for multiple languages and scripts (but cannot be easily applied to image processing in other tasks). The current lack of datasets that meet all the requirements for handwritten task recognition presents an opportunity for the schema generator to create all the characters from schemes. Our results are encouraging.

Both tested autoencoder solutions need to be trained on the existing, non-labelled datasets, they can be easily applied to nearly every project with image processing. It is the most generic and the best of the tested solutions.

Our experiments confirm that inquiries into the possibilities of data augmentation by generating synthetic data, which can replace the need for authentic handwritten data to train deep learning models, is a direction researchers need to focus on.

References

1. Aaref, A.M.: English character recognition algorithm by improving the weights of MLP neural network with dragonfly algorithm. Period. Eng. Nat. Sci. **9**, 616–628 (2021)
2. Alonso, E., Moysset, B., Messina, R.: Adversarial generation of handwritten text images conditioned on sequences. In: ICDAR (2019)
3. T.E. de Campos, B.B.: Chars74k dataset (2009)
4. Chaudhuri, A., et al.: Optical Character Recognition Systems for Different Languages with Soft Computing. Springer, Heidelberg (2017). https://doi.org/10.1007/978-3-319-50252-6
5. Chen, Q.: HIT-OR3C dataset. corpus for Chinese characters (2011)
6. Geetha, R.E.A.: Effective offline handwritten text recognition model based on a sequence-to-sequence approach with CNN-RNN networks. Neural Comput. Appl. **33**, 10923–10934 (2021)
7. Ghosh, D., Shivaprasad, A.: An analytic approach for generation of artificial hand-printed character database from given generative models. Pattern Recognit **32**, 907–920 (1999)
8. Jaderberg, M., Simonyan, K., Vedaldi, A., Zisserman, A.: Synthetic data and artificial neural networks for natural scene text recognition (2014)
9. Krishnan, P., Jawahar, C.V.: Generating synthetic data for text recognition (2016)
10. LeCun, Y.: The mnist database of handwritten digits (2019)
11. Mouchère, H.: Crohme: Competition on recognition of online handwritten mathematical expressions (2014)
12. NIST: Nist handprinted forms and characters - nist special database 19 (2019)
13. Santosh, K.: Devanagari character dataset (2011)
14. Simonyan, K., Zisserman, A.: Very deep convolutional networks for large-scale image recognition. In: ICLR (2015)
15. Slimane, F.: Arabic printed text image database (2009)
16. Taufique, M., et al.: Handwritten Bangla character recognition using inception convolutional neural network. Int. J. Comput. Appl. **181**, 48–59 (2018)
17. Tensmeyer, C., Wigington, C.: Training full-page handwritten text recognition models without annotated line breaks. In: ICDAR (2019)
18. Veit, A., Matera, T., Neumann, L., Matas, J., Belongie, S.: Coco-text: dataset and benchmark for text detection and recognition in natural images. In: ICDAR (2017)
19. Zhang, J.: C-Bézier curves and surfaces. Graphical Models and Image Processing (1999)

PCMO: Partial Classification from CNN-Based Model Outputs

Jiarui Xie[1]($\boxtimes$) (iD), Violaine Antoine[1] (iD), and Thierry Chateau[2,3] (iD)

[1] Université Clermont-Auvergne, CNRS, Mines de Saint-Étienne,
Clermont-Auvergne-INP, LIMOS, 63000 Clermont-Ferrand, France
{jiarui.xie,violaine.antoine}@uca.fr
[2] Logiroad.ai, Chevroliere, France
[3] Institut Pascal, UMR6602, CNRS, University of Clermont Auvergne,
Clermont-Ferrand, France
thierry.chateau@uca.fr

Abstract. The partial classification can assign a sample to a class subset when this sample has similar probabilities for multiple classes. However, the extra information for making such predictions usually comes at the cost of retraining the model, changing the model architecture, or applying a new loss function. In an attempt to alleviate this computational burden, we fulfilled partial classification only based on pre-trained CNN-based model outputs (PCMO), by transforming the model outputs to beliefs for predicted sets under the Dempster-Shafer theory. The PCMO method has been executed on six prevalent datasets, four classical CNN-based models, and compared with three existing methods. For instance, experiments with MNIST and CIFAR10 datasets show the superiority of PCMO in terms of average discounted accuracy (0.23% and 7.71% improvement, respectively) when compared to other methods. The performance demonstrated that the PCMO method makes it possible to improve classification accuracy and to make cautious decisions by assigning a sample to a class subset. Moreover, the PCMO method is simple to implement compared to the existing methods, as the PCMO method does not need to retrain the model or conduct any further modifications.

Keywords: CNN-based model · Decision making · Dempster-Shafer theory · Partial classification

1 Introduction

The precise or certainty classification [17,34] is a well-known issue in which a sample is classified into one and only one of the training classes. Unfortunately, such a strict classification sometimes results in misclassification when the input sample does not contain sufficient evidence to identify a certain class. The partial classification [9,21,24] is one of the more practical ways to solve this problem. It is defined as the assignment of a sample into a class subset. For example, let us consider a class set $\Omega = \{\omega_1, \omega_2, \omega_3\}$. Here, we cannot manage to reliably classify

© Springer Nature Switzerland AG 2021
T. Mantoro et al. (Eds.): ICONIP 2021, LNCS 13110, pp. 150–163, 2021.
https://doi.org/10.1007/978-3-030-92238-2_13

a sample into a single class, but it is almost sure that it does not belong to ω_1. Consequently, it is more reasonable to assign it to the subset $\{\omega_2, \omega_3\}$. In practice, high ambiguity emerges in numerous applications, and large-scale datasets contain a fair amount of confusing samples, these are the bedrock of the usage of partial classification. For instance, the goal of road surfaces classification [38] is to produce a prediction with almost null error which can be expected from partial classification.

A considerable amount of literature has been published on partial classification and has always led to different classification strategies. On the one hand, researchers attempted to predict a subset with prior fixed cardinality [28] or with a rejection option [12,14,18]. They can be seen as a special case of partial classification by classifying the sample into one specific class subset. On the other hand, a number of authors attempted to modify the loss function [3,9,24] or build a new classifier [29,33,36] to provide beliefs for predicted sets. Usually, such algorithms are time-consuming. To this end, it is essential to reduce the computation and time complexity by efficiently and sufficiently leveraging the information provided by the pre-trained neural network.

In this paper, we proposed a new partial classification method based on pre-trained CNN-based model outputs (PCMO). Different from the existing methods, the PCMO method simply and efficiently fulfilled partial classification only based on pre-trained CNN-based model outputs, and provided beliefs to predicted sets for further prediction. As manifested in Fig. 1, at first, the CNN-based model extracts features from the input layer through the combination of the feature extraction process and the fully connected layer between the last hidden layer and the output layer. Second, the received features are converted into beliefs under the Dempster-Shafer theory (DST) [30] through the output to possibility and the possibility to belief processes. Finally, the PCMO method performs partial classification based on the produced beliefs by choosing the maximum belief and generating the corresponding class subset as the prediction.

The contributions of our work can be summarized as follows:

- The most striking achievement is that the proposed method is fulfilled only based on model outputs that can be applied to any pre-trained CNN-based model without any demand to retrain the model or conduct any further modifications.
- By considering good features of log function and analyzing the regular pattern of model outputs, a novel and reasonable transformation from model outputs to possibility distribution is proposed.

2   Related Work

Partial Classification. Partial classification also known as set-valued classification becomes prevalent recently imputable to its capability dealing with ambiguity. At the first glance partial classification seems to be linked to multi-label

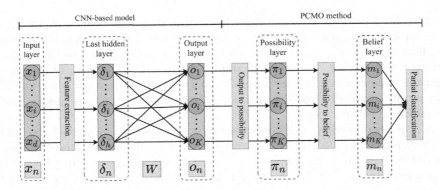

Fig. 1. Framework of the proposed method. The feature extraction process is demonstrated simply, it can be any kind of CNN-based architecture, e.g., fully connected layer, LeNet [15], GoogLeNet [31], or ResNet [10]. The detailed output to possibility, possibility to belief, and partial classification processes are presented in Sect. 4.

classification [4,32]. The confusion comes from the fact that both methods produce a class subset as the prediction. However, the crucial dissimilarity comes in that an input sample is labeled by a subset of classes for multi-label classification, whereas for partial classification, only a single class.

The straightforward way to fulfill partial classification is to always predict a fixed number of classes such as the top five most proper classes [28]. However, there is no reason to predict exactly five or any other a prior fixed number of classes all the time. Classification with rejection option [12,14,18] is another plain strategy that concerns the treatment of outliers that are not defined by any of the training classes. Depending on this strategy, such samples are assigned to the empty set, or the entire set, reflecting maximum uncertainty. Both the top five and rejection strategies can be seen as a special case of partial classification by giving belief for one specific class subset. Apart from the above methods, there are two directions that aim at modifying loss function or building the new classifier to provide beliefs for predicted sets. On the one hand, Ha [9] introduces a loss function consist of the sum of two terms, one reflecting the loss of missing the ground-truth labels, and the other penalizing imprecision. A similar loss function used in [24] is composed of the uncertainty quantified by conditional class probabilities, and the quality of the predicted set measured by a utility function. Besides, Coz et al. [3] propose a loss function inspired by aggregating precision and recall. On the other hand, Vovk et al. [33] proposed an approach to learn a partial classifier with finite sample confidence guarantees. In the same context, Sadinle et al. [29] designed a classifier that guarantees user-defined levels of coverage while minimizing ambiguity. As we can see, the weakness of the above methods is the time-consuming nature, e.g., retraining model, changing model architecture, and applying the new loss function.

Dempster-Shafer Theory. Dempster-Shafer theory (DST) [30] is a general framework for reasoning with uncertainty which is proposed by Arthur P. Dempster [5] then refined by Glenn Shafer [30] also know as evidence theory.

The existing works related to CNN have the following three main directions. The first one is classifier fusion, in which the outputs of several classifiers are transformed into belief functions and aggregated by suitable combination rules [1,19,40]. Another direction is evidential calibration, the decisions of classifiers are converted into belief functions with some frequency calibration property [20,22,23]. The last approach is to design evidential classifiers [6,7], which transformed the evidence of the input sample into beliefs and combine them by appropriate combination rules.

3   Background

3.1   The Pattern of the CNN-Based Model Outputs

The convolutional neural network (CNN) [15] is a machine learning method that uses multiple layers to progressively extract features from raw data as sample representation. Define a training dataset $\mathcal{D}^{train} = \{\boldsymbol{x}_n, y_n\}_{n=1}^N$ has K classes, where $\boldsymbol{x}_n \in \mathbb{R}^d$, and $y_n \in \{1, \ldots, i, \ldots, K\}$. A CNN-based model $f(\boldsymbol{x}; \boldsymbol{\theta})$, with the entire model parameter $\boldsymbol{\theta}$. From the last hidden layer $\boldsymbol{\delta}_n = \{\delta_1, \ldots, \delta_i, \ldots, \delta_h\}$ to the output layer $\boldsymbol{o}_n = \{o_1, \ldots, o_i, \ldots, o_K\}$, the weight $\boldsymbol{W} \in \mathbb{R}^{h \times K}$ defines a transformation, i.e., $\boldsymbol{o}_n = \boldsymbol{W}\boldsymbol{\delta}_n$. In general, the empirical loss $\mathcal{L}(\boldsymbol{\theta})$ over $\mathcal{D}^{train}$ has the following form:

$$\mathcal{L}(\boldsymbol{\theta}) = \sum_{n=1}^N \ell(f(\boldsymbol{x}_n; \boldsymbol{\theta}), y_n) \tag{1}$$

where $\ell(\cdot)$ is the specified loss function, e.g., logistic loss, exponential loss, or cross-entropy loss.

The CNN-based model mentioned in this article respects two usual and reasonable assumptions. The model loss Eq. (1) converges to zero when iteration t approaches infinity, i.e., $\lim_{t \to \infty} \mathcal{L}(\boldsymbol{\theta}_t) = 0$, and the model's last hidden layer and the output layer are fully connected. Based on the two assumptions, [39] demonstrates, both theoretically and empirically, that the last weight layer $\boldsymbol{W}$ of a neural network converges to a support vector machine (SVM) trained on the last hidden layer output $\boldsymbol{\delta}$ with the commonly used cross-entropy loss.

Since $\boldsymbol{W}$ represents a hyperplane, the farther the input sample is from the hyperplane, the greater the corresponding class output, i.e., $max(\boldsymbol{o}_n)$ will be. As illustrated in Fig. 2, the model output contours of the CNN-based model are radiated, becoming higher as the distance from the hyperplane increases.

3.2   Dempster-Shafer Theory

The Dempster-Shafer theory [30] (or evidence theory) is a mathematical framework that enables the reflection of partial and uncertain knowledge. Let $\Omega = \{\omega_1, \ldots, \omega_i, \ldots, \omega_K\}$ be the finite class sets. The belief function $m : 2^\Omega \to [0, 1]$

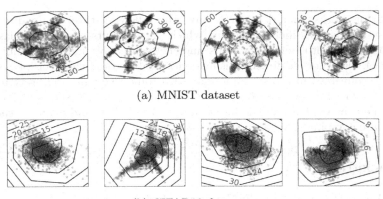

(a) MNIST dataset

(b) CIFAR10 dataset

Fig. 2. The model output contours on MNIST [16] and CIFAR10 [13] datasets. Different prevalent CNN-based models are verified. From left to right are LeNet [15], GoogLeNet [31], ResNet [10], and MobileNet [11]. For visualization purposes, the h that appears in the last hidden layer is set as two. Meanwhile, according to the minimum and maximum column values of $\delta \in \mathbb{R}^{N \times 2}$ a 2D mesh can be generated. Feed this mesh into the last hidden layer to get the outputs which can be regarded as contours. As we can see, the pattern of the CNN-based model is that a sample far from the training dataset can bring high outputs and lead to high probabilities for several classes. Under this context, partial classification rather than precise classification should be used.

applied on x_n measures the degree of belief that the ground-truth label of x_n belongs to a subset $A_i \subseteq \Omega$. It satisfies the following equation:

$$\sum_{A_i \subseteq \Omega} m(A_i) = 1 \tag{2}$$

The subset A_i such that $m(A_i) > 0$ is called the focal set of m. When the focal set is nested, m is said to be *consonant*. As we can see the maximum quantity of beliefs is 2^K which is a significant difference from K for probability. Since the maximum quantity of subsets of classes is also 2^K, belief instead of probability is inherently more suitable for partial classification.

4   Proposed Method

As can be seen from Sect. 3.1, a sample that is far from the training dataset occupies high outputs for several classes leading to high probabilities for the corresponding classes, resulting in the improper execution of precise classification. Consider, from another angle, the high outputs for multiple classes can be regarded as evidence to classify a sample into a class subset. From this point, we proposed to calculate beliefs only based on pre-trained CNN-based model outputs to fulfill partial classification. Moreover, we chose the possibility as the bridge between model outputs and beliefs, then proposed the following transformations.

Sorting o_n by descending order to get $o'_n = \{o'_1 \geq \cdots \geq o'_i \geq \cdots \geq o'_K\}$, where o'_i is the ith largest element in o_n. Then, a prerequisite step is to prepare a temporary vector $v_n = \{v_1, \ldots, v_i, \ldots, v_K\}$ based on Eq. (3) that coordinates with Eq. (4) to calculate the target possibility distribution.

$$v_i = \frac{1}{|A_i|} \sum_{k=1}^{i} \log_2(1 + \max(0, o'_k)) \tag{3}$$

where $\frac{1}{|A_i|}$ is used to penalize the ambiguity caused by classifying x_n to A_i. If we consider a reasonable assumption that the desired possibility transformation should keep the original pattern of outputs, escalating the difference for small values while narrowing the difference for bigger values. The $\log_2$ function should be chosen, which tends to be flat after the initial rapid growth. At the same time, in order to avoid the negative possibility, use $\max(0, o'_k)$ to clamp the outputs and move the $\log_2$ to the left by one unit.

After min-max normalization by Eq. (4), we can get the possibility distribution $\pi_n = \{\pi_1, \ldots, \pi_i, \ldots, \pi_K\}$. Following the theory proposed in [2] that any possibility distribution is a plausibility function corresponding to a consonant m. Our possibility distribution π_n can be transformed to belief function m according to Eq. (5), the detailed calculation is presented in Fig. 3. In our case, π_K equals zero, which implies that $m(\Omega)$ equals zeros.

$$\pi_n = \frac{v_n - \min(v_n)}{\max(v_n) - \min(v_n)} \tag{4}$$

$$m(A_i) = \begin{cases} \pi_j - \pi_{j+1} & \text{if } A_i = \{\omega_1, \cdots, \omega_j\} \text{ for some } j \in \{1, \cdots, K-1\} \\ \pi_K & \text{if } A_i = \Omega \\ 0 & \text{otherwise} \end{cases} \tag{5}$$

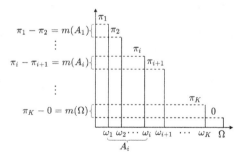

Fig. 3. Calculation of the belief function m [2].

The PCMO classification algorithm is demonstrated in Algorithm 1. Based on the beliefs calculated through Eqs. (3), (4), and (5), we chose the subset with the maximum belief as the prediction. Suppose, the maximum belief is $m(A_i)$, the PCMO method will generate the predicted set $\{\omega_1, \ldots, \omega_i\}$ corresponding to the top i maximum outputs. In addition, the usage is flexible, for example, $\overline{m(A_1)} = 1 - m(A_1)$ can be regarded as the uncertainty for the classification as used in Sect. 5.4.

Algorithm 1: Classification process for a sample x_n

Data: Model outputs $o_n \in \mathbb{R}^{1 \times K}$
Result: Predicted set $predSet$
Sort o_n in descending order to get the sorted index $index$ and sorted outputs o'_n;
Calculate vector v_n base on o'_n according to Eq. (3);
Calculate possibility distribution π_n base on v_n according to Eq. (4);
Calculate belief m_n base on π_n according to Eq. (5);
Obtain the maximum belief index $idx = argmax(m_n)$ for the sample x_n;
Generate the predicted set $predSet = list(index[0 : i + 1])$, which contains the candidate classes;
return $predSet$;

---

5   Experiments

5.1   Experiment Protocol

There are six datasets involved, a Road Surface dataset manually generated and five prevalent datasets, i.e., modified national institute of standards and technology (MNIST) [16], canadian institute for advanced research 10 (CIFAR10) [13], street view house number (SVHN) [27], large-scale scene understanding challenge (LSUN) [35], and canadian institute for advanced research 100 (CIFAR100) [13]. The characteristics of the datasets are shown in Table 1. Four classical CNN-based models, i.e., LeNet [15], GoogLeNet [31], residential energy services network (ResNet) [10], and MobileNet [11], are adopted to prove the efficiency of the PCMO method. We used the cross-entropy loss as the loss function and the rectified linear unit (ReLU) as the activation function. For method comparison, we chose energy score (based on pre-trained model outputs) [18], dropout score (based on several executions of the pre-trained model on the testing dataset) [12], and ensemble score (based on executions of several pre-trained models on the testing dataset) [14].

5.2   Criteria

Traditional accuracy becomes improper when partial classification is allowed. In this case, Zaffalon [37] proposed the following discounted accuracy:

$$a = \frac{1}{|A_i|} I(y_n \in A_i) \tag{6}$$

where $I(\cdot)$ is the indicator function.

Table 1. A quick view of datasets involved.

Name	# Classes	# Training samples	# Testing samples
MNIST [16]	10	55000	10000
CIFAR10 [13]	10	50000	10000
SVHN [27]	10	4000	1000
LSUN [35]	10	2400	600
CIFAR100 [13]	10	8000	2000
Road surface	3	2040	780

For a dataset, the accuracy is evaluated by the average discounted accuracy (ADA). The ADA is a single value with the requirement that the better the prediction, the larger the ADA is.

$$\text{ADA} = \frac{1}{N} \sum_{n=1}^{N} a_n \tag{7}$$

In addition, to approximately measure the goodness of calibration, expected calibration error (ECE) [26] defined by Eq. (8) was adopted. This groups the probability interval into B bins with n_b samples inside and assigns each predicted probability to the bin that encompasses it. The calibration error is the difference between the fraction of predictions in the bin that are correct (accuracy) and the mean of the probabilities in the bin (confidence).

$$\text{ECE} = \sum_{b=1}^{B} \frac{n_b}{N} |\,\text{acc}(b) - \text{conf}(b)| \tag{8}$$

where $\text{acc}(b)$ and $\text{conf}(b)$ are the accuracy and confidence of bin b, respectively.

5.3 Evaluation of the PCMO Method

The PCMO method performs partial classification by choosing the predicted set that occupies the maximum belief. Naturally, the bigger cardinality of the predicted set indicates a more confusing input sample. Thus, in order to verify the efficiency of partial classification and the capacity of reducing the classification risk under the PCMO method. We rejected the most confusing samples according to different rejection rates [25].

On the one hand, we executed the PCMO method for different CNN-based models when rejection rates change from 0.0 to 1.0. Figure 4 is quite revealing in two ways. First, the ADA increases along with the increase of rejection rates. Second, the selected four classical CNN-based models achieved good ADA values, except for the slightly worse initial accuracy of LeNet and MobileNet due to its simple model architecture. This indicates that the PCMO method performed partial classification based on the calculated beliefs. The performance on the

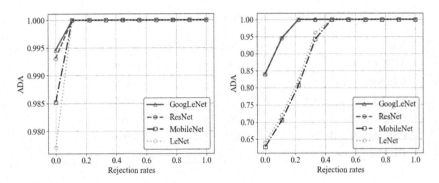

Fig. 4. The performance in terms of ADA values of different CNN-based models with different rejection rates based on MNIST (left) and CIFAR10 (right) datasets.

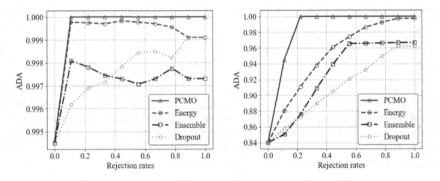

Fig. 5. The performance in terms of ADA values of different methods with different rejection rates based on MNIST (left) and CIFAR10 (right) datasets.

different CNN-based models further proves that partial classification can be achieved only based on the CNN-based model outputs.

On the other hand, we verified different methods based on GoogLeNet (the best performing model in the previous step), as demonstrated in Fig. 5. The ADA of the PCMO method increases significantly when the rejection rate increases from 0.0 to 0.1 or 0.2. In contrast, the ADA of the other methods performed fluctuation or insensitivity when the rejection rate increased from 0.0 to 1.0. The striking performance is evidence that the PCMO method makes a well-distributed partial classification while the others only classified samples to a class subset when the rejection rate is large.

To manifest the efficiency of different methods and models when against a small rejection rate, we set rejection rate equals 0.1 and received Table 2. The PCMO method achieves the highest ADA and comparable ECE values among all the methods. Compared to other methods, there is a 0.23% and 7.71% improvement in terms of ADA for MNIST and CIFAR10, respectively, based on GoogelNet. The detailed statistics are demonstrated in Table 3. It is also clear, for each

Table 2. Comparative experimental results on five datasets for four CNN-based models and four methods when the rejection rate equals 0.1.

Datasets		MNIST		CIFAR10		SVHN		LSUN		CIFAR100	
Criteria		ADA	ECE	ADA	ECE	ADA	ECE	ADA	ECE	ADA	ECE
PCMO based on models	GoogLeNet	1.000	0.004	0.933	0.046	1.000	0.014	0.811	0.096	0.961	0.091
	ResNet	1.000	0.003	0.934	0.040	1.000	0.018	0.924	0.078	0.994	0.082
	MobileNet	1.000	0.006	0.712	0.019	0.790	0.071	0.333	0.077	0.406	0.178
	LeNet	1.000	0.009	0.698	0.031	0.209	0.015	0.135	0.041	0.144	0.037
Methods based on GoogLeNet	PCMO	**1.000**	**0.004**	**0.933**	**0.046**	**1.000**	**0.014**	**0.811**	**0.096**	**0.961**	**0.091**
	Energy	1.000	0.001	0.877	0.041	0.987	0.008	0.765	0.102	0.912	0.072
	Ensemble	0.998	0.001	0.847	0.039	0.967	0.016	0.732	0.098	0.856	0.085
	Dropout	0.995	0.003	0.844	0.043	0.964	0.014	0.736	0.102	0.873	0.079
Methods based on ResNet	PCMO	**1.000**	**0.003**	**0.934**	**0.040**	**1.000**	**0.018**	**0.924**	**0.078**	**0.994**	**0.082**
	Energy	1.000	0.002	0.878	0.033	0.991	0.008	0.898	0.054	0.939	0.053
	Ensemble	0.998	0.001	0.846	0.034	0.967	0.015	0.824	0.075	0.884	0.071
	Dropout	0.993	0.003	0.840	0.037	0.956	0.019	0.811	0.080	0.878	0.081
Methods based on MobileNet	PCMO	**1.000**	**0.006**	**0.712**	**0.019**	**0.790**	**0.071**	**0.333**	**0.077**	**0.406**	**0.178**
	Energy	0.989	0.004	0.668	0.018	0.743	0.073	0.312	0.077	0.370	0.181
	Ensemble	0.998	0.016	0.671	0.028	0.752	0.079	0.325	0.080	0.365	0.166
	Dropout	0.976	0.006	0.639	0.016	0.708	0.061	0.298	0.075	0.354	0.174
Methods based on LeNet	PCMO	**1.000**	**0.009**	**0.698**	**0.031**	**0.209**	**0.015**	**0.135**	**0.041**	**0.144**	**0.037**
	Energy	0.999	0.003	0.662	0.030	0.188	0.018	0.144	0.039	0.155	0.037
	Ensemble	0.998	0.003	0.667	0.016	0.206	0.013	0.144	0.039	0.166	0.050
	Dropout	0.984	0.009	0.627	0.032	0.192	0.011	0.137	0.032	0.149	0.033

Table 3. The detailed performance improvement statistics. This represents the subtraction between the ADA value produced by the PCMO method and the averaged ADA value of the other three methods.

Models	Datasets				
	MNIST	CIFAR10	SVHN	LSUN	CIFAR100
GoogLeNet	0.23%	7.71%	2.74%	2.74%	2.74%
ResNet	0.31%	7.90%	2.89%	2.89%	2.89%
MobileNet	1.20%	5.24%	5.56%	5.56%	5.56%
LeNet	0.63%	4.60%	1.36%	1.36%	1.36%

CNN-based model, that no matter what dataset is involved, the PCMO method performs better than other methods. This proved that the PCMO method is effective in reducing the misclassification of the confusing sample with the beliefs calculated only based on pre-trained CNN-based model outputs.

5.4   Practical Usage of the PCMO Method

The PCMO method has a wide range of application contexts, e.g., autonomous driving [38]. To prove the PCMO method is effective in reducing classification risk practically, we applied it to the road surface classification, which is an essential part of autonomous driving. The Road Surface dataset is mixed manually from the Crack [38] and Pothole [8] datasets, which contains three classes, i.e., crack, pothole, and normal. Several samples are shown in Fig. 6

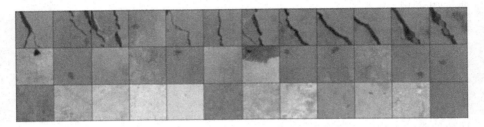

Fig. 6. The visualization of the Road Surface dataset, which contains cracks (the first row), potholes (the second row), and normals (the third row). As we can see, each class contains several confusing samples, e.g., the fourth one, the third one, and the second one for each class, respectively. The PCMO method aims to detect confusing samples, reducing the classification risk.

Figure 7(a) shows a certain crack sample, in which the crack area is obvious and clean. Thus, the PCMO method produced high belief $m(crack) = 0.85$ and low uncertainty $\overline{m(crack)} = 0.15$ to indicate the certain classification. Similarly, Fig. 7(b) and Fig. 7(c) show a certain pothole and normal sample, respectively, which contains sufficient evidence (obvious class characteristics) to support a certain prediction.

Inversely, Fig. 7(d) demonstrated a confusing crack sample that is difficult to identify from a pothole sample due to its circle-shaped crack. Figure 7(e) manifested a confusing pothole sample whose pothole is too shallow to identify from a normal sample. And Fig. 7(f) demonstrated a confusing normal sample whose crack indicates it is a crack sample rather than a normal sample. Correspondingly, the PCMO method produces low belief and high uncertainty as illustrated in Fig. 7. Based on the PCMO method, we can reduce classification risk and enhance the quality of the target system by rejecting confusing samples.

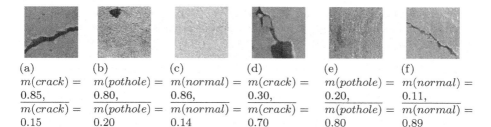

(a)	(b)	(c)	(d)	(e)	(f)
$m(crack) =$ 0.85,	$m(pothole) =$ 0.80,	$m(normal) =$ 0.86,	$m(crack) =$ 0.30,	$m(pothole) =$ 0.20,	$m(normal) =$ 0.11,
$\overline{m(crack)} =$ 0.15	$\overline{m(pothole)} =$ 0.20	$\overline{m(normal)} =$ 0.14	$\overline{m(crack)} =$ 0.70	$\overline{m(pothole)} =$ 0.80	$\overline{m(normal)} =$ 0.89

Fig. 7. The beliefs calculated based on the proposed method for different types of samples. (a) Certain crack sample, (b) certain pothole sample, (c) certain normal sample, (d) confusing crack sample, (e) confusing pothole sample, and (f) confusing normal sample.

6   Conclusion

In this paper, we proposed a new partial classification method named PCMO, which is fulfilled based on pre-trained CNN-based model outputs. At first, we theoretically and empirically proved our hypothesis that a sample far from the training dataset can provide high outputs and lead to high probabilities for several classes. Second, we adopted possibility as the bridge fulfilling the transformation from model outputs to beliefs for the predicted sets. Then, we verified the PCMO method with different CNN-based models, as well as different methods based on five datasets. Finally, to demonstrate the practical usage of the PCMO method, we conducted experiments based on a manually generated road surface dataset. From the production of ADA and ECE criteria, we can tell that the PCMO method performs better than the existing methods, as it can provide a high belief to a certain sample, as well as a high uncertainty to a confusing sample. The PCMO method proved effective in increasing prediction accuracy and ultimately reducing the classification risk. In the future, we plan to explore the feasibility of PCMO methods on all types of pre-trained models and try to develop a method to generate beliefs for all subsets of the entire set.

Acknowledgments. This work was funded by the Auvergne Rhône Alpes region: project AUDACE2018. The authors would like to thank the reviewers for their very insightful comments that helped to improve the article.

References

1. Bi, Y.: The impact of diversity on the accuracy of evidential classifier ensembles. Int. J. Approximate Reasoning **53**(4), 584–607 (2012)
2. Dubois, D., Prade, H.: On several representations of an uncertainty body of evidence. Fuzzy Inf. Decis. Process. 167–181 (1982)
3. Del Coz, J.J., Diez, J., Bahamonde, A.: Learning nondeterministic classifiers. J. Mach. Learn. Res. **10**(10), 2273–2293 (2009)

4. Dembczyński, K., Waegeman, W., Cheng, W., Hüllermeier, E.: On label dependence and loss minimization in multi-label classification. Mach. Learn. 5–45 (2012). https://doi.org/10.1007/s10994-012-5285-8
5. Dempster, A.P.: Upper and lower probabilities induced by a multivalued mapping. In: Classic Works of the Dempster-Shafer Theory of Belief Functions, pp. 325–339 (1967)
6. Denœux, T.: A neural network classifier based on Dempster-Shafer theory. IEEE Trans. Syst. Man Cybern. Part A: Syst. Hum. 30(2), 131–150 (2000)
7. Denœux, T., Kanjanatarakul, O., Sriboonchitta, S.: A new evidential K-nearest neighbor rule based on contextual discounting with partially supervised learning. Int. J. Approximate Reasoning 113, 287–302 (2019)
8. Fan, R., Ai, X., Dahnoun, N.: Road surface 3D reconstruction based on dense subpixel disparity map estimation. IEEE Trans. Image Process. 27(6), 3025–3035 (2018)
9. Ha, T.: The optimum class-selective rejection rule. IEEE Trans. Patt. Anal. Mach. Intell. 19(6), 608–615 (1997)
10. He, K., Zhang, X., Ren, S., et al.: Deep residual learning for image recognition. In: Proceedings of the IEEE Conference on Computer Vision and Pattern Recognition, pp. 770–778 (2016)
11. Howard, A.G., et al.: MobileNets: efficient convolutional neural networks for mobile vision applications. CoRR abs/1704.04861 (2017)
12. Kingma, D.P., Salimans, T., Welling, M.: Variational dropout and the local reparameterization trick. Adv. Neural Inf. Process. Syst. 28, 2575–2583 (2015)
13. Krizhevsky, A.: Learning multiple layers of features from tiny images. University of Toronto, May 2012
14. Lakshminarayanan, B., Pritzel, A., Blundell, C.: Simple and scalable predictive uncertainty estimation using deep ensembles. In: Advances in Neural Information Processing Systems, pp. 6402–6413 (2017)
15. LeCun, Y., Bottou, L., Bengio, Y., Haffner, P.: Gradient-based learning applied to document recognition. Proc. IEEE 86(11), 2278–2324 (1998)
16. LeCun, Y., Cortes, C., Burges, C.: MNIST handwritten digit database. ATT Labs 2 (2010)
17. Leng, B., Liu, Y., Yu, K., Zhang, X., Xiong, Z.: 3D object understanding with 3D convolutional neural networks. Inf. Sci. 366, 188–201 (2016)
18. Liu, W., Wang, X., Owens, J., Li, Y.: Energy-based out-of-distribution detection. Adv. Neural Inf. Process. Syst. (2020)
19. Liu, Z., Pan, Q., Dezert, J., Han, J.W., He, Y.: Classifier fusion with contextual reliability evaluation. IEEE Trans. Cybern. 48(5), 1605–1618 (2017)
20. Ma, H., Xiong, R., Wang, Y., Kodagoda, S., Shi, L.: Towards open-set semantic labeling in 3D point clouds: analysis on the unknown class. Neurocomputing 275, 1282–1294 (2018)
21. Ma, L., Denœux, T.: Partial classification in the belief function framework. Knowl.-Based Syst. 214, 106742 (2021)
22. Minary, P., Pichon, F., Mercier, D., Lefevre, E., Droit, B.: Face pixel detection using evidential calibration and fusion. Int. J. Approximate Reasoning 91, 202–215 (2017)
23. Minary, P., Pichon, F., Mercier, D., Lefevre, E., Droit, B.: Evidential joint calibration of binary SVM classifiers. Soft Comput. 23(13), 4655–4671 (2019)
24. Mortier, T., Wydmuch, M., Dembczyński, K., Hüllermeier, E., Waegeman, W.: Efficient set-valued prediction in multi-class classification. Data Mining Knowl. Disc. 35(4), 1435–1469 (2021). https://doi.org/10.1007/s10618-021-00751-x

25. Nadeem, M.S.A., Zucker, J.D., Hanczar, B.: Accuracy-rejection curves (ARCs) for comparing classification methods with a reject option, vol. 8, pp. 65–81 (2009)
26. Naeini, M.P., Cooper, G., Hauskrecht, M.: Obtaining well calibrated probabilities using Bayesian binning. In: Proceedings of the AAAI Conference on Artificial Intelligence, vol. 29 (2015)
27. Netzer, Y., Wang, T., Coates, A., Bissacco, A., Wu, B., Ng, A.Y.: Reading digits in natural images with unsupervised feature learning. In: NIPS Workshop on Deep Learning and Unsupervised Feature Learning (2011)
28. Russakovsky, O., et al.: ImageNet large scale visual recognition challenge. Int. J. Comput. Vis. **115**(3), 211–252 (2015)
29. Sadinle, M., Lei, J., Wasserman, L.: Least ambiguous set-valued classifiers with bounded error levels. J. Am. Stat. Assoc. **114**(525), 223–234 (2019)
30. Shafer, G.: A Mathematical Theory of Evidence, vol. 42. Princeton University Press, Princeton (1976)
31. Szegedy, C., et al.: Going deeper with convolutions. In: Proceedings of the IEEE Conference on Computer Vision and Pattern Recognition, pp. 1–9 (2015)
32. Vovk, V.: Conditional validity of inductive conformal predictors. In: Asian Conference on Machine Learning, pp. 475–490 (2012)
33. Vovk, V., Gammerman, A., Shafer, G.: Algorithmic learning in a random world. Springer Science & Business Media (2005). https://doi.org/10.1007/b106715
34. Wang, J., Ju, R., Chen, Y., Liu, G., Yi, Z.: Automated diagnosis of neonatal encephalopathy on aEEG using deep neural networks. Neurocomputing **398**, 95–107 (2020)
35. Yu, F., Zhang, Y., Song, S., Seff, A., Xiao, J.: LSUN: construction of a large-scale image dataset using deep learning with humans in the loop (2015)
36. Zaffalon, M.: The Naive Credal classifier. J. Stat. Plann. Inference **105**(1), 5–21 (2002)
37. Zaffalon, M., Corani, G., Mauá, D.: Evaluating Credal classifiers by utility-discounted predictive accuracy. Int. J. Approximate Reasoning **53**(8), 1282–1301 (2012)
38. Zhang, L., Yang, F., Zhang, Y.D., Zhu, Y.J.: Road crack detection using deep convolutional neural network. In: 2016 IEEE International Conference on Image Processing (ICIP), pp. 3708–3712 (2016)
39. Zhang, Y., Liao, S.: A Kernel perspective for the decision boundary of deep neural networks. In: 2020 IEEE 32nd International Conference on Tools with Artificial Intelligence, pp. 653–660 (2020)
40. Zhou, C., Lu, X., Huang, M.: Dempster-Shafer theory-based robust least squares support vector machine for stochastic modelling. Neurocomputing **182**, 145–153 (2016)

Multi-branch Fusion Fully Convolutional Network for Person Re-Identification

Shanshan Ji🆔, Te Li$^{(\boxtimes)}$, Shiqiang Zhu, Qiwei Meng, and Jianjun Gu

Intelligent Robotics Research Center, Zhejiang Lab, Hangzhou, Zhejiang, China
{jiss,lite,zhusq,mengqw,jgu}@zhejianglab.com

Abstract. Building effective CNN architectures with light weight has become an increasing application demand for person re-identification (Re-ID) tasks. However, most of the existing methods adopt large CNN models as baseline, which is complicated and inefficient. In this paper, we propose an efficient and effective CNN architecture named Multi-branch Fusion Fully Convolutional Network (MBF-FCN). Firstly, multi-branch feature extractor module focusing on different receptive field sizes is designed to extract low-level features. Secondly, basic convolution block units (CBU) are used for constructing candidate network module to obtain deep-layer feature presentation. Finally, head structures consisted of multi-branches will be adopted, combining not only global and local features but also lower-level and higher-level features with fully convolutional layer. Experiments demonstrate our superior trade-off among model size, speed, computation, and accuracy. Specifically, our model trained from scratch, only has 2.1 million parameters, 0.84 GFLOPs and 384-dimensional features, reaching the state-of-the-art result on Market-1501 and DuckMTMCreID dataset of Rank-1/mAP = 94.5%/84.3%, Rank-1/mAP = 86.6%/73.5% without re-ranking, respectively.

Keywords: Person Re-Identification · Lightweight network · Multi-branch fusion · Fully convolutional network

1 Introduction

Person re-identification (Re-ID) [1,2] is a popular research topic in the field of computer vision, which aims to match and recognize a query person among all the gallery pedestrian images taken from different cameras or the same camera in different occasions. With the rapid development of the deep convolutional network, several deep Re-ID methods [3–5] have achieved a high-level performance on some public datasets. However, it is still challenging work to distinguish different persons on subtle changes on clothing, body characteristics, as well as confirm the same person from significant changes in viewpoint, illumination, occlusion and so on.

Most person Re-ID models utilize large deep networks such as VGG [6], Inception [7] and ResNet [8] as backbones. Taking advantage of these standard

© Springer Nature Switzerland AG 2021
T. Mantoro et al. (Eds.): ICONIP 2021, LNCS 13110, pp. 164–175, 2021.
https://doi.org/10.1007/978-3-030-92238-2_14

baselines, by further exploiting local features together with global features, the robust feature representations can be learned and the performance can be further improved for person Re-ID. PCB [9] and MGN [10] methods take the advantage of the body parts and extract local-region features for learning partial discriminative feature representations. Aligned Re-ID [11], Relation Network for Re-ID [12], CAN [13] and HPM [14] design special network structures to further study the relationship of global and local features of pedestrian body parts. Besides, auxiliary feature representation learning methods such as semantic segmentation and human body pose are also used for person Re-ID [15,16]. Despite these methods, researchers also focus on how to build effective CNN backbones specifically suitable for Re-ID tasks. [17] proposes an automatic search strategy named Auto-ReID to find an efficient and effective CNN architecture for Re-ID task. [18] proposes a Multi-Level Factorisation Net (MLFN) which is composed of multiple stacked blocks for learning identity-discriminative and view-invariant visual factors at multiple semantic levels. An efficient small-scale network named OSNet is proposed in [19]. In OSNet work, some basic building blocks consists of multiple convolutional streams are designed for omni-scale feature learning. To improve the performance in person Re-ID tasks, most methods adopt complicate net structures, which is lower efficiency both in training and testing procedure. Furthermore, the datasets for Re-ID are often of moderate size, it may lead to overfitting if the network structure is too deep and complicate. Moreover, large dimensional output features also greatly reduce the feature matching speed of the retrieval process.

In this paper, we propose a multi-branch fusion fully convolutional network (MBF-FCN) for person Re-Identification. It is a lightweight net framework with less computation cost but higher performance. In summary, our contributions are as follows:

1) We develop the multi-branch feature extractor module for extracting the most discriminative features by different receptive field sizes. This design obtains multi-scale body information and make best use of low-level visual features.

2) Based on Inception and ResNet blocks, we design several basic feature extract blocks called convolution block unit (CBU), These basic blocks enable us to construct more optimal structures for extracting deeper-layer features with less parameters and fewer computation.

3) The embedding head of the structure also consisted of multi-branches, combining not only global and local features but also lower-level and higher-level features. Furthermore, we use fully convolutional layer instead of fully connect layer, which is lightweight and has a further consistent increase in accuracy.

2   Related Works

Lightweight CNN Architectures for Person Re-ID. Although CNN models with large backbones can achieve high performance [20,21], these methods also have disadvantages of increasing overall model complexity and higher computational costs. One straightforward way is to use SqueezeNet and MobileNet series, which are lightweight and perform well in image classification tasks, as strong baselines for person Re-ID. Another important research direction is developing specific network architectures for person Re-ID task [22]. Harmonious Attention Convolution Neural Network (HA-CNN), a lightweight yet deep CNN architecture is proposed in [23]. It learns hard region-level and soft pixel-level attention within arbitrary person bounding boxes along with re-id feature representations for maximizing the correlated complementary information between attention selection and feature discrimination. [19] designs an efficient small-scale network named OSNet (Omni-Scale Network). OSNet is lightweight and is capable of detecting the small-scale discriminative features at a certain scale. [24] propose a branch-cooperative architecture based on OSNet, namely BC-OSNet, enriching the feature representation by stacking four cooperative branches. It is well known that development of new network takes substantial efforts, hard work and computation cost to acquire excellent results. Therefore, we design some basic convolution block units (CBUs) for constructing various Re-ID candidate network structures and use multi-loss function and training strategy to adapt to Re-ID task. In addition, our net is fairly lightweight and can achieve competitive performance when training from scratch.

Instance Batch Normalization (IBN). To improve the generalization capacities of CNN networks, a combination of batch normalization (BN) and Instance Normalization (IN) method has been adopted to build CNN blocks in recent work. According to [25], an IBN-Net, which integrates Instance Normalization (IN) and Batch Normalization (BN) in a certain manner can improve the robustness and achieve better results. Therefore, we refer to the IBN-Net block, replacing the original BN layers of our CNN blocks by half IN and BN for the other half.

Re-ranking. In addition to the development of CNN architectures, re-ranking is an effective post-processing technique for boosting the performance of Re-ID. Generally, Euclidean or cosine distances in the feature space are used for similarity criteria in the retrieval stage of Re-ID. By taking the neighbor features of nearest samples into consideration, a k-reciprocal re-ranking method, which inherits the advantages of k-reciprocal nearest neighbors to improve the initial ranking list by combining the original distance and Jaccard distance, is proposed in [26]. Considering the high computation efficiency of most existing re-ranking methods, recent work [27] designs a hardware-friendly acceleration approach by re-formulating the re-ranking process as a graph neural network (GNN), significantly reducing the running time of re-ranking processing.

3   Proposed Method

3.1   Network Architecture

Our model (MBF-FCN) is a novel deep network architecture which is more effective for extracting features for person Re-ID. The architecture of our model is shown in Fig. 1. In General, the main framework can be separated into four parts: low-layer feature extractor module, multi-branch feature extractor module, feature fusion and deep-layer feature extractor module, feature embedding and the loss function module.

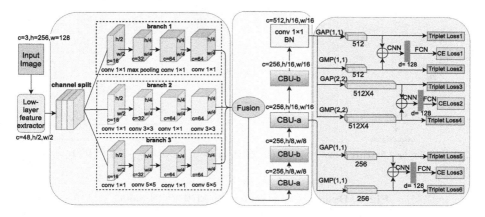

Fig. 1. Architecture of MBF-FCN model, the main framework can be separated into four parts: low-layer feature extractor module, multi-branch feature extractor module, feature fusion and deep-layer feature extractor module, feature embedding and the loss function module.

Firstly, the base features of the input image are extracted by the low-layer feature extractor module. The detail of proposed structure is shown in Fig. 2 (a), which is consisted of two streams that includes an average pooling layer and two 3 × 3 convolutional layers. This structure is designed to process the input images and realize channel expansion and feature map transformation, therefore extracting a useful low-level visual feature map.

Secondly, as shown in Fig. 1, features obtained by the first module are split into three branches by the channel split layer. The channel number of each branch is set to the same. Each branch can obtain different levels of features with less computation. For branch 1, it mainly includes 1 × 1 convolution layer and a max pooling layer. For branch 2 and branch3, 3 × 3 convolution layers and 5 × 5 convolution layers are used, respectively. The three branches showed in Fig. 1 are designed refering to Inception module. The first layer and the third layer of each branch with a 1 × 1 convolution layer is designed for transformation in the channel space of the feature map. With mixture of 1 × 1 and 3 × 3 convolutions, 1 × 1 and 5 × 5 convolutions, receptive field of different sizes can be captured for locally spatial information. The basic building block of each branch consisted of convolution layers of different kernel sizes, which aims to get the most discriminative features by different receptive field sizes, can make the net 'got wide' to generate a network with less layers and less computation.

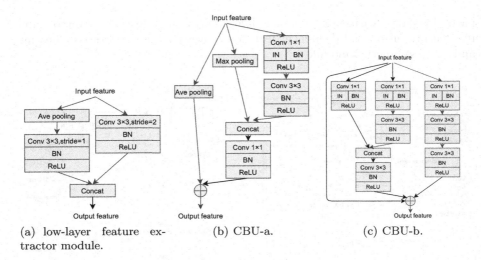

(a) low-layer feature extractor module.

(b) CBU-a.

(c) CBU-b.

Fig. 2. Basic CNN blocks for the MBF-FCN model.

Thirdly, the feature maps extracted from the multi-branch feature extractor module are fused by channels using a concatenate layer, and a 1 × 1 conv-bn-relu module is also used to obtain information of cross-channel. After fusing, convolution block units (CBU) are used to extract deeper features. We design two kinds of convolution block units named CBU-a and CBU-b respectively. As shown in Fig. 2(b), CBU-a contains an average-pooling layer and a max-pooling layer to reduce the feature dimensions, and 3 × 3 convolution layers to obtain suitable deeper features. For example, the input feature map for the first CBU-a step is [N, C, H, W], then the output feature map will be [N, C, H/2, W/2]. CBU-b shown in Fig. 2(c) is inspired by Xception and ResNet block, which consists of multiple streams with 1 × 1 convolution layers and 3 × 3 convolution layers, and shortcut connection is also used. For the first 1 × 1 convolution blocks of CBU-a and CBU-b, we replacing the original BN layers by half IN and BN for the other half. We proposed these structures to reduce computation complexity and it can provide better extensibility for the model. By stacking multiple CBU-a and CBU-b (Two couples applied in this paper, and it can be combined in different ways according to specific tasks), we can construct different feature extraction units for efficient feature extraction.

Finally, for embedding head and the loss function module, we follow the structure in [28] and employ a multi-branch network structure to combine global and local features. For each branch, global average pooling (GAP) and global max pooling (GMP) are adopted for information extraction of the feature map. 1 × 1 conv-bn layer is used for extract output feature of the last CBU-b, and then obtain a global feature map of 512 dimensions. For the first branch, we set the target output size of 1 × 1 for both GAP and GMP to get global features, and for the second branch, we set the target output size of 2 × 2 for GAP and GMP, which divides the feature map into horizontal and vertical parts for obtaining partial features. In the third branch, the output feature map with 256 dimensions of the last CBU-a is used as the input, and the target output size of GAP and GMP is set to 1 × 1, thus, the lower-level feature of the pedestrian can also be exploited. These three branches have fused global-and-local features as well as lower-and-higher features together, so that the overall information of multi-grained for the pedestrians can be acquired. For each branch, we combine features of GAP and GMP by summing up operations and use a 1 × 1 convolution layer with BN to get 128-dimensional features. We use a fully convolutional net structure (FCN, i.e. using conv-bn layer to realize classification from the input of [N, 128, 1, 1] to the out-put of [N, 751, 1, 1] for Market1501 dataset, instead of a fully connected layer), combined with CE loss function to optimize identity loss. For triplet loss (TP loss), embedding obtained in each branch after GAP and GMP is individually used for optimization with respect to the loss function.

The main framework of the proposed net is a global-and-local multi-branch structure that combines global features and local features utilize a fully convolutional network, while keeps the overall number of parameters and embeddings low.

3.2 Training and Loss Functions

To train our model for person Re-ID task, we use a typical strategy of combining Cross Entropy (CE) loss and Triplet (TP) Loss. For CE loss, we further use label smoothing. The overall multi-loss function is the sum of each output branch including three for CE losses and six for TP losses, it can be formulated as:

$$L = \sum_{i=0}^{n} W_{CE}^{(i)} * L_{CE}^{(i)} + \sum_{j=0}^{m} W_{TP}^{(j)} * L_{TP}^{(j)} \tag{1}$$

Where we denote by $W_{CE}^{(i)}$ and $W_{TP}^{(j)}$ the weights of CE loss and TP loss, respectively. The balanced weights of the multi-loss function depended on the branch numbers of i and j. The corresponding values are $i = 0, 1, 2$ and $j = 0, 1, 2, 3, 4, 5$, where $W_{CE}^{(i)} = 1/n$, $W_{TP}^{(j)} = 1/m$, of which $n = 3$ and $m = 6$.

4 Experiments

4.1 Datasets and Evaluation Protocols

Datasets. The experiments are conducted on the following mainstream person Re-ID datasets: Market-1501 [29], DukeMTMC-reID [30]. The Market1501 dataset is composed of 32668 images captured by six cameras, a total of 1501 labeled person identities. There are 12936 images of 751 identities for training set and 3368 query images and 19732 gallery images for testing set. The DukeMTMC-reID dataset contains 16522 images of 702 identities for training set, taken by 8 cameras. The testing set includes 2228 query images and 17661 gallery images. The details of the datasets are shown in Table 1.

Table 1. Information of image-based person Re-ID datasets.

Datasets	Training		Testing			Cameras
	IDs	Images	IDs	Query	Gallery	
Martket-1501	751	12936	750	3368	19732	6
DuckMTMC-reID	702	16522	702	2228	17661	8

Evaluation Protocols. Generally, Cumulative Matching Characteristics (CMC) and mean Average Precision (mAP) are two widely used evaluation metrics. In our experiments, Floating-point Operations Per second (FLOPs) and the network parameter are also used to evaluate the performance. These two metrics are more comprehensive measurements to evaluate the Re-ID models considering the efficiency and complexity in training/testing with limited computational resources. For a fair comparison with other existing methods, the input images are set to the same size.

4.2   Implementation Details

In training stage, the input images are resized to 256×128 and augmented by horizontal flip, random crop and random erasing [31]. Since we apply Cross-Entropy Loss with label smoothing and Triplet Loss as the multi-loss function, the method with each batch includes 32 images from 16 different person IDs, where each ID includes 4 images is used. The network is optimized by SGD with momentum 0.9. The initial learning rate is set to 0.03 and decayed by 0.1 at 80 and 150 epochs. A Warming-up strategy is also used with a start learning rate 0.0003 for 10 epochs. The total training procedure lasts for 250 epochs. It takes about 1.5 h to conduct a completely training using a single NVIDIA RTX 2080Ti GPU with 11 GB memory for Market1501. We implemented our model in the Pytorch framework. All of our models are trained from scratch, that is to say, we do not use any pre-training models. Existing Re-ID models boost the performance from model tuning, but it is widely known, obtaining the original model is typically difficult and the subsequent operation is time-consuming.

4.3   Comparison with State-of-the-Art Methods

We compare our method against state-of-the-art Re-ID methods on Market-1501 and DukeMTMC-reID. Comparisons between the proposed model and existing models are shown in Table 2. Our model has the smallest model size (only 2.1 million parameters) and 384-dimensional output features. Moreover, all of our models are trained from scratch. The results on Market-1501 dataset show that our model achieves accuracy of rank1 = 94.5% and mAP = 84.3%, which outperforms the best of previous works. For DuckMTMC-ReID dataset, our method achieves the best performance of rank1 = 86.6% and mAP = 73.5%. Compared with other Re-ID methods, our model offers a good compromise in terms of Floating-point Operations Per second (FLOPs), network parameters, the dimension of features.

Table 2. Performance of our Re-ID model compared with other deep methods on Market1501 and DuckMTMC-reID datasets.

Trained from scratch					Market1501		DukeMTMC	
Model	Input	Params (10^6)	GFLOPs	F-dim	R1(%)	mAP(%)	R1(%)	mAP(%)
Resnet50	(256, 128)	23.5	2.7	2048	87.9	70.4	78.3	58.9
MLFN	(256, 128)	32.5	2.8	1024	90.1	74.3	81.1	63.2
HACNN	(160, 64)	4.5	0.5	1024	91.2	75.6	80.1	63.2
MobileNetV2_x1_0	(256, 128)	2.2	0.2	1280	87.1	69.5	74.2	54.7
OSNet_x1_0	(256, 128)	2.2	0.98	512	93.6	81.0	84.7	68.6
MBF-FCN	**(256, 128)**	**2.1**	**0.84**	**384**	**94.5**	**84.3**	**86.6**	**73.5**

4.4   Further Analysis and Discussions

Model Complexity. We compare our MBF-FCN model with popular Re-ID CNN architectures PCB, PGFA, MLFN, MGN, MDRS. Our model has the smallest model size (only 2.1 million parameters), much less than the ResNet-based methods. Moreover, training our model from scratch on Market1501 datasets for 250 epochs with batchsize = 32 only takes 1.5 h, which requires fewer computing resources and takes less time. The comparison of complexity between the proposed model and existing models is shown in Table 3.

Table 3. Comparison of complexity with some existing models on Market1501 dataset.

Models	GPU(s)	Parameters	Batch size	Epochs	Training time
PCB	Two NVIDIA TITAN XP GPUs	24.033M	64	60	50 min
PGFA	–	25.082M	32	60	–
MLFN	A K80 GPU	32.473M	256	307	–
MGN	Two NVIDIA TITAN XP GPUs	68.808M	64	80	2 h
MDRS	A single NVIDIA GTX 1080Ti GPU	113.637M	32	440	12 h
MBF-FCN	**A single NVIDIA GTX 2080Ti GPU**	**2.1M**	**32**	**250**	**1.5 h**

Effect of Branches. In our model, we design a multi-branch feature extractor module by fusion three branches. The basic building block of each branch consisted of convolution layers of different kernel sizes, which can get the most discriminative features by different receptive field sizes. We further evaluate the effect of different branch combinations ways. The results in Table 4 suggest that by using all three branches together, our model achieves the best results on both datasets. This structure benefits from extracting features by different receptive field sizes of multi-scale body information. Different branches combination methods have their respective advantages but our integration of multi-branch method gets the superior performance of feature learning. All above experiments do not use pre-training models.

Table 4. Ablation study of branches.

Branches	Market-1501		DuckMTMC	
	Rank1(%)	mAP(%)	Rank1(%)	mAP(%)
[1, 1, 1]	92.8	81.5	85.1	71.6
[2, 2, 2]	94.0	83.6	86.4	**73.7**
[3, 3, 3]	93.9	83.5	86.3	73.5
[1, 1, 2]	93.9	83.4	86.4	73.2
[1, 1, 3]	93.8	83.2	86.6	72.9
[2, 2, 1]	93.7	83.2	86.6	73.3
[2, 2, 3]	93.4	83.6	86.4	73.6
[3, 3, 1]	93.5	83.2	86.1	72.7
[3, 3, 2]	94.1	83.5	86.6	73.5
[1, 2, 3]	**94.5**	**84.3**	**86.6**	73.5

Effect of Instance Normalization. According to the idea of IBN-Net, we trained our model with IBN (half IN and BN for the other half) instead of BN. Table 5 gives performance of models with IBN layers added to different parts of our net. BN means we don't use any IBN layers in our net; IBN-part1 means using IBN in the first 1×1 conv layer of each branch for multi-branch feature extractor module parts; IBN-part2 means using IBN in the CUB-a and CBU-b parts; IBN-part1,2 means using IBN in both multi-branch feature extractor module parts and CUB parts. The results show that using IBN can increase the rank-1 and mAP performance of both datasets. It can be seen that the performance is further improved with IBN layers. Among all methods, adding IBN layer to both multi-branch parts and CUB parts has the best performance, but when IBN layers are added only to multi-branch parts, the performance is lowest. This indicates that our multi-branch and CBU modules help to preserve important feature information, and the choice of IBN among layers of the framework can be important for improving the model performance.

Table 5. Comparison of our model with IBN methods.

	Market-1501		DuckMTMC	
	Rank1(%)	mAP(%)	Rank1(%)	mAP(%)
BN	92.8	80.8	83.5	69.2
IBN-part1	93.0	81.1	84.8	70.3
IBN-part2	93.9	83.9	86.5	73.5
IBN-part1,2	**94.5**	**84.3**	**86.6**	**73.6**

Effect of Re-ranking. With the help of re-ranking, the accuracy of the model can be further improved. Table 6 gives performance of the two re-ranking methods. It can be seen that after implementing re-ranking, the rank-1 and mAP accuracy further improves approximately 1.0%–3.5%, 10%–14%, respectively, which outperforms previous works. However, it is time consuming when searching in a large gallery as well as the model has a high feature dimension. We recommend using a more effective method proposed in [22], namely GNN re-ranking. We evaluate these two re-ranking methods on NVIDIA RTX 2080Ti GPU platform. By comparing two kinds of re-ranking methods, we observe that GNN re-ranking method achieve the best mAP performs than k-reciprocal re-ranking method, but the second-best in terms of Rank1. Besides, the running time cost is also been tested. We can see that the GNN-based method achieves a competitive speed compared with k-reciprocal re-ranking methods.

Table 6. Comparison of our model with re-ranking methods.

	Market-1501			DuckMTMC		
	Rank1(%)	mAP(%)	Time cost	Rank1(%)	mAP(%)	Time cost
Without RE	94.5	84.3	–	86.6	73.5	–
+ k-reciprocal RE	**95.3**	92.6	55 s	**89.6**	85.5	41 s
+ GNN-based RE	95.1	**93.4**	**1.07 s**	89.1	**87.4**	**0.59 s**

5   Conclusion

In this paper, we present a novel Multi-branch Fusion Fully Convolutional Network (MBF-FCN), a lightweight CNN architecture for person Re-ID. Our model outperforms other Re-ID methods such as ResNet-based models and several lightweight CNN models, while being much smaller and without requiring a good pre-trained model. We provide some basic CNN blocks for building lightweight person Re-ID architecture, with less parameters and fewer computation. Experiments demonstrate the effectiveness of the proposed methods both in training and testing.

References

1. Zheng, L., Yang, Y., Hauptmann, A.G.: Person re-identification: past, present and future. arXiv preprint arXiv:1610.02984 (2016)
2. Wu, L., Shen, C., van den Hengel, A.: PersonNet: person re-identification with deep convolutional neural networks. arXiv preprint arXiv:1601.07255 (2016)
3. Ye, M., Shen, J., Lin, G., Xiang, T., Shao, L., Hoi, S.C.: Deep learning for person re-identification: a survey and outlook. IEEE Trans. Patt. Anal. Mach. Intell. (2021)
4. He, L., Liao, X., Liu, W., Liu, X., Cheng, P., Mei, T.: FastReID: a Pytorch toolbox for general instance re-identification. arXiv preprint arXiv:2006.02631 (2020)
5. Luo, H., et al.: A strong baseline and batch normalization neck for deep person re-identification. IEEE Trans. Multimedia **22**(10), 2597–2609 (2019)
6. Simonyan, K., Zisserman, A.: Very deep convolutional networks for large-scale image recognition. arXiv preprint arXiv:1409.1556 (2014)
7. Szegedy, C., et al.: Going deeper with convolutions. In: Proceedings of the IEEE Conference on Computer Vision and Pattern Recognition, pp. 1–9 (2015)
8. He, K., Zhang, X., Ren, S., Sun, J.: Deep residual learning for image recognition. In Proceedings of the IEEE Conference on Computer Vision and Pattern Recognition, pp. 770–778 (2016)
9. Sun, Y., Zheng, L., Yang, Y., Tian, Q., Wang, S.: Beyond part models: person retrieval with refined part pooling (and a strong convolutional baseline). In: Proceedings of the European Conference on Computer Vision (ECCV), pp. 480–496 (2018)
10. Wang, G., Yuan, Y., Chen, X., Li, J., Zhou, X.: Learning discriminative features with multiple granularities for person re-identification. In: Proceedings of the 26th ACM International Conference on Multimedia, pp. 274–282 (2018)
11. Zhang, X., et al.: AlignedReID: surpassing human-level performance in person re-identification. arXiv preprint arXiv:1711.08184 (2017)
12. Liang, J., Zeng, D., Chen, S., Tian, Q.: Related attention network for person re-identification. In: 2019 IEEE Fifth International Conference on Multimedia Big Data (BigMM), pp. 366–372. IEEE (2019)
13. Li, W., et al.: Collaborative attention network for person re-identification. J. Phys. Conf. Ser. **1848**, 012074. IOP Publishing (2021)
14. Yang, F., et al.: Horizontal pyramid matching for person re-identification. In: Proceedings of the AAAI Conference on Artificial Intelligence, vol. 33, pp. 8295–8302 (2019)
15. Huang, J., Liu, B., Lihua, F.: Joint multi-scale discrimination and region segmentation for person Re-ID. Patt. Recogn. Lett. **138**, 540–547 (2020)
16. Gao, S., Wang, J., Lu, H., Liu, Z.: Pose-guided visible part matching for occluded person ReID. In: Proceedings of the IEEE/CVF Conference on Computer Vision and Pattern Recognition, pp. 11744–11752 (2020)
17. Quan, R., Dong, X., Wu, Y., Zhu, L., Yang, Y.: Auto-ReID: searching for a part-aware convnet for person re-identification. In: Proceedings of the IEEE/CVF International Conference on Computer Vision, pp. 3750–3759 (2019)
18. Chang, X., Hospedales, T.M., Xiang, T.: Multi-level factorisation net for person re-identification. In: Proceedings of the IEEE Conference on Computer Vision and Pattern Recognition, pp. 2109–2118 (2018)
19. Zhou, K., Yang, Y., Cavallaro, A., Xiang, T.: Learning generalisable omni-scale representations for person re-identification. IEEE Trans. Patt. Anal. Mach. Intell. (2021)

20. Zheng, Z., Zheng, L., Yang, Y.: Unlabeled samples generated by GAN improve the person re-identification baseline in vitro. In: Proceedings of the IEEE International Conference on Computer Vision, pp. 3754–3762 (2017)
21. Zheng, Z., Zheng, L., Yang, Y.: A discriminatively learned CNN embedding for person reidentification. ACM Trans. Multimedia Comput. Commun. Appl. (TOMM) **14**(1), 1–20 (2017)
22. Xiong, F., Xiao, Y., Cao, Z., Gong, K., Fang, Z., Zhou, J.T.: Towards good practices on building effective CNN baseline model for person re-identification. arXiv preprint arXiv:1807.11042 (2018)
23. Li, W., Zhu, X., Gong, S.: Harmonious attention network for person re-identification. In: Proceedings of the IEEE Conference on Computer Vision and Pattern Recognition, pp. 2285–2294 (2018)
24. Zhang, L., Wu, X., Zhang, S., Yin, Z.: Branch-cooperative OSNet for person re-identification (2020)
25. Pan, X., Luo, P., Shi, J., Tang, X.: Two at once: enhancing learning and generalization capacities via ibn-net. In: Proceedings of the European Conference on Computer Vision (ECCV), pp. 464–479 (2018)
26. Zhong, Z., Zheng, L., Cao, D., Li, S.: Re-ranking person re-identification with k-reciprocal encoding. In: Proceedings of the IEEE Conference on Computer Vision and Pattern Recognition, pp. 1318–1327 (2017)
27. Zhang, X., Jiang, M., Zheng, Z., Tan, X., Ding, E., Yang, Y.: Understanding image retrieval re-ranking: a graph neural network perspective. arXiv preprint arXiv:2012.07620 (2020)
28. https://github.com/douzi0248/Re-ID
29. Zheng, L., Shen, L., Tian, L., Wang, S., Wang, J., Tian, Q.: Scalable person re-identification: a benchmark. In: Proceedings of the IEEE International Conference on Computer Vision, pp. 1116–1124 (2015)
30. Ristani, E., Solera, F., Zou, R., Cucchiara, R., Tomasi, C.: Performance measures and a data set for multi-target, multi-camera tracking. In: Hua, G., Jégou, H. (eds.) ECCV 2016. LNCS, vol. 9914, pp. 17–35. Springer, Cham (2016). https://doi.org/10.1007/978-3-319-48881-3_2
31. Zhong, Z., Zheng, L., Kang, G., Li, S., Yang, Y.: Random erasing data augmentation. In: Proceedings of the AAAI Conference on Artificial Intelligence, vol. 34, pp. 13001–13008 (2020)

Fast Organization of Objects' Spatial Positions in Manipulator Space from Single RGB-D Camera

Yangchang Sun[1,2], Minghao Yang[1,2(✉)], Jialing Li[3], Baohua Qiang[3],
Jinlong Chen[3], and Qingyu Jia[3]

[1] School of Artificial Intelligence, University of Chinese Academy of Sciences,
Beijing 100049, China
[2] Research Center for Brain-inspired Intelligence (BII), Institute of Automation,
Chinese Academy of Sciences (CASIA), Beijing 100190, China
`sunyangchang2020@ia.ac.cn`, `mhyang@nlpr.ia.ac.cn`
[3] School of Computer Science and Information Security,
Guilin University of Electronic Technology, Guilin 541004, China

Abstract. For the grasp task in physical environment, it is important for the manipulator to know the objects' spatial positions with as few sensors as possible in real time. This work proposed an effective framework to organize the objects' spatial positions in the manipulator 3D workspace with a single RGB-D camera robustly and fast. It mainly contains two steps: (1) a 3D reconstruction strategy for objects' contours obtained in environment; (2) a distance-restricted outlier point elimination strategy to reduce the reconstruction errors caused by sensor noise. The first step ensures fast object extraction and 3D reconstruction from scene image, and the second step contributes to more accurate reconstructions by eliminating outlier points from initial result obtained by the first step. We validated the proposed method in a physical system containing a Kinect 2.0 RGB-D camera and a Mico2 robot. Experiments show that the proposed method can run in quasi real time on a common PC and it outperforms the traditional 3D reconstruction methods.

Keywords: 3D reconstruction · Real-time system · Robot grasping

1 Introduction

For the grasp task in physical environment, the manipulator needs to know the objects' spatial positions before picking or grasping. Basically, this procedure belongs to the eye-to-hand calibration task [29]. While demanding reliable accuracy, the system is supposed to construction task space as fast as possible. It usually contains two steps: (1) detect objects in image and (2) project them to manipulator task space fast and accurately. However, in spite of various sensors and 3D reconstruction (eye-to-hand) methods, there is still lack of an effective framework to organize the objects' spatial positions in manipulator task space

© Springer Nature Switzerland AG 2021
T. Mantoro et al. (Eds.): ICONIP 2021, LNCS 13110, pp. 176–187, 2021.
https://doi.org/10.1007/978-3-030-92238-2_15

robustly and fast in real time. The challenges come from the sensor noise and computational complexity.

This work proposed an effective framework to organize the objects' spatial positions in the manipulator 3D workspace with a single RGB-D camera robustly and fast. It mainly contains two steps: (1) a 3D reconstruction strategy for objects' contours obtained in environment; (2) a distance-restricted outlier point elimination strategy to reduce the reconstruction errors caused by sensor noise. The first step ensures fast object extraction and 3D reconstruction from scene image, and the second step contributes to more accurate reconstructions by eliminating outlier points from initial result obtained by the first step.

The remainders of this work are organized as following: related works are introduced in Sect. 2; details of the proposed method are presented in Sect. 3; and experiments, discussion and conclusions are given in Sect. 4 and 5.

2   Related Work

2.1   Object Extraction from Cluster Environment

Various methods have been proposed for object extraction, such as image patches matching [11], points descriptors similarity [5,19,28], regression objects pose from their 3D point clouds (obtained by RGB and depth information) directly [7, 27,30], etc. In physical environment, these methods are not suitable for real-time object extraction, since object templates are needed for shape matching or feature points retrieving in these methods [4,5,11,19,28]. There are some other methods proposed for object extraction in cluster environment [3,10,11,17,20, 26]. However, these methods need markers placed around operating plane or previously modeled objects' meshes or 3D points [7,27,30].

Recently, with the development of deep learning techniques, the task of object detection from cluster environment has a series of solutions such as Faster RCNN [23], SSD [15], YOLO-v1 to Yolo-v4 [2,21,22], etc. In addition, deep-architecture-based instance segmentation methods, such as Mask-RCNN, has the abilities of extracting objects from cluster environment at pixel level [8,8]. In this work, we adopt Mask-RCNN to extract objects contours from RGB-D camera image, the contours were then reconstructed in manipulator 3D workspace.

2.2   Objects Contours Reconstructed (Eye-to-Hand Calibration)

Once the objects are extracted in RGB images, their contour points can be obtained from RGB image. At the same time, the depth information of contour points is obtained from depth images. There are mainly two kinds of eye-to-hand methods proposed to reconstruct the contour points from 2D RGB image to 3D manipulator workspace. The first one is geometrical method [1,6,16]. The geometrical methods have the advantages of high accuracy. However, they usually demand the axis of camera perpendicular to the plane of workspace [6], or markers are needed for accurate computation of camera rotations and translations [1]. These demands are very strict in physical environment.

The second eye-to-hand reconstruction strategy is neural network method [9, 13,29]. Compared with geometry methods, the camera position setting and camera intrinsic parameters are not explicit parameters in the neural network method [9,13,29]. The neural network methods are more flexible and convenient than geometrical methods. We prefer neural network method to geometrical method in this work. However, the neural network method costs more time than geometrical methods. Considering that an object image contains tens and hundreds of pixels, a challenge of this work is to reduce the time cost on objects' pixels reconstruction using neural network. In this work, we prefer to reconstructing the objects contours instead of all pixels. Once the objects contours are reconstructed, the objects' shapes could be presented in 3D workspace.

2.3   Reconstruction Error Elimination

Quite a few RGB-D cameras obtain depth information from infrared reflection. Hollows and speckle burrs exist in the depth data because of strong backlight, objects out of reflection range and measurement errors caused by multiple infrared reflections [12], especially those on the objects' contour points. These sensor noises bring errors to the objects reconstruction in manipulator 3D workspace.

Quit a few methods have been proposed to evaluate outliers from cluster points, such as semilocal constraints [25], density peaks [24], average relative error (AVErr) [14], etc. The semilocal constraints and AVErr methods need the geometric relations on source and target images, while density peaks method evaluates the outliers only rely on the points' inner distances. Inspired by the density peaks outliers evaluation method, we proposed a density peaks based distance-restricted outlier point adjustment (DP-DROPA) method for reconstruction error elimination. As far as we know, this method has not been discussed in the field of real time objects reconstruction from grasp environment.

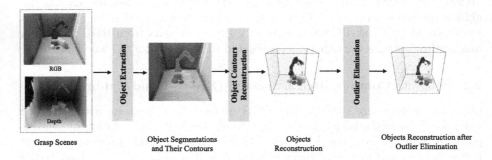

Fig. 1. The pipeline of the proposed method.

3   The Proposed Method

The outline of the proposed framework is given in Fig. 1, which contains two steps: object contours reconstruction from the extracted contour points, and the reconstruction errors elimination. Just as what we introduced in last paragraph, we adopt Mask-RCNN [8] in object extraction from RGB image, and used neural network method [13] to reconstruct the object contours. After the object contours reconstructed in 3D workspace, there are unavoidable factors that cause 3D reconstruction errors: 1) Contour depth errors: the contour points close to the boundaries are easily located at the hollow and burr areas of depth images. These points are outliers which have abnormal depth values. 2) Reconstruction errors: despite high performance of various reconstruction methods, there are still unavoidable errors between the reconstruction points and the physical object 3D position caused by measurement and calculation errors.

For the contour depth errors, if the depth value of a point is 0, or twice times larger or less than the average values of other points, we consider this point an outlier and remove it. For the reconstruction errors, we proposed a density peaks based distance-restricted outlier point adjustment (DP-DROPA) method to reduce the objects contours reconstruction errors through the distances between the reconstructed points. We first introduce the principle of DP-DROPA, and then discuss how it is used for reconstruction errors elimination.

3.1   DP-DROPA (Density Peaks Based Distance-Restricted Outlier Point Adjustment)

Definition. Supposing that P is a point set for an object contours, which contains m points: $p_1,...,p_i,...,p_m$, where $p_i \in R^S$ and their center point p_0 is obtained by $p_0 = \sum_{i=1}^{m} p_i/m$. For each point p_i in P, we construct a new item set X, its element $x_i = \|p_i - p_0\| (1 \leq i \leq m)$. Then according to article [24], the local density ρ_i and distance δ_i for each x_i are given as Eq. (1) and Eq. (2):

$$\rho_i = \sum_{j=1,j\neq i}^{m} \chi(d_{ij} - d_c) \qquad (1)$$

$$\delta_i = min_{j:\rho_j > \rho_i}(d_{ij}) \qquad (2)$$

In Eq. (1), d_{ij} is the distance between item x_i and x_j. Function $\chi(x) = 1$ if $x<0$ and $\chi(x) = 0$ otherwise. d_c is a cutoff distance, which is suggested to be the values that ensure the average number of neighbors at least 1 to 2.0% of the total number of points [24]. In Eq. (2), δ_i is measured by the minimum distance between the item x_i and any other item with higher density.

Then according to the definition of DP-DROPA [24], the items which have the max values of δ are the cluster center points of good items, and the items with low values of ρ are outliers. After we obtain the good item set with $X_{in} = X - X_{out}$, where X_{out} are the outlier items in X, then we obtain the corresponding good contour point set P_{in}.

3.2   Object Extraction from Cluster Environment and Object Contours Reconstruction

Mask-RCNN has the abilities of instances segmentation for objects in cluster environment [8]. A pretrained model of Mask-RCNN on MS COCO dataset was applied, and the FCN layer was fine-tuned for our task. Supposing (u_i, v_i) and d_i as the $i^{th}(1 \leq i \leq m)$ contour point obtained by Mask-RCNN in RGB and depth images respectively. Let $p_i = (u_i, v_i, d_i)$, then the point set $P_{RGB-D}(p_i \in P_{RGB-D})$ presents the object contour in RGB-D presentation.

After the contour points obtained, we use neural network (NN) to transfer contour points in RGB-D presentation to the manipulator workspace [29]. The transfer procedure could be written as Eq. (3), where M is the transfer matrix and P_{Recon} is the reconstructed point set in manipulator workspace. As the idea presented in [29], we adopt a full connection neuronal network structure to transfer RGB-D points to 3D space. Then we obtain the 3D reconstruction points P_{Recon} from P_{RGB-D}.

$$P_{Recon} = M \times P_{RGB-D} \tag{3}$$

3.3   DP-DROPA Based 3D Contour Outliers Elimination

Despite high accuracy of neural network method on eye-to-hand calibration, in physical environment, there are still reconstructions errors because of distortion of camera information, pixels distance noise in RGB images and depth images. We adopt DP-DROPA to adjust the reconstruction of P_{Recon} in this work.

Supposing that there are m reconstruction points $(x_i, y_i, z_i, 0 \leq i \leq m)$ for an extracted object in P_{Recon}, $x_{max}, x_{min}, y_{max}, y_{min}, z_{max}, z_{min}$ are the max and min values in the coordinates of manipulator workspace for P_{Recon}. Then the contour points set P_{Recon} could be transferred as point set $\overline{P}_{Recon}$ with values in the range of (0, 1] for extracted objects using Eq. (4)–Eq. (6).

$$\overline{x}_i = (x_i - x_{min})/(x_{max} - x_{min}) \tag{4}$$

$$\overline{y}_i = (y_i - y_{min})/(y_{max} - y_{min}) \tag{5}$$

$$\overline{z}_i = (z_i - z_{min})/(z_{max} - z_{min}) \tag{6}$$

Continually, DP-DROPA algorithm is used to remove the possible outliers in $\overline{P}_{Recon}$, and obtained the rectified contour collection $\widetilde{P}_{Recon}$.

4   Experiments

In this section, we first introduced the experiment environment and then we compared the reconstruction results obtained by neural network (NN) and DP-DROPA-NN method respectively, where the item DP-DROPA-NN means the proposed DP-DROPA based 3D contour outlier elimination method with the neural network reconstruction results. Finally, we discussed the time cost for each step in the proposed method in detail and presented some reconstruction examples in physical environment.

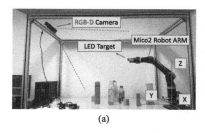 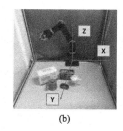 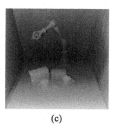

(a)   (b)   (c)

Fig. 2. The experiment environment and RGB-D camera configuration in workspace. (Color figure online)

4.1 Experiment Setting

The environment is presented as Fig. 2, where an RGB-D Kinect is placed in the upper left corner of the Kinova Mico2 manipulator workspace. Depth information of RGB-D camera ranges from 500 mm to 2500 mm. Both the Mico2 and Kinect 2 are connected to a computer with 2.60 GHz CPU, 8.0G RAM and NVidia 1080 GTX 8G GPU. Figure 2(b) and Fig. 2(c) present the RGB and depth images captured from Kinect 2.0.

4.2 Object Extraction and Contour Points Selection

We constructed 42 objects in 12 categories (box, mouse, cup, bottle, keyboard, stapler, toothbrush, glove, ball, scissors, screwdriver, pen) with different sizes in the reconstruction experiments. Figure 3(a) presents several samples in experiment.

We used a mobilenet based Mask-RCNN to extract object contours from RGB images. The Mask-RCNN runs on the MindSpore AI framework [18] deployed on Ascend 910. The original RGB-D images are in the resolution ratio of 480×270 and are resized to 256×256 before input to Mask-RCNN. Supposing the number of contour points obtained by Mask-RCNN for an object is w, the values of w for an object with size from 30×30 to 100×100 resolutions is about from 100 to 500. It is time-consuming to reconstruct all contour points. Therefore, we only reconstruct the selected k contour points, where $k = \sqrt{w}$. And the point set P_{Recon} is obtained by selecting every k interval point for the w contour points.

The images in Fig. 3(b) and Fig. 3(c) present the m object contour points obtained by Mask-RCNN and the k selected ones. The values of w and k for the white box, blue box, mouse, dark cup and white cup are 260, 220, 70, 172, 85 and 16, 15, 8, 13, 9 respectively. We can see from the images in Fig. 3(c) that in spite of the contour points decreased from w to k, the object contours are still well maintained in Fig. 3(c) compared with those in Fig. 3(b).

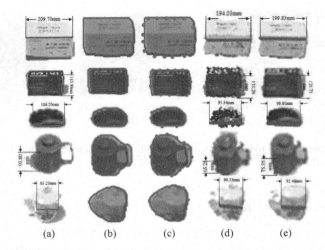

(a)   (b)   (c)   (d)   (e)

Fig. 3. The extraction and reconstruction results for several objects in the experiments. (a) objects extracted from RGB image by Mask-RCNN. (b) objects masks obtained by Mask-RCNN. (c) selected contour points. (d) reconstruction results of NN. (e) reconstruction results of DP-DROPA-NN. (Color figure online)

4.3 Object Reconstruction

Once the k contour points are determined, they are reconstructed in the manipulator 3D work space using Eq. (3) by neural network. Each voxel is labeled by an LED on the end-effector, just like the method presented in [9] and [29].

The images in Fig. 3(d) and Fig. 3(e) present the object reconstructions obtained by neural network (NN) method, and the proposed DP-DROPA-NN strategy is shown in Fig. 3(a), where the red lines indicate the ground truth distance between two contour points, and the corresponding ones in Fig. 3(d) and Fig. 3(e) present the reconstructed distances by NN and DP-DROPA-NN respectively. Supposing that the length of ground truth distance is L_j^{GT}, and the reconstructed length are L_j^{NN} and L_j^{DP} for NN and DP-DROPA-NN respectively, where $j (0 \leq j \leq J)$ is the number of reconstructed lines used for comparison. We calculated the reconstruction errors for NN and DP-DROPA-NN using Eq. (7) and Eq. (8) respectively.

Table 1 presents the performance of the proposed method. The values in the column of accuracy present the absolute and relative errors for the object given in Fig. 3(a), where Err^{NN} and Err^{DP} are relative errors. For each object, at least twelve lines and their distances are calculated, namely $j \geq 12 \times 5$ for each object. We can see from Table 1 that the absolute reconstruction errors obtained by DP-DROPA-NN are obviously smaller than those of neural network, about 6.1 mm, especially for the white cup, about 5.06 mm. A reason for this is that there are loops and irregular curves in cups, and the loops and irregular curves easily cause the contours partly located at the background. Figure 4(a) and Fig. 4(b) give an example of the cup reconstruction, especially the loop part reconstruction in detail. We can see from Fig. 4 that DP-DROPA-NN has better reconstruction performance at cup handle part.

Table 1. The performance of each step of the proposed method.

	Contour		Accuracy				Time cost (ms)		
	w	k	$\frac{\sum_j (L_j^{NN} - L_j^{GT})}{J}$	$\frac{\sum_j (L_j^{DP} - L_j^{GT})}{J}$	$Err^{NN}(\%)$	$Err^{DP}(\%)$	T_{MRCNN}	T_{Recon}^{NN}	T_{Align}
WhiteBox	260	16	14.20 mm	7.40 mm	6.67	3.53		24	5
BlueBox	220	15	12.10 mm	5.89 mm	10.66	5.19		20.6	4.6
Mouse	70	8	7.03 mm	4.77 mm	6.62	4.49	**200**	15.6	3.8
DarkCup	172	13	7.79 mm	5.76 mm	8.3	6.14		16.2	3.8
WhiteCup	85	9	8.10 mm	5.06 mm	9.74	6.07		31.4	6.4
Average	/	/	**9.84 mm**	**5.77 mm**	**8.42**	**5.08**		**21.56**	**4.72**

In addition, for the convex objects such as boxes and mouse, Mask-RCNN could archive convex contours for them. In spite of similar reconstruction performance obtained by DP-DROPA-NN and neural network, the proposed DP-DROPA-NN obtains smaller relative errors than neural network. The average error of DP-DROPA-NN is also less than that of neural network, about 3.34%.

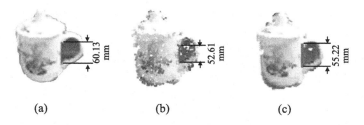

(a)   (b)   (c)

Fig. 4. The reconstruction details at the handle part of the white cup. (a) original RGB image; (b) the reconstruction result obtained by NN, and (c) the reconstruction result obtained by DP-DROPA-NN.

4.4   Time Cost

The proposed framework contains two steps: object extraction and reconstruction, where the reconstruction step contains two inner phases: contour reconstruction and outliers elimination. Therefore, there are total three phases in the proposed method: objects extraction from RGB using Mask-RCNN (T_{MRCNN}), contour points reconstruction (T_{Recon}^{NN}) and contour outliers elimination (T_{Align}). "Time Cost" in Table 1 present the time cost for each phase in detail. It is worth noting that the Mask-RCNN time cost for a given RGB image is about 200 ms, which has no relation to the number of objects in the scene. This is because that Mask-RCNN always evaluates each pixel in image for segmentation. The average time costs for T_{Recon}^{NN} and T_{Align} are about 21.56 ms and 4.72 ms.

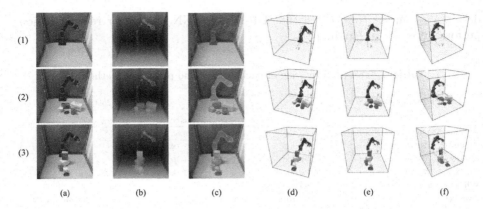

Fig. 5. Reconstruction results from a single RGB-D images of increasing complex scenes. (a) RGB presentation; (b) depth information; (c) objects extraction result obtained by Mask RCNN, (d-f) 3D reconstruction results from different views. (Color figure online)

4.5 Organization of the Objects' Spatial Positions from a Single RGB-D Image

Figure 5 presents the objects' spatial positions generated by the proposed model from RGB-D camera, the images in the column of Fig. 5(a) and Fig. 5(b) are RGB and depth images. The images in the column of Fig. 5(c) are the results of objects extraction obtained by Mask-RCNN, and the images in the columns of Fig. 5(d), Fig. 5(e) and Fig. 5(f) are the 3D reconstructions from different views.

Figure 5(1), Fig. 5(2) and Fig. 5(3) present three scenes where the scenes are increasingly complex from row (1) to row (3). In the scene presented in the images of row (1), there is no object in this scene. While in the images of row (2) and row (3), there are six objects. In the image of row (2), these objects are placed around a circle in the workspace, and there are at least 5.0 mm distance among them. While in the scene of row (3), they are placed together compactly, and obvious overlaps exist in the view of RGB-D image.

With these settings, we can see from Fig. 5(2) that when the objects are scattered in the workspace, they are well reconstructed with the proposed method. The object pixels could be well distinguished in the work space (as shown in Fig. 5(2e)). It indicates that the proposed method provides effective reconstruction for the objects in manipulator space when these objects are scattered.

In the scene of Fig. 5(3), these objects are placed together compactly, and there are obviously overlaps and occlusions. However, these objects' reconstruction points are still correctly arranged in their positions (as shown in Fig. 5(3e)). It indicates that even the objects are placed compactly in RGB-D images, their 3D reconstruction points are distributed correctly in the space and their spatial relationships in the robot manipulator workspace are well presented. In addition, the reconstructed 3D points of Mico2 in Fig. 5(d), Fig. 5(e), and Fig. 5(f) help to provide space relationship between end-effector and objects.

4.6   Discussion

The reconstruction results in Table 1 and Fig. 5 indicate that the proposed method provide an effective framework to reconstruct and organize the vision information to their 3D presentation from a single RGB-D image. With the average absolute reconstruction errors about 5.7 mm and relative errors 5.08%, the proposed method contributes to the objects accurate reconstruction in manipulator 3D work space, even these objects are placed compactly in the scene in stacked or overlap style.

5   Conclusion

In this work, we proposed an effective framework to organize the objects spatial positions in the manipulator 3D workspace with a single RGB-D camera. With the help of Mask-RCNN, the objects in RGB image could be extracted from the cluster environment. Furthermore, through the contour point reconstruction with neural network and the density peaks based distance-restricted outlier point adjustment, namely the proposed DP-DROPA-NN method, the objects contour points are well reconstructed in the manipulator 3D workspace. The proposed method is validated in a physical cluster 3D space. Experiments show that the proposed method runs quasi-real-time on a common PC and can obtain more completive reconstruction results from physical environment compared to the traditional neural network reconstruction methods.

In addition, one advantage of our method is that the reconstructed objects in the 3D space are automatically organized. And with the labels obtained by Mask-RCNN, the objects' reconstruction points are also automatically labeled in 3D space synchronously. Finally, with 3FPS time cost in real practical environments, the proposed method has latent capacity in the further real grasp applications in physical environment with a single RGB-D camera configuration.

Acknowledgments. This work is supported by the National Key Research & Development Program of China (No. 2018AAA0102902), the National Natural Science Foundation of China (NSFC) (No.61873269), the Beijing Natural Science Foundation (No: L192005), the CAAI-Huawei MindSpore Open Fund (CAAIXSJLJJ-20202-027A), the Guangxi Key Research and Development Program (AB18221011, AB21075004, AD18281002, AD19110137), the Natural Science Foundation of Guangxi of China (No: 2020GXNSFAA297061, 2019GXNSFDA185006, 2019GXN SFDA185007), Guangxi Key Laboratory of Intelligent Processing of Computer Images and Graphics (No GIIP201702) and Guangxi Key Laboratory of Trusted Software (NO kx201621,kx201715).

References

1. Boby, R.A., Saha, S.K.: Single image based camera calibration and pose estimation of the end-effector of a robot. In: 2016 IEEE International Conference on Robotics and Automation (ICRA), pp. 2435–2440. IEEE (2016)
2. Bochkovskiy, A., Wang, C.Y., Liao, H.Y.M.: Yolov4: optimal speed and accuracy of object detection (2020). arXiv preprint arXiv:2004.10934
3. Brachmann, E., Michel, F., Krull, A., Yang, M.Y., Gumhold, S., et al.: Uncertainty-driven 6d pose estimation of objects and scenes from a single rgb image. In: Proceedings of the IEEE Conference on Computer Vision and Pattern Recognition, pp. 3364–3372 (2016)
4. Cao, Z., Sheikh, Y., Banerjee, N.K.: Real-time scalable 6D of pose estimation for textureless objects. In: 2016 IEEE International Conference on Robotics and Automation (ICRA), pp. 2441–2448. IEEE (2016)
5. Collet, A., Martinez, M., Srinivasa, S.S.: The moped framework: object recognition and pose estimation for manipulation. Int. J. Rob. Res. 30(10), 1284–1306 (2011)
6. Durović, P., Grbić, R., Cupec, R.: Visual servoing for low-cost scara robots using an rgb-d camera as the only sensor. Automatika: časopis za automatiku, mjerenje, elektroniku, računarstvo i komunikacije 58(4), 495–505 (2017)
7. Gao, G., Lauri, M., Wang, Y., Hu, X., Zhang, J., Frintrop, S.: 6D object pose regression via supervised learning on point clouds. In: 2020 IEEE International Conference on Robotics and Automation (ICRA), pp. 3643–3649. IEEE (2020)
8. He, K., Gkioxari, G., Dollár, P., Girshick, R.: Mask r-cnn. In: Proceedings of the IEEE International Conference on Computer Vision, pp. 2961–2969 (2017)
9. Jones, M., Vernon, D.: Using neural networks to learn hand-eye co-ordination. Neural Comput. Appl. 2(1), 2–12 (1994)
10. Kehl, W., Manhardt, F., Tombari, F., Ilic, S., Navab, N.: SSD-6D: Making rgb-based 3D detection and 6D pose estimation great again. In: Proceedings of the IEEE International Conference on Computer Vision, pp. 1521–1529 (2017)
11. Kehl, W., Milletari, F., Tombari, F., Ilic, S., Navab, N.: Deep learning of local RGB-D patches for 3D object detection and 6D pose estimation. In: Leibe, B., Matas, J., Sebe, N., Welling, M. (eds.) ECCV 2016. LNCS, vol. 9907, pp. 205–220. Springer, Cham (2016). https://doi.org/10.1007/978-3-319-46487-9_13
12. Kuan, Y.W., Ee, N.O., Wei, L.S.: Comparative study of intel r200, kinect v2, and primesense rgb-d sensors performance outdoors. IEEE Sens. J. 19(19), 8741–8750 (2019)
13. Levine, S., Pastor, P., Krizhevsky, A., Ibarz, J., Quillen, D.: Learning hand-eye coordination for robotic grasping with deep learning and large-scale data collection. Int. J. Rob. Res. 37(4–5), 421–436 (2018)
14. Li, E., Mo, H., Xu, D., Li, H.: Image projective invariants. IEEE Trans. Pattern Anal. Mach. Intell. 41(5), 1144–1157 (2018)
15. Liu, W., et al.: SSD: single shot multibox detector. In: Leibe, B., Matas, J., Sebe, N., Welling, M. (eds.) ECCV 2016. LNCS, vol. 9905, pp. 21–37. Springer, Cham (2016). https://doi.org/10.1007/978-3-319-46448-0_2
16. Meng, Y., Zhuang, H.: Self-calibration of camera-equipped robot manipulators. Int. J. Rob. Res. 20(11), 909–921 (2001)
17. Michel, F., et al.: Global hypothesis generation for 6D object pose estimation. In: Proceedings of the IEEE Conference on Computer Vision and Pattern Recognition, pp. 462–471 (2017)

18. Mindspore: Mask-rcnn-mobilenetv1. Website (2020). https://gitee.com/mindspore/mindspore/blob/r1.1/model_zoo/official/cv/maskrcnn_mobilenetv1/src/maskrcnn_mobilenetv1/mobilenetv1.py
19. Pavlakos, G., Zhou, X., Chan, A., Derpanis, K.G., Daniilidis, K.: 6-dof object pose from semantic keypoints. In: 2017 IEEE International Conference on Robotics and Automation (ICRA), pp. 2011–2018. IEEE (2017)
20. Rad, M., Lepetit, V.: Bb8: a scalable, accurate, robust to partial occlusion method for predicting the 3D poses of challenging objects without using depth. In: Proceedings of the IEEE International Conference on Computer Vision, pp. 3828–3836 (2017)
21. Redmon, J., Divvala, S., Girshick, R., Farhadi, A.: You only look once: unified, real-time object detection. In: Proceedings of the IEEE Conference on Computer Vision and Pattern Recognition, pp. 779–788 (2016)
22. Redmon, J., Farhadi, A.: Yolov3: an incremental improvement (2018). arXiv preprint arXiv:1804.02767
23. Ren, S., He, K., Girshick, R., Sun, J.: Faster r-cnn: towards real-time object detection with region proposal networks. Adv. Neural Inf. Process. Syst. **28**, 91–99 (2015)
24. Rodriguez, A., Laio, A.: Clustering by fast search and find of density peaks. Science **344**(6191), 1492–1496 (2014)
25. Schmid, C., Mohr, R.: Local grayvalue invariants for image retrieval. IEEE Trans. Pattern Anal. Mach. Intell. **19**(5), 530–535 (1997)
26. Tekin, B., Sinha, S.N., Fua, P.: Real-time seamless single shot 6D object pose prediction. In: Proceedings of the IEEE Conference on Computer Vision and Pattern Recognition, pp. 292–301 (2018)
27. Wang, C., Xu, D., Zhu, Y., Martín-Martín, R., Lu, C., Fei-Fei, L., Savarese, S.: Densefusion: 6D object pose estimation by iterative dense fusion. In: Proceedings of the IEEE/CVF Conference on Computer Vision and Pattern Recognition, pp. 3343–3352 (2019)
28. Wohlhart, P., Lepetit, V.: Learning descriptors for object recognition and 3D pose estimation. In: Proceedings of the IEEE Conference on Computer Vision and Pattern Recognition, pp. 3109–3118 (2015)
29. Wu, H., Tizzano, W., Andersen, T.T., Andersen, N.A., Ravn, O.: Hand-eye calibration and inverse kinematics of robot arm using neural network. In: Kim, J.-H., Matson, E.T., Myung, H., Xu, P., Karray, F. (eds.) Robot Intelligence Technology and Applications 2. AISC, vol. 274, pp. 581–591. Springer, Cham (2014). https://doi.org/10.1007/978-3-319-05582-4_50
30. Zeng, A., et al.: Multi-view self-supervised deep learning for 6D pose estimation in the amazon picking challenge. In: 2017 IEEE International Conference on Robotics and Automation (ICRA), pp. 1386–1383. IEEE (2017)

EvoBA: An Evolution Strategy as a Strong Baseline for Black-Box Adversarial Attacks

Andrei Ilie[(✉)], Marius Popescu, and Alin Stefanescu

University of Bucharest, Bucharest, Romania
{cilie,marius.popescu,alin}@fmi.unibuc.ro

Abstract. Recent work has shown how easily white-box adversarial attacks can be applied to state-of-the-art image classifiers. However, real-life scenarios resemble more the black-box adversarial conditions, lacking transparency and usually imposing natural, hard constraints on the query budget.

We propose **EvoBA** (All the work is open source: https://github.com/andreiilie1/BBAttacks A full paper version is available at https://arxiv.org/abs/2107.05754), a black-box adversarial attack based on a surprisingly simple evolutionary search strategy. **EvoBA** is query-efficient, minimizes L_0 adversarial perturbations, and does not require any form of training.

EvoBA shows efficiency and efficacy through results that are in line with much more complex state-of-the-art black-box attacks such as **AutoZOOM**. It is more query-efficient than **SimBA**, a simple and powerful baseline black-box attack, and has a similar level of complexity. Therefore, we propose it both as a new strong baseline for black-box adversarial attacks and as a fast and general tool for gaining empirical insight into how robust image classifiers are with respect to L_0 adversarial perturbations.

There exist fast and reliable L_2 black-box attacks, such as **SimBA**, and L_∞ black-box attacks, such as **DeepSearch**. We propose **EvoBA** as a query-efficient L_0 black-box adversarial attack which, together with the aforementioned methods, can serve as a generic tool to assess the empirical robustness of image classifiers. The main advantages of such methods are that they run fast, are query-efficient, and can easily be integrated in image classifiers development pipelines.

While our attack minimises the L_0 adversarial perturbation, we also report L_2, and notice that we compare favorably to the state-of-the-art L_2 black-box attack, **AutoZOOM**, and of the L_2 strong baseline, **SimBA**.

1 Introduction

With the increasing performance and applicability of machine learning algorithms, and in particular of deep learning, the safety of such methods became more relevant than ever. There has been growing concern over the course of the last few years regarding adversarial attacks, i.e., algorithms which are able to fool machine learning models with minimal input perturbations, as they were shown to be very effective.

Ideally, theoretical robustness bounds should be obtained in the case of critical software involving image classification components. There has been important recent research in this direction [5,9,16], but most often the algorithms generating these

© Springer Nature Switzerland AG 2021
T. Mantoro et al. (Eds.): ICONIP 2021, LNCS 13110, pp. 188–200, 2021.
https://doi.org/10.1007/978-3-030-92238-2_16

bounds work for limited classes of models, do not scale well with larger neural networks, and require complete knowledge of the target model's internals. Therefore, complementary empirical robustness evaluations are required for a better understanding of how robust the image classifiers are. For this, one has to use effective adversarial attacks that resemble real-life conditions, such as in the black-box query-limited scenario.

In general, adversarial attacks are classified as either white-box or black-box. White-box adversarial attacks, where the attacker has complete knowledge of the target model, were shown to be particularly successful, most of them using gradient-based methods [2,7,19]. In the case of black-box adversarial attacks, the attacker can only query the model, and has no access to the model internals and to the data used for training. These restrictions make the black-box adversarial setup resemble more real-life scenarios. Furthermore, the attacker usually has to minimise the number of queries to the model, either due to time or monetary constraints (such as in the case of some vision API calls).

Previous state-of-the-art black-box adversarial attacks focused on exploiting the transferability phenomenon, which allowed the attackers to train substitute models imitating the target one, and perform white-box attacks on these [14,15]. More recently, a class of black-box adversarial attacks, called Zeroth Order Optimization (**ZOO**) [4], has gained momentum, providing one of the current state-of-the art attacks, **AutoZOOM** ([20]). Interestingly, a much simpler algorithm, **SimBA** (Simple Black-box Attack) [8], achieves a similar, slightly lower success rate than state-of-the-art attacks, including **AutoZOOM**, and is more query-efficient. Therefore, **SimBA** is proposed as a default baseline for adversarial attacks, but is itself an unexpectedly powerful algorithm.

We propose **EvoBA**, an untargeted black-box adversarial attack that makes use of a simple evolutionary search strategy. **EvoBA** only requires access to the output probabilities of the target model for a given input and needs no extra training. **EvoBA** is more query-efficient than **SimBA** and **AutoZOOM**, and has a perfect success rate, surpassing **SimBA** and being aligned with **AutoZOOM** from this point of view. We designed **EvoBA** to minimize the L_0 adversarial perturbation, however we also report the L_2 norms of the perturbations it generates and compare them with the L_2 norms of **Auto-ZOOM** and **SimBA**, which are L_2 adversarial attacks. Despite our algorithm aiming to minimize a different metric than these methods, we achieve similar L_2 norms in the perturbations we generate, and a significantly better L_0.

While L_2 and L_∞ adversarial attacks are more commonly studied, L_0 adversarial attacks can fit better real-life, physical settings, such as in the well-known cases of graffiti perturbations on stop-signs ([6]) and of adversarial eyeglass frames ([17]). **EvoBA** focuses on the L_0 norm, is fast, query-efficient, and effective. Therefore, we propose using it together with similarly fast and efficient methods that focus on different norms, such as **SimBA**, which focuses on L_2, and **DeepSearch** ([21]), which focuses on L_∞, to empirically evaluate the robustness of image classifiers. These methods can act together as a fast and general toolbox used along the way of developing systems involving image classification models.

To wrap it up, we propose **EvoBA** as a strong baseline for black-box adversarial attacks and as a tool that can provide empirical insight into how robust image classifiers are. Its main advantages are that it is as effective as state-of-the-art black-box attacks, such as **AutoZOOM**, and more query-efficient. While it is an L_0 adversarial attack, it achieves L_2 perturbations of similar magnitudes with state-of-the-art L_2 black-box attacks and requires no training. Furthermore, **EvoBA** is highly parallelisable and runs significantly faster than **SimBA**, the other powerful baseline black-box attack.

2   Related Work

One general approach for black-box attacks exploits the transferability property of adversarial examples [14,15]. The attacker can train their own substitute model and perform white-box attacks on it, which usually yield good adversarial samples for the target model as well. However, [18] showed the limitations of this method, highlighting how not all white-box attacks and not all architectures transfer well.

A class of approaches that do not rely on transferability is based on Zeroth Order Optimization (ZOO), which tries to estimate the gradients of the target neural network. The early **ZOO** [4] managed to reach similar success rates with state-of-the-art white-box attacks. The main disadvantage of traditional ZOO-type attacks is that they usually require plenty of queries for approximating the coordinate-wise gradients. **AutoZOOM** [20] solved this by performing a dimensionality reduction on top of the target image, and then using the ZOO approach in the reduced space. It achieved results in line with previous ZOO state-of-the-art methods, but reduced the query count by up to 93%. Our method is comparable to **AutoZOOM** in terms of performance, achieving a similar success rate with a slightly lower query count, but is considerably simpler. Our approach does not need to estimate gradients at all and, compared to **AutoZOOM-AE** - the more powerful attack from [20], it does not require any form of training or knowledge about the training data. In addition, **AutoZOOM** demands access to the output probabilities over all classes, unlike our method, which only requires the output class and its probability. While **AutoZOOM** is an attack that minimizes the L_2 perturbations norm, **EvoBA** is focused on minimising the L_0 norm. However we also report the L_2 and notice that it is comparable to **AutoZOOM**'s results.

SimBA ([8]) is a very simple greedy strategy that has comparable performance to the significantly more complex **AutoZOOM**. **EvoBA** has a similar complexity to **SimBA**, both of them being good candidates for strong adversarial attack baselines. **SimBA** greedily attacks random pixels one by one, while **EvoBA** strikes a better balance between exploration and exploitation, which is common for evolutionary algorithms. This makes **EvoBA** achieve a slightly better success rate than **SimBA**, with a lower query budget. Similarly to **AutoZOOM**, **SimBA** is an L_2 adversarial attack, while **EvoBA** focuses on L_0. Nevertheless, the average L_2 perturbation norm of **EvoBA** is slightly higher, but comparable to **SimBA**'s results. We also run **SimBA** and remark that the L_0 perturbation norms that **EvoBA** achieves are significantly better. Furthermore, **SimBA** does not allow for any kind of parallelisation, while the evolution strategy we use in **EvoBA** is highly parallelisable and, accordingly, faster.

DeepSearch ([21]) is another simple, yet very efficient black-box adversarial attack, which achieves results in line with much more complex methods. It is an L_∞ attack, perturbing with very high probability all the pixels (maximal L_0 norm of the perturbations), which makes it incomparable with the L_0 attack **EvoBA**. Similarly, while

EvoBA optimizes the L_0 norm of the perturbations, it most often produces high L_∞ distortions, with sometimes near-maximal perturbations for the few chosen pixels. While both **DeepSearch** and **EvoBA** have non-complex implementations, **EvoBA** is conceptually simpler. **DeepSearch** is based on the idea of linear explanations of adversarial examples ([7]), and exploits three main aspects: it devises a mutation strategy to perturb images as fast as possible, it performs a refinement on top of the earliest adversarial example in order to minimise the L_∞ norm, and adapts an existing hierarchical-grouping strategy for reducing the number of queries. Furthermore, **EvoBA** has the advantage of being highly parallelisable, while **DeepSearch** is inherently sequential.

As black-box adversarial attacks are ultimately search strategies in obscured environments, it has been natural to also explore the path of evolutionary algorithms. One example is **GenAttack** [1], an approach that follows the classic pattern of genetic algorithms. While it was developed at the same time with **AutoZOOM**, the authors report similar results for the targeted versions of the two methods, without providing untargeted attack results. We focused on the untargeted scenario, and our results are also in line with **AutoZOOM**. **GenAttack** focuses on minimising the L_∞ perturbation norm, and, in expectation, the L_0 it achieves is equal to the count of all pixels in the image. In comparison, **EvoBA** is an L_0 attack, achieving considerably small perturbations under this norm. In addition, **EvoBA** is less complex and more suitable for a strong baseline.

A related approach to ours, which also makes use of evolution strategies, is [13], which tries to minimise the L_∞ norm of adversarial perturbations. It proposes different evolution strategies applied on top of a tiling approach inspired by [11], where the authors use a Bandits approach. The attacks they propose focus on minimising the L_∞ norm of the perturbations and the authors do not report any other results regarding different norms. The L_2 cost of this approach is not clear, one of the main issues being that it can become rather high. The L_0 is equal in general to the number of pixels in the entire image, in comparison with **EvoBA**, which is L_0-efficient. Qualitatively, the applied adversarial tiles it generates are easily perceivable by a human, yielding grid-like patterns on top of the target image, while the samples produced by **EvoBA** are imperceptible (Fig. 6, ImageNet) or look like benign noise (Fig. 3, CIFAR-10).

3   The Method

3.1   Notation and Threat Model

We work under black-box adversarial settings, with limited query budget and L_0 perturbation norm. We consider the untargeted attack scenario, where an adversary wants to cause perturbation that changes the original, correct prediction of the target model for a given image to any other class.

We denote by $\mathbf{F}$ the target classifier. By a slight abuse of notation, we let $\mathbf{F}(\mathbf{x})$ be the output distribution probability of model $\mathbf{F}$ on input image $\mathbf{x}$ and $\mathbf{F}_k(\mathbf{x})$ be the output probability for class k. Then, $\mathbf{F}$ can be seen as a function $\mathbf{F} : \mathbb{I} \mapsto \mathbb{R}^K$, where $\mathbb{I}$ is the image space (a subset of $\mathbb{R}^{h \times w \times c}$) and K is the number of classes. As we are working under black-box conditions, we have no information about the internals of $\mathbf{F}$, but we have query access to it, i.e., we can retrieve $\mathbf{F}(\mathbf{x})$ for any $\mathbf{x} \in \mathbb{I}$. We will see that for our method we just need access to $\arg\max_k \mathbf{F}_k(\mathbf{x})$ and to its corresponding probability.

Let us consider an image $\mathbf{x}$, which is classified correctly by $\mathbf{F}$. The untargeted attack goal is to find a perturbed version $\tilde{\mathbf{x}}$ of $\mathbf{x}$ that would make

$$\underset{k}{\arg\max}\, \mathbf{F}_k(\tilde{\mathbf{x}}) \neq \arg\max_k \mathbf{F}_k(\mathbf{x}), \tag{1}$$

constrained by the query and L_0 bounds.

3.2   Proposed Algorithm

It is usual for black-box attacks to deal with a surrogate optimization problem that tries to find a perturbed version $\tilde{\mathbf{x}}$ of $\mathbf{x}$ that minimizes $\mathbf{F}(\tilde{\mathbf{x}})$. This is clearly not equivalent to the formulation at (1), but it often yields good adversarial examples and is easier to use in practice. In loose terms, this surrogate optimization problem can be formulated as follows for an image $\mathbf{x}$ with true label y:

$$\min_{\delta \in \mathbb{R}^{h \times w \times c}} \mathbf{F}_y(\mathbf{x} + \boldsymbol{\delta}), \text{w.r.t. queries} \leq Q, \|\delta\|_0 \leq \epsilon. \tag{2}$$

Algorithm 1: EvoBA

Data: black-box model $\mathbf{F}$, image $\mathbf{x}$, correct class k, query budget Q, L_0 threshold ϵ, pixel batch size B, generation size G

1   PARENT $\leftarrow \mathbf{x}$, PREDICTION $\leftarrow \mathbf{F}_k(\mathbf{x})$, QUERY_CNT $\leftarrow 1$
2   **while** $\|$PARENT $- \mathbf{x}\|_0 < \epsilon$ *and* QUERY_CNT $< Q$ **do**
3   OFFSPRING $\leftarrow [\,]$, FITNESSES $\leftarrow [\,]$, PIXELS $\leftarrow$ SAMPLE_PIXELS(PARENT, B)
4   **for** IDX $\leftarrow 1 \dots G$ **do**
5   CHILD $\leftarrow$ PARENT
6   **for** PIXEL $\leftarrow$ PIXELS **do**
7   CHILD[PIXEL] $\leftarrow$ SAMPLE_VALUES()
8   OFFSPRING $\leftarrow$ OFFSPRING $+$ [CHILD]
9   PRED_CHILD $\leftarrow \mathbf{F}($CHILD$)$, QUERY_CNT $\leftarrow$ QUERY_CNT $+ 1$
10   **if** *arg max* PRED_CHILD $\neq k$ **then**
11   **return** CHILD
12   FITNESSES $\leftarrow$ FITNESSES $+ [1 - $ PRED_CHILD$_k($CHILD$)]$
13   BEST_CHILD $\leftarrow$ arg max(FITNESSES)
14   PARENT $\leftarrow$ OFFSPRING[BEST_CHILD]
15 **return** PERTURBATION FAILED

We adopt a simple evolution strategy for (2) that yields results in line with state-of-the-art black-box attacks. The algorithm works by iteratively creating generations of perturbed images as follows: it selects the fittest individual in each generation (with lowest probability to be classified correctly), starting from the unperturbed image, then samples small batches of its pixels and randomly perturbs them, stopping when the fittest individual is either no longer classified correctly or when one of the constraints no longer holds (when the query count or the distance becomes too large).

The pseudocode of **EvoBA** is given as Algorithm 1. The sample function (line 3) does a random, uniform sample over the pixels of PARENT, and returns a list of size at most B (the sampling is done with repetition) containing their coordinates. Its purpose is to pick the pixels that will be perturbed. The function SAMPLE_VALUES generates pixel perturbed values. For our L_0 objective, we let it pick uniformly a random value in the pixel values range. The algorithm follows a simple and general structure of evolution strategies. The mutation we apply on the best individual from each generation is selecting at most B pixels and assigning them random, uniform values. The fitness of an individual is just the cumulative probability of it being misclassified.

One important detail left out of the pseudocode, which allowed us to get the best results, is how we deal with multi-channel images. If the target image has the shape $h \times w \times c$ (height, width, channels), then we randomly sample a position in the $h \times w$ grid and add all of its channels to the PIXELS that will be perturbed. We hypothesise that this works well because of an "inter-channel transferability" phenomenon, which allows **EvoBA** to perturb faster the most sensible zones in all the channels. Note that this yields a cost of c for every grid-sample to the L_0 perturbation norm, so in the case of ImageNet or CIFAR-10 images it counts as 3, and in the case of MNIST as 1.

The query budget and L_0 constraints impose a compromise between the total number of generations in an **EvoBA** run and the size G of each generation. The product of these is approximately equal to the number of queries, so for a fixed budget we have to strike the right balance between them. The bigger G is, the more we favour more exploration instead of exploitation, which should ultimately come at a higher query cost. The smaller G is, the search goes in the opposite direction, and we favour more the exploitation. As each exploitation step corresponds to a new perturbation, this will result in bigger adversarial perturbations. Furthermore, lacking proper exploration can even make the attack unsuccessful in the light of the query count and L_0 constraints. The batch size B allows selecting multiple pixels to perturb at once in the mutation step. It is similar to the learning rate in general machine learning algorithms: the higher it is, the fewer queries (train steps, in the case of machine learning algorithms) we need, at the cost of potentially missing local optimal solutions.

The space complexity of Algorithm 1 is $\mathcal{O}(G \times \text{size}(\mathbf{x}))$. As the size of $\mathbf{x}$ is in general fixed, or at least bounded for specific tasks, we can argue that the space complexity is $\mathcal{O}(G)$. However, **EvoBA** can be easily modified to only store two children at a moment when generating new offspring, i.e., the currently generated one and the best one so far, which makes **EvoBA**'s space complexity $\mathcal{O}(1)$. The dominant component in the time complexity is given by $\mathbf{F}$ queries. In the form of Algorithm 1, they could be at most $\min(\frac{\epsilon}{B} \times G, Q)$. Assuming that the budget Q will generally be higher, the time complexity would roughly be $\mathcal{O}(\frac{\epsilon}{B} \times G \times (\text{query cost of } \mathbf{F}))$. This is merely the sequential time complexity given by the unoptimised pseudocode, however the G $\mathbf{F}$-queries can be batched, yielding much faster, parallelised runs.

4   Experiments

4.1   Experimental Setup

We used TensorFlow/Keras for all our experiments. All the experiments were performed on a MacBook with 2,6 GHz 6-Core Intel Core i7, without a GPU.

We ran locally **SimBA**, the strong and simple L_2 black-box adversarial attack, and compared **EvoBA** to both our local **SimBA** results and reported results from the paper introducing it ([8]). We also monitored and compared against the **AutoZOOM** results, but for these we used different target models, as the main focus was on comparing to **SimBA** (which already achieves results in line with state-of-the-art approaches, such as **AutoZOOM**, being more query-efficient), so we adopted their models.

We don't do a head-to-head comparison with the L_∞ baseline **DeepSearch**, as it is a direct consequence of their approach that they get near-maximal L_0 perturbations, while **EvoBA** aims to optimize the L_0 norm. Similarly, **EvoBA** creates high L_∞ perturbations, as it modifies very few pixels with random, possibly large quantities.

We ran multiple experiments over three datasets: MNIST, CIFAR-10, and ImageNet. On MNIST we ran the experiments on a classic target LeNet architecture [12], while **SimBA** does not report any results on this dataset, and **AutoZOOM** uses a similar architecture to ours, with additional dropout layers (taken from [3]). For comparing with **SimBA** on MNIST, we ran the attack ourselves.

We initially used MNIST to validate **EvoBA** against a completely random black-box adversarial attack, similar to the one introduced in [10]. The purely random strategy iterates by repeatedly sampling a bounded number of pixels from the original target image and changing their values according to a random scheme. While the purely random method introduced in [10] achieves surprising results for such a simple approach, it does a very shallow form of exploration, restarting the random perturbation process with each miss (i.e., with each perturbed image that is still classified correctly). We will refer to the completely random strategy as **CompleteRandom** in the experiments.

For CIFAR-10, we use a ResNet-50 target model, similarly to **SimBA**, and we compare to their reported results and to our local run of their attack. For ImageNet, we use a similar target ResNet-50 to **SimBA**, while **AutoZOOM** used InceptionV3.

We will use the following shorthands in the results below: **SR** (Success Rate), **QA** (Queries Average), **L0** (Average of L0 successful perturbations), **L2** (Average L2 norm of successful perturbations). We will refer to **EvoBA** that perturbs at most B pixels at once and that has generation size G as **EvoBA(B,G)**. We will explicitly mention in each section which thresholds were used for the experiments.

For all the local runs of **SimBA**, we used $\epsilon = 0.2$ (a hyperparameter specific to **SimBA**), which was also used in the paper [8]. We only replicated locally the results of the Cartesian Basis version of **SimBA**, which resembles more an L_0 adversarial attack, but which is less efficient than the Discrete Cosine Transform (DCT) version from the paper. We will use **SimBA-LCB** to refer to the local run of **SimBA** on top of the exact same target models as **EvoBA**. We will use **SimBA-CB** to refer to the results of the Cartesian Basis paper results, and **SimBA-DCT** for the DCT paper results.

In the cases where the **AutoZOOM** paper [20] provides data, we will only compare to **AutoZOOM-BiLIN (AutoZ-BL)**, the version of the attack which requires no additional training and data, being closer to our and **SimBA**'s frameworks.

4.2   Results on MNIST

We only experiment with **EvoBA**(B,G) with $B > 1$ on MNIST, and focus on $B = 1$ for subsequent experiments. We impose an L_0 perturbation limit of 100 and a query threshold of 5000. Running **SimBA (SimBA-LCB)** on a local machine with the mentioned specifications, requires an average of **93 s** per MNIST sample. In comparison, all the **EvoBA** experiments on MNIST took between **1.94 s** and **4.58 s** per sample (Table 1).

Table 1. MNIST results. **SimBA-LCB** has a very low success rate in the case of a LeNet architecture. If we take a look at the 56% images perturbed by **EvoBA(1,10)** for example, the QA becomes 192.8, which is not apparent from the table, but which is more efficient than **SimBA-LCB**.

	SR	QA	L0	L2
EvoBA(1,10)	100%	301.4	29.32	3.69
EvoBA(1,20)	100%	549.4	26.88	3.58
EvoBA(1,40)	100%	894.2	21.92	3.33
EvoBA(2,20)	100%	312.6	30,72	3.65
EvoBA(2,30)	100%	265.4	38,6	3.89
SimBA-LCB	56%	196.86	48.16	2.37
AutoZ-BL	100%	98.82	–	3.3
CompleteRandom	60.5%	576.1	93.98	5.59

Table 2. CIFAR-10 results. **AutoZOOM**, **SimBA-CB**, and **SimBA-DCT** do not report the L_0 metrics. However, we have discussed already why it is very likely that **AutoZOOM** perturbs most of the pixels.

	SR	QA	L0	L2
EvoBA(1,30)	100%	178.56	17.67	1.82
SimBA-LCB	100%	206.5	99.46	1.73
SimBA-CB	100%	322	–	2.04
SimBA-DCT	100%	353	–	2.21
AutoZ-BL	100%	85.6	–	1.99
CompleteRandom	69.5%	161.2	97.17	3.89

We also report the **CompleteRandom** results, for which we impose an L_0 perturbation limit of 100 and a query threshold of 5000 (similarly to **EvoBA**).

We randomly sampled 200 images from the MNIST test set and ran **SimBA-LCB**, **EvoBA**, and **CompleteRandom** against the same LeNet model [12]. For reference, we also add the results of **AutoZOOM**, which are performed on a different architecture. Therefore, the results are not directly comparable with them. L_0 data is not available for **AutoZOOM**, but it is usual for ZOO methods to perturb most of the pixels, so it is very likely that the associated L_0 is very high.

All the **EvoBA** configurations have a 100% success rate, and **SimBA-LCB** achieves 56%. While **SimBA** achieves near-perfect success rates on other tasks, one could argue that attacking a simple target model such as LeCun on a relatively easy task such as MNIST is harder than performing attacks on more intricate target models and tasks.

This is a natural trade-off between the complexity of a model and its robustness. If we restrict **EvoBA(1,10)** to its top 56% perturbed images in terms of query-efficiency, it achieves an average of 192.79 queries, which is below **SimBA-LCB**'s queries average of 196.86. Similarly, if we restrict **EvoBA(1,10)** to its top 56% perturbed images in terms of L_2-efficiency, it achieves an L_2 of 3.07, which is higher, but closer to **SimBA-LCB**'s L_2 result of 2.37.

CompleteRandom achieves a success rate of 60.5%, far below **EvoBA**'s 100%, but surprisingly above **SimBA-LCB**'s 56%. However, the nature of **CompleteRandom**'s perturbations is to lie at high distances, achieving L_0 and L_2 distances that are significantly higher than the distances achieved by the other methods.

4.3   Results on CIFAR-10

We impose an L_0 perturbation limit of 100, and a query threshold of 2000 for both **EvoBA** and **CompleteRandom**. We randomly sample 2000 images for **EvoBA** and 50 images for **SimBA-LCB**. **EvoBA** and **SimBA-LCB** are run on the exact same target ResNet-50 model, while **SimBA-DCT** and **SimBA-CB** also run on a target ResNet-50 model. **AutoZOOM-BiLIN** targets an Inception V3 model. **SimBA-LCB** required an average of **26.15 s** per CIFAR-10 sample, while **EvoBA** required **1.91s** per sample.

As shown in Table 2, all the attacks achieved 100% success rate in the

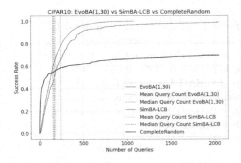

Fig. 1. The success rate (ratio of perturbed images) as a function of the max query budget. We compare **EvoBA** with **SimBA** and **CompleteRandom** on CIFAR-10.

CIFAR-10 experiments, with the sole exception of **CompleteRandom** (69.5%). **EvoBA(1,30)** has a better query average than all **SimBA** approaches, which targeted the same ResNet-50 architecture. While **EvoBA(1,30)** targeted a different architecture than **AutoZOOM-BiLIN**, we still remark how the latter is twice more query efficient. However, **EvoBA(1,30)** surprisingly achieves an L_2 metric which is better than the reported numbers of **SimBA-CB** and **SimBA-DCT**, which are L_2 adversarial attacks.

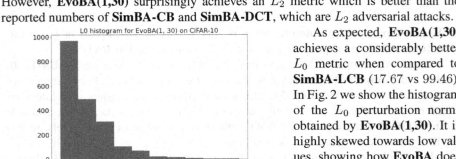

Fig. 2. L_0 perturbation norms obtained by **EvoBA** (1,30) on CIFAR-10 with target model ResNet-50.

As expected, **EvoBA(1,30)** achieves a considerably better L_0 metric when compared to **SimBA-LCB** (17.67 vs 99.46). In Fig. 2 we show the histogram of the L_0 perturbation norms obtained by **EvoBA(1,30)**. It is highly skewed towards low values, showing how **EvoBA** does well in finding quick small perturbations with respect to the L_0 norm. The success rate of

CompleteRandom (69.5%) and its low average query count (161.2) are surprisingly good results for the trivial nature of the method. However, these come at the cost of an average L_0 that is roughly 5.5 times higher and of an average L_2 that is roughly 1.4 times higher in comparison with **EvoBA(1,30)**'s average results. In Fig. 1 we compare **EvoBA(1,30)** with **SimBA-LCB**, while also providing the **CompleteRandom** results. We plot the success rate as a function of the number of queries in order to understand how each method behaves for different query budgets. **EvoBA(1,30)** has a better success rate than **SimBA-LCB** for any query budget up to 2000. For very low query budgets (under 112 queries), **CompleteRandom** has a better success rate than **EvoBA(1,30)**, but it starts converging fast after their intersection point to the success rate of 69.5%. It is natural for the **CompleteRandom** strategy to find quick perturbations for the least robust images, as it performs bulk perturbations of many pixels at once, while **EvoBA** does all perturbations sequentially. This illustrates the trade-off between exploration and exploitation that any black-box optimization problem encounters. We present qualitative results in Fig. 3.

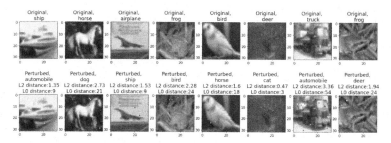

Fig. 3. The first row contains original CIFAR-10 samples, which are classified correctly by ResNet-50. The second row contains adversarial examples created by **EvoBA**, and are labelled with the corresponding ResNet-50 predictions.

4.4   Results on ImageNet

We adopt a similar framework to **AutoZOOM**: we randomly sample 50 correctly classified images and run **EvoBA** on top of them. For **EvoBA**, similarly to **SimBA**, we use a ResNet50 model, while **AutoZOOM** uses an InceptionV3. We impose an L_0 perturbation limit of 1000, and a query threshold of 10000.

The median number of queries for **EvoBA(1,15)** is surprisingly low: 728.5. Its median L_0 is 200, and its median L_2 is 5.69. **EvoBA(1,15)** achieves the best average query metric among the given experiments. Surprisingly, its L_2 is almost equal to the one of **AutoZOOM-BiLIN**. **EvoBA(1,15)** also achieves a 100% success rate, which is in line with **AutoZOOM-BiLIN**, and better than **SimBA**'s results (Figs. 4, 5 and Table 3).

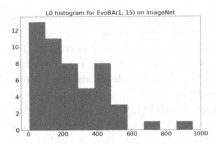

Fig. 4. L_0 perturbation norms obtained by **EvoBA(1,15)** on ImageNet with target model ResNet-50.

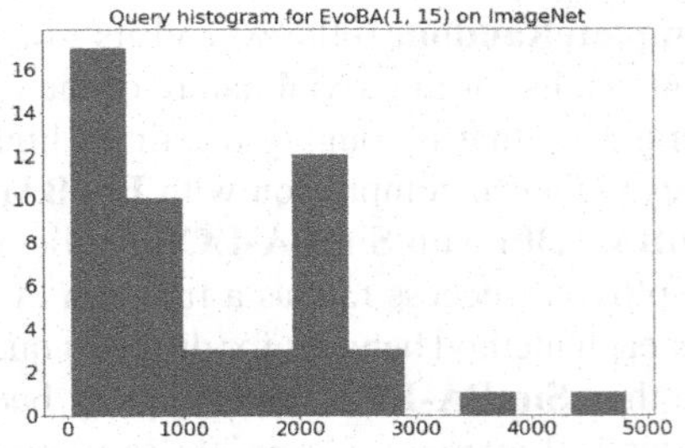

Fig. 5. Query counts obtained by **EvoBA(1,15)** on ImageNet with target model ResNet-50.

Table 3. ImageNet results. **Auto-ZOOM**, **SimBA-CB**, and **SimBA-DCT** do not report the L_0 metrics. We discussed already why it is very likely that **AutoZOOM** perturbs most of the pixels.

	SR	QA	L0	L2
EvoBA(1,15)	100%	1242.4	247.3	6.09
EvoBA(1,20)	100%	1412.51	211.03	5.72
SimBA-CB	98.6%	1665	–	3.98
SimBA-DCT	97.8%	1283	–	3.06
AutoZ-BL	100%	1695.27	–	6.06

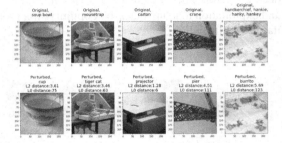

Fig. 6. The first row contains original ImageNet samples, which are classified correctly by ResNet-50. The second row contains adversarial examples created by **EvoBA**, and are labelled with the corresponding ResNet-50 predictions.

5   Conclusion

We proposed **EvoBA**, an L_0 black-box adversarial attack based on an evolution strategy, which serves as a powerful baseline for black-box adversarial attacks. It achieves results in line with state-of-the-art approaches, such as **AutoZOOM**, but is far less complex. Simple yet efficient methods, such as **EvoBA**, **SimBA**, **DeepSearch**, and even **CompleteRandom** shed a light on the research potential of the black-box adversarial field, outlining the inherent security issues in many machine learning applications.

Our further work consists of aggregating **EvoBA** with the other strong black-box baseline attacks we have seen for different norms (**SimBA** for L_2 and **DeepSearch** for L_∞) and creating a generic open-source framework to empirically assess the robustness of image classifiers and to help with their development process. Such a toolbox could also be used to assess the quality of adversarial training methods, as many focus on improving the robustness of the target models with respect to L_p norms.

Acknowledgement. This work was partially supported by the Romanian Ministry of Research and Innovation UEFISCDI 401PED/2020 and PN-III-P2-2.1-PTE-2019-0820.

References

1. Alzantot, M., Sharma, Y., Chakraborty, S., Zhang, H., Hsieh, C.J., Srivastava, M.B.: GenAttack: practical black-box attacks with gradient-free optimization. In: Proceedings of the Genetic and Evolutionary Comp. Conf. (GECCO'18), pp. 1111–1119 (2019)
2. Athalye, A., Carlini, N., Wagner, D.: Obfuscated gradients give a false sense of security: circumventing defenses to adversarial examples. In: Proceedings of International Conference on Machine Learning (ICML'18), pp. 274–283 (2018)
3. Carlini, N., Wagner, D.: Towards evaluating the robustness of neural networks. In: IEEE Symposium on Security and Privacy (SP'17), pp. 39–57. IEEE (2017)
4. Chen, P.Y., Zhang, H., Sharma, Y., Yi, J., Hsieh, C.J.: Zoo: zeroth order optimization based black-box attacks to deep neural networks without training substitute models. In: Proceedings of the 10th ACM Workshop on Artificial Intelligence and Security, pp. 15–26 (2017)
5. Dvijotham, K., Stanforth, R., Gowal, S., Mann, T.A., Kohli, P.: A dual approach to scalable verification of deep networks. In: UAI, vol. 1, p. 3 (2018)
6. Eykholt, K., et al.: Robust physical-world attacks on deep learning visual classification. In: Proceedings of the IEEE Conference on Computer Vision and Pattern Recognition, pp. 1625–1634 (2018)
7. Goodfellow, I., Shlens, J., Szegedy, C.: Explaining and harnessing adversarial examples. In: International Conference on Learning Representations (2015)
8. Guo, C., Gardner, J.R., You, Y., Wilson, A.G., Weinberger, K.: Simple black-box adversarial attacks. In: Proceedings of International Conference on Machine Learning, pp. 2484–2493 (2019)
9. Huang, X., Kwiatkowska, M., Wang, S., Wu, M.: Safety verification of deep neural networks. In: Majumdar, R., Kunčak, V. (eds.) CAV 2017. LNCS, vol. 10426, pp. 3–29. Springer, Cham (2017). https://doi.org/10.1007/978-3-319-63387-9_1
10. Ilie, A., Popescu, M., Stefanescu, A.: Robustness as inherent property of datapoints. In: AISafety Workshop, IJCAI (2020)
11. Ilyas, A., Engstrom, L., Madry, A.: Prior convictions: black-box adversarial attacks with bandits and priors (2018). arXiv:1807.07978
12. LeCun, Y., et al.: LeNet-5, convolutional neural networks (2015). http://yann.lecun.com/exdb/lenet
13. Meunier, L., Atif, J., Teytaud, O.: Yet another but more efficient black-box adversarial attack: tiling and evolution strategies (2019). arXiv:1910.02244
14. Papernot, N., McDaniel, P., Goodfellow, I.: Transferability in machine learning: from phenomena to black-box attacks using adversarial samples (2016). arXiv:1605.07277
15. Papernot, N., McDaniel, P., Jha, S., Fredrikson, M., Celik, Z.B., Swami, A.: The limitations of deep learning in adversarial settings. In: IEEE European Symposium on Security and Privacy (EuroS&P'16), pp. 372–387. IEEE (2016)
16. Ruan, W., Huang, X., Kwiatkowska, M.: Reachability analysis of deep neural networks with provable guarantees (2018). arXiv:1805.02242
17. Sharif, M., Bhagavatula, S., Bauer, L., Reiter, M.: Accessorize to a crime: real and stealthy attacks on state-of-the-art face recognition. In: Proceedings of ACM SIGSAC Conference on Computer and Communication Security (CCS'16), pp. 1528–1540 (2016)
18. Su, D., Zhang, H., Chen, H., Yi, J., Chen, P.Y., Gao, Y.: Is robustness the cost of accuracy? - a comprehensive study on the robustness of 18 deep image classification models. In: Proceedings of the European Conference on Computer Vision (ECCV'18), pp. 631–648 (2018)

19. Szegedy, C., et al.: Intriguing properties of neural networks (2013). arXiv:1312.6199
20. Tu, C.C., et al.: AutoZOOM: Autoencoder-based zeroth order optimization method for attacking black-box neural networks. In: Proceedings of the AAAI Conference, vol. 33, pp. 742–749 (2019)
21. Zhang, F., Chowdhury, S.P., Christakis, M.: Deepsearch: a simple and effective blackbox attack for deep neural networks. In: Proceedings of the 28th ACM Joint Meeting on European Software Engineering Conference and Symposium on the Foundations of Software Engineering, pp. 800–812 (2020)

A Novel Oversampling Technique for Imbalanced Learning Based on SMOTE and Genetic Algorithm

Juan Gong[✉]

College of Science, University of Shanghai for Science and Technology,
Shanghai 200093, China

Abstract. Learning from imbalanced datasets is a challenge in machine learning, oversampling is an effective method to solve the problem of class imbalance, owing to its easy-to-go capability of achieving the balance by synthesizing new samples. However several problems still exist such as noise samples, selection of boundary samples and the diversity of synthetic samples. To solve these problems, this paper proposes a new improved oversampling method based on SMOTE and genetic algorithm (GA-SMOTE). The main steps of GA-SMOTE are as follows. Firstly GA-SMOTE uses genetic algorithm to find an optimal noise processing scheme. Then GA-SMOTE assigns different sampling weight to each sample and the sample closer to the boundary is assigned greater weight. Finally, GA-SMOTE divides raw dataset into multiple sub-clusters by K-means clustering and intra-cluster neighborhood triangular sampling method is used in each sub-cluster to improve the diversity of synthetic samples. A large number of experiments have proved that GA-SMOTE is superior to the other five comparison methods in dealing with imbalanced data classification.

Keywords: Machine learning · Data mining · Oversampling · Genetic algorithm

1 Introduction

Class imbalance is an important reason for the performance degradation of classifiers. In practical applications, there are many datasets with highly skewed distribution, such as medical diagnosis [1], text classification [2], etc. This paper mainly studies binary classification imbalanced datasets, which the class with a large number of samples are called majority class and the class with a small number of samples are called minority class.

In imbalanced datasets, although the prediction ability of classifiers are impaired due to class imbalance, the classifier can still achieve high prediction accuracy because the prediction ability of traditional classifiers tend to majority samples. However, minority samples usually contain more useful information and require more attention. For example, in medical diagnosis [1], patients are

© Springer Nature Switzerland AG 2021
T. Mantoro et al. (Eds.): ICONIP 2021, LNCS 13110, pp. 201–212, 2021.
https://doi.org/10.1007/978-3-030-92238-2_17

regarded as minority samples and the healthy people are regarded as majority samples. If a patient is misclassified as a healthy people, it will pay a huge price. Therefore, in order to solve the class imbalance problem, researchers have proposed many methods, which can generally be divided into three categories: cost sensitive methods [3], ensemble learning [4] and data-level methods [5, 6].

Cost sensitive methods and ensemble learning mainly modify the classification algorithms so that they are only suitable for specific classifiers. The data-level methods change the structure of dataset to balance the number of samples of minority and majority classes so that they are suitable for more classifiers and are used more widely. The data-level methods include undersampling and oversampling methods, the former balance dataset by removing majority samples and the latter balance dataset by copying original minority samples or synthesizing new samples.

Oversampling methods are used commonly because these methods not only retain more information but also generate new information. Synthetic minority over-sampling technique (SMOTE) [6] is the most classic oversampling method. Researchers put forward a variety of improved methods for SMOTE, such as Borderline-SMOTE (BSMOTE) [7], adaptive synthetic sampling (ADASYN) [8] and majority weighted minority oversampling technique (MWMOTE) [9], these methods solve the class imbalance problem mainly by identifying minority samples that are difficult to learn at the decision boundary.

All these oversampling methods are based on the k-nearest neighbor method when considering the noise samples problem. In this paper, an improved oversampling method based on genetic algorithm is proposed to deal with noise samples (GA-SMOTE). In brief, the main contributions of this paper are as follows.

(1) GA-SMOTE proposes a new noise removing method that finds an optimal noise processing scheme by genetic algorithm.

(2) GA-SMOTE proposes an intra-cluster neighborhood triangular sampling method, which can effectively improve the diversity of synthetic samples and avoid noise generation.

(3) Empirical experiments on public datasets by different classifiers are used to show the effectiveness of GA-SMOTE. The experiment results show that GA-SMOTE has a better performance than other comparison methods.

The extensive experiments design is described as follows. We use 12 publicly datasets from the UCI machine learning repository [15] to validate the performance of GA-SMOTE, which is compared with 5 benchmark approaches. We regard F-measure, G-Mean and area under ROC graph (AUC) [19] as the evaluation measures, and K Nearest Neighbor (KNN) [17], Decision Tree (DT) [18] classifiers are employed. Moreover, in order to relieve the influence of randomness, by using 5-fold stratified cross validation each experiment are repeated five times to report the average as result.

The remainder of this paper is organized as follows. Section 2 presents the related work. Section 3 introduces the proposed oversampling algorithm in detail. Section 4 describes the experimental design and results. Conclusions and future works are provided in Sect. 5.

2   Related Work

In this section, we will review oversampling methods and genetic algorithm.

2.1   Oversampling Methods

The most classic method in oversampling is SMOTE [6], which generates synthetic samples by linear interpolation for minority samples. Following are details about the procedure of the generation of samples. For minority samples x_i and x_j where x_j is randomly selected from the neighbours of x_i according to k-nearest neighbor approach, a new synthetic sample x_{new} is generated between x_i and x_j by linear interpolation. It can be expressed as follow:

$$x_{new} = x_i + \alpha \times (x_j - x_i) \tag{1}$$

where α is a random number $\in [0, 1]$. SMOTE method does not simply copy minority samples but uses linear interpolation method to synthesize new samples. SMOTE effectively alleviates the overfitting problem. In recent years, scholars have further proposed a variety of improved SMOTE algorithms based on the consideration of strengthening the decision boundary and improving the diversity of new samples.

BSMOTE, ADASYN, MWMOTE are improved methods for strengthening boundary areas. Specifically, BSMOTE [7] strengthens the decision boundary by finding boundary minority samples as target samples to synthesize new samples. ADASYN [8] method adaptively allocates the weight to each minority sample and pays attention to samples with learning difficulties. MWMOTE [9] method assigns weights to minority samples with hard-to-learn information. Considering the diversity of synthetic samples, three minority samples are randomly selected to synthesize a new sample by Random SMOTE (RSMOTE) [10].

2.2   Genetic Algorithm

John Holland proposes genetic algorithm and the algorithm has been widely studied and applied [11]. Genetic algorithm is also used in imbalanced learning and applied to the SMOTE model. For example [12–14], compare with GA-SMOTE, whose difference is the way of applying Genetic algorithm to the process. We use genetic algorithm to find an optimal noise processing scheme and remove noise samples.

In addition, the basic process of a standard genetic algorithm is as follows. Firstly genetic algorithm determines the fitness function and then establishes an initial population composed of multiple chromosomes. Each solution of the optimization problem corresponds to one chromosome, in which binary coding is mainly used to code chromosomes. Combined with the size of fitness value, a series of genetic operations including selection, crossover and mutation are carried out on the population. After multiple generations of evolution, the individual with the best fitness value is obtained, which is the optimal solution of the optimization problem.

3   The Proposed GA-SMOTE Method

From the previous oversampling methods, it is easy to find that the determination of noise samples is usually based on k-nearest neighbor. If k nearest neighbors of a sample all belongs to its different class, the sample will be defined as a noise sample. Therefore, it exists a certain blindness that easy to regard the boundary samples with important information as noise samples or regard the true noisy samples as normal samples.

As shown in Fig. 1(a), N1, N2 and N3 will be regarded as noise samples by k-nearest neighbor method. N1 is a minority sample that falls in majority samples area and it is indeed a true noise sample. N3 is a majority sample that falls in minority samples area and it is also a true noise sample. However, N2 is a minority sample that closes to the decision boundary so that N2 is a boundary sample with hard-to-learn information and is needed more attention. To solve the problem, this paper proposes a noise removing method based on genetic algorithm.

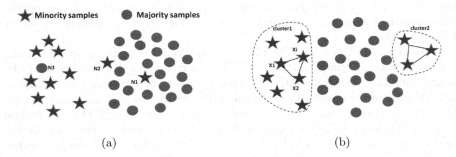

(a)   (b)

Fig. 1. (a) Schematic diagram of noise samples, (b) Intra-cluster neighborhood triangular sampling method.

In this paper, we regard each suspicious noise based on the k nearest neighbor as gene in a chromosome, and code each suspicious noise as a binary bit (0 or 1). Let the number of the suspicious noise is r, then each chromosome is an r-dimension vector. In a chromosome, 1 means that the suspicious noise is a false noise to be retained in the dataset, while 0 represents that the suspicious noiset is a true noise to be eliminated. Chromosome individual in GA-SMOTE is represented as follow:

$$X = (x_1, x_2, ..., x_r). \tag{2}$$

In order to evaluate and sequence the chromosomes of each generation, an appropriate fitness function needs to be designed. The classification performance on the original dataset can be used to reflect the reasonable degree of noise removing. Therefore, in GA-SMOTE, the fitness function is measured by G-Mean. First we remove the true noise samples according to each chromosome that is each noise process scheme. Then after noise removing, SMOTE method is applied for the dataset to obtain a balanced dataset. Finally, Support Vector Machine (SVM) [16] is used to classify the dataset and obtain corresponding G-Mean, SVM is provided by Python sklearn package and the parameters of the classifier are the default values [22]. The calculation method of G-Mean is as follow:

$$G - Mean = \sqrt{\frac{TP}{TP + FN}} \times \sqrt{\frac{TN}{TN + FP}}. \tag{3}$$

After a series of genetic operations and multiple generations of evolution, according to fitness function G-Mean, the individual with the best fitness value is obtained. The individual is an optimal noise process scheme in this paper, we will remove noise by this scheme.

The pseudo-code of noise removing algorithm can be described as follow Algorithm 1 and the pseudo-code of GA-SMOTE can be described as follow Algorithm 2.

Algorithm 1. Noise Removing

Input: S: set of all samples; k: number of nearest neighbors; P: population size; G maximum iteration times.

Output: S_{nr}: set of samples after noise processing.

1: We can get a set of suspicious noise r by using the k-nearest neighbor method to identify suspicious noise of S;
2: **for** $i = 1$ to P **do**
3:     Initialize chromosome X_i for samples in r by Eq. (2) and join in population P_{g-1};
4:     Calculate the fitness function value of X_i by Eq. (3);
5: **end for**
6: **for** $g = 1$ to G **do**
7:     Selection operation: select the first P chromosomes from the rank of chromosomes $P_{g-1} \cup P_g$ by fitness function value as P_g;
8:     Crossover operation: randomly select paternal chromosomes from P_g and part of the genes exchange to generate two progeny chromosomes;
9:     Mutation operation: randomly invert certain genes in chromosome X_i, for example, invert 0 to 1 or invert 0 to 1;
10:     The chromosomes with maximum fitness function value as optimal solution X_o;
11: **end for**
12: We can get S_{nr} after noise processing by global optimal solution X_o.

Algorithm 2. GA-SMOTE

Input: S_{min}: set of minority samples from S_{nr}; k: number of nearest neighbors; A: number of attributes. N: number of synthetic samples needed to be generated.

Output: S_{syn}: set of generated synthetic samples.

1: Noise removing by algorithm 1;

2: For $i = 1$ to S_{min}, calculate the k nearest neighbors of sample x_i;

3: Adaptively calculate the sampling weight of x_i: $r_i = m_i/k$, where m_i is the number of corresponding majority samples in k;

4: Normalize r_i to $R_i = r_i/\sum_{i=1}^{n} r_i$, where n represents the number of all minority samples in dataset;

5: Find the clusters of minority samples by K-means clustering algorithm;

6: For $i = 1$ to S_{min}, calculate the number of synthetic samples for x_i as g_i, $g_i = [R_i * N]$, where $[\cdot]$ represents the floor function;

7: For $j = 1$ to g_i, first randomly select two minority neighbor samples x_1 and x_2 of x_i, from the cluster of x_i, then the steps of synthetic sample s_i as following steps;

   (a) For $a = 1$ to A, each attribute of synthetic sample t_{ia}: $t_{ia} = x_{1a} + \alpha \times (x_{2a} - x_{1a})$;

   (b) For $a = 1$ to A, each attribute of synthetic sample s_{ia}: $s_{ia} = x_{ia} + \alpha \times (t_{ia} - x_{ia})$;

   (c) Add s_{ia} into s_i;

8: Add s_i into S_{syn}, we can get the set of generated synthetic samples S_{syn}.

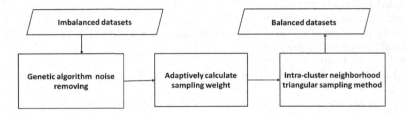

Fig. 2. GA-SMOTE method flow chart.

The GA-SMOTE method first uses the genetic algorithm to find an optimal noise process scheme. Then GA-SMOTE calculates the sampling weight adaptively and pays attention to samples that near the decision boundary. Finally GA-SMOTE uses intra-cluster neighborhood triangular sampling method to generate synthetic samples. To be specific, the intra-cluster neighborhood triangular sampling method uses K-means algorithm to cluster minority samples into multiple sub-clusters and generate synthetic samples in intra-cluster, which effectively prevents the generation of noise samples. Besides, the synthetic samples will fall in the triangle region formed by x_i and its two minority neighbor samples x_1 and x_2, which improves the diversity of the synthesized samples, and the sketch map is shown in Fig. 1(b). GA-SMOTE flow chart is shown in Fig. 2.

4   Experiments

4.1   Datasets Analysis

To investigate the performance of GA-SMOTE, we selected 12 public imbalanced datasets with different features or sizes from the UCI machine learning repository [15]. Table 1 demonstrates the detailed information of the 12 public datasets, including the names of datasets, the number of features, the size of the whole dataset, the number of minority samples, the number of majority samples and imbalanced ratio (IR).

Table 1. Information of the UCI datasets.

Data	Attributes	Size	Minority	Majority	IR
wdbc	30	569	212	357	1.68
Ionosphere	34	351	126	225	1.79
Haberman	3	306	81	225	2.78
Vehicle	18	846	199	647	3.25
Leaf	16	340	70	270	3.86
Parkinsons	23	195	48	147	3.87
Libra	90	360	72	288	4
Ecoli	7	336	52	284	5.46
Balance	4	625	49	576	11.76
Page-blocks	10	5473	329	5144	15.64
Yeast	8	1484	51	1433	28.1
Abalone	7	4177	99	4078	41.19

4.2   Evaluation Metrics

Table 2. Confusion matrix.

	Observed positive	Observed negative
Predicted positive	TP	FP
Predicted negative	FN	TN

To test the performance of GA-SMOTE, we should use suitable evaluation measures. Referring to [7,19], we use F-measure, G-Mean and AUC as the criteria to evaluate the performance of the different methods, these evaluation measures are based on confusion matrix, following the suggestions of [20,21], as is illustrated in Table 2, positive is the minority class and negative is the majority class. These evaluation measures that we used can be described as follows:

$$F - measure = (2 \times \frac{TP}{TP+FP} \times \frac{TP}{TP+FN})/(\frac{TP}{TP+FP} + \frac{TP}{TP+FN}) \quad (4)$$

$$AUC = \frac{1}{2} \times (\frac{TP}{TP+FN} + \frac{TN}{TN+FP}) \quad (5)$$

where the values of these evaluation measures that F-measure, G-Mean (G-Mean has been introduced in Sect. 3) and AUC are closer to 1, the classification performance is better. In addition, mean ranking of the 6 methods averaged over the 12 datasets is calculated and Friedman test [23] is applied to compare GA-SMOTE with other methods.

4.3 Compare Methods and Parameter Settings

In this experiment, we compared with 5 other popular oversampling methods to verify the effectiveness of GA-SMOTE. The 5 algorithms include: 1) SMOTE, 2) RSMOTE, 3) BSMOTE, 4) ADASYN, 5) MWMOTE. The implementation of these algorithms was based on Python, and for a fair comparison, k (number of nearest neighbors) value was set to 5 for all oversampling methods.

The classifiers include K Nearest Neighbor (KNN) [17] and Decision Tree (DT) [18] which were provided by Python sklearn package, the parameters of all classifiers were all the default values [22].

Besides, GA-SMOTE needs a category of parameters to be set. In genetic algorithm, the population size was set in 5 and maximum iteration times was set in 50.

4.4 Algorithm Comparison and Result Analysis

The detailed results of 'F-measure', 'G-Mean' and 'AUC' on 12 datasets are shown in Tables 3-4. The bold indexes represent the best measure among different algorithms. Apparently, compared GA-SMOTE method with other methods in all classifiers and all datasets, GA-SMOTE method is the highest in the times of manifestation best. Therefore, GA-SMOTE is superior to other methods on many imbalanced datasets.

Then, Table 5 shows the mean ranking to each method on all imbalanced datasets and in terms of F-measure, G-Mean and AUC. Among them, the algorithm with the best performance ranked 1 and the worst ranked 6. Clearly, according to Table 5, the mean rating index of GA-SMOTE is the best.

In addition, the results of the Friedman test [23] for each algorithm are given in Table 5. The null hypothesis of this test is that the all algorithms perform similarly in each classifier. It is easy to find that for all classifiers and evaluation measures in Table 5, at a significance level $\alpha = 0.05$, the all null hypothesis are rejected. In addition, the mean ranking results in Table 5 indicate that GA-SMOTE performs better than the other methods.

On the whole, according to experimental results, it is not difficult to find that the comprehensive performance of GA-SMOTE is better than other algorithms.

Table 3. Results for these methods on the 12 datasets classified using KNN.

F-measure dataset	SMOTE	RSMOTE	BSMOTE	ADASYN	MWMOTE	GA-SMOTE
wdbc	0.9335	0.9337	0.9276	0.9171	0.9206	**0.9419**
Ionosphere	0.9184	0.927	0.9443	0.9394	0.8964	**0.9627**
Haberman	0.708	0.7262	0.7351	0.6751	0.7002	**0.7839**
Vehicle	0.9508	**0.9535**	0.9533	0.9491	0.9388	0.9502
Leaf	0.867	0.8587	0.8512	0.8482	0.8565	**0.8904**
Parkinsons	0.8832	0.8845	0.8707	0.8589	0.8193	**0.9059**
Libra	0.9966	0.9914	0.9948	0.9968	0.9885	**0.9979**
Ecoli	0.9525	0.9485	0.9703	0.9337	0.9569	**0.9741**
Balance	0.8564	0.8484	0.9158	0.8675	0.8252	**0.9669**
Page-blocks	0.9844	0.9837	0.9886	0.981	0.9818	**0.9934**
Yeast	0.9559	0.9566	0.9717	0.9538	0.945	**0.9809**
Abalone	0.9344	0.9299	0.9643	0.933	0.8892	**0.9682**
G-Mean dataset	SMOTE	RSMOTE	BSMOTE	ADASYN	MWMOTE	GA-SMOTE
wdbc	0.934	0.9335	0.9311	0.9217	0.921	**0.9431**
Ionosphere	0.9183	0.9267	0.9439	0.9393	0.8961	**0.9623**
Haberman	0.7336	0.7501	0.7621	0.7166	0.7098	**0.8206**
Vehicle	0.9557	**0.9581**	0.9575	0.955	0.9453	0.9553
Leaf	0.8884	0.8737	0.875	0.8795	0.8769	**0.9072**
Parkinsons	0.8934	0.8928	0.8876	0.8776	0.8176	**0.9212**
Libra	0.9966	0.9914	0.9949	0.9969	0.9885	**0.9979**
Ecoli	0.9545	0.9503	0.9711	0.9433	0.9585	**0.9755**
Balance	0.8871	0.8879	0.919	0.8936	0.8685	**0.9682**
Page-blocks	0.9846	0.9839	0.9888	0.9816	0.982	**0.9935**
Yeast	0.9597	0.9603	0.9733	0.9577	0.9501	**0.9816**
Abalone	0.9429	0.9393	0.9664	0.942	0.9032	**0.9701**
AUC dataset	SMOTE	RSMOTE	BSMOTE	ADASYN	MWMOTE	GA-SMOTE
wdbc	0.9334	0.9335	0.9299	0.92	0.9215	**0.9436**
Ionosphere	0.9152	0.9244	0.9443	0.9401	0.8898	**0.9601**
Haberman	0.7299	0.7469	0.7568	0.7099	0.7098	**0.8113**
Vehicle	0.953	**0.9556**	0.9554	0.9517	0.9418	0.9527
Leaf	0.8789	0.8686	0.8653	0.865	0.8699	**0.8997**
Parkinsons	0.8919	0.8923	0.8822	0.8724	0.8204	**0.9148**
Libra	0.9965	0.9912	0.9947	0.9968	0.9885	**0.9979**
Ecoli	0.9537	0.9499	0.9707	0.9381	0.9579	**0.9749**
Balance	0.8726	0.8682	0.9182	0.881	0.8485	**0.9681**
Page-blocks	0.9845	0.9838	0.9887	0.9813	0.982	**0.9934**
Yeast	0.9576	0.9584	0.9724	0.9557	0.9475	**0.9812**
Abalone	0.9383	0.9342	0.9653	0.9371	0.8971	**0.9691**

Table 4. Results for these methods on the 12 datasets classified using DT.

F-measure dataset	SMOTE	RSMOTE	BSMOTE	ADASYN	MWMOTE	GA-SMOTE
wdbc	0.9473	0.9346	0.9322	0.9311	0.9374	**0.9521**
Ionosphere	0.8946	0.875	0.8856	0.8654	0.8822	**0.9075**
Haberman	0.6788	0.7417	0.7108	0.6861	0.7085	**0.7789**
Vehicle	0.957	0.9593	0.9631	**0.9679**	0.9531	0.9618
Leaf	0.8711	0.8691	0.8634	0.8462	0.8687	**0.8836**
Parkinsons	0.9127	0.8969	0.9058	0.8811	0.8643	**0.9269**
Libra	0.9295	0.9395	0.9363	**0.9512**	0.9163	0.9472
Ecoli	0.9442	0.9462	0.9502	0.9353	0.9448	**0.9715**
Balance	0.9056	0.8964	0.9198	0.9024	0.8978	**0.9614**
Page-blocks	0.9885	0.9868	0.9905	0.9869	0.9871	**0.9949**
Yeast	0.9559	0.9543	0.9695	0.9543	0.9479	**0.9778**
Abalone	0.9482	0.9461	0.9652	0.9469	0.9008	**0.9732**
G-Mean dataset	SMOTE	RSMOTE	BSMOTE	ADASYN	MWMOTE	GA-SMOTE
wdbc	0.9479	0.9364	0.9356	0.9326	0.9387	**0.952**
Ionosphere	0.8928	0.8752	0.8856	0.8675	0.8845	**0.9066**
Haberman	0.6938	0.7397	0.717	0.7036	0.7062	**0.7773**
Vehicle	0.9577	0.9596	0.9638	**0.9686**	0.9537	0.9618
Leaf	0.8727	0.8705	0.8696	0.8546	0.873	**0.8875**
Parkinsons	0.9157	0.8999	0.9094	0.8865	0.87	**0.9273**
Libra	0.9322	0.941	0.9368	**0.9513**	0.9168	0.9479
Ecoli	0.9453	0.9473	0.9509	0.9366	0.946	**0.9724**
Balance	0.9139	0.897	0.9225	0.9139	0.8985	**0.961**
Page-blocks	0.9885	0.9868	0.9905	0.9869	0.9871	**0.9949**
Yeast	0.9566	0.9548	0.9699	0.9549	0.9486	**0.978**
Abalone	0.9493	0.9467	0.9657	0.9479	0.9018	**0.9734**
AUC dataset	SMOTE	RSMOTE	BSMOTE	ADASYN	MWMOTE	GA-SMOTE
wdbc	0.9481	0.936	0.9337	0.9331	0.9385	**0.9529**
Ionosphere	0.8914	0.8762	0.886	0.8685	0.8836	**0.906**
Haberman	0.6917	0.7405	0.7167	0.7028	0.7068	**0.7791**
Vehicle	0.9577	0.9595	0.9636	**0.9686**	0.9535	0.9624
Leaf	0.8725	0.8705	0.8702	0.8532	0.8734	**0.887**
Parkinsons	0.9133	0.8986	0.9101	0.8865	0.8677	**0.9276**
Libra	0.9298	0.9391	0.9365	**0.9522**	0.9168	0.9482
Ecoli	0.9444	0.9463	0.9509	0.9369	0.9452	**0.9717**
Balance	0.9109	0.8973	0.9218	0.9084	0.899	**0.961**
Page-blocks	0.9885	0.9868	0.9905	0.9869	0.9871	**0.9949**
Yeast	0.9564	0.9547	0.9698	0.9548	0.9485	**0.9779**
Abalone	0.949	0.9466	0.9656	0.9476	0.9018	**0.9734**

Table 5. Results for mean ranking of the 6 methods averaged over the 12 datasets and results for Friedman test.

Meas.	SMOTE	RSMOTE	BSMOTE	ADASYN	MWMOTE	GA-SMOTE	P-value
Classification method: KNN							
F-measure	3.417	3.5	2.75	4.75	5.333	1.25	8.70019E-07
G-Mean	3.333	3.833	2.75	4.333	5.5	1.25	1.0144E-06
AUC	3.417	3.583	2.75	4.667	5.333	1.25	1.1826E-06
Classification method: DT							
F-measure	3.417	4.125	2.917	4.458	4.833	1.25	2.11794E-05
G-Mean	3.542	4.25	2.917	4.375	4.667	1.25	3.95365E-05
AUC	3.5	4.25	2.917	4.417	4.667	1.25	3.63988E-05

5   Conclusion

GA-SMOTE considering the problem of noise, the decision boundary and the diversity of synthetic samples. The algorithm finds an optimal noise process scheme and calculates the sampling weight of each minority sample, which effectively removes noise samples and reinforces decision boundary. Then GA-SMOTE uses intra-cluster neighborhood triangular sampling method, which improves the diversity of synthetic samples and avoids the generation of noise samples. Besides, it should be noticed that class imbalance problem not only exists in binary datasets, but also exists in multi-class datasets, and the application of GA-SMOTE to solve the problem of multi-class imbalance classification will become the focus in our future research.

References

1. Han, W.H., Huang, Z.Z., Li, S.D., Jia, Y.: Distribution-sensitive unbalanced data oversampling method for medical diagnosis. J. Med. Syst. **43**(10), 39 (2019)
2. Zheng, Z., Wu, X., Srihari, R.K.: Feature selection for text categorization on imbalanced data. Sigkdd Explor. **6**(1), 80–89 (2004)
3. Fan, W., Stolfo, S., Zhang, J., Chan, P.: Adacost: misclassification cost-sensitive boosting. In: International Conference on Machine Learning, pp. 97–105 (1999)
4. Galar, M., Fernandez, A., Barrenechea, E., Bustince, H., Herrera, F.: A review on ensembles for the class imbalance problem: bagging-, boosting-, and hybrid-based approaches. IEEE Trans. Syst. Man Cybern. Part C Appl. Rev. **42**(4), 463–484 (2012)
5. Tomek, I.: Two modifications of CNN. IEEE Trans. Syst. Man Cybern. **6**, 769–772 (1976)
6. Chawla, N.V., Bowyer, K.W., Hall, L.O., Kegelmeyer, W.F.: SMOTE: synthetic minority over-sampling technique. J. Artif. Intell. Res. **16**, 321–357 (2002)
7. Han, H., Wang, W.-Y., Mao, B.-H.: Borderline-SMOTE: a new over-sampling method in imbalanced data sets learning. In: Huang, D.-S., Zhang, X.-P., Huang, G.-B. (eds.) ICIC 2005. LNCS, vol. 3644, pp. 878–887. Springer, Heidelberg (2005). https://doi.org/10.1007/11538059_91
8. He, H., Bai, Y., Garcia, E. A., Li, S.: ADASYN: adaptive synthetic sampling approach for imbalanced learning. In: IJCNN, Hong Kong, China, pp. 1322–1328 (2008)

9. Barua, S., Islam, M.M., Yao, X., Murase, K.: MWMOTE- majority weighted minority oversampling technique for imbalanced data set learning. IEEE Trans. Knowl. Data Eng. **26**(2), 405–425 (2014)
10. Dong, Y., Wang, X.: A new over-sampling approach: random-SMOTE for learning from imbalanced data sets. In: Xiong, H., Lee, W.B. (eds.) KSEM 2011. LNCS (LNAI), vol. 7091, pp. 343–352. Springer, Heidelberg (2011). https://doi.org/10.1007/978-3-642-25975-3_30
11. Holland, J.: Adaptation in Natural and Artificial Systems. University of Michigan press, Ann Arbor (1975)
12. Jiang, K., Lu, J., Xia, K.: A novel algorithm for imbalance data classification based on genetic algorithm improved SMOTE. Arab. J. Sci. Eng. **41**(8), 3255–3266 (2016)
13. Gu, Q., Wang, X.M., Wu, Z., Ning, B., Xin, C.S.: An improved SMOTE algorithm based on genetic algorithm for imbalanced data classification. J. Dig. Inf. Manag. **14**(2), 92–103 (2016)
14. Tallo, T.E., Musdholifah, A.: The implementation of genetic algorithm in smote (synthetic minority oversampling technique) for handling imbalanced dataset problem. In 2018 4th International Conference on Science and Technology (ICST), pp. 1–4. IEEE (2018)
15. Lichman, M.: UCI Machine Learning Repository (2016). http://archive.ics.uci.edu/ml
16. Suykens, J.A.K., Gestel, T.V., Brabanter, J.D., Moor, B.D., Vandewalle, J.: Least squares support vector machines. Int. J. Circ. Theory Appl. **27**, 605–615 (2002)
17. Fix, E., Hodges, J.L.: Discriminatory analysis-nonparametric discrimination: Consistency properties. Technical Report 4, USAF School of Aviation Medicine, Randolph Field 57(3) (1951)
18. Wang, W., Xie, Y.B., Yin, Q.: Decision tree improvement method for imbalanced data. J. Comput. Appl. **39**(3), 623–628 (2019)
19. Guo, H., Viktor, H.L.: Learning from imbalanced data sets with boosting and data generation: the databoost-IM approach. ACM Sigkdd Explor. Newsl. **6**(1), 30–39 (2004)
20. Kubat, M., Matwin, S.: Addressing the curse of imbalanced training sets: one-sided selection. In: Proceedings of International Conference on Machine Learning, pp. 179–186 (1997)
21. Dunn, J.C.: A fuzzy relative of the ISODATA process and its use in detecting compact well-separated clusters. J. Cybern. **3**(3), 32–57 (1973)
22. Pedregosa, F., et al.: Scikit-learn: machine learning in python. J. Mach. Learn. Res. **12**, 2825–2830 (2011)
23. Friedman, M.: The use of ranks to avoid the assumption of normality implicit in the analysis of variance. J. Am. Stat. Assoc. **32**(200), 675 (1937)

Dy-Drl2Op: Learning Heuristics for TSP on the Dynamic Graph via Deep Reinforcement Learning

Haojie Chen, Jiangting Fan, Yong Liu[✉], and Xingfeng Lv[✉]

Department of Computer Science and Technology, Heilongjiang University,
Harbin, China
{liuyong123456,lvxingfeng}@hlju.edu.cn

Abstract. In recent years, learning effective algorithms for combination optimization problems based on reinforcement learning has become a popular topic in artificial intelligence. In this paper, we propose a model Dy-Drl2Op that combines the multi-head attention mechanism with hierarchical reinforcement learning to solve the traveling salesman problem on a dynamic graph. In the Dy-Drl2Op model, we design a policy network to process the feature vector of each node, with the help of distributed reinforcement learning algorithm to quickly predict the probobility of each node being selected at the next step. Dy-Drl2Op also utilizes the beam search algorithm to explore the optimal solution in the whole solution space. The trained model will generate the action decision sequence that meets the specific target reward function in real time. Dy-Drl2Op is evaluated on a series of dynamic traveling salesman problems. The experimental results show that Dy-Drl2Op is better than several baseline models in terms of solution quality and efficiency.

Keywords: Combination optimization problems · Reinforcement learning · Attention mechanism · Beam search

1 Introduction

Many decision-making problems belong to combination optimization problems [9] and belong to np-hard problems [6]. The scale of these problems increases exponentially with the increase of the number of nodes. For the large-scale dynamic traveling salesman problem with weak generalization ability, the algorithm cannot find the optimal solution.

This paper uses the reinforcement learning [14] algorithm to solve the problem of node sequence [17] decision-making by an end-to-end approach in the dynamic combinatorial optimization problems. Aiming at the traveling salesman problem and its variants, we propose a framework Dy-Drl2Op that combines a policy network based on variant transformer [5] and distributed reinforcement learning [11]. Regarding the dynamic traveling salesman problem as a decision problem of a node sequence, we use Markov decision process [4] modeling to

© Springer Nature Switzerland AG 2021
T. Mantoro et al. (Eds.): ICONIP 2021, LNCS 13110, pp. 213–223, 2021.
https://doi.org/10.1007/978-3-030-92238-2_18

design a probability distribution function of an optimal strategy to achieve the purpose of dynamically outputting feasible solutions.

2   Related Work

Using reinforcement learning [25] to solve the combination optimization problems has been the development trend of artificial intelligence in recent years. Vinyals et al. [22], Bello [8] tried to use reinforcement learning algorithms to deal with the knapsack problem with two-dimensional coordinates as node information, and solved the problem of labeling required for reinforcement learning. Since the design of the model is only applicable to the Euclidean space without considering the graph structure, it has not been well extended and applied.

Dai [10] proposed a single model based on graph network [12] representation. The model inserts the output nodes into the local route by fitting the Q-learning [13] training model. Compared with the previous Bello's framework, the graph convolutional neural network Structure2Vec [16] is more suitable for solving graph-based combination optimization problems. The disadvantage is that the model requires artificial design of auxiliary functions, and the generalization ability is poor. Nazari [20] proposed to replace the encoder of the pointer network with a one-dimensional convolutional layer to directly perform the node representation process. Using this method to update the representation vector can reduces a lot of unnecessary calculations after the state change.

Based on this, Kool et al. [1] trained a model based on the Transformer [18] architecture. The model outputs a fractional solution to the multi-people traveling salesman problems. The result can be seen as a solution to the linear relaxation of the problem [7]. However, the model has fixed requirements for the dimensions of input and output, and its scalability is low.

3   The Proposed Neural Network Model

Aiming at the process of obtaining the optimal solution of the dynamic traveling salesman problem, the paper designs a lightweight model Dy-Drl2Op, which construction includes three stages: input preprocessing, policy network parametrization, training procedure, as shown in Fig. 1.

3.1   Input Preprocessing

The graph can be defined as $G(V, E)$, where V is the set of nodes, E is the set of edges. $X_v^{(t)}$ represents the representation vector of node v at time t and the initial setting of $X_v^{(t)}$ is $X_v^{(t)} = \{[V_v^{(t)}, F_v, I_v^{(t)}]\}$, where $[\cdot, \cdot, \cdot]$ is the concatenation operation. $V_v^{(t)}$ represents whether the node is selected at time t. The initial value of $V_v^{(t)}$ is 0. In the subsequent process, if the node is selected, the value of $V_v^{(t)}$ will be set to 1. F_v represents the current position of the node v. $I_v^{(t)}$ represents the customer demand of the node v at time t. Its update method

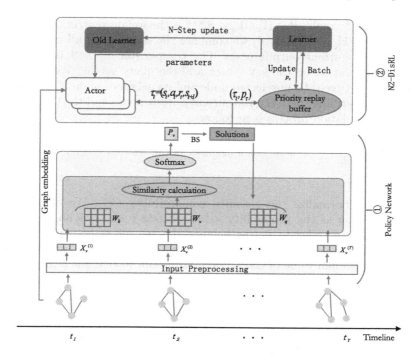

Fig. 1. The Model diagram of Dy-Drl2Op

is in the Sect. 3.2. At time $t \in T$, a series of representation vectors $X_v^{(t)} \in \mathbb{R}^d$ are obtained by one-dimensional convolution transformation on all nodes $v \in V$ in graph G.

3.2 Policy Network Parametrization

In this module, we propose a policy network, which can predict the target node sequence. The policy network inputs the representation vector of the nodes whose $V_v^{(t)}$ are 0 at time t and the representation vector of the nodes whose $V_v^{(t-1)}$ are 1 at the previous time into the variant transformer [24], which only contains the multi-head context attention mechanism and the softmax [2] layer.

The h heads of the multi-head attention respectively perform low-dimensional projections on the node representations $X_v^{(t)}$ and $X_u^{(t-1)}$ to obtain the query vector q^t key vector, $k_{(v)}^t$ and value vector $v_{(v)}^t$. $X_u^{(t-1)}$ is the node representation vector that has been selected at the previous moment. The calculation process is as follows:

$$k_{(v)}^t = W_k X_v^{(t)} \tag{1}$$

$$v_{(v)}^t = W_v X_v^{(t)} \tag{2}$$

$$q^t = W_q X_u^{(t-1)} \tag{3}$$

In the above formula, $W_k, W_v, W_q \in \mathbb{R}^{d \times d}$ is the weight matrix. After obtaining the value of $k_{(v)}^t$, $v_{(v)}^t$ and q^t, we compare the sequence of all target output nodes with the current output node. The comparison method is to perform dot product operation on the key vector and the query vector. We can get the compatibility between the unselected node and its neighbor nodes. Then, we calculate the compatibility between the current output nodes and the target output nodes. The calculation process is as follows:

$$
C_{(v)} = \begin{cases} \dfrac{(q^t)^T k_{(v)}^t}{\sqrt{d_k}}, & \text{if node } v \text{ is a neighbor of the selected node } u \text{ recently} \\ -\infty, & \text{otherwise} \end{cases} \tag{4}
$$

In order to normalize the obtained compatibility $C_{(v)}$. We use the value obtained by Eq. 5 as the probability value of node v and neighbor nodes. As shown in Eq. 6, in order to better balance the two parts of exploration and utilization, when selecting nodes, we add random exploration degree entropy noise which a random value of Gaussian distribution, so that the policy network can explore more information of node in the initial stage.

$$
P_v = softmax(tanh(C_{(v)})) \tag{5}
$$

During the prediction process, we use the beam search algorithm [19]. According to Eq. 5 and Eq. 6, the first b nodes with the largest probability value and noise value are selected to join the set S until the output path covers all the nodes V. After the policy network outputs the feasible solution sequence, we get the reward r_t which depends on the distance between the node v and its neighbors.

$$
v_i = \pi(a_{t+1}|s_t, a_t) = \underset{V_i \notin S}{\arg\max}(P_v + entropy) \tag{6}
$$

3.3   Computational Experiment

In order to better deal with the traveling salesman problem in a dynamic environment, we set up a clearly defined Markov state transition function. In the process of predicting each node by the policy network [15] in Sect. 3.2, the vehicle will select all undelivered customer nodes or warehouses to proceed to the next node state. We use the following function to update the dynamic process of customer demand and vehicle load in the traveling salesman problem. Because the traveling salesman problem we studied is a classical traveling salesman problem, we mainly introduce the others.

Distribution Collection Traveling Salesman Problem (DCTSP). The dispatcher starts from the distribution center node v_0 in graph G, and a limited-capacity vehicle is responsible for the distribution of customers whose demand exceeds 0. Compared with the traveling salesman problem in the classical combination optimization problems, this problem adds the constraint condition of limited capacity vehicle load $L_{(t)}$.

$$I_v^{(t)} = 0, \text{ if } L_{(t)} \geq I_v^{(t-1)} \tag{7}$$

In Eq. 7, if the remaining load $L_{(t)}$ of the vehicle at the current moment t is large enough to meet the demand of the current customer node v, the demand $I_v^{(t)}$ of the customer v at the current moment t is set to 0. At the same time, the customer node v is placed in the visited solution sets and no longer dispatched. Otherwise, the customer will be blocked.

$$L_{(t)} = \max(L_{(t-1)} - I_v^{(t-1)}, \ 0) \tag{8}$$

In Eq. 8, if the demand of the customer node at the previous time exceeds the remaining load of the vehicle at the previous time, the remaining load $L_{(t)}$ of the vehicle after delivery is 0. Otherwise, the remaining load $L_{(t)}$ of the vehicle at the current moment is the subtraction between the vehicle load and the demand of the customer node at the previous moment.

Split Delivery Traveling Salesman Problem (SDTSP). Each customer's demand $L_v^{(t)}$ can be split into multiple parts, and we allow dispatchers to deliver to customers whose demand $L_v^{(t)}$ exceeds the current vehicle load. The solution can distribute the needs of a given customer to multiple routes.

$$I_v^{(t)} = \max(I_v^{(t-1)} - L_{(t-1)}, \ 0) \tag{9}$$

In Eq. 9, if the remaining load of the vehicle at the previous moment is sufficient to deliver the customer's demand, the current moment demand $v_{(t)}$ is set to 0. Otherwise, the customer demand is the subtraction between the customer demand the remaining load of the vehicle at the previous moment, which causes the current customer's demand to be met multiple times by dispatchers and there are multiple deliveries.

$$L_{(t)} = \max(L_{(t-1)} - I_v^{(t-1)}, \ 0) \tag{10}$$

In Eq. 10, if the remaining load of the vehicle at time t is enough to exceed the customer's demand, the remaining load of the vehicle is the subtraction between the vehicle load at time $t-1$ and the customer's demand $I_v^{(t-1)}$ that has been met. Otherwise, the remaining load L_t of the vehicle is less than the number required by the customer, and the remaining load is updated to 0 after delivery to the customer.

3.4   Training Procedure

We propose N2-DisRL to train the policy network, which allows the agent to autonomously learn to perform actions through feedback of reward values in a given environment. The goal of training is to learn an interactive strategy to maximize the cumulative reward.

We introduce a distributed reinforcement learning [3] method which use the parallel agent strategy. That is to say, we copy n threads and run an agent in each thread to interact with the given environment to speed up the data collection time in parallel.

After the actor exploring the environment, we need define the total reward value obtained by the agent as R_t at time t, and the agent chooses action a_t that maximizes R_t each time. The total reward R_t is updated using a multi-step reward method [23]. The attenuation factor γ is used to incorporate multiple future reward r_t with varying degrees of attenuation into the total value R_t. The calculation process is as follows:

$$R_t = r_{t+1} + \gamma r_{t+2} + \cdots + \gamma^{n-1} r_{t+n} + \gamma^n Q_{t+n} \tag{11}$$

In order to increase the probability of sampling data that is difficult to train, we consider adopting the experience replay system and propose to calculate the TD-error of the step from the exploration environment as the priority p_τ to measure the difference in state estimation value at different times, and store it in the experience replay in the form of (τ, p_τ). In Eq. 12, the subtraction between the value of the real state and the value of the target evaluation indicates that the greater the priority p_τ, the greater the error of the policy network, indicating that the accuracy needs to be retrained and adjusted.

$$p_\tau = r_t + \gamma Q_t - Q_{t-1} \tag{12}$$

Later, we should takes a batch of high-priority data from the experience replay to update the parameters. We create a baseline network for the Learner network which is called old Learner. The old Learner will obtain the new Learner network parameters to update. If the loss of the old Learner network is smaller after T steps, the parameters of Learner network will be covered to stabilize the training process. The loss function is as follows:

$$L_t(\tau; W) = (r_{t+1} + \gamma r_{t+2} + \cdots + \gamma^{n-1} r_{t+n} + \gamma^n Q_{t+n} - Q_t)^2 \tag{13}$$

The loss function is the square subtraction between the total reward value R_t fed back by the exploration environment and the label Q function output from the training model. The stochastic gradient descent method is used to train and update the parameters of the model, while using the old Learner network to adjust the update direction. In order to prevent invalid data from occupying too much cache, the system will periodically update the data of experience replay. After the Learner cyclically updates at every interval, it removes all batches of data to reduce the possibility of data overflow.

Algorithm 1. Training model

Input: Steps per epoch T, batch size B.

Output: Update weight matrix W.

1: Initialize $W = (W_1, W_2, \ldots, W_g)$, replay memory D to capacity N
2: $W_{old} \leftarrow W$
3: Get s_0 and choose a_0 randomly;
4: **for** $t = 1$ to $T - 1$ **do**
5:    Perform a_t according to policy $\pi(a_{t+1}|S_t, a_t) = \underset{V_i \notin S}{\arg\max}(P_v + entropy)$;
6:    Receive reward r_t and new state s_{t+1};
7:    Store transition $\tau_t = (s_t, a_t, r_t, s_{t+1})$ in replay memory D with priority $p_\tau = r_t + \gamma Q_t - Q_{t-1}$ in such form (τ_t, p_τ);
8:    **if** D.size() $> B$ **then**
9:       Sample B transitions with maximal p_τ;
10:       Training W by using SGD on equation 13;
11:       $W \leftarrow$ Update(parameters());
12:       $W_a \leftarrow$ Learner.parameters();
13:    **end if**
14:    **if** $L_t(\tau_t; W) < L_t(\tau_t; W_{old})$ **then**
15:       Copy parameters into new network $W_{old} \leftarrow W$;
16:    **end if**
17: **end for**
18: Replay.delete();
19: Return W;

---

4   Performance Evaluation

4.1   Hyperparameters

At any time $t \in T$, where $T > 0$, the number of customer nodes of different traveling salesman problems is set to 20, 50 and 100. The data set composed of 1000 Euclide graphs serves as the input of this framework Dy-Drl2Op. In order to keep the training time controlled and the total number of parameter updates consistent, each epoch has a total of 0.64M training instances and is set to process 256 batches.

The number of iterations in the experiment was all set to 10 times, and the maximum time T_{max} was set to 128. In the process of training model, the learning rate is set to 0.0001 and the number of training examples sampled from Replay Memory M is set to 16. The parameter θ of the Learner network is initialized by the Xavier initializer [26] to keep the gradient of each layer consistent. Later, we use the Adam [21] optimizer to perform stochastic gradient descent on the loss function Eq. 13 to update the parameter.

4.2   Comparison of Optimal Performance

Table 1. Optional length of different models

Optional Length	LKH3	S2V	Our model	DyRL	AM
TSP20	3.86	3.89	3.86	–	3.86
TSP50	5.69	5.99	5.75	–	5.76
TSP100	7.97	8.31	7.98	–	8.01
DCTSP20	6.14	–	6.27	6.41	6.29
DCTSP50	10.38	–	10.55	11.25	10.61
DCTSP100	15.65	–	15.93	16.99	16.33
SDTSP20	–	–	6.28	6.36	6.26
SDTSP50	–	–	10.58	11.1	10.58
SDTSP100	–	–	16.2	16.88	16.27

By comparing our framework Dy-Drl2Op with S2V [10], AM [1], DyRL [20] and LKH3, the following results are obtained by averaging more than 8 training results. Table 1 shows the optimal length of all Traveling Salesman Problem with different number of nodes for different models. − means that the model does not support this type of problem. Based on the experimental results in Table 1, it can be found that this model can obtain better results than other comparative models. Especially after the number of nodes up to 100, our model is obviously better than other models in different problems, and outputs the route closest to the optimal solution within a limited time.

For the problem of DCTSP and SDTSP, DyRl and AM models were applied to this problem, and there was a difference of 0.01 to 0.4 between the different models, and our model was superior to AM by 0.1. For different customer nodes, our model can obtain the optimal path length from 3.86 to 16.2, which is relatively close to the AM model, and can reach the optimal path length of LKH3 at 20 nodes.

4.3   Visualization

In order to explain the process of generating solutions more clearly, we will visualize the construction process of the dispatch route. As shown in Fig. 2, we show solutions for the distribution collection traveling salesman problem and the split delivery traveling salesman problem [27] with different numbers of nodes. In the visualization process, the five-pointed star represents the warehouse node, the dot represents the customer node, and the line segment represents the delivery route. The number of the node $1, 2, \ldots$ in the legend indicates the customer delivery order.

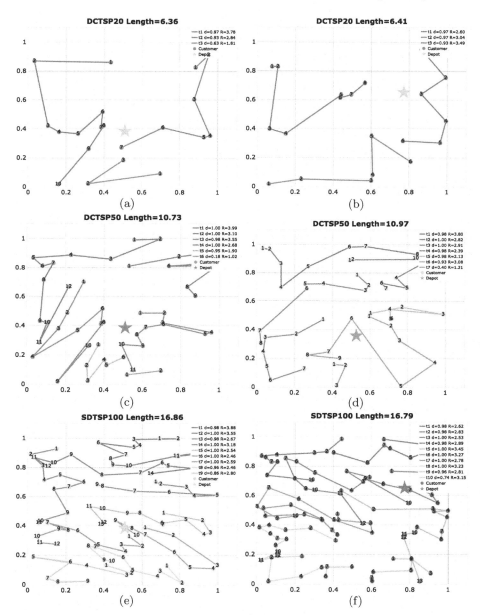

Fig. 2. Route of DCTSP and SDTSP

Since the load of the dispatched vehicle is fixed at 1, the vehicle need to return to the warehouse multiple times to fill up the capacity. The legend on the right indicates the order of the delivery route, the vehicle capacity used for that route, and the total reward value earned for each route.

In the case of the split delivery traveling salesman problem, some customer nodes will be visited multiple times. The delivery vehicle reduces the cost by dispatching the needs of a single customer node in batches. In this example, we observe that the blue route t1 and red route t2 in Fig. 2 (e) jointly deliver a customer node, and the dispatcher only meets the current customer node 11 (or the second delivery of node 1) part of the demand during each delivery process, which can reduce the number of deliveries required to meet all needs.

5   Conclusion

We propose a distributed reinforcement learning framework Dy-Drl2Op. The policy network is combined with the N2-DisRL distributed network to train and adjust the policy network parameters to dynamically generate solutions. The model can explore environmental features from multiple aspects and synthesize multiple solutions in a very short time. Through many experiments with a large number of different node numbers, Dy-Drl2Op is significantly faster and higher quality than state-of-the-art models.

Acknowledgment. This work was supported by the National Natural Science Foundation of China (No. 61972135), the Natural Science Foundation of Heilongjiang Province in China (No. LH2020F043), the Innovation Talents Project of Science and Technology Bureau of Harbin in China (No. 2017RAQXJ094), and the Postgraduate Innovative Scientific Research Project of Heilongjiang University in China (No.YJSCX2021-197HLJU).

References

1. Kool, W., Van Hoof, H., Welling, M.: Attention, Learn to Solve Routing Problems! (2018)
2. Salakhutdinov, R., Hinton, G.E.: Replicated softmax: an undirected topic model. In: Advances in Neural Information Processing Systems 22: Conference on Neural Information Processing Systems (2009)
3. Sutton, R.S., Barto, A.G.: Reinforcement learning. Bradford Book **15**(7), 665–685 (1998)
4. White, C.C., et al.: Markov decision processe. Eur. J. Oper. Res. (1989)
5. Vaswani, A., et al.: Attention is all you need. In: Advances in Neural Information Processing Systems, pp. 6000–6010 (2017)
6. Hochba, D.S.: Approximation algorithms for NP-hard problems. ACM SIGACT News **28**(2), 40–52 (1997)
7. Jordan, M.I., Rumelhart, D.E.: Forward models: supervised learning with a distal teacher. Cogn. Sci. **16**(3), 307–354 (2010)
8. Bello, I., Pham, H., Le, Q.V., Norouzi, M., Bengio, S.: Neural combinatorial optimization with reinforcement learning. In: International Conference on Learning Representations (ICLR) (2017)

9. Colorni, A., Dorigo, M., Maffioli, F., et al.: Heuristics from nature for hard combinatorial optimization problems. Int. Trans. Oper. Res. **3**(1), 1–21 (2010)
10. Khalil, E., Dai, H., Zhang, Y., Dilkina, B., Song, L.: Learning combinatorial optimization algorithms over graphs. In: Advances in Neural Information Processing Systems, pp. 6351–6361 (2017)
11. Liang, E., Liaw, R., Moritz, P., et al.: RLlib: abstractions for distributed reinforcement learning (2017)
12. Joshi, C.K., Laurent, T., Bresson, X.: An efficient graph convolutional network technique for the travelling salesman problem (2019)
13. Watkins, C., Dayan, P.: Technical note: q-learning. Mach. Learn. **8**(3–4), 279–292 (1992)
14. Kulkarni, T.D., Narasimhan, K.R., Saeedi, A., et al.: Hierarchical deep reinforcement learning: integrating temporal abstraction and intrinsic motivation (2016)
15. Zaremba, W., Sutskever, I., Vinyals, O.: Recurrent Neural Network Regularization (2014). Eprint
16. Edwards, M., Xie, X.: Graph convolutional neural network. In: British Machine Vision Conference (2016)
17. Sutskever, I., Vinyals, O., Le, Q.V.: Sequence to sequence learning with neural networks. In: Advances in Neural Information Processing Systems (2014)
18. Garg, S., Peitz, S., Nallasamy, U., et al.: Jointly learning to align and translate with transformer models (2019)
19. Wiseman, S., Rush, A.M.: Sequence-to-sequence learning as beam-search optimization. In: Proceedings of the 2016 Conference on Empirical Methods in Natural Language Processing (2016)
20. Nazari, M.R., Oroojlooy, A., Snyder, L., Takac, M.: Reinforcement learning for solving the vehicle routing problem. In: Advances in Neural Information Processing Systems, pp. 9860–9870 (2018)
21. Kingma, D., Ba, J.: Adam: a method for stochastic optimization. Comput. Sci. (2014)
22. Vinyals, O., Fortunato, M., Jaitly, N.: Pointer networks. In: Advances in Neural Information Processing Systems, pp. 2692–2700 (2015)
23. Osband, I., Blundell, C., Pritzel, A., et al.: Deep exploration via bootstrapped DQN (2016)
24. Nassar, K.: Transformer-based language modeling and decoding for conversational speech recognition (2020)
25. Barrett, T., Clements, W., Foerster, J., et al.: Exploratory combinatorial optimization with reinforcement learning. Proc. AAAI Conf. Artif. Intell. **34**(4), 3243–3250 (2020)
26. Glorot, X., Bengio, Y.: Understanding the difficulty of training deep feedforward neural networks. J. Mach. Learn. Res. **9**, 249–256 (2010)
27. Helsgaun, K.: An extension of the Lin-Kernighan-Helsgaun TSP solver for constrained traveling salesman and vehicle routing problems. Technical report (2017)

Multi-label Classification of Hyperspectral Images Based on Label-Specific Feature Fusion

Jing Zhang[1], PeiXian Ding[1], and Shuai Fang[1,2(✉)]

[1] School of Computer and Information, Hefei University of Technology,
Hefei 230601, China
fangshuai@hfut.edu.cn
[2] Key Laboratory of Industrial Safety and Emergency Technology, Hefei 230601,
Anhui, China

Abstract. For hyperspectral classification, the existence of mixed pixels reduces the classification accuracy. To solve the problem, we apply the multi-label classification technique to hyperspectral classification. The focus of multi-label classification is to construct label-specific features. However, some algorithms do not consider the construction of label-specific features from multiple perspectives, resulting in that useful information is not selected. In this paper, we propose a new hyperspectral image multi-label classification algorithm based on the fusion of label-specific features. The algorithm constructs label-specific features from the three perspectives: distance information and linear representation information between instances, clustering information between bands, and then merges three feature subsets to obtain a new label feature space, making each label has highly discriminative features. Comprehensive experiments are conducted on three hyperspectral multi-label data sets. Comparison results with state-of-the-art algorithms validate the superiority of our proposed algorithm.

Keywords: Hyperspectral classification · Multi-label classification · Label-specific features

1 Introduction

In recent years, hyperspectral image classification has been widely used in ground objects component analysis, precision forestry, water quality detection, and other fields. Hyperspectral image classification is usually based on the characteristics of the spectral curve to classify each pixel. To improve classification accuracy, a series of classification methods combining spectral information and spatial information have been formed [1], such as FGF-SVM-SNI [2]. Classification methods based on spectral-neighborhood features have also been produced, such as JSRC [3], SC-MK [4], and so on. In addition, methods based on deep learning [5] have also achieved good results in hyperspectral classification, such as Zhang et al.

© Springer Nature Switzerland AG 2021
T. Mantoro et al. (Eds.): ICONIP 2021, LNCS 13110, pp. 224–234, 2021.
https://doi.org/10.1007/978-3-030-92238-2_19

[6] used 3-DGAN as a spectral-spatial classifier. These algorithms have greatly improved the classification accuracy of hyperspectral images. How-ever, there is still a major difficulty in the classification of hyperspectral images: the existence of mixed pixels causes a decrease in the classification accuracy.

The spatial resolution of hyperspectral images is usually low, and there are a large number of mixed pixels in the image. Within the mixed pixel, the spectral curves of different ground objects are fitted together, which makes it difficult to determine the classification boundary between different ground objects and in-creases the difficulty of hyperspectral single-label classification. If different ground objects are regarded as multiple labels, the method of multi-label can be applied to mixed pixels. The multi-label classification method can not only reduce the negative effects of mixed pixels on hyperspectral single-label classification, but also retain more ground objects information. Therefore, this paper will adopt the multi-label classification method to study hyperspectral image classification.

Generally speaking, there are two solutions to the multi-label classification problem: problem transformation method and algorithm adaptation method. The problem transformation method transforms the multi-label classification problem into one or more single-label classification problems, among which the three most commonly used algorithms are BR [7], CC [8], and LP [9]. The algorithm adaptation method is to adapt to the classification of multi-label data by adjusting the existing learning algorithms, such as ML_KNN [10]. The traditional multi-label classification strategy uses the original feature space to operate on labels and does not consider that different features have different distinguishing capabilities for labels. Using the same feature for all labels may even cause feature redundancy and interference. Based on this problem, the LIFT [11] algorithm first proposed the concept of label-specific features and constructed features with exclusive distinguishing capabilities for each label based on the distance mapping mechanism. Some other algorithms were inspired by LIFT and improved the construction of label-specific features through various technologies, such as ML_DFL [12], SLMLC [13].

The above-mentioned multi-label algorithms have achieved better classification performance after constructing the label-specific features. However, most algorithms do not construct label-specific features from multiple perspectives, which are far from fully mining the label information in hyperspectral images. According to the characteristics of rich spectral information and wideband range of hyperspectral images, we expand the label-specific features from three perspectives: distance information and linear representation information between instances, clustering information between bands. Multi-label classification of hyperspectral images is performed by fusing label-specific feature subsets. Our main contributions are as follows:

1) Extract the weight coefficients of the cluster center to linearly represent the instances to expand the specific characteristics of the labels between the instances.

2) Combining the clustering results of the original spectral bands, extract the distance information between the instance and the central feature of the band cluster to expand the specific features of the label between the bands.
3) Fuse the label-specific features from three perspectives, so that the features of each label have high discrimination.

2   The Proposed Algorithm

In multi-label learning, $\mathbf{X} = \{x_1, x_2, \cdots, x_n\}^T$ represents the instance space, $\mathbf{L} = \{l_1, l_2, \cdots, l_q\}$ represents the label space, $\mathbf{D} = \{\{x_i, y_i\} | 1 \leq i \leq n\}$ represents the training data set with n samples, where $x_i \in R^d$ represents the d-dimensional feature vector, and $y_i \in \{0,1\}^q$ represents the label vector related to x_i. The value of y_i is 1 or 0 indicating whether the label is associated with the instance x_i, respectively.

2.1   Construction of Distance Mapping Feature Between Instances

Firstly, for each label $l_k \in \mathbf{L}$, the examples in the training set $\mathbf{D}$ are divided into a positive instance set $\mathbf{P}_k$ containing the label l_k and a negative instance set $\mathbf{N}_k$ that does not contain the label l_k. $\mathbf{P}_k$ and $\mathbf{N}_k$ are respectively denoted as:

$$\mathbf{P}_k = \{x_i | \{x_i, y_i\} \in \mathbf{D}, \ l_k \in y_i\} \tag{1}$$

$$\mathbf{N}_k = \{x_i | \{x_i, y_i\} \in \mathbf{D}, \ l_k \notin y_i\} \tag{2}$$

Secondly, use the K-means algorithm [14] to perform cluster analysis on the two sets of $\mathbf{P}_k$ and $\mathbf{N}_k$ respectively. To reduce the adverse effects of class imbalance in multi-label data, the two sets are set to the same number of clusters. The formula of cluster number is as follows:

$$m_k = \lceil \lambda \cdot min \lceil |\mathbf{P}_k|, |\mathbf{N}_k| \rceil \rceil \tag{3}$$

For Eq. (3), $\lambda \in [0,1]$ is a parameter used to adjust the number of clusters, $|\mathbf{P}_k|$ and $|\mathbf{N}_k|$ respectively represent the number of examples contained in the set of positive and negative instances, $\lceil \cdot \rceil$ indicates rounding up. The cluster center set of the instances is denoted as $\mathbf{C}_k = [\mathbf{CP}_k, \mathbf{CN}_k]$, where $\mathbf{CP}_k$ is the cluster center set of positive instances, and $\mathbf{CN}_k$ is the cluster center set of negative instances.

To use the label information represented by the cluster center, the distance between the instance and the cluster center is used as label-specific features. Intuitively speaking, the essence of extracting the label l_k distance mapping feature is to transform the original d-dimensional instance space $\mathbf{X}$ into the $2m_k$-dimensional feature space $\mathbf{Z}_k$. The mapping function is $\phi_k : \mathbf{X} \rightarrow \mathbf{Z}_k$, which is expressed as follows:

$$\phi_k(x) = [d[x_i, cp_1^k], \cdots, d[x_i, cp_{m_k}^k], d[x_i, cn_1^k], \cdots, d[x_i, cn_{m_k}^k]] \tag{4}$$

For Eq. (4), $d\left(\cdot,\cdot\right)$ returns the Euclidean distance between two examples.

In hyperspectral images, the instances and cluster centers are represented by spectral curves, and $d\left(\cdot,\cdot\right)$ obtains the Euclidean distance between the two spectral curves. Figure 1(a) illustrates the features of distance mapping between instances to cluster centers, where the green dot is an original instance, and the red and blue triangles are the cluster centers of positive and negative instances, respectively. The line segments between the instance and the cluster centers represent the distance between them.

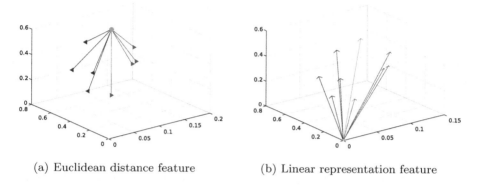

(a) Euclidean distance feature   (b) Linear representation feature

Fig. 1. Schematic diagram of features between instances (Color figure online)

2.2 Construction of Linear Representation Mapping Features Between Instances

Hyperspectral images have very rich spectral information. Merely measuring the distance information between instances is not enough to capture all the label-specific features in the spectrum. Therefore, we linearly represent the instance by positive and negative clustering centers [15], and extract the weight information of clustering centers as a supplement to the distance mapping features. Figure 1(b) illustrates the linear representation of feature mapping in hyperspectral images, where the green vector is an original instance, and the red and blue vectors are the cluster centers of the positive and negative instances, respectively.

The essence of linear representation is to map the original d-dimensional instance eigenspace $\mathbf{X}$ to the new $2m_k$-dimensional eigenspace $\mathbf{Z}_{\mathbf{k}}'$. The mapping function is $\phi_k' : \mathbf{X} \rightarrow \mathbf{Z}_{\mathbf{k}}'$, which is expressed as follows:

$$\phi_k'\left(x\right) = \left[\omega_1^k, \omega_2^k, \cdots, \omega_{2mk}^k\right] \tag{5}$$

For Eq. (5), $\omega^k = \left[\omega_1^k, \omega_2^k, \cdots, \omega_{2m_k}^k\right]$ is the linear weight coefficient of the positive and negative cluster centers of the k-th label. Generally, ω^k is normalized, that is, $\sum_{j=1}^{2m_k} \omega_j^k = 1$.

For the solution of ω^k, we can use the idea of locally linear embedding [16]. Suppose that the instance x_i can be linearly represented by positive and negative clustering centers of label l_k : $x_i = \omega_1^k cp_1^k + \cdots + \omega_m^k cp_{m_k}^k + \omega_{m+1}^k cn_1^k + \cdots + \omega_{2m_k}^k cn_{m_k}^k$. The above problem can be transformed into the problem of solving the loss function:

$$min \sum_{i=1}^{n} \left\| x_i - \sum_{j=1}^{2m_k} \omega_j^{ki} C_j^k \right\|_2^2 \tag{6}$$

According to the treatment of the above equation by the Lagrange multiplier method, we can obtain the following linear equation:

$$CM_{ki} \cdot \omega_{ki} = 1 \tag{7}$$

For Eq. (7), CM_{ki} is the covariance matrix of the instance x_i about positive and negative clustering center $\mathbf{C}_k$ in each label l_k, and 1 is a $2m_k$-dimensional column vector.

To obtain ω_{ki}, the irreversible phenomenon of CM_{ki} needs to be avoided, so regularize CM_{ki} to CM'_{ki}. Therefore:

$$\omega_{ki} = CM_{ki}'^{-1} 1 \tag{8}$$

Finally, by normalizing ω_{ki}, the weight coefficients of the positive and negative cluster centers are obtained:

$$\omega_{ki} = \frac{\omega_{ki}}{\sum_{j=1}^{2m_k} \omega_j^{ki}} \tag{9}$$

ω_{ki} is the linear representation mapping feature of label l_k between instances.

2.3   Construction of Band Clustering Mapping Feature

Hyperspectral images can continuously record hundreds of spectral bands, and a large number of bands can provide an irreplaceable role in hyperspectral classification. To expand the useful information of label-specific features, we obtain discriminative label-specific features from the angle of spectral bands.

Firstly, for each label l_k, we call the spectral bands in the positive and negative examples positive band set and negative band set respectively, which are represented by $\overline{\mathbf{P}_k}$ and $\overline{\mathbf{N}_k}$, and $\overline{\mathbf{P}_k} = \mathbf{P}_k^T$, $\overline{\mathbf{N}_k} = \mathbf{N}_k^T$. The specific definition is as follows:

$$\overline{\mathbf{P}_k} = \{f_i | \{x_i, y_i\} \in \mathbf{P}_k, \ \exists x_{ij} \neq 0\} \tag{10}$$

$$\overline{\mathbf{N}_k} = \{f_i | \{x_i, y_i\} \in \mathbf{N}_k, \ \exists x_{ij} \neq 0\} \tag{11}$$

Secondly, we quote the idea of dividing the feature set into blocks in [17], and adopt the spectral clustering method [18] to divide $\overline{\mathbf{P}_k}$ and $\overline{\mathbf{N}_k}$ into band subsets.

Similar to instance clustering, set $\overline{\mathbf{P}_k}$ and $\overline{\mathbf{N}_k}$ the same number of clusters. The formula for the number of clusters is:

$$n_k = \left\lceil \mu \cdot min \left\lceil \left|\overline{\mathbf{P}_k}\right|, \left|\overline{\mathbf{N}_k}\right| \right\rceil \right\rceil \tag{12}$$

For Eq. (12), $\mu \in [0,1]$ is the ratio parameter used to control the number of clusters, $\left|\overline{\mathbf{P}_k}\right|$ and $\left|\overline{\mathbf{N}_k}\right|$ are the number of bands contained in the set of positive and negative band sets respectively, $\lceil \cdot \rceil$ indicates rounding up.

In each band subset obtained after spectral clustering, we calculate the mean value of the examples as the central feature of the band subset. Finally, the feature subset corresponding to the central feature of the band set is extracted in the whole training set, and the distance from the examples in each feature subset to the central features of the positive and negative band set is calculated as the label l_k label-specific features at the band level.

The essence of the band clustering mapping feature is to map the original d-dimensional instance space $\mathbf{X}$ to the $2n_k$-dimensional feature space $\mathbf{Z}_k''$, the mapping function is $\phi_k'' : \mathbf{X} \to \mathbf{Z}_k''$, specifically as follows:

$$\phi_k''(x) = \left[d\left[x_i, \overline{cp_1^k}\right], \cdots, d\left[x_i, \overline{cp_{n_k}^k}\right], d\left[x_i, \overline{cn_1^k}\right], \cdots, d\left[x_i, \overline{cn_{n_k}^k}\right] \right] \tag{13}$$

For Eq. (13), $d(\cdot, \cdot)$ returns the distance between two examples, $\left|\overline{\mathbf{CP}_k}\right|$ is the central feature set of the positive band set, and $\left|\overline{\mathbf{CN}_k}\right|$ is the central feature set of the negative band set.

For a clearly understanding, the positive band set of label l_1 is taken as an example to demonstrate the process of band clustering feature mapping in Fig. 2. $[x1, \cdots, x8]$ denotes 8 pixels in l_1, the value of l_1 is 1 or 0, which denotes positive and negative instances respectively. $[f1, \cdots, f8]$ denotes that each pixel has 8 bands, cp- denotes the central feature of the positive band set, and d denotes the distance vector between the pixel and the central feature. Figure 2 shows that the original 8-dimensional instance space is mapped to the 3-dimensional feature space in the positive band set of label l_1.

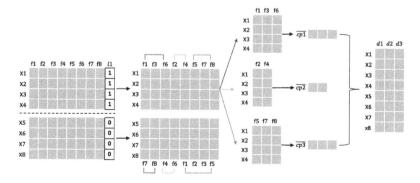

Fig. 2. Schematic diagram of band clustering feature.

2.4   Construction of Classification Model

For each label $l_k \in L$, the integration of $\phi_k(x)$, $\phi_k'(x)$ and $\phi_k''(x)$ makes the original d-dimensional feature space $\mathbf{X}$ map to the $(2m_k + 2m_k + 2n_k)$-dimensional label-specific feature space $\psi_k(x)$:

$$\psi_k(x) = \left[\phi_k[x], \phi_k'[x], \phi_k''[x]\right] \tag{14}$$

Then, for each label l_k, using the reconstructed feature space $\psi_k(x)$ to create a new training set B_K with n instances from the original training set $\mathbf{D}$:

$$B_K = \{\psi_k\{x_i\}, y_i | \{x_i, y_i\} \in D\} \tag{15}$$

When $l_k \in y_i$ the value of y_i is 1; when $l_k \notin y_i$, the value of y_i is 0.

Finally, the classification model $g(k)$ is constructed according to the training set B_K of the label l_k, and q classifier models can be obtained. Given a test instance $x \in X$, the prediction of its related label set can get the following results:

$$y = \{f_k\{g_k\{\psi_k\{x\}\}\} > 0, 1 \le k \le q\} \tag{16}$$

3   Experimental Results and Analysis

3.1   The Experimental Setup

The Data Set
Due to the lack of a hyperspectral image multi-label data set, the multi-label data set in the experiment is made by combining the endmember set and the pixel abundance coefficient set provided in the literature [19–21]. The data sets are Jasper Ridge, Samson, and Urban. If the abundance coefficient of the label is greater than the threshold θ, we set y = 1; Otherwise, y = 0. Table 1 summarizes the detailed characteristics of the multi-label dataset. The label base represents the average number of relevant labels an example has.

Table 1. Description of the dataset

Data set	Number of samples	Characteristics of several	Tag number	Label base
Urban	94 249	162	6	1.4728
Jasper ridge	10 000	198	4	1.4035
Samson	9 025	156	3	1.3835

The Evaluation Index
To test the classification effect on mixed pixels, the verification indicators used in this paper are accuracy, precision, recall, F_β, and mixed pixel classification accuracy SA (the ratio of the number of correctly classified mixed pixels to the

total number of mixed pixels). For each label, TP (true positive) represents the number of instances where the prediction is positive and the actual is positive; FP (false positive) represents the number of instances where the prediction is positive but the actual is negative; TN (true negative) represents the number of instances where the prediction is negative and the actual is negative; FN (false negative) represents the number of instances where the prediction is negative and the actual is positive. The accuracy, precision, recall, and F_β are defined as:

$$acc = \frac{TP + TN}{TP + TN + FP + FN} \tag{17}$$

$$pre = \frac{TP}{TP + FP} \tag{18}$$

$$rec = \frac{TP}{TP + FN} \tag{19}$$

$$F_\beta = \frac{\left(1 + \beta^2\right) \cdot TP}{(1 + \beta^2) \cdot TP + \beta^2 \cdot FN + FP} \tag{20}$$

The Experiment Design
This paper compares our proposed algorithm LIFT_LR_BC with 6 other multi-label classification algorithms, including the traditional multi-label classification algorithms: BR [7], ECC [8] and ML-KNN [10], and the algorithms that construct the label-specific features: LIFT [11], ML_DEL [12] and LLSF [22]. The results of the ablation experiment are also given in the article, where LIFT_LR adds linear representation features based on distance mapping features, and LIFT_BC adds band clustering features based on distance mapping features.

In the experiment, the ratio parameter λ of instance clustering is set to 0.2, and the ratio parameter μ of band clustering is also set to 0.2. For a fair comparison, all comparison algorithms use LIBSVM with linear kernel [23] as the basic binary classifier. For the conventional data set Jasper Ridge and Samson, we conducted three 10-fold cross-validation; for the large-scale dataset Urban, we conducted three 50-fold cross-validation.

3.2 Multi-label Algorithm Comparison

Table 2, Table 3, and Table 4 respectively give detailed experimental results of all comparison algorithms using six evaluation indicators on three data sets. For each evaluation index "↑" means "the larger the value is, the better", and the bold value indicates the best performance among all comparison algorithms.

Table 2. Experimental results on the Jasper Ridge data set

Method	sa↑	acc↑	pre↑	rec↑	$f_\beta(\beta = 1)$↑
BR	0.9457	0.9860	0.9806	0.9627	0.9714
ECC	0.9479	0.9862	0.9817	0.9621	0.9716
ML-KNN	0.9271	0.9796	0.9644	0.9584	0.9609
LLSF	0.9459	0.9866	0.9810	0.9623	0.9715
LIFT	0.9523	0.9872	0.9842	0.9627	0.9732
ML_DFL	0.9534	0.9875	0.9834	0.9656	0.9743
LIFT_LR	0.9537	0.9877	**0.9852**	0.9659	0.9752
LIFT_BC	0.9527	0.9874	0.9848	0.9636	0.9739
LIFT_LR_BC	**0.9540**	**0.9878**	0.9850	**0.9660**	**0.9753**

Table 3. Experimental results of the Samson data set

Method	sa↑	acc↑	pre↑	rec↑	$f_\beta(\beta = 1)$↑
BR	0.9502	0.9833	**0.9856**	0.9843	0.9849
ECC	0.9493	0.9831	0.9853	0.9836	0.9844
ML-KNN	0.9525	0.9841	0.9841	0.9859	0.9850
LLSF	0.9572	0.9838	0.9843	0.9861	0.9852
LIFT	0.9623	0.9874	0.9840	0.9898	0.9868
ML_DFL	0.9643	0.9881	0.9851	**0.9908**	**0.9879**
LIFT_LR	0.9650	0.9883	0.9847	0.9907	0.9876
LIFT_BC	0.9625	0.9875	0.9841	0.9898	0.9869
LIFT_LR_BC	**0.9652**	**0.9884**	0.9847	**0.9908**	0.9877

Table 4. Experimental results of the Urban data set

Method	sa↑	acc↑	pre↑	rec↑	$f_\beta(\beta = 1)$↑
BR	0.8084	0.9636	0.9332	0.8671	0.8970
ECC	0.8122	0.9640	0.9352	0.8693	0.8993
ML-KNN	0.8116	0.9601	0.9207	0.8437	0.8748
LLSF	0.8728	0.9675	0.9477	0.8761	0.9049
LIFT	0.8779	0.9766	0.9583	0.8953	0.9225
ML_DFL	0.8809	0.9772	0.9549	0.9044	0.9272
LIFT_LR	0.8782	0.9769	0.9581	0.8966	0.9233
LIFT_BC	0.8782	0.9767	0.9579	0.9007	0.9265
LIFT_LR_BC	**0.8820**	**0.9776**	**0.9584**	**0.9075**	**0.9307**

As shown in the table, the overall effect of traditional multi-label algorithms such as ML-KNN on the three hyperspectral multi-label data sets is average. LLSF uses raw spectral data for classification, and the effect is not as good as LIFT. ML_DFL further distinguishes the boundary between positive and negative classes, and the classification effect is better than LIFT. The performance of LIFT_LR and LIFT_BC exceeds LIFT, which proves that the linear representation mapping feature and band clustering mapping feature are effective. LIFT_LR_BC achieved the best results in the three data sets, indicating that the strategy of fusing distance mapping features, linear representation mapping features, and band clustering mapping features can improve the hyperspectral multi-label classification.

4 Conclusion

Since the previous multi-label classification algorithms did not make full use of the label-specific feature information, we combine the characteristics of hyperspectral images and constructs label-specific features from the three perspectives: distance information and linear representation information between in-stances, clustering information between bands. Through the fusion of the three feature subsets, a better feature space is obtained, and the feature information with high discriminability is constructed for each label. Experimental results verify the effectiveness of the proposed algorithm.

References

1. Ghamisi, P., Maggiori, E., Li, S., et al.: New frontiers in spectral-spatial hyperspectral image classification: the latest advances based on mathematical morphology, Markov random fields, segmentation, sparse representation, and deep learning. IEEE Geosci. Remote Sens. Mag. **6**(3), 10–43 (2018)
2. Hu, C., Xv, M., Fan, Y.: Hyperspectral image classification method combining fast guided filtering and spatial neighborhood information. In: 2020 International Conference on Computer Communication and Network Security (CCNS), pp. 55–58. IEEE (2020)
3. Hsu, P.H., Cheng, Y.Y.: Hyperspectral image classification via joint sparse representation. In: IGARSS 2019–2019 IEEE International Geoscience and Remote Sensing Symposium, pp. 2997–3000. IEEE (2019)
4. Fang, L., Li, S., Duan, W., et al.: Classification of hyperspectral images by exploiting spectral-spatial information of superpixel via multiple Kernels. IEEE Trans. Geosci. Remote Sens. **53**(12), 6663–6674 (2015)
5. Zhu, X.X., Tuia, D., Mou, L., et al.: Deep learning in remote sensing: a comprehensive re-view and list of resources. IEEE Geosci. Remote Sens. Mag. 5(4), 8–36 (2017)
6. Zhang, M., Gong, M., Mao, Y., et al.: Unsupervised feature extraction in hyperspectral images based on Wasserstein generative adversarial network. IEEE Trans. Geosci. Remote Sens. **57**(5), 2669–2688 (2018)

7. Tsoumakas, G., Katakis, I., Vlahavas, I.: Mining multi-label data. In: Data Mining and Knowledge Discovery Handbook, pp. 667–685. Springer, Boston, MA (2009). https://doi.org/10.1007/978-0-387-09823-4_34
8. Read, J., Pfahringer, B., Holmes, G., et al.: Classifier chains for multi-label classification. In: Joint European Conference on Machine Learning and Knowledge Discovery in Databases, pp. 254–269. Springer, Berlin, Heidelberg (2009). https://doi.org/10.1007/s10994-011-5256-5
9. Boutell, M.R., Luo, J., Shen, X., et al.: Learning multi-label scene classification. Patt. Recogn. **37**(9), 1757–1771 (2004)
10. Zhang, M.L., Zhou, Z.H.: ML-KNN: a lazy learning approach to multi-label learning. Patt. Recogn. **40**(7), 2038–2048 (2007)
11. Zhang, M.L., Wu, L.: Lift: multi-label learning with label-specific features. IEEE Trans. Pattern Anal. Mach. Intell. **37**(1), 107–120 (2014)
12. Zhang, J.J., Fang, M., Li, X.: Multi-label learning with discriminative features for each label. Neurocomputing **154**, 305–316 (2015)
13. He, Z.F., Yang, M.: Sparse and low-rank representation for multi-label classification. Appl. Intell. **49**(5), 1708–1723 (2019)
14. Jain, A.K., Murty, M.N., Flynn, P.J.: Data clustering: a review. ACM Comput. Surveys (CSUR) **31**(3), 264–323 (1999)
15. Guo, Y., Chung, F., Li, G., et al.: Leveraging label-specific discriminant mapping features for multi-label learning. ACM Trans. Knowl. Disc. Data (TKDD) **13**(2), 1–23 (2019)
16. Roweis, S.T., Saul, L.K.: Nonlinear dimensionality reduction by locally linear embed-ding. Science **290**(5500), 2323–2326 (2000)
17. Guan, Y., Li, W., Zhang, B., et al.: Multi-label classification by formulating label-specific features from simultaneous instance level and feature level. Appl. Intell. **51**(6), 3375–3390 (2021)
18. Shi, J., Malik, J.: Normalized cuts and image segmentation. IEEE Trans. Patt. Anal. Mach. Intell. **22**(8), 888–905 (2000)
19. Zhu, F., Wang, Y., Fan, B., et al.: Effective spectral unmixing via robust representation and learning-based sparsity. arXiv preprint arXiv:1409.0685 (2014)
20. Zhu, F., Wang, Y., Fan, B., et al.: Spectral unmixing via data-guided sparsity. IEEE Trans. Image Process. **23**(12), 5412–5427 (2014)
21. Zhu, F., Wang, Y., Xiang, S., et al.: Structured sparse method for hyperspectral unmixing. ISPRS J. Photogramm. Remote. Sens. **88**, 101–118 (2014)
22. Huang, J., Li, G., Huang, Q., et al.: Learning label specific features for multi-label classification. In: 2015 IEEE International Conference on Data Mining. IEEE, pp. 181–190 (2015)
23. Chang, C.C., Lin, C.J.: LIBSVM: a library for support vector machines. ACM Trans. Intell. Syst. Technol. (TIST) **2**(3), 1–27 (2011)

A Novel Multi-scale Key-Point Detector Using Residual Dense Block and Coordinate Attention

Li-Dan Kuang⬭, Jia-Jun Tao, Jianming Zhang(✉)⬭, Feng Li, and Xi Chen

School of Computer and Communication Engineering,
Changsha University of Science and Technology, Changsha 410114, China
jmzhang@csust.edu.cn

Abstract. Object detection, one of the core missions in computer vision, plays a significant role in various real-life scenarios. To address the limitations of pre-defined anchor boxes in object detection, a novel multi-scale key-point detector is proposed to achieve rapid detection of natural scenes with high accuracy. Compared with the method based on key-point detection, our proposed method has fewer detection points which are the sum of pixels on four-layer compared to one-layer. Furthermore, we use feature pyramids to avoid ambiguous samples. Besides, in order to generate feature maps with high quality, a novel residual dense block with coordinate attention is proposed. In addition to reducing gradient explosion and gradient disappearance, it can reduce the number of parameters by 5.3 times compared to the original feature pyramid network. Moreover, a non-key-point suppression branch is proposed to restrain the score of bounding boxes far away from the center of the target. We conduct numerous experiments to comprehensively verify the real-time, effectiveness, and robustness of our proposed algorithm. The proposed method with ResNet-18 and resolution of 384×384 achieves 77.3% mean average precision at a speed of 87 FPS on the VOC2007 test, better than CenterNet under the same settings.

Keywords: Object detection · Anchor-free · Key-point detector · Residual dense block · Coordinate attention

1 Introduction

Object detection is an important, complicated, and challenging mission, which is widely used in facial analysis [1], autonomous driving cars [2], medical, and other scenes. The main tasks are locating where it is and identifying what it is. The traditional method first selects the region of interest, and then uses sliding windows with multi-scale or multi-aspect ratios to scan the entire image. It involves complicated calculations and redundant windows. Thanks to the advantages of convo-lutional neural networks, deep learning technology has become a powerful method of object detection.

© Springer Nature Switzerland AG 2021
T. Mantoro et al. (Eds.): ICONIP 2021, LNCS 13110, pp. 235–246, 2021.
https://doi.org/10.1007/978-3-030-92238-2_20

In the past few years, mainstream detectors use a backbone network to extract features for an input image and then use the feature pyramid network (FPN) [3] to fuse the different feature layers. Finally, the detection head is divided into several branches for prediction according to different mission requirements. The most representative method above is RetinaNet [4]. RetinaNet uses a set of pre-defined possible bounding boxes and undoubtedly achieves good performance, but it still has some limitations. Cleverly, the pre-defined boxes have been sidestepped with the emergence of anchor-free methods which are more robust to bounding box distributions than anchor-based methods. The anchor-free methods could be divided into two categories: key-point prediction and pixel-wise prediction. Key-point detectors, such as CornerNet [5], ExtremeNet [6], and CenterNet [7] predict the regression of several key points within the bounding box. Pixel-wise detectors, such as Densebox [8], FCOS [9], FSAF [10], SAPD [11], and FoveaBox [12] predict per-pixel on the feature pyramid maps. FoveaBox proposed a fovea area which can predict the points in the center area of the objects. This has an enlightening effect on reducing the number of points in the pixel-wise prediction. The anchor-free detectors mentioned above are simpler and faster. In addition, since anchor boxes are no longer used, anchor-free models have higher requirements for feature ability, especially in lightweight models.

The high precision of key-point detectors relied on a single feature map with a stride of 2 or 4 rather than feature pyramid maps, thus suffering from high memory consumption, long training, and testing time [11]. Pixel-wise detectors use multi-level prediction with FPN to avoid intractable ambiguity. Can we combine the advantages of the two methods? We show that the answer is affirmative. A simple and lightweight anchor-free detector that uses key-point detection, multi-level prediction, and residual dense block with coordinate attention module are proposed in the paper to balance both precision and computing pressure. The main contributions of this paper are summarized as follows:

1. Instead of the original FPN, we propose a novel feature fusion module that uses residual dense block with coordinate attention (shorted as RDBCA). It can reduce the number of parameters in the network, and further increase the performance.
2. We add multi-scale prediction in key-point detector to address ambiguous samples problem. Our proposed method has fewer detection points which are the sum of pixels on four feature maps with strides of 8, 16, 32, and 64 compared to one feature map with a stride of 4.
3. We propose non-key-point suppression (shorted as NKS) to reduce the impact of non-key points on detection so that non-key points can be eliminated by confidence threshold and non-maximum suppression (NMS) [13] process.

The rest of this paper is arranged as follows. Section 2 describes the structure of the entire network and the composition of each module. In Sect. 3, we conduct the experimental results to verify the proposed method. Finally, the conclusion of the paper is given in Sect. 4.

2   Proposed Method

In this section, we first describe the entire network of proposed model. Then, we detail how to generate feature fusion module, and how NKS works. Finally, we introduce the loss function and process of inference.

2.1   Network Architectures

Backbone Network. As shown in Fig. 1, the backbone network of this paper uses ResNet-18 [14] as the baseline. The backbone network was originally designed for image classification. The final fully-connected layer of the backbone network outputs a fixed-length one-dimensional vector and discards the spatial information. We modify the last fully-connected layer and classification layer of the backbone network to construct a fully convolutional network. After each down-sampling of the backbone network, the size becomes one-half of the original input size. Finally, a three-layer image pyramid containing C_3, C_4, and C_5 with different sizes and channels from the backbone network is obtained, and the strides of each layer are 8, 16, and 32, respectively. We do not include C_1 and C_2 into the pyramid due to its large memory footprint.

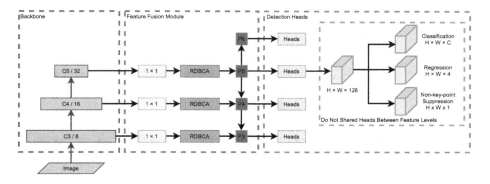

Fig. 1. The framework of the proposed multi-scale key-point detector using RDBCA. We use an input image with resolution of 384 × 384 as an example. 8, 16, and 32 are the down-sampling ratios for each feature map layer.

Detection Heads. We only use one 3 × 3 depth-wise separable convolution with the number of channels 128 for each final layer. The parameters of this convolution are shared by three different branches, but we do not share heads among different feature layers.

Classification Branch. The classification branch predicts the probability of objects at each positive point location for each object class. Following [4], we train C binary classifiers instead of training a multi-class classifier, where C is the number of categories of natural scenes, which is 20 for the Pascal VOC dataset [15].

Regression Branch. The regression branch predicts a 4D vector of point-to-boundary distance. Through the exponential function (denoted as $\exp(\cdot)$), the real number is mapped to $(0, \infty)$ on the regression branch [9].

Proposed Non-Key-Point Suppression Branch. As the densely clustered targets may cause multiple fake key points that exceed the threshold in the heatmap, a target with multiple boxes matches and cannot be eliminated by NMS (as shown in Fig. 4A(1)), we propose a simple but effective NKS branch to suppress detected bounding boxes which is far away from the key-point. The score of non-key-point suppression branch ranges from 0 to 1 and is trained with binary cross entropy loss. The final classification score with NKS branch is defined as:

$$p_{cls\_nks} = (p_{cls})^{(2-p_{nks})}, \tag{1}$$

where p_{cls} is classification score, p_{nks} is NKS score.

2.2   The Feature Fusion Module with Proposed RDBCA

Now we present details about our proposed feature fusion module as displayed in Fig. 2. Our feature fusion module contains top-down pathway, lateral connections and proposed RDBCA.

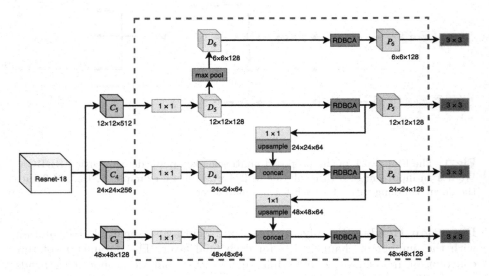

Fig. 2. The structure of feature fusion module with proposed RDBCA.

Top-Down Pathway and Lateral Connections. A three-layer image pyramid is constructed from the backbone network, which is used to extract features at different layers. Firstly, as shown in Fig. 2, D_3, D_4, and D_5 are generated by the output of backbone network C_3, C_4, and C_5 followed by a 1×1 convolutional layer to adjust the number of channels to 128, 64, and 64. Secondly, P_5 generated from D_5 is sent into RDBCA to integrate and utilize local feature information. Thirdly, we use a 1×1 convolution to reduce the number of channels one-half of the original, and use bilinear interpolation to up-sample the feature map to the same size as the previous layer. The up-sampled map merges with the corresponding bottom-up map by concatenating operation. Finally, the way of producing P_3 and P_4 is similar to the second step mentioned above. P_6 is directly generated by a max-pooling layer with the stride being 2 from D_5, and followed by RDBCA. As a result, the feature layers P_3, P_4, P_5, and P_6 have strides 8, 16, 32, and 64, respectively. We use concatenation instead of element-wise addition. The element-wise addition focuses on reusing features, and the concatenation operation benefits from the discovery of new features.

Proposed RDBCA. As shown in Fig. 3, we first use three 3×3 depth-wise separable convolutions and preserve the information for the purpose of accessing to all the subsequent layers and passing on information [16]. Each data after 3×3 depth-wise separable convolution will be skip-connected with the original input to minimize gradient explosion and gradient disappearance. Next, we concatenate the original input and all outputs from three convolutions, and the channel of the output feature is 4 times the original input. Then, we use a 1×1 convolutional layer to reduce the number of channels matching the original input for local feature fusion. Finally, we use coordinate attention [17] and add original input features and output from coordinate attention together for another local feature fusion. It should be noted that the original residual dense block uses concatenate operation similarly to the dense block to combine former residual dense block, which leads to extremely large computation. Therefore, we use element-wise addition instead of concatenating operation except the last one. The residual dense block with coordinate attention is defined as follows:

$$F_{1_1} = \delta(\varphi_{3\times3}(F_0) + F_0), F_{1_2} = \delta(\varphi_{3\times3}(F_{1_1}) + F_0), F_1 = \delta(\varphi_{3\times3}(F_{1_2}) + F_0), \quad (2)$$

$$F_2 = \delta(\varphi_{1\times1}(concat([F_0, F_{1_1}, F_{1_2}, F_1])) + F_0), \quad (3)$$

$$F_3 = \varphi_{1\times1}(concat([HAvgPool(F_2), WAvgPool(F_2)])), \quad (4)$$

$$F_{3_1}, F_{3_2} = split(F_3), F_4 = F_2 \times \sigma(\varphi_{3\times1}(F_{3_1})) \times \sigma(\varphi_{3\times1}(F_{3_2})), \quad (5)$$

$$F_{out} = \delta(F_0 + F_4). \quad (6)$$

Here F_0 is the original input, F_{out} is output feature map, $\varphi_{1\times1}$ is a 1×1 convolutional operation, $\varphi_{3\times3}$ is a 3×3 depth-wise separable convolution, δ is a non-linear activation function which is ReLU function [18], "concat" is concatenate operation, $HAvgPool(\cdot)$ and $WAvgPool(\cdot)$ denotes 2D adaptive average pooling at height and width respectively, and $split(\cdot)$ can split the tensor into chunks which are the views of the original tensor.

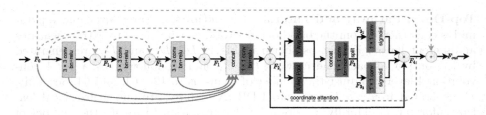

Fig. 3. The structure of proposed RDBCA.

2.3 Scale Assignment

FPN is a divide-and-conquer method to address the problem of ambiguous samples caused by overlapping objects of different sizes. Different scales are equivalent to different focal lengths, thus focusing on objects of different sizes.

Let W and H be width and height of input image, and P_l be the feature maps at layer l, where $l \in 3, 4, 5, 6$. The size of P_l is $(W/2^l) \times (H/2^l) \times 128$. In Sect. 2.2, we have introduced that we use four feature maps with different scales $\{P_3, P_4, P_5, P_6\}$. In order to make targets with different scales fall into different layers, we set a value corresponding to the maximum distance for each layer. For layer P_l, the maximum distance m_l is computed by,

$$m_l = A \times 2^{l-1}, l = 3, 4, 5, 6. \tag{7}$$

Since P_2 does not used in our model, we set m_2 to 0, and A to 12. In this paper, m_3, m_4, m_4, and m_6 are set as 48, 96, 192, and 384, respectively. If the shorter side of the bounding box satisfies $m_{l-1} < min(x_{max} - x_{min}, y_{max} - y_{min}) \leq m_l$, it is considered the target falls into P_l (e.g., the minimum of the target is 64 for P_4). Unless specified, the resolution used 384×384 in this paper.

2.4 Label Assignment

The ground truth can be defined as $B^k = [x_{min}^k, y_{min}^k, x_{max}^k, y_{max}^k]$. k is the number of the bounding boxes. (x_{min}^k, y_{min}^k) and (x_{max}^k, y_{max}^k) denote the coordinates of top-left and bottom-right corners. c^k is the category of the object.

For any object which satisfies the conditions mentioned in Sect. 2.3 falls into the layer l, we calculate the center point of the bounding box k which is the only one positive sample marked as $(x_{l,pos}^k, y_{l,pos}^k)$ for this ground truth. Here $x_{l,pos}^k = \lfloor (x_{min}^k + x_{max}^k)/(2 \times 2^l) \rfloor$ and $y_{l,pos}^k = \lfloor (y_{min}^k + y_{max}^k)/(2 \times 2^l) \rfloor$, and other non-central points in the bounding box become negative samples naturally marked as $(x_{l,neg}^k, y_{l,neg}^k)$.

For each positive sample, the distances from the center point to the four edges are recorded by a 4-dimension vector $\mathbf{V}^k = (d_l^k, d_t^k, d_r^k, d_b^k)$, and uses as the regression targets for the location. Here d_l^k, d_t^k, d_r^k, and d_b^k are the original image distances of the bounding box k. The training regression targets can be expressed as,

$$\begin{cases} d_l^k = \frac{x_{min}^k + x_{max}^k}{2} - x_{min}^k, d_t^k = \frac{y_{min}^k + y_{max}^k}{2} - y_{min}^k, \\ d_r^k = x_{max}^k - \frac{x_{min}^k + x_{max}^k}{2}, d_b^k = y_{max}^k - \frac{y_{min}^k + y_{max}^k}{2}. \end{cases} \tag{8}$$

We put all positive samples on a heatmap, and let category $Y_{l,x,y,c}^{gt} = 1$ and non-key-point prediction $Q_{l,x,y}^{gt} = 1$. For negative samples, we use a Gaussian kernel as background and set non-key-point prediction $Q_{l,x,y}^{gt} = 0$. The Gaussian kernel is computed by,

$$Y_{l,x,y,c}^{gt} = \exp\left(-\frac{(x_{l,pos} - x_{l,neg})^2 + (y_{l,pos} - y_{l,neg})^2}{2\sigma^2}\right), \tag{9}$$

here σ is an adaptive standard deviation based on the bounding box [5,7]. If two Gaussian values of the same class conflict in the same feature layer, the element-wise maximum will be taken.

2.5 Loss Function

Referring [7], the loss function for classification defined as,

$$L_{cls} = \frac{-1}{N}\sum_{l,x,y,c}\begin{cases} \left(1 - Y_{l,x,y,c}^{pred}\right)^\alpha \log\left(Y_{l,x,y,c}^{pred}\right), & \text{if } Y_{l,x,y,c}^{gt} = 1, \\ \left(1 - Y_{l,x,y,c}^{gt}\right)^\beta \left(Y_{l,x,y,c}^{pred}\right)^\alpha \log\left(1 - Y_{l,x,y,c}^{pred}\right), & \text{otherwise,} \end{cases} \tag{10}$$

where x and y are pixels of feature maps, c is the number of classes, α and β are hyper-parameters of the focal loss, and N is the number of positive points in image. We use $\alpha = 2$ and $\beta = 4$ in all our experiments [7].

We use IoU loss [19] as the regression loss function. We only regress distances of bounding box for positive samples. The regression loss function defined as,

$$L_{iou} = \frac{1}{N}\sum_{l,x,y}\begin{cases} 1 - \frac{\text{Intersection}(\mathbf{v}^{pred}, \mathbf{v}^{gt})}{\text{Union}(\mathbf{v}^{pred}, \mathbf{v}^{gt})}, & \text{if } Y_{l,x,y}^{gt} = 1, \\ 0, & \text{otherwise,} \end{cases} \tag{11}$$

where $\mathbf{v}^{pred}$ and $\mathbf{v}^{gt}$ are point-to-boundary distances for positive samples and ground truth boxes respectively. N is the number of positive points in image.

We use binary cross entropy loss as the NKS loss function. The NKS loss function defined as,

$$L_{nks} = -\frac{1}{M}\sum_{l,x,y}\left[Q_{l,x,y}^{gt}\log\left(Q_{l,x,y}^{pred}\right) + \left(1 - Q_{l,x,y}^{gt}\right)\log\left(1 - Q_{l,x,y}^{pred}\right)\right], \tag{12}$$

where M is the total number of samples of all feature layers. Thus, the loss for whole network L is the summation of L_{cls}, L_{iou}, and L_{nks}.

$$L = L_{cls} + L_{iou} + L_{nks} \tag{13}$$

2.6   Inference

During inference, given an input image, we first use NKS branch to suppress the score from non-key-point with Eq. 1. Then, we select the top 100 scoring predictions from all prediction layers. Next, we use a confidence threshold of 0.05 to filter out predictions with low confidence. Finally, NMS with a threshold of 0.5 is applied for each class. The predicted bounding boxes are obtained by inverting Eq. 8.

3   Experiments and Results

Pascal VOC is a popular object detection dataset [15]. In this section, we train our model on the Pascal VOC 2007 and Pascal VOC 2012 which contains 16551 training images, and evaluate our performance on Pascal VOC 2007 test set which contains 4952 testing images.

3.1   Training Details

In this paper, we build our experimental environments under Windows Server 2019 system, Intel(R) Xeon(R) Gold 6226R CPU, one NVIDIA GeForce 2060s GPU, PyTorch 1.8.1, CUDA 10.2.89, and CUDNN 7.6.5. Our detection model is trained with Adam optimizer. We use one GPU with 24 images per batch. All models are trained for 150K iterations. We use 10^{-5} to warm up at the first 500 iterations [14], then go back to 10^{-3} and continue training for 15K iterations. Then we decay the learning rate with a cosine annealing [20]. We initialize our backbone network with the weights pre-trained on ImageNet. Data augmentation includes resizing the input image with unchanged ratios using padding, random expanding, random mirroring, and random sample cropping.

We experiment with our modified ResNet-18, ResNet-50, and MobileNet v2 [21]. We train an input resolution of 384×384 and 512×512 with all backbone mentioned above. Furtherly, for resolution of 512×512, we add one more feature layer P_7 which is directly generated by a max-pooling layer with the stride being 2 from D_6, and followed by RDBCA. The scales m_3 to m_7 are set as 48, 96, 192, 384, and 768 respectively.

We use widely-used mean average precision (mAP) and Frames Per Second (FPS) to evaluate the proposed method. We measure the speed with batch size to 1. We set the confidence threshold to 0.05, NMS with a threshold of 0.5, and report the mAP at IoU thresholds 0.5 (shorted as mAP@0.5). Code for evaluating is downloaded from [22]. FPS is tested on GeForce 2060s GPU for each model.

3.2   Without NKS vs. with NKS

We first compare the proposed method without and with NKS as shown in Fig. 4. We set confidence threshold to 0.3 and NMS threshold to 0.5. When without NKS, although there is only one horse in Fig. 4A(1), multiple areas which denote high scores are predicted as key points, resulting in repeated and conflicting bounding boxes that cannot be eliminated by NMS (see Fig. 4A(3)). There are also four fake key points of person are detected in Fig. 4A(4). While our proposed NKS branch can simply address these issues. As shown in Fig. 4B(1), all key points fall into the center of the targets. The point on the edge of the target is suppressed so that the score cannot reach the confidence threshold. Moreover, thanks to the feature pyramid used in our method, the horse is assigned to larger feature maps, and persons are assigned to smaller ones, so we can easily use Eq. 1 to suppress the fake key, and will not affect other key points in the same location (see Figs. 4B(3)–(4)).

Figure 5 further verifies this conclusion. After NKS, points shifted to the left and distributed more evenly after NKS. Points below the red line with a high classification score but low IoU are suppressed, and could be filtered out by confidence threshold or NMS. Consequently, NKS could down-weight the classification score of points which far away from the key points.

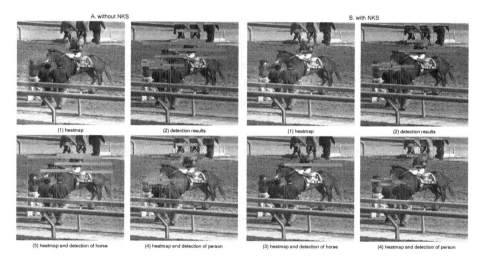

Fig. 4. Qualitative comparison results of without NKS (A) and with NKS (B). The heatmaps, detection results, heatmap and detection results of horse and heatmap and detection results of person are respectively displayed in (1)–(4).

3.3 Proposed RDB vs. Original FPN

Table 1 shows our proposed method greatly reduces the number of parameters by 5.3 times compared to the original FPN with 256 output channels. For the sake of fairness and consistency, we also compare with original FPN with 128 output channels which is 1.9 times the number of parameters of our method. It can be concluded from Table 1 the proposed method has fewer parameters and higher mAP than the original FPN.

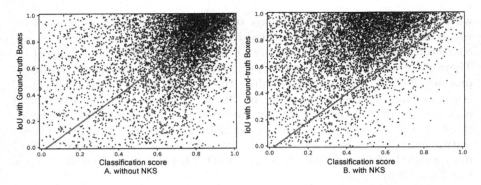

Fig. 5. Classification score of proposed method without NKS (A) and with NKS (B). Each point represents a box predicted from detector. Points below the red line have high classification score but low IoU, which make detector results far away from center. (Color figure online)

Table 1. Ablation study for different feature fusion modules. "RDB" is proposed feature fusion module residual dense block. "CA" is coordinate attention. "Channels" is the number of output channels. "Params(M)" is the parameters of the feature fusion module in millions.

Methods	RDB	CA	Channels	Params(M)	mAP@0.5
Original FPN			128	1.148	74.7
		√	128	1.161	75.1
			256	3.180	75.8
		√	256	3.207	76.3
Proposed	√		128	0.594	76.7
	√	√	128	0.607	77.3

3.4 Comparison Results on VOC2007

We compare our proposed detector with other SOTA detectors in VOC2007 test in Table 2. Our proposed method with ResNet-18 achieves the highest mAP and FPS values, compared with the same backbone of CenterNet. We deliberate that the proposed method has fewer detecting points than CenterNet which uses one feature map with the stride of 4. Besides, our proposed detector with ResNet-50 achieves higher mAP and FPS values than CenterNet with Resnet-101 and FCOS with ResNet-50. Compared with FCOS with ResNet-101 at a resolution of 512 × 512, we use ResNet-50 to achieve similar accuracy and to show 4.1 times faster FPS. We also tested on lightweight backbone such as MobileNetV2. The proposed method can also be combined with MobileNetV2 to achieve mAP at 76.2% and 79.0% at two different resolutions, which are higher than FCOS with the same backbone.

Table 2. Performance on different backbones. "mAP@0.5" with superscript * is copied from the original publications. CenterNet does not use NMS when inference.

Methods	Resolution	mAP@0.5	FPS
FCOS-ResNet50 [9]	384×384	74.6	27
	512×512	76.6	25
FCOS-ResNet101 [9]	384×384	81.1	17
	512×512	82.4	14
FCOS-MobileNetV2 [9]	384×384	68.9	32
	512×512	72.3	30
CenterNet-ResNet18 [7]	384×384	72.6*	63
	512×512	75.7*	55
CenterNet-ResNet101 [7]	384×384	77.6*	26
	512×512	78.7*	20
Proposed-ResNet18	384×384	77.3	87
	512×512	79.3	77
Proposed-ResNet50	384×384	80.5	65
	512×512	82.2	60
Proposed-MobileNetV2	384×384	76.2	70
	512×512	79.0	63

4   Conclusion

In this work, we propose a novel multi-scale key-point detector using RDBCA. We add feature pyramid maps to reduce ambiguous samples, and reduce the number of detection points compare to one feature map. The proposed RDBCA in feature fusion module not only can extract convolutional features from previous features, but also can fuse features for residual learning significantly. Furthermore, we proposed NKS branch to reduce score of classification where is far away from real key point. Experiment results demonstrate that our proposed method can easily combine with different backbone networks, and show the advantage of our proposed method. More specifically, proposed method with Resnet-18 and resolution of 384×384 achieves 77.3% mAP at a speed of 87 FPS on VOC2007 test, better than CenterNet under the same settings. Code is available at https://github.com/Tao-JiaJun/MSKPD-RDBCA.

Acknowledgments. This work was supported by National Natural Science Foundation of China under Grants 61901061, 61972056, Natural Science Foundation of Hunan Province of China under Grant 2020JJ5603, the Scientific Research Fund of Hunan Provincial Education Department under Grant 19C0031, 19C0028, the Young Teachers' Growth Plan of Changsha University of Science and Technology under Grant 2019QJCZ011.

References

1. Zheng, Y., Pal, D.K., Savvides, M.: Ring loss: convex feature normalization for face recognition. In: 2018 CVPR, pp. 5089–5097. IEEE, Salt Lake City (2018)
2. Wang, D., Devin, C., Cai, Q., et al.: Deep object-centric policies for autonomous driving. In: ICRA, pp. 8853–8859. IEEE, Montreal (2019)
3. Lin, T.Y., Dollár, P., Girshick, R., Hariharan, B., Belongie, S., et al.: Feature pyramid networks for object detection. In: CVPR, pp. 2117–2125. IEEE (2017)
4. Lin, T.Y., Goyal, P., Girshick, R., et al.: Focal loss for dense object detection. In: CVPR, pp. 2980–2988. IEEE (2017)
5. Law, H., Deng, J.: CornerNet: detecting objects as paired keypoints. Int. J. Comput. Vis. **128**(3), 642–656 (2019). https://doi.org/10.1007/s11263-019-01204-1
6. Xingyi, Z., Jiacheng, Z., Krahenbuhl, P.: Bottom-up object detection by grouping extreme and center points. In: CVPR, pp. 850–859. IEEE (2019)
7. Zhou, X., Wang, D., Philipp K.: Objects as points. arXiv:1904.07850 (2019)
8. Huang, L., Yang, Y., Deng, Y., et al.: DenseBox: unifying landmark localization with end to end object detection. arXiv:1509.04874 (2015)
9. Tian, Z., Shen, C., Chen, H., et al.: FCOS: fully convolutional one-stage object detection. In: ICCV, pp. 9627–9636. IEEE (2019)
10. Zhu, C., He, Y., Savvides, M.: Feature selective anchor-free module for single-shot object detection. In: CVPR, pp. 840–849. IEEE (2019)
11. Zhu, C., Chen, F., Shen, Z., et al.: Soft anchor-point object detection. In: ECCV, vol. 12354. Springer, Cham (2020). https://doi.org/10.1007/978-3-030-58545-7_6
12. Kong, T., Sun, F., Liu, H., et al.: FoveaBox: beyound anchor-based object detection. IEEE Trans. Image Process. **29**, 7389–7398. IEEE (2020)
13. Bodla, N., Singh, B., Chellappa, R., et al.: Soft-NMS-improving object detection with one line of code. In: ICCV, pp. 5561–5569. IEEE (2017)
14. He, K., Zhang, X., Ren, S., et al.: Deep residual learning for image recognition. In: CVPR, pp. 770–778. IEEE (2016)
15. Everingham, M., Van Gool, L., Williams, C.K., et al.: The pascal visual object classes (VOC) challenge. Int. J. Comput. Vis. **88**(2), 303–338 (2010)
16. Zhang, Y., Tian, Y., Kong, Y., et al.: Residual dense network for image super-resolution. In: CVPR, pp. 2472–2481. IEEE (2018)
17. Hou, Q., Zhou, D., Feng, J.: Coordinate attention for efficient mobile network design. In: CVPR. IEEE (2021)
18. Glorot, X., Bordes, A., Bengio, Y.: Deep sparse rectifier networks. In: Proceedings of the AISTATS, pp. 315–323 (2011)
19. Yu, J., Jiang, Y., Wang, Z., et al.: UnitBox: an advanced object detection network. In: Proceedings of the 24th ACM International Conference on Multimedia, pp. 516–520. Elsevier (2016)
20. Ilya, L., Frank, H.: SGDR: Stochastic gradient descent with warm restarts. In: 5th International Conference on Learning Representations. Elsevier (2017)
21. Sandler, M., Howard, A., Zhu, M., et al.: MobileNetV2: inverted residuals and linear bottlenecks. In: CVPR, pp. 4510–4520. IEEE (2018)
22. https://github.com/Cartucho/mAP , 29 May 2020

Alleviating Catastrophic Interference in Online Learning via Varying Scale of Backward Queried Data

Gio Huh[(✉)]

Shanghai American School, Shanghai, China
gio02px2022@saschina.org

Abstract. In recent years, connectionist networks have become a staple in real world systems due to their ability to generalize and find intricate relationships and patterns in data. One inherent limitation to connectionist networks, however, is catastrophic interference, an inclination to lose retention of previously formed knowledge when training with new data. This hindrance has been especially evident in online machine learning, where data is fed sequentially into the connectionist network. Previous methods, such as rehearsal and pseudo-rehearsal systems, have attempted to alleviate catastrophic interference by introducing past data or replicated data into the data stream. While these methods have proven to be effective, they add additional complexity to the model and require the saving of previous data.

In this paper, we propose a comprehensive array of low-cost online approaches to alleviating catastrophic interference by incorporating three different scales of backward queried data into the online optimization algorithm; more specifically, we averaged the gradient signal of the optimization algorithm with that of the backward queried data of the classes in the data set. Through testing our method with online stochastic gradient descent as the benchmark, we see improvements in the performance of the neural net-work, achieving an accuracy approximately seven percent higher with significantly less variance.

Keywords: Catastrophic interference · Online learning · Backward query · Stochastic gradient descent

1 Introduction

A connectionist network is able to effectively generalize to data due to its ability to optimize itself through epochs of training cases. Weights within the network are tuned by an optimization algorithm, and the information learned from these training cases is distributed throughout the weights within the network.

In training connectionist networks, the size and quality of datasets are generally considered vital factors. Under ideal circumstances, datasets are elaborate and large enough to improve the generalization of the neural network. Realistically, however, obtaining such datasets may not be practical or feasible, especially

© Springer Nature Switzerland AG 2021
T. Mantoro et al. (Eds.): ICONIP 2021, LNCS 13110, pp. 247–256, 2021.
https://doi.org/10.1007/978-3-030-92238-2_21

when the dataset may be given as a function of time or is too large to run at once.

Online machine learning is employed to overcome these impediments by sequentially utilizing the dataset. If a dataset is limited by time, the neural network is initially optimized with the original dataset and incrementally trained as more data become available. If a dataset is too large, a stream-based online learning algorithm is used [1,2]. In these cases, stochastic gradient descent (SGD) is normally chosen as the online machine learning algorithm because of its simplicity, increased update frequency, and low cost-perliteration [3,4]. This method, however, is susceptible to noisy gradient signals and variance, which diminish its convergence speed. Additionally, SGD gives priority to the most recent training cases, which reduces the value that previous training cases provide, and hence becomes prone to catastrophic forgetting [5]. Due to these factors, SGD is not an ideal learning algorithm; there is a need for a more streamlined approach to online/stream-based machine learning.

Prior papers have attempted to overcome the aforementioned limitations of SGD by improving SGD with rehearsal/pseudo-rehearsal methods [6–11] and regularization [12–14]. For example, one approach is locally weighted regression (LWR), a modelling technique centered around producing local models from subsets of the dataset and combining them into a single global model in order to reduce catastrophic forgetting. Another approach is rehearsal [6], which merges old data with newly received data for online learning. A variant of this is pseudo-rehearsal [7–11], which combines randomly generated data with the online machine learning process. The advantage of this approach is that the neural network can avoid saving old data. All of these approaches do show improvements in online machine learning; however, these approaches, by no means, are perfect, each having its limitations.

Recently, backward queried data have been shown to possess the potential of alleviating catastrophic interference [20]. The extent of the research, however, has been introductory and requires further investigation and development.

In this paper, we suggest three low-cost variants of online SGD that overcome the catastrophic inference found in online machine learning when a subset of the dataset is made accessible initially, and the remaining data are made available through a data stream. We propose an improvement to SGD that incorporates different scales of backward queried samples during the training of the remaining data. Each backward queried data can be seen as a "representative" training case of its respective classes. The proposed variants include online SGD incorporated with the backward queried data of the class, online SGD incorporated with the backward queried data of all classes, and online SGD incorporated with backward queried data of all classes excluding that of the training case. Through these three variants, we aim to reduce catastrophic interference of online SGD while improving its ability to converge to minima. We benchmarked our learning algorithm using the Fashion- MNIST and Kuzushiji-MNIST datasets and then evaluated our results with those of the online SGD, using 10,000 unknown test cases.

2   Related Works

The two foremost methods we would like to mention are rehearsal-based: rehearsal and pseudo-rehearsal. Rehearsal is a method which saves old data for the online training of the neural network [6], and pseudo-rehearsal is a method which pairs the data stream with replicated training cases generated using the current neural network [7–10]. The aim of both of these methods is to improve the neural network's ability to retain previous data and its faculty to learn different concepts through a data stream. Recently, there has been success using rehearsal [15] and pseudo-rehearsal based methods for vision tasks [16–19]. Specific variants of rehearsal-based method [11] include Locally Weighted Regression Pseudo-Rehearsal (LW-PR2), a version of LWR, which attempts to mitigate catastrophic interference and improve SGD by retaining information through local models and combining them into a sin-gle global model; and Deep Generative Replay, which uses generative adversarial networks (GAN) to replicate past data with desired outputs. Unfortunately, all of these approaches are susceptible to the common limitations of rehearsal and pseudo-rehearsal due to the fact that rehearsal methods increasingly have to store training data while pseudo-rehearsal is computationally complex, making both methods relatively difficult to implement into varying problems. Moreover, a unique drawback of LWR, which LW-PR2 employs, is that it is computationally taxing and requires thousands of local models to be accurate, which also poses a problem when the data set is relatively large as the task becomes time intensive.

Another way that SGD has been attempted to be improved is through regularization, which reduces the overfitting and catastrophic forgetting of the neural network through parameters [12–14].

All of the previous works mentioned above can be classified into two categories: improving SGD in a parametric way (regularization) or retaining information from the previous data set (rehearsal-based). While it is difficult to assess which method is superior in online training, each has its own advantages and drawbacks. Since training cases provide more information than parameters, in some sense, rehearsal-based methods may be more beneficial. Having said that, saving old data or generating representative data can be computationally costly. In this paper, we try to capitalize on the advantages of utilizing the previous data set during online training while reducing the computational cost of saving such data: we propose a simple but effective approach of using representative data by using backward queried image per class and feeding this image with newly received data during online learning.

3   Conceptualization

Our method comes in the form of combining regular SGD with backward queried data. In this section, we explain SGD, then backward query, and finally our proposed method in relation to the previous two concepts.

3.1 Stochastic Gradient Descent

SGD is preferred in online learning because of its characteristics of tuning the neural network for every single sample as in (1).

$$\theta_{t+1} = \theta_t - \eta \cdot \nabla_\theta \mathbb{J}\left(\theta; x^{(i)}; y^{(i)}\right) \tag{1}$$

where θ denotes weights, η denotes learning rate, and $x^{(i)}$ denotes each training sample of label $y^{(i)}$. In this formula, we observe two specific characteristics of SGD: its simplicity and updating frequency. The single training case per iteration makes SGD relatively simple compared to other gradient descents, such as mini-batch or batch gradient descent (GD), which averages out the gradients of different training cases within a batch or the whole dataset. Similarly, SGD possesses a relatively fast updating frequency since it optimizes the neural network for every training case instead of batches.

These characteristics make SGD distinct from other gradient descent. It is also the source behind SGD's drawbacks. Since SGD is updated for every training case, training cases could generate noise within the neural network, especially when the quality of training cases is poor. Moreover, as SGD frequently updates with new training cases, it is likely to forget the changes made by previous training cases (catastrophic interference).

3.2 Backward Query

In generating backward queried samples, the process of calculating values in a feed-forward neural network is reversed. Instead of initially computing from the input layer, labels that represent a specific class are inputted into the output layer. It is then computed with the inverse of the activation function and multiplied with the weights to obtain new values corresponding with the previous layer. This process repeats until the values have backwardly propagated to the input layer. The final values become the backward queried sample of the aforementioned class.

The significance behind backward query is that the samples, in essence, become the representative data of the class and provides insight into the trained neural network because the values provide an imprint of how the weights and activate function save and compute the input value internally. This is advantageous for online training since we can effectively capture the value of the entire dataset into a single case, which can be later used to retain previous information. In Fig. 1, the 10 images represent the backward queried sample of each class in the Fashion-MNIST dataset. The implementation of backward query can be seen in Algorithm 1.

Algorithm 1. Backward Query

1: **Input:**
      float O [10] = {...}
2: **Output:**
      int I [784] = {...}
3:   weights [4] = {$W_{256,784}$, $W_{128,256}$, $W_{100,128}$,$W_{10,100}$}
4:   $hidden$ =O
5: **for** $i = 4$ to 1 **do**
6:     $W = weights[i]$
7:     $hidden = W^T \cdot inversion\ of\ activation(hidden)$
8:     $hidden = normalize(hidden)$
9: **end for**
10: $I = hidden$

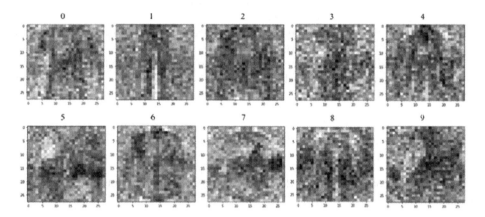

Fig. 1. Backward queried sample of Fashion-MNIST dataset

3.3   Proposed Concept

In our approach, three different scales of backward query data were incorporated with SGD. In the first variant, we paired every training case with their corresponding backward queried data from the initial dataset. We then averaged the gradient of the two samples before updating the network. The proposed approach can be modelled using the formula in (2).

$$\theta_{t+1} = \theta_t - \eta \cdot \frac{1}{2} \left(\nabla_\theta \mathbb{J} \left(\theta; x^{(i)}; y^{(i)} \right) + \nabla_\theta \mathbb{J} \left(\theta; b^{(y)}; y^{(i)} \right) \right) \tag{2}$$

where θ denotes weights, η denotes learning rate, $x^{(i)}$ denotes each training sample of label $y^{(i)}$, and $b^{(y)}$ denotes the backward queried data of label y. In this variant, we expect faster convergence speeds due to the smaller scale of backward queried data but at the cost of higher susceptibility to noise.

In the second variant, we averaged the gradient of the training case with the gradients of all of the backward queried of the dataset. The proposed approach is modelled using the formula in (3).

$$\theta_{t+1} = \theta_t - \eta \cdot \frac{1}{(\kappa+1)} \left(\nabla_\theta \mathbb{J}\left(\theta; x^{(i)}; y^{(i)}\right) + \sum_{j=1}^{\kappa} \nabla_\theta \mathbb{J}\left(\theta; b^{(j)}; y^{(j)}\right) \right) \quad (3)$$

where θ denotes weights, η denotes learning rate, $x^{(i)}$ denotes each training sample of label $y^{(i)}$, κ denotes the number of classes within the dataset, and $b^{(j)}$ denotes the backward queried data of label y. Compared to the first variant, we expect this variant to better alleviate catastrophic interference because more backward queried images allow for the neural network to optimize itself based on what it has learned in the past, retaining more of its previous knowledge and reducing the negative effects that new training cases can have on weights which are proponents of the neural network's understanding of the dataset; however, this may result in slower convergence.

The final variant averages the gradient of the training case with the gradients of the backward queried data excluding that of the training case. The final variant is modelled in (4).

$$\theta_{t+1} = \theta_t - \eta \cdot \frac{1}{\kappa} \left(\nabla_\theta \mathbb{J}\left(\theta; x^{(i)}; y^{(i)}\right) + \sum_{j=1,\ j\neq i}^{\kappa} \nabla_\theta \mathbb{J}\left(\theta; b^{(j)}; y^{(j)}\right) \right) \quad (4)$$

where θ denotes weights, η denotes learning rate, $x^{(i)}$ denotes each training sample of label $y^{(i)}$, κ denotes the number of classes within the dataset, and $b^{(j)}$ denotes the backward queried data of label y. In comparison to the first and second variants, we designed this variant to possess an equal representation of all classes within the dataset before each update; each class is represented by either a training case or backward queried data. We expect this variant to fluctuate significantly less than the first variant and converge slightly faster than the second variant due to its intermediate scale of backward queried data.

4   Experiment

In evaluating the performance of SGD and our method, we establish our experimental settings and benchmark our proposed variants with online SGD over an online data stream. We recorded the accuracy of SGD and the proposed methods with 60,000 training samples, which are divided into the initial training set and subsequent streaming data. To measure the performance, 10,000 unknown samples are used.

4.1   Experimental Settings

- Training & Test Data: We use two datasets, Fashion-MNIST and Kuzushiji-MNIST, both of which consist of 60,000 training cases and 10,000 test cases of 28 by 28 grey-scale pixel maps with ten classes.
- Neural Network Architecture: We implemented a multilayer perceptron (MLP) with Python and NumPy. The layers of the MLP are [784, 256, 128, 100, 10], where 784 and 10 are input and output layers, respectively. We use sigmoid as the activation function, 0.01 as the learning rate, and the Mean Squared Error (MSE) as the loss function.
- Initial Training Dataset and Data Stream: To emulate an online environment, we allocated the first 10,000 training cases to the initial dataset and the final 50,000 training cases to the subsequent data stream. To measure the performance, we used a separate 10,000 unknown test samples.

4.2   Experiment with SGD and the Proposed Method

As known, mini-batch is favorable when the entire dataset is given, so we use it as the optimization algorithm for the initial 10,000 samples, which will be optimized for 100 epochs. After the initial training, we use the remaining 50,000 samples as the stream data and update the model using SGD or the proposed method. The results of the performance of SGD and the variants can be seen in the following figures.

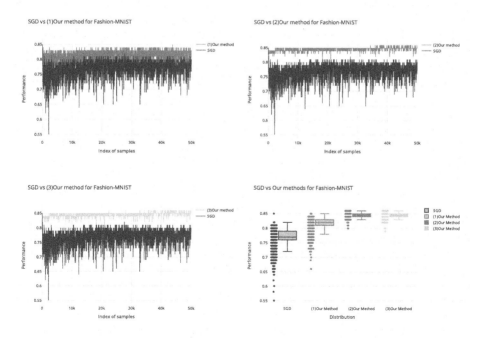

Fig. 2. Performance of SGD and three methods over Fashion-MNIST

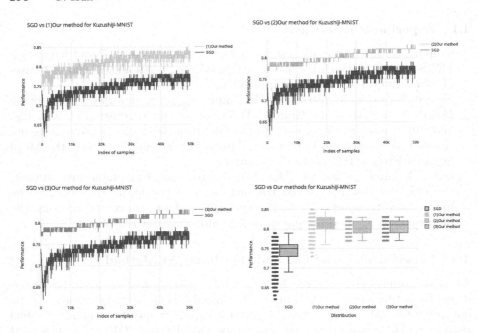

Fig. 3. Performance of SGD and three variants over Kuzushiji-MNIST

From assessing the performance rate of the perceptron over the data stream, our proposed variants were able to perform at a higher standard than SGD while maintaining a smaller variance. For the data stream of Fashion-MNIST seen in Fig. 2, when the performance of online SGD dropped significantly after the first 5,000 samples, all of our variants were able to retain more of the original performance of the neural network from the initial dataset. Interestingly, the second and the third variants performed more favorably than the first variant in all statistical categories, both achieving higher accuracies for all quartiles with significantly less fluctuation. For the data stream of Kuzushiji-MNIST in Fig. 3, the proposed variants similarly performed more favorably than SGD. Within the first few thousand samples, the perceptron optimized by SGD underwent a drop of more than ten percent and recovered by the end of the data stream; in contrast, our proposed variants suffered a smaller drop in performance and was able to gradually improve upon the original performance rate of the perceptron by the end of the data stream. One notable observation is that the first variant achieves the highest performance rate during the data stream despite having higher fluctuations in performance.

While there is no standard metric for the susceptibility of a neural network to catastrophic interference, the reduction of catastrophic interference can be seen through the retention and improvement of the performance rate with using the proposed variants; the neural network is able to retain more knowledge by using backward queried data to reduce the superimposition of gradient corresponding to newer training cases on vital weights that hold previously formed data.

5   Conclusion

In this paper, we proposed three online learning algorithms that incorporate different scales of backward queried data, and, from our results, we were able to validate the significance of our method for the application of online training. This is because we saw a reduction in the standard deviation of SGD and an improvement in its performance for both Fashion-MNIST and Kuzushiji-MNIST. The results also demonstrate that our variants are a viable option for real world applications of online machine learning due to the fact that we were able to enhance the faculty of the neural network with relatively low cost and easy implementation. Still, it should be cautioned that our approach was exclusively benchmarked with the task of image classification. In addition, our method includes a new heuristic with three different variants, which have their own respective advantages and drawbacks. Nevertheless, our paper provides a novel low-cost approach to overcoming catastrophic interference and demonstrates its potential in online machine learning.

References

1. Cesa-Bianchi, N., Long, P.M., Warmuth, M.: Worst-case quadratic loss bounds for prediction using linear functions and gradient descent. IEEE Trans. Neural Networks **7**(3), 604–619 (1996)
2. Kivinen, J., Warmuth, M.K.: Exponentiated gradient versus gradient descent for linear predictors. Inf. Comput. **132**(1), 1–63 (1997)
3. Bottou, L., LeCun, Y.: Large scale online learning. In: Advances in Neural Information Processing Systems 16. MIT Press, Cambridge, MA (2004). Location (1999)
4. Zhang, T.: Solving large scale linear prediction problems using stochastic gradient descent algorithms. In: Proceedings of the 21st International Conference on Machine Learning (ICML), Banff, Alberta, Canada (2004)
5. McCloskey, M., Cohen, N.J.: Catastrophic interference in connectionist networks: the sequential learning problem. Psychol. Learn. Motivation **24**, 109–165 (1989)
6. Ratcliff, R.: Connectionist models of recognition memory: constraints imposed by learning and forgetting functions. Psychol. Rev. **97**(2), 285 (1990)
7. Robins, A.: Catastrophic forgetting in neural networks: the role of rehearsal mechanisms. In: First New Zealand International Two-Stream Conference on Artificial Neural Networks and Expert Systems, pp. 65–68. IEEE (1993)
8. Robins, A.: Catastrophic forgetting, rehearsal and pseudorehearsal. J. Neural Comput. Artif. Intell. Cogn. Res. **7**, 123–146 (1995)
9. Robins, A.: Sequential learning in neural networks: a review and a discussion of pseudore- hearsal based methods. Intell. Data Anal. **8**(3), 301–322 (2004)
10. French, R.M.: Using pseudo-recurrent connectionist networks to solve the problem of sequential learning. In: Proceedings of the 19th Annual Cognitive Science Society Conference, vol. 16 (1997)
11. Williams, G.R., Goldfain, B., Lee, K., Gibson, J., Rehg, J.M., Theodorou, E.A: Locally weighted regression pseudo-rehearsal for adaptive model predictive control. In: Proceedings of the Conference on Robot Learning, PMLR, vol. 100, pp. 969–978 (2020)

12. Tibshirani, R.: Regression shrinkage and selection via the lasso. J. Roy. Stat. Soc. B **58**(1), 267–288 (1996)
13. Xiao, L.: Dual averaging method for regularized stochastic learning and online optimization. In: Advanced in Neural Information Processing Systems 11 (2009)
14. Langford, J., Li, L., Zhang, T.: Sparse online learning via truncated gradient. In: Advanced in Neural Information Processing Systems 21 (2008)
15. Rebuffi, S.-A., Kolesnikov, A., Sperl, G., Lampert, C.H.: Incremental classifier and representation learning. In: Conference on Computer Vision and Pattern Recognition (CVPR), pp. 5533–5542. IEEE (2017)
16. Shin, H., Lee, J.K., Kim, J., Kim, J.: Continual learning with deep generative replay. In: Advances in Neural Information Processing Systems, pp. 2990–2999 (2017)
17. Atkinson, C., McCane, B., Szymanski, L., Robins, A.V.: Pseudo-rehearsal: achieving deep reinforcement learning without catastrophic forgetting. CoRR, abs/1812.02464 (2018)
18. Mellado, D., Saavedra, C., Chabert, S., Salas, R.: Pseudorehearsal approach for incremental learning of deep convolutional neural networks. In: Barone, D.A.C., Teles, E.O., Brackmann, C.P. (eds.) LAWCN 2017. CCIS, vol. 720, pp. 118–126. Springer, Cham (2017). https://doi.org/10.1007/978-3-319-71011-2_10
19. Kemker, R., Kanan, C.: Fearnet: brain-inspired model for incremental learning. CoRR, abs/1711.10563 (2017)
20. Huh, G.: Enhanced stochastic gradient descent with backward queried data for online learning. In: IEEE International Conference on Machine Learning and Applied Network Technologies (2020)

Construction and Reasoning for Interval-Valued EBRB Systems

Ji-Feng Ye and Yang-Geng Fu(✉)

College of Computer and Data Science, Fuzhou University,
Fuzhou, People's Republic of China
fu@fzu.edu.cn

Abstract. Due to various uncertain factors, sometimes it is difficult to obtain accurate data. In comparison, interval-valued data can better represent uncertain information. However, most of the existing theoretical researches on Extended Belief Rule-Based (EBRB) system are aimed at real-valued data and lack methods applied to interval-valued data. Based on the theory of interval evidence reasoning, this paper proposes a new method of constructing and reasoning for interval-valued EBRB systems. This model can not only be applied to interval-valued data but also real-valued data. And after analyzing the problems of the conventional EBRB similarity formula, we use the method of normalized Euclidean distance. In addition, because of the ignorance of attribute weights by conventional methods, this paper combines mutual information methods to determine attribute weights. Some case studies about interval-valued datasets and real-valued datasets are provided in the last. The experimental results have proven the feasibility and efficiency of the proposed method.

Keywords: Extended belief rule-based system · Interval data · Interval evidence reasoning

1 Introduction

With the industrial process becoming more and more refined, intelligent and complex, industrial production is facing more and more uncertainty, such as fuzzy, incomplete, inaccurate and so on. Because of various noise interference, the data collected by the monitoring equipment often contains errors. In this case, data in interval format may be more suitable for describing uncertain data. However, the classic evidence theory requires a precise belief structure.

Lee and Zhu [1] first studied evidence combination with interval belief structures and gave the interval evidence combination formula based on the generalized addition and multiplication method. Yager [2] combines interval evidence

---

This research was supported by the National Natural Science Foundation of China (No. 61773123) and the Natural Science Foundation of Fujian Province, China (No. 2019J01647).

© Springer Nature Switzerland AG 2021
T. Mantoro et al. (Eds.): ICONIP 2021, LNCS 13110, pp. 257–268, 2021.
https://doi.org/10.1007/978-3-030-92238-2_22

based on interval intersection and union. However, these methods do not standardize the results, resulting in combined interval-valued results that may be invalid. Thus, Denoeux [3,4] systematically studied the combination and standardization of interval evidence, and innovatively proposed a quadratic programming model. Wang et al. [5] modified Denocux's interval evidence combination formula and proposed the optimal interval evidence combination formula. Wang's interval evidence combination formula implements the interval evidence combination operation and interval result standard operation with a unified optimization model, thus avoiding the shortcomings of insufficient correlation between the two operations. Then, based on the theory of interval evidence combination, Wang et al. [6] proposed the interval evidence reasoning method.

Belief Rule-Based [7] (BRB) system is a production expert system logical reasoning method under uncertain information. Since then, Liu et al. [8] proposed an Extended Belief Rule-Based system to address the problem of "combination explosion". How to extend interval belief and interval evidence reasoning methods to the theory of belief rule bases has also been studied. Li et al. [9] proposed a forward reasoning algorithm for belief rules under the input of interval information. Gao et al. [10] introduced the interval belief structure into the BRB system and gave the corresponding construction and inference methods. However, this method is limited by expert knowledge. Zhu et al. [11] introduced the interval belief structure into the EBRB system and proposed an interval-valued belief rule inference method based on evidential reasoning (IRIMER). However, it does not maintain the interval-valued property when calculating the activation weights, which leads to the lack of sufficient correlation between the similarity calculation of the antecedent attributes and the belief composition of the consequent attributes.

Based on Wang's interval evidence reasoning model and the traditional EBRB system, this paper proposes the Interval Extended Belief Rule-Based (IEBRB) system. The contributions of this paper are summarized as follows.

1. Firstly, by improving the structure of the belief rule, the similarity operation of antecedent and the evidence reasoning operation of consequent are unified in the optimization model, which enhances the relevance of two operations.
2. Then, the similarity formula is improved to the normalized Euclidean distance formula and the mutual information method is used to determine attribute weight, which improves the interpretability and accuracy of system.
3. Finally, experiments on interval-valued data and real-valued data show that the proposed method is effective.

2   Problem Formulation

IRIMER combines the ability of EBRB to handle uncertain information and the ability of the interval evidence reasoning model to handle interval-valued data very well. For interval belief structures, IRIMER calculates the distance between the antecedents of input data and the antecedents of rule as follows.

$$d_i^k = \sqrt{\frac{1}{2J_i} \sum_{j=1}^{J_i} \left(\alpha_{i,j}^{k-} - \alpha_{i,j}^{-}\right)^2 + \left(\alpha_{i,j}^{k+} - \alpha_{i,j}^{+}\right)^2} \tag{1}$$

where $[\alpha_{i,j}^{k-}, \alpha_{i,j}^{k+}](i = 1, 2, \ldots, M; k = 1, 2, \ldots, L)$ represents the interval belief degree of jth reference value of ith antecedent attribute in kth rule. $[\alpha_{i,j}^{-}, \alpha_{i,j}^{+}]$ represents the interval belief degree of jth reference value of ith antecedent attribute in input data. But this formula only uses the upper and lower bounds of the interval belief of the antecedent, and the calculated distance becomes real-valued again, which is not sufficient for the use of interval-valued information. There are two problems with this approach.

Problem 1. The distance between interval-valued cannot be measured well.

For example, if the antecedent belief of the rule is $\alpha(x^{(R)}) = [0.4, 0.6]$, the antecedent belief of two data is $\alpha(x^{(1)}) = [0.5, 0.5]$, $\alpha(x^{(2)}) = [0.5, 0.7]$, the distance obtained by using Eq. 1 is $d(x^{(R)}, x^{(1)}) = 0.1$, $d(x^{(R)}, x^{(2)}) = 0.1$, which can be found the two results are the same. This is the problem caused by only considering the boundary of the interval.

Problem 2. The similarity operation of the antecedent and the evidence reasoning operation of the consequent do not reflect the interval.

EBRB system uses activation weights to associate the similarity operation of the antecedent with the evidence reasoning operation of the consequent. The activation weight calculated using Eq. 1 is real-valued, and most of interval-valued information of the antecedent will be lost. Particularly when IRIMER applies to the interval-valued classification problem, it can not transfer the interval-valued information of the antecedent well. Since the activation weights and the consequence of the classification problems are real-valued, the belief results obtained by evidential reasoning operation on the results are also real-valued.

3   Interval Extended Belief Rule-Based System

3.1   Construction of IEBRB

If the belief of the antecedent is directly used as the parameter constraint of the optimization model, the number of parameters of the optimization model will increase by multiples of the number of parameter values. Therefore, we directly introduce the interval-valued of the training data in the rule's antecedent.

$$
\begin{aligned}
R_k : &IF\ U_1\ is\ [x_1^{k-}, x_1^{k+}] \wedge U_2\ is\ [x_2^{k-}, x_2^{k+}] \wedge \ldots \wedge U_M\ is\ [x_M^{k-}, x_M^{k+}] \\
&THEN\ \left\{ (D_1, [\beta_1^{k-}, \beta_1^{k+}]), (D_2, [\beta_2^{k-}, \beta_2^{k+}]), \ldots, (D_N, [\beta_N^{k-}, \beta_N^{k+}]) \right\}, \\
&with\ rule\ weight\ \theta_k\ and\ attribute\ weights\ \{\delta_1, \delta_2, \ldots, \delta_M\}, \\
&s.t.\quad 0 \leqslant \beta_n^{k-}, \beta_n^{k+}, \delta_i, \theta_k \leqslant 1,\quad \sum_{n=1}^{N} \beta_n^{k-} \leqslant 1 \leqslant \sum_{n=1}^{N} \beta_j^{k+},
\end{aligned} \tag{2}
$$

where $[x_i^{k-}, x_i^{k+}](i = 1, 2, \ldots, M; k = 1, 2, \ldots, L)$ represents the interval-valued of the ith antecedent attribute of the kth rule. $D_n (n = 1, 2, \ldots, N)$ represents the n reference value of the consequent attribute. $[\beta_1^{k-}, \beta_1^{k+}]$ represents the interval belief degree that D_n is the final result. θ_k and δ_i represents rule weight and attribute weight respectively

For classification problem, according to the category y^k, we set the belief of the reference value corresponding to the category to 1, and the other reference values to 0:

$$E\left(y^k\right) = \left\{(D_n, [\beta_n^{k-}, \beta_n^{k+}]), n = 1, \ldots, N\right\}, \text{ with } \beta_n^{k-} = \beta_n^{k+} = \begin{cases} 1, & D_n \text{ is } y^k \\ 0, & otherwise \end{cases} \quad (3)$$

For regression problems, we use Wang's method [6] to calculate the belief distribution form of the consequent interval-valued $[y^{k-}, y^{k+}]$:

$$E\left([y^{k-}, y^{k+}]\right) = \left\{\left(D_n, [\beta_n^{k-}, \beta_n^{k+}]\right), n = 1, \ldots, N\right\}. \quad (4)$$

Suppose interval-valued $[y^{k-}, y^{k+}]$ surrounds multiple evaluation levels, and is surrounded by two evaluation D_n, D_M. We regard interval-valued as the value range of the variable $y^k \in [y^{k-}, y^{k+}]$. Then y^k will only be located between two adjacent reference values at a time. In response to this special case, 0–1 integer variables are introduced:

$$I_{n+m-1,n+m}^k = \begin{cases} 1, & u(D_{n+m-1}) < y^k \leqslant u(D_{n+m}) \\ 0, & otherwise \end{cases}, \text{ with } \sum_{m=1}^M I_{n+m-1,n+m}^k = 1. \quad (5)$$

Then, the interval-valued changing to belief distribution is as follows:

$$\begin{cases} \beta_n^{k-} = I_{n,n+1}^k \cdot \max\left(0, \frac{u(D_{n+1})-y^{k+}}{u(D_{n+1})-u(D_n)}\right) \quad \text{and} \quad \beta_n^{k+} = I_{n,n+1}^k \cdot \frac{u(D_{n+1})-y^{k-}}{u(D_{n+1})-u(D_n)} \\ \beta_{n+m}^{k-} = I_{n+m-1,n+m}^k \cdot \max\left(0, \frac{y^{k-}-u(D_{n+m-1})}{u(D_{n+m})-u(D_{n+m-1})}\right) \\ \quad + I_{n+m,n+m+1}^k \cdot \max\left(0, \frac{u(D_{n+m+1})-y^{k+}}{u(D_{n+m+1})-u(D_{n+m})}\right) \\ \quad \text{and} \quad \beta_{n+m}^{k+} = I_{n+m-1,n+m}^k + I_{n+m,n+m+1}^k, \quad m = 2, \ldots, M-1 \\ \beta_{n+M}^{k-} = I_{n+M-1,n+M}^k \cdot \max\left(0, \frac{y^{k-}-u(D_{n+M-1})}{u(D_{n+M})-u(D_{n+M-1})}\right) \\ \quad \text{and} \quad \beta_{n+M}^{k+} = I_{n+M-1,n+M}^k \cdot \frac{y^{k+}-u(D_{n+M-1})}{u(D_{n+M})-u(D_{n+M-1})} \\ \beta_j^{k-} = 0 \quad \text{and} \quad \beta_j^{k+} = 0, \quad j = 1, 2, \ldots, n-1, n+M+1, \ldots, N \end{cases}$$
$$(6)$$

3.2 Reasoning of IEBRB

This paper extends the idea of nonlinear optimization model proposed by Denoeux [3,4] and Wang [6] to the antecedent part of EBRB. By taking the interval-valued of the antecedent attribute and the interval-valued of the belief of the consequent as constraints, the reasoning process of IEBRB is regarded as the following optimization model:

$$\text{Max/Min} \quad \beta_n$$

$$\text{s.t.} \quad x_i^- \leqslant x_i \leqslant x_i^+, \tag{7a}$$

$$x_i^{k-} \leqslant x_i^k \leqslant x_i^{k+}, \tag{7b}$$

$$\beta_n^{k-} \leqslant \beta_n^k \leqslant \beta_n^{k+} \text{ and } \sum_{n=1}^{N} \beta_n^k \leqslant 1, \tag{7c}$$

$$E(x_i) = \{(A_{i,j}, \alpha_{i,j}), j = 1, 2, \ldots, J_i\} \tag{7d}$$

$$E(x_i^k) = \{(A_{i,j}, \alpha_{i,j}^k), j = 1, 2, \ldots, J_i\} \tag{7e}$$

$$S_i^k = 1 - \sqrt{\sum_{j=1}^{J_i} (\alpha_{ij} - \alpha_{ij}^k)^2 \Bigg/ \left[\sum_{j=1}^{J_i} \alpha_{ij}^2 + \sum_{j=1}^{J_i} \alpha_{ij}^{k\,2} \right]} \tag{7f}$$

$$\omega_k = \theta_k \times \prod_{i=1}^{T_k} \left(S_i^k \right)^{\delta_i} \Bigg/ \sum_{l=1}^{L} \left[\theta_l \times \prod_{i=1}^{T_l} \left(S_i^l \right)^{\delta_i} \right] \tag{7g}$$

$$\mu = \left[\sum_{j=1}^{N} \prod_{k=1}^{L} \left(\omega_k \beta_j^k + 1 - \omega_k \sum_{i=1}^{N} \beta_i^k \right) - (N-1) \prod_{k=1}^{L} \left(1 - \omega_k \sum_{i=1}^{N} \beta_i^k \right) \right]^{-1} \tag{7h}$$

$$\beta_n = \frac{\mu \times \left[\prod_{k=1}^{L} \left(\omega_k \beta_n^k + 1 - \omega_k \sum_{i=1}^{N} \beta_i^k \right) - \prod_{k=1}^{L} \left(1 - \omega_k \sum_{i=1}^{N} \beta_i^k \right) \right]}{1 - \mu \times \left[\prod_{k=1}^{L} \left(1 - \omega_k \right) \right]} \tag{7i}$$

After constructing this optimization model, you can use the "fmincon" function or intelligent algorithms to solve it. This paper uses differential evolution algorithm to solve it. The obtained maximum and minimum values are regarded as the upper and lower bounds of the interval, thus making up the interval belief output. Finally, it can be expressed as:

$$f(x) = \left\{ \left(D_1, [\beta_1^-, \beta_1^+] \right), \left(D_2, [\beta_2^-, \beta_2^+] \right), \ldots, \left(D_N, [\beta_N^-, \beta_N^+] \right) \right\} \tag{8}$$

For regression problems, we can extend the optimization model to the utility value calculation part. The optimization model is:

$$\text{Max} \quad u_{\max} = \sum_{n=1}^{N} u(D_n) \beta_n + u(D_N) \left(1 - \sum_{n=1}^{N} \beta_n \right)$$

$$\text{Min} \quad u_{\min} = \sum_{n=1}^{N} u(D_n) \beta_n + u(D_1) \left(1 - \sum_{n=1}^{N} \beta_n \right) \tag{9}$$

$$\text{s.t.} \quad \text{use } Eqs.\,(7a) - -(7i)$$

3.3 New Individual Matching Degree

In addition, it needs to be particularly emphasized that we have improved the formula for calculating the degree of individual match. This is because the traditional EBRB [8] system uses Euclidean distance to measure the similarity between the input data and the kth rule on the ith attribute, but the individual matching degree S_i^k will have a negative value. In response to this problem, Yang et al. [12,13] proposed the following standardized Euclidean distance method.

$$d_i^k = \sqrt{\frac{1}{2}\sum_{j=1}^{Ji}(\alpha_{ij} - \alpha_{ij}^k)^2} \qquad (10)$$

But the standardized Euclidean distance formula still cannot satisfy the monotonicity. For intuitive convenience, we only consider a single attribute $M = 1$, the value range of the sample is $[1,5]$, and the reference value is $A = \{1,2,3,4,5\}$. Figure 1(a) shows the individual matching degree s_i^k between a single training data x_i^k and a single input data x_i obtained using Eq. 10. It can be seen that the individual matching degree curve is not a monotonously changing relationship. With the distance between x_i^k and x_i gets farther, the similarity does not continue to decrease until it is zero.

To address this problem, this paper using the normalized Euclidean distance Eq. 7f. Figure 1(b) shows the individual matching degree s_i^k between a single training data x_i^k and a single input data x_i obtained by using normalized Euclidean distance. It can be found that the similarity curve satisfies the monotonicity requirement. As the distance between x_i^k and x_i gets farther, the similarity will continue to decrease until it reaches zero.

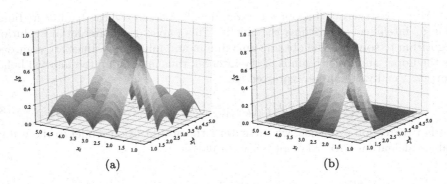

Fig. 1. Analyze the monotonicity of the individual match degree formula

3.4 Determine Attribute Weights

The attribute weight will affect the inference result of the EBBR system, but the traditional method is to determine the attribute weight through expert knowledge. Therefore, this paper proposes a method to determine attribute weights based on mutual information. If the antecedent data or the consequent data are continuous, we need to convert to discrete data first. Assuming that the system has a total of T interval data, the t-th data has M antecedent attributes and 1 consequent attribute, expressed as $\{x_1^t, x_2^t, \ldots, x_M^t; y^t\}, x_i^t = [x_i^{t-}, x_i^{t+}]$. According to the data value range of the i-th attribute, Q level areas can be divided evenly into $\{R_{i,1}, R_{i,2}, \ldots, R_{i,Q}\}$, then according to this Q interval, discretize the attribute U_i from continuous data:

$$D(x_i^t) = \{R_{i,j} | R_{i,j} \cap x_i^t \neq \emptyset, \quad \forall j = 1, 2, \ldots, Q\} \tag{11}$$

For classification problem, suppose there are a total of N categories in the consequent attributes, so according to the categories, the consequents are divided into N subsets $y(t) \in \{C_1, C_2, \ldots, C_N\}$. For regression problem, the continuous data can be discretized in the same way as Eq. 11. Then calculate the mutual information and conditional entropy of the discrete feature of the antecedent attribute U_i and consequent attribute Y:

$$MI(U_i, Y) = \sum_{j=1}^{Q} \sum_{n=1}^{N} \frac{|R_{i,j} \cap C_n|}{T_i'} \log \left(\frac{|R_{i,j} \cap C_n|}{|R_{i,j}||C_n|} \times T_i' \right), \quad \text{with } T_i' = \sum_{j=1}^{Q} |R_{i,j}| \tag{12}$$

$$H(U_i) = -\sum_{j=1}^{Q} \frac{|R_{i,j}|}{T_i'} \log \frac{|R_{i,j}|}{T_i'} \tag{13}$$

$$H(Y) = -\sum_{n=1}^{N} \frac{|C_n|}{T_i'} \log \frac{|C_n|}{T_i'} \tag{14}$$

According to this mutual information, it is used to measure the relevance of the antecedent attribute data to the consequent attribute. In order to make the value range of the mutual information the same as the value range of the weight, it is within the interval of $[0, 1]$, we adopt standardization Mutual information as attribute weight:

$$\delta(U_i) = NMI(U_i, Y) = \frac{2 \times MI(U_i, Y)}{H(U_i) + H(Y)} \tag{15}$$

3.5 Framework

Figure 2 is the framework of the IEBRB system. In the rule construction stage, we use training set to construct interval belief rules, and use the mutual information to determine attribute weights. In the rule reasoning stage, we conduct interval evidence reasoning. In the model evaluation stage, the test set is applied to verify IEBRB system.

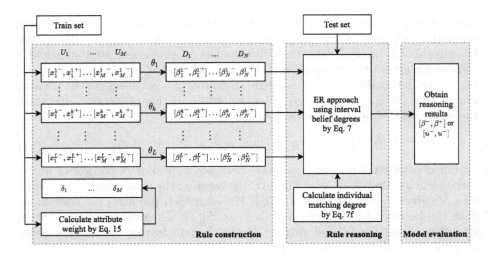

Fig. 2. The framework of the IEBRB system

4 Experiment and Analysis

IEBRB can be regarded as a more general improvement of EBRB, which can be applied not only to the problem of interval-valued datasets, but also to the problem of real-valued datasets. In order to verify the effectiveness of the proposed method, several experiments will be performed. For the regression problem, this paper uses two cases: graphite detection [14] and oil pipeline leak detection [15]. For classification problems, this paper uses 4 datasets from the well-known UCI machine-learning repository.

4.1 Interval-Valued Data Problem Cases

UCI datasets are all real-valued. Therefore, this paper uses the statistical method [16–18] to convert the antecedent attribute data from real-value to interval-value. Assume that the t-th sample has M antecedent attributes and 1 consequent attributes $(x_1^t, x_2^t, \ldots, x_M^t; y^t)$, and a total of N classes $y^t \in \{C_1, C_2, \ldots, C_N\}$. Converting real-valued data x_i^t into interval-valued data $[x_i^{t-}, x_i^{t+}]$ is as follows:

$$\begin{cases} x_i^{t-} = x_i^t - rand()\tau\sigma_{i,n} \\ x_i^{t+} = x_i^t + rand()\tau\sigma_{i,n} \end{cases} \quad , \quad \text{if } y^t == C_n \qquad (16)$$

where τ is the adjustment parameter of the interval value length, generally set to 2. In order to increase randomness, we introduce the random function. $\sigma_{i,n}$ represents the standard deviation of x_i in the same class C_n. That is,

$$\sigma_{i,n} = \sqrt{\frac{1}{|C_n|-1}\sum_{x_i^t\in C_n}(x_i^t - \bar{a}_{i,n})^2}, \quad \text{with } \bar{a}_{i,n} = \frac{1}{|C_n|}\sum_{x_i^t\in C_n}x_i^t. \quad (17)$$

We specify $[x_i^{t^-}, x_i^{t^+}]$ as the attribute value of the attribute a_i. Through this method, an interval dataset can be constructed.

Table 1 shows the classification effect of EBRB-IM, IRIMER [11], and IEBRB on self-constructed interval dataset. In this experiment, 5-CV is adopted, and take the average of classification accuracy of 5 duplicated experiments as result. The EBRB-IM method is to take the midpoint of interval and convert interval value to a real value $x_i^{t'} = (x_i^{t^-} + x_i^{t^+})/2$, then apply it to the traditional EBRB [8] method.

Table 1. Accuracy (%) in classification problems with interval-valued

Dataset	EBRB-IM	IRIMER	IEBRB
Ecoli	85.47	67.16	**86.31**
Glass	63.18	74.32	**78.98**
Iris	90.29	93.33	**95.33**
Seeds	90.28	89.91	**91.43**

From Table 1, it can be found that the classification effect of the proposed method is optimal on the four datasets. Comparing IEBRB method and IRIMER method, the effect is much more outstanding on Ecoli and Glass, with an increase of 19.15% and 4.66%, respectively, and it also has a good effect on Iris and Seeds. Comparing IEBRB method and EBRB-IM method, there are outstanding advantages in Glass and Iris, increasing by 15.8% and 5.04%, respectively. There is a slight advantage in Ecoli and Seeds.

In order to demonstrate the applicability of the IEBRB method in regression problems, it is applied to graphite detection. The interval dataset comes from the literature [11]. This is a hierarchical Extended Belief Rule-Based problem, which consists of five sub-expanded belief rule bases in total. Table 2 shows the prediction utility values.

It can be seen from Table 2 that IEBRB method and IRIMER method are relatively consistent in average utility value, and there is not much difference between the utility value of EBRB-IM, which shows that IEBRB's reasoning method is reasonable. Moreover, the interval utility value obtained by the IEBRB method includes the interval utility value obtained by the IRIMER method. The reason for this result is that the IRIMER method is calculated by using Eq. 1 in the antecedent, which only considers the upper and lower bounds of the interval and has a huge loss of interval information, so the interval utility value obtained is shorter. The IEBRB method treats the interval as a range of values and makes full use of the interval information. Although the interval is longer, it can more

Table 2. Utility values in regression problems with interval-valued

	EBRB-IM	IRIMER		IEBRB	
		Interval	Average	Interval	Average
Sub-EBRB-1	0.9296	(0.7794, 0.8711)	0.8253	(0.5792, 0.9898)	0.7845
Sub-EBRB-2	0.9569	(0.8996, 0.9295)	0.9146	(0.8531, 0.9843)	0.9187
Sub-EBRB-3	0.5059	(0.3066, 0.5841)	0.4454	(0.2927, 0.6126)	0.4526
Sub-EBRB-4	0.8918	(0.7143, 0.8856)	0.8000	(0.6957, 0.9740)	0.8349
Sub-EBRB-5	0.6312	(0.4626, 0.6839)	0.5733	(0.5010, 0.8656)	0.6833

fully consider each situation. For example, the results of EBRB-IM on Sub-EBRB-1, Sub-EBRB-2, and Sub-EBRB-4 are outside the interval utility range predicted by IRIMER, while the interval utility value of the IEBRB method in the five sub-rule bases all includes the results of EBRB-IM, which also proves our analysis.

4.2 Real-Valued Data Problem Cases

This paper treats the interval-valued as a range of values, so we can also treat real-valued as a range of values, but this range of values will be special, and its upper and lower bounds are equal. Only one value can be taken in this interval, i.e., $x_i^- = x_i^+$. Therefore, IEBRB can be applied to real-valued data problems.

Table 3 shows the predictive effect of IEBRB and other EBRB methods in the detection of oil pipeline leaks. It can be seen that although both IRIMER and IEBRB can be applied to interval data, the performance of IRIMER is very poor in the real-valued dataset. Among the four methods, IEBRB's MAE ranks first and MSE ranks second. This is because IEBRB inherits the EBRB calculation method and improves the individual matching formula, which is more effective than the traditional EBBR method.

Table 3. Comparison with other method in oil pipe leak detection

	EBRB [8]	DEA-EBRB [13]	IRIMER	IEBRB
MAE	0.2734	0.2688	1.4397	**0.2523**
MSE	0.5506	**0.4495**	4.1718	0.5404

Table 4 shows the classification effect of IEBRB and other methods on the four datasets from UCI repository. Each dataset performs a 10-fold cross-validation test. It can be found that IEBRB can be well applied to real-valued classification data, and through the improvement of the individual matching degree formula and the determination of attribute weights, it can be found that the obtained IEBRB has the best results on Ecoli. In Iris and Seeds are also

Table 4. Compare the accuracy (%) for classification datasets with real-valued

Dataset	Ecoli	Glass	Iris	Seeds	Avg. Rank	Avg. Acc
EBRB [8]	81.16 (7)	67.85 (7)	95.33 (4)	91.33 (7)	6.25	83.92
DEA-EBRB [13]	84.82 (6)	71.5 (3)	95.26 (5)	92.38 (4)	4.5	85.99
VP-EBRB [19]	84.87 (5)	71.75 (2)	95.13 (7)	92.57 (3)	4.25	86.08
MVP-EBRB [19]	85.61 (4)	72.06 (1)	95.87 (2)	92.38 (4)	2.75	**86.48**
KDT-EBRB [12]	86.93 (2)	69.86 (6)	95.2 (6)	93.71 (1)	3.75	86.42
BT-EBRB [20]	86.28 (3)	70.01 (5)	95.99 (1)	92.2 (6)	3.75	86.12
IRIMER	65.12 (8)	62.52 (8)	94.8 (8)	88 (8)	8	77.61
IEBRB	87.21 (1)	70.38 (4)	95.47 (3)	92.86 (2)	**2.5**	**86.48**

ranked in the top three. In the Glass, although IEBRB ranks fourth, the gap with the first is not very large. This can be seen from the fact that the average ranking of IEBRB is the first and average accuracy of IEBRB is tied for the first place with the MVP-EBRB method. In addition, notice that compared with the IRIMER method that also uses interval-valued data, the classification effect of the IEBRB method far exceeds that of the four datasets, which proves that the IEBRB method has good adaptability on real-valued datasets.

5   Conclusion

Aiming at the problem that EBRB cannot be applied to interval-valued data, this paper proposes Interval Extended Belief Rule-Based system. Most of the previous methods for interval-valued datasets only consider the upper and lower bounds of the interval values, but the proposed method regards the interval as the range of values and constructs an optimization model to solve the interval result, which can more fully consider the interval values information. The inference result is also more reasonable. Moreover, the proposed method can well inherit the calculation method of traditional EBRB, not only for interval-valued datasets but also for real-valued datasets. Through the improvement of the calculation formula of the individual matching degree and the determination of the attribute weight, the effect of EBRB has been improved to a certain extent. The case studies and experimental results demonstrate that the IEBRB method is feasible and effective in interval-valued datasets and real-valued datasets.

References

1. Stanley Lee, E., Zhu, Q.: An interval dempster-Shafer approach. Comput. Math. Appl. **24**(7), 89–95 (1992)
2. Yager, R.R.: Dempster-Shafer belief structures with interval valued focal weights. Int. J. Intell. Syst. **16**(4), 497–512 (2001)

3. Denœux, T.: Reasoning with imprecise belief structures. Int. J. Approxim. Reason. **20**(1), 79–111 (1999)
4. Denoeux, T.: Modeling vague beliefs using fuzzy-valued belief structures. Fuzzy Sets Syst. **116**(2), 167–199 (2000)
5. Wang, Y.-M., Yang, J.-B., Xu, D.-L., Chin, K.-S.: On the combination and normalization of interval-valued belief structures. Inf. Sci. **177**(5), 1230–1247 (2007)
6. Wang, Y.-M., Yang, J.-B., Dong-Ling, X., Chin, K.-S.: The evidential reasoning approach for multiple attribute decision analysis using interval belief degrees. Eur. J. Oper. Res. **175**(1), 35–66 (2006)
7. Yang, J.-B., Liu, J., Wang, J., Sii, H.-S., Wang, H.-W.: Belief rule-base inference methodology using the evidential reasoning approach-RIMER. IEEE Trans. Syst. Man Cybern. Part A Syst. Hum. **36**(2), 266–285 (2006)
8. Liu, J., Martinez, L., Calzada, A., Wang, H.: A novel belief rule base representation, generation and its inference methodology. Knowl.-Based Syst. **53**, 129–141 (2013)
9. Li, B., Wang, H.-W., Yang, J.-B., Guo, M., Qi, C.: A belief-rule-based inventory control method under nonstationary and uncertain demand. Expert Syst. Appl. **38**(12), 14997–15008 (2011)
10. Gao, F., Zhang, A., Bi, W.: Weapon system operational effectiveness evaluation based on the belief rule-based system with interval data. J Intell. Fuzzy Syst. **39**, 1–15 (2020). Preprint
11. Zhu, H., Zhao, J., Xu, Y., Du, L.: Interval-valued belief rule inference methodology based on evidential reasoning-IRIMER. Int. J. Inf. Technol. Decis. Making **15**(06), 1345–1366 (2016)
12. Yang, L.-H., Wang, Y.-M., Qun, S., Fu, Y.-G., Chin, K.-S.: Multi-attribute search framework for optimizing extended belief rule-based systems. Inf. Sci. **370**, 159–183 (2016)
13. Yang, L.-H., Wang, Y.-M., Lan, Y.-X., Chen, L., Fu, Y.-G.: A data envelopment analysis (DEA)-based method for rule reduction in extended belief-rule-based systems. Knowl.-Based Syst. **123**, 174–187 (2017)
14. Hodges, J., Bridge, S., Yie, S.Y.: Preliminary results in the use of fuzzy logic for radiological waste characterization expert system. Mississippi State University, MS, Technical report. MSU-960626 (1996)
15. Dongling, X., et al.: Inference and learning methodology of belief-rule-based expert system for pipeline leak detection. Expert Syst. Appl. **32**(1), 103–113 (2007)
16. Leung, Y., Fischer, M.M., Wu, W.-Z., Mi, J.-S.: A rough set approach for the discovery of classification rules in interval-valued information systems. Int. J. Approximate Reasoning **47**(2), 233–246 (2008)
17. Zhang, X., Mei, C., Chen, D., Li, J.: Multi-confidence rule acquisition and confidence-preserved attribute reduction in interval-valued decision systems. Int. J. Approximate Reasoning **55**(8), 1787–1804 (2014)
18. Zhang, Y., Li, T., Luo, C., Zhang, J., Chen, H.: Incremental updating of rough approximations in interval-valued information systems under attribute generalization. Inf. Sci. **373**, 461–475 (2016)
19. Lin, Y.-Q., Fu, Y.-G., Qun, S., Wang, Y.-M., Gong, X.-T.: A rule activation method for extended belief rule base with VP-tree and MVP-tree. J Intell Fuzzy Syst **33**(6), 3695–3705 (2017)
20. Fu, Y.-G., Zhuang, J.-H., Chen, Y.-P., Guo, L.-K., Wang, Y.-M.: A framework for optimizing extended belief rule base systems with improved ball trees. Knowl.-Based Syst. **210**, 106–484 (2020)

Theory and Applications of Natural Computing Paradigms

Brain-mimetic Kernel: A Kernel Constructed from Human fMRI Signals Enabling a Brain-mimetic Visual Recognition Algorithm

Hiroki Kurashige[1,2]([✉]), Hiroyuki Hoshino[3,4], Takashi Owaki[4], Kenichi Ueno[5], Topi Tanskanen[6], Kang Cheng[5], and Hideyuki Câteau[2,7]

[1] School of Information and Telecommunication Engineering, Tokai University, Tokyo, Japan
[2] RIKEN BTCC, RIKEN, Saitama, Japan
[3] Faculty of Engineering, Aichi Institute of Technology, Toyota, Aichi, Japan
[4] TOYOTA Central R & D Labs, Inc, Nagakute, Aichi, Japan
[5] RIKEN CBS, RIKEN, Saitama, Japan
[6] University of Helsinki, Helsinki, Finland
[7] University of Tsukuba, Ibaraki, Japan

Abstract. Although the present-day machine learning algorithm sometimes beats humans in visual recognition, we still find significant differences between the brain's and the machine's visual processing. Thus, it is not guaranteed that the information sampled by the machines is the one used in the human brain. To overcome this situation, we propose a novel method for extracting the building blocks of our brain information processing and utilize them in machine learning algorithms. The visual features used in our brain are identifiable by applying kernel canonical correlation analysis (KCCA) to paired data of visual stimuli (images) and evoked functional magnetic resonance imaging (fMRI) activity. A machine learning algorithm incorporating the identified visual features represented in the brain is expected to inherit the characteristics of brain information processing. In the proposed method, the features are incorporated into kernel-based algorithms as a positive-definite kernel. Applying the method to fMRI data measured from a participant seeing natural and object images, we constructed a support vector machine (SVM) working on the visual features presumably used in the brain. We showed that our model outperforms the SVM equipped with a conventional kernel, especially when the size of the training data is small. Moreover, we found that the performance of our model was consistent with physiological observations in the brain, suggesting its neurophysiological validity.

Keywords: Brain-mimetic visual recognition algorithms · Kernel canonical correlation analysis · Functional magnetic resonance imaging

This work was supported by the Japan Society for the Promotion of Science (JSPS) KAKENHI Grant Number 23500267 and the Leading Initiative for Excellent Young Researchers (MEXT, Japan).

© Springer Nature Switzerland AG 2021
T. Mantoro et al. (Eds.): ICONIP 2021, LNCS 13110, pp. 271–283, 2021.
https://doi.org/10.1007/978-3-030-92238-2_23

1   Introduction

This study introduces and evaluates a novel method to design a brain-mimetic computational algorithm. Our brains interact with the world by selectively sampling the information using intrinsic methods, and then work on the sampled information and show significant adaptability and creativity. Therefore, to realize computer algorithms that have human-like adaptability and creativity, setting ourselves on the information selected by our brains should be a good starting point. Moreover, analyzing the computer algorithm that is ensured similarity with the human brain provides good opportunities to understand our cognitive natures emerged in the brain.

Hence, we attempt to develop a method to identify information about the external world used by the human brain using functional magnetic resonance imaging (fMRI) data and to utilize it in a visual recognition algorithm. The identification of the information used in the brain is realized by applying a machine learning method to fMRI data. This study provides a brain-mimetic model that is usable in neuroscientific inspections of human cognitive abilities and gives a policy to develop brain-like AI systems.

Theoretically, the scope of our methodological framework includes action generation, as well as recognition (see Discussion). However, in the present study, we narrowed our focus on visual recognition.

1.1   Related Work

There is a long history of the development of brain-mimetic algorithms. In the earliest stages, Rosenblatt proposed a pattern recognition algorithm, perceptron, that can learn pattern separation based on error signals [28]. Inspired by Hubel and Wiesel's neurophysiological observations [13], Fukushima developed a visual recognition algorithm, neocognitron [8], that is the source of the current deep neural network [23] via LeCun's learnable multi-layer network, LeNet [24]. Rumelhart et al. proposed an effective learning algorithm called back-propagation [29]. Back-propagation has been broadly applied to neural network learning, including multi-layer perceptron, recurrent neural networks, and deep convolutional neural networks (CNNs). On the one hand, back-propagation is a highly useful tool for engineering purposes. On the other hand, the resultant neural networks learned using back-propagation can provide rich insights into psychological and psychiatric phenomena. These efforts have converged to recent large-scale deep CNNs [11, 20, 35].

From a neuroscientific perspective, one of the most noteworthy facts about CNNs is that we can find the units in the CNNs whose representations are similar to the representations of the visual neurons in the biological brain [9, 18, 41, 42]. Therefore, CNNs may be considered a good model of the brain. However, more recent findings have addressed significant differences in visual representations between the brain and the CNN that have been revealed by more detailed investigations [15, 39, 40]. At the training of the CNNs, the parameters are not directly fit to reproduce biological observations but to satisfy an achievement in an engineering purpose. This results in difficulty to relate a large proportion of units in the CNN to biological neurons. Additionally, the biological brain adopts population coding, where the information is represented by the multivariate pattern of neural activity, not by single neuronal activity. Although the representation similarity analysis has shown that some aspects of the population behavior are observable in the CNNs [5], any mechanism to utilize it is not explicitly incorporated. Thus, it is not guaranteed that the information sampled by CNNs from the external world is the one used in the human brain. Hence, although the CNNs mimic the biological brains in some senses, we cannot regard the information sampled by CNNs as the starting point for our brain-mimetic algorithms.

1.2   Contributions

All brain-mimetic algorithms mentioned above were constructed on the basis of purified qualitative knowledge of neurophysiology and/or neuroanatomy. Between the wild behaviors of recognition performed by our brain and such neuroscientific knowledge, the purification processes are interleaved to make them interpretable. The processes are executed by ignoring the nature of biological recognition that prevents interpretation because of the complexity. This results in distorted knowledge about how our brains use the information about the external world. To avoid the distortion inevitably occurring as a byproduct of the researcher's interpretation, we study an alternative way.

Here, we propose a method in which brain activity is more directly associated with the external world to identify the information used in the biological brain (Fig. 1). To this end, we assumed that the information about the world used in the brain is represented well by neural activity. Thus, we tried to identify the visual features represented well by some neural activity patterns and to identify the neural activity patterns that represent some visual features well. To solve this dual identification problem simultaneously, we applied the nonlinear variant of canonical correlation analysis, kernel CCA (KCCA) [1, 3, 22] to paired data of the images (including object images and natural scene images) and the human brain activities evoked by the images that were recorded using fMRI. Subsequently, we constructed a kernel function from the identified visual features. It is important that the resultant kernel function should be brain-mimetic because the kernel function is the inner product of the Hilbert space in which the distances between the mapped images correspond to those in the space of the neural representations in the brain. Therefore, algorithms with the kernel function become brain-mimetic. In the present study, we tested the recognition performance of a support vector machine (SVM) incorporating a brain-mimetic kernel constructed using this method in the pedestrian detection problem. We observed that our algorithm outperformed the SVM with a Gaussian kernel, particularly when the training dataset was small and the test images were masked. Additionally, we observed that our kernel worked well when it was based on data recorded from neurophysiologically relevant areas. Finally, we discuss the expandability of the proposed method and its future directions.

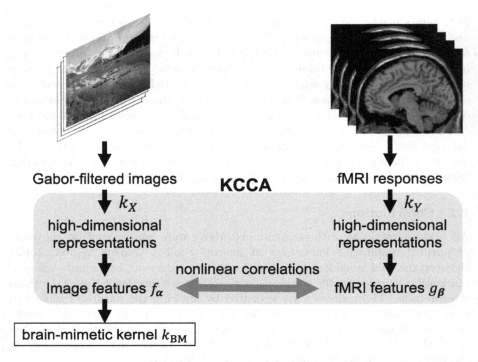

Fig. 1. Proposed Method. While object and natural visual images were presented to the participant, the evoked fMRI signals were recorded using fMRI. After applying the Gabor filters to images and extraction of fMRI response intensities, KCCA was applied to the data pair. As a result, a transformation that maps visual images onto the visual features used in the brain was identified. Incorporating the transformation into machine learning algorithms as kernels, brain-mimetic machines were constructed.

Note that the method presented in this study was originally proposed in our preliminary report [21], in which we showed the results for simple simulation data of the *artificial* neural networks to provide an intuitive understanding of our method. There, we evaluated the method and observed that the characteristics of the artificial neural networks can be inherited by an SVM through a kernel constructed on the basis of visual features identified by the KCCA. In contrast, here, we validated the method using real *biological neurophysiological* data acquired from the human brain using fMRI.

The most novel and innovative point in the present study is to propose a method to construct a brain-mimetic system that is ensured similarity with the human brain in the sense that it is based on maximizing the canonical correlation between population-level neural activity and sensory features representing the external world. Therefore, this study should provide, especially to the neuroscience community, a useful tool to investigate the cognitive natures of the brain and, especially to the AI community, hints to develop more brain-like AI.

2 Materials and Methods

2.1 Participant

One Japanese male undergraduate student (22 years old, right-handed, normal vision) without a history of neurological or psychiatric diseases participated in this study. Written informed consent was obtained in accordance with the protocol approved by the ethics committees of RIKEN. We conducted an extensive sampling from an individual to cover as much of the diversity of natural visual experience as possible, which is a strategy that has been recognized the effectiveness in the recent studies of modeling brain functions under naturalistic conditions using neurophysiological data [26,31].

2.2 MRI Data Acquisition

We conducted MRI experiments using a 4T MRI (Agilent Technologies). T1-weighted structural images and T2*-weighted functional images were acquired. For functional data, 836 or 824 four-shot echo planar images covering the ventral visual cortex were acquired: 21 transverse slices, slice thickness 3 mm, slice gap 0.5 mm, field-of-view 192 mm × 192 mm, matrix size 64 × 64, and efficient repetition time (TR) of 766 ms under TSENSE reconstruction. During the functional scans, 1020 picture images (including object images and natural scene images) were presented to a participant using a headcoil mounted mirror reflecting a display screen. For structural data, whole-brain gradient-echo images were acquired with a matrix size of 220 × 256, 256 slices, voxel size 1 mm × 1 mm × 1 mm.

2.3 MRI Preprocessing

Slice timing correction was conducted using a custom-made program. To correct the head motions and fit the structural image onto the mean functional image, we used SPM12 (Welcome Trust Centre for Human Neuroimaging, England, UK). Smoothing was not performed. To make the participant's brain atlas, a recon-all script in Freesurfer (Freesurfer Version 5.3.0) was executed. In this study, we used the Desikan-Killiany atlas [6]. The structural image generated by recon-all was fitted to the original structural image using flirt in FSL (FSL Version 5.0.6; http://www.fmrib.ox.ac.uk/fsl/). We removed noise signals from functional volumes by regressing the mean signals from the white matter and ventricles.

2.4 Estimating Activation Coefficients of image-Induced fMRI Signals

We applied an alternate iteration of estimating the responses to visual images and estimating the hemodynamic response curves for each voxel. Serial correlations of errors were removed using the Cochrane-Orcutt procedure. In total, this model fitting procedure is a variant of the basis-restricted separable model method [16,17].

2.5   Image Preprocessing

Although the images presented to the participant during fMRI recording had various sizes, for the analysis using KCCA (see below), we reduced their pixels by shrinking the images to be 60-pixel heights and then cropping 60×30 pixels. Then, these images were filtered by convolving the set of Gabor functions with directions of $-60, -30, 0, 30, 60,$ and $90°$ and receptive field sizes of 8×8 pixels. Similarly, the set of Daimler images [25], which were used in SVM tasks, were exchanged to 60×30 pixels and were Gabor-filtered. Masked Daimler images were generated by adding a vertical 60×7 bar to a random position of each image.

2.6   KCCA-SVM

Here, we propose a procedure to identify the visual features used in the brain from visual fMRI data and to construct a positive-definite kernel function that represents an inner product in the space spanned by the identified visual features. Using the kernel in kernel-based machine learning algorithms, we process images implicitly in the space in which the distance (generally defined based on an inner product) between images inherits the one used in the brain. The procedure shown in this section was proposed in our preliminary study, in which we provided a proof of concept using computer simulations of artificial neural networks [21], and in our published patent [12].

The strategy adopted in the present study was to explore the visual features that were highly correlated with the appearance of some patterns in brain activity measured with fMRI. This is because the visual features that are used in the brain are assumed to be represented as neural activities evoked in the brain while seeing the images including them. In contrast, visual features that are not important in information processing in the brain and should be ignored are poorly correlated with any neural representation. Therefore, we sought the pairs of visual features in images and the patterns of fMRI signals induced by the images using a nonlinear variant of the CCA method, KCCA, introduced right below (Fig. 1).

Here, we consider a set of N image vectors, $X = \{\mathbf{x}^1, \cdots, \mathbf{x}^N\}$. By measuring the fMRI brain activity of a participant watching the images, a set of N fMRI signal vectors, $Y = \{\mathbf{y}^1, \cdots, \mathbf{y}^N\}$, is obtained. The image and fMRI signal vectors that have the same superscripts correspond to each other. Applying the KCCA to these data, the pairs of the brain-used visual features and the corresponding patterns of fMRI activity that are nonlinearly correlated are simultaneously identified.

In the original linear version of CCA, a projection of $\mathbf{x}$ onto $\mathbf{u}$ (defined as $f_\mathbf{u}(\mathbf{x}) = \mathbf{u}^T\mathbf{x}$) and a projection of $\mathbf{y}$ onto $\mathbf{v}$ (defined as $g_\mathbf{v}(\mathbf{y}) = \mathbf{v}^T\mathbf{y}$) are considered. CCA is a method to identify pairs of vectors $\{(\mathbf{u}^{(j)}, \mathbf{v}^{(j)})\}_j$ ($j = 1, 2, \cdots$) that are highly correlated with each other. A KCCA is a nonlinear extension of the CCA using positive-definite kernels [1,3,22]. This can identify nonlinearly correlated pairs of the projections, each of which is implicitly represented in the (generally) high-dimensional spaces defined by the kernel (shown as k_X and k_Y in Fig. 1). Because we consider high-dimensional visual and neural spaces, the possible nonlinear correlations between the preprocessed visual images and fMRI activities are calculated, resulting in the highly correlated pairs of nonlinear projections shown as image features f_α and fMRI features g_β in Fig. 1.

Based on the theoretical result of the reproducing kernel Hilbert space [32], we can consider a positive-definite kernel $k_X(\mathbf{x}, \mathbf{x}')$ as an inner product in a high-dimensional space. Thus, $k_X(\mathbf{x}, \mathbf{x}') = \phi(\mathbf{x})^T \phi(\mathbf{x}')$, where $\phi(\mathbf{x})$ is a representation of $\mathbf{x}$ in the high-dimensional space implicitly defined by $k_X(\mathbf{x}, \mathbf{x}')$. Because the Gaussian kernel is universal in function approximation [10] and most standard in the field of pattern recognition, in the present study, the Gaussian kernel was used for k_X and k_Y in the KCCA. The parameter values were set using a cross-validation procedure on the separated data.

Similar to the case of the linear CCA, we consider a projection of $\phi(\mathbf{x})$ onto $\mathbf{u}$ in this space, that is, $f_\mathbf{u}(\mathbf{x}) = \mathbf{u}^T \phi(\mathbf{x})$. From the reproducing property of the kernel [32], we can express $\mathbf{u}$ as a weighted sum of $\phi(\mathbf{x}^n)$: $\mathbf{u} = \sum_{n=1}^{N} \alpha_n \phi(\mathbf{x}^n)$. Therefore, the projection $f_\mathbf{u}(\mathbf{x})$ can be expressed in a high-dimensional visual space as $f_\alpha(\mathbf{x}) = \sum_{n=1}^{N} \alpha_n \phi(\mathbf{x}^n)^T \phi(\mathbf{x}) = \sum_{n=1}^{N} \alpha_n k_X(\mathbf{x}, \mathbf{x}^n)$. Similarly, when $k_Y(\mathbf{y}, \mathbf{y}') = \psi(\mathbf{y})^T \psi(\mathbf{y}')$, the projection in the high-dimensional neural space, $g_\mathbf{v}(\mathbf{y}) = \mathbf{v}^T \psi(\mathbf{y})$ can be expressed as $g_\beta(\mathbf{y}) = \sum_{n=1}^{N} \beta_n k_Y(\mathbf{y}, \mathbf{y}^n)$.

A correlation between the projections with L2 regularization is following:

$$\rho_{\text{KCCA}} = \frac{\text{Cov}(f_\mathbf{u}(\mathbf{x}), g_\mathbf{v}(\mathbf{y}))}{\sqrt{\text{Var}(f(\mathbf{x})) + \zeta_x \|\mathbf{u}\|^2} \sqrt{\text{Var}(g(\mathbf{y})) + \zeta_y \|\mathbf{v}\|^2}}$$
$$= \frac{\alpha^T K_X K_Y \beta}{\sqrt{\alpha^T (K_X^2 + \zeta_x K_X) \alpha} \sqrt{\beta^T (K_Y^2 + \zeta_y K_Y) \beta}}$$

where K_X and K_Y are the Gram matrices corresponding to $k_X(\mathbf{x}, \mathbf{x}')$ and $k_Y(\mathbf{y}, \mathbf{y}')$, and ζ_x and ζ_y are parameters for controlling the regularization.

Because the correlation is scaling invariant with respect to α and β, we impose the following constraints: $\alpha^T (K_X^2 + \zeta_x K_X) \alpha = \beta^T (K_Y^2 + \zeta_y K_Y) \beta = 1$. Our task is to find (α, β) that maximizes ρ_{KCCA} under these constraints. Considering the Lagrange multiplier method, we maximize the following function:

$$L(\alpha, \beta) = \alpha^T K_X K_Y \beta - \frac{\lambda_x}{2} \alpha^T (K_X^2 + \zeta_x K_X) \alpha - \frac{\lambda_y}{2} \beta^T (K_Y^2 + \zeta_y K_Y) \beta.$$

This results in the following generalized eigenvalue problem:

$$\begin{pmatrix} 0 & K_X K_Y \\ K_Y K_X & 0 \end{pmatrix} \begin{pmatrix} \alpha \\ \beta \end{pmatrix} = \lambda \begin{pmatrix} K_X^2 + \zeta_x K_X & 0 \\ 0 & K_Y^2 + \zeta_y K_Y \end{pmatrix} \begin{pmatrix} \alpha \\ \beta \end{pmatrix}. \quad (1)$$

We have $2N$ eigenvalues and the corresponding eigenvectors, $\{(\alpha^{(j)T}, \beta^{(j)T})^T\}$, $j = 1, \cdots, 2N$. The jth pair of visual and neural canonical variables is defined as $f_{\alpha^{(j)}}(\mathbf{x}) = \sum_{n=1}^{N} \alpha_n^{(j)} k_X(\mathbf{x}, \mathbf{x}^n)$ and $g_{\beta^{(j)}}(\mathbf{y}) = \sum_{n=1}^{N} \beta_n^{(j)} k_Y(\mathbf{y}, \mathbf{y}^n)$ based on the

jth eigenvector $(\boldsymbol{\alpha}^{(j)^T}, \boldsymbol{\beta}^{(j)^T})^T$. The canonical variables with different suffixes j are orthogonal, which means that the variables are mutually uncorrelated.

These canonical variables are candidates for visual features and neural representations to construct our brain-mimetic visual recognition systems. We selected the pairs of $(f_{\boldsymbol{\alpha}^{(j)}}(\mathbf{x}), g_{\boldsymbol{\beta}^{(j)}}(\mathbf{y}))$ with the d highest correlation values. In this study, we set $d = 100$. In this process, the correlation values were evaluated based on the data that were not used to solve the generalized eigenvalue problem mentioned above. We consider the resultant d pairs of visual and neural canonical variables, $\{(f_{\boldsymbol{\alpha}^{(j)}}(\mathbf{x}), g_{\boldsymbol{\beta}^{(j)}}(\mathbf{y}))\}_{j=1,\cdots,d}$, to be highly correlated pairs of visual features and neural representations.

Naturally, we assume that the visual features $\{(f_{\boldsymbol{\alpha}^{(j)}}(\mathbf{x})\}_{j=1,\cdots,d}$ that are well represented in the brain, which is suggested by high correlation values, are used in the brain. Hereafter, these features are called *brain-used visual features*. We expect that visual recognition algorithms working on brain-used visual features behave brain-like. Such algorithms perform visual processing not within the entire feature space, but in the restricted space spanned by the brain-used visual features. This restriction possibly plays a role resembling a prior probability in Bayesian statistics [4]. We incorporate the restriction in the restricted space using a specially designed kernel representing the similarity between images $\mathbf{x}$ and $\mathbf{x}'$ by the similarity (i.e. inner product) in the space spanned by d brain-used visual features, $\{f_{\boldsymbol{\alpha}^{(j)}}(\mathbf{x})\}_{j=1,\cdots,d}$. We call this the *brain-mimetic kernel*. Formally, the brain-mimetic kernel is defined as

$$k_{\mathrm{BM}}(\mathbf{x}, \mathbf{x}') \equiv k(q(\mathbf{x}), q(\mathbf{x}')) \tag{2}$$

where $q(\mathbf{x}) = (f_{\boldsymbol{\alpha}^{(1)}}(\mathbf{x}), \cdots, f_{\boldsymbol{\alpha}^{(d)}}(\mathbf{x}))^T$ is the mapping of images $\mathbf{x}$ to the space that is spanned by the visual features of the brain. k is an arbitrary positive-semidefinite kernel (e.g., linear kernel, Gaussian kernel, polynomial kernel, etc.). In the present study, we used the Gaussian kernel as k. Therefore,

$$k_{\mathrm{BM}}(\mathbf{x}, \mathbf{x}') = \exp(-\frac{|q(\mathbf{x}) - q(\mathbf{x}')|^2}{2\sigma^2}). \tag{3}$$

The brain-mimetic kernel is usable in any kernel-based algorithm (e.g., SVM, KCCA, and kernel principal component analysis) and makes them brain-mimetic by introducing the similarity between images used in the brain.

3 Results

To evaluate the method on real neurophysiological data, we applied it to a set of pairs of visual images and evoked fMRI activity. We presented 1020 images (including object images and natural scene images) to a participant while measuring brain activity using fMRI. After extracting the neural response intensity for each image and filtering each image using the Gabor functions, we identified the visual features multivariately well-represented in the brain using KCCA. We studied five visual areas as regions of interests: fusiform gyrus (FG), lateral

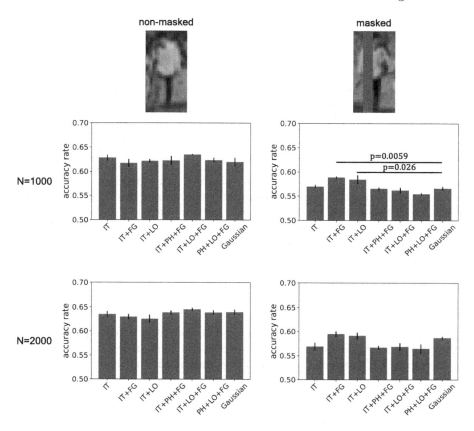

Fig. 2. Experimental Result. Performance of the SVM with Gaussian and brain-mimetic kernels for the pedestrian detection task. The upper row shows the case of a small training data size ($N = 1000$), and the lower row shows the case of a large training data size ($N = 2000$). The left column shows the case when the test images are not masked, and the right column shows the case when the test images are partially masked. Note that no masked image was used either to train the SVMs or to fit the KCCA. The kernels from the IT+FG and IT+LO data outperform the Gaussian kernel when the training data are small and the test images are masked.

occipital cortex (LO), inferior temporal cortex (IT), parahippocampal gyrus (PH), and isthmus cingulate gyrus (IC). For the five areas and their doublet and triplet combinations, we surveyed KCCA parameters and explored the areas whose KCCA correlation values between fMRI signals and Gabor-filtered images were significantly positive. Under the multiple comparisons controlling false discovery rate, IT, IT+FG, IT+LO, IT+LO+FG, IT+FG+PH, and LO+FG+PH were significant. Therefore, we analyzed only those six areas.

For these areas, we identified the visual features using the KCCA and constructed brain-mimetic kernels from them (see Materials and Methods). To evaluate these kernels, we incorporated them into SVMs and executed pedestrian detection tasks (Fig. 2). Here, we trained SVMs to discriminate images with pedestrians from images without pedestrians. We then applied the SVMs to the new test datasets. In our daily life, we can detect pedestrians even if masking objects partially cover them. This is a typical situation where human visual recognition works superiorly. To mimic the situations, we gave the partially masked images in the test sessions and not in the training sessions. In such cases, we found that the SVM with brain-mimetic kernels constructed from IT+FG and IT+LO outperformed the SVM with a Gaussian kernel when the size of the training data is small ($N = 1000$) (IT+FG: $t = 3.348, p = 0.0059$, IT+LO: $t = 2.722, p = 0.026$ using Dunnett's multiple comparison method). When the size is large ($N = 2000$), the difference became non-significant. On the other hand, no brain-mimetic kernel significantly outperformed the Gaussian kernel in either $N = 1000$ or 2000 when images were non-masked.

Since our brains utilize preexisting knowledge and work well, especially when information is partial, it is considered that these observations rely on the inheritance of the superior characteristics of the brain into the SVM through the brain-mimetic kernels.

4   Discussion

In the present study, we propose a method to construct a brain-mimetic algorithm based on visual fMRI data recorded from a human watching object and natural images. The novelty of the proposed method is that it utilizes the information in our brain more directly, in which we incorporate the quantitative and nonlinear characteristics latent in our brain into machine learning algorithms by using a nonlinear multivariate technique for data analysis. As our method can identify the relationship between brain activity representing various things (e.g. speeches and motions) and the features of them, it should not be restricted to visual recognition.

On applying the proposed KCCA-SVM framework to visual fMRI data, we found that the visual features represented in two combined regions of interests, IT+FG and IT+LO, were useful for the pedestrian detection task. This result is relevant from a neurophysiological perspective. Many studies have suggested that the IT is the most vital area for general object vision [14,19,36]. Because the visual features represented in the IT have mid-level abstraction, it is considered that the SVM with the kernel constructed from IT activity discriminates the pedestrian images from non-pedestrian images in a moderately abstract subspace. The FG probably plays a role more specific to our pedestrian detection task, because the FG includes the area that processes the visual images of body parts, called the fusiform body area [27,30,37]. As the LO includes another area for visual processing of body parts called the extrastriate body area [2,7,38], LO might also provide useful visual features for detecting pedestrians. Thus, the performance of our model resembles that of the biological brain. This suggests the neurophysiological validity of the proposed method. The KCCA could identify the visual features used in the human brain. Moreover, we could successfully and easily incorporate them into the SVM through the kernel.

As we explained in Materials and Methods, our strategy of extensive sampling from an individual to cover the diversity of natural visual experience is reasonable. However, it is important to extend the study to the larger population of participants to capture diversities of visual recognitions across individuals. To this end, use of open fMRI datasets [33, 34] may be useful in future study.

Since our method was based on the direct maximization of the canonical correlation between fMRI activity and visual features, this kind of similarity of the visual recognition algorithm and the brain is originally ensured. On the other hand, there are other kinds of similarities to be investigated, such as the correspondence of the representational dissimilarity matrices [5] and mutual predictability of brain activity from machine responses [41]. Therefore, comparing the proposed method with various brain-inspired machine learning methods including deep CNNs with respect to such similarities is an important next step.

The most important future challenge is to apply the proposed methodology to studies of various neural phenomena to understand information processing in the brain. For instance, comparing properties of the kernel algorithms (e.g. KCCA, kernel PCA, kernel LDA) incorporating the brain-mimetic kernels constructed using activity data from a brain area (e.g. inferior temporal cortex) with functional and anatomical properties of its downstream brain areas (e.g. medial temporal lobe). Such efforts will lead deep understanding of the specificities of human information processing realized by the brain and the development of AI systems that possess more human-like characteristics.

References

1. Akaho, S.: A kernel method for canonical correlation analysis. arXiv preprint arXiv:cs/0609071 (2006)
2. Astafiev, S.V., Stanley, C.M., Shulman, G.L., Corbetta, M.: Extrastriate body area in human occipital cortex responds to the performance of motor actions. Nat. Neurosci. **7**(5), 542–548 (2004)
3. Bach, F.R., Jordan, M.I.: Kernel independent component analysis. J. Mach. Learn. Res. **3**, 1–48 (2002)
4. Bishop, C.M.: Pattern Recognition and Machine Learning. Springer, New York (2006)
5. Cichy, R.M., Khosla, A., Pantazis, D., Torralba, A., Oliva, A.: Comparison of deep neural networks to spatio-temporal cortical dynamics of human visual object recognition reveals hierarchical correspondence. Sci. Rep. **6**, 27755 (2016)
6. Desikan, R.S., et al.: An automated labeling system for subdividing the human cerebral cortex on MRI scans into gyral based regions of interest. NeuroImage **31**(3), 968–980 (2006)
7. Downing, P.E., Jiang, Y., Shuman, M., Kanwisher, N.: A cortical area selective for visual processing of the human body. Science **293**(5539), 2470–2473 (2001)
8. Fukushima, K., Miyake, S.: Neocognitron: A self-organizing neural network model for a mechanism of visual pattern recognition. In: Amari, S., Arbib, M.A. (eds.) Competition and Cooperation in Neural Nets, Lecture Notes in Biomathematics, vol. 45, pp. 267–285, Springer, Heidelberg (1982). https://doi.org/10.1007/978-3-642-46466-9_18

9. Güçlü, U., van Gerven, M.A.: Deep neural networks reveal a gradient in the complexity of neural representations across the ventral stream. J. Neurosci. **35**(27), 10005–10014 (2015)
10. Hammer, B., Gersmann, K.: A note on the universal approximation capability of support vector machines. Neural Process. Lett. **17**(1), 43–53 (2003)
11. He, K., Zhang, X., Ren, S., Sun, J.: Deep residual learning for image recognition. In: 2016 IEEE Conference on Computer Vision and Pattern Recognition (CVPR), pp. 770–778 (2016)
12. Hoshino, H., Owaki, T., Kurashige, H., Kato, H.: Sensory data identification apparatus and program (September 2015)
13. Hubel, D.H., Wiesel, T.N.: Receptive fields and functional architecture of monkey striate cortex. J. Physiol. **195**(1), 215–243 (1968)
14. Hung, C.P., Kreiman, G., Poggio, T., DiCarlo, J.J.: Fast readout of object identity from macaque inferior temporal cortex. Science **310**(5749), 863–866 (2005)
15. Jacob, G., Pramod, R.T., Katti, H., Arun, S.P.: Qualitative similarities and differences in visual object representations between brains and deep networks. Nat. Commun. **12**(1), 1872 (2021)
16. Kay, K.N., David, S.V., Prenger, R.J., Hansen, K.A., Gallant, J.L.: Modeling low-frequency fluctuation and hemodynamic response timecourse in event-related fMRI. Hum. Brain Mapp. **29**(2), 142–156 (2008)
17. Kay, K.N., Naselaris, T., Prenger, R.J., Gallant, J.L.: Identifying natural images from human brain activity. Nature **452**(7185), 352–355 (2008)
18. Khosla, M., Ngo, G.H., Jamison, K., Kuceyeski, A., Sabuncu, M.R.: Cortical response to naturalistic stimuli is largely predictable with deep neural networks. Sci. Adv. **7**(22), eabe7547 (2021)
19. Kriegeskorte, N., et al.: Matching categorical object representations in inferior temporal cortex of man and monkey. Neuron **60**(6), 1126–1141 (2008)
20. Krizhevsky, A., Sutskever, I., Hinton, G.E.: Imagenet classification with deep convolutional neural networks. In: Advances in Neural Information Processing Systems, pp. 1097–1105 (2012)
21. Kurashige, H., Câteau, H.: A method to construct visual recognition algorithms on the basis of neural activity data. In: Lu, B.-L., Zhang, L., Kwok, J. (eds.) ICONIP 2011. LNCS, vol. 7064, pp. 485–494. Springer, Heidelberg (2011). https://doi.org/10.1007/978-3-642-24965-5_55
22. Lai, P.L., Fyfe, C.: Kernel and nonlinear canonical correlation analysis. Int. J. Neural Syst. **10**(05), 365–377 (2000)
23. LeCun, Y., Bengio, Y., Hinton, G.: Deep learning. Nature **521**(7553), 436–444 (2015)
24. LeCun, Y., Bottou, L., Bengio, Y., Haffner, P.: Gradient-based learning applied to document recognition. Proc. IEEE **86**(11), 2278–2324 (1998)
25. Munder, S., Gavrila, D.M.: An experimental study on pedestrian classification. IEEE Trans. Pattern Anal. Mach. Intell. **28**(11), 1863–1868 (2006)
26. Naselaris, T., Allen, E., Kay, K.: Extensive sampling for complete models of individual brains. Curr. Opin. Behav. Sci. **40**, 45–51 (2021)
27. Peelen, M.V., Downing, P.E.: Selectivity for the human body in the fusiform gyrus. J. Neurophysiol. **93**(1), 603–608 (2005)
28. Rosenblatt, F.: The perceptron: a probabilistic model for information storage and organization in the brain. Psychol. Rev. **65**(6), 386 (1958)
29. Rumelhart, D.E., Hinton, G.E., Williams, R.J.: Learning representations by back-propagating errors. Nature **323**, 533–536 (1986)

30. Schwarzlose, R.F., Baker, C.I., Kanwisher, N.: Separate face and body selectivity on the fusiform gyrus. J. Neurosci. **25**(47), 11055–11059 (2005)
31. Seeliger, K., Sommers, R.P., Güçlü, U., Bosch, S.E., van Gerven, M.A.J.: A large single-participant fMRI dataset for probing brain responses to naturalistic stimuli in space and time. bioRxiv 687681 (2019)
32. Shawe-Taylor, J., Cristianini, N.: Kernel Methods for Pattern Analysis. Cambridge University Press, Cambridge (2004)
33. Shen, G., Horikawa, T., Majima, K., Kamitani, Y.: Deep image reconstruction from human brain activity. PLOS Comput. Biol. **15**(1), 1–23 (2019)
34. Shen, G., Horikawa, T., Majima, K., Kamitani, Y.: Deep image reconstruction. OpenNeuro (2020). https://doi.org/10.18112/openneuro.ds001506.v1.3.1
35. Simonyan, K., Zisserman, A.: Very deep convolutional networks for large-scale image recognition. arXiv preprint arXiv:1409.1556 (2014)
36. Tanaka, K.: Inferotemporal cortex and object vision. Ann. Rev. Neurosci. **19**(1), 109–139 (1996)
37. Taylor, J.C., Wiggett, A.J., Downing, P.E.: Functional MRI analysis of body and body part representations in the extrastriate and fusiform body areas. J. Neurophysiol. **98**(3), 1626–1633 (2007)
38. Urgesi, C., Candidi, M., Ionta, S., Aglioti, S.M.: Representation of body identity and body actions in extrastriate body area and ventral premotor cortex. Nat. Neurosci. **10**(1), 30–31 (2007)
39. Xu, Y., Vaziri-Pashkam, M.: Examining the coding strength of object identity and nonidentity features in human occipito-temporal cortex and convolutional neural networks. J. Neurosci. **41**(19), 4234–4252 (2021)
40. Xu, Y., Vaziri-Pashkam, M.: Limits to visual representational correspondence between convolutional neural networks and the human brain. Nat. Commun. **12**(1), 2065 (2021)
41. Yamins, D.L., Hong, H., Cadieu, C.F., Solomon, E.A., Seibert, D., DiCarlo, J.J.: Performance-optimized hierarchical models predict neural responses in higher visual cortex. Proc. Nat. Acad. Sci. **111**(23), 8619–8624 (2014)
42. Zhuang, C., Yan, S., Nayebi, A., Schrimpf, M., Frank, M.C., DiCarlo, J.J., Yamins, D.L.K.: Unsupervised neural network models of the ventral visual stream. Proc. Nat. Acad. Sci. **118**(3), e2014196118 (2021)

Predominant Sense Acquisition with a Neural Random Walk Model

Attaporn Wangpoonsarp(iD) and Fumiyo Fukumoto(✉)(iD)

Graduate Faculty of Interdisciplinary Research, University of Yamanashi,
4-3-11, Takeda, Kofu, Yamanashi 400-8510, Japan
{g16dhl01,fukumoto}@yamanashi.ac.jp

Abstract. Domain-Specific Senses (DSS) acquisition has been one of
the major topics in Natural Language Processing (NLP). However, most
results from unsupervised learning methods are not effective. This paper
addresses the problem and proposes an approach for improving performance based on deep learning. To obtain DSS, we utilize Approximate
Personalized Propagation of Neural Predictions (APPNP) consisting of
Graph Convolutional Networks (GCN) and PageRank. GCN is a neural
network that performs on graphs to learning sense features from neighbors' senses and using Personalized PageRank for propagation. For constructing sense features, we collect glosses from WordNet and obtained
sense embedding by using Bidirectional Encoder Representations from
Transformers (BERT). Our experimental results show that the approach
works well and attain at 0.614 Macro F1-score. In addition, to demonstrate the efficacy that DSS can work well in the NLP task, we apply
the results on DSS to text categorization and gain a macro F1-score at
0.920, while the CNN baseline method is 0.776.

Keywords: Domain-specific senses · Neural random walk model ·
Word sense disambiguation · Text categorization · Natural language
processing

1 Introduction

Domain-Specific Senses (DSS) is the task to detect which sense of the word is
appropriate in a given context. It is curious because a word has multiple senses
and each sense is used in a different context, thus many NLP researchers are
attracted to solve this problem. DSS can be applied in a variety of applications
such as information retrieval, question answering, machine translation, text categorization. The simple way to identify a proper sense of the word is the First
Sense Heuristic (FSH) that is an approach to select the first sense of a word in
the dictionary and it is a strong influence, particularly for words with highly
skewed sense distributions [1]. However, the weakness of the FSH is a small corpus, consequently, we can not utilize the sense that does not appear in SemCor.
Moreover, FSH is based on the frequency of SemCor instead of the domain. For

© Springer Nature Switzerland AG 2021
T. Mantoro et al. (Eds.): ICONIP 2021, LNCS 13110, pp. 284–295, 2021.
https://doi.org/10.1007/978-3-030-92238-2_24

instance, the noun word, "book" has eleven senses in the WordNet [2]. The first sense of "book" is "a written work or composition that has been published", and it is often used in the "publishing" domain rather than the "religion" domain. In contrast, the ninth sense of "book", i.e. "the sacred writings of the Christian religions" is more likely to be used in the "religion" domain. From the above example, it has been observed that DSS influences the proper sense of words. DSS can be obtained from three approaches including supervised learning, unsupervised learning and semi-supervised learning. The supervised learning has been very successful. However, it requires a lot of sense annotated data and it is expensive, while an unsupervised learning is ineffective and it does not use labeled training data. Semi-supervised learning utilizes a small portion of sense annotated data to predict labels for a lot of sense unannotated data. The goal of semi-supervised learning is to attain performance similar to supervised learning.

In this paper, we focus on DSS based on semi-supervised learning and propose a method for identifying predominant sense from a corpus. We first select only noun and verb words, extracting them from Reuters news corpus (RCV1) [3], and then collect their senses and gloss text from the WordNet. Gloss texts are utilized to build sense embedding using Bidirectional Encoder Representations from Transformers (BERT) [4]. Next, we create an adjacency matrix of sense from RCV1 to constructing a graph structure. Both of the sense embeddings and adjacency matrix are applied for training neural networks called Approximate Personalized Propagation of Neural Predictions (APPNP) [5] to classify sense category and evaluated them with the Subject Field Codes (SFC) [6]. Lastly, to conduct the extrinsic evaluation of our method, we apply the DSS with text categorization to examine the effectiveness of our method.

2   Related Work

The early attempt to determine the correct sense in a context was presented by Gale et al. [7]. They reported a word that appeared in each discourse tended to share the same sense extremely. Magnini et al. [8] presented a method to Word Sense Disambiguation (WSD) based on the assumption that domains are properties used to link words together as context with SFC in WordNet Domain. However, matching the domain to a target word was semi-automated and required manual annotation. Fukumoto and Suzuki [9] focus on the problem and proposed an automatic method to detect DSS based on Support Vector Machines (SVMs) and they tested only noun words for twelve categories. The average precision of the results achieved to 0.661. Yuan [24] proposed LSTM based method to perform WSD. Their method includes two parts. The first part is to predict the appropriate sense of sentence perform by supervised WSD with the LSTM model and the second part is semi-supervised learning with label propagation to predict labels for unlabeled sentences using several labeled sentences. The results demonstrate the state-of-art performance in several benchmarks. However, it relies on predictive results from the LSTM model which is supervised learning, and a large number of sentences and labels are required in the training stage. Pasini and Navigli [10] presented Train-O-Matic, Supervised WSD that was a method for sense labeling automatically and it only required a minimum knowledge, WordNet-like resource. They exploited the domain of senses and make a graph to connect senses together using Personalized PageRank (PPR) as a propagation scheme. The motivation for their work is similar to ours, which is to identify predominant senses in different domains by training senses within a graph.

There has been plenty of works that attempted to map a word into a semantic vector space. For example, a Bag-of-Words approach that represents the occurrence of words within a document as a vector. Nevertheless, it encountered some problems such as sparse representations and word semantics discarding. Mikolov et al. [11] addressed the problem and proposed Word2Vec that is a neural network to learn individual words and generated dense representation. However, the drawback of the Word2Vec model is incapable of the handle a word with multiple senses that mean Word2Vec represented a word with only the same vector, regardless of how many senses a word have. Recently, a celebrated technique was proposed by Devlin et al. [4] called BERT that tackled the Word2Vec problem. It can produce different word vectors of a word from a given context.

Currently, there has been a variety of graph-based models, for example, Deep-Walk proposed by Perozzi et al. [12] is the technique for learning node embedding capturing the contextual information surrounding nodes. Agirre et al. [13] developed the variant of Personalize PageRank for knowledge-based WSD. Graph Convolutional Networks (GCN) [15] is another technique that can generate useful features of nodes on the graph. However, the propagation of neural networks with message passing loses its focus on the local neighborhood [16] when many layers are applied. Klicpera et al. [5] addressed the problem and proposed the PPR as a propagation scheme instead to solve the problem.

Text categorization is an approach for assigning a label to the text according to the content of the text. Convolutional Neural Network (CNN) is a famous technique presented by Kim [17]. Their results from tuning with little hyperparameter are better than the state of art by 4 out of 7 datasets. Most text categorization frequently emphasizes single-label, however, Liu et al. [18] presented a CNN model that can distinguish multi-label problems effectively. While Wang et al. [19] proposed a method for short text categorization by fusion an explicit approach with an implicit approach that would enable short text to obtain useful information and reduce ambiguity. Their approach applied fine-grained and large-scale semantic knowledge according to the domain in which the word occurred in documents. In contrast, our approach uses only the senses of WordNet that are tagged with the domain. The contributions of our paper are as follows:

1. We propose a semi-supervised method for detecting domain-specific senses which leverage word embedding and deep learning technique. The method requires a few sense-annotated data for a training stage only.
2. The method is automated and required only documents from the given category and thesaurus such as WordNet.
3. We show that the result obtained by DSS improves the performance of text categorization through extrinsic evaluation.

3   Framework of the System

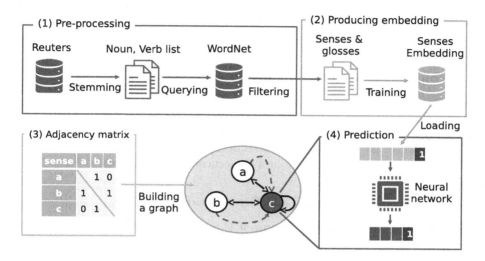

Fig. 1. System framework

The objective of our work is to identify the predominant sense for each domain by learning the features of each sense through the relationship of senses in the graph structure (Fig. 1).

3.1   Pre-processing

The objective of the pre-processing is to extract senses and glosses in the Word-Net from given categories. We initially gather the documents from RCV1. Each word is annotated for Part Of Speech (POS) and is lemmatized using Stanford CoreNLP [20]. Noun and verb words are chosen and used to find their senses and gloss texts from the WordNet. For each category, noun and verb words are extracted.

3.2   Producing Embedding

The sense embedding is learned with BERT that is a type of neural net-work model for pre-training language embeddings developed by Devlin et al. [4]. We apply BERT to learns the feature representation of gloss texts $S_i = \{w_1, \cdots, w_m\}$ where each $w_i, (1 \leq i \leq m)$ denotes a word in gloss texts to build sense embeddings as an input for the prediction stage. In this work, we use The pre-trained BERT model as BERT-Base, Uncased (L = 12, H = 768, A = 12) which represents a model consisting of 12 layers, 768 hidden units, and 12 atten-tion heads. BERT's input formatting has two important special tokens consisting of [SEP] and [CLS]. Both tokens are inserted into a sentence after that we sum token embedding, the segment embedding, and the position embedding to build input embeddings. We use input embeddings which have four dimensions including the number of layers, the number of batches, the number of tokens, and the number of features for creating sense embeddings. We select the pooling strategy as the average pooling on the second-to-last layer to obtained sense embeddings.

3.3 Building a Graph with an Adjacency Matrix

We begin to create a co-occurrence matrix between senses, each of which meets three criteria: firstly, target senses, and their POS are found in RCV1 documents. Secondly, they have the same category. Thirdly, each document contains more than one sense. By utilizing sense relationships, we create an adjacency matrix by converting the non-zero value in a co-occurrence matrix equal to one.

3.4 Predicting Categories and Propagation

We utilize the APPNP model because it is based on GCN model [15] that is a very powerful neural network even 2 layers of GCN can generate useful feature representation of nodes on graphs and it also solves the lost focus issue with PPR. We use sense embedding which is the result from the second step as an input and then training with the APPNP model that predicts a proper category for each sense.

We build a graph. $G_d = (V, E)$ is a graph that represents the relationships between senses in all domains by adjacency matrix $A \in \mathbb{R}^{n \times n}$ where n is a node set. $\tilde{A} = A + I_n$ represents the adjacency matrix with added self-loops. V is the vertices set consisting vertex v_i that is a sense gloss texts. E is a edge set. Each edge e_{ij} denotes co-occurrence relationship of v_i and v_j in documents. The sense gloss texts v_n are represented by the feature matrix $X \in \mathbb{R}^{n \times f}$ where f denotes the number of features, and the category matrix $Y \in \mathbb{R}^{n \times c}$ where c denotes the number of categories. GCN model with two layers is defined by

$$Z_{\text{GCN}} = \text{softmax}\left(\hat{\tilde{A}} \text{ ReLu}\left(\hat{\tilde{A}} X W^{(0)}\right) W^{(1)}\right), \tag{1}$$

where $Z_{GCN} \in \mathbb{R}^{n \times c}$ is the prediction of each node, $\hat{\tilde{A}} = \tilde{D}^{-1/2} \tilde{A} \tilde{D}^{-1/2}$ is a normalized adjacency matrix with self-loops, $\tilde{D}_{ij} = \Sigma_k \tilde{A}_{ik} \delta_{ij}$ is the diagonal matrix, and $W^{(0)}$ and $W^{(1)}$ are weight matrix.

The propagation in this graph model derived from PageRank [14] that is defined by $\pi_{pr} = A_{rw} \pi_{pr}$ with $A_{rw} = AD^{-1}$ where D denotes the diagonal matrix. The Personalized PageRank adapts from PageRank for recognizing the connection between nodes and is defined by $\pi_{ppr}(i_x) = \alpha \left(I_n - (1-\alpha)\hat{\tilde{A}}\right)^{-1} i_x$ where x is root node and i_x denotes teleport vector with teleport probability $\alpha \in [0, 1]$.

APPNP applies the above ideas and produces the first prediction for each node and then propagate it with PPR to produce the final prediction. It is defined by

$$Z^{(0)} = H, \quad H = f_\theta(X),$$
$$Z^{(k+1)} = (1 - \alpha)\hat{\tilde{A}}Z^{(k)} + \alpha H, \tag{2}$$
$$Z^{(K)} = \text{softmax}\Big((1 - \alpha)\hat{\tilde{A}}Z^{(K-1)} + \alpha H\Big),$$

where $Z \in \mathbb{R}^{n \times c}$ is the prediction of each node. H is the prediction for each node and represents both the starting vector and teleport set, f_θ is a neural network with the number of parameters θ. K denotes the number of power iteration steps for approximate topic-sensitive PageRank and $k \in [0, K - 2]$.

4   Applying DSS to Text Categorization

For this work, we choose to apply DSS to text categorization among NLP application because recently it has been widely used, such as spam detection to filter spam email from a mailbox. Our assumption on text categorization corresponds to Magnini's work [8], which is word sense correlates with any document only if it has the same domain as word sense. Therefore, we perform an experiment by combining the DSS results of our work with each document to measure the effectiveness of the DSS on text categorization. From Fig. 2, we apply the CNN model for text categorization. We replace the target word in the document with the gloss text of sense which has a domain that matches the document domain and then every word including gloss text is transformed into word embedding that enables to learn the features of both document and the semantic of the target word at the same time.

5   Experiments

We evaluate our method using 6-months RCV1 corpus and WordNet 3.0 with SFC resource [6] that is a gold standard domain of word senses. We choose 14 categories of documents out of 126 categories that appear in the SFC. We obtain nouns and verbs list appearing 14 categories of SFC and their corresponding sense from the WordNet.

Table 1 shows data statistics. "#doc" refers to the number of documents and "#sense" shows the number of senses. The total number of senses is 6,082 senses. Of these, we remove senses which do not have any relationship with other senses. As a result, we use 4,567 senses and create a graph. In the prediction step, the total senses in a graph are divided into a visible and a test set. The 3,000 nodes are sampling for the visible set and then we split a training set size per category vary from 5 to 35. In the second part of the visible set, we split 500 nodes for an early stopping set and the rest nodes are the number of the validation set. A test set of 1,567 nodes were sampled. For the parameter settings

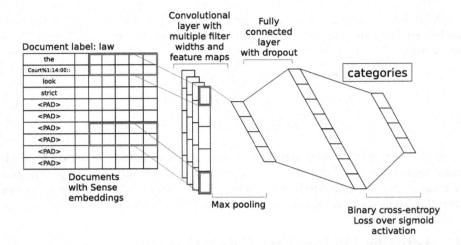

Fig. 2. CNN model for text categorization

in our model, We set the number of hidden units as 64 and the dropout rate as 0.5. We also set patience to 300 for early stopping, the maximum epochs for training as 10,000. We select Adam optimizer with a learning rate as 0.01 and the teleport probability (α) as 0.2.

For the performance comparison of DSS, we compare an unsupervised approach (WMD-DSS), a semi-supervised approach with LSTM (LSTM-DSS) and a semi-supervised approach (APPNP-DSS). For the acquisition of WMD-DSS, we firstly pre-process for RCV1 corpus. We select the only noun and verb words to fetch senses and glosses from WordNet and then create word embedding with Word2Vec by applying the skip-gram model as a training algorithm and using hierarchical softmax as model training and dimension of word vector of 100 and window size of 5. We measure the similarity of senses with Word Mover's Distance (WMD) algorithm [21]. Finally, the similarity results are applied with a simple Markov Random Walk (MRW) to ranking senses within a graph order by the most predominant senses with the high WMD value.

To obtained LSTM-DSS based on [24], we assume a predicted target sense of a sentence with LSTM that has a strong relationship with a category. We begin to extract sample sentences from sense glosses that have a target word within a sentence and then used them as input for the LSTM model that results in context vectors and proper domains of the target words. The total of sample sentences are obtained from sense glosses equal to 2,023 sentences. We configure 2,048 hidden units, 512 dimension context layer and 512 dimensional word embeddings to LSTM model whereas batch size equal to 32. Before training label propagation, we specified 20 sample set sizes with labels per category corresponding to the context vector as seeds to the unlabeled remain set. LabelSpreading function from scikit-learn [25] is used to perform label propagation and calculate similarity measure using inner product.

Table 1. The subject field codes (SFC) and reuters category correspondences.

SFC/Reuters	& Doc	& Sense
Admin/Management	5,830	401
Art/Arts	1,906	317
Economy/Economics	59,888	741
Fashion/Fashion	194	420
Finance/Funding	20,760	75
Industry/Production	12,156	394
Law/Legal	6,607	785
Meteorology/Weather	2,164	108
Military/War	15,864	696
Politics/Politics	28,668	624
Publishing/Advertising	1,230	259
Religion/Religion	1,478	627
Sports/Sports	18,410	371
Tourism/Travel	291	264

Table 2 shows the test dataset results obtained from our method "APPNP-DSS F-score" when using labeled training size equals to 20. "WMD-DSS F-score" denotes the results obtained by the topmost 20% senses from an unsupervised approach that computed sense similarity with WMD that relies on Word2Vec and ranked with PageRank. "LSTM-DSS F-score" denotes the result obtained by LSTM model with label propagation when using labeled training size equals to 20.

We can see from Table 2 that the overall performance obtained by our model attains at the macro average F-score at 0.614. The best F-score is "Meteorology" while the worst is "Finance". One of the reasons that "Finance" is less effective than other categories is that the number of neighbor nodes of the "Finance" category is lower than other categories. Another reason is the number of senses in the "Finance" category is too small compared to other categories. The number of labeled training sets consists of 20 nodes However, the number of validation sets of the "Finance" category is only 17 nodes. Consequently, it affects to tuning hyperparameter of the model in this category which deteriorates the overall performance.

To examine how the number of training data affects the overall performance, we have an experiment. The results are illustrated in Fig. 3. The training size of 20 gain the best F-score at 0.614 whereas the smallest training size of 5 attained at 0.463 F-score. We note that when the number of training data increases, the F-score increases accordingly. For acquiring DSS with WMD, it is an unsupervised method, thus we can obtain the predominant senses per category with only the topmost 20%. At the topmost 30%, we obtain the F-score of 0.480, and the F-score drops gradually until the topmost 100% which obtains a 0.248 F-score.

Table 2. The results of sense assignment

SFC	WMD-DSS F-score	LSTM-DSS F-score	APPNP-DSS F-score
Admin	0.367	0.636(+.269)	0.406(−.023)
Art	0.489	0.638(+.149)	0.561(−.077)
Economy	0.500	0.692(+.192)	0.421(−.271)
Fashion	0.719	0.676(−.043)	0.755(+.079)
Finance	0.600	0.162(−.438)	**0.215(+.053)**
Industry	0.744	0.670(−.074)	0.637(−.033)
Law	0.470	0.714(+.244)	0.635(−.079)
Meteorology	0.889	0.160(−.729)	**0.807(+.647)**
Military	0.581	0.670(+.089)	0.765(+.095)
Politics	0.356	0.700(+.344)	0.527(−.173)
Publishing	0.543	0.662(+.119)	0.676(+.014)
Religion	0.630	0.680(+.050)	0.775(+.095)
Sports	0.508	0.692(+.184)	0.804(+.112)
Tourism	0.714	0.544(−.017)	0.616(+.072)
Macro F-score	0.579	0.592(+.013)	**0.614(+0.022)**

We apply the results of our method (APPNP-DSS) as an input to text categorization and compare the performance with CNN, WMD-DSS and LSTM-DSS approach. The dataset used for text categorization, 6-month RCV1 corpus is divided into two folds consisting of 80% for the training sets, and 20% for the test sets. We then divide the training sets once, 80% for the training sets, and 20% for the validation sets. All three methods have the same CNN model configurations as shown in Table 3. We use the Optuna framework [22] for optimizing the best settings of the CNN model.

For categorizing using the CNN model, we replace the target words in the document with glosses from our prediction method. The target word is replaced only if the category of sense and category of document match. Another condition is POS of sense and POS of the target word match.

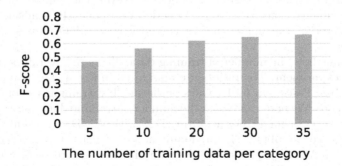

Fig. 3. The number of training data against F-score

Table 3. CNN model configurations

Description	Values	Description	Values
Input size	Maximum length × 100	A number of output categories	14
Input word vectors	Word2Vec	Filter region size	(2,3,4)
Stride size	1	Feature maps (m)	256
Filters	256 × 3	Activation function	ReLu
Pooling	1-max pooling	Dropout	Randomly selected
Dropout rate1	0.25	Dropout rate2	0.5
Hidden layers	512	Batch sizes	100
Learning rate	Predicted by Adam	Epoch	40 with early stopping
Loss function	BCE loss over sigmoid activation	Threshold value for MSF	0.5

Table 4. Classification performance

Category	CNN	WMD-DSS 20%	LSTM-DSS	APPNP-DSS
Law	0.843	0.911(+.068)	0.908(−.003)	0.961(+.053)
Finance	0.904	0.945(+.041)	0.956(+.011)	0.967(+.011)
Industry	0.793	0.899(+.106)	0.881(−.018)	0.901(+.020)
Publishing	0.723	0.819(+.096)	0.778(−.041)	0.831(+.053)
Admin	0.864	0.913(+.049)	0.948(+.035)	0.974(+.026)
Economy	0.927	0.973(+.046)	0.961(−.012)	0.978(+.017)
Art	0.730	0.773(+.043)	0.786(+.013)	**0.911(+.125)**
Fashion	0.666	0.775(+.109)	0.956(+.181)	**0.896(−.06)**
Politics	0.818	0.926(+.108)	0.931(+.005)	0.981(+.050)
Religion	0.655	0.855(+.200)	0.825(−.030)	0.916(+.091)
Sports	0.988	0.992(+.004)	0.993(+.001)	0.997(+.004)
Tourism	0.246	0.493(+.247)	0.643(+.150)	0.715(+.072)
Military	0.871	0.933(+.062)	0.907(−.026)	0.973(+.066)
Meteorology	0.842	0.885(+.043)	0.914(+.029)	0.936(+.022)
Micro F-score	0.886	0.945(+.059)	0.941(−.004)	0.970(+.029)
Macro F-score	0.776	0.864(+.088)	0.885(+.021)	**0.920(+.035)**

Table 4 shows F-score of categorization obtained by CNN, an unsupervised method, and a semi-supervised method. Overall, the result obtained by DSS shows better performance than a normal CNN model. APPNP-DSS performs better than LSTM-DSS. The best improvement is "Art" (+.125) and the worst is "Fashion" (−.06). APPNP-DSS is better than LSTM-DSS as the Macro F1-score is 0.920. One possible reason why category "Fashion" of APPNP-DSS have a lower F-score than LSTM-DSS is that the number of times the target word is replaced with the gloss text of APPNP-DSS is lower than that of LSTM-DSS. Also, if the number of words consisting of the gloss text is large, it affects categorization as well.

6   Conclusion

We proposed a semi-supervised method for acquiring the DSS based on BERT embedding and deep learning techniques. We also compare the results of our method with the unsupervised method which works well at the topmost 20% and the semi-supervised method (LSTM with label propagation). Our method can reach an F-score of 0.614 for 1,567 senses, whereas the semi-supervised LSTM with label propagation attain F-score of 0.592 for 1,743 senses and the unsupervised method at the topmost 20% obtains an F-score of 0.579 for 892 senses. Moreover, we apply our results with text categorization to examine the performance of DSS. The results of this experiment showed that our method can improve text categorization performance as it achieved a 0.920 macro F-score and 0.144 improvements compared with the CNN baseline model.

For future work, there are several approaches that can be further study. We are going to apply DSS to other NLP applications such as machine translation, question answering, and sentiment analysis. We also apply our method to other part-of-speech e.g. adjective and adverb as well as other datasets and thesaurus for quantitative evaluation of our method. Comparison to the state-of-art WSD technique [23] by using the same datasets, SemEval Check whether this dataset is correct or not is also necessary to examine the effectiveness of the method.

References

1. Koeling, R., McCarthy, D., Carroll, J.: Domain-specific sense distributions and predominant sense acquisition. In: Proceedings of the Human Language Technology Conference and Conference on Empirical Methods in Natural Language Processing, pp. 419–426 (2005)
2. Miller, G.A.: WordNet: a lexical database for English. J. Commun. ACM **38**, 39–41 (1995)
3. Rose, T., Stevenson, M., Whitehead, M.: The Reuters corpus volume 1 - from yesterday's news to tomorrow's language resources. In: Proceedings of the 3rd International Conference on Language Resources and Evaluation (LREC 2002), pp. 29–31 (2002)
4. Devlin, J., Chang, M-W, Lee, K., Toutanova, K.: BERT: pre-training of deep bidirectional transformers for language understanding. Journal of CoRR, arXiv:1810.04805, pp. 4171–4186 (2018)
5. Klicpera, J., Bojchevski, A., Günnemann, S.: Personalized embedding propagation: combining neural networks on graphs with personalized page rank. Journal of CoRR, arXiv:1810.05997 (2018)
6. Magnini, B., Cavaglià, G.: Integrating subject field codes into WordNet. In: Proceedings of the Second International Conference on Language and Evaluation (LREC 2000), pp. 1413–1418 (2000)

7. Gale, W.A., Church, K.W., Yarowsky, D.: One sense per discourse. In: Proceedings of the workshop and Speech and Natural Language, pp. 233–237 (1992)
8. Magnini, B., Strapparava, C., Pezzulo, G., Gliozzo, A.: Using domain information for word sense disambiguation. In: Proceedings of the the Second International Workshop on Evaluating Word Sense Disambiguation Systems, pp. 111–114 (2001)
9. Fukumoto, F., Suzuki, Y.: Identifying domain-specific senses and its application to text classification. In: Proceedings of the International Conference on Knowledge Engineering and Ontology Development, pp. 263–268 (2010)
10. Pasini, T., Navigli, R.: Train-o-Matic: supervised Word Sense Disambiguation with no (manual) effort. J. Artif. Intell. **279**, 103215 (2019)
11. Mikolov, T., Chen K., Corrado, G., Dean, J.: Efficient estimation of word representations in vector space. In: Proceedings of 1st International Conference on Learning Representations, ICLR 2013 (2013)
12. Perozzi, B., Al-Rfou, R., Skiena, S.: DeepWalk: Online learning of social representations. Journal of CoRR, arXiv:1403.6652 (2014)
13. Agirre, E., Lacalle, O.L., Soroa, A.: Random walks for knowledge-based word sense disambiguation. J. Comput. Linguist. **40**, 57–84 (2014)
14. Brin, S., Page, L.: The anatomy of a large-scale hypertextual web search engine. J. Comput. Netw. **30**, 107–117 (1998)
15. Kipf, T. N., Welling, M.: Semi-Supervised Classification with Graph Convolutional Networks. Journal of CoRR, arXiv:1609.02907 (2016)
16. Li, Q., Han, Z., Wu, X-M: Deeper Insights into Graph Convolutional Networks for Semi-Supervised Learning. Journal of CoRR, arXiv:1801.07606 (2018)
17. Kim, Y.: Convolutional Neural Networks for Sentence Classification. Journal of CoRR, arXiv:1408.5882 (2014)
18. Liu, J., Chang, W.C., Wu, Y., Yang, Y.: Deep learning for extreme multi-label text classification. In: Proceedings of 40th International ACM SIGIR conference on Research and Development in Information Retrieval (2017)
19. Wang, J., Wang, Z., Zhang, D., Yan, J.: Combining knowledge with deep convolutional neural networks for short text classification. In: Proceedings of the 26th International Joint Conference on Artificial Intelligence IJCAI-17 (2017)
20. Manning, C., Surdeanu, M., Bauer, J., Finkel, J., Bethard, S., McClosky, D.: The Stanford CoreNLP natural language processing toolkit. In: Proceedings of 52nd Annual Meeting of the Association for Computational Linguistics: System Demonstrations, pp. 55–60 (2014)
21. Kusner, M., Sun, Y., Kolkin, N., Weinberger, K.: From word embeddings to document distances. In: Proceedings of Machine Learning Research, pp. 957–966 (2015)
22. Akiba, T., Sano, S, Yanase, T., Ohta, T., Koyama, M.: Optuna: a next-generation Hyperparameter Optimization Framework. Jounal of CoRR, arXiv:1907.10902 (2019)
23. Bevilacqua, M., Navigli, R.: Breaking through the 80% glass ceiling: raising the state of the art in word sense disambiguation by incorporating knowledge graph information. In: Proceedings of the 58th Annual Meeting of the Association for Computational Linguistics, pp. 2854–2864 (2020)
24. Yuan, D., Richardson, J., Doherty R. Evans, C., Altendorf, E.: Semi-supervised Word Sense Disambiguation with Neural Models. In: Proceedings of COLING 2016, the 26th International Conference on Computational Linguistics: Technical Papers, pp. 1374–1385 (2016)
25. Zhou, D., Bousquet, O., Lal, T.N., Weston, J., Schölkopf, B.: Learning with local and global consistency. In: Proceedings of Advances in Neural Information Processing Systems, vol. 16, pp. 321–328 (2004)

Processing-Response Dependence on the On-Chip Readout Positions in Spin-Wave Reservoir Computing

Takehiro Ichimura, Ryosho Nakane, and Akira Hirose[✉]

Department of Electrical Engineering and Information Systems,
The University of Tokyo, Tokyo 153-8656, Japan
`ahirose@ee.t.u-tokyo.ac.jp`

Abstract. This paper reports and discusses the processing response dependence on a spin-wave reservoir chip, a natural computing device, to present one of the important steps to design a spin-wave reservoir computing hardware. As an example, we deal with a sinusoidal-square wave distinction task, where signals with a certain duration switch to each other at random. Observation of the transient response provides us with information useful for determining chip size and other parameters. Accumulation of this type of investigations will elucidate how we should design a spin-wave reservoir chip.

Keywords: Natural computing · Reservoir computing · Spin-wave reservoir chip

1 Introduction

Reservoir computing (RC) is a framework of processing time-serial data stably with low calculation cost [9,13,27], and is realizable by using physical phenomena directly [23]. Many proposals have been made so far such as the "bucket" RC [2], soft matter robots [14,15] as well as circuit-based one [29]. Physical RC will realize low-power artificial intelligence (AI) devices for, e.g., real-time applications such as sensing and imaging in the near future. Among others, optical RC is one of the most promising directions [1,12,21]. For example, an optical RC is found showing the "edge of chaos" characteristics, which enhance the RC performance, in a certain range of physical parameters without any special treatment [28].

Spin is another promising physics. In a wide scope of information processing, spin has been attracting many researchers for logic processing [10]. However, it is rather new for spin phenomena to target RC [20,22], including the use of spin torque [24,25]. Recently, the authors proposed a spin-wave reservoir chip

A. Hirose—This work was supported in part by the New Energy and Industrial Technology Development Organization (NEDO) under Project JPNP16007, and in part by Tohoku University RIEC Cooperative Research Project.

ⓒ Springer Nature Switzerland AG 2021
T. Mantoro et al. (Eds.): ICONIP 2021, LNCS 13110, pp. 296–307, 2021.
https://doi.org/10.1007/978-3-030-92238-2_25

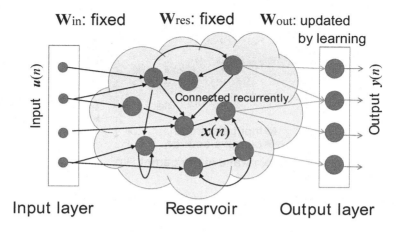

Fig. 1. Basic construction of a reservoir computing network.

[16,18]. The spin-wave chip utilizes not only propagation but also its nonlinearity, hysteresis, and other specific characteristics of spin waves to map time-serial input signals into a higher dimensional information space [5,17,19]. The use of waves also leads to low-power consumption [3,6,11]. Another advantage is the "non-elaboration," which means no need of exceedingly precise elaboration in the chip design and fabrication stages [4].

A spin-wave reservoir chip is based on the spin-wave spatiotemporal dynamics. Its features are explored actively with various scopes. The authors investigated its characteristics in exclusive OR (XOR) learning and processing [7] and effective readout electrode arrangements for sinusoidal/square-wave discrimination [8]. As the results, it was shown that a chip can map input signals into sufficiently higher information space with a realistic number of readout electrodes.

This paper investigates further the influence of on-chip readout positions on the processing response in a spin-wave reservoir computing chip. First, we check the spin waveform in a reservoir chip as well as the instantaneous output signal when we prepare a sufficient but realistic number of readout electrodes. In particular, we examine when the errors occur. We find that the errors appear at the transient of sinusoidal-to-square and square-to-sinusoidal input waveforms. The errors of this type are unavoidable in time-serial processing in general. We investigate the mechanisms of the transient errors occur experimentally. The result will be a design building block for spin-wave RC chips practically workable in the near future.

2   Spin-Wave Reservoir Chip

Figure 1 shows the basic construction of reservoir computing processing networks. Time-serial signals $u(n)$ as a function of discrete time n are weighted by

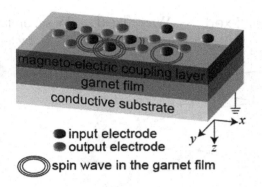

Fig. 2. Basic structure of the reservoir chip [18].

$\mathbf{W}_{\text{in}}$ and fed to neurons in the reservoir. The neurons, connected to each other with weights $\mathbf{W}_{\text{res}}$, process the signals recurrently to generate reservoir-internal signals at the neurons as

$$x(n) = f\left((1 - \gamma_{\text{leak}})x(n-1) + \gamma_{\text{leak}}(\mathbf{W}_{\text{in}}u + \mathbf{W}_{\text{res}}x(n-1))\right) \qquad (1)$$

where γ_{leak}, μ_0 and f are leak rate, permeability of free space and nonlinear activation function of reservoir neurons, respectively. The output neurons read the signals through the readout weights $\mathbf{W}_{\text{out}}$ to generate their outputs as

$$y(n) = f_{\text{out}}\left(\mathbf{W}_{\text{out}}x(n)\right) \qquad (2)$$

where f_{out} is another activation function of the output neurons, which is the identical function or a nonlinear function. The functions work in an element-wise manner.

Figure 2 presents the structure of the reservoir chip [18] where we utilize the spin waves propagating in the garnet film. There are several types of transducers to convert electrical signals into spin-wave signals, with which the chip exchange information at the input and output (readout) electrodes. The input signals propagates in the chip, being mapped into a higher dimensional information space through nonlinearity, hysteresis, dispersion, anisotropy, and so on. This is a heavy signal processing part in conventional neural networks. However, the physical RC performs this processing with physical mechanisms, resulting in the reduction of computational cost and power consumption. After this main processing, the learning process can be very light but the total performance is very high. This is the major advantage of the physical RC.

The spin wave is a wave carried by magnetization vector M. The dynamics is represented by Landau-Lifshitz-Gilbert equation as

$$\frac{dM}{dt} = -\gamma\mu_0 M \times H_{\text{eff}} + \frac{\alpha}{M_{\text{s}}}\left(M \times \frac{dM}{dt}\right) \qquad (3)$$

where γ and α are the material-specific gyromagnetic constant and the damping constant, respectively, and the effective magnetic field H_{eff} is the sum of exter-

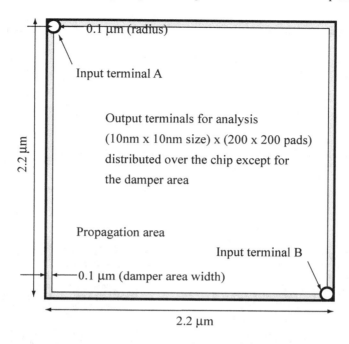

Fig. 3. Dimensions of the garnet film, damper region and input electrodes used in the numerical analysis [7].

nal magnetic field H_{ext}, demagnetization field H_d, uniaxial magnetocrystalline anisotropy H_k related to saturation field M_s, and exchange field H_{ex} determined by M_s and exchange stiffness constant A_{ex}.

3 Numerical Analysis of Reservoir Response

3.1 Learning in Sinusoidal and Square Wave Distinction and Its Processing Results

Figure 3 shows the chip structure analyzed here. The total chip area is $2.2 \times 2.2\ \mu m^2$, thickness is 100 nm, and the damping constant is $\alpha = 0.001$ in the propagation area (inner part) while $\alpha = 1$ in the damping area (edge), the saturation magnetic field is $M_s = 100$ kA/m and the exchange stiffness is $A_{EX} = 3.6 \times 10^{-12}$. We prepare two input electrodes at the two corners. For the numerical analysis, we use MuMax3 [26] with a 50 nm mesh in the three directions and a 0.01 ns time step.

To both of the input terminals, we feed the input signals shown in Fig. 4, namely, sinusoidal wave and square wave as a development up to the fourth. For the both, the (fundamental) frequency is 2.5 GHz. We modulate the uniaxial magnetic anisotropy with the input signals between a lower value $K_U^L = 1$ kJ/m^3 and a higher value $K_U^H = 10$ kJ/m^3.

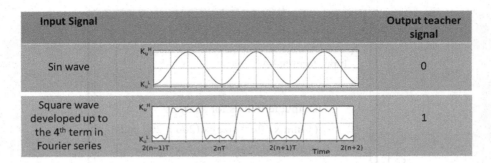

Fig. 4. Sinusoidal and square input signal waveforms and corresponding output teacher signals.

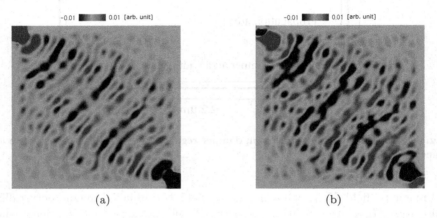

Fig. 5. A capture of the spatial representation of the spin-wave x-component when (a) sinusoidal or (b) square signal is input.

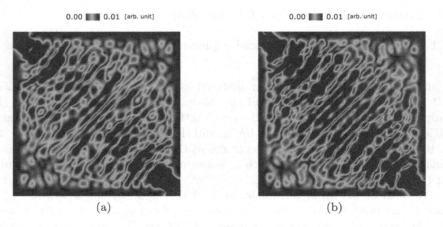

Fig. 6. A capture of the spatial representation of the spin-wave amplitude when (a) sinusoidal or (b) square signal is input.

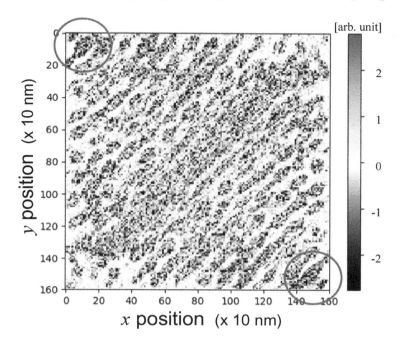

Fig. 7. Spatial representation of the output neuron weights when we virtually prepare a readout electrodes at every calculation mesh and input sinusoidal/square waves switched at random [8]. (Color figure online)

Figure 5 shows a distribution example of the x component of the spin wave. Spin waves evoked at the input electrode propagate around. The damper area on the edges suppresses the reflections. Though the waveform for the sinusoidal input is similar to that for the square wave, we find the details different from each other.

Figure 6 presents the corresponding amplitude distribution. In the present numerical experiment, an output neuron reads the amplitude values through the weights to generate a time-serial output signal. We find the details are different again.

The output signal of the single neuron decides sinusoidal or square wave with a simple threshold. As shown in the right column in Fig. 4, the teacher signal is 0 for the sinusoidal and 1 for the square wave. We employ a minibatch learning process by generating a random sinusoidal/square wave sequences, each of which has duration of 1,280 time steps = 12.8 ns time duration corresponding to 12.8 ns × 2.5 GHz = 32 wave crests.

Figure 7 is the readout weight distribution when we put an electrode virtually at every calculation mesh, which was published in Ref. [8] but shown here again for need in the following analysis. The weights show a scattered distribution, and their absolute values are within a modest range. There are almost-zero regions

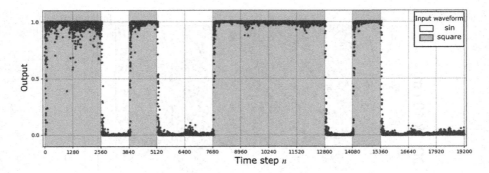

Fig. 8. Output neuron waveform when we prepare realistic number of electrodes (289) and input sinusoidal/square waves switched at random [8].

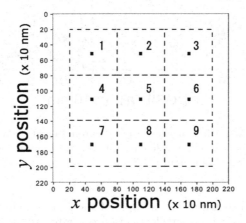

Fig. 9. Nine tiles and its number labels to read spin-wave signals at the centers of the respective tiles.

extending in belts, which reflects the spin-wave propagation. In a close observation, we find the absolute values are a little larger near the input electrodes marked by red circles.

Figure 8 presents an example of the output signals after learning when a random sinusoidal/square sequence is fed to the reservoir [8]. Here, the number of the readout electrodes is 289 realistically and located at random. The learning is performed in the same way. We find a good performance of the distinction. We also find that errors occur at the transient region, that is, at the switching time region from sinusoidal to square and vice versa. This error is inevitable in a RC to a certain extent, reflecting the RC essence of time-serial processing by utilizing a memory effect inside. At the same time, the task to distinguish sinusoidal and square waves also requires an observation time of, at least, one cycle of the wave essentially. However, in Fig. 8, we find a latency of about 100 steps = 1 ns and

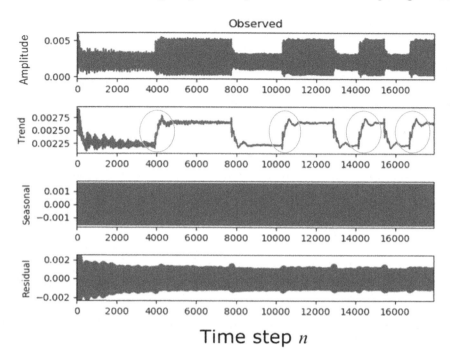

Time step n

Fig. 10. Output spin-wave amplitude, trend, seasonal and residue at the center of the chip (= center of Tile 5) when we input sinusoidal/square waves switched at random.

a time lag of first order with a time constant of about 2 ns. They are a little longer than a unit cycle of 2.5 GHz wave, i.e., 0.4 ns.

3.2   Origins of the Latency and the Time Lag

In this section, we discuss the origins of the latency and the time lag. As shown in Fig. 9, we divide the chip into nine tiles numbered from 1 to 9.

First, we check the response at the center (black dot) in Tile 5. Figure 10 shows (top) the amplitude of the spin wave and its decomposition as the "trend" component, the "seasonal (steady oscillation)" component, and (bottom) the residue, which are obtained by using a default Python function for rough estimation. In the trend component curve, the low value time regions show the sinusoidal inputs while the high regions correspond to square ones. We observe the transient at the switching time points.

Figure 11 shows averaged waveforms observed at the nine points at the tile centers for rise (sinusoidal to square), marked by ovals, or falling (square to sinusoidal) transient regions observed in the time serial response. Among them, Tile 1 and Tile 9 present a quick rise and fall.

In the response, the latency (100–200 time steps, 1–2 ns depending on tile) is considered to reflect the propagation time of the spin wave, generated at

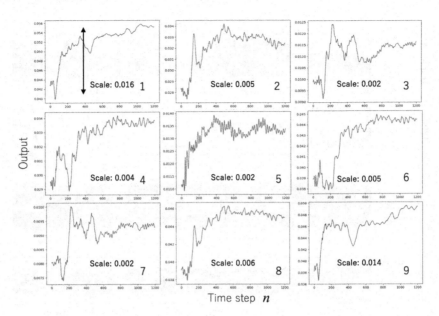

Fig. 11. Averaged rise response (sinusoidal to square wave) with different vertical scales indicated respectively.

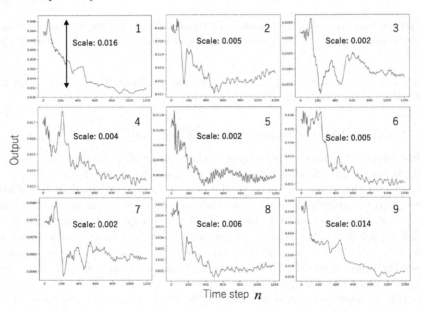

Fig. 12. Averaged falling response (square to sinusoidal wave) with different vertical scales indicated respectively.

the input electrodes, traveling to an observation point. This is determined by the chip size, electrode distance and wave velocity. In the same way, the time lag (about 200 time steps, 2 ns) is determined by the wave velocity, dispersion and other dynamics. They are different from the time essentially required for sinusoidal/square wave distinction. In this sense, this task is a complicated one including the two different time constants. However, the spin-wave RC learned the waves to reduce the errors to the minimum in the given condition.

This result is consistent with the weight distribution presented in Fig. 7. That is, a larger readout weights are needed at around the input electrodes in order to obtain a switching information without delay. In the present chip configuration and the task, the amplitude is large around the input electrodes, and simultaneously the weights are also large there. Figure 12 shows the same for the falling transient. We find that the latency and the time lag show the same tendency as those of the rise in Fig. 11.

A spin-wave RC learns the task to realize a target waveform with transient response by including the latency and the time lag. The design of a spin-wave chip should roughly include such temporal response determined by the size and electrode positions.

4   Conclusion

This paper presented a set of numerical experiment results to discuss a possible chip design of a spin-wave RC device, a natural computing hardware. A RC chip does not need elaboration since it works after learning. Nonetheless, however, we need to roughly determine its size and construction. Such discussion will lead to a meaningful design procedure needed in the fabrication of near-future RC hardware.

Acknowledgment. The authors thank Dr. T.Yamane, Dr. J.B.Heroux, Dr. H.Numata, and D.Nakano of IBM–Research Tokyo and Dr. G.Tanaka of the University of Tokyo for their helpful discussion.

References

1. Bueno, J., et al.: Reinforcement learning in a large-scale photonic recurrent neural network. Optica 5(6), 756–760 (2018)
2. Fernando, C., Sojakka, S.: Pattern recognition in a bucket. In: Banzhaf, W., Ziegler, J., Christaller, T., Dittrich, P., Kim, J.T. (eds.) ECAL 2003. LNCS (LNAI), vol. 2801, pp. 588–597. Springer, Heidelberg (2003). https://doi.org/10.1007/978-3-540-39432-7_63
3. Hirose, A.: Physical reservoir computing and complex-valued neural networks. In: International Conference on Neuromorphic Systems. Oak Ridge National Laboratory, Oak Ridge (online) (July 2020)
4. Hirose, A., Nakane, R., Tanaka, G.: Keynote speech: Information processing hardware, physical reservoir computing and complex-valued neural networks. In: Kimura, M. (ed.) IEEE International Meeting for Future of Electron Devices, Kansai 2019 (IMFEDK) Kyoto, pp. 19–24 (November 2019)

5. Hirose, A., et al.: Physical reservoir computing: possibility to resolve the inconsistency between neuro-AI principles and its hardware. Aust. J. Intell. Inf. Process. Syst. (AJIIPS) **16**(4), 49–54 (2019)
6. Hirose, A., et al.: Proposal of carrier-wave reservoir computing. In: Cheng, L., Leung, A.C.S., Ozawa, S. (eds.) ICONIP 2018. LNCS, vol. 11301, pp. 616–624. Springer, Cham (2018). https://doi.org/10.1007/978-3-030-04167-0_56
7. Ichimura, T., Nakane, R., Tanaka, G., Hirose, A.: Spatial distribution of information effective for logic function learning in spin-wave reservoir computing chip utilizing spatiotemporal physical dynamics. In: 2020 International Joint Conference on Neural Networks (IJCNN), pp. 1–8 (2020). https://doi.org/10.1109/IJCNN48605.2020.9207629
8. Ichimura, T., Nakane, R., Tanaka, G., Hirose, A.: A numerical exploration of signal detector arrangement in a spin-wave reservoir computing device. IEEE Access **9**, 72637–72646 (2021). https://doi.org/10.1109/ACCESS.2021.3079583
9. Jaeger, H.: The "echo state" approach to analysing and training recurrent neural networks-with an erratum note. Bonn, Germany: German Natl. Res. Cent. Inf. Technol. GMD Tech. Rep. **148**(34), 13 (2001)
10. Kanazawa, N., et al.: Demonstration of a robust magnonic spin wave interferometer. Sci. Rep. **6**(1), 30268 (2016). https://doi.org/10.1038/srep30268
11. Katayama, Y., Yamane, T., Nakano, D., Nakane, R., Tanaka, G.: Wave-based neuromorphic computing framework for brain-like energy efficiency and integration. IEEE Trans. Nanotechnol. **15**(5), 762–769 (2016)
12. Larger, L., Baylón-Fuentes, A., Martinenghi, R., Udaltsov, V.S., Chembo, Y.K., Jacquot, M.: High-speed photonic reservoir computing using a time-delay-based architecture: million words per second classification. Phys. Rev. X **7**, 011015 (2017)
13. Maass, W., Natschläger, T., Markram, H.: Real-time computing without stable states: a new framework for neural computation based on perturbations. Neural Comput. **14**(11), 2531–2560 (2002)
14. Nakajima, K., Hauser, H., Li, T., Pfeifer, R.: Information processing via physical soft body. Sci. Rep. **5**, 10487 (2015)
15. Nakajima, K., Li, T., Hauser, H., Pfeifer, R.: Exploiting short-term memory in soft body dynamics as a computational resource. J. R. Soc. Interface **11**, 20140437 (2018)
16. Nakane, R.: On-chip reservoir computing device utilizing spin waves. J. IEICE **102**(2), 140–146 (2019). (in Japanese)
17. Nakane, R., Tanaka, G., Hirose, A.: Demonstration of spin-wave-based reservoir computing for next-generation machine-learning devices. In: International Conference on Magnetism (ICM) 2018 San Francisco, pp. 26–27 (July 2018)
18. Nakane, R., Tanaka, G., Hirose, A.: Reservoir computing with spin waves excited in a garnet film. IEEE Access **6**, 4462–4469 (2018)
19. Nakane, R., Tanaka, G., Hirose, A.: Numerical analysis on wave dynamics in a spin-wave reservoir for machine learning. In: IEEE/INNS International Joint Conference on Neural Networks (IJCNN) 2019 Budapest, N-20170 (2019). https://doi.org/10.1109/IJCNN.2019.8852280
20. Nomura, H., et al.: Reservoir computing with dipole-coupled nanomagnets. Japan. J. Appl. Phys. **58**(7), 070901 (2019). https://doi.org/10.7567/1347-4065/ab2406
21. Paquot, Y., et al.: Optoelectronic reservoir computing. Sci. Rep. **2**, 287 (2012)
22. Prychynenko, D., et al.: Magnetic skyrmion as a nonlinear resistive element: a potential building block for reservoir computing. Phys. Rev. Appl. **9**, 014034 (2018)
23. Tanaka, G., et al.: Recent advances in physical reservoir computing: a review. Neural Netw. **115**, 100–123 (2019)

24. Torrejon, J., et al.: Neuromorphic computing with nanoscale spintronic oscillators. Nature **547**(7664), 428 (2017)
25. Tsunegi, S., et al.: Physical reservoir computing based on spin torque oscillator with forced synchronization. Appl. Phys. Lett. **114**(16), 164101 (2020)
26. Vansteenkiste, A., Leliaert, J., Dvornik, M., Helsen, M., Garcia-Sanchez, F., Van Waeyenberge, B.: The design and verification of mumax3. AIP Adv. **4**(10), 107133 (2014)
27. Verstraeten, D., Schrauwen, B., D'Haene, M., Stroobandt, D.: An experimental unification of reservoir computing methods. Neural Netw. **20**(3), 391–403 (2007). https://doi.org/10.1016/j.neunet.2007.04.003, http://www.sciencedirect.com/science/article/pii/S089360800700038X, echo State Networks and Liquid State Machines
28. Yamane, T., et al.: Simulation study of physical reservoir computing by nonlinear deterministic time series analysis. In: International Conference on Neural Information Processing (ICONIP) 2017 Guangzhou, pp. 639–647 (2017)
29. Yi, Y., et al.: FPGA based spike-time dependent encoder and reservoir design in neuromorphic computing processors. Microprocess. Microsyst. **46**, 175–183 (2016)

Advances in Deep and Shallow Machine Learning Algorithms for Biomedical Data and Imaging

A Multi-task Learning Scheme for Motor Imagery Signal Classification

Rahul Kumar$^{(\boxtimes)}$ and Sriparna Saha

Indian Institute of Technology, Patna 801103, Bihar, India
{rahul_1911mt11,sriparna}@iitp.ac.in

Abstract. Motor imagination is an act of thinking about body motor parts and motor imagery signals are the brain activities generated while performing motor imagination. Electroencephalogram (EEG) is a non-invasive way by which brain activities can be recorded with electrodes placed on the scalp. EEG measurements recorded for every other individual are different even if they think about same moving body parts. And this makes motor imagery signal classification difficult. Many authors have proposed various machine learning and deep learning approaches for motor imagery signal classification. To the best knowledge of authors, in most of the studies, subject specific model is trained as EEG signature is subject specific. However, even when EEG measurments from different individuals are recorded, if they are for same motor imagination, then, there must be some hidden common features for a specific motor imagination across all individuals. With subject specific models, those features can't be learnt. We have proposed yet another deep learning approach for motor imagery signal classification where a single deep learning system is trained with a multi-task learning approach. The results illustrate that a single mult-task learnt model performs even better than subject specific trained models.

Keywords: EEG · Motor imagery · Deep learning · Multi-task learning

1 Introduction

A Brain Computer Interface (BCI) is a system that measures Central Nervous System (CNS) activity and converts it into artificial output such that it can replace or improve natural CNS output [1]. In other words, BCI is a neuro-imaging system which helps in mapping of neural signal of an individual to his/her cognitive state. Signal acquisition, feature extraction, and signal classification are the basic building blocks of a BCI system. Motor imagery signal classification plays a key role for a BCI system to work properly.

BCI can be categorized into two main categories namely invasive BCI system and non-invasive BCI system with respect to different signal acquisition methods. Electrodes are implanted inside the brain to acquire brain signals in invasive

© Springer Nature Switzerland AG 2021
T. Mantoro et al. (Eds.): ICONIP 2021, LNCS 13110, pp. 311–322, 2021.
https://doi.org/10.1007/978-3-030-92238-2_26

BCI system while in non-invasive BCI system, brain signals are recorded from the scalp itself. EEG is the most popular non-invasive method for acquiring brain signals as it is economical, portable, reliable and has excellent temporal resolution[1].

Brain waves recorded via EEG are categorized in 5 major frequency bands as follows : delta waves(0.5 Hz–3 Hz), theta waves(4 Hz–7 Hz), alpha waves(8 Hz–13 Hz), beta waves(14 Hz–30 Hz) and gamma waves(> 30 Hz) [2]. And according to various studies it is found that neural signal while performing motor imagery mostly consists of alpha and beta waves. So feature extraction in frequency domain prior to classification becomes an important preprocessing step for motor imagery signal classification.

Common Spatial Filter (CSP) [3,4] is one of the most common feature extraction methods which is used in BCI field for extracting features from motor imagery data. Winner of BCI IV 2008 competition used a variant of CSP to attain the state-of-the-art results. Other than CSP, feature extraction methods like Principal Component Analysis(PCA), Independent Component Analysis (ICA) [5] etc. are also being used. Earlier, conventional machine learning methods were mostly employed for motor imagery signal classification. Conventional machine learning algorithms like Support Vector Machine(SVM), Bayesian Classifier, Nearest neighbour classifier, Linear Discriminant Analysis (LDA) etc. were used for motor imagery signal classification [6–8]. With the recent development in deep learning algorithms, those are being applied for solving motor imagery classification problems also. Deep learning algorithms like Convolutional Neural Network(CNN) [9–11], Recurrent Neural Networks(RNN) [12], Restricted Boltzman Machine(RBM) [13] etc. are being used in recent studies for motor imagery signal classification. We have proposed a yet another deep learning technique which is a multi-task learning based approach.

A multi-task model is preferred over single task model architecture when two or more tasks are correlated to each other. It leverages correlation among multiple tasks and learns better features from the shared representation as compared to single task model. Multi-task learning has been proved effective in various domains. In the field of natural language processing, learning similar tasks like emotion, sentiment and depression with a multi-task model have been already proved to produce better results as compared to single task models [14,15]. A multi-task framework achives better generalization, improves the performance of each task and requires only one unified model for all the tasks. We have tried to implemented multi-task learning for classification of motor imagery signal where each subject's motor imagery has been considered a separate task.

In this study, BCI IV-2b dataset (described in detail in Sect. 4.1) consisting of motor imagery data for 9 different subjects is used. The main issue with motor imagery signal classification is that different individuals have different EEG signatures even for the same motor imagination. That is, two different persons performing same motor imagery tasks will have slightly different EEG recordings, which makes its classification difficult. And to address this problem, in

---

[1] (Source code avaliable at https://github.com/RahulnKumar/EEG-Multi-Tasking).

most of the studies, separate models are trained for different subjects. Although
EEG signatures are different for different subjects but they are highly correlated
tasks as they are for same motor imagination. Therefore, we have proposed a
single multi-task learnt model where we do not need to train 9 different models
for 9 subjects but only one unified multi-task model is capable of handling multiple subjects. In our multi-task setup, motor imagery classification of different
subjects is considered as different tasks. We have experimented with three different kinds of multi-task architectures, namely, fully-shared multi-task model,
private-shared multi-task model and adversarial multi-task model architectures.
Subject specific models, i.e., single task models were also trained and compared
with results of multi-task models. The results we obtained for single multi-task
model were even better than 9 subject specific trained single task models.

The remainder of the paper is organised as follows: Sect. 2 discusses related
works and studies motor imagery classification. The proposed multi-task learning
approach is described in Sect. 3. Section 4 describes experimental results and
details of the datasets used. Section 5 summarizes the results of this work and
draws conclusions.

2   Related Works

Various approaches for motor imagery signal classification have been described
in this section. Before advancements in deep learning algorithms, mostly conventional machine learning algorithms like support vector machine, Bayesian
classifier, nearest neighbour classifiers etc. were used for motor imagery signal
classification. In 2008, BCI-IV competition was held where motor imagery signal classification has to be carried out for 9 subjects. Various conventional and
machine learning methods were proposed to solve this problem. Filter Bank
Common Spatial Pattern (FBCSP) is a novel machine learning approach proposed by Ang et al. [3] which won that competition. In FBCSP, common spatial
filtering was applied on the band pass filtered raw EEG data. And post feature
extraction, classification was carried out using machine learning algorithms like
Bayesian classifier, K-nearest neighbour, SVM, LDA etc.

Deep learning based algorithms for motor imagery signal classification have
also been proposed in studies in recent years. Authors in [13] adopted a deep
learning scheme based on Restricted Boltzmann Machines(RBMs) for motor
imagery classification. They first converted time domain EEG signal into frequency domain using Fast Fourier Transform (FFT) and then used Wavelet
Package Decomposition to extract features. A combined CNN and SAE model
for motor imagery classification has been proposed in [9]. In this work they
first considered Short Time Fourier Transform (STFT) of the band-pass filtered
raw EEG signals to get image representation of the EEG data. Then they used
resulting EEG data in image form to train their model. Ping et al. [12] proposed a Long Short Term Memory (LSTM) framework where they have used
one dimension-aggregate approximation for EEG feature extraction. And further they employed channel weighing technique to improve their model. Ko et al.
[16] introduced a novel Recurrent Spatio-Temporal Neural Network (RSTNN)

framework for motor imagery signal classification. With RSTNN, EEG feature extraction is being done in two parts, namely, temporal feature extractor and spatial feature extractor, and three neural networks are used for classification.

As discussed earlier, EEG signature is subject specific even for the same motor imagination among different subjects. To solve this problem, most of the studies had proposed solution where they simply train separate model for each subject. But some hidden features which are common to all subjects are not being learnt by subject specific trained model. Thus, in this study, we have experimented with three different types of multi-task architectures and compared their results with subject specific trained single task models.

3   Proposed Methodology

In order to classify motor imagery signals, we have implemented three different kinds of multi-task architectures. Motor imagery data of 9 different subjects has been used from BCI IV 2008 competition. This dataset is described in detail in Sect. 4.1. In the following subsections, motor imagery signal preprocessing steps and different multi-task architectures used have been explained.

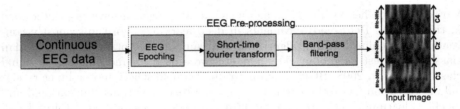

Fig. 1. Preprocessing pipeline

3.1   Preprocessing

We have designed our preprocessing pipeline similar to Tabar et al. [9]. EEG data from BCI competition IV-2b dataset has been used. It consisted of 3 channel EEG recordings (C3, Cz, C4) of 9 different subjects who were performing motor imagination of their left and right hands. At first, EEG epoching was employed, i.e., we extracted all the motor imagery trials from continuous raw EEG time series data. It has been proved in different studies, that frequency domain features provide better results than time domain features in case of motor imagery signal classification [13]. So, we considered STFT such that both time domain and frequency domain features can be leveraged. We considered the STFT of each motor imagery trials with window size of 0.128 s and time lapse of 0.028 s. The resulting spectrogram was band pass filtered for alpha waves (8 Hz–13 Hz) and beta waves(14 Hz–30 Hz). We also normalized the band pass filtered spectrogram which is in image form so that it can be trained properly with various neural network models. Finally, the three STFT outputs corresponding to three electrodes EEG recordings were stacked on the top of each other as it can be seen in Fig. 1. In this way, input image that we received for training is having time, frequency and location information of the EEG signals.

3.2 Learning Architectures

In this section, we've described different learning architectures which we used for motor imagery signal classification. As discussed earlier, EEG signature is different for different individuals even for the same motor imagination, so at first we trained 9 subject specific models for each subject. And then different multi-task models were implemented to train motor imagery signals of all the 9 subjects with one model. The multi-task models implemented are fully-shared, private-shared and adversarial-shared multi-task models.

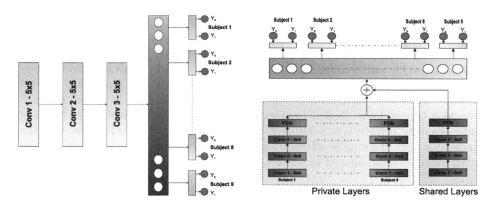

Fig. 2. Conventional multi-task model

Fig. 3. Private-shared multi-task model

Single Task Model. Subject specific models, i.e., separate single task models were trained for each of the 9 subjects. Single task model consisted of 3 convolution-pooling layers, 1 fully connected layer and a softmax output layer. To allow each layer to learn more independently, batch normalization is used. Raw EEG signals were transformed into image forms with the preprocessing pipeline as described in Sect. 3.1 and then these images were given as inputs to CNN. CNNs were used in the model as they are capable of extracting state-of-the-art learning features from the image representation. Outputs from CNN were fed to a fully connected layer and then to a softmax classifier for motor imagery classification.

Conventional-Multitask Models. These can also be called fully-shared multi-task models. Conventional multi-task model consisted of a fully shared three convolution-pooling layers and a fully connected layer as shown in Fig. 2. In place of one softmax output layer as in the case of single task model, it consisted of 9 separate softmax output layers where each output corresponds to a particular subject motor imagery classification label. Fully shared layer comprised of three convolution-pooling layers and one fully connected layer which were common for all the subjects. And the softmax output layers were task specific layers. When several tasks are highly correlated (here motor imagery data for all the subjects), then it is better to train a single multi-task model rather than several single task models. However, the caveat with this conventional multi-task model is that it is not able to learn subject specific features properly as the task specific layers are not private for each subject. These task specific layers are directly connected to shared layers and hence affect in learning subject specific traits. So, to address this issue, private-shared multi-task model was trained.

Private-Shared Multitask Model. It consists of dedicated separate task specific network for each subject and a shared network common for all subjects as shown in Fig. 3. Unlike fully shared multi-task network, where task specific layers were originating from shared layers, it has completely independent task specific private network for each subject. Input EEG images are fed to private and shared layers separately and their outputs are concatenated and then fed to a fully connected and softmax layers. The caveat with this model is that it became a relatively complex model with 9 different private layers. It is almost equivalent to training nine subject specific models with a shared layer. And the sole purpose of these 9 private layers is to learn subject specific features. A simple adversarial network can also be employed to learn subject specific features. So, instead of training nine separate private layers, subject specific traits can also be learned with the generator part of a simple adversarial network. Therefore, we also experimented with an adversarial-shared multitask model.

Adversarial Multitask Model. This architecture consists of a shared network and an adversarial network as shown in Fig. 4. It is very simple architecture as compared to the private-shared multi-task architecture described above. In place of nine different private layers, it has one adversarial network whose task is to learn task specific features. The generator in adversarial network tries to learn subject specific features and the discriminator does subject classification. So, while training, adversarial network tries to learn subject specific features and shared network tries to learn some hidden features which are common to all the subjects. And finally outputs of the generator network (subject specific features) and shared network are fused together to make the final prediction. This type of network does ensure that task specific layers and shared layers learn different sets of parameters.

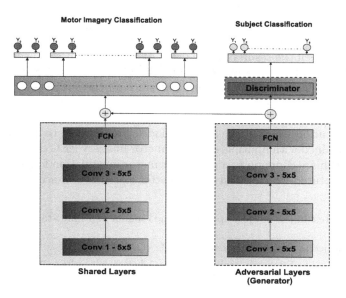

Fig. 4. Adversarial-shared multi-task model

4   Dataset, Results and Discussion

4.1   Dataset

We have conducted our experiments with BCI IV-2b dataset [17]. This dataset consisted of 3 channel EEG recordings (C3, Cz, C4) with a sampling rate 250 Hz of 9 subjects while they were performing motor imagination. All the 9 subjects were performing two different motor imagery tasks, viz. right hand and left hand motor imaginations. Motor imagery data was recorded in 5 sessions. Each session comprised of 6 runs separated by short breaks. And each run comprised of 20 trials (10 trial for left hand motor imagination and 10 for right hand motor imagination). Out of five sessions, first two sessions were conducted without feedback while the last three with feedback. And the first three sessions were provided with labels. In our study we have used first three sessions for training and testing. First two sessions and 50% of third session are used for training and rest 50% of third session are used as testing dataset for all the subjects except for subject 1. For subject 1, second session EEG data[2] was missing and therefore, only first session and 50% of third session are used for training and rest 50% of third session is used for testing.

4.2   Comparison of Results of Single Task and Multi-task Models

Results for single task models, i.e., subject specific trained models and multi-task model are presented in Table 1. Results presented illustrate that motor imagery

---

[2] http://www.bbci.de/competition/iv/.

signals are classified better with a model trained with multiple subject's EEG data rather than a single subject. A paired t-test revealed that there is significant difference between the results (p = 0.016). At first it might seem suspicions that if EEG data is different for different subjects, then in that case subject specific trained model should perform better. But since all the subjects are performing same motor imagination, i.e., they are thinking of moving their either left or right hand, hence, there must be some common features associated with similar motor imagination which single subject models are not able to decode. So, when a model is trained with multiple subjects' motor imagery EEG data, some hidden features which are common to a particular motor imagination for all the subjects are being captured which in turn gives better classification accuracy.

Table 1. Results for subject specific trained model and multi-task learnt model

Subjects	Classification accuracy %	
	Explicitly trained model	Adversarial multi-task model
1	76.25	80.17
2	60.00	56.77
3	48.75	56.79
4	96.25	98.80
5	67.50	78.32
6	70.00	76.13
7	80.00	87.21
8	86.25	89.11
9	78.75	82.33
Mean	76.75	78.40

4.3   Comparison of Different Multi-task Models

As discussed in Sect. 3.2, we have implemented three different multi-task model architectures. Results presented in Table 2 illustrate that conventional multi-task model, i.e., fully shared multi-task model is not able to perform as good as the other multi-task architectures. The reason is that these fully-shared multi-task models are not able to learn task specific features for highly correlated tasks. Although these multi-task models are having task specific layers but they do not have dedicated private layers. Weights in the task specific layers are influenced by other subjects' incoming weights from the fully shared layers which are common for all the subjects. With fully shared layers, conventional multi-task model is able to learn features which are common to all the subjects but task specific layers are not designed well enough that can distinguish between correlated tasks.

To overcome this issue, dedicated private layers were designed in the multi-task architecture. So, in case of private shared multi-task architecture, it was able to learn subject specific features which were not influenced by the shared features. It can be seen in the Table 2, that private shared multi-task model performed better than fully-shared multi-task model. However, caveat with this model is that it became a comparatively complex model. As it can be seen from Fig. 3, private-shared multi-task architecture consists of 10 different sub-networks (9 are subject specific and 1 shared network). This is almost same as training 10 different models and out of these, 9 sub-networks are purposed to capture subject specific features.

In adversarial multi-task model, instead of those 9 sub-networks, we train only one adversarial network whose generator's task is to generate subject specific features. With adversarial network, we are leveraging all the 9 sub-networks with just one network. Although paired t-test revealed that there is no significant difference between private shared and adversarial multi-task architectures (p = 0.183), but it is evident from results presented in Table 2, that adversarial multi-task network outperformed private-shared multi-task model architecture and attained better results.

Table 2. Comparison of different multi-task models

Subjects	Classification accuracy % (Mean ± Standard deviation)		
	Fully-shared	Private-shared	Adversarial-shared
1	76.7 ± 2.7	78.6 ± 2.5	80.17 ± 2.9
2	53.2 ± 3.1	55.8 ± 3.3	56.77 ± 3.5
3	52.1 ± 3.7	56.5 ± 2.4	56.79 ± 4.1
4	98.8 ± 1.0	98.8 ± 1.0	98.8 ± 1.0
5	75.2 ± 3.5	78.2 ± 4.1	78.32 ± 3.4
6	75.4 ± 3.1	77.3 ± 3.2	76.13 ± 4.3
7	83.8 ± 3.3	85.2 ± 3.1	87.21 ± 3.3
8	87.1 ± 2.9	90.1 ± 2.1	89.11 ± 2.8
9	79.3 ± 3.7	79.6 ± 3.2	82.33 ± 3.4
Mean	75.7 ± 3.0	77.8 ± 2.8	78.4 ± 3.2

4.4 Comparison of Multi-task Model with State-of-the-Art Methods

We have also compared the results of our multi-task model with the state-of-the-art model, i.e., CNN-SAE model [9] and BCI IV 2008 competition winner's algorithm, i.e., FBCSP [4]. While training CNN-SAE model, authors have used first 3 sessions for training and testing. Authors in [4] have used different sessions for different subjects for training their FBCSP model based on some exhaustive search. Most of researchers have used kappa as evaluation metric for motor imagery signal classification. Kappa is used as evaluation metric in classification problems as it removes the effect of random classification performed by the model. Kappa value is defined as follows:

$$\kappa = \frac{A_o - A_e}{1 - A_e} \tag{1}$$

Here, A_o is actual accuracy and A_e is expected accuracy by chance.

Table 3 presents kappa values attained by CNN-SAE, FBCSP and proposed multi-task learnt models. It is clear from the results that our proposed multi-task learnt model outperforms winner's algorithm and state-of-the-art model as well. Although our multi-task learnt model surpasses the CNN-SAE model by a small margin, but our proposed model is a very simple model as compared to the CNN-SAE model. Moreover, CNN-SAE model used in [9] is trained explicitly for all the subjects while ours is a single unified model trained for all the subjects (and the main takeaway is that some common hidden features are learnt by a multi-task model which are not learnt with subject specific trained model).

Table 3. Comparisons of FBCSP, CNN-SAE and Multi-task learnt models

Subjects	Mean kappa value		
	FBCSP	CNN-SAE	Multi-task learnt model
1	0.546	0.517	0.603
2	0.208	0.324	0.135
3	0.244	0.496	0.136
4	0.888	0.905	0.976
5	0.692	0.655	0.576
6	0.534	0.579	0.522
7	0.409	0.488	0.744
8	0.413	0.494	0.782
9	0.583	0.463	0.647
Mean	0.502	0.547	0.568

4.5 Effect of Different Hyperparameters

We tried different numbers of hidden layers, i.e., 2 layers (1 shared and 1 task specific), 3 layers (2 shared and 1 task-specific), 4 layers (2 shared and 2 task-specific), 5 layers (3 shared and 2 task-specific) and 6 layers (3 shared and 2 task-specific). Conventional multi-task model performed better with 5 layers. Private-shared and adversarial multi-task model performed better with 4 layers in their task-specific and shared network. We also experimented with different learning rates and found that variable learning rate attained better results for motor imagery signal classification. Learning rates of both motor imagery classification model and subject classification (discriminator) model, were set to vary with classification accuracy.

5   Conclusion

In this paper, we have proposed a multi-task learning approach for motor imagery signal classification. Although EEG signature for same motor imagination is different for different persons but they are highly correlated and there are some hidden features which can not be learnt with subject specific trained model. We have shown that motor imagery classification of a particular subject can leverage from concurrent learning of motor imagery of other subjects. Raw EEG time series data was first converted into image form with the help of STFT and then fed into different neural networks. We have experimented with three multi-task architectures, viz. fully shared multi-task model, private-shared multi-task model and adversarial-shared multi-task model. Our experiments showed that multi-task learnt model performs better as compared to subject specific trained model. Among various multi-task model architectures, adversarial-shared multi-task model attains the best classification accuracy. Motor imagery signal classification model can be improved further as we only experimented with vanilla CNN and performed simple concatenation of shared and private layers before prediction. Future studies could fruitfully explore this area further by implementing different model architectures and improve motor imagery classification accuracy.

Acknowledgments. Dr. Sriparna Saha gratefully acknowledges the Young Faculty Research Fellowship (YFRF) Award, supported by Visvesvaraya Ph.D. Scheme for Electronics and IT, Ministry of Electronics and Information Technology (MeitY), Government of India, being implemented by Digital India Corporation (formerly Media Lab Asia) for carrying out this research.

References

1. Wolpaw, J., Wolpaw, E.W.: Brain-Computer Interfaces: Principles and Practice. Oxford University Press, USA (2012)
2. Teplan, M., et al.: Fundamentals of EEG measurement. Meas. Sci. Rev. **2**(2), 1–11 (2002)
3. Ang, K.K., Chin, Z.Y., Zhang, H., Guan, C.: Filter bank common spatial pattern (FBCSP) in brain-computer interface. In: 2008 IEEE International Joint Conference on Neural Networks (IEEE World Congress on Computational Intelligence), pp. 2390–2397. IEEE (2008)
4. Ang, K.K., Chin, Z.Y., Wang, C., Guan, C., Zhang, H.: Filter bank common spatial pattern algorithm on BCI competition iv datasets 2a and 2b. Front. Neurosci. **6**, 39 (2012)
5. Stewart, A.X., Nuthmann, A., Sanguinetti, G.: Single-trial classification of EEG in a visual object task using ICA and machine learning. J. Neurosci. Methods **228**, 1–14 (2014)
6. Lotte, F., Congedo, M., Lécuyer, A., Lamarche, F., Arnaldi, B.: A review of classification algorithms for EEG-based brain-computer interfaces. J. Neural Eng. **4**(2), R1 (2007)

7. Vaid, S., Singh, P., Kaur, C.: EEG signal analysis for BCI interface: a review. In: 2015 Fifth International Conference on Advanced Computing & Communication Technologies, pp. 143–147. IEEE (2015)
8. Nicolas-Alonso, L.F., Gomez-Gil, J.: Brain computer interfaces, a review. Sensors 12(2), 1211–1279 (2012)
9. Tabar, Y.R., Halici, U.: A novel deep learning approach for classification of EEG motor imagery signals. J. Neural Eng. 14(1), 016003 (2016)
10. Dai, M., Zheng, D., Na, R., Wang, S., Zhang, S.: EEG classification of motor imagery using a novel deep learning framework. Sensors 19(3), 551 (2019)
11. Tang, Z., Li, C., Sun, S.: Single-trial EEG classification of motor imagery using deep convolutional neural networks. Optik 130, 11–18 (2017)
12. Wang, P., Jiang, A., Liu, X., Shang, J., Zhang, L.: LSTM-based EEG classification in motor imagery tasks. IEEE Trans. Neural Syst. Rehabil. Eng. 26(11), 2086–2095 (2018)
13. Lu, N., Li, T., Ren, X., Miao, H.: A deep learning scheme for motor imagery classification based on restricted boltzmann machines. IEEE Trans. Neural Syst. Rehabil. Eng. 25(6), 566–576 (2016)
14. Akhtar, S., Ghosal, D., Ekbal, A., Bhattacharyya, P., Kurohashi, S.: All-in-one: emotion, sentiment and intensity prediction using a multi-task ensemble framework. IEEE Trans. Affect. Comput. (2019)
15. Qureshi, S.A., Dias, G., Hasanuzzaman, M., Saha, S.: Improving depression level estimation by concurrently learning emotion intensity. IEEE Comput. Intell. Mag. 15(3), 47–59 (2020)
16. Ko, W., Yoon, J., Kang, E., Jun, E., Choi, J.S., Suk, H.I.: Deep recurrent spatio-temporal neural network for motor imagery based BCI. In: 2018 6th International Conference on Brain-Computer Interface (BCI), pp. 1–3. IEEE (2018)
17. Leeb, R., Brunner, C., Müller-Putz, G., Schlögl, A., Pfurtscheller, G.: BCI competition 2008-Graz data set b, pp. 1–6. Graz University of Technology, Austria (2008)

An End-to-End Hemisphere Discrepancy Network for Subject-Independent Motor Imagery Classification

Li Nie, Huan Cai, Yihan Wu, and Yangsong Zhang[✉]

School of Computer Science and Technology, Southwest University of Science and Technology, Mianyang, China

Abstract. Left-hand and right-hand motor imagery(MI) demonstrated discrepant neurological patterns between two brain hemispheres. This mechanism has not been considered in the existing deep learning model for MI classification. In this paper, we first proposed a novel end-to-end deep learning model termed as Hemisphere Discrepancy Network(HDNet), to improve electroencephalogram(EEG) based MI classification in the cross-subject scenario. Concretely, a temporal layer was designed to explore temporal information by employing a convolution along the time points of EEG data to learn temporal representations. Then, a bi-hemispheric spatial discrepancy block composing three parallel layers was constructed to obtain the deep spatial representations, not only keeping intrinsic spatial dependence but also extracting discrepancy information between the left and right brain hemispheres. After these operations, a classifier was followed. Evaluation experiments were conducted on a two-class MI EEG dataset, which contained 54 subjects. The results exhibited that HDNet outperformed the baseline methods. This study demonstrated that exploring the discrepancy information between brain hemispheres holds the promise for MI classification.

Keywords: Brain-computer interfaces · Electroencephalogram · Motor imagery · Hemispheric discrepancy

1 Introduction

A majority of studies have developed brain-computer interface (BCI) for different applications with the goals ranging from providing means of communication to functional rehabilitation over the past decades [1,2]. BCI based on non-invasive electroencephalogram (EEG) sensorimotor rhythms is termed as motor imagery(MI), referring to a thinking activity in which one can imagine to perform a specific movement without actually executing it [3,4]. Researches on MI-BCI have been gaining more attention as it can provide help for strokes and epilepsy patients to communicate, as well as be used to control external devices [5,6].

© Springer Nature Switzerland AG 2021
T. Mantoro et al. (Eds.): ICONIP 2021, LNCS 13110, pp. 323–333, 2021.
https://doi.org/10.1007/978-3-030-92238-2_27

In the past, researchers have employed traditional method to extract features from preprocessed EEG signals firstly, and then to classify the extracted features by a number of linear and nonlinear algorithms. One popular method is the filter bank common spatial patterns (FBCSP) [7,8]. This method utilized spectral power modulations of EEG signal and linear discriminant analysis (LDA)or support vector machine (SVM) as a classifier, which has achieved the satisfactory performance. Although the conventional machine learning methods have been successful up to a certain extent in classifying the EEG MI data, they have had to choose weights carefully and mainly concentrated on training models for each subject individually in intra-subject scenario, which was time-consuming and labor-intensive, and they have not been able to reach good decoding accuracies with hand-crafted features in the inter-subject manner [9].

The recent success of deep learning methods has driven researchers to apply them for EEG classification. Schirrmeister et al. proposed several end-to-end convolutional neural networks(CNNs) [10], in which models named DeepNet and ShallowNet have the capabilities of feature extraction and classification to process raw MI data. Lawhern et al. [11] proposed EEGNet, a compact CNN that uses depthwise and separable convolutions to construct an EEG-specific model with lightweight parameters. Filter-Bank Convolutional Network (FBC-Net) recently has been proposed by Mane et al. [12], which effectively aggregated the EEG time-domain information using variance layer. On the other hand, many researches made their efforts to explore more spatial features. For instance, Zhao et al. introduced a new 3D representation of EEG and a corresponding multi-branch 3D CNN [13]. Zhang et al. proposed a Graph-based Convolutional Recurrent Attention Model (G-CRAM) to learn spatial information with the aid of graph representations of EEG nodes and extract attentional temporal dynamics using a recurrent attention network [14]. These methods mentioned above adopted a convolutional kernel whose length is the same as the number of EEG channels to extract the global spatial pattern. However, previous studies have demonstrated that motor imagery can cause differences in the spatial distribution of event-related desynchronization(ERD) or event-related synchronization (ERS) observed from the EEG signal [15,16]. Specifically, when imagining the movement of the left hand, the ERD phenomenon in the right motor cortex is more prominent, while when imagining the movement of the right hand, the area where the ERD phenomenon is prominent is in the left brain area.

In the field of emotion recognition, a considerable amount of studies have proved that taking advantage of discriminative information from the hemispheres can improve the performance in terms of classification accuracy [17,18]. Huang et al. proposed BiDCNN model [19] with a three-input and single-output network structure to effectively learn the different response patterns between the left and right hemispheres. Recently, Ding et al. introduced a multi-scale CNN (termed as TSception) to learn temporal dynamics and spatial asymmetry from affective EEG emotion data [20]. Nonetheless, the asymmetric EEG patterns of the cerebral hemispheres were ignored in the previous studies on MI classification.

Inspired by these studies, in this paper, we first proposed a novel a hemisphere discrepancy network (HDNet) to learn the bi-hemispheric information for cross-subject motor imagery classification. HDNet architecture is composed of three modules: a temporal layer, a bi-hemispheric spatial discrepancy block and a final classifier. We utilized a one-dimension CNN layer to learn the temporal features of each EEG channel, and then learned discrepant spatial asymmetry features based on the temporal embedding features. In addition, these spatial features were concatenated and finally fed into classifier. A dataset with fifty-four subjects was adopted to evaluate the effectiveness of the proposed method. The experimental results demonstrated the HDNet yielded superior performance than the comparison methods.

2   Methods

This section first briefly describes the dataset as well as its preprocessing. Secondly, we formally define the research problems and introduce the design of the proposed model.

2.1   Data Description and Preprocessing

The KU dataset [21], provided by Korea University, was used to evaluate the proposed method and other baseline methods. Data acquisition process was conducted with 54 healthy subjects, using 64 Ag/AgCl electrodes in 10–20 international system, with a sampling rate 1000 Hz. Subjects were asked to imagine left- and right-hand movements (two-class MI task). There were 400 trials per subject obtained from two sessions. Each session provides 100 trials of the recorded EEG without feedback, defined as an offline condition and 100 trials with feedback, so-called on-line condition. In this paper, only the first session of all subjects (200 × 54) was selected to evaluate the proposed method.

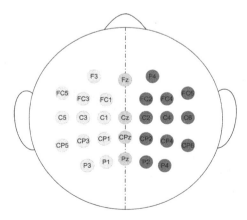

Fig. 1. The placement of the selected 28 EEG electrodes.

During the preprocessing stage, a third-order Butterworth band-pass filter from 0-40Hz was applied to the original EEG data, and then the data was downsampled 250 Hz. In each trial, the data in 4s time interval after stimulus onset was extracted for following analysis. In this study, 28 electrodes from the motor cortex region were selected (F-5/z/4, FC-5/3/1/2/4/6, C-5/3/1/z/2/4/6, CP-5/3/1/z/2/4/6 and P-3/1/z/2/4), as shown in Fig. 1 colourful circles.

In fact, instead of removing the electrodes on the central line of the brain at the very beginning, we maintained and used these electrodes in the temporal layer and global convolution layer while dropping them in the hemisphere convolution layer and bi-hemisphere subtraction layer. Furthermore, to ensure the following hemispheric operations were meaningful, we rearranged the channels of each 2D matrix trial according to electrode pairs corresponding to the left and right hemispheres, about which more details can be found in Subsect. 2.2 and Fig. 2.

2.2   Proposed Hemisphere Discrepancy Network

The framework of the HDNet model is presented in Fig. 2, which is composed of three main modules, i.e., a Temporal layer, a Bi-hemispheric Spatial Discrepancy Block, and a Classifier. The details are given in the following subsections.

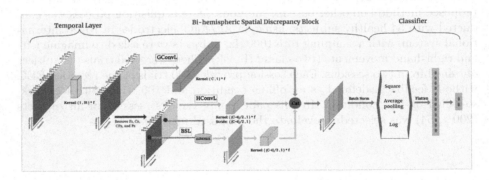

Fig. 2. Overall visualization of HDNet Architecture. The temporal layer takes in the rearranged data. And then fed the output into bi-hemispheric spatial discrepancy block with three parallel branches. Finally, apply a classifier on the concatenated output of previous block. (GConvL: global convolution layer, HConvL: hemisphere convolution layer, BSL: bi-hemisphere subtraction layer, f: number of filters)

Temporal Layer. In this layer, each filter performed a convolution over time dimension without activation function. As described in [10], the transformation performed by the temporal layer is analogous to the bandpass step in FBCSP. The temporal layer was realized with 2D convolution whose kernel was set to (1, 31) for exploring long-term temporal dynamics. In the following sections, we denote a single two-dimensional(2D) raw trial as matrix $X_I \in \mathbb{R}^{C \times P}$ and its corresponding label as $y \in [0,1]$. Here, C and P represent number of the EEG channels and time points respectively, and I is an abbreviating representation of input data. The final output of this layer is computed as below:

$$\mathcal{M}_T = TConv(X_I^k), \mathcal{M}_T \in \mathbb{R}^{n \times f \times C \times P'}, k \in [1,n] \qquad (1)$$

where P' is the length of time points, n is the number of EEG samples, $TConv(\cdot)$ is the temporal convolution operation and f is the number of filters.

To ensure that the following operations performed on the hemispheres make sense, the middle electrodes(Fz, Cz, CPz, and Pz) will be eliminated. Thus each sample output of temporal layer without middle electrodes can be defined as $M_T^* = [T^l, T^r]$, where T^l and T^r denote the output of the left(in yellow) and right(in blue) hemisphere channels as shown in Fig. 1, respectively.

Bi-Hemispheric Spatial Discrepancy Block. The neuroscience study proves that the left and right hemispheres of human brains have different ERD/ERS phenomena while the subject images left or right movements. According to that, we tried to investigate whether the features from two different hemispheres can help us improve the MI decoding performance or not. Therefore, in this bi-hemispheric spatial discrepancy block, we constructed three parallel layers: global convolution layer($GConvL$), hemisphere convolution layer($HConvL$) and bi-hemisphere subtraction layer(BSL).

Global Convolution Layer(GConvL). In this layer, we used a depthwise 2D convolution layer of kernel size (C,1) to span across all the channels for learning the global spatially discriminative information termed as global convolution layer($GConvL$). Here C is the total number of channels.

However, the kernel of global convolution($GConv$) is (C,1), which means that we did not remove the central electrodes in this layer as well as the temporal convolution layer. Because the electrodes located on the central line may preserve some inherent relation between two hemispheres. The final output of this layer is expressed as:

$$\mathcal{M}_G = GConv(M_T^k), \mathcal{M}_G \in \mathbb{R}^{n \times f \times 1 \times P'}, k \in [1,n] \qquad (2)$$

Hemisphere Convolution Layer(HConvL). To explore the asymmetrical information for left- and right-hand MI tasks, inspired by [20], which combined the brain asymmetry into the kernel design, we introduced a hemisphere convolution layer ($HConvL$) which was used to extract the discrepancy pattern between the left and right hemispheres by sharing the convolutional kernel. As mentioned above, the middle electrodes(Fz, Cz, CPz, and Pz) have been eliminated in this layer and the kernel was shared by two hemispheres without overlapping. Thus the kernel size of the hemisphere convolution($HConv$) was ($\frac{C-4}{2}$, 1) and the step was ($\frac{C-4}{2}$, 1), where C is the total number of channels. Thus, the output of $HConvL$ can be defined as:

$$\mathcal{M}_H = HConv(M_T^{*k}) = HConv([T^l, T^r]^k), \mathcal{M}_H \in \mathbb{R}^{n \times f \times 2 \times P'}, k \in [1,n] \qquad (3)$$

Bi-hemisphere Subtraction Layer(BSL). To further explore the asymmetrical information, inspired by [17], we proposed a bi-hemisphere subtraction layer (*BSL*) which first performed a subtraction operation on the paired channels referring to the symmetric positions on the brain scalp and then applied a 2D spatial convolution on the channel dimension. Specifically, we did not only remove the middle electrodes and divide the electrodes into two groups (left and right) coarsely, but also carefully rearrange the sequence of electrodes according to the location of the symmetrical electrode pairs as shown in Fig. 2. Therefore, the subtraction operation was performed as $e_s^{lj} = (e_t^{lj} - e_t^{rj}), j = 0, \ldots, \frac{C-4}{2}$, where t and s are abbreviating representations of the data in the corresponding operation stages of temporal convolution and subtraction respectively. After that, a 2D spatial convolution*(SConv)* whose kernel is $(\frac{C-4}{2}, 1)$ was conducted on the output of subtraction operation to integrate the asymmetric information to a high-level feature map. The final output of this layer is calculated as below:

$$\mathcal{M}_B = SConv(M_T^{*k}) = SConv(\ominus(e_t^{lj}, e_t^{rj})), \mathcal{M}_B \in \mathbb{R}^{n \times f \times 1 \times P'}, k \in [1, n] \quad (4)$$

where the $\ominus$ is subtraction operation.

After these operations, we concatenated the three components($\mathcal{M}_G$, $\mathcal{M}_H$ and $\mathcal{M}_B$) in the manner of the channel dimension, which is the final output of this spatial block, defining as follows:

$$\mathcal{M}_S = BN[Cat(\mathcal{M}_G, \mathcal{M}_H, \mathcal{M}_B)], \mathcal{M}_S \in \mathbb{R}^{n \times f \times 4 \times P'} \quad (5)$$

where $Cat(\cdot)$ stands for concatenation operation along the channel dimension and $BN(\cdot)$ denotes the 2D batch normalization. After concatenation, the channel dimension is 4.

Classifier. After the bi-hemispheric spatial discrepancy block of HDNet, the outputs were transformed by a nonlinear squaring operation, a mean pooling layer and a logarithmic activation function, which aimed to learn high-level fusion representations. Together, these steps were analogous to the trial log-variance computation in FBCSP as describing in [10] . The final learned representations were generated by the following formula:

$$\mathcal{M}_F = \phi_{Log}\{AvgPool[Square(\mathcal{M}_S)]\}, \mathcal{M}_F \in \mathbb{R}^{n \times f \times 4 \times P''} \quad (6)$$

where P'' is the length of time points after these operations.

Finally, $\mathcal{M}_F$ was flattened to 1-D vector, and a dropout function $\psi_{dp}(\cdot)$ was added in fully connected layer to prevent overfitting. The final output was activated by the softmax function $\phi_{Softmax}$, which can be calculated by:

$$\mathcal{O}utput = \phi_{Softmax}\{FC[\psi_{dp}(\mathcal{M}_F)]\}, \mathcal{O}utput \in \mathbb{R}^{n \times 2} \quad (7)$$

Then the cross-entropy loss function was evaluated .

$$\mathcal{L}_{cross-entropy} = -\sum_{i=1}^{c} y_i \log \widehat{y}_i \quad (8)$$

where y_i and $\widehat{y_i}$ are the true label and the predicted classification probability, and c is the total number of classes.

3   Experiments and Results

3.1   Implementation Details

To evaluate the performance of HDNet, the inter-subject scenario was adopted, because this setting is more valuable and can make a new subject use the BCI system without the time-consuming and labor-intensive calibration procedure. A 9-fold cross-validation analyse was conducted on the KU dataset acquired from 54 subjects. Specifically, the dataset was divied into 9 subsets. Each subset has 1200 trials from 6 subjects (6 × 200 trials). Each subset (1200 trials) was served as the testing dataset, the remaining subsets (9600 trials) were used as the training dataset. As for the HDNet, the number of filters f was set to 10.

During the training process, the stochastic gradient descent(SGD) was used as the optimizer to minimize the cross-entropy loss function with a learning rate of 0.001; the dropout rate was set to 0.5. In addition, the training batch size was 64 and the epoch was 30. The entire experiments were conducted on the *Geforce RTX 3090* platform by making use of the *PyTorch* framework.

3.2   Baseline Models

For a fair comparison, the same training procedure of HDNet was set for all comparison models except CRAM [22] of which optimizer was Adam, as it got a lower accuracy using SGD optimizer. The proposed model was firstly compared with the recently published RNN model name CRAM [22], which splits the raw EEG trials into temporal slices and feed them into a spatio-temporal encoder, after that, a recurrent attention mechanism with LSTM units and self-attention mechanism is introduced. In addition, we compared our model with a compact CNN named EEGNet [11]. It encapsulates feature extraction designed for classifying raw EEG when only limited amounts of data are available. Finally, a comparison with the ShallowNet and DeepNet models proposed by [10] is involved. The DeepNet architecture is made up of five convolutional layers following by a softmax layer as a classifier and more details can be found in Fig. 1 of [10]. The ShallowNet consists of a temporal and a spatial convolution layers without activation function between them, which is analogous to the mechanism principle of FBCSP.

3.3   Results

Table 1 shows the results of average accuracies and F1-scores for the HDNet and baseline models using nine fold cross-validation method for subject-independent classification. Among deep learning methods, HDNet achieves the best performance, with the average accuracy and F1-score being 77.11% and 76.48% respectively, which achieves 12.74% higher accuracy and 11.56% higher F1-score than

330   L. Nie et al.

Table 1. Average accuracies and F1-scores of 9-fold cross-validation in subject-independent scenario.

Criterion	Model	Fold1	Fold2	Fold3	Fold4	Fold5	Fold6	Fold7	Fold8	Fold9	Average
Accuracy	CRAM	0.7131	0.6252	0.6217	0.5408	0.7250	0.7196	0.6059	0.6416	0.6002	0.6437
	DeepNet	0.7608	0.6725	0.6725	05533	0.7750	0.7550	0.6466	0.7108	0.6558	0.6891
	EEGNet	0.7850	0.6633	0.6983	0.5716	0.7950	0.7641	0.6891	0.6983	0.6700	0.7038
	ShallowNet	0.8033	0.6658	0.6791	0.7066	0.8033	**0.8133**	0.7091	0.7316	0.6400	0.7280
	HDNet	**0.8550**	**0.7016**	**0.7025**	**0.7275**	**0.8650**	0.8008	**0.7558**	**0.8416**	**0.6900**	**0.7711**
F1-Score	CRAM	0.7429	0.5737	0.6229	0.5721	0.7337	0.7279	0.6311	0.6639	0.5745	0.6492
	DeepNet	0.7691	**0.6727**	0.7078	0.5663	0.7890	0.7745	0.6569	0.7243	0.6136	0.6971
	EEGNet	0.7813	0.6424	**0.7176**	0.5673	0.7980	0.7635	0.6960	0.7346	**0.6597**	0.7067
	ShallowNet	0.8217	0.6364	0.6565	0.7188	0.7972	**0.8136**	0.7084	0.7659	0.5924	0.7234
	HDNet	**0.8625**	0.6715	0.6826	**0.7390**	**0.8634**	0.8036	**0.7596**	**0.8460**	0.6549	**0.7648**

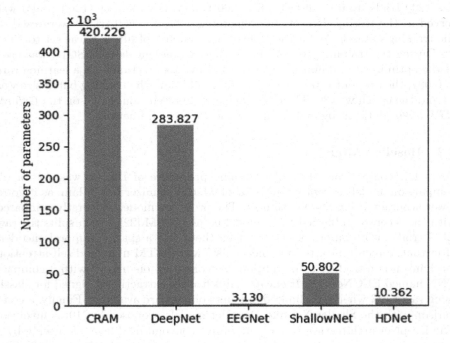

Fig. 3. The number of trainable parameters of models.

CRAM. ShallowNet achieves second place compared with the other methods (72.80% for accuracy and 72.34% for F1-score), yet it's accuracy is 4.31% lower than HDNet. DeepNet and EEGNet achieve similar accuracies and F1-scores.

To further compare these models, Fig. 3 illustrates the number of trainable parameters of all models in Table 1. CRAM has the highest amount of parameters and the lowest accuracy, while HDNet's parameters are lower than other models, except EEGNet, with the highest accuracy. And and training runtime of the HDNet in 30 epoches is nearly 164 min, which is shorter than three hours.

Although HDNet achieved relatively higher classification accuracies than the baselines, the limitations of this work should also be considered. In this work, only KU dataset was adopted, and only the first session data from this dataset was used, the generalization ability on the two-session KU dataset and other different datasets should be evaluated in the future. Besides, we should further investigate the performance of HDNet on multiple-class classification tasks of MI, such as left-hand, right-hand, foot and tongue movements. Although the training time of HDNet is less than 3 h, we will optimize the architecture of HDNet so that its running time is still less with more channels and data volume.

3.4   Ablation Study

To further verify that the asymmetry information of hemispheres plays a dominant role in the performance improvement, ablation studies were further conducted on bi-hemispheric spatial discrepancy block of the HDNet by examining the contribution of each component in this block on the nine fold cross-validation dataset. The results about the accuracies are summarized in Table 2.

Table 2. The accuracy obtained with ablation study on the proposed HDNet model

Method	Fold1	Fold2	Fold3	Fold4	Fold5	Fold6	Fold7	Fold8	Fold9	Average
withoutS&H	0.8175	0.6850	0.6633	**0.7433**	0.7383	0.7333	0.7300	0.7325	0.6458	0.7210
withoutHConvL	0.7925	0.6991	**0.7025**	0.6133	0.8175	**0.8100**	0.7375	0.7791	0.6783	0.7366
withoutBSL	0.8366	0.6908	0.6941	0.7216	0.8466	0.8000	0.7541	0.8058	0.6650	0.7572
HDNet	**0.8550**	**0.7016**	**0.7025**	0.7275	**0.8650**	0.8008	**0.7558**	**0.8416**	**0.6900**	**0.7711**

According to the table, the most significant drop of accuracy were observed when both hemispheric convolution layer and bi-hemisphere subtraction layer were ablated from the model(termed as "withoutS&H") with the decrement being 4.61%. To an extent, this results indicates that when the convolutional network only extracts time information and global brain spatial information, the contribution to classification accuracy is limited. Furthermore, the model settings "withoutHConvL" and "withoutBSL" stand for removing the hemispheric convolution layer and removing the bi-hemisphere subtraction layer repectively. The results show that the former layer contributes more than the latter layer. The accuracy drops from 77.11% to 73.66% if the hemispheric convolution layer is removed, while the accuracy decreases form 77.11% to 75.72% when removing bi-hemisphere subtraction layer. From the neurological perspective, the drop shows that the *HConvL* has a higher ability (2.06%) than *BSL* to capture hemispheric difference information. According to the above ablation studies, the proposed HDNet incorporating bi-hemispheric spatial discrepancy block outperforms the baseline models, demonstrating the effectiveness of each component of the proposed model.

3.5   Limitations and Future Work

Although HDNet achieved relatively higher classification accuracies than the baselines, the limitations of this work should also be considered. In this work, only KU dataset was adopted, and only the first session data from this dataset was used, the generalization ability on the two-session KU dataset and other different datasets should be evaluated in the future. Besides, we should further investigate the performance of HDNet on multiple-class classification tasks of MI, such as left-hand, right-hand, foot and tongue movements. Although the training time of HDNet is less than 3 h, we will optimize the architecture of HDNet so that its running time is still less with more channels and data volume.

4   Conclusion

This study proposed a novel end-to-end deep learning method HDNet, which was inspired by the neuroscience findings of the different response patterns on brain hemispheres to the left- and right-hand MI. The proposed model yielded 77.11% classification accuracy and 76.48% F1-score in the subject-independent scenario, and achieved better performance comparing with the state-of-the-art methods. To our best knowledge, this is the first study that extracting the discrepant feature of hemispheres for EEG-based MI task. Moreover, we showed that the performance of the deep learning-based model in EEG MI does not completely couple with the total amount of model parameters, which is different with computer vision(CV). This study demonstrated that integrating the brain mechanisms of BCI task into the DL models is a potential way to achieve satisfactory performance for the BCI system.

References

1. Ang, K.K., Guan, C.: EEG-based strategies to detect motor imagery for control and rehabilitation. IEEE Trans. Neural Syst. Rehabil. Eng. **25**(4), 392–401 (2017)
2. Lazarou, I., Nikolopoulos, S., Petrantonakis, P.C., Kompatsiaris, I., Tsolaki, M.: EEG-based brain-computer interfaces for communication and rehabilitation of people with motor impairment: a novel approach of the 21st century. Front. Human Neurosci. **12**, 14 (2018)
3. He, B., Baxter, B., Edelman, B.J., Cline, C.C., Wenjing, W.Y.: Noninvasive brain-computer interfaces based on sensorimotor rhythms. Proc. IEEE **103**(6), 907–925 (2015)
4. Jiao, Y., et al.: Sparse group representation model for motor imagery EEG classification. IEEE J. Biomed. Health Inf. **23**(2), 631–641 (2019)
5. Chaudhary, U., Birbaumer, N., Ramos-Murguialday, A.: Brain-computer interfaces for communication and rehabilitation. Nature Rev. Neurol. **12**(9), 513–525 (2016)
6. Li, Y., et al.: Multimodal BCIs: target detection, multidimensional control, and awareness evaluation in patients with disorder of consciousness. Proc. IEEE **104**(2), 332–352 (2016)

7. Ang, K.K., Chin, Z.Y., Zhang, H., Guan, C.: Filter bank common spatial pattern (FBCSP) in brain-computer interface. In: 2008 IEEE International Joint Conference on Neural Networks (IEEE World Congress on Computational Intelligence), pp. 2390–2397. IEEE (2008)

8. Ang, K.K., Chin, Z.Y., Wang, C., Guan, C., Zhang, H.: Filter bank common spatial pattern algorithm on BCI competition IV datasets 2a and 2b. Front. Neurosci. **6**, 39 (2012)

9. Al-Saegh, A., Dawwd, S.A., Abdul-Jabbar, J.M.: Deep learning for motor imagery EEG-based classification: a review. Biomed. Signal Process. Control **63**, 102172 (2021)

10. Schirrmeister, R.T., et al.: Deep learning with convolutional neural networks for EEG decoding and visualization. Hum. Brain Mapp. **38**(11), 5391–5420 (2017)

11. Lawhern, V.J., Solon, A.J., Waytowich, N.R., Gordon, S.M., Hung, C.P., Lance, B.J.: EEGNet: a compact convolutional neural network for EEG-based brain-computer interfaces. J. Neural Eng. **15**(5), 056013 (2018)

12. Mane, R., et al.: FBCNet: A multi-view convolutional neural network for brain-computer interface. arXiv preprint arXiv:2104.01233 (2021)

13. Zhao, X., Zhang, H., Zhu, G., You, F., Kuang, S., Sun, L.: A multi-branch 3d convolutional neural network for EEG-based motor imagery classification. IEEE Trans. Neural Syst. Rehabil. Eng. **27**(10), 2164–2177 (2019)

14. Zhang, D., Chen, K., Jian, D., Yao, L.: Motor imagery classification via temporal attention cues of graph embedded EEG signals. IEEE J. Biomed. Health Inf. **24**(9), 2570–2579 (2020)

15. Yuan, H., He, B.: Brain-computer interfaces using sensorimotor rhythms: current state and future perspectives. IEEE Trans. Biomed. Eng. **61**(5), 1425–1435 (2014)

16. Blankertz, B., Tomioka, R., Lemm, S., Kawanabe, M., Muller, K.R.: Optimizing spatial filters for robust EEG single-trial analysis. IEEE Signal Process. Mag. **25**(1), 41–56 (2007)

17. Cui, H., Liu, A., Zhang, X., Chen, X., Wang, K., Chen, X.: EEG-based emotion recognition using an end-to-end regional-asymmetric convolutional neural network. Knowl.-Based Syst. **205**, 106243 (2020)

18. Li, Y., et al.: A novel bi-hemispheric discrepancy model for EEG emotion recognition. IEEE Trans. Cogn. Dev. Syst. **13**(2), 354–367 (2021)

19. Huang, D., Chen, S., Liu, C., Zheng, L., Tian, Z., Jiang, D.: Differences first in asymmetric brain: a bi-hemisphere discrepancy convolutional neural network for EEG emotion recognition. Neurocomputing **448**, 140–151 (2021)

20. Ding, Y., Robinson, N., Zeng, Q., Guan, C.: Tsception: Capturing temporal dynamics and spatial asymmetry from EEG for emotion recognition. arXiv preprint arXiv:2104.02935 (2021)

21. Lee, M.H., et al.: EEG dataset and OpenBMI toolbox for three BCI paradigms: an investigation into BCI illiteracy. GigaScience **8**(5), giz002 (2019)

22. Zhang, D., Yao, L., Chen, K., Monaghan, J.: A convolutional recurrent attention model for subject-independent EEG signal analysis. IEEE Signal Process. Lett. **26**(5), 715–719 (2019)

Multi-domain Abdomen Image Alignment Based on Joint Network of Registration and Synthesis

Yizhou Chen[1], Zhengwei Lu[1], Xu-Hua Yang[1], Haigen Hu[1], Qiu Guan[1(✉)], and Feng Chen[2(✉)]

[1] College of Computer Science and Technology, Zhejiang University of Technology, Hangzhou, China
gq@zjut.edu.cn
[2] The First Affiliated Hospital, Zhejiang University School of Medicine, Hangzhou, China
chenfenghz@zju.edu.cn

Abstract. Multi-domain abdominal image alignment is a valuable and challenging task for clinical research. Normally, with the assistance of the image synthesizer, the register can perform well. However, the deviation of the synthesizer is likely to be the bottleneck of the register performance. In this case, using the registration training information to guide the synthesizer optimization is meaningful. Therefore, we propose the Joint Network of Registration and Synthesis (RSNet). The network calculates the loss caused only by the synthetic deviation according to the registration training, which effectively improves the performance of the synthesizer. Meanwhile, the network constructs an adaptive weighting factor based on the similarity measure of the synthetic image, which significantly generalizes the performance of the register. The real-world datasets are collected and processed with the help of a cooperative hospital. Our experiments demonstrate that RSNet can achieve state-of-the-art performance.

Keywords: Multi-domain abdominal image alignment · Registration · Synthesis · Joint network

1 Introduction

Multi-domain abdominal images include multi-modality images (e.g., CT/ MR) and multi-phase images (e.g., CT plain/venous phase images). When the anatomical content of the multi-domain image is aligned (i.e., the shape and position of the organ and tissue in the image are the same), the doctor can make

This work is supported in part by the National Natural Science Foundation of China (61802347, 61972347, 61773348, U20A20171), and the Natural Science Foundation of Zhejiang Province (LGF20H180002, LY21F020027, LSD19H180003).

T. Mantoro et al. (Eds.): ICONIP 2021, LNCS 13110, pp. 334–344, 2021.
https://doi.org/10.1007/978-3-030-92238-2_28

a more appropriate diagnosis and treatment plan for the patient [1]. However, the anatomical structure of the abdomen will inevitably be deformed during image acquisition, thus the obtained multi-domain images are usually unaligned [2].

The register can calculate the spatial deformation function between two images, which is able to align the two images [3,4]. However, for abdominal images, when the image styles of the two domains differ significantly (i.e., the pixel intensity distribution of multi-domain images differs significantly), the accurate calculation of the spatial deformation function often requires auxiliary information from labels [5–7]. However, labeling is usually expensive, making it difficult to have many labels to assist the training of the register [8].

Therefore, the researchers introduced image synthesis [9,10] to the multi-domain alignment. Ideally, the deformation function calculation between multi-domain real images is simplified to the function calculation between mono-domain real/synthetic images [11,12]. At this time, the deformation function can be accurately calculated without labels, which can align the multi-domain real image (note: the spatial deformation operation will not be affected by the pixel intensity value [4]). However, synthesizers usually have deviations. The smaller the deviation, the smaller the interference to the calculation of the deformation function [1,13]. Therefore, researchers try to use the registration training loss to guide the optimization of the synthesizer to reduce the deviation [14,15]. Unfortunately, when this strategy is implemented on multi-domain abdominal images, the alignment performance decreases [14].

It is because the registration training loss usually contains synthesis deviation and registration deviation. The latter one can interfere with the optimization of the synthesizer. Thus, if the loss term is caused only by the synthesis deviation, it can provide effective constraints for optimizing the synthesizer.

Meanwhile, if the register can adapt to the synthesis deviation, the spatial deformation function between the real/synthetic image calculated by the register can be more accurate. Even if the synthesizer is further optimized, the synthesis deviation cannot be eliminated. Therefore, it is essential to build a mechanism for the register to adapt to the synthesis deviation.

In this paper, we propose a multi-domain abdominal image alignment method based on the registration and synthesis joint network (RSNet). First, the register performs pre-training on the real image in the same domain. Then, the register and the synthesizer are optimized jointly. The contribution of the proposed method is concluded as follows.

1) A two-stage training RSNet is proposed. Through the first stage, the register can perform an ideal mono-domain alignment. Through the second stage, RSNet can calculate the loss term caused only by the synthetic deviation according to the pre-trained register. As a result, the performance of the synthesizer is improved.

2) In the joint optimization stage, an adaptive generalization mechanism for optimizing the register is proposed. Thus, for the same synthetic image, the register can perform relatively accurate abdominal image alignment.

2   Related Work

The multi-domain registration method performed with the assistance of image synthesis is briefly reviewed.

Wei *et al.* [11] and Qin *et al.* [12] train the synthesizer on unaligned data [9,10]. Ideally, the synthesizer can translate the real image from one domain to another domain, and the synthetic image's anatomical structure content is consistent with the original image. Then, the register can be trained on the real/synthetic image of the mono-domain. However, due to the lack of sufficient training constraints, the synthesis deviation is usually significant.

Xu *et al.* [13] tried to compromise the mono-domain deformation function and the multi-domain deformation function to dilute the synthesis deviation. However, when the multi-domain image styles differ significantly, the method will be invalid. Liu *et al.* [1] and Zhou *et al.* [14] used segmentation labels to provide constraints for the training of the synthesizer and the register. However, when the label is lacking, these methods will no longer be tractable.

Without using segmentation labels, Zhou *et al.* [14] used the registration training loss as a constraint of the synthesizer. But they found that the alignment performance worse than that of the traditional registration network [16]. Arar *et al.* [15] also fed back the loss to the training of the register and achieved success in the field of natural image alignment. However, the structural content complexity of medical images is usually higher than that of natural images. Overall, the alignment performance of this method on multi-domain abdominal images is open to question.

3   Proposed Method

3.1   The Overview of RSNet

Let X and Y be the two domains of this research, x and y are the real images in X and Y, respectively; m pairs of paired but unaligned x/y constitute the training set of this research. Note that x/y images from the same anatomical location of the same patient. They are paired images. The task of this research is to align the x to its paired y.

The register pre-training of RSNet is shown in Fig. 1(a). First, the real images y_i/y_j are randomly sampled. Where, $1 \leq i \neq j \leq m$. In other word, the two samples are different. Then, the register R calculates the deformation function φ between y_i/y_j, which can be given as:

$$\varphi_{j \to i} = R(y_i, y_j) \tag{1}$$

where, the former item in $R(.)$ is a fixed image, and the latter item is a moving image. The calculated $\varphi_{j \to i}$ can align the content of the anatomical structure of y_j to y_i and get moved y_j. The $\mathcal{L}_{ncc}$ between the image and the y_i is used to optimize R (note: $\mathcal{L}_{ncc}$ is the mainstream registration training loss, and can refer to [4]), which can be given as:

$$\mathcal{L}_{R\_real} = |\mathcal{L}_{ncc}(y_i, T(\varphi_{j \to i}, y_j))| \tag{2}$$

For the implementation details of the space transform "T", please refer to [2]. By minimizing $\mathcal{L}_{R\_real}$, the performance of R is optimized.

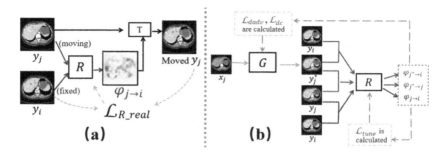

Fig. 1. Framework diagram of RSNet. Where, (a) is the register pre-training; (b) is the joint optimization training. "T" represents the spatial transform. The real images x_j/y_j are paired but unaligned; y_j^* is translated from x_j. The subscript of φ represents the alignment direction.

Subsequently, R and synthesizer G are jointly optimized, as shown in Fig. 1(b). At this time, the real image x_j is also sampled (note: x_j and y_j are paired but unaligned). G translates x_j into a synthetic image y_j^*. R calculates the φ of y_i/y_j^* and y_j/y_j^*, respectively, which can be given as:

$$\varphi_{j^*\to i} = R(y_i, y_j^*); \quad \varphi_{j^*\to j} = R(y_j, y_j^*) \tag{3}$$

Based on the φ, the deformation function adversarial loss $\mathcal{L}_{dadv}$ and the deformation consistency loss $\mathcal{L}_{dc}$ are constructed. These two loss terms and $G's$ own loss term jointly constrain the training of G. At the same time, the adaptive generalization loss $\mathcal{L}_{tune}$ is constructed and used for R generalization.

In the application stage, when the paired but unaligned real images x_t/y_t are input to RSNet (note: t represents the test sample), x_t is translated to y_t^* by G, the deformation function $\varphi_{t^*\to t}$ between y_t/y_t^* is calculated by R, and $\varphi_{t^*\to t}$ align x_t to y_t.

3.2 The Optimization of the Synthesizer in the Joint Training

When G is ideal, y_j^* can be regarded as the real image in Y, and the alignment of y_i/y_j^* and y_j/y_j^* by R can reach the pre-training level. Conversely, when there is a deviation in y_j^*, the alignment performance will decrease. The loss caused by this can guide the training of G.

1) $\mathcal{L}_{dadv}$. When G is ideal, y_j/y_j^* is extremely similar, so the $\varphi_{j^*\to i}/\varphi_{j\to i}$ is also extremely similar. As shown in Fig. 2(a), we designed a deformation function discriminator D_φ to improve the G in the form of adversarial learning [17], which can be given as:

$$\mathcal{L}_{dadv} = |log(D_\varphi(\varphi_{j\to i})) + log(1 - D_\varphi(\varphi_{j^*\to i}))| \tag{4}$$

The output value range of D_φ is $[0,1]$. When D_φ training, $\mathcal{L}_{dadv}$ is expected to be minimized. When G training, $\mathcal{L}_{dadv}$ is expected to be maximized. After the iterative training of D_φ and G, the performance of G is improved.

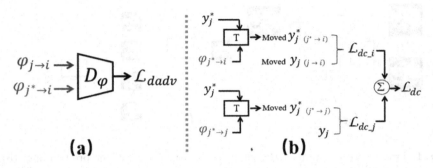

Fig. 2. Diagram of the losses integrates into the synthesizer. Where, (a) shows the calculation of $\mathcal{L}_{dadv}$; (b) show the calculation of $\mathcal{L}_{dc}$. The method of obtaining moved y_j is the same as Fig. 1(a).

2) $\mathcal{L}_{dc}$. When G is ideal, the moved y_j^*/y_j obtained by deforming y_j^*/y_j to y_i should be aligned. At the same time, the styles of the two images should also be consistent. Therefore, the difference in $\mathcal{L}_1$ between the two images can guide the optimization of G, which can be given as:

$$\mathcal{L}_{dc\_i} = \|T(\varphi_{j\to i}, y_j) - T(\varphi_{j^*\to i}, y_j^*)\|_1 \tag{5}$$

Similarly, when G is ideal, the moved y_j^* deformed from y_j^* to y_j should be consistent with y_j. At this time, the difference between the two images can also be fed back to G training, which can be given as:

$$\mathcal{L}_{dc\_j} = \|y_j - T(\varphi_{j^*\to j}, y_j^*)\|_1 \tag{6}$$

$\mathcal{L}_{dc\_i}$ and $\mathcal{L}_{dc\_j}$ are added in the same weight to get $\mathcal{L}_{dc}$. By minimizing $\mathcal{L}_{dc}$, the performance of G is optimized.

3) The Overall Optimization of the Synthesizer. CycleGAN [9] is adopted as G in this research because it can perform image synthesis training on unaligned data sets, and its own optimization goal is $\mathcal{L}_{CycleGAN} = \mathcal{L}_{adv} + 10\mathcal{L}_{cyc}$, please refer to [9]. In the joint training, $\mathcal{L}_{dadv}$ and $\mathcal{L}_{dc}$ are integrated into the training of G, which can be given as:

$$\mathcal{L}_{Syn} = \mathcal{L}_{CycleGAN} + \lambda \times \mathcal{L}_{dadv} + \beta \times \mathcal{L}_{dc} \tag{7}$$

By minimizing $\mathcal{L}_{Syn}$, the performance of G is improved. Among them, λ is set to 1 (both $\mathcal{L}_{dadv}$ and $\mathcal{L}_{adv}$ are adversarial learning losses); β is set to 10 (both $\mathcal{L}_{dc}$ and $\mathcal{L}_{cyc}$ are $\mathcal{L}_1$ loss).

3.3   The Optimization of the Register in the Joint Training

R aligns y_i/y_j^*, y_j/y_j^*, and y_i/y_j, respectively, and the registration training loss is integrated, as shown in Fig. 3. As a result, while $R's$ alignment performance for y_i/y_j^* and y_j/y_j^* is improved, y_i/y_j can still be accurately aligned (that is, the $\mathcal{L}_{dadv}$ and $\mathcal{L}_{dc}$ will not be invalidated with R parameter updates).

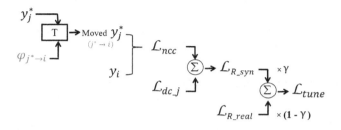

Fig. 3. Diagram of registration generalization loss in the joint training. Here, $\mathcal{L}_{R\_syn}$ is the registration loss of the real/synthetic image. The $\mathcal{L}_{R\_real}$ and $\mathcal{L}_{dc\_j}$ calculations are the same as Fig. 1(a) and Fig. 2(b), respectively.

We constructed an adaptive weighting factor based on Frechet Inception Distance (FID) [18] to prevent low-quality synthetic images from interfering with R training. The index can be calculated without aligning the images. In the joint training process, the similarity fid_y between batch-level y/y^* is calculated, $fid\_y \in [0, +\infty)$. The smaller the value, the more realistic the synthetic image (note: there are several fixed images and moving images in the sampled image batch). According to the obtained fid_y, the "adaptive" weight γ is calculated, which can be given as:

$$\gamma = \begin{cases} \delta/(\ fid\_y + \delta) & s.t.\ fid\_y \leq \delta \\ 0 & s.t.\ \delta < fid\_y \end{cases} \qquad (8)$$

where, δ is the artificially set threshold. According to experience, this value is set to 300. The better the quality of the synthetic image, the closer γ is to 1. The generalization loss of R is formulated as:

$$\mathcal{L}_{tune} = \gamma \times [|\mathcal{L}_{ncc}(y_i, T(\varphi_{j^* \to i}, y_j^*))| + \mathcal{L}_{dc\_j}] + (1 - \gamma) \times \mathcal{L}_{R\_real} \qquad (9)$$

By minimizing $\mathcal{L}_{tune}$, the performance of R is generalized.

4   Experiments

4.1   Datasets and Implementation Details

Datasets. Since there is currently no publicly available abdominal data set with paired images, we used 1020 pairs of unaligned CT plain/ CT venous/MR venous phase samples from 51 patients in the cooperative hospital (i.e., CTP/CTV/MRV). We divide the training set and the test set at 4:1. The length/width of the sample is 512/512 pixels. For the test sample, the outlines of the liver and spleen are manually labeled.

Parameter Settings. Experiments are performed on a gtx2080TI GPU with 11 GB of memory. For reducing operating costs, the batch size in the training phase is set to 2, the epoch of pre-training and joint training is set to 500. The pixel intensity values of the input samples are normalized. The structure of G and D_φ is the same as Wei et al. [11]. The R adopts the mainstream VoxelMorph network [16]. The structure of the network used to calculate FID are the same as Szegedy et al. [19].

Comparison Experiments. We compare the RSNet with five state-of-the-art registration methods that do not use labels [11–13,15,20]. Where, Kim et al. [20] do not use image synthesis. Due to the lack of contour labeling in the training set (Note: this study only manually labeled the test set), the comparison method [1,6,7,14] that requires labels during the training process has not been implemented.

Ablation Study. When "$\mathcal{L}_{dadv}$", "$\mathcal{L}_{dc}$", "$\mathcal{L}_{tune}$", or the adaptive weight is ablated (i.e., the "Fixed weight" in Table 2, at this time, $\gamma = 0.5$), the performance of the RSNet is tested.

Evaluation Strategy. We refer to [1,11,13,14] and use the Dice ratios and Average surface distance (ASD) of moved/fixed image organ contours as quantitative evaluation metrics. The higher the Dice ratios, or the lower the ASD, the better the alignment performance.

4.2   Comparison Analysis

The qualitative comparison of multi-modality and multi-phase alignment is shown in Fig. 4 and Fig. 5, respectively. The alignment performance of the CTV image to the MRV image and the opposite direction is tested; the two-way alignment of the CTP/CTV image is also tested. It can be seen that the organ contours of the moved image obtained by RSNet and that of fixed image are closer than other methods. In addition, in the multi-modality alignment, the moved image obtained by the comparison method has distortion, which indicates that the multi-modality alignment is challenging.

The quantitative results are counted in Table 1. For multi-modality alignment, the result of RSNet is significantly better than that of the five comparison methods. It shows that, for CTV and MRV with significant differences in style, cross-modality synthesis and multi-modality real image alignment are more challenging. Furthermore, the corresponding quantitative results of Xu et al. [13] and Kim et al. [20] are lower than that of Wei et al. [11] and Qin et al. [12], indicating that the φ calculated by the R for multi-modality real images is not accurate enough. For Arar et al. [15], the quantitative result is low, indicating that the registration deviation will interfere with the training of the G.

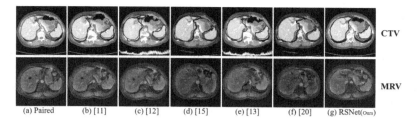

Fig. 4. The results of the multi-modality alignment comparison experiment. Where, the two images in (a) are paired and to be aligned; the red and blue lines are the outlines of the organs. Image in (b–g) is the moved image; the yellow and green lines are the outlines of the organs. (Color figure online)

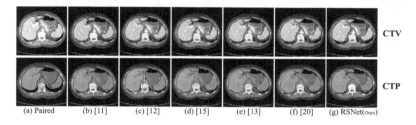

Fig. 5. The results of the multi-phase alignment comparison experiment. (Color figure online)

For multi-phase alignment, the alignment performance of the five comparison methods is relatively ideal. The reason may be that the style difference between CTP and CTV is relatively tiny. Therefore, the challenge of cross-phase image synthesis and multi-phase real image alignment is relatively tiny. Nevertheless, the multi-phase alignment performance of the RSNet is still slightly better than that of the five comparison methods.

Table 1. Quantitative results of multi-domain alignment comparison experiment [the average of Dice ratios (%)↑/ASD↓]. Where, "CTV→MRV" represents the deformation of CTV image to MRV image, and so on.

	CTV→MRV	MRV→CTV	CTV→CTP	CTP→CTV
Wei *et al.* [11]	83.98/3.94	82.67/4.70	89.71/2.31	90.33/2.22
Qin *et al.* [12]	83.82/3.99	82.14/5.12	89.24/2.40	89.90/2.29
Arar *et al.* [15]	80.49/7.71	79.66/8.78	89.26/2.40	89.45/2.37
Xu *et al.* [13]	83.44/4.16	81.91/5.23	90.49/2.13	91.12/1.95
Kim *et al.* [20]	78.86/10.03	79.42/8.85	90.13/2.26	91.03/1.99
RSNet (ours)	**86.76 /3.22**	**85.94 /3.49**	**91.62 /1.88**	**92.08 /1.71**

4.3   Ablation Studies

Figure 6 and Fig. 7 are qualitatively demonstrated the ablation study of multi-modality and multi-phase alignment, respectively. From these figures, when a loss or adaptive setting in RSNet is ablated, the organ contour distance between the moved image and the fixed image increases, and the moved image may have distortion.

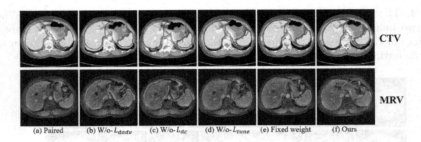

(a) Paired   (b) W/o-L_{dadv} (c) W/o-L_{dc} (d) W/o-L_{tune} (e) Fixed weight   (f) Ours

Fig. 6. The qualitative results of multi-modality abdominal alignment ablation study. (Color figure online)

Quantitative results are statistically in Table 2. For multi-modality alignment, when the $\mathcal{L}_{tune}$ is lacking (i.e., "Without-$\mathcal{L}_{tune}$"), the decrease of quantitative results in the corresponding method is slighter than that of "Without-$\mathcal{L}_{dadv}$" or "Without-$\mathcal{L}_{dc}$". The possible reason is that the styles of multi-modality images are quite different, and it is difficult for G to generate an ideal synthetic image by relying on its training constraints. When the synthetic image quality is limited, the adjustment of $\mathcal{L}_{tune}$ to R is limited. Therefore, RSNet will first enhance the alignment performance by lifting the quality of the synthetic image; then, the $\mathcal{L}_{tune}$ is used to further enhance the alignment performance. In other words, the mentioned $\mathcal{L}_{dadv}$ and $\mathcal{L}_{dc}$ are of great significance to the performance improvement of multi-modality alignment.

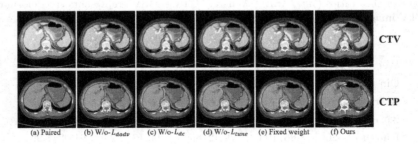

(a) Paired   (b) W/o-L_{dadv} (c) W/o-L_{dc} (d) W/o-L_{tune} (e) Fixed weight   (f) Ours

Fig. 7. The qualitative results of multi-phase abdominal alignment ablation study. (Color figure online)

For multi-phase alignment, the decline of quantitative results of "Without-$\mathcal{L}_{tune}$" is more significant than that of "Without-$\mathcal{L}_{dadv}$" or "Without-$\mathcal{L}_{dc}$". The possible reason is that cross-phase abdominal image synthesis is easy to achieve relatively ideal performance. The demand for improving the quality of the synthetic image is lower than that of generalizing.

In addition, for each group of ablation studies, the quantitative results of "Fixed weight" are lower than that of "Ours", which proves the effectiveness of the proposed weight "adaptive" mechanism.

Table 2. The quantitative results of the ablation study of multi-domain alignment.

	CTV→MRV	MRV→CTV	CTV→CTP	CTP→CTV
Without-$\mathcal{L}_{dadv}$	84.13/3.90	82.87/4.76	90.52/2.10	91.30/1.92
Without-$\mathcal{L}_{dc}$	83.95/3.95	82.26/5.02	90.38/2.21	91.16/1.97
Without-$\mathcal{L}_{tune}$	85.30/3.55	83.45/4.15	90.04/2.32	90.43/2.18
Fixed weight	85.87/3.51	83.85/4.04	91.15/1.97	91.56/1.90
RSNet (ours)	**86.76 /3.22**	**85.94 /3.49**	**91.62 /1.88**	**92.08 /1.71**

5   Conclusions

In this research, we propose a novel RSNet. The loss term caused only by synthesis deviation is constructed and used to optimize the synthesizer. The adaptive generalization loss is constructed and use to generalize the register. Experiments show that RSNet can reach state-of-the-art performance, especially for multi-modality alignment.

The quality evaluation threshold of the synthetic image of this research relies on experience. In future work, we will use optimization algorithms and deep network feedback mechanisms based on the initial threshold setting. Thus, the threshold can be flexibly adjusted by itself to adapt to the RSNet training process.

References

1. Liu, F., Cai, J., Huo, Y., Cheng, C., et al.: JSSR: A joint synthesis, segmentation, and registration system for 3D multi-domain image alignment of large-scale pathological CT scans (ECCV), July 2020
2. Huijskens, S.C., et al.: Abdominal organ position variation in children during image-guided radiotherapy. In: Radiat Oncol 13, vol. 173 (2018)
3. Zhang, Y., Jiang, F., Shen, R.: Region-based face alignment with convolution neural network cascade. In: Liu, D., Xie, S., Li, Y., Zhao, D., El-Alfy, E.S. (eds.) ICONIP 2017. LNCS, vol. 10636, pp. 300–309. Springer, Cham (2017). https://doi.org/10.1007/978-3-319-70090-8_31
4. Jaderberg, M., Simonyan, K., et al.: Spatial transformer networks. In: Advances in Neural Information Processing Systems, pp. 2017–2025 (2015)

5. Cao, X., Yang, J., Gao, Y., Guo, Y., Wu, G., Shen, D.: Dual-core steered non-rigid registration for multi-domain images via bi-directional image synthesis. In: Medical Image Analysis, vol. 41, pp. 18–31 (2017)
6. Fan, J., Cao, X., Wang, Q., et al.: Adversarial learning for mono- or multi-domain registration. Med. Image Anal. **58**, 101545 (2019)
7. Dubost, F., Bruijne, M.D., Nardin, M., et al.: Multi-atlas image registration of clinical data with automated quality assessment using ventricle segmentation. Med. Image Anal. **63**, 101698 (2020)
8. Blendowski, M., Hansen L., Heinrich, M.P.: Weakly-supervised learning of multi-domain features for regularised iterative descent in 3D image registration. Med. Image Anal. **67**, 101822 (2021)
9. Zhu, J., Park, T., Isola, P., Efros, A.: Unpaired image-to-image translation using cycle-consistent adversarial networks. In: IEEE International Conference on Computer Vision, pp. 2242–2251 (2017)
10. Liu, M., Thomas, B., Jan, K.: Unsupervised image-to-image translation networks. In: Annual Conference on Neural Information Processing Systems, pp. 701–709 (2017)
11. Wei, D., Ahmad, S., Huo, J., et al.: SLIR: synthesis, localization, inpainting, and registration for image-guided thermal ablation of liver tumors. Med. Image Anal. **65**, 101763 (2020)
12. Qin, C., Shi, B., Liao, R., Mansi, T., Rueckert, D., Kamen, A.: Unsupervised deformable registration for multi-modal images via disentangled representations. In: Chung, A.C.S., Gee, J.C., Yushkevich, P.A., Bao, S. (eds.) IPMI 2019. LNCS, vol. 11492, pp. 249–261. Springer, Cham (2019). https://doi.org/10.1007/978-3-030-20351-1_19
13. Xu, Z., et al.: Adversarial uni- and multi-modal stream networks for multimodal image registration. In: Martel, A.L., et al. (eds.) MICCAI 2020. LNCS, vol. 12263, pp. 222–232. Springer, Cham (2020). https://doi.org/10.1007/978-3-030-59716-0_22
14. Zhou, B., Augenfeld, Z., Chapiro, J., Zhou, S.K., Liu, C., Duncan, J.S.: Anatomy-guided multidomain registration by learning segmentation without ground truth: application to intraprocedural CBCT/MR liver segmentation and registration. Med. Image Anal. **71**, 102041 (2021)
15. Arar, M., Ginger, Y., et al.: Unsupervised multi-domain image registration via geometry preserving image-to-image translation (CVPR), March 2020
16. Balakrishnan, G., Zhao, A., Sabuncu, M.R., Guttag, J., Dalca, A.V.: VoxelMorph: a learning framework for deformable medical image registration. IEEE Trans. Med. Imaging **38**(8), 1788–1800 (2019)
17. Goodfellow, I., et al.: Generative adversarial nets. In: Proceedings of the International Conference on Neural Information Processing Systems, Montr šSal, Canada, pp. 2672–2680 (2014)
18. Heusel, M., Ramsauer, H., et al.: GANs trained by a two time-scale update rule converge to a local Nash equilibrium (ANIPS), June 2017
19. Szegedy, C., Vanhoucke, V., Ioffe, S., Shlens, J., Wojna, Z.: Rethinking the inception architecture for computer vision. In: Proceedings of the IEEE Conference on Computer Vision and Pattern Recognition, pp. 2818–2826 (2016)
20. Kim, B., Kim, D.H., Park, S.H., Kim, J., Lee, J., Ye, J.C.: CycleMorph: cycle consistent unsupervised deformable image registration. Med. Image Anal. **71**, 102036 (2021)

Coordinate Attention Residual Deformable U-Net for Vessel Segmentation

Cong Wu, Xiao Liu$^{(\boxtimes)}$, Shijun Li, and Cheng Long

Hubei University of Technology, Wuhan, China

Abstract. The location information of features is essential for pixel-level segmentation tasks such as retinal vessel segmentation. In this study, we proposed the CARDU-Net (Coordinate Attention Gate Residual Deformable U-Net) model based on coordinate attention mechanism for the segmentation task, which can extract effective features by accurately locating feature location information and enhance the accuracy of segmentation. The deformable convolution and residual structure with Dropblock are also introduced to refine the encoder structure of U-Net. The model is applied to DRIVE, CHASE_DB1, and LUNA (2017) datasets, and the experimental results on the three public datasets demonstrate the superior segmentation capability of CARDU-Net, and the modified part is reflected by ablation experiments in this work. The results show that the CARDU-Net model performs better compared to other network models and can segment medical images accurately.

Keywords: Retinal vessel segmentation · U-Net · Coordinate Attention

1 Introduction

A range of ocular diseases brought about by glaucoma, diabetes and hypertension are the main causes of blindness [1], and these diseases are inextricably linked to the state of the retinal vasculature. Doctors can further diagnose the condition by observing the morphology of the retinal vessels, therefore, the segmentation of retinal vessels plays an important supporting role in the diagnosis and treatment of ocular diseases [2]. However, the large number of vessels in fundus images with complex backgrounds and different morphologies makes segmentation difficult. Traditionally, experienced radiologists and ophthalmologists perform manual segmentation, which is not only time and energy consuming, but also requires a high level of professionalism. To speed up the diagnostic efficiency and reduce manual bias, it is necessary to develop an automated and accurate segmentation method.

In earlier studies, the images were mainly analyzed with the help of some methods of machine learning, but eventually, the segmentation still needs to be

© Springer Nature Switzerland AG 2021
T. Mantoro et al. (Eds.): ICONIP 2021, LNCS 13110, pp. 345–356, 2021.
https://doi.org/10.1007/978-3-030-92238-2_29

done manually, which is relatively inefficient [3,4]. In recent years, deep learning has achieved good results in natural language processing and image classification, which provides novel solutions to researchers in the field of medical image segmentation [5]. However, the task of medical image segmentation is based on the pixel level, and the segmentation accuracy has not been high enough.

To address this problem, Ronneberger et al. stated the U-Net [6] structure for semantic segmentation of biological images. It is mainly composed of encoder and decoder modules with skip connection. Due to its great performance, researchers have improved upon it, such as Zhuang reported LadderNet [7], Wu et al. presented MS-NFN [8], and Khan et al. [9] used a structure similar to U-Net++ for the retinal vessel segmentation. Although these networks performs well by stacking U-Net or encoders, but they all ignore the position information of the feature in the image. Meanwhile overlaying the encoder-decoder structure to extract features undoubtedly sacrifices the image resolution and requires a larger computational cost.

Only a small proportion of the pixel information in the retinal vessel image is a valid feature [10], and U-Net is easily disturbed by the retinal background information when extracting features. Wu et al. [11] integrated attentional mechanisms into Dense U-Net to withdraw retinal vessel information. Wang et al. proposed DEU-Net [12], which uses channel-specific attention to select the necessary features and combines them with a dual encoder. Guo et al. [13] introduced a modified channel attention in the U-Net structure to enhance the interdependence between feature channels. Guo et al. presented RSAN [14], which introduces spatial attention into the skip connection of U-Net to infer the attention graph along the spatial dimension to achieve feature refinement. As can be seen, none of the above attentions provide the location information of the features to the network well. Hou et al. proposed coordinate attention [15], which can generate feature location attention maps by encoding channel relationships and long-term dependencies with precise location information. The coordinate attention has never been applied to medical image segmentation before, and this study introduces it to the U-Net structure for the first time to perform segmentation of medical images.

It was observed that in addition to the location information, the variable morphology of retinal vessels also limits the accuracy of vessel segmentation. The common convolution approach to extract this unknown image morphology has certain drawbacks because of the fixed geometry of the convolution kernel and can not sample the feature map better, and the feature information extraction is not sufficient. Dai et al. [16] proposed a deformable convolution approach, the sampling points of the convolution kernel learn a certain offset according to the position of the target so that the effective perceptual field increases and the network boosts the feature extraction capability.

To address the above problems and inspired by coordinate attention and deformable convolution, this paper proposes CARDU-Net, a network model based on U-Net, which introduces coordinate attention to enable the network to accurately learn the location information of features while suppressing the interference of invalid background information. The residual blocks with Drop-block [18] is also added to the encoder and decoder respectively and the normal convolution is replaced by deformable convolution to improve the extraction capability of the model. The experimental results show that our network has state-of-the-art performance. The specific contributions of this paper are as fol-lows:

(1) A novel model (CARDU-Net) is proposed to accomplish the image segmen-tation task, employing deformable convolution to extract retinal vessels with complex morphology.
(2) Coordinate Attention is introduced to medical image segmentation for the first time to obtain location information of features for enhanced feature extraction.
(3) CARDU-Net has excellent segmentation results on DRIVE, CHASE_DB1, and LUNA (2017) datasets with good generalization, which can be applied to other segmentation tasks.

2   Method

Our goal is to establish a deep learning model to segment retinal vessels accu-rately. Inspired by Coord Attention and Deformable conv, we propose a new net-work model CARDU-Net. Figure 1 introduces the architecture of it. CARDU-Net is composed of four downsampling blocks and four upsampling blocks contain-ing skip connection between them. After downsampling, a coordinate attention is followed to provide the location information of the features. Each downsam-pling module includes two 3 × 3 convolution layers with residual structure and a maxpooling layer. In the convolution layer, the common convolution module is replaced by deformable convolution. After each convolution layer, there is a BN layer and a ReLU layer.

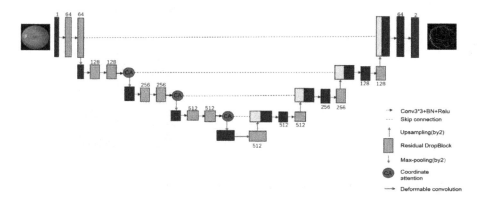

Fig. 1. CARDU-Net network architecture.

2.1 CARDU-Net

Attention mechanism has been widely applied in the field of computer vision. In recent years, SEblock, CBAM, etc. have been outstanding, many scholars have integrated them into U-Net to undertake segmentation tasks. However, when faced with the segmentation task, they just pay attention to the channel where the feature is located, but not the position of the feature. Especially for retinal blood vessel images, the feature information is complex and there are many background noise, so the positional relationship is undoubtedly crucial for retinal vessel segmentation. The coordinate attention presented recently can solve this problem well, which makes the background noise in the network segmentation removed more comprehensively, and inhibits the extraction of invalid features, while the effective information is still preserved.

The network architecture of CARDU-Net is shown in Fig. 1, we put coordinate attention between the downsampling operations in the U-Net model. The structure of coordinate attention is shown in Fig. 2, the core of which is to decompose the global pooling into two one-dimensional feature extractions, then spliced to form a perceptual attention map. With the constant adjustment of the weight of coordinate attention, this perceptual attention tries to make the important information in the whole image clearer, so the coordinate attention will be maximized to assist feature extraction immediately after downsampling. Therefore, our network can achieve better segmentation results by learning more accurate attention weight matrix before output.

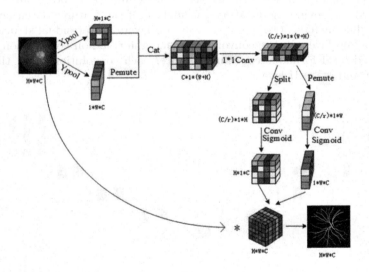

Fig. 2. Structure of coordinate attention

After inputting a tensor with the size of H × W × C, the global average pooling operation is performed from H and W dimensions respectively:

$$F_c(h) = \frac{1}{W} \sum_{0 \le i < W} x_c(h, i) \tag{1}$$

$$F_c(w) = \frac{1}{H} \sum_{0 \le j < H} x_c(j, w) \tag{2}$$

At this time, the obtained tensor sizes are C × W × 1 and C × 1 × H respectively. Use shared 1 × 1 convolution operation after splicing:

$$F = (([F^h, F^w])C)\sigma \tag{3}$$

where σ is the ReLU function, which splits F along the spatial dimension and then restores the two tensor to the same number of channels as the original image using a 1 × 1 convolution:

$$G^h = \delta\left(C_h\left(F^h\right)\right) \tag{4}$$

$$G^w = \delta\left(C_w\left(F^w\right)\right) \tag{5}$$

Here δ is sigmoid function, the objective is to enable valid features to continue to be maintained while the probability of invalid features being extracted tends to zero. At this time, we get a three-dimensional weight matrix, which is multiplied by the original input:

$$Y_c(i, j) = X_c(i, j) * G_c{}^h(i) * G_c{}^w(j) \tag{6}$$

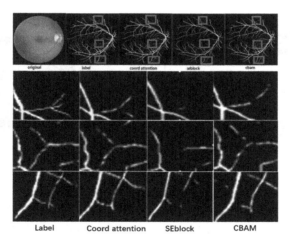

Fig. 3. Comparison of details. (In the detail diagram below, from front to back are the results of label, coordinate attention, SEblock and CBAM)

There are several differences with SEblock. Coordinate attention uses two-dimensional pooling to obtain information separately, instead of directly using global average pooling. This has the advantage of knowing the position of feature information more accurately and suppressing noise. Coordinate attention does not use fully-connected layer to reduce dimension and enhance dimension, but uses 1 × 1 convolution to reduce channel dimension, and then uses two 1 × 1 convolutions to enhance dimension respectively. Cross-channel dependence is realized by applying nonlinear activation function. Coordinate attention pays attention not only to the channel but also to the specific location of features, which is obviously better than other attention mechanisms in experiments, with less information loss and higher segmentation accuracy. It can be seen from Fig. 3 that the segmentation effect of coordinated attention is the best, and it extracts more vessel information than SEblock and CBAM.

2.2   Residual Blocks with Dropblock

In this work, due to the limited amount of data, gradient disappearance and performance degradation problems occur during the training process. He, et al. proposed ResNet [17], in which the residual structure can alleviate the problems caused by a larger number of network layers and enable multi-scale fusion of features. Ghiasi et al. proposed DropBlock [18], a structured Dropout which randomly discards local block regions to achieve more efficient regularization, as an effective module to prevent overfitting of convolutional networks [13,14]. Inspired by the above work, we added RDB (residual blocks with Dropblock) to the encoder and decoder to optimize the network that ease the training difficulty of our network.

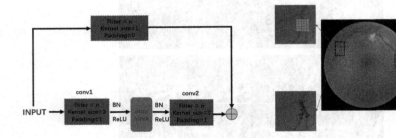

Fig. 4. Diagram of residual block

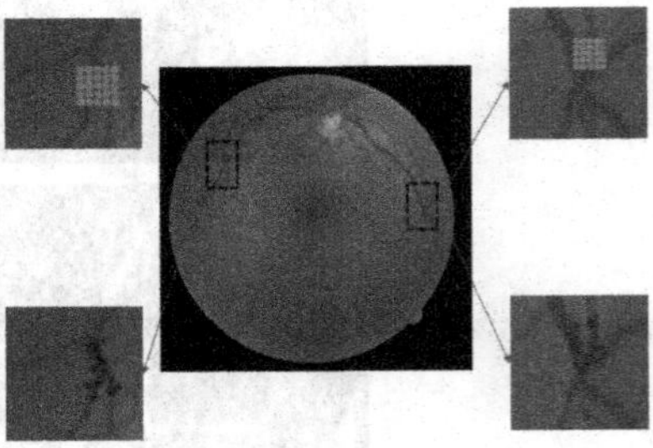

Fig. 5. Comparison of deformable convolution and ordinary convolution

The structure of the residual block is illustrated in Fig. 4. Weightless short-cuts become the path from input to output, which increases the spread of information flow, reduces the loss of features and deepens feature extraction. The residual block consists of residual mapping and identity mapping. The identity mapping is shortcut, which transfers the input directly to the back to reduce the loss of feature information. In CARDU-Net model, the double-layer ordinary convolution in encoder and decoder is changed to convolution with residual structure, and the number of convolution cores in each layer in encoder is 64, 128, 256, 512 respectively. A DropBlock immediately follows the BN and ReLU, which can effectively ease overfitting by regularizing convolutional layers. The residual structure makes the training of the network less difficult and facilitates the extraction of retinal blood vessel images, hence improving the segmentation effect.

2.3   Deformable Convolution

Retinal vessels have different morphologies and complex variations, and a single convolution method cannot better take into account the coarse or unconventional vessel features, which leads to poor segmentation results, so we use a deformable convolution method to solve this problem. As shown in Fig. 5, compared with the conventional convolution kernel, the deformable convolution can change the offset of the convolution block according to the target morphology, which makes the effective perceptual field larger and leads to better extraction of vessel features and excellent segmentation results.

As shown in Fig. 6, we conducted a control experiment with deformable convolution and normal convolution and dilated convolution on DRIVE. It can be seen that the deformable convolution performs better when facing task of retinal vessel segmentation, mainly at complex deformations and too large or too small vessels. The reason for the lack of the dilated convolution may be the Gridding Effect, which may be effective for segmentation of large objects, but is a fatal problem for pixel-by-pixel segmentation tasks.

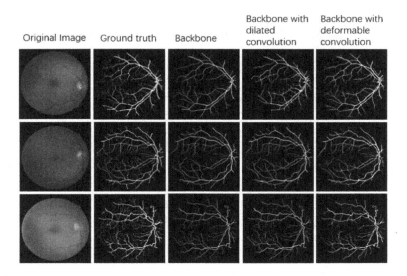

Fig. 6. Comparison of several convolution methods on DRIVE

3   Experiments and Results

The most characteristic features of the fundus retinal dataset images are non-uniform background illumination, poor contrast between vessels and background, and many tiny vessels. In this work, we use the DRIVE, CHASE_DB1 and LUNA (2017) datasets. For the DRIVE dataset, we use 20 samples as the training set and another 20 samples as the test set. And for the CHASE_DB1 dataset, we divide 20 samples for network training and another 8 samples for testing. We use data augument to make up for the lack of image data. In the training stage, we select the RMSPprop optimization algorithm, and the learning rate is 0.0001. We have trained 600 epochs in total, and batchsize is 4. Using the BCEWithLogitsLoss loss function which can be expressed as:

$$Loss = \{l_1, ..., l_N\}, l_n = -[y_n \cdot log(\sigma(x_n)) + (1 - y_n) \cdot log(1 - \sigma(x_n))] \quad (7)$$

In the field of image segmentation, there are several indicators to measure the segmentation effect, including SE, SP, ACC, F1, AUC and so on.

$$SE = \frac{TP}{TP + FN} \quad (8)$$

$$ACC = \frac{TP + TN}{TP + FP + TN + FN} \quad (9)$$

$$F1 = 2 * \frac{PR * SE}{PR + SE} \quad (10)$$

Table 1. The results of different models on DRIVE.

Method	SE	SP	F1	ACC	AUC
U-Net [6]	0.7537	0.9820	0.8142	0.9531	0.9755
Liskowski [19]	0.7726	0.9820	0.8149	0.9553	0.9722
R2U-Net [20]	0.7792	0.9813	0.8171	0.9556	0.9784
RSAN [14]	**0.8149**	0.9839	0.8222	0.9691	**0.9855**
LadderNet [7]	0.7856	0.9810	0.8202	0.9561	0.9793
MF-NFN [8]	0.7844	0.9819	N.A	0.9567	0.9807
SF^{3}N [9]	0.8112	0.9795	N.A	0.9651	0.9830
DEU-Net [12]	0.7952	0.9816	**0.8270**	0.9568	0.9770
CARDU-Net	0.8123	**0.9862**	0.8205	**0.9699**	0.9829

Table 1 and Table 2 show the working effect of CARDU-Net model proposed in this paper, and how it compares with the segmentation results of other network model. The results show that our CARDU-Net does better in segmentation accuracy and performs well in AUC and F1 compared to other good networks.

Table 2. The results of different models on CHASE_DB1.

Method	SE	SP	F1	ACC	AUC
U-Net [6]	0.8288	0.9701	0.7783	0.9578	0.9772
Liskowski [19]	0.7507	0.9797	N.A	0.9581	0.9716
R2U-Net [20]	0.7756	0.9820	0.8171	0.9634	0.9815
RSAN [14]	**0.8486**	0.9836	0.8111	0.9751	**0.9894**
LadderNet [7]	0.7978	0.9818	0.8031	0.9656	0.9839
MF-NFN [8]	0.7538	0.9847	N.A	0.9637	0.9825
SF³N [9]	0.8107	0.9811	N.A	0.9689	0.9815
DEU-Net [12]	0.8074	0.9821	0.8037	0.9661	0.9812
CARDU-Net	0.8312	**0.9868**	**0.8184**	**0.9763**	0.9854

Table 3. Effect of RDB on the model.

Method	SE	ACC	F1
U-Net [6]	0.7537	0.9531	0.8142
CARDU-Net without RDB	0.7839	0.9648	0.8176
CARDU-Net with RDB	**0.8123**	**0.9699**	**0.8205**

To demonstrate the role of RDB in this network, we designed ablation experiments on DRIVE, as shown in Table 3, when the RDB module is included in CARDU-Net, the ACC improves by about 0.5% and the F1-score improves by about 0.3%.

We have conducted a comparison experiment between deformable convolution and normal convolution and null convolution on DRIVE. As shown in Table 4, Backbone is the network in CARDU-Net when the convolution method is normal convolution. The experimental results show that the deformable convolution performs best.

Table 4. Results of convolution methods.

Method	SE	ACC	F1
Backbone	0.7726	0.9652	0.8157
Backbone with dilated convolution	0.7757	0.9659	0.8169
Backbone with deformable convolution	**0.8123**	**0.9699**	**0.8205**

We also compared the effect of the coordinate attention at different locations in the network structure on segmentation performance, respectively. As in Fig. 7. we individually place coordinate attention at four different locations:

Table 5. Results of coord attention in different positions

position	DRIVE		CHASE_DB1	
	ACC	F1_score	ACC	F1_score
A0	0.9568	0.8113	0.9591	0.7702
A1	0.9673	0.8107	0.9705	0.7867
A2	**0.9699**	**0.8133**	**0.9843**	**0.8213**
A3	0.9597	0.8110	0.9682	0.7923

(1) A0: without coordinate attention.
(2) A1: after all down sampling layers.
(3) A2: after all down sampling layers, but before the first down sampling layer.
(4) A3: after all up sampling layers.

Table 5 shows a comparison of the performance impact of coordinating attention to different locations in the network. The model performs best when the coordination attention is placed at position A2. At the same time, it can be found that if the coordinate attention is added to the first down sampling layer, the performance does not change better instead. The main reason may be that there are too many low-level features and the feature map is too close to the input layer.

Table 6. Comparison of partial lung segmentation results.

Method	SE	ACC
U-Net [6]	0.938	0.975
CE-Net [21]	0.980	0.990
CARDU-Net	0.988	0.992

To demonstrate the good generalization ability of CARDU-Net in the face of different segmentation tasks, we use LUNA(2017) for testing. LUNA(2017) includes 534 lung images and labels with a size of 512×512, we use 80% of the data for training and the rest for testing. We show the experimental results in Table 6 and can see that the ACC improves from 0.975 to 0.992 compared to U-Net. In Fig. 8, we show a comparative example of partial lung segmentation.

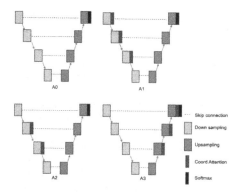

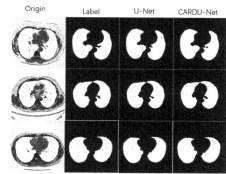

Fig. 7. Coordinated attention position

Fig. 8. Partial lung segmentation results (compared with U-Net [6])

4 Conclusion

In this work, we propose CARDU-Net for enhancing the accuracy of retinal vessel segmentation to achieve end-to-end automatic segmentation. We adopt U-Net as the backbone, introduce a coordinate attention to obtain the location information of features while inhibiting the background noise, and change the normal convolutional block in the encoder to deformable convolution while using the improved residual block to alleviate the network degradation problem. Excellent segmentation performance on DRIVE and CHASE_DB1 eye datasets and lung dataset LUNA (2017), and it can be seen through experiments that our proposed method achieves excellent segmentation results. In addition, when segmenting the lung structure, we noticed that the bordered parts of some results are not smooth enough, so in the next work, some recent loss functions can be imported to improve this problem, and also generative adversarial networks can be used to alleviate this problem.

References

1. Smart, T.J., Richards, C.J., Bhatnagar, R., Pavesio, C., Agrawal, R., Jones, P.H.: A study of red blood cell deformability in diabetic retinopathy using optical tweezers. In: Optical Trapping and Optical Micromanipulation XII, vol. 9548, p. 954825. International Society for Optics and Photonics (2015)
2. Cheung, CYl., et al.: Retinal vascular tortuosity, blood pressure, and cardiovascular risk factors. Ophthalmology **118**(5), 812–818 (2011)
3. Sinthanayothin, C., et al.: Automated detection of diabetic retinopathy on digital fundus images. Diabet. Med. **19**(2), 105–112 (2002)
4. Gardner, G.G., Keating, D., Williamson, T.H., Elliott, A.T.: Automatic detection of diabetic retinopathy using an artificial neural network: a screening tool. Br. J. Ophthalmol. **80**(11), 940–944 (1996)
5. Litjens, G., et al.: A survey on deep learning in medical image analysis. Med. Image Anal. **42**, 60–88 (2017)

6. Ronneberger, O., Fischer, P., Brox, T.: U-Net: convolutional networks for biomedical image segmentation. In: Navab, N., Hornegger, J., Wells, W.M., Frangi, A.F. (eds.) MICCAI 2015. LNCS, vol. 9351, pp. 234–241. Springer, Cham (2015). https://doi.org/10.1007/978-3-319-24574-4_28
7. Zhuang, J.: Laddernet: multi-path networks based on u-net for medical image segmentation. arXiv preprint arXiv:1810.07810 (2018)
8. Wu, Y., Xia, Y., Song, Y., Zhang, Y., Cai, W.: Multiscale network followed network model for retinal vessel segmentation. In: Frangi, A.F., Schnabel, J.A., Davatzikos, C., Alberola-López, C., Fichtinger, G. (eds.) MICCAI 2018. LNCS, vol. 11071, pp. 119–126. Springer, Cham (2018). https://doi.org/10.1007/978-3-030-00934-2_14
9. Khan, T.M., Robles-Kelly, A., Naqvi, S.S.: A semantically flexible feature fusion network for retinal vessel segmentation. In: Yang, H., Pasupa, K., Leung, A.C.-S., Kwok, J.T., Chan, J.H., King, I. (eds.) ICONIP 2020. CCIS, vol. 1332, pp. 159–167. Springer, Cham (2020). https://doi.org/10.1007/978-3-030-63820-7_18
10. Badar, M., Haris, M., Fatima, A.: Application of deep learning for retinal image analysis: a review. Comput. Sci. Rev. **35**, 100203 (2020)
11. Wu, C., Zou, Y., Zhan, J.: DA-U-Net: densely connected convolutional networks and decoder with attention gate for retinal vessel segmentation. In: IOP Conference Series: Materials Science and Engineering, vol. 533, p. 012053. IOP Publishing (2019)
12. Wang, B., Qiu, S., He, H.: Dual encoding U-Net for retinal vessel segmentation. In: Shen, D., et al. (eds.) MICCAI 2019. LNCS, vol. 11764, pp. 84–92. Springer, Cham (2019). https://doi.org/10.1007/978-3-030-32239-7_10
13. Guo, C., Szemenyei, M., Hu, Y., Wang, W., Zhou, W., Yi, Y.: Channel attention residual u-net for retinal vessel segmentation. In: ICASSP 2021–2021 IEEE International Conference on Acoustics, Speech and Signal Processing (ICASSP), pp. 1185–1189. IEEE (2021)
14. Guo, C., Szemenyei, M., Yi, Y., Zhou, W., Bian, H.: Residual spatial attention network for retinal vessel segmentation. In: Yang, H., Pasupa, K., Leung, A.C.-S., Kwok, J.T., Chan, J.H., King, I. (eds.) ICONIP 2020. LNCS, vol. 12532, pp. 509–519. Springer, Cham (2020). https://doi.org/10.1007/978-3-030-63830-6_43
15. Hou, Q., Zhou, D., Feng, J.: Coordinate attention for efficient mobile network design. In: Proceedings of the IEEE/CVF Conference on Computer Vision and Pattern Recognition, pp. 13713–13722 (2021)
16. Dai, J., et al.: Deformable convolutional networks. In: Proceedings of the IEEE International Conference on Computer Vision, pp. 764–773 (2017)
17. He, K., Zhang, X., Ren, S., Sun, J.: Deep residual learning for image recognition. In: Proceedings of the IEEE Conference on Computer Vision and Pattern Recognition, pp. 770–778 (2016)
18. Ghiasi, G., Lin, T.Y., Le, Q.V.: Dropblock: a regularization method for convolutional networks. arXiv preprint arXiv:1810.12890 (2018)
19. Liskowski, P., Krawiec, K.: Segmenting retinal blood vessels with deep neural networks. IEEE Trans. Med. Imaging **35**(11), 2369–2380 (2016)
20. Alom, M.Z., Hasan, M., Yakopcic, C., Taha, T.M., Asari, V.K.: Recurrent residual convolutional neural network based on U-Net (R2U-Net) for medical image segmentation. arXiv preprint arXiv:1802.06955 (2018)
21. Gu, Z., et al.: CE-Net: context encoder network for 2D medical image segmentation. IEEE Trans. Med. Imaging **38**(10), 2281–2292 (2019)

Gated Channel Attention Network for Cataract Classification on AS-OCT Image

Zunjie Xiao[1], Xiaoqing Zhang[1], Risa Higashita[1,2(✉)], Yan Hu[1], Jin Yuan[3], Wan Chen[3], and Jiang Liu[1,4,5,6(✉)]

[1] Department of Computer Science and Engineering, Southern University of Science and Technology, Shenzhen, China
[2] Tomey Corporation, Nagoya, Japan
[3] Zhongshan Ophthalmic Center, Sun Yat-sen University, Guangzhou, China
[4] Cixi Institute of Biomedical Engineering, Ningbo Institute of Materials Technology and Engineering, Chinese Academy of Sciences, Ningbo, China
[5] Guangdong Provincial Key Laboratory of Brain-inspired Intelligent Computation, Department of Computer Science and Engineering, Southern University of Science and Technology, Shenzhen, China
[6] Research Institute of Trustworthy Autonomous Systems, Southern University of Science and Technology, Shenzhen, China

Abstract. Nuclear cataract (NC) is the leading cause of blindness and vision impairment globally. Accurate NC classification is significant for clinical NC diagnosis. Anterior segment optical coherence tomography (AS-OCT) is a non-contact, high-resolution, objective imaging technique, which is widely used in diagnosing ophthalmic diseases. Clinical studies have shown that there is a significant correlation between the pixel density of the lens region on AS-OCT images and NC severity levels; however, automatic NC classification on AS-OCT images has not been seriously studied. Motivated by clinical research, this paper proposes a gated channel attention network (GCA-Net) to classify NC severity levels automatically. In the GCA-Net, we design a gated channel attention block by fusing the clinical priority knowledge, in which a gated layer is designed to filter out abundant features and a Softmax layer is used to build the weakly interacting for channels. We use a clinical AS-OCT image dataset to demonstrate the effectiveness of our GCA-Net. The results showed that the proposed GCA-Net achieves 94.3% in accuracy and outperformed strong baselines and state-of-the-art attention-based networks.

Keywords: AS-OCT · Nuclear cataract · Gated channel attention · Deep learning

1 Introduction

Cataract is the leading cause of reversible blindness and vision impairment worldwide [5]. Early treatment can address vision impairment and restore vision to

---

Z. Xiao and X. Zhang—Equal contribution.

© Springer Nature Switzerland AG 2021
T. Mantoro et al. (Eds.): ICONIP 2021, LNCS 13110, pp. 357–368, 2021.
https://doi.org/10.1007/978-3-030-92238-2_30

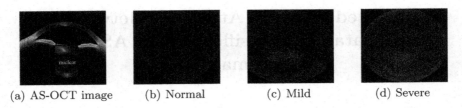

(a) AS-OCT image　　(b) Normal　　　(c) Mild　　　(d) Severe

Fig. 1. The whole AS-OCT image shown in (a), and the center area is the nucleus. (b) Normal nucleus image; (c) Mild NC image where the nuclear opacity is asymptomatic;(d) Severe NC image where the nuclear opacity is symptomatic.

improve the cataract patient's quality of life. According to the location of the opacities, cataracts can be generally classified into three types: nuclear cataract (NC), cortical cataract (CC), and posterior subcapsular cataract (PSC). NC is the most common type of cataract, characterized by the increase of light scattering in the nucleus region of the crystalline lens area. In clinical practice, slit-lamp image is routinely used to diagnose NC based on standard cataract classification systems. Lens opacity classification system III (LOCS III) [4] is a well-accepted slit lamp image-based cataract classification system. With the development of nuclear opacity pathology, nuclear cataract can be divided into three stages [15]. (1) Normal: healthy or without nuclear opacity in the slit-lamp image; (2) mild (grade=1 or 2 in LOCS III): the nuclear opacity is asymptomatic; (3) severe (grade≤3 in LOCS III): the nuclear opacity is symptomatic. Mild NC can be relieved by clinical intervention, while severe NC needs to prepare for surgery as soon. Figure 1 shows the representative figures of AS-OCT nuclear areas at the three stages.

Anterior segment optical coherence tomography (AS-OCT) is a non-contact, high-resolution tomography technique, which can objectively and quickly obtain overall information of the entire lens. AS-OCT images have gradually been used in the diagnosis of various anterior segment ocular diseases such as glaucoma, cataracts, and keratitis [5]. For NC diagnosis, AS-OCT image can capture the nucleus region clearly while other ophthalmology images like fundus images cannot. The clinical study has shown that the average lens density (ALD) has a strong linear relationship with the nucleus region of AS-OCT images based on the LOCS III [21], which provided clinical support for automatic cataract classification on AS-OCT images. Following [21], clinical research [3,14,16,20] further got the similar statistics results. Motivated by the preliminary works, [27] studied NC classification based on AS-OCT image, which uses the convolutional neural network (CNN), but they achieved poor performance.

Average nucleus density (AND) is a clinical indicator on AS-OCT image for nuclear cataract diagnosis, which is defined as the average pixel density in the nucleus region [21]. Figure 2 shows the distribution of AND in different stages of nuclear cataract. It can be seen that there are significant differences in the AND distribution among different NC stages, while many images are difficult to

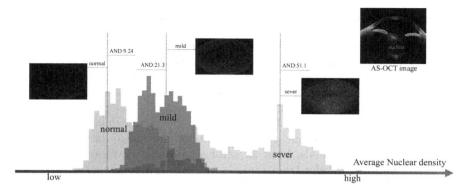

Fig. 2. The different nuclear opacity stages are reflected in the OCT image; the histogram reflects the sample distribution of three nuclear cataract severity levels (different colors mean different NC stages).

classify the severity of cataracts simply by AND (the overlap area as shown in Fig. 2).

In recent years, channel attention mechanism has become one of the most popular attention mechanisms due to its simplicity and effectiveness, which directly learns importance weights of each channel. In channel attention block, global average pooling (GAP) is used for integrating channel-wise information, which calculates the mean value of each channel. GAP collects the global mean value, which enhances the representation ability for global information, especially AND. Inspired by this relationship, we propose a simple yet effective gated channel attention network (GCA-Net) for NC classification automatically. In the GCA-Net, this paper designs a novel gated channel attention block, where a gating operator is used to mask and applies a weakly-interacting operator to model the global channel information.

The main contributions of this paper are as follows: (1) We develop a novel convolutional neural network (CNN) model named GCA-Net to discriminate opacity information for classifying NC levels into three severity levels. (2) This paper designs a simple yet effective channel attention (GCA) block comprised of three stages: gating, squeezing, and interacting, to capture the global information. (3) The results on a clinical AS-OCT image show that our GCA-Net surpasses state-of-the-art attention-based networks.

2   Related Work

2.1   Cataract Classification

In recent years, research scholars have proposed many advanced machine learning and deep learning methods for automatic cataract classification on different ophthalmology image modalities [26]. [12] proposes an automatic NC classification system that contains three stages (region detection, pixel feature extraction, and level prediction) based on the ACHIKO-NC slit-lamp dataset, and achieves an average error of 0.36. Xu et al. also performed NC classification on the ACHIKO-NC dataset, using the group sparse regression (GSR) method and achieved 83.4% accuracy [25]; [24] proposed the semantic similarity method for slit lamp image-based NC classification and obtained better performance than GSR. [1] achieves an accuracy of 95% using support vector machines (SVM) to classify NC on ultrasound images, but the ultrasound image data sets used for their work are from animals. Li et al. achieved accurate cataract screening by improving the Haar wavelet transform algorithm on fundus images [2].

Compared with machine learning methods, deep learning methods are skilled at capturing useful feature representations. Gao et al. proposed a hybrid model of convolutional neural network (CNN) and recurrent neural network (RNN) based on slit-lamp images and achieved 82.5% accuracy for NC classification [6]. A team of Sun Yat-sen University proposed a congenital cataract screening platform based on deep learning [13]. Xu et al. proposed a global-local hybrid CNN network by fusing different parts of pathological information that achieves better performance than previous methods on fundus images [23,26].

There are relatively few NC classification studies on AS-OCT images. Some clinical studies have verified its reliability on NC classification based on LOCS III [3,16,21]. [27] tried preliminary NC classification using deep learning methods on AS-OCT images. We combine clinical and methodological research to propose our own method.

2.2 Attention Mechanism

Attention mechanisms have empowered CNN models and achieved state-of-the-art results on various learning tasks [19]. In general, attention mechanisms can be mainly summarized into two groups, channel attention mechanism and spatial attention mechanism. SENet [10] firstly proposed the channel attention mechanism. It performs the GAP for channel squeeze, then reconstructs inter-dependencies of the channels through fully-connected (fc) layers, finally a *Sigmoid* layer is applied to generate channel weights for each channel. GENet [9] introduces a learnable layer for better exploiting the context feature, and FcaNet [19] increases the diversification of extracted features by extracting multi-band information. Bottleneck Attention Module (BAM) [17] and Convolutional Block Attention Module (CBAM) [22] combine the two attention mechanisms for getting the fused attention weights. To improve efficiency, ECANet [18] uses one-dimensional convolution layers to replace the original fully-connected layers in SENet.

3   Method

In this section, we first revisit the classical channel attention mechanism. Then we elaborate our GCA block in detail.

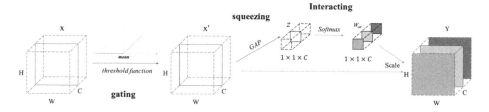

Fig. 3. A gated channel attention block.

3.1 Revisiting of Channel Attention

Channel attention is one of the most widely used attention module in CNNs. It uses a learnable block to adjust the importance of each channel and enhance the feature representation ability of the model. Given $X \in \mathbb{R}^{C \times W \times H}$ is the input feature tensor, where C denotes the number of channels, H and W denote the height and width of the feature map, respectively. The output $Y \in \mathbb{R}^{C \times W \times H}$ has the same shape of X with re-weighting of each channel. SENet [10] is the most classic channel attention mechanism consist of squeeze and excitation operation. The formula can be written as:

$$Y = \mathbf{F}_{scale}(W_{att}, X), \qquad (1)$$

$$W_{att} = \mathbf{F}_{ex}(\mathbf{F}_{sq}(X)), \qquad (2)$$

where $W_{att} \in \mathbb{R}^C$ is the channel attention weight, $\mathbf{F}_{scale}$ refers to channel-wise multiplication, $\mathbf{F}_{sq}$ represents the squeeze function GAP, and $\mathbf{F}_{ex}$ is the excite function to transform the squeeze info to attention weights. Generally, the squeeze step compresses channel information, and excitation step calculates the channel weights W_{att}. For the first step, it usually use parameter-free function like global average pooling (GAP) [10] or global max pooling (GMP) [22] to compute channel-statistics information. For the second step, it adopts fc layers for inter-channel dependency reconstruction.

In this paper, we found that the dependency among channel-statistics information is weak, and fc layers do not work well for AS-OCT image-based NC classification. This is because AND is an important indicator for NC diagnosis on AS-OCT images. Hence, we design a simple yet effective channel attention block named gated channel attention (GCA) block and will be introduced in the next section.

3.2 Gated Channel Attention Block

Figure 3 shows the diagram of the structure of a gated channel attention (GCA) block, which comprises three stages: gating, squeezing, and interacting.

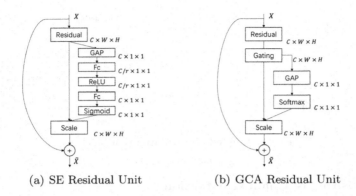

(a) SE Residual Unit   (b) GCA Residual Unit

Fig. 4. The schema of the SE residual unit (left) and the GCA residual unit (right).

Gating: To suppress the redundant features in a feature map, we devise a gated unit to mask the irrelevant features. According to the clinical studies in Sect. 2.1, the higher density region has higher relevance with cataract. To this end, we proposed a high-value gate for masking the low-value influence. It is an adaptive *threshold* function in which we use the global average value from each feature map as the threshold value. This is because [11] demonstrated that pooling value below average suppressed neuron activations in a CNN model. Formally, the gated tensor $X' \in \mathbb{R}^{C \times W \times H}$ is generated by masking the low-value of input tensor $X \in \mathbb{R}^{C \times W \times H}$, such that the *c-th* channel is formulated by:

$$(X'_c)_{ij} = \mathbf{F}_{gating}(X_c)_{ij} = Max(Mean(X_c), (X_c)_{ij}), \qquad (3)$$

where *Mean* function calculates the mean value of the feature map, *Max* function returns the largest item of input.

Squeezing: We use a squeezing operator to follow the gating operator, which is used to compute the channel-statistics feature information from each channel. This paper uses global average pooling (GAP) as squeezing operator, equivalent to the *AND* indicator for NC diagnosis. It can be written as follows:

$$z_c = \mathbf{F}_{GAP}(X'_c) = \frac{1}{W \times H} \sum_{i=1}^{W} \sum_{j=1}^{H} (X'_c)_{ij}, \qquad (4)$$

where z_c denotes the output of GAP in *c-th* channel.

In the experiments, we test the effects of different pooling operators.

Interacting: In the third stage, we propose a weakly interacting operator to construct weak dependencies of inter-channel and set the relative weights for channels. The fully-connection operator is the first proposed method for channel interacting in channel attention block. However, it brings higher model complexity, and [18] simplifies the interacting stage using local-connection. We further reduce the interacting complexity, and achieve channel interacting base on a

Softmax function. This paper uses the following formulation to get attention weights:

$$(W_{att})_c = Softmax(z)_c = \frac{e^{z_c}}{\sum_{i=1}^{C} e^{z_i}}, \tag{5}$$

where W_{att} is the channel attention weight same as formula 2.

As shown in the formula 5, the attention weight $(W_{att})_c$ of each channel can be obtained through the dependencies between a single channel (z_c) and all channels (z). Thus, *Softmax* function can be regarded as a weakly-connection among channels. On the contrary, *Sigmoid* obtain the channel weights independently with a lack of interaction. In the experiments, we will make a comparison between these two interaction methods.

The final output of the GCA block is obtained by rescaling X' with the channel weights W_{att}:

$$Y_c = F_{scale}(X'_c, (W_{att})_c) = (W_{att})_c X'_c, \tag{6}$$

where Y_c is the *c-th* channel of final output, $\mathbf{F}_{scale}(X'_c, (W_{att})_c)$ is a channel-wise multiplication between the weight W_{attc} and the feature map X'_c.

Discussion: To demonstrate the effectiveness of our GCA block, we use ResNet18 and ResNet34 as the backbone networks. We use them based on two reasons: 1) ResNet is a universal backbone, and ResNet18 and ResNet34 have low computational cost. 2) Most attention mechanism blocks have been verified to be effective on the ResNet backbone. The final GCA-Net is stacked by repeated GCA units shown in Fig. 4(b).

4   Experiments

4.1   Dataset and Evaluation Measures

We use a clinical AS-OCT images dataset, which is collected through the CASIA2 ophthalmology device (Tomey Corporation, Japan). The original AS-OCT image is shown as Fig. 1(a). However, only the nucleus area is associated with NC classification [21], and we extracted the nucleus part of the whole AS-OCT image manually as shown in Fig. 1(b)(c)(d).

The AS-OCT image dataset contains 17200 AS-OCT images from 543 participants with the average age of 61.3±18.7 (range: 14~95) years old, and there are 135 males and 335 females among the participants with gender information. The participants were asked to collect images of one eye or both eyes, and the total number of collected eyes is 860 (440 left eyes and 420 right eyes). Each eye has 20 AS-OCT images, and We discarded 999 images without complete nucleus region due to the occlusion of the eyelids during collection. Finally, we use 16201 AS-OCT images for NC classification.

We divide the dataset based on participants into three disjoint subsets: training dataset, validation dataset, and testing dataset. Table 2 summarizes the distribution of three NC stages on the three datasets.

Table 1. The AS-OCT image distribution of NC stages on different datasets.

Dataset	Normal	Mild	Severe
Training	896	3219	5504
Validation	317	793	2331
Testing	390	830	1921
Total	1603	4842	9756

We resize the nucleus images to 224*224 and perform the random rotation and random horizontal flipping for data augmentation. All models are implemented on the Pytorch platform and trained on a TITAN-V GPU with 12GB memory. We use the stochastic gradient descent (SGD) optimizer with the batch size of 64. The initial learning rate is set to 0.0015 and decreased by a factor of 10 every 10 epochs after 100 epochs.

We use three commonly-used evaluation metrics: Acc, $F1$ and $Kappa$ value to evaluate the performance of the model [7]. The calculation formulas are as follows:

$$Acc = \frac{TP + TN}{TP + FP + TN + FN}, \tag{7}$$

$$Recall = \frac{TP}{TP + FN}, \tag{8}$$

$$Precision = \frac{TP}{TP + FP}, \tag{9}$$

$$F1 = 2 \times \frac{Recall \times Precision}{Recall + Precision}, \tag{10}$$

where TP, FP, TN, and FN denote the numbers of true positives, false positives, true negatives, and false negatives, respectively.

$$Kappa = \frac{p_0 - p_e}{1 - p_0}, \tag{11}$$

where p_0 is the relative observed agreement among raters, and p_e is the hypothetical probability of chance agreement. Furthermore, we use $\#P$ to denote the number of parameters and $GFLOPs$ [10] to measure the computation.

4.2   Comparison with State-of-art Attention Attention Blocks

Table 2 compares the proposed GCA block with state-of-art attention blocks on ResNet18 and ResNet34. Our GCA-Net achieves the best NC classification results among all methods. It obtains the accuracies of **94.24%** and **94.31%**, respectively, and outperforms state-of-art attention blocks by more than **3%** accuracy. Furthermore, It also consistently improves performance over other methods on $F1$ and $Kappa$ value, demonstrating the effectiveness of the proposed GCA-Net. Moreover, compared with ResNets and comparative attention-based CNN models, the GCA-Net parameters are equal to ResNets and are

Table 2. Comparison with state-of-the-art attention blocks.

Method	Backbone	Acc	F1	Kappa	#P	GFLOPs
ResNet [8]		91.02	90.98	83.43	11.18M	1.82
SENet [10]	ResNet18	90.61	90.88	82.44	11.27M	1.82
CBAM [22]		89.18	89.03	79.21	11.27M	1.82
GCA-Net		**94.24**	**94.76**	**89.48**	11.18M	1.82
ResNet [8]		88.57	88.57	78.27	21.29M	3.67
SENet [10]	ResNet34	88.35	88.39	77.75	21.44M	3.67
CBAM [22]		91.21	90.93	83.54	21.44M	3.67
GCA-Net		**94.31**	**94.55**	**89.45**	21.29M	3.67

smaller than SENet and CBAM. Furthermore, our GCA-Net does not add additional GFlops through comparisons to other state-of-the-art attention methods. In general, Our GCA-Net works better between accuracy and complexity.

4.3 Ablation Study

Table 3. Effects of pooling operators in GCA based on ResNet18 (✓ denotes using gating operator before squeezing and ✗ denotes not).

Squeeze	Gating	Acc	F1	Kappa
Global max pooling	✓	93.51	93.80	85.95
Global std pooling	✓	93.25	93.59	87.84
Global average pooling	✗	93.64	93.67	88.31
Global average pooling	✓	**94.24**	**94.76**	**89.48**

Effects of Different Pooling Operators. Table 3 shows the classification results of three different pooling operators in the GCA block based on ResNet18. Compared with global max pooling and global std pooling, the GAP achieves the best results on three evaluation measures. This is because GAP can be taken as another representation of average nucleus density (AND) from the nucleus region. Furthermore, the results also demonstrate that the gating operator significantly improves the classification results for the GCA block.

Table 4. Classification results of channel interaction operators in GCA block based on ResNet18.

Squeezing operator	Acc	F1	Kappa
Fully-connection [10]	91.27	91.08	83.41
Local-connection [18]	93.31	93.34	87.77
Non-connection	92.04	92.12	85.44
Weakly-connection	**94.24**	**94.76**	**89.48**

Effect of Different Channel Interaction. Table 4 presents the classification results of four interaction operations: fully-connection, local-connection, non-connection($Sigmoid$) and weakly-connection ($Softmax$). Our weakly-connection interaction operation obtains the best classification results among four interaction operations. Two reasons can explain these: 1) $Softmax$ operation not only sets the relative weights for channels, but also suppresses the unimportant channels. 2) Inter-channel dependencies are weak, and it is difficult to build good dependencies among channels in training.

5   Conclusion

This paper proposes a simple yet effective gated channel attention network named GCA-Net to classify severity levels of nuclear cataract automatically on AS-OCT images. In the GCA-Net, we design a gated channel attention (GCA) block to mask redundant features and use the $Softmax$ layer to set relative weights for all channels, which is motivated by the clinical study of average nucleus density (AND). The results on a clinical AS-OCT image dataset demonstrate that our GCA-Net achieves the best classification performance and outperforms advanced attention-based CNN models. Moreover, the computation complexity of our GCA-Net is equal to previous methods, which indicates that it has the potential to deploy our method on the real machine.

In the future, we will collect more AS-OCT images to verify the overall performance of the GCA-Net and plug the GCA block in other CNN models to test its effectiveness.

Acknowledgment. This work was supported in part by Guangdong Provincial Department of Education (2020ZDZX3043, SJJG202002), Guangdong Provincial Key Laboratory (2020B121201001), Shenzhen Natural Science Fund (JCYJ202 00109140820699 and the Stable Support Plan Program 20200925174052004).

References

1. Caixinha, M., Amaro, J., Santos, M., Perdigão, F., Gomes, M., Santos, J.: In-Vivo automatic nuclear cataract detection and classification in an animal model by ultrasounds. IEEE Trans. Biomed. Eng. **63**(11), 2326–2335 (2016)

2. Cao, L., Li, H., Zhang, Y., Zhang, L., Xu, L.: Hierarchical method for cataract grading based on retinal images using improved Haar wavelet. Inf. Fusion **53**, 196–208 (2020)
3. Chen, D., Li, Z., Huang, J., Yu, L., Liu, S., Zhao, Y.E.: Lens nuclear opacity quantitation with long-range swept-source optical coherence tomography: correlation to LOCS III and a Scheimpflug imaging-based grading system. Br. J. Ophthalmol. **103**(8), 1048–1053 (2019)
4. Chylack, L.T., et al.: The lens opacities classification system iii. Arch. Ophthalmol. **111**(6), 831–836 (1993)
5. Gali, H.E., Sella, R., Afshari, N.A.: Cataract grading systems: a review of past and present. Curr. Opin. Ophthalmol. **30**(1), 13–18 (2019)
6. Gao, X., Lin, S., Wong, T.Y.: Automatic feature learning to grade nuclear cataracts based on deep learning. IEEE Trans. Biomed. Eng. **62**(11), 2693–2701 (2015)
7. Hao, H., et al.: Open-Appositional-Synechial anterior chamber angle classification in AS-OCT sequences. In: Martel, A.L., et al. (eds.) MICCAI 2020. LNCS, vol. 12265, pp. 715–724. Springer, Cham (2020). https://doi.org/10.1007/978-3-030-59722-1_69
8. He, K., Zhang, X., Ren, S., Sun, J.: Deep residual learning for image recognition. In: Proceedings of the IEEE Conference on Computer Vision and Pattern Recognition, pp. 770–778 (2016)
9. Hu, J., Shen, L., Albanie, S., Sun, G., Vedaldi, A.: Gather-excite: exploiting feature context in convolutional neural networks. arXiv preprint arXiv:1810.12348 (2018)
10. Hu, J., Shen, L., Sun, G.: Squeeze-and-excitation networks. In: Proceedings of the IEEE Conference on Computer Vision and Pattern Recognition, pp. 7132–7141 (2018)
11. Kobayashi, T.: Global feature guided local pooling. In: Proceedings of the IEEE/CVF International Conference on Computer Vision, pp. 3365–3374 (2019)
12. Li, H., Lim, J.H., Liu, J., Wong, T.Y.: Towards automatic grading of nuclear cataract. In: 2007 29th Annual International Conference of the IEEE Engineering in Medicine and Biology Society, pp. 4961–4964. IEEE (2007)
13. Long, E., et al.: An artificial intelligence platform for the multihospital collaborative management of congenital cataracts. Nat. Biomed. Eng. **1**(2), 1–8 (2017)
14. Makhotkina, N.Y., Berendschot, T.T., van den Biggelaar, F.J., Weik, A.R., Nuijts, R.M.: Comparability of subjective and objective measurements of nuclear density in cataract patients. Acta Ophthalmol. **96**(4), 356–363 (2018)
15. Ozgokce, M., et al.: A comparative evaluation of cataract classifications based on shear-wave elastography and B-mode ultrasound findings. J. Ultrasound **22**(4), 447–452 (2019)
16. Panthier, C., Burgos, J., Rouger, H., Saad, A., Gatinel, D.: New objective lens density quantification method using swept-source optical coherence tomography technology: Comparison with existing methods. J. Cataract Refract. Surg. **43**(12), 1575–1581 (2017)
17. Park, J., Woo, S., Lee, J.Y., Kweon, I.S.: Bam: bottleneck attention module. arXiv preprint arXiv:1807.06514 (2018)
18. Qilong, W., Banggu, W., Pengfei, Z., Peihua, L., Wangmeng, Z., Qinghua, H.: ECA-Net: Efficient channel attention for deep convolutional neural networks (2020)
19. Qin, Z., Zhang, P., Wu, F., Li, X.: Fcanet: frequency channel attention networks. arXiv preprint arXiv:2012.11879 (2020)
20. Wang, W., et al.: Objective quantification of lens nuclear opacities using swept-source anterior segment optical coherence tomography. Br. J. Ophthalmol. (2021)

21. Wong, A.L., et al.: Quantitative assessment of lens opacities with anterior segment optical coherence tomography. Br. J. Ophthalmol. **93**(1), 61–65 (2009)

22. Woo, S., Park, J., Lee, J.Y., Kweon, I.S.: CBAM: convolutional block attention module. In: Proceedings of the European Conference on Computer Vision (ECCV), pp. 3–19 (2018)

23. Xu, X., Zhang, L., Li, J., Guan, Y., Zhang, L.: A hybrid global-local representation CNN model for automatic cataract grading. IEEE J. Biomed. Health Inform. **24**(2), 556–567 (2019)

24. Xu, Y., Duan, L., Wong, D.W.K., Wong, T.Y., Liu, J.: Semantic reconstruction-based nuclear cataract grading from slit-lamp lens images. In: Ourselin, S., Joskowicz, L., Sabuncu, M.R., Unal, G., Wells, W. (eds.) MICCAI 2016. LNCS, vol. 9902, pp. 458–466. Springer, Cham (2016). https://doi.org/10.1007/978-3-319-46726-9_53

25. Xu, Y., et al.: Automatic grading of nuclear cataracts from slit-lamp lens images using group sparsity regression. In: Mori, K., Sakuma, I., Sato, Y., Barillot, C., Navab, N. (eds.) MICCAI 2013. LNCS, vol. 8150, pp. 468–475. Springer, Heidelberg (2013). https://doi.org/10.1007/978-3-642-40763-5_58

26. Zhang, X., Fang, J., Hu, Y., Xu, Y., Higashita, R., Liu, J.: Machine learning for cataract classification and grading on ophthalmic imaging modalities: a survey. arXiv preprint arXiv:2012.04830 (2020)

27. Zhang, X., et al.: A novel deep learning method for nuclear cataract classification based on anterior segment optical coherence tomography images. In: 2020 IEEE International Conference on Systems, Man, and Cybernetics (SMC), pp. 662–668. IEEE (2020)

Overcoming Data Scarcity for Coronary Vessel Segmentation Through Self-supervised Pre-training

Marek Kraft[1]([envelope]), Dominik Pieczyński[1], and Krzysztof 'Kris' Siemionow[2]

[1] Institute of Robotics and Machine Intelligence, Poznań University of Technology, Piotrowo 3A, 60-965 Poznań, Poland
marek.kraft@put.poznan.pl
[2] Kardiolytics Inc., 1415 W37th Street, Chicago, IL 60609, USA

Abstract. Cardiovascular diseases affect a significant part of the population, leading to deterioration in life quality, health degradation, and even premature death. One of the most effective diagnostic methods for the disease is based on medical imaging, specifically Computed Tomography Angiography, from which the complete 3D image of the coronary vessels can be reconstructed. Manual annotation and reconstruction is a tedious process, so a range of automated methods have been proposed over the years, with the methods based on deep neural networks achieving the best results recently. On the downside, such methods require extensive datasets for training. To overcome the problems with data scarcity, we propose a method for self-supervised pre-training of neural networks performing the task of coronary vessel segmentation. The method is based on a vesselness filter and significantly boosts performance, reducing the training time and boosting the accuracy without additional annotated data.

Keywords: Coronary vessels · Segmentation · Deep learning · Self-supervised learning

1 Introduction

Health problems associated with cardiovascular diseases (CVDs) are affecting a significant part of the population. According to the recent World Health Organisation statistics, CVDs are one of the most common non-communicable diseases [27]. Moreover, they are the leading cause of premature deaths on a global scale. Even if not lethal, it can potentially lead to severe degradation of overall health and the quality of life [24]. Furthermore, the course of the disease (especially in its early stages) is often latent, with the subjects appearing seemingly healthy [1]. The leading cause of CVDs is myocardial ischemia, hypoxia or necrosis that results from gradual narrowing even complete obstruction of blood vessel lumen due to the gradual buildup of plaque [20].

© Springer Nature Switzerland AG 2021
T. Mantoro et al. (Eds.): ICONIP 2021, LNCS 13110, pp. 369–378, 2021.
https://doi.org/10.1007/978-3-030-92238-2_31

Multiple diagnostics methods can be used to diagnose the problem, including baseline and exercise electrocardiography. Still, methods based on medical imaging are applied to get a complete insight into the patient's condition and the state of the coronary vessels. One of the most commonly used diagnostic methods for relatively noninvasive CVD diagnostics is medical imaging using Computed Tomography Angiography (CTA). Introducing the contrast agent into the blood vessels makes them more visible, enabling the capture of their structure, along with the possible stenoses and other malformations, using standard computed tomography techniques. Although this method offers good imaging accuracy and is a highly valuable diagnostics tool for stenosis qualification, manual mapping and tracing of coronary vessels is a lengthy process and requires domain expertise. Moreover, manual annotations may present significant inter-observer variation [17].

In this paper, we present a method for self-supervised pre-training of a neural network performing the 3D segmentation of coronary vessels from CTA images. The method is based on the use of the Jerman vesselness filter [13,14]. We use a relatively large validation dataset to verify how the number of samples (complete 3D CTA scans) in the dataset for self-supervision and the segmentation training dataset and their proportion influences the training process and the achieved metrics. We demonstrate a notable improvement in accuracy by using self-supervised pre-training, especially in the scenarios with severe deficiency of data used for segmentation training. We hope the presented solution will facilitate more accurate reconstruction of coronary vessels using deep learning methods – an application in which the annotated data is typically scarce, but the raw CTA scan data might be readily available.

2   Related Work

Relevance of the CVD-related problems for public health and general well-being has spurred significant research interest in the field. The classical approaches involved centerline extraction [22], region growing [21] or active contours [6]. However, most of the more recent approaches take advantage of the deep learning approach. A similar approach followed by the use of level sets for result refinement is presented in [5].

A direct approach based on a convolutional neural network using an encoder-decoder architecture is presented in [9]. The authors also explore a range of approaches to input data representation and acknowledge that annotated data scarcity is a problem in this application.

In [3], the authors take advantage of the vesselness filter and use the output of the Frangi filter [7] as output, an additional input channel for a 3D U-Net [25] aside from the CTA scan. The introduction of an additional channel improves the segmentation results. The approach presented in [26] uses a graph neural network to extract the vessel lumen represented as a set of points on the contour. The points are collected along the centerline and used to construct the surface mesh of the vessels. The method assumes that the centerline is already extracted. An approach using a combination of convolutional and recurrent neural networks is presented in [18]. The authors claim that such a combination is well suited for coronary vessel segmentation due to their tubular, elongated, continuous shape and tree-like structure. The results are reported on a large dataset (916 scans overall), but it was not released to the public. Overall, the use of machine learning methods in the field of cardiac imaging, including CTA processing, experiences notable growth [28]. It is worth noting that all the presented segmentation methods rely on hand-annotated data for training.

As mentioned, the application of deep learning techniques in the medical image processing field is oftentimes hindered by the lack of expert-generated labels. Fortunately, this can be alleviated to some degree by suing a technique called self-supervised training [16]. The training is composed of two tasks:

- Upstream (pretext) task enables learning of visual features are learned by learning objective functions of pretext tasks. It does not require hand-generated labels, as it relies on characteristics solely derived from the data attributes alone.
- Downstream task, which is essentially supervised training with hand-generated labels, using the upstream task training result as the starting point.

Successful applications of self-supervised training in the context of vessel segmentation has been reported in the case of 2D retinal images. A solution utilising complementary imaging modality as the label image is presented in [8]. The images require prior alignment to ensure image-label coincidence. Another solution is presented in [11] – it is based on gradient-based flow feature extraction.

3   Methods

3.1   The Jerman Vesselness Filter

The method used to generate self-supervision training data is the Jerman multi-scale vesselness filter [13,14]. The method uses Hessian filtering to enhance tubular and spherical structures in 2D and 3D images, which makes it a very useful tool for medical applications related to blood vessels. The method compares favourably to other popular methods such as the Frangi filter [7] in terms of accuracy across multiple benchmarks. It is also easier to tune since it features less tuneable parameters requiring user input. The formulation of the Jerman filter for the 3D case is given in Eq. 1

$$
V_p = \begin{cases} 0, & \text{if } \lambda_2 \leq 0 \vee \lambda_p \leq 0, \\ 1, & \text{if } \lambda_2 \geq \lambda_p/2 > 0, \\ \lambda_2^2(\lambda_p - \lambda_2)[\frac{3}{\lambda_2+\lambda_p}]^3 & \text{otherwise.} \end{cases} \tag{1}
$$

The value of λ_p is computed as shown in Eq. 2.

$$
\lambda_p(s) = \begin{cases} \lambda_3, & \text{if } \lambda_3 > \tau \, max_{\mathbf{x}}\lambda_3(\mathbf{x}, s), \\ \tau max_{\mathbf{x}}\lambda_3(\mathbf{x}, s), & \text{if } 0 < \lambda_3 \leq \tau \, max_{\mathbf{x}}\lambda_3(\mathbf{x}, s), \\ 0 & \text{otherwise.} \end{cases} \tag{2}
$$

In the above equations, the λ_i stand for the i-th Hessian eigenvalue. The Hessian operator is assumed to be applied to the 3D volume with second order partial image derivative filter convolutions. The $\mathbf{x}$ stands for a set of 3D coordinates in the volume, τ stands for a threshold value, and s denotes the scale.

To make an informed selection of Jerman filter settings, a Bayesian optimisation over hyperparameters is used. The process is applied to 10 randomly selected scans with manually segmented coronary arteries and maximises the area under the receiver operating characteristic curve measured between filter output and ground truth. The explored parameters are scale range, number of equally spaced scale steps in that range and threshold. Resulting from the described Bayesian parameter search, the selected values of the parameters are $s_{range} = [0.05, 5.70]$, $s_{steps} = 15$ and $\tau = 0.3$.

3.2   Volume Masking

The whole volume containing the heart and captured during the CTA procedure contains multiple elongated, vessel-like objects aside from the coronary vessels (ribs, pulmonary arteries). To mask out the irrelevant portions of the volume, a binary segmentation neural network that segments the pericardium part was used. The neural network is an ACS U-Net with a ResNet50 encoder as presented in [29], which enabled taking advantage of Imagenet pre-training in a 3D segmentation setting. Sample results of the masking process are given in Fig. 1. A detailed description of the neural network is beyond the scope of this paper – it is used merely as a tool to facilitate more accurate coronary vessel segmentation results and speed up the training process.

Fig. 1. The raw volume with additional distractors (left) and the binary pericardium segmentation masks (right, top row), along with the masking results (right, bottom row).

3.3   Segmentation Neural Network

A version of the standard U-Net network modified for segmenting 3D data [4] is used to ascertain the positive impact of self-supervised pretraining. The architecture is broadly used for medical images segmentation [19] and has therefore proven to be applicable to a variety of tasks. While the state of the art approaches to specific segmentation problems utilise different methods, those are oftentimes developed by extending the U-Net base [2,15,30]. Because of the limited amount of labelled data, neural network models with fewer parameters can be considered. However, such approaches may not result in the improvement of general network performance, as the problem itself is very difficult and impossible to solve with overly simplified architectures.

An existing implementation of 3D U-Net [25] is used to train the coronary arteries segmentation. The network uses the ReLU activation function and Batch Normalisation layers [10] after convolutions. As the segmentation problem is binary, sigmoid is used as the final activation function.

3.4   Self-supervised Training and Validation

The segmentation network is trained in two stages. The task of the first stage is to learn to reconstruct the original Jerman filtering output from raw data. The loss function utilised for this task is presented in Eq. 3.

$$LOSS = MAE(y, \hat{y}) + 0.1 \cdot MAE(y_{grad}, \hat{y_{grad}}) + 0.1 \cdot (1 - SSIM(y, \hat{y})) \quad (3)$$

In the equation above, MAE is mean absolute error, SSIM is structural similarity index measure, y is the ground truth, $\hat{y}$ is the prediction, y_{grad} is a gradient image of the ground truth and $\hat{y_{grad}}$ is a gradient image of the prediction.

The purpose of including structural similarity index measure and mean absolute error between gradient images is to mitigate an issue of the blurred results when using loss functions such as mean absolute error or mean squared error directly. The aforementioned components should guide the neural network to learn producing structure-preserving reconstructions of the Jerman filtering results.

The task during the second stage of the training process is to learn to segment the coronary arteries. The loss function used is Dice loss calculated between ground truth and predicted volumes.

Adam optimiser with the learning rate value set to $3e - 4$ is used for both stages. The learning rate is automatically reduced by half if no improvement in training loss value is for at least 3 epochs. The training is stopped if the validation loss does not improve for 20 epochs.

The CT data is prepared by limiting the range of Hounsfield values to a $[0, 1500]$ window and scaling it using min-max normalisation.

To facilitate the training process, both the CT data and the labels are split into $64 \times 64 \times 64$ cubes, as three-dimensional neural networks use a significant amount of GPU memory. The cubes are subject to training-time data augmentation containing common transformations, such as random translation, random rotation, random shearing, coarse dropout and addition of Gaussian noise.

The validation dataset size is 50 samples (cases) and remains constant for all tests to provide a common ground for the evaluation.

4 Results and Discussion

Overall, all combinations of pre-training dataset sizes (0, 50, 100, 200) and training dataset sizes (10, 20, 50) were tested. The values of final metrics achieved for each combination are given in Table 1.

Table 1. Metric values for the validation dataset for all combinations of pre-training and training dataset sizes. PTD stands fot the number of pre-training datasets, HLD for the number of hand-labeled datasets used in final downstream training. DSC stands for dice score.

F1			
[PTD]	[HLD]		
	10	20	50
0	0.733	0.777	0.807
50	0.738	0.785	0.806
100	0.764	0.791	0.820
200	0.752	0.798	0.816

Precision			
[PTD]	[HLD]		
	10	20	50
0	0.810	0.830	0.853
50	0.835	0.840	0.861
100	0.851	0.851	0.844
200	0.844	0.857	0.873

Recall			
[PTD]	[HLD]		
	10	20	50
0	0.722	0.770	0.797
50	0.714	0.776	0.788
100	0.728	0.773	0.825
200	0.741	0.777	0.793

DSC			
[PTD]	[HLD]		
	10	20	50
0	0.730	0.774	0.803
50	0.733	0.781	0.803
100	0.759	0.786	0.816
200	0.761	0.794	0.814

The results clearly show that using self-supervised pre-training helps to achieve better metric results. The gain is especially visible for the cases where only 10 labeled datasets were used for final training. However, using a combination of the largest pre-training and training set results in degradation of the recall metric. The precision metric is still boosted, but the resulting F1 metric and dice score are a bit lower as a result. This is an interesting phenomenon requiring further investigation.

To gain better insight into how the upstream pre-training affects the final downstream training process, we also registered the course of the training. Validation dataset learning curves for the tested combinations of pre-training dataset and training dataset sizes are given in Fig. 2.

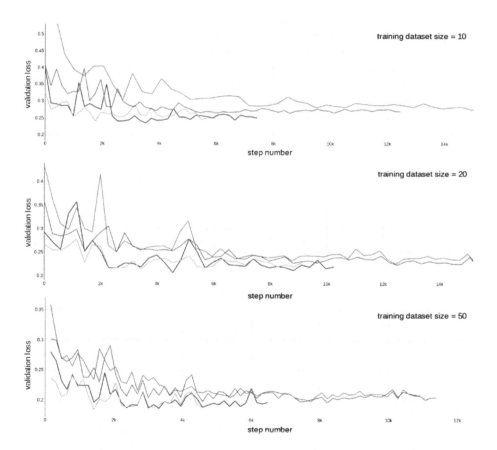

Fig. 2. Validation dataset learning curves for the tested combinations of pre-training dataset and training dataset sizes. Red, green, orange and blue colour curves represent the pre-training dataset sizes of 0, 50, 100 and 200, respectively. The difference in the number of steps is a result of using the early stopping training policy. (Color figure online)

It can be seen that all the pre-trained versions start with encoded knowledge of the problem by forming internal representations since the loss value is closer to zero for all of the cases, as opposed to the versions that don't take advantage of pre-training. The difference becomes less evident as the number of samples in the downstream task training dataset becomes larger. Using larger number of samples for the upstream task results in faster completion of training and facilitates achieving lower validation loss. However, please note, that there is an additional upfront cost associated with the completion of the upstream pre-training.

Sample segmentation reults are given in Fig. 3. The raw output from the neural network was binarised and filtered with a $3 \times 3 \times 3$ morphological closing operator. Then, all the consistent volumes aside from the largest two, corresponding to the left and right coronary artery, were filtered out. As shown, the pre-training has a positive impact on the final results.

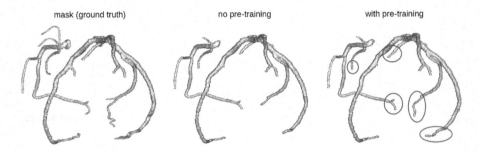

Fig. 3. Sample segmentation result (200 samples used for upstream pre-training, 10 for downstream training). The results after pre-training managed to retrieve a larger portion of the coronary vessels (see the highlighted regions).

5   Conclusions and Future Work

A method for overcoming data scarcity for coronary vessel segmentation using deep neural networks was presented in this paper. The method utilises self-supervised learning by neural network pre-training with the data generated with a vesselness filter as an upstream task. Pre-training data does not require any expert annotation. The operation creates internal feature representations useful for the downstream task of supervised coronary vessel segmentation training using expert-generated ground truth masks. We successfully demonstrated that the procedure makes the training faster, and the segmentation quality metrics are improved. The evaluation was performed for a variety of combinations of dataset sizes for self-supervised training and supervised training. The improvement is visible, especially in the cases of small hand-labelled dataset sizes, emphasising the presented method's usefulness whenever the labelled training data is scarce.

With such promising results, we think the approach has potential for future extensions. The plans include performing the tests of other downstream tasks such as inpainting [23], imaging artefact detection [12], or combining such tasks. Using the techniques for other tasks involving vascular imaging, such as segmentation of cerebral or liver blood vessels, is also planned.

References

1. Benjamin, E.J., et al.: Heart disease and stroke statistics-2018 update: a report from the American heart association. Circulation **137**(12), e67–e492 (2018)
2. Cao, H., et al.: Swin-Unet: Unet-like pure transformer for medical image segmentation. arXiv preprint arXiv:2105.05537 (2021)
3. Chen, Y.C., et al.: Coronary artery segmentation in cardiac CT angiography using 3D multi-channel U-net. arXiv preprint arXiv:1907.12246 (2019)
4. Çiçek, Ö., Abdulkadir, A., Lienkamp, S.S., Brox, T., Ronneberger, O.: 3D U-net: learning dense volumetric segmentation from sparse annotation. In: Ourselin, S., Joskowicz, L., Sabuncu, M.R., Unal, G., Wells, W. (eds.) MICCAI 2016. LNCS, vol. 9901, pp. 424–432. Springer, Cham (2016). https://doi.org/10.1007/978-3-319-46723-8_49
5. Duan, X., Li, Y., Wang, J.: Coronary artery CTA image segmentation and three-dimensional visualization based on U-Net. In: Proceedings of the 2020 International Symposium on Artificial Intelligence in Medical Sciences, pp. 64–68 (2020)
6. Florin, C., Paragios, N., Williams, J.: Globally optimal active contours, sequential Monte Carlo and on-line learning for vessel segmentation. In: Leonardis, A., Bischof, H., Pinz, A. (eds.) ECCV 2006. LNCS, vol. 3953, pp. 476–489. Springer, Heidelberg (2006). https://doi.org/10.1007/11744078_37
7. Frangi, A.F., Niessen, W.J., Vincken, K.L., Viergever, M.A.: Multiscale vessel enhancement filtering. In: Wells, W.M., Colchester, A., Delp, S. (eds.) MICCAI 1998. LNCS, vol. 1496, pp. 130–137. Springer, Heidelberg (1998). https://doi.org/10.1007/BFb0056195
8. Hervella, Á.S., Rouco, J., Novo, J., Ortega, M.: Self-supervised deep learning for retinal vessel segmentation using automatically generated labels from multimodal data. In: 2019 International Joint Conference on Neural Networks (IJCNN), pp. 1–8. IEEE (2019)
9. Huang, W., et al.: Coronary artery segmentation by deep learning neural networks on computed tomographic coronary angiographic images. In: 2018 40th Annual International Conference of the IEEE Engineering in Medicine and Biology Society (EMBC), pp. 608–611. IEEE (2018)
10. Ioffe, S., Szegedy, C.: Batch normalization: accelerating deep network training by reducing internal covariate shift. In: Bach, F., Blei, D. (eds.) Proceedings of the 32nd International Conference on Machine Learning. Proceedings of Machine Learning Research, vol. 37, pp. 448–456. PMLR, Lille, 07–09 July 2015
11. Jena, R., Singla, S., Batmanghelich, K.: Self-supervised vessel enhancement using flow-based consistencies. arXiv preprint arXiv:2101.05145 (2021)
12. Jenni, S., Favaro, P.: Self-supervised feature learning by learning to spot artifacts. In: Proceedings of the IEEE Conference on Computer Vision and Pattern Recognition, pp. 2733–2742. IEEE (2018)
13. Jerman, T., Pernuš, F., Likar, B., Špiclin, Ž.: Beyond Frangi: an improved multiscale vesselness filter. In: Medical Imaging 2015: Image Processing, vol. 9413, p. 94132A. International Society for Optics and Photonics (2015)

14. Jerman, T., Pernuš, F., Likar, B., Špiclin, Ž: Enhancement of vascular structures in 3D and 2D angiographic images. IEEE Trans. Med. Imaging **35**(9), 2107–2118 (2016)
15. Jha, D., Riegler, M.A., Johansen, D., Halvorsen, P., Johansen, H.D.: DoubleU-net: a deep convolutional neural network for medical image segmentation. In: 2020 IEEE 33rd International Symposium on Computer-Based Medical Systems (CBMS), pp. 558–564. IEEE (2020)
16. Jing, L., Tian, Y.: Self-supervised visual feature learning with deep neural networks: a survey. IEEE Trans. Pattern Anal. Mach. Intell. **43**(11), 4037–4058 (2021)
17. Joskowicz, L., Cohen, D., Caplan, N., Sosna, J.: Inter-observer variability of manual contour delineation of structures in CT. Eur. Radiol. **29**(3), 1391–1399 (2018). https://doi.org/10.1007/s00330-018-5695-5
18. Kong, B., et al.: Learning tree-structured representation for 3D coronary artery segmentation. Comput. Med. Imaging Graph. **80**, 101688 (2020)
19. Litjens, G., et al.: A survey on deep learning in medical image analysis. Med. Image Anal. **42**, 60–88 (2017)
20. Mendis, S., Puska, P., Norrving, B., World Health Organization, et al.: Global atlas on cardiovascular disease prevention and control. World Health Organization (2011)
21. Öksüz, İ., Ünay, D., Kadıpaşaoğlu, K.: A hybrid method for coronary artery stenoses detection and quantification in CTA images. In: Proceedings of MICCAI Workshop on 3D Cardiovascular Imaging: A MICCAI Segmentation Challenge (2012)
22. Schaap, M., et al.: Standardized evaluation methodology and reference database for evaluating coronary artery centerline extraction algorithms. Med. Image Anal. **13**(5), 701–714 (2009)
23. Singh, S., et al.: Self-supervised feature learning for semantic segmentation of overhead imagery. In: Proceedings of 2018 British Machine Vision Conference, vol. 1, p. 4 (2018)
24. Waltz, M., Badura, B.: Subjective health, intimacy, and perceived self-efficacy after heart attack: predicting life quality five years afterwards. Soc. Indic. Res. **20**(3), 303–332 (1988)
25. Wolny, A., et al.: Accurate and versatile 3D segmentation of plant tissues at cellular resolution. eLife **9**, e57613 (2020)
26. Wolterink, J.M., Leiner, T., Išgum, I.: Graph convolutional networks for coronary artery segmentation in cardiac CT angiography. In: Zhang, D., Zhou, L., Jie, B., Liu, M. (eds.) GLMI 2019. LNCS, vol. 11849, pp. 62–69. Springer, Cham (2019). https://doi.org/10.1007/978-3-030-35817-4_8
27. World Health Organization, et al.: World health statistics 2019: monitoring health for the SDGs, sustainable development goals (2019)
28. Yang, G., Zhang, H., Firmin, D., Li, S.: Recent advances in artificial intelligence for cardiac imaging. Comput. Med. Imaging Graph. **90**, 101928 (2021)
29. Yang, J., et al.: Reinventing 2D convolutions for 3D images. IEEE J. Biomed. Health Inform. (2021)
30. Zhou, Z., Rahman Siddiquee, M.M., Tajbakhsh, N., Liang, J.: UNet++: a nested U-net architecture for medical image segmentation. In: Stoyanov, D., et al. (eds.) DLMIA/ML-CDS -2018. LNCS, vol. 11045, pp. 3–11. Springer, Cham (2018). https://doi.org/10.1007/978-3-030-00889-5_1

Self-Attention Long-Term Dependency Modelling in Electroencephalography Sleep Stage Prediction

Georg Brandmayr[1,2]([✉]) [ID], Manfred Hartmann[2] [ID], Franz Fürbass[2] [ID], and Georg Dorffner[1] [ID]

[1] Section for Artificial Intelligence, Medical University of Vienna, Vienna, Austria
[2] Center for Health and Bioresources, AIT Austrian Institute of Technology GmbH, Vienna, Austria
georg.brandmayr@ait.ac.at

Abstract. Complex sleep stage transition rules pose a challenge for the learning of inter-epoch context with Deep Neural Networks (DNNs) in ElectroEncephaloGraphy (EEG) based sleep scoring. While DNNs were able to overcome the limits of expert systems, the dominant bidirectional Long Short-Term Memory (LSTM) still has some limitations of Recurrent Neural Networks. We propose a sleep Self-Attention Model (SAM) that replaces LSTMs for inter-epoch context modelling in a sleep scoring DNN. With the ability to access distant EEG as easily as adjacent EEG, we aim to improve long-term dependency learning for critical sleep stages such as Rapid Eye Movement (REM). Restricting attention to a local scope reduces computational complexity to a linear one with respect to recording duration. We evaluate SAM on two public sleep EEG datasets: MASS-SS3 and SEDF-78 and compare it to literature and an LSTM baseline model via a paired t-test. On MASS-SS3 SAM achieves $\kappa = 0.80$, which is equivalent to the best reported result, with no significant difference to baseline. On SEDF-78 SAM achieves $\kappa = 0.78$, surpassing previous best results, statistically significant, with +4% F1-score improvement in REM. Strikingly, SAM achieves these results with a model size that is at least 50 times smaller than the baseline.

Keywords: Attention · Sleep scoring · Inter-epoch context.

1 Introduction

The visual scoring of sleep stages based on polysomnography (PSG) is essential for the diagnosis of many sleep disorders, but the instrumentation burden and time effort limit it's application. While expert systems could solve sleep scoring of PSG already in 2005 [1] with human level agreement, limitations were in high development effort and low flexibility for reuse. End-to-end learning via Convolutional Neural Networks (CNNs) enabled sleep scoring based on electroencephalography (EEG) only (the brain signal subset of PSG) and promises

© Springer Nature Switzerland AG 2021
T. Mantoro et al. (Eds.): ICONIP 2021, LNCS 13110, pp. 379–390, 2021.
https://doi.org/10.1007/978-3-030-92238-2_32

to reduce both burden and time effort [22]. Figure 1 shows an example of nightly EEG with a hypnogram according to the rules of the American Association of Sleep Medicine (AASM). It depicts a time series of sleep stages, i.e., brain-states, based on 30 s segments termed *epochs* (not to be confused with training epochs). Soon after the first CNN solutions it was realized that sleep scoring is not only a pattern recognition problem, but also a sequence transduction problem. Complex sleep stage transition rules [4] pose a challenge for the learning of sleep EEG inter-epoch context with DNNs. The inset in Fig. 1 shows an example of a prolonged Rapid Eye Movement (REM) stage. Sequence transduction tasks are typically found in Natural Language Processing (NLP) and can be solved with encoder-decoder architectures based on Recurrent Neural Networks (RNNs). However, it was not until the solution of the vanishing gradient problem with LSTM, that many sequence tasks such as translation were improved dramatically [9]. Automatic sleep scoring was no exception, and today the modeling of inter-epoch context is dominated by bidirectional LSTMs [12,20,21]. However, their sequential nature makes them harder to train than feed-forward networks and the fixed size state may still limit representation of distant sequence elements. NLP research demonstrated that direct access to encoded sequence elements via the attention mechanism improves performance [3]. The Transformer model drops RNNs completely and uses a pure feed-forward, self-attention based, encoder-decoder mechanism for sequence transduction [23].

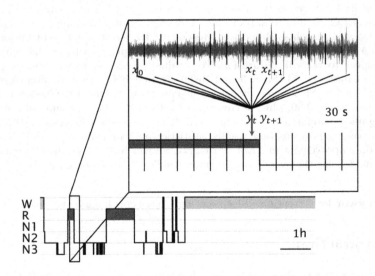

Fig. 1. Hypnogram sleep scoring from EEG with sleep stages W awake, R rapid eye movement (REM) sleep (red), N1 transitional sleep, N2 normal sleep and N3 deep sleep. The inset shows an inter-epoch context of 14 epochs (30 s long) for prolonged R scoring.

2 Motivation

In this paper we seek to replace LSTMs with a self-attention-based sequence encoder for epoch features in a sleep-scoring DNN, demonstrating an application to time series classification. We argue that directly accessing epoch features reduces the long-term dependency challenge, since no encoding into states is necessary. This could aid especially stage R (REM sleep), which can depend on distant epochs in the past and future. Since attention relates every output element with every input element, the computational complexity is quadratic in sequence length. This becomes an issue when scoring an entire night of EEG, opposed to NLP with short sequences [23]. As solution we propose to restrict the attention context L to a fixed size, moving scope. The inset in Fig. 1 shows an example of a restricted context of $L = 14$ epochs, where the prediction of sleep stage y_l depends not only on $\mathbf{X}_l$ but also on $\mathbf{X}_0$. With this approach, the computational complexity is reduced to linear, however at the price of a limited attention scope. Due to the biological limitation of sleep cycle length we conjecture that a full night scope is not required. Thus, it remains a tradeoff to choose a scope size L. We hypothesize, that the direct access to distant epochs adds more representative capacity, than what may be lost by limited scope.

3 Method

Competitive sleep scoring models typically use, with only few exceptions [18], a sequence to sequence approach [19–21] based on LSTMs. The proposed model SAM also uses this approach, but replaces bidirectional LSTMs with a sequence encoder stack, based on the encoder part of the Transformer. Figure 2 left shows the main building blocks. The model estimates from a length L sequence of EEG epochs $\boldsymbol{X} = [\mathbf{X}^{(1)}, \mathbf{X}^{(2)}, \ldots, \mathbf{X}^{(L)}]$ a sequence of sleep stages $y \in V^L$ with the set of sleep stages V. While some other work uses transformed versions of the EEG [6,19], here we process raw EEG. The input $\boldsymbol{X} \in \mathbb{R}^{f_s \times T \times C}$ is a high-resolution signal with sampling frequency f_s, C channels and T seconds duration. Overall, the output probability sequence $\mathbf{P} \in \mathbb{R}^{L \times K}$ with $K = |V|$ is estimated from the input $\boldsymbol{X}$ via the non-linear, parameterized function g:

$$\mathbf{P} = g(\boldsymbol{X}; \theta) = f_s([f_e(\mathbf{X}_0; \theta_e), \ldots, f_e(\mathbf{X}_L; \theta_e)]; \theta_s) \tag{1}$$

with parameters $\theta = (\theta_e, \theta_s)$ for embedder f_e and sequence encoder f_s.

3.1 Embedder

The epoch embedder operates on each sequence element individually, and thus treats the sequence like a batch, indicated by the dashed line in Fig. 2. It is a CNN designed to reduce a raw EEG epoch $\mathbf{X}_l$ to a vector $\mathbf{e}_l \in \mathbb{R}^{N_F}$ of high level, representative features. Every CNN layer has the same kernel size and is followed by batch normalization [10] and ReLU activation, with padding and stride set

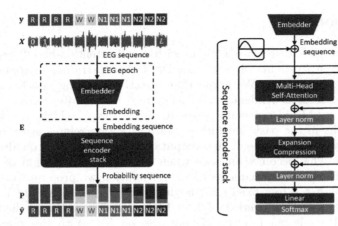

Fig. 2. Left: Model overview. The model predicts from the length-L EEG sequence X the sleep probability sequence $\mathbf{P}$ (W white, R red, N1–3 grayscale). **Right: Encoder.** Five layers of MHSA and feed-forward net model relational representations.

for identical in- and output resolution. The actual reduction of resolution gets done via identical max pooling layers with kernel size two. We found identical resolution reduction performing better than aggressive resolution reduction [18]. Final max pooling concludes the embedder and reduces the finally N_{F} feature maps to dimension $1 \times N_{\mathrm{F}}$. Skip connections [8] after every second max pooling layer facilitate deep training and diversify the receptive field [14].

3.2   Sequence Encoder Stack

The embeddings, stacked to the sequence $\mathbf{E} \in \mathbb{R}^{L \times N_{\mathrm{F}}}$, are the input for inter-epoch context modelling in the sequence encoder stack. It uses, unlike most state-of-the-art solutions based on RNNs, only self-attention to model the inter-epoch dependencies in the encoded sequence $\mathbf{Z} \in \mathbb{R}^{L \times N_{\mathrm{F}}}$.

Attention. It solves the problem of access to past information without regard of the distance and avoids the bottleneck of squeezing the past into a single state vector. It computes relational representations of sequences via a form of content-addressed retrieval. Attention is a function $f(\mathbf{q}_l, \mathbf{k}, \mathbf{v})$ that maps a query vector $\mathbf{q}_l$ via key-value pairs to a retrieved context vector $\mathbf{c}_l$. With a query of dimension Q, $L \times Q$ key matrix $\mathbf{K}$ and $L \times V$ value matrix $\mathbf{V}$ the l-th context $\mathbf{c}_l$ of dimension V is:

$$\mathbf{c}_l = \boldsymbol{\alpha}_l \mathbf{V} \qquad (2)$$

with $\boldsymbol{\alpha}_l = \mathrm{softmax}(f_{\mathrm{score}}(\mathbf{q}_l, \mathbf{K})$ and the scoring function f_{score}. Among different attention functions [16], we consider dot-product attention $f_{\mathrm{score}}(\mathbf{q}_l, \mathbf{K}) = \mathbf{q}_l^{\mathsf{T}} \mathbf{K}^{\mathsf{T}}$ for its favorable grouping to matrix operations. With input scaling and application to a sequence of L queries at once:

$$f_{\text{SDPA}}(\mathbf{Q}, \mathbf{K}, \mathbf{V}) = \text{softmax}\left(\frac{\mathbf{Q}\mathbf{K}^{\mathsf{T}}}{\sqrt{d_k}}\right) \tag{3}$$

scaled dot-product attention forms the basic attention operation. Attention is, due to its definition (3), agnostic to element permutation. To supply element ordering information a solution is to add positional embeddings to the input E. We use fixed sinusoids with position dependent frequency $u_{lk} = \sin l$ according to [23].

Multi-headed Self-attention. The relational representation of different positions of a single sequence is obtained via $\mathbf{Q} = \mathbf{K} = \mathbf{V} = \mathbf{E}$ and called self-attention or intra-attention. It originates in NLP and has been successfully applied in tasks such as reading comprehension [5]. Multi-headed self-attention splits the attention operation along the model dimension into N_{H} attention heads applied in parallel to linear projections of the input $\mathbf{E}$:

$$f_{\text{MHSA}}(\mathbf{E}; \mathbf{W}) = [\mathbf{C}_1, \mathbf{C}_2, \ldots, \mathbf{C}_{N_{\text{H}}}]\mathbf{W}_{\text{O}} \tag{4}$$

with head context $\mathbf{C}_i = f_{\text{SDPA}}(\mathbf{E}\mathbf{W}_{\mathbf{Q}i}, \mathbf{E}\mathbf{W}_{\mathbf{K}i}, \mathbf{E}\mathbf{W}_{\mathbf{V}i})$, projection weights $\mathbf{W}_{\mathbf{Q}i}$, $\mathbf{W}_{\mathbf{K}i} \in \mathbb{R}^{M \times Q}, \mathbf{W}_{\mathbf{V}i} \in \mathbb{R}^{M \times V}$ and $\mathbf{W}_{\text{O}} \in \mathbb{R}^{N_{\text{H}}V \times M}$ reshaping inputs from and the output to the same model dimension M.

Architecture. According Fig. 2 right the sequence encoder stack is comprised of N_{s}-layers (i.e., parameters are duplicated) of f_{MHSA} followed by a 2-layer fully connected feed-forward net. This net $f_{\text{EC}}(\mathbf{z}_l) = \max(0, \mathbf{z}\mathbf{W}_1 + \mathbf{b}_1)\mathbf{W}_2 + \mathbf{b}_1$ performs expansion and compression for each encoded sequence element. After the last encoder layer class probabilities are obtained with the same linear-softmax projection for each sequence element. Residual connections [8] facilitate model convergence during parameter optimization and layer normalization [2] avoids overfitting.

4 Experiments

We conduct experiments with the public Sleep-EDF Database Expanded sleep cassette study (SEDF-78) from Physionet [7] and the Montreal Archive of Sleep Studies subset 3 (MASS-SS3). In the following, we compare SAM with existing approaches on the single EEG sleep scoring task.

4.1 Datasets

Table 1 provides an overview of both datasets. Both cover healthy subjects and span a total age range from 20 to 101 years. While MASS-SS3 [17] is scored with the actual 5 class AASM standard, SEDF-78 [11] is scored with the older Rechtschaffen and Kales (R&K) standard. R&K has 6 sleep stages, but it is possible to merge the 2 stages S3 and S4 for close resemblance of N3. Thus, we evaluated both datasets with $V = \{\text{W}, \text{N1}, \text{N2}, \text{N3}, \text{R}\}$.

Table 1. Datasets.

Dataset	SEDF-78	MASS-SS3
N	78	62
F:M	41:37	34:28
Mean age (range)	59.0 yrs (25–101)	42.5 yrs (20–69)
Sleep disorders	None	AHI < 10
Scoring standard	R& K	AASM
Epoch duration	30 s	30 s
Records	153	62
Derivation	Fpz-Cz	EOG-F4
Sampling rate	100 Hz	256 Hz

4.2 Preparation

MASS signals were resampled 256 Hz to the model sampling rate $f_S = 100$ Hz with a polyphase filter with up conversion 25 and down conversion 64 was used. The $C = 1$ input channels were bipolar derivations, with derivation Fpz-Cz for SEDF-78 and EOG-F4 for MASS-SS3. Both datasets were scored with epoch duration $T = 30$ s. To remove drifts and low frequency artifacts the data were filtered with a forward-backward, i.e., zero phase, Butterworth high pass filter of 5-th order with 0.1 Hz cutoff. According to this specification the l-th input epoch is $\mathbf{X}_l \in \mathbb{R}^{3000 \times 1}$.

4.3 Model Setup

According to the experimental EEG size of 3000 samples we chose 9 max pooling layers followed by final max pooling in the embedder. The resulting 20 CNN layers had kernel size 5. This contrasts other work, such as the successful Deep-SleepNet with kernels up to size 400 [21]. After the first convolution feature maps increased super-linear from 8 to $N_F = 64$. A complete specification is provided in the appendix Table 3. The sequence encoder layer was specified with model dimension $M = 64$, $N_H = 4$ attention heads and feed forward expansion 200. We stacked $N_s = 5$ identical layers (with different parameters) and chose an attention scope $L = 30$. This resulted in a total number of 360 k parameters. For SEDF-78 we reimplemented the DeepSleepNet [21] LSTM model as a baseline. Notably, this model has 22 M parameters.

4.4 Training

While some work applies separated and subsequent training of embedder and sequence encoder [21], we train SAM jointly, via cross-entropy loss:

$$\mathcal{L}\left(\theta; \boldsymbol{X}, \mathbf{y}\right) = -\frac{1}{L} \sum_{l=1}^{L} \sum_{k=1}^{K} \delta_{y_l k} \log g\left(\boldsymbol{X}; \theta\right)_{lk} \tag{5}$$

with the label y_l one hot coded and the model 3. Unlike many other solutions we use uniform class weights. We optimize (3) with AdamW [15] with $\beta_1 = 0.9$, $\beta_2 = 0.999$, $\epsilon = 10^{-8}$, a batch size of 32 and weight decay of 10^{-4}. The learning rate was set to 10^{-3} and followed a fixed schedule with 4 epochs ramp up, 4 epochs ramp down and a total duration of 20 epochs. Since no validation set was used, we always used the final model for testing. During training we use 10% dropout probability after MHSA and expansion-compression. The model is implemented in PyTorch with a proprietary data loader. Training and testing were performed on a 64 GB workstation with an NVIDIA 3090 GPU with 24 GB RAM. The complete CV protocol on SEDF-78 required 1.7 h for SAM, while the reimplemented DeepSleepNet required 12.2 h.

4.5   Protocol

Both datasets were evaluated with k-fold cross-validation (CV) with randomly split *subjects*. In accordance with literature we chose $k = 10$ for SEDF-78 and $k = 31$ for MASS-SS3. In all training and test runs, a single subject was treated atomic, i.e., a single subject's data, on record or epoch level, was never split over test and training. This avoided over-fitting due to correlated test- and training data. We randomly chose 4 folds (2 per dataset) to find optimal hyperparameters from a set of combinations.

We report Cohen's κ, specific F1-scores and the macro F1-score (MF1) from pooled subjects. Thus, every result is from a single confusion matrix. In comparison with our baseline LSTM, we can compare results at the subject level. We use a paired-sample t-test to show statistical significance at the $\alpha = 0.05$ level. Note that we do not compare to work that does not treat subjects atomic, uses a different k in CV, does not report comparable metrics or uses different electrode derivations.

Table 2. Results for the proposed SAM and the LSTM baseline DeepSleepNet compared to literature. Our work is indicated by *. The best results are boldfaced.

Dataset	Model	Overall scores		Sleep stage F1-scores				
		κ	MF1	W	N1	N2	N3	R
MASS-SS3	SAM* (0.36 M)	**0.80**	**82%**	87%	56%	**91%**	85%	88%
EOGL-F4	DeepSleepNet*	**0.80**	**82%**	**88%**	58%	**91%**	84%	88%
	DeepSleepNet [21] (22 M)	**0.80**	**82%**	87%	**60%**	90%	82%	**89%**
	IIT [20]	0.79	81%	85%	54%	**91%**	**87%**	85%
SEDF-78	SAM*	**0.78**	**79%**	**93%**	49%	**86%**	**82%**	**84%**
Fpz-Cz	DeepSleepNet*	0.76	77%	92%	48%	84%	80%	79%
	CNN-LSTM [12]	0.77	–	–	–	–	–	–
	U-Time [18]	0.75	76%	92%	**51%**	83%	75%	80%

5   Results and Discussion

Table 2 shows overall agreement scores and sleep stage specific agreement for all datasets. On MASS-SS3 SAM achieves $\kappa = 0.80$ and MF1 = 82%, which is on par with the best reported result. While our model is less accurate in N1 it is more accurate in the clinically important N3 stage. The reimplemented LSTM baseline DeepSleep-Net achieves comparable results to the published version. On SEDF-78 our results surpass the DeepSleepNet and the literature with $\kappa = 0.78$ and MF1 = 79%. While all sleep stages except N1 improve between 1% and 2%, the largest improvement occurs in R with 4%.

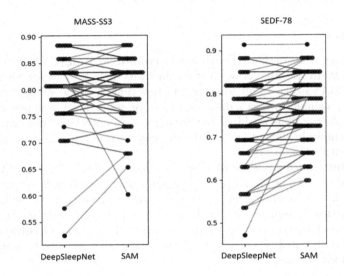

Fig. 3. Paired samples plot of binned subject-wise Cohen's κ for the proposed model SAM and the baseline LSTM DeepSleepNet for datasets MASS-SS3 and SEDF-78, with binned average in red. (Color figure online)

Figure 3 shows paired samples plots of binned subject level κ. For MASS-SS3 there is no difference on average (horizontal red line), analogous to the pooled result (cf. Table 2). The paired-sample t-test confirms this expectation statistically with $p = 0.7 > \alpha$. For SEDF-78 most subjects show an increase by at least one bin. Also, on average, there is an increase of 1 bin, which has a size of 0.025. The t-test confirms a statistically significant difference with $p = 4 \times 10^{-5} < \alpha$.

Our reimplemented DeepSleepNet* achieves higher agreement on SEDF-78 (MF1 = 0.77) than U-Time, although their DeepSleepNet reimplementation (MF1 = 0.73) was inferior to U-Time. We carefully reimplemented the original layer details (e.g., CNN padding) and adapted learning rates for the single, joint training session. In accordance with comparable published and reimplemented results on MASS-SS3 we assume that our higher baseline result on SEDF-78 is representative. Results show twice as large database-differences for DeepSleep-Net ($\Delta\kappa = 0.04$) than for SAM ($\Delta\kappa = 0.02$), caused by two different reasons. First, we consider MASS-SS3 a simpler problem than SEDF-78 (EEG only), since the former provides the network with EEG and EOG— the most important information human experts use to score sleep—reflected by higher scores on MASS-SS3. Second, DeepSleepNet results are particularly high on MASS-SS3 compared to SEDF (both to SEDF-78 here and SEDF-20 in [21]). According the authors the DeepSleepNet architecture was optimized only on a MASS-SS3 subset, which may cause an architectural bias towards MASS-SS3.

While these are promising results, we acknowledge that they should be solidified with more subjects from an independent dataset. A question out of this work's scope are clinical benefits of the proposed method. Since raw agreement has no direct clinical relevance, improvements must be interpreted cautiously. However, since stage R is clinically important (e.g., for time-to-REM) our improvements could be relevant. Considered that SAM is more than 50 times smaller than DeepSleepNet it's parity on MASS-SS3 is remarkable. On the harder EEG only SEDF-78 experiment SAM could achieve a considerable REM improvement ($+4\%$ F1-R), albeit the small size. The results support the introductory hypothesis that REM accuracy benefits most from attention. We attribute this to the direct (i.e., not state encoded) access to distant embeddings and the distant-indifferent (i.e., constant) maximum path length of attention. Although we conceived SAM for sleep scoring, other tasks such as abnormality detection or movement intention detection as well may fit as well.

6   Conclusion

This contribution introduced SAM, a simple, and lightweight EEG local attention model. We showed that SAM with 360 k parameters is on par with the 22 M parameter state-of-the-art on the MASS-SS3 PSG sleep scoring task. On the harder SEDF-78 EEG sleep scoring task SAM achieves the new state-of- the-art performance and proves long-term dependency modelling benefits with a considerable improvement in the practically important REM sleep stage. On top of its effectiveness SAM is also efficient on EEG of arbitrary length since computational complexity scales linear with EEG length. SAM may be an important step towards reduced-instrumentation, but PSG-quality sleep scoring and we look forward to investigations in clinical use.

Acknowledgments. Asan Agibetov helped with LaTeX help and Kluge Tilmann provided computing infrastructure. This work was supported by the Austrian Research Promotion Agency (FFG) grant number 867615.

Appendix

Table 3 shows the embedder layer specification.

Table 3. Embedder layer specification based on CNN blocks (FCi) and residual blocks (FRi). Conv BN layers comprise CNN, batch norm and ReLU activation and are specified by kernel size, feature maps C_o, stride and padding.

Layer			Out dim	Conv./Pooling			
ID	Group	Type		Size	C_O	Stride	Padd.
1	FC1	Input	3000 × 1				
2		Conv BN	3000 × 8	5	8	1	2
3	FR2	Conv BN	3000 × 18	5	18	1	2
4		Max Pool	1500 × 18	2		2	0
5		Conv BN	1500 × 18	5	18	1	2
6		Conv BN	1500 × 18	5	18	1	2
7	FC3	Conv BN	1500 × 21	5	21	1	2
8		Max Pool	750 × 21	2		2	0
9	FR4	Conv BN	750 × 25	5	25	1	2
10		Max Pool	375 × 25	2		2	0
11		Conv BN	375 × 25	5	25	1	2
12		Conv BN	375 × 25	5	25	1	2
13	FC5	Conv BN	375 × 29	5	29	1	2
14		Max Pool	187 × 29	2		2	0
15	FR6	Conv BN	187 × 34	5	34	1	2
16		Max Pool	93 × 34	2		2	0
17		Conv BN	93 × 34	5	34	1	2
18		Conv BN	93 × 34	5	34	1	2
19	FC7	Conv BN	93 × 40	5	40	1	2
20		Max Pool	46 × 40	2		2	0
21	FR8	Conv BN	46 × 47	5	47	1	2
22		Max Pool	23 × 47	2		2	0
23		Conv BN	23 × 47	5	47	1	2
24		Conv BN	23 × 47	5	47	1	2
25	FC9	Conv BN	23 × 54	5	54	1	2
26		Max Pool	11 × 54	2		2	0
27	FR10	Conv BN	11 × 64	5	64	1	2
28		Max Pool	5 × 64	2		2	0
29		Conv BN	5 × 64	5	64	1	2
30		Conv BN	5 × 64	5	64	1	2
31	F11	Max Pool	1 × 64	5		5	0
32		Flatten	64				
33	C	Linear	5				

References

1. Anderer, P., et al.: An E-health solution for automatic sleep classification according to Rechtschaffen and kales: validation study of the Somnolyzer 24 x 7 utilizing the Siesta database. Neuropsychobiology **51**(3), 115–133 (2005). https://doi.org/10.1159/000085205
2. Ba, J.L., Kiros, J.R., Hinton, G.E.: Layer normalization. arXiv preprint arXiv:1607.06450, July 2016
3. Bahdanau, D., Cho, K.H., Bengio, Y.: Neural machine translation by jointly learning to align and translate. In: 3rd International Conference on Learning Representations, ICLR 2015 - Conference Track Proceedings, pp. 1–15 (2015)
4. Berry, R.B., et al.: The AASM manual for the scoring of sleep and associated events. Rules, Terminology Tech. Specifications, Darien, Ill., Am. Acad. Sleep Med. **176**, 2012 (2012)
5. Cheng, J., Dong, L., Lapata, M.: Long short-term memory-networks for machine reading. In: EMNLP 2016 - Conference on Empirical Methods in Natural Language Processing, Proceedings, pp. 551–561 (2016). https://doi.org/10.18653/v1/d16-1053
6. Dong, H., Supratak, A., Pan, W., Wu, C., Matthews, P.M., Guo, Y.: Mixed neural network approach for temporal sleep stage classification. IEEE Trans. Neural Syst. Rehabil. Eng. **26**(2), 324–333 (2018)
7. Goldberger, A.L., et al.: Physiobank, physiotoolkit, and physionet. Circulation **101**(23) (2000)
8. He, K., Zhang, X., Ren, S., Sun, J.: Deep residual learning for image recognition. In: Proceedings of the IEEE Computer Society Conference on Computer Vision and Pattern Recognition 2016-December, pp. 770–778 (2016). https://doi.org/10.1109/CVPR.2016.90
9. Hochreiter, S., Schmidhuber, J.: Long short-term memory. Neural Comput. **9**(8), 1735–1780 (1997)
10. Ioffe, S., Szegedy, C.: Batch normalization: accelerating deep network training by reducing internal covariate shift. In: International Conference on Machine Learning, pp. 448–456. PMLR (2015)
11. Kemp, B., Zwinderman, A.H., Tuk, B., Kamphuisen, H.A., Oberyé, J.J.: Analysis of a sleep-dependent neuronal feedback loop: the slow-wave microcontinuity of the EEG. IEEE Trans. Biomed. Eng. **47**(9), 1185–1194 (2000). https://doi.org/10.1109/10.867928
12. Korkalainen, H., et al.: Accurate deep learning-based sleep staging in a clinical population with suspected obstructive sleep apnea. IEEE J. Biomed. Health Inform. **24**(7), 2073–2081 (2019)
13. Krizhevsky, A., Sutskever, I., Hinton, G.E.: Imagenet classification with deep convolutional neural networks. Adv. Neural. Inf. Process. Syst. **25**, 1097–1105 (2012)
14. Li, W., et al.: On the compactness, efficiency, and representation of 3D convolutional networks: brain parcellation as a pretext task. In: Niethammer, M., et al. (eds.) IPMI 2017. LNCS, vol. 10265, pp. 348–360. Springer, Cham (2017). https://doi.org/10.1007/978-3-319-59050-9_28
15. Loshchilov, I., Hutter, F.: Decoupled weight decay regularization. arXiv preprint arXiv:1711.05101 (2017)
16. Luong, M.T., Pham, H., Manning, C.D.: Effective approaches to attention-based neural machine translation. In: Conference Proceedings - EMNLP 2015: Conference on Empirical Methods in Natural Language Processing, pp. 1412–1421 (2015)

17. O'Reilly, C., Gosselin, N., Carrier, J., Nielsen, T.: Montreal archive of sleep studies: an open-access resource for instrument benchmarking and exploratory research. J. Sleep Res. **23**(6), 628–635 (2014)
18. Perslev, M., Jensen, M.H., Darkner, S., Jennum, P.J., Igel, C.: U-time: a fully convolutional network for time series segmentation applied to sleep staging. In: Proceedings of the 33rd International Conference on Neural Information Processing Systems, pp. 4415–4426 (2019)
19. Phan, H., Andreotti, F., Cooray, N., Chen, O.Y., De Vos, M.: Seqsleepnet: end-to-end hierarchical recurrent neural network for sequence-to-sequence automatic sleep staging. IEEE Trans. Neural Syst. Rehabil. Eng. **27**(3), 400–410 (2019)
20. Seo, H., Back, S., Lee, S., Park, D., Kim, T., Lee, K.: Intra- and inter-epoch temporal context network (IITNET) using sub-epoch features for automatic sleep scoring on raw single-channel EEG. Biomed. Signal Process. Control **61**, 102037 (2020)
21. Supratak, A., Dong, H., Wu, C., Guo, Y.: DeepSleepNet: a model for automatic sleep stage scoring based on raw single-channel EEG. IEEE Trans. Neural Syst. Rehabil. Eng. **25**(11), 1998–2008 (2017)
22. Tsinalis, O., Matthews, P.M., Guo, Y., Zafeiriou, S.: Automatic sleep stage scoring with single-channel EEG using convolutional neural networks. arXiv preprint arXiv:1610.01683, October 2016, http://arxiv.org/abs/1610.01683
23. Vaswani, A., et al.: Attention is all you need. In: Advances in Neural Information Processing Systems, pp. 5998–6008 (2017)

ReCal-Net: Joint Region-Channel-Wise Calibrated Network for Semantic Segmentation in Cataract Surgery Videos

Negin Ghamsarian[1]([⊠]), Mario Taschwer[1], Doris Putzgruber-Adamitsch[2], Stephanie Sarny[2], Yosuf El-Shabrawi[2], and Klaus Schöffmann[1] [iD]

[1] Department of Information Technology, Alpen-Adria-Universität Klagenfurt, Klagenfurt, Austria
{negin,mt,ks}@itec.aau.at
[2] Department of Ophthalmology, Klinikum Klagenfurt, Klagenfurt, Austria
{doris.putzgruber-adamitsch,stephanie.sarny,Yosuf.El-Shabrawi}@kabeg.at

Abstract. Semantic segmentation in surgical videos is a prerequisite for a broad range of applications towards improving surgical outcomes and surgical video analysis. However, semantic segmentation in surgical videos involves many challenges. In particular, in cataract surgery, various features of the relevant objects such as blunt edges, color and context variation, reflection, transparency, and motion blur pose a challenge for semantic segmentation. In this paper, we propose a novel convolutional module termed as *ReCal* module, which can calibrate the feature maps by employing region intra-and-inter-dependencies and channel-region cross-dependencies. This calibration strategy can effectively enhance semantic representation by correlating different representations of the same semantic label, considering a multi-angle local view centering around each pixel. Thus the proposed module can deal with distant visual characteristics of unique objects as well as cross-similarities in the visual characteristics of different objects. Moreover, we propose a novel network architecture based on the proposed module termed as *ReCal-Net*. Experimental results confirm the superiority of ReCal-Net compared to rival state-of-the-art approaches for all relevant objects in cataract surgery. Moreover, ablation studies reveal the effectiveness of the ReCal module in boosting semantic segmentation accuracy.

Keywords: Cataract surgery · Semantic segmentation · Feature map calibration

1 Introduction

Cataract surgery is the procedure of restoring the eye's clear vision by removing the occluded natural lens and implanting an intraocular lens (IOL). Being one of the most frequently performed surgeries, enhancing the outcomes of cataract surgery and diminishing its potential intra-operative and post-operative risks is of great importance.

This work was funded by the FWF Austrian Science Fund under grant P 31486-N31.

T. Mantoro et al. (Eds.): ICONIP 2021, LNCS 13110, pp. 391–402, 2021.
https://doi.org/10.1007/978-3-030-92238-2_33

Accordingly, a large body of research has been focused on computerized surgical work-flow analysis in cataract surgery [7,9,16,17,23], with a majority of approaches relying on semantic segmentation. Hence, improving semantic segmentation accuracy in cataract surgery videos can play a leading role in the development of a reliable computerized clinical diagnosis or surgical analysis approach [6,18,19].

Semantic segmentation of the relevant objects in cataract surgery videos is quite challenging due to (i) transparency of the intraocular lens, (ii) color and contextual variation of the pupil and iris, (iii) blunt edges of the iris, and (iv) severe motion blur and reflection distortion of the instruments. In this paper, we propose a novel module for joint Region-channel-wise Calibration, termed as *ReCal* module. The proposed module can simultaneously deal with the various segmentation challenges in cataract surgery videos. In particular, the ReCal module is able to (1) employ multi-angle pyramid features centered around each pixel position to deal with transparency, blunt edges, and motion blur, (2) employ cross region-channel dependencies to handle texture and color variation through interconnecting the distant feature vectors corresponding to the same object. The proposed module can be added on top of every convolutional layer without changing the output feature dimensions. Moreover, the ReCal module does not impose a significant number of trainable parameters on the network and thus can be used after several layers to calibrate their output feature maps. Besides, we propose a novel semantic segmentation network based on the ReCal module termed as *ReCal-Net*. The experimental results show significant improvement in semantic segmentation of the relevant objects with ReCal-Net compared to the best-performing rival approach (85.38% compared to 83.32% overall IoU (intersection over union) for ReCal-Net vs. UNet++).

The rest of this paper is organized as follows. In Sect. 2, we briefly review state-of-the-art semantic segmentation approaches in the medical domain. In Sect. 3, we first discuss two convolutional blocks from which the proposed approach is inspired, and then delineate the proposed ReCal-Net and ReCal module. We detail the experimental settings in Sect. 4 and present the experimental results in Sect. 5. We finally conclude the paper in Sect. 6.

2   Related Work

Since many automatic medical diagnosis and image analysis applications entail semantic segmentation, considerable research has been conducted to improve medical image segmentation accuracy. In particular, U-Net [21] achieved outstanding segmentation accuracy being attributed to its skip connections. In recent years, many U-Net-base approaches have been proposed to address the weaknesses of the U-Net baseline [5,12,14,25,26]. UNet++ [26] is proposed to tackle automatic depth optimization via ensembling varying-depth U-Nets. MultiResUNet [14] factorizes large and computationally expensive kernels by fusing sequential convolutional feature maps. CPFNet [5] fuses the output features of parallel dilated convolutions (featuring different dilation rates) for scale awareness. The SegSE block [20] and scSE block [22], inspired by Squeeze-and-Excitation (SE) block [13], aim to recalibrate the channels via extracting inter-channel dependencies. The scSE block [22] further enhances feature

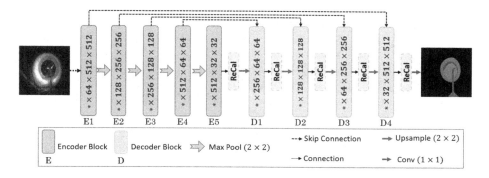

Fig. 1. The overall architecture of ReCal-Net containing five ReCal blocks.

representation via spatial channel pooling. Furthermore, many approaches are proposed to enhance semantic segmentation accuracy for particular medically relevant objects, including but not limited to liver lesion [2], surgical instruments [18,19], pulmonary vessel [3], and lung tumor [15].

3 Methodology

Notations. Everywhere in this paper, we show convolutional layer with the kernel-size of $(m \times n)$, P output channels, and g groups as $*_{(m \times n)}^{P,g}$ (we consider the default dilation rate of 1 for this layer). Besides, we show average-pooling layer with a kernel-size of $(m \times n)$ and a stride of s pixels as $\sum_{(m \times n)}^{s}$, and global average pooling as $\sum^{G}$.

Feature Map Recalibration. The Squeeze-and-Excitation (SE) block [13] was proposed to model inter-channel dependencies through squeezing the spatial features into a channel descriptor, applying fully-connected layers, and rescaling the input feature map via multiplication. This low-complexity operation unit has proved to be effective, especially for semantic segmentation. However, the SE block does not consider pixel-wise features in recalibration. Accordingly, scSE block [22] was proposed to exploit pixel-wise and channel-wise information concurrently. This block can be split into two parallel operations: (1) spatial squeeze and channel excitation, exactly the same as the SE block, and (2) channel squeeze and spatial excitation. The latter operation is conducted by applying a pixel-wise convolution with one output channel to the input feature map, followed by multiplication. The final feature maps of these two parallel computational units are then merged by selecting the maximum feature in each location.

ReCal-Net. Figure 1 depicts the architecture of ReCal-Net. Overall, the network consists of three types of blocks: (i) encoder blocks that transform low-level features to semantic features while compressing the spatial representation, (ii) decoder blocks that are responsible for improving the semantic features in higher resolutions by employing the symmetric low-level feature maps from the encoder blocks, (iii) and *ReCal* modules that account for calibrating the semantic feature maps. We use the VGG16 network as

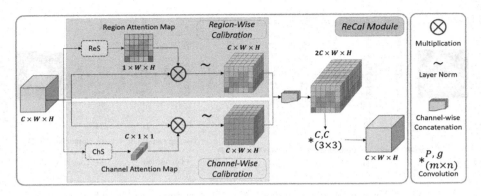

Fig. 2. The detailed architecture of ReCal block containing regional squeeze block (ReS) and channel squeeze block (ChS).

the encoder network. The ith encoder block (Ei, $i \in \{1, 2, 3, 4\}$) in Fig. 1 correspond to all layers between the i-1th and ith max-pooling layers in the VGG16 network (max-pooling layers are indicated with gray arrows). The last encoder block (E5) corresponds to the layers between the last max-pooling layer and the average pooling layer. Each decoder block follows the same architecture of decoder blocks in U-Net [21], including two convolutional layers, each of which being followed by batch normalization and ReLU.

ReCal Module. Despite the effectiveness of SE and scSE blocks in boosting feature representation, both fail to exploit region-wise dependencies. However, employing region-wise inter-dependencies and intra-dependencies can significantly enhance semantic segmentation performance. We propose a joint region-channel-wise calibration (ReCal) module to calibrate the feature maps based on joint region-wise and channel-wise dependencies. Figure 2 demonstrates the architecture of the proposed ReCal module inspired by [13,22]. This module aims to reinforce a semantic representation considering inter-channel dependencies, inter-region and intra-region dependencies, and channel-region cross-dependencies. The input feature map of ReCal module $\mathcal{F}_{In} \in \mathbb{R}^{C \times W \times H}$ is first fed into two parallel blocks: (1) the Region-wise Squeeze block (*ReS*), and (2) the Channel-wise Squeeze block (*ChS*). Afterward, the region-wise and channel-wise calibrated features ($\mathcal{F}_{Re} \in \mathbb{R}^{C \times W \times H}$ and $\mathcal{F}_{Ch} \in \mathbb{R}^{C \times W \times H}$) are obtained by multiplying ($\otimes$) the input feature map to the region-attention map and channel-attention map, respectively, followed by the layer normalization function. In this stage, each particular channel $\mathcal{F}_{In}(C_j) \in \mathbb{R}^{W \times H}$ in the input feature map of a ReCal module has corresponding region-wise and channel-wise calibrated channels ($\mathcal{F}_{Re}(C_j) \in \mathbb{R}^{W \times H}$ and $\mathcal{F}_{Ch}(C_j) \in \mathbb{R}^{W \times H}$). To enable the utilization of cross-dependencies between the region-wise and channel-wise calibrated features, we concatenate these two feature maps in a depth-wise manner. Indeed, the concatenated feature map ($\mathcal{F}_{Concat}$) for each $p \in [1, C]$, $x \in [1, W]$, and $y \in [1, H]$ can be formulated as (1).

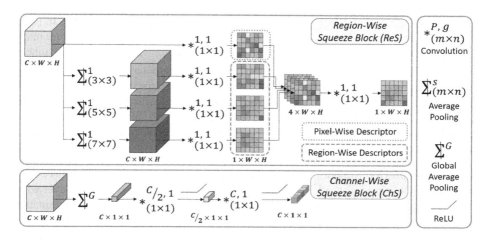

Fig. 3. Demonstration of regional squeeze block (ReS) and channel squeeze block (CS).

$$\begin{cases} \mathcal{F}_{Concat}(2p, x, y) = \mathcal{F}_{Re}(p, x, y) \\ \mathcal{F}_{Concat}(2p - 1, x, y) = \mathcal{F}_{Ch}(p, x, y) \end{cases} \tag{1}$$

The cross-dependency between region-wise and channel-wise calibrated features is computed using a convolutional layer with C groups. More concretely, every two consecutive channels in the concatenated feature map undergo a distinct convolution with a kernel-size of (3×3). This convolutional layer considers the local contextual features around each pixel (a 3×3 window around each pixel) to determine the contribution of each of region-wise and channel-wise calibrated features in the output features. Using a kernel size greater than one unit allows jointly considering inter-region dependencies.

Region-Wise Squeeze Block. Figure 3 details the architecture of the ReS block, which is responsible for providing the region attention map. The region attention map is obtained by taking advantage of multi-angle local content based on narrow to wider views around each distinct pixel in the input feature map. We model multi-angle local features using average pooling layers with different kernel sizes and the stride of one pixel. The average pooling layers do not any number of impose trainable parameters on the network and thus ease using the ReS block and ReCal module in multiple locations. Besides, the stride of one pixel in the average pooling layer can stimulate a local view centered around each distinctive pixel. We use three average pooling layers with kernel-sizes of (3×3), (5×5), and (7×7), followed by pixel-wise convolutions with one output channel ($*_{(1\times1)}^{1,1}$) to obtain the region-wise descriptors. In parallel, the input feature map undergoes another convolutional layer to obtain the pixel-wise descriptor. The local features can indicate if some particular features (could be similar or dissimilar to the centering pixel) exist in its neighborhood, and how large is the neighborhood of each pixel containing particular features. The four attention maps are then concatenated and fed into a convolutional layer ($*_{(1\times1)}^{1,1}$) that is responsible for determining the contribution of each spatial descriptor in the final region-wise attention map.

Channel-Wise Squeeze Block. For ChS Block, we follow a similar scheme as in [13]. At first, we apply global average pooling ($\sum^G$) on the input convolutional feature map. Afterward, we form a bottleneck via a pixel-wise convolution with C/r output channels ($*_{(1\times1)}^{C/r,1}$) followed by ReLU non-linearity. The scaling parameter r can curbs the computational complexity. Besides, it can act as a smoothing factor that can yield a better-generalized model by preventing the network from learning outliers. In experiments, we set $r = 2$ as it is proved to have the best performance [22]. Finally, another pixel-wise convolution with C output channels ($*_{(1\times1)}^{C,1}$) followed by ReLU non-linearity is used to provide the channel attention map.

Module Complexity. Suppose we have an intermediate layer in the network with convolutional response map $\mathcal{X} \in \mathbb{R}^{C \times H \times W}$. Adding a ReCal module on top of this layer with its scaling parameter being equal to 2, amounts to "$C^2 + 22C + 4$" additional trainable weights. More specifically, each convolutional layer $*_{(m \times n)}^{P,g}$ applied to C input channels amounts to $((m \times n) \times C \times P)/g$ trainable weights. Accordingly, we need "$4C + 4$" weights for the ReS block, "C^2" weights for the ChS block, and "$18C$" weights for the last convolution operation of the ReCal module. In our proposed architecture, adding five ReCal modules on convolutional feature maps with 512, 256, 128, 64, and 32 channels sums up to $371K$ additional weights, and only $21K$ more trainable parameters compared to the SE block [13] and scSE block [22].

4   Experimental Settings

Datasets. We use four datasets in this study. The iris dataset is created by annotating the cornea and pupil from 14 cataract surgery videos using "supervisely" platform. The iris annotations are then obtained by subtracting the convex-hull of the pupil segment from the cornea segment. This dataset contains 124 frames from 12 videos for training and 23 frames from two videos for testing[1]. For lens and pupil segmentation, we employ the two public datasets of the LensID framework [8], containing the annotation of the intraocular lens and pupil. The lens dataset consists of lens annotation in 401 frames sampled from 27 videos. From these annotations, 292 frames from 21 videos are used for training, and 109 frames from the remaining six videos are used for testing. The pupil segmentation dataset contains 189 frames from 16 videos. The training set consists of 141 frames from 13 videos, and the testing set contains 48 frames from three remaining videos. For instrument segmentation, we use the instrument annotations of the CaDIS dataset [11]. We use 3190 frames from 18 videos for training and 459 frames from three other videos for testing.

Rival Approaches. Table 1 lists the specifications of the rival state-of-the-art approaches used in our evaluations. In "Upsampling" column, "Trans Conv" stands for *Transposed Convolution*. To enable direct comparison between the ReCal module and scSE block, we have formed scSE-Net by replacing the ReCal modules in ReCal-Net with scSE modules. Indeed, the baseline of both approaches are the same, and the only difference is the use of scSE blocks in scSE-Net at the position of ReCal modules in ReCal-Net.

---

[1] The dataset will be released with the acceptance of this paper.

Table 1. Specifications of the proposed and rival segmentation approaches.

Model	Backbone	Params	Upsampling	Reference	Year
UNet++ (/DS)	VGG16	24.24 M	Bilinear	[26]	2020
MultiResUNet	✗	9.34 M	Trans Conv	[14]	2020
BARNet	ResNet34	24.90 M	Bilinear	[19]	2020
PAANet	ResNet34	22.43 M	Trans Conv & Bilinear	[18]	2020
CPFNet	ResNet34	34.66 M	Bilinear	[5]	2020
dU-Net	✗	31.98 M	Trans Conv	[25]	2020
CE-Net	ResNet34	29.90 M	Trans Conv	[12]	2019
scSE-Net	VGG16	22.90 M	Bilinear	[22]	2019
U-Net	✗	17.26 M	Bilinear	[21]	2015
ReCal-Net	VGG16	22.92 M	Bilinear	Proposed	

Data Augmentation Methods. We use the Albumentations [1] library for image and mask augmentation during training. Considering the inherent features of the relevant objects and problems of the recorded videos [10], we apply motion blur, median blur, brightness and contrast change, shifting, scaling, and rotation for augmentation. We use the same augmentation pipeline for the proposed and rival approaches.

Neural Network Settings. We initialize the parameters of backbones for the proposed and rival approaches (in case of having a backbone) with ImageNet [4] training weights. We set the input size of all networks to $(3 \times 512 \times 512)$.

Training Settings. During training with all networks, a threshold of 0.1 is applied for gradient clipping. This strategy can prevent the gradient from exploding and result in a more appropriate behavior during learning in the case of irregularities in the loss landscape. Considering the different depths and connections of the proposed and rival approaches, all networks are trained with two different initial learning rates ($lr \in \{0.005, 0.002\}$) for 30 epochs with SGD optimizer. The learning scheduler decreases the learning rate every other epoch with a factor of 0.8. We list the results with the highest IoU for each network.

Loss Function. To provide a fair comparison, we use the same loss function for all networks. The loss function is set to a weighted sum of binary cross-entropy (BCE) and the logarithm of soft Dice coefficient as follows.

$$\mathcal{L} = (\lambda) \times BCE(\mathcal{X}_{true}(i,j), \mathcal{X}_{pred}(i,j))$$
$$- (1-\lambda) \times (\log \frac{2 \sum \mathcal{X}_{true} \odot \mathcal{X}_{pred} + \sigma}{\sum \mathcal{X}_{true} + \sum \mathcal{X}_{pred} + \sigma}) \tag{2}$$

Soft Dice refers to the dice coefficient computed directly based on predicted probabilities rather than the predicted binary masks after thresholding. In (2), $\mathcal{X}_{true}$ refers to the ground truth mask, $\mathcal{X}_{pred}$ refers to the predicted mask, $\odot$ refers to Hadamard product

Table 2. Quantitative comparisons among the semantic segmentation results of Recal-Net and rival approaches based on IoU(%).

Network	Lens	Pupil	Iris	Instruments	Overall
U-Net	61.89 ±20.93	83.51 ±20.24	65.89 ±16.93	60.78 ±26.04	68.01 ±21.03
CE-Net	78.51 ±11.56	92.07 ± 4.24	71.74 ± 6.19	69.44 ±17.94	77.94 ± 9.98
dU-Net	60.39 ±29.36	68.03 ±35.95	70.21 ±12.97	61.24 ±27.64	64.96 ±26.48
scSE-Net	86.04 ±11.36	96.13 ± 2.10	78.58 ± 9.61	71.03 ±23.25	82.94 ±11.58
CPFNet	80.65 ±12.16	93.76 ± 2.87	77.93 ± 5.42	69.46 ±17.88	80.45 ± 9.58
BARNet	80.23 ±14.57	93.64 ± 4.11	75.80 ± 8.68	69.76 ±21.29	79.86 ±12.16
PAANet	80.30 ±11.73	94.35 ± 3.88	75.73 ±11.67	68.01 ±22.29	79.59 ±12.39
MultiResUNet	61.42 ±19.91	76.46 ±29.43	49.99 ±28.73	61.01 ±26.94	62.22 ±26.25
UNet++/DS	84.53 ±13.42	96.18 ± 2.62	74.01 ±13.13	65.99 ±25.66	79.42 ±14.75
UNet++	85.74 ±11.16	96.50 ± 1.51	81.98 ± 6.96	69.07 ±23.89	83.32 ±10.88
ReCal-Net	**87.94** ±10.72	**96.58** ± 1.30	**85.13** ± 3.98	**71.89** ±19.93	**85.38** ± 8.98

Table 3. Quantitative comparisons among the semantic segmentation results of Recal-Net and rival approaches based on Dice (%).

Network	Lens	Pupil	Iris	Instruments	Overall
U-Net	73.86 ±20.39	89.36 ±15.07	78.12 ±13.01	71.50 ±25.77	78.21 ±18.56
CE-Net	87.32 ± 9.98	95.81 ± 2.39	83.39 ± 4.25	80.30 ±15.97	86.70 ± 8.15
dU-Net	69.99 ±29.40	73.72 ±34.24	81.76 ± 9.73	71.30 ±27.62	74.19 ±25.24
scSE-Net	91.95 ± 9.14	98.01 ± 1.10	87.66 ± 6.35	80.18 ±21.49	89.45 ± 9.52
CPFNet	88.61 ±10.20	96.76 ± 1.53	87.48 ± 3.60	80.33 ±15.85	88.29 ± 7.79
BARNet	88.16 ±10.87	96.66 ± 2.30	85.95 ± 5.73	79.72 ±19.95	87.62 ± 9.71
PAANet	88.46 ± 9.59	97.05 ± 2.16	85.62 ± 8.50	78.15 ±21.51	87.32 ±10.44
MultiResUNet	73.88 ±18.26	82.45 ±25.49	61.78 ±25.96	71.35 ±26.88	72.36 ±24.14
UNet++/DS	90.80 ±11.41	98.03 ± 1.41	84.38 ± 9.06	75.64 ±25.38	87.21 ±11.81
UNet++	91.80 ±11.16	98.26 ± 0.79	89.93 ± 4.51	78.54 ±22.76	89.63 ± 9.80
ReCal-Net	**93.09** ± 8.56	**98.26** ± 0.68	**91.91** ± 2.47	**81.62** ±17.75	**91.22** ± 7.36

(element-wise multiplication), and σ refers to the smoothing factor. In experiments, we set $\lambda = 0.8$ and $\sigma = 1$.

5   Experimental Results

Table 2 and Table 3 compare the segmentation performance of ReCal-Net and ten rival state-of-the-art approaches based on the average and standard deviation of IoU and Dice coefficient, respectively[2]. Overall, ReCal-Net, UNet++, scSE-Net, and CPFNet have

---

[2] The "Overall" column in Table 2 and Table 3 is the average of the other four values.

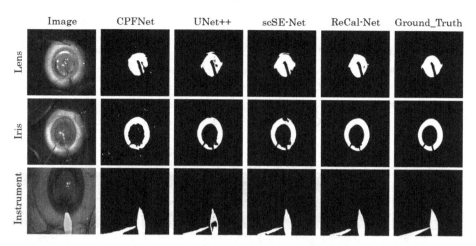

Fig. 4. Qualitative comparisons among the top four segmentation approaches.

shown the top four segmentation results. Moreover, the experimental results reveal that ReCal-Net has achieved the highest average IoU and Dice coefficient for all relevant objects compared to state-of-the-art approaches. Considering the IoU report, ReCal-Net has gained considerable enhancement in segmentation performance compared to the second-best approach in lens segmentation (87.94% vs. 86.04% for scSE-Net) and iris segmentation (85.13% vs. 81.98% for UNet++). Having only 21k more trainable parameters than scSE-Net (0.08% additive trainable parameters), ReCal-Net has achieved 8.3% relative improvement in iris segmentation, 2.9% relative improvement in instrument segmentation, and 2.2% relative improvement in lens segmentation in comparison with scSE-Net. Regarding the Dice coefficient, ReCal-Net and UNet++ show very similar performance in pupil segmentation. However, with 1.32M fewer parameters than UNet++ as the second-best approach, ReCal-Net shows 1.7% relative improvement in overall Dice coefficient (91.22% vs. 89.63%). Surprisingly, replacing the scSE blocks with the ReCal modules results in 4.25% higher Dice coefficient for iris segmentation and 1.44% higher Dice coefficient for instrument segmentation.

Figure 4 provides qualitative comparisons among the top four segmentation approaches for lens, iris, and instrument segmentation. Comparing the visual segmentation results of ReCal-Net and scSE-Net further highlights the effectiveness of region-wise and cross channel-region calibration in boosting semantic segmentation performance.

Table 4 reports the ablation study by comparing the segmentation performance of the baseline approach with ReCal-Net considering two different learning rates. The baseline approach refers to the network obtained after removing all ReCal modules of ReCal-Net in Fig. 1. These results approve of the ReCal module's effectiveness regardless of the learning rate.

Table 4. Impact of adding ReCal modules on the segmentation accuracy based on IoU(%).

Learning rate	Network	Lens	Iris	Instrument
0.002	Baseline	84.83 ±11.62	81.49 ± 6.82	70.04 ±23.94
	ReCal-Net	85.77 ±12.33	83.29 ± 5.82	71.89 ±19.93
0.005	Baseline	86.13 ±11.63	81.00 ± 8.06	67.16 ±24.67
	ReCal-Net	87.94 ±10.72	85.13 ± 3.98	70.43 ±21.17

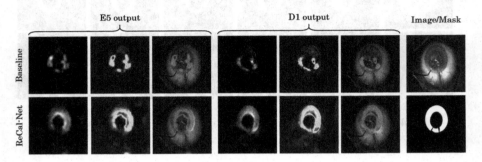

Fig. 5. Visualizations of the intermediate outputs in the baseline approach and ReCal-Net based on class activation maps [24]. For each output, the figures from left to right represent the grayscale activation maps, heatmaps, and heatmaps on images.

To further investigate the impact of the ReCal modules on segmentation performance, we have visualized two intermediate filter response maps for iris segmentation in Fig. 5. The E5 output corresponds to the filter response map of the last encoder block, and the D1 output corresponds to the filter response map of the first decoder block (see Fig. 1). A comparison between the filter response maps of the baseline and ReCal-Net in the same locations indicated the positive impact of the ReCal modules on the network's semantic discrimination capability. Indeed, employing the correlations between the pixel-wise, region-wise, and channel-wise descriptors can reinforce the network's semantic interpretation.

6   Conclusion

This paper presents a novel convolutional module, termed as ReCal module, that can adaptively calibrate feature maps considering pixel-wise, region-wise, and channel-wise descriptors. The ReCal module can effectively correlate intra-region information and cross-channel-region information to deal with severe contextual variations in the same semantic labels and contextual similarities between different semantic labels. The proposed region-channel recalibration module is a very light-weight computational unit that can be applied to any feature map $\mathcal{X} \in \mathbb{R}^{C \times H \times W}$ and output a recalibrated feature map $\mathcal{Y} \in \mathbb{R}^{C \times H \times W}$. Moreover, we have proposed a novel network architecture based on the ReCal module for semantic segmentation in cataract surgery videos, termed as ReCal-Net. The experimental evaluations confirm the effectiveness of the proposed ReCal module and ReCal-Net in dealing with various segmentation challenges in cataract surgery. The proposed ReCal module and ReCal-Net can be adopted for various medical image segmentation and general semantic segmentation problems.

References

1. Buslaev, A., Iglovikov, V.I., Khvedchenya, E., Parinov, A., Druzhinin, M., Kalinin, A.A.: Albumentations: Fast and flexible image augmentations. Information **11**(2), 125 (2020). https://doi.org/10.3390/info11020125
2. Chen, X., Zhang, R., Yan, P.: Feature fusion encoder decoder network for automatic liver lesion segmentation. In: 2019 IEEE 16th International Symposium on Biomedical Imaging (ISBI 2019), pp. 430–433 (2019). https://doi.org/10.1109/ISBI.2019.8759555
3. Cui, H., Liu, X., Huang, N.: Pulmonary vessel segmentation based on orthogonal fused U-Net++ of chest CT images. In: Shen, D., et al. (eds.) MICCAI 2019. LNCS, vol. 11769, pp. 293–300. Springer, Cham (2019). https://doi.org/10.1007/978-3-030-32226-7_33
4. Deng, J., Dong, W., Socher, R., Li, L.J., Li, K., Fei-Fei, L.: ImageNet: a large-scale hierarchical image database. In: 2009 IEEE Conference on Computer Vision and Pattern Recognition, pp. 248–255. IEEE (2009)
5. Feng, S., et al.: CPFNet: context pyramid fusion network for medical image segmentation. IEEE Trans. Med. Imaging **39**(10), 3008–3018 (2020). https://doi.org/10.1109/TMI.2020.2983721
6. Ghamsarian, N.: Enabling relevance-based exploration of cataract videos. In: Proceedings of the 2020 International Conference on Multimedia Retrieval, ICMR 2020, pp. 378–382. Association for Computing Machinery, New York (2020). https://doi.org/10.1145/3372278.3391937
7. Ghamsarian, N., Amirpourazarian, H., Timmerer, C., Taschwer, M., Schöffmann, K.: Relevance-based compression of cataract surgery videos using convolutional neural networks. In: Proceedings of the 28th ACM International Conference on Multimedia, MM 2020, pp. 3577–3585. Association for Computing Machinery, New York (2020). https://doi.org/10.1145/3394171.3413658
8. Ghamsarian, N., Taschwer, M., Putzgruber-Adamitsch, D., Sarny, S., El-Shabrawi, Y., Schoeffmann, K.: LensID: a CNN-RNN-based framework towards lens irregularity detection in cataract surgery videos. In: de Bruijne, M., et al. (eds.) MICCAI 2021. LNCS, vol. 12908, pp. 76–86. Springer, Cham (2021). https://doi.org/10.1007/978-3-030-87237-3_8
9. Ghamsarian, N., Taschwer, M., Putzgruber-Adamitsch, D., Sarny, S., Schoeffmann, K.: Relevance detection in cataract surgery videos by spatio-temporal action localization. In: 2020 25th International Conference on Pattern Recognition (ICPR), pp. 10720–10727 (2021). https://doi.org/10.1109/ICPR48806.2021.9412525
10. Ghamsarian, N., Taschwer, M., Schoeffmann, K.: Deblurring cataract surgery videos using a multi-scale deconvolutional neural network. In: 2020 IEEE 17th International Symposium on Biomedical Imaging (ISBI), pp. 872–876 (2020). https://doi.org/10.1109/ISBI45749.2020.9098318
11. Grammatikopoulou, M., et al.: CaDIS: cataract dataset for image segmentation (2020)
12. Gu, Z., et al.: CE-net: context encoder network for 2D medical image segmentation. IEEE Trans. Med. Imaging **38**(10), 2281–2292 (2019). https://doi.org/10.1109/TMI.2019.2903562
13. Hu, J., Shen, L., Sun, G.: Squeeze-and-excitation networks. In: 2018 IEEE/CVF Conference on Computer Vision and Pattern Recognition, pp. 7132–7141 (2018). https://doi.org/10.1109/CVPR.2018.00745

14. Ibtehaz, N., Rahman, M.S.: MultiResUNet: rethinking the U-net architecture for multimodal biomedical image segmentation. Neural Netw. **121**, 74–87 (2020). https://doi.org/10.1016/j.neunet.2019.08.025.   https://www.sciencedirect.com/science/article/pii/S089360801930 2503

15. Jiang, J., et al.: Multiple resolution residually connected feature streams for automatic lung tumor segmentation from CT images. IEEE Trans. Med. Imaging **38**(1), 134–144 (2019). https://doi.org/10.1109/TMI.2018.2857800

16. Jin, Y., et al.: Multi-task recurrent convolutional network with correlation loss for surgical video analysis. Med. Image Anal. **59**, 101572 (2020). https://doi.org/10.1016/j.media.2019. 101572. https://www.sciencedirect.com/science/article/pii/S1361841519301124

17. Marafioti, A., et al.: CataNet: predicting remaining cataract surgery duration. In: de Bruijne, M., et al. (eds.) MICCAI 2021. LNCS, vol. 12904, pp. 426–435. Springer, Cham (2021). https://doi.org/10.1007/978-3-030-87202-1_41

18. Ni, Z.L., et al.: Pyramid attention aggregation network for semantic segmentation of surgical instruments. In: Proceedings of the AAAI Conference on Artificial Intelligence, vol. 34, no. 07, pp. 11782–11790, April 2020. https://doi.org/10.1609/aaai.v34i07.6850. https://ojs.aaai. org/index.php/AAAI/article/view/6850

19. Ni, Z.L., et al.: BARNet: bilinear attention network with adaptive receptive fields for surgical instrument segmentation. In: Bessiere, C. (ed.) Proceedings of the Twenty-Ninth International Joint Conference on Artificial Intelligence, IJCAI-20, pp. 832–838. International Joint Conferences on Artificial Intelligence Organization, July 2020. https://doi.org/10.24963/ ijcai.2020/116. Main track

20. Pereira, S., Pinto, A., Amorim, J., Ribeiro, A., Alves, V., Silva, C.A.: Adaptive feature recombination and recalibration for semantic segmentation with fully convolutional networks. IEEE Trans. Med. Imaging **38**(12), 2914–2925 (2019). https://doi.org/10.1109/TMI.2019. 2918096

21. Ronneberger, O., Fischer, P., Brox, T.: U-net: convolutional networks for biomedical image segmentation. In: Navab, N., Hornegger, J., Wells, W.M., Frangi, A.F. (eds.) MICCAI 2015. LNCS, vol. 9351, pp. 234–241. Springer, Cham (2015). https://doi.org/10.1007/978-3-319- 24574-4_28

22. Roy, A.G., Navab, N., Wachinger, C.: Recalibrating fully convolutional networks with spatial and channel "squeeze and excitation" blocks. IEEE Trans. Med. Imaging **38**(2), 540–549 (2019). https://doi.org/10.1109/TMI.2018.2867261

23. Twinanda, A.P., Shehata, S., Mutter, D., Marescaux, J., de Mathelin, M., Padoy, N.: EndoNet: a deep architecture for recognition tasks on laparoscopic videos. IEEE Trans. Med. Imaging **36**(1), 86–97 (2017). https://doi.org/10.1109/TMI.2016.2593957

24. Wang, H., et al.: Score-CAM: score-weighted visual explanations for convolutional neural networks. In: 2020 IEEE/CVF Conference on Computer Vision and Pattern Recognition Workshops (CVPRW), pp. 111–119 (2020). https://doi.org/10.1109/CVPRW50498. 2020.00020

25. Zhang, M., Li, X., Xu, M., Li, Q.: Automated semantic segmentation of red blood cells for sickle cell disease. IEEE J. Biomed. Health Inform. **24**(11), 3095–3102 (2020). https://doi. org/10.1109/JBHI.2020.3000484

26. Zhou, Z., Siddiquee, M.M.R., Tajbakhsh, N., Liang, J.: UNet++: redesigning skip connections to exploit multiscale features in image segmentation. IEEE Trans. Med. Imaging **39**(6), 1856–1867 (2020). https://doi.org/10.1109/TMI.2019.2959609

Enhancing Dermoscopic Features Classification in Images Using Invariant Dataset Augmentation and Convolutional Neural Networks

Piotr Milczarski$^{(\boxtimes)}$ ⓘ, Michał Beczkowski ⓘ, and Norbert Borowski ⓘ

Faculty of Physics and Applied Informatics, University of Lodz,
Pomorska str. 149/153, 90-236 Lodz, Poland
{piotr.milczarski,michal.beczkowski,norbert.borowski}@uni.lodz.pl
https://www.wfis.uni.lodz.pl

Abstract. The Invariant Dataset Augmentation (IDA) method shows its advantages in image classification using CNN networks. In the paper, several pretrained neural networks have been used e.g. Inception-ResNet-v2, VGG19, Xception, etc., trained on the PH2 and Derm7pt dermoscopic datasets in different scenarios. One of them relies on the Invariant Dataset Augmentation not only for training and but also for validation and testing. That original research and method show that the classification characteristics e.g. values of the weighted accuracy and sensitivity (true positive rate), and precision (positive predictive rate) tests F1 and MCC are much higher. That general approach provides better results with higher classification results e.g. sensitivity and precision (positive predictive rate) and can be used in other disciplines. In the paper, the results are shown in the research of the assessment of the skin lesions on two distinct dermoscopic datasets, PH2 and Derm7pt, on the classification of the presence or its absence of the feature called blue-white veil within the lesion.

Keywords: Invariant Dataset Augmentation · Data augmentation · Dermoscopic images · Classification of skin lesions features · Blue-white veil · Pretrained CNN

1 Introduction

According to the European Cancer Information System [1] and the American Cancer Society [2] statistics melanoma, as well as skin diseases, are very high on the mortality lists due to cancer. The defined Invariant Dataset Augmentation [3][4] used in the dermoscopic images classification allow to obtain much higher values of the classification characteristics e.g. values of the weighted accuracy and sensitivity (true positive rate), and precision (positive predictive rate) tests F1 and MCC, in the case when one expands the original image dataset. While examing the lesions in the direction of the search for the blue-white veil feature,

© Springer Nature Switzerland AG 2021
T. Mantoro et al. (Eds.): ICONIP 2021, LNCS 13110, pp. 403–417, 2021.
https://doi.org/10.1007/978-3-030-92238-2_34

the image taken by the dermatology specialist might be taken from different angles. Nonetheless, if the feature is present in reality it must be present on each of the original images because the conditions e.g. light, time exposure, etc. are the same in the dermoscopic scan.

In the paper, several neural networks have been built based on the pre-trained CNN networks such as Inception-ResNet-v2, VGG19, Xception, etc. Then, they have been trained on specially prepared images provided by the PH2 [5] and Derm7pt [6,7] datasets with image augmentation described by the method named the Invariant Dataset Augmentation. The motivation of the research is to enhance screening methods by raising the classification probability of the positive features that point to the possible health problems. The original author's approach, i.e. using Invariant Dataset Augmentation (IDA) of the images in much higher values of the classification results if we take into account classification characteristics like weighted accuracy and true positive rate as well as the F1 test and Matthews correlation coefficient (MCC) than based only on the original images and their random augmentation rather than targeted one as in IDA case. To show the advantages of the IDA method we have prepared the images from two distinct datasets. The Derm7pt is provided by Argenziano et al. [6] and used in Kawahara et al. in [7]. The PH2 dataset is provided by Mendonca et al. [5]. The results vary for different CCN networks but they can be used in a support system for the general practitioners [15–17].

The CNN networks used in the research usually classify the images from the PH2 of the lesions quite well if we take into account the accuracy. But the true positive rate (the blue-white veil is present) was rather low i.e. 65–80% in comparison to the true negative rate (the blue-white veil is absent) that varied from 95 to 98% [24,25]. The reason for that is the fact that there are only 18–20% of the images pointed out as the blue-white veil (is present) in the PH2 and Derm7pt datasets.

As a dermoscopic feature, the blue-white veil or blue-whitish veil (BWV) is used in dermoscopic methodologies. It has been examined using two distinct machine learning methods i.e. the feature- and appearance-based ones. In appearance-based ones, some CNN networks have been used to classify the image dataset [7,22,23]. Kawahara et al. [7] show in their research that the InceptionV3 pretrained network provided them with 87.6% of unbalanced and 85.8% balanced accuracy, more than 96.5% sensitivity, but only with 49% specificity and 51% false positive rate. For that research, the values of precision were 89% and 0.87 AUC ROC. The research was done for the blue-white veil classification of the dermoscopic images derived from the Derm7pt [7,22].

In the second machine learning approach using feature-based methods [18–20,24–27], there is a problem with how to extract the BWV as a feature. In one of the research [25], the authors also used Derm7pt to train and test on a subset of 179 its images. Their research resulted in an accuracy of around 85% with help of the C4.5 method.

In our previous paper, we tested some pretrained networks (e.g. VGG19) and the results showed more than 85% accuracy, reaching even 98% [3]. In the current paper, we do not limit the research to only one CNN network and one dataset. We have taken into account Inception-ResNet-v2, VGG19, and Xception pretrained CNNs, trained on the set of images provided by the PH2 and Derm7pt datasets with the help of the authors' method called the Invariant Dataset Augmentation (IDA) [3]. The IDA is used not only for data augmentation but also as a proliferation of validation and test datasets. That results in the eightfold classification of the same image as a set of bytes but different due to its topology. The achieved in the research average accuracy is around 94 % but the balanced accuracy is around 90% using the PH2 dataset, while the sensitivity is even 20% higher than in a standard approach and reached even 98–100%.

The paper is organized as follows. In Sect. 2 the Invariant Dataset Augmentation (IDA) is shown. The blue-white veil and its dermatological importance and screening methods are discussed in Sect. 3. The chosen pretrained convolutional neural networks and some of their parameters are shown in Sect. 4. The research and its results are presented in Sect. 4. General method description using CNN classification with the invariant dataset augmentation is described in detail in the next section. Then, the final results are provided in Sect. 6. Finally, Sect. 7 shows the conclusions.

2   Importance of the Invariant Dataset Augmentation

The Invariant Dataset Augmentation [3] and its foundation are based on the idea of how to make the available image dataset bigger as much as possible not changing the data within (i.e. pixels) and its order and relation to other data saved in pixels. The conclusion was to transform the images with the help of the geometrical invariant and reversible transformations. These transformations will not have any impact on the features of the skin lesions. In the Three-Point Checklist of Dermoscopy, the asymmetry of shape, hue, and presence of some structures and their distributions are taken into account in the lesion assessment. While data augmentation authors often use a rotation of scaling the image, but while looking for very fragile features in the images, these features can vanish after that procedure e.g. blue-whitish veil.

After some research, we choose seven geometrical operations: rotation by 90°, 180° and 270°, mirror reflection by a vertical and horizontal axis of the images and their rotations by 90°. Figure 1 shows the original image IMD168, taken from PH2 [5] dataset and its invariant geometrical transformations described above. Figure 2 presents the same operations for the original image FAL094.jpg from the Derm7pt dataset. Altogether, we achieve eight images from one initial image not changing a single pixel in them and their relation towards other pixels. The method can be used in a case where the angle of the image acquisition is random as in dermoscopy, where the doctor can take an image of the same lesion from a different angle.

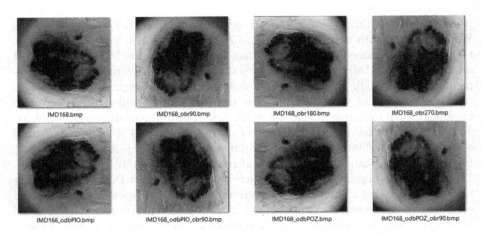

Fig. 1. The image (top-left corner) IMD168 and its invariant copies provided by the PH2 dataset [5].

3   Dermoscopic Screening Methodologies and Image Datasets

In dermoscopy, the Three-Point Checklist of Dermoscopy (3PCLD) [9–11,13,14] is defined and regarded as a sufficient screening method. The dermatology experts proved it is a sufficient screening method of skin lesions assessments. There are three criteria in the 3PLCD method: asymmetry of shape, hue, and structure distribution within the lesion with discrete value either 0, 1, or 2; blue-white structures and atypical pigment network. The Seven-Point Check-list (7PCL) defined by Soyer et al. in [8] is another screening method used in skin lesions assessment. In the 7PCL case, dermatology experts examine pigment network, blue-white veil, streaks,   pigmentation, regression structures, dots and globules, and vascular structures as the important assessment factors. The ABCD and ABCDE rules [12] are examples of the screening methods that take into account the symmetry/asymmetry, border, color, diameter, and evolving of the lesion.

Several dermoscopic images datasets with skin lesions images are available for the researchers. In our research, we use Derm7pt [6,7] consisting of 1011 images with 195 with a blue-white veil feature. With IDA methodology we expanded it to 8088 and 1560 corresponding cases. The second image dataset used in our research is PH2 [5]. It consists of 200 images, but as in Derm7pt, there is a small number of blue-white veil cases i.e. 36. Using IDA we achieve 1600 images altogether with 288 having a blue-white veil. The next dermoscopic dataset is the ISIC dataset provided by the International Skin Imaging Collaboration initiative [21]. It contains more than 24000 images. In the ISIC image dataset, the description lacks whether there is present or not the blue-white veil as well as other features. That is why we are using Derm7pt and PH2 datasets only.

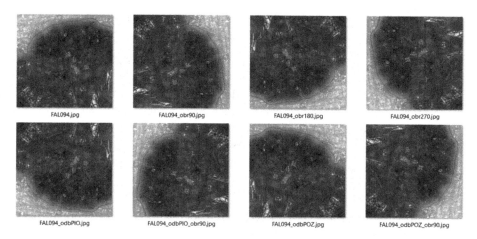

Fig. 2. Original image FAL094 (top-left corner) from the Derm7pt dataset [7] and its invariant copies.

There are also some printed versions with the images of the lesions e.g. [6] but the scans should not be used as a train set.

4 Available Pretrained CNN Networks and Their Parameters

The pretrained CCN networks differ from each other due to their architecture, the number of parameters, depth, image input size, speed, the hardware needed, time of execution, etc. These facts the researchers need to examine choosing a network as well as the initial problem they want to solve. Table 1. summarizes some of the CNN networks' feature. The input images to the networks are RGB images.

5 General Method Description

In the paper, we use the PH2 and Derm7pt datasets as the train, validation and test sets as two independent input image collections. A detailed description of the method can be found in the previous paper [3]. Let us summarize it in the following steps:

1. The preprocessing of the image set to fulfill the CNN networks requirements (see Table 1) for the image resolution:
   (a) Cropp the images into a square.
   (b) Scale the images to 299 × 299 px (see Table 1) results in 200 images for PH2 and 1011 ones from Derm7pt datasets scaled to 299 × 299 px or according to the demanded by the network respectively (see Tables 3 and 4).

Table 1. Pretrained CNNs and their chosen parameters

Network	Depth	Size [MB]	Parameters [millions]	Image input size	Average accuracy [%]
VGG19	19	535	144	224 × 224	~70
Xception	71	85	22.9	299 × 299	80<
Inception-ResNet-v2	164	209	55.9	299 × 299	~80
NASNet-Large	*	360	88.9	331 × 331	>80
InceptionV3	48	89	23.9	299 × 299	~77

(c) Apply the Invariant Dataset Augmentation, as the results, there are 1600 images for the PH2 dataset and 8088 for Derm7pt.

2. The setting up the convolutional neural network, e.g. VGG19, Xception, etc.
3. Network training for each dataset on:
   (a) The dataset from Step 2c of the procedure (eight copies).
   (b) The dataset from Step 2b (one copy).

   The datasets are divided into training sets with 65% of the images, validation sets out with 10% of the images, and 25% of the testing ones. The trained networks achieved in Steps 3.a and 3.b are saved.
4. Testing the networks.

   Steps 3 and 4 are conducted five times for each set of training, validation, and test sets. We achieve tests results with twenty confusion matrices (see Tables 5, 6, 7 and 8).
5. Calculating the confusion matrix parameters using Eqs. 1–7 and their average values, variance, minimum and maximum ones.

The whole procedure might be used to classify other features e.g. asymmetry [4]. If we take into account the dermoscopic methodologies or screening methods and implement them in the research we need to be sure that our operations on the images do not make the fragile features vanish. The Invariant Dataset Augmentation guarantees that the features will not vanish. It also provides us with an expanded eightfold image dataset in the case of dermoscopic features classification. The invariant transformations do not change that features.

Next point is that during validation and testing the researchers usually use only one original image. In our method, we use eight of them. Table 2 shows the results of the classification probability for the chosen exemplary image from the Derm7pt dataset and its seven copies. The original image and the copies were tested on the same network that was trained on the different image set for each of the CNN networks. But the train and validation sets for each network were the same. The results from the table underline our idea to classify not only original images but also its invariant copies. As we can see, the probability of the classification in all chosen networks varies from almost 0 to 0.86 (VGG19), 0.96 (Xception), and 0.88 (IRN2).

It can be explained by the properties of the convolution given by the example equation:

$$I(x,y) = \sum_{i=0}^{n} \sum_{j=0}^{m} k(i,j) I(x+i, \; y+j),$$

where the kernel of the operation is $k(I,j)$ is of the size $n x m$. The image size is $N x M$, where $N \geq n$ and $M \geq m$. The classification probability of the invariant images can vary from 0.0 to 1.0. That is why the IDA method can be successfully used in automatic screening methods based on CNN networks.

Table 2. The classification probability of the image Fbl040 (Derm7pt [7]).

Fbl040 and its version	CNN network		
	VGG19	XN	IRN2
Original	0.11258	0.00322	0.88311
Rotated by 90	0.00020	0.00107	0.07039
Rotated by 180	0.00004	0.08929	0.13821
Rotated by 270	0.00001	0.96155	0.02221
Mirrored vertically	0.85665	0.00014	0.00022
Mirrored vertically and rotated by 90	0.00000	0.22520	0.38297
Mirrored horizontally	0.00019	0.68363	0.81576
Mirrored horizontally & rotated by 90	0.00002	0.00724	0.00227

The results also mean that how the camera is positioned relative to the lesion has a great impact on the classification results. Rotating the camera above the lesion provides us with the same lesion. The same does for the Invariant Dataset Augmentation for a train set. From the other point of view, fragile features like the blue-whitish veil can be obscured by the image transformation as the rotation by a random angle. That is why using augmentation by random angle usually impacts the training process while classifying fragile features. The Invariant Dataset Augmentation is used not only in the training and validation phases but also in the test one. The confusion matrix parameters are defined according to the following formulas:

$$ACC = (TP + TN)/N \tag{1}$$

$$TPR = TP/(TP + FN) \tag{2}$$

$$FPR = FN/(FP + TN) = 1 - TNR \tag{3}$$

$$w. \; ACC = (TPR + TNR)/2 \tag{4}$$

$$Prec = TP/(TP + FP) \tag{5}$$

$$F1 = 2TP/(2TP + FP + FN) \tag{6}$$

$$MCC = \frac{(TP * TN - FP * FN)}{\sqrt{(TP + FP)(TP + FN)(TN + FP)(TN + FN)}} \tag{7}$$

where the number is given in the Eqs. 1–7 mean: N – the sum of all cases; TP, true positive – correctly classified items; TN, true negatives – correctly classified items of the opposite class; FP, false positives – wrongly classified items as positive ones; FN – false negative – wrongly classified cases as negative ones; ACC – accuracy; TPR also called Recall - true positive rate; FPR – false-positive rate; Prec. – precision; F1 score test; AUC ROC – the area under curve for the receiver operating curve; MCC – Matthews correlation coefficient.

To show the advantage of the Invariant Dataset Augmentation we have always compared the results of the methodology and trained, validated, and tested on the same image subsets. The BW1 is using the IDA images i.e. images and their copies. The BW2 uses only original images for training and validation. The same divisions of the images were made for the BW1 and BW2 approach (Table 4).

Table 3. The division of the PH2 dataset using IDA.

Number of images	PH2 dataset			IDA dataset		
	Total	Train	Test	Total	Train	Test
BWV present	36	27	9	288	216	72
BWV absent	164	123	41	1312	984	328
Total	200	150	50	1600	1200	400

Table 4. The division of the Derm7pt dataset using IDA.

Number of images	Derm7pt dataset			IDA dataset		
	Total	Train	Test	Total	Train	Test
BWV present	195	146	49	1560	1168	392
BWH absent	816	613	203	6528	4904	1624
Total	1011	759	252	8088	6072	2016

6   Results

The research has been conducted using Matlab 2019b with Deep Learning Toolbox^{TM} v. 12. With the that toolbox, one can perform transfer learning with pretrained CNN as shown in Table 1. The Microsoft Windows 10 Pro operating system was used on the computer with the configuration parameters are as follows: Processor: Intel® Core™, i7-8700K CPU, 3.70 GHz, RAM memory: 64 GB, Graphics Card: NVIDIA GTX 1080Ti with 11 GB of GDRAM.

The chosen confusion matrix factors are shown in Tables 5 and 6 for the BW1 and the BW2 approaches and the PH2 dataset and Tables 7 and 8 respectively for the Derm7pt dataset. It can be derived from Tables 5, 6, 7 and 8 that the balanced accuracy, the true positive rate (TPR) are higher in the BW2 approach and it is contrary to the accuracy. In the columns of Tables 5, 6, 7 and 8, the results in the columns with ID T1 and T8 refer to the results of the tests on a single original image (out of eight), on eight images, and results from the authors' IDA procedure.

The worst-case approach used in the IDA columns means that after the original image and its invariant copies are classified we choose the positive result if at least one of the classifications shows a positive result. Taking into account the Table 2 example result we could also define a measure that could show the probability of how many images out of eight have been classified as a positive one.

All the networks were trained with softmax function and showed the validation accuracy >90% and the validation loss close to 0.5. The initial research

Table 5. The classification results for the PH2 dataset using IDA for the BW1 approach.

CM factor	Name	VGG19			Xception			IRN2		
		T1	T8	IDA	T1	T8	IDA	T1	T8	IDA
ACC [%]	AVG	94.2	94.1	93.8	93.0	93.2	92.8	92.2	92.6	91.5
	VAR	2.8	2.2	2.6	1.3	1.4	1.6	1.9	1.3	1.9
	Min	86.0	89.0	88.0	92.0	91.5	90.0	88.0	89.5	86.0
	Max	98.0	97.8	98.0	96.0	96.5	96.0	96.0	94.5	96.0
w.ACC [%]	AVG	87.4	87.7	89.5	84.9	84.1	89.5	82.9	82.9	85.5
	MAX	94.4	94.3	98.8	92.0	91.3	93.2	88.9	88.0	93.2
TPR [%]	AVG	76.7	77.6	82.7	72.2	69.9	84.4	68.3	67.8	76.1
	VAR	12.1	9.6	10.23	7.5	7.9	6.5	8.1	9.3	8.8
	Min	55.6	62.5	66.7	55.6	58.3	66.7	44.4	47.2	66.7
	Max	88.9	88.9	100	88.9	83.3	88.9	77.8	79.2	88.9
FPR [%]	AVG	2.0	2.3	3.8	2.4	1.7	5.4	2.6	1.9	5.1
	VAR	2.3	1.8	1.6	1.7	1.0	1.7	1.4	1.4	3.2
	Min	0.0	0.3	2.4	0.0	0.3	2.4	0.0	0.3	2.4
	Max	7.3	5.2	7.3	4.9	3.4	7.3	7.3	5.8	12.2
Test	F1	0.94	0.93	0.95	0.84	0.90	0.89	0.88	0.81	0.89
	MCC	0.93	0.91	0.94	0.81	0.88	0.86	0.86	0.78	0.86

Table 6. The classification results for the PH2 dataset using IDA for the BW2 approach.

CM factor	Name	VGG19			Xception			IRN2		
		T1	T8	IDA	T1	T8	IDA	T1	T8	IDA
ACC [%]	AVG	93.2	93.6	94.0	92.1	91.9	92.1	91.0	90.8	90.3
	VAR	2.9	3.0	3.0	1.5	0.7	2.9	3.0	1.2	1.9
	Min	88.0	89.6	90.0	88.0	90.3	86.0	86.0	85.5	88.0
	Max	98.0	98.5	100	94.0	93.0	98.0	96.0	93.3	96.0
w.ACC [%]	AVG	87.2	88.1	92.2	82.2	80.2	88.0	79.8	78.2	83.9
	Max	98.8	97.5	100	87.7	85.3	94.4	88.9	86.1	93.2
TPR [%]	AVG	77.8	79.6	89.4	66.7	61.8	81.7	62.2	58.4	73.9
	VAR	11.7	10.2	8.2	7.9	7.1	9.5	13.3	16.1	10.1
	Min	55.6	66.7	77.8	55.6	51.4	66.7	33.3	23.6	55.6
	Max	100	97.2	100	77.8	73.6	88.9	77.8	76.4	88.9
FPR [%]	AVG	3.4	3.3	5.0	2.3	1.5	5.6	2.7	2.1	6.1
	VAR	2.2	2.0	2.2	1.6	1.1	3.9	1.7	1.3	2.7
	Min	0.0	0.0	0.0	0.0	0.0	0.0	0.0	0.6	2.4
	Max	7.3	7.0	7.3	4.9	3.0	14.6	4.9	5.2	12.2
Test	F1	0.95	0.96	1.0	0.82	0.79	0.94	0.88	0.80	0.89
	MCC	0.94	0.95	1.0	0.79	0.74	0.94	0.86	0.76	0.86

Table 7. The classification results for the Derm7pt dataset using IDA for the BW1 approach.

CM factor	Name	VGG19			Xception			IRN2		
		T1	T8	IDA	T1	T8	IDA	T1	T8	IDA
ACC [%]	AVG	87.4	87.9	86.9	88.0	87.9	85.7	86.1	86.1	84.0
	VAR	2.8	2.4	2.5	2.2	1.6	1.8	1.0	1.1	1.3
	Min	83.3	84.3	83.3	84.9	85.0	82.9	85.3	84.5	82.1
	Max	92.9	92.5	91.7	92.1	90.3	88.9	87.7	87.4	85.7
w. ACC [%]	AVG	78.3	79.1	82.2	78.4	78.1	82.8	74.7	75.2	80.8
	MAX	87.8	86.1	88.6	85.8	82.9	89.2	76.1	78.2	83.4
TPR [%]	AVG	63.6	64.7	74.4	62.9	62.1	78.1	56.1	57.3	75.5
	VAR	6.8	5.2	5.5	7.7	6.9	8.3	2.3	3.8	4.6
	Min	53.1	57.7	65.3	53.1	50.0	65.3	53.1	53.6	69.4
	Max	79.6	75.8	83.7	75.5	71.2	89.8	59.2	63.3	81.6
FPR [%]	AVG	6.9	6.5	10.0	6.0	5.9	12.5	6.7	6.9	13.9
	VAR	2.4	2.3	2.4	1.6	0.8	2.0	1.0	1.3	2.0
	Min	3.0	2.8	5.9	3.9	4.6	9.4	4.9	5.4	11.3
	Max	11.8	11.4	14.8	8.9	8.0	16.7	7.4	9.1	16.7
Test	F1	0.81	0.80	0.80	0.79	0.73	0.76	0.62	0.66	0.67
	MCC	0.77	0.75	0.75	0.74	0.70	0.70	0.54	0.59	0.59

Table 8. The classification results for the Derm7pt dataset using IDA for the BW2 approach.

CM factor	Name	VGG19			Xception			IRN2		
		T1	T8	IDA	T1	T8	IDA	T1	T8	IDA
ACC [%]	AVG	88.0	87.9	85.9	87.2	86.8	82.9	87.2	84.3	80.6
	VAR	2.1	1.8	2.4	2.3	1.3	2.4	1.4	0.9	1.6
	Min	83.3	85.0	81.0	83.7	85.1	78.6	82.1	82.1	77.8
	Max	92.1	90.8	89.7	90.9	88.8	88.5	87.7	85.6	83.3
w.ACC [%]	AVG	79.9	79.4	83.9	76.9	76.5	80.6	68.1	71.8	77.3
	MAX	84.2	84.9	89.7	81.2	78.8	87.4	76.9	74.1	80.9
TPR [%]	AVG	66.5	65.5	80.5	60.1	59.5	76.8	50.4	51.3	71.8
	VAR	6.0	5.7	6.4	5.2	3.9	4.7	4.6	2.7	4.6
	Min	44.9	47.7	61.2	51.0	51.3	71.4	42.9	46.2	65.3
	Max	73.5	76.5	89.8	67.3	64.0	85.7	59.2	55.9	77.6
FPR [%]	AVG	6.8	6.7	12.8	6.3	6.6	15.7	7.8	7.7	17.2
	VAR	2.6	2.0	2.9	2.4	1.3	2.3	1.5	1.1	1.8
	Min	2.5	3.1	8.9	2.5	3.9	10.8	5.4	6.3	13.8
	Max	13.3	11.1	20.2	9.4	8.1	20.2	11.3	11.0	22.2
Test	F1	0.78	0.75	0.77	0.74	0.69	0.74	0.65	0.59	0.64
	MCC	0.73	0.69	0.72	0.69	0.62	0.68	0.58	0.51	0.55

to trim the training parameters was made with: more than thirty epochs in the BW1 cases for all networks; more than sixty epochs for InceptionResNetv2 networks; from sixty to hundred and more epochs for Xception and VGG19 in the BW2 cases.

The validation accuracy, the loss, and the confusion matrix parameters have not changed while increasing the number of epochs. The learning rate was examined with the values varying from 5e−5 to 1e−2.

The time of training for the BW1 approach has varied from around twelve minutes (VGG19), forty minutes (Xception), and sixty minutes (IRN2) for the PH2 dataset; from eighty to a hundred twenty minutes for VGG19, around a hundred forty minutes for XN, and 300 min for IRN2 for the Derm7pt dataset.

The times of the training of the CNNs in the BW2 case varied from 5 min (VGG19 with 60 epochs) to 14 min (XN with 100 epochs), and 21 min (IRN2 with 60 epochs) for the PH2 dataset. For the Derm7pt dataset, they were correspondingly around 4–5 times higher.

For the BW1 and BW2, the same confusion matrix parameters are used for the chosen CNN network and they are calculated according to Eqs. 1–7. Tables 5, 6, 7 and 8 columns and rows are described as follows: T1 - test results for an original image set (T1); T8 – the results for eight copies treated as a separate file; IDA classification. The other confusion matrix parameters are noted as

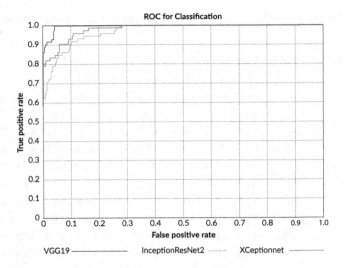

Fig. 3. The AUC ROC for the three chosen CNNs for the PH2 dataset

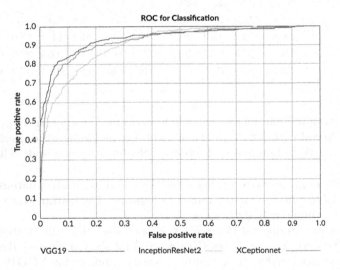

Fig. 4. The AUC ROC for the three chosen CNNs for the Derm7pt dataset

in the description of the Eqs. 1–7. If we take into account the dermoscopic methodologies or screening methods and implement them in the research we need to be sure that our dataset operations do not make the fragile features vanish.

The Invariant Dataset Augmentation guarantees that the features will not vanish. It also provides us with an expanded eightfold image dataset in the case of dermoscopic features classification. The invariant transformations do not change that features. What is more, we obtain seven new copies out of one each, altogether eight images. Next point is that during validation and testing the researchers usually use only one original image. In our method, we use eight of them.

Figures 3 and 4 show the values of ROC with the highest value of AUC area for the best CNNs for the PH2 dataset and Derm7pt. For the BW1 approach, the AUC ROC achieved for BW1 the average values 0.966, 0.969, and 0.957, as well as the maximum values 0.996 for VGG19, 0.985 for Xception, and 0.971 for InceptionResNetv2. For the BW2 approach, both values of AUC for each network were even higher reaching 1.0 or 0.999 as its maximum value in the case of T8.

7 Conclusions

In the research, three CNN networks, Inception-ResNet-v2, VGG19, and Xception, were used to classify the dermoscopic images from PH2 and Derm7pt datasets in search of a blue-white veil. The results achieved in the research show that the classification parameters are higher for the proposed method. The weighted accuracy was above 90 reaching even 94% percent for the IDA approach that can be seen in Tables 5, 6, 7 and 8. Table 2 shows the classification probabilities of one chosen image and its invariant copies (the image Fbl040 Derm7pt dataset [7]) by the CNN networks.

The test F1 and MCC test have higher values from 5 to 20% than using only original images. In the paper, the VGG19 network reached 98.8% of the weighted accuracy, 100% of the true positive rate, 2.3% false positive rate, the test F1 equal to 0.95, MCC test equal to 0.95, and AUC close to 1.0.

References

1. European Cancer Information System (ECIS). https://ecis.jrc.ec.europa.eu. Accessed 14 Jul 2021
2. ACS - American Cancer Society. https://www.cancer.org/research/cancer-facts-statistics.html. Accessed 14 Jul 2021
3. Milczarski, P., Beczkowski, M., Borowski, N.: Blue-white veil classification of dermoscopy images using convolutional neural networks and invariant dataset augmentation. In: Barolli, L., Woungang, I., Enokido, T. (eds.) AINA 2021. LNNS, vol. 226, pp. 421–432. Springer, Cham (2021). https://doi.org/10.1007/978-3-030-75075-6_34
4. Beczkowski, M., Borowski, N., Milczarski, P.: Classification of dermatological asymmetry of the skin lesions using pretrained convolutional neural networks. In: Rutkowski, L., Scherer, R., Korytkowski, M., Pedrycz, W., Tadeusiewicz, R., Zurada, J.M. (eds.) ICAISC 2021. LNCS (LNAI), vol. 12855, pp. 3–14. Springer, Cham (2021). https://doi.org/10.1007/978-3-030-87897-9_1
5. Mendoncca, T., Ferreira, P.M., Marques, J.S., Marcal, A.R.S., Rozeira, J.: PH2 - a dermoscopic image database for research and benchmarking. In: 35th Annual International Conference of the IEEE Engineering in Medicine and Biology Society (EMBC), Osaka, pp. 5437–5440 (2013)

6. Argenziano, G., Soyer, H.P., De Giorgi, V., et al.: Interactive Atlas of Dermoscopy. EDRA Medical Publishing & New Media, Milan (2002)
7. Kawahara, J., Daneshvar, S., Argenziano, G., Hamarneh, G.: Seven-point checklist and skin lesion classification using multitask multimodal neural nets. IEEE J. Biomed. Health Inform. **23**(2), 538–546 (2019)
8. Soyer, H.P., Argenziano, G., Zalaudek, I., et al.: Three-point checklist of dermoscopy. A new screening method for early detection of melanoma. Dermatology **208**(1), 27–31 (2004)
9. Argenziano, G., Soyer, H.P., et al.: Dermoscopy of pigmented skin lesions: results of a consensus meeting via the Internet. J. Am. Acad. Dermatol. **48**(9), 679–693 (2003)
10. Milczarski, P.: Symmetry of hue distribution in the images. In: Rutkowski, L., Scherer, R., Korytkowski, M., Pedrycz, W., Tadeusiewicz, R., Zurada, J.M. (eds.) ICAISC 2018. LNCS (LNAI), vol. 10842, pp. 48–61. Springer, Cham (2018). https://doi.org/10.1007/978-3-319-91262-2_5
11. Argenziano, G., Fabbrocini, G., et al.: Epiluminescence microscopy for the diagnosis of doubtful melanocytic skin lesions. Comparison of the ABCD rule of dermatoscopy and a new 7-point checklist based on pattern analysis. Arch. Dermatol. **134**, 1563–1570 (1998)
12. Nachbar, F., Stolz, W., Merkle, T., et al.: The ABCD rule of dermatoscopy. High prospective value in the diagnosis of doubtful melanocytic skin lesions. J. Am. Acad. Dermatol. **30**(4), 551–559 (1994)
13. Milczarski, P., Stawska, Z., Maslanka, P.: Skin lesions dermatological shape asymmetry measures. In: Proceedings of the IEEE 9th International Conference on Intelligent Data Acquisition and Advanced Computing Systems: Technology and Applications, IDAACS, pp. 1056–1062 (2017)
14. Menzies, S.W., Zalaudek, I.: Why perform dermoscopy? The evidence for its role in the routine management of pigmented skin lesions. Arch. Dermatol. **142**, 1211–1222 (2006)
15. Simonyan, K., Zisserman, A.: Very deep convolutional networks for large-scale image recognition. In: Conference Track Proceedings of 3rd International Conference on Learning Representations (ICRL), San Diego, USA, (2015)
16. Was, L., Milczarski, P., Stawska, Z., Wiak, S., Maslanka, P., Kot, M.: Verification of results in the acquiring knowledge process based on IBL methodology. In: Rutkowski, L., Scherer, R., Korytkowski, M., Pedrycz, W., Tadeusiewicz, R., Zurada, J.M. (eds.) ICAISC 2018. LNCS (LNAI), vol. 10841, pp. 750–760. Springer, Cham (2018). https://doi.org/10.1007/978-3-319-91253-0_69
17. Celebi, M.E., Kingravi, H.A., Uddin, B.: A methodological approach to the classification of dermoscopy images. Comput. Med. Imaging Graph. **31**(6), 362–373 (2007)
18. Was, L., et al.: Analysis of dermatoses using segmentation and color hue in reference to skin lesions. In: Rutkowski, L., Korytkowski, M., Scherer, R., Tadeusiewicz, R., Zadeh, L.A., Zurada, J.M. (eds.) ICAISC 2017. LNCS (LNAI), vol. 10245, pp. 677–689. Springer, Cham (2017). https://doi.org/10.1007/978-3-319-59063-9_61
19. Milczarski, P., Stawska, Z., Was, L., Wiak, S., Kot, M.: New dermatological asymmetry measure of skin lesions. Int. J. Neural Netw. Adv. Appl. **4**, 32–38 (2017)
20. Milczarski, P., Stawska, Z.: Classification of skin lesions shape asymmetry using machine learning methods. In: Barolli, L., Amato, F., Moscato, F., Enokido, T., Takizawa, M. (eds.) WAINA 2020. AISC, vol. 1150, pp. 1274–1286. Springer, Cham (2020). https://doi.org/10.1007/978-3-030-44038-1_116

21. The International Skin Imaging Collaboration: Melanoma Project. http://isdis. net/isic-project/. Accessed 14 Jul 2021

22. Esteva, A., Kuprel, B., Novoa, R.A., et al.: Dermatologist-level classification of skin cancer with deep neural networks. Nature **542**, 115–118 (2017)

23. He, K., Zhang, X., Ren S. and Sun, J.: Deep residual learning for image recognition. In: Proceedings of IEEE Conference on Computer Vision and Pattern Recognition, pp. 770–778 (2016)

24. Madooei, A., Drew, M.S., Sadeghi, M., Atkins, M.S.: Automatic detection of blue-white veil by discrete colour matching in dermoscopy images. In: Mori, K., Sakuma, I., Sato, Y., Barillot, C., Navab, N. (eds.) MICCAI 2013. LNCS, vol. 8151, pp. 453–460. Springer, Heidelberg (2013). https://doi.org/10.1007/978-3-642-40760-4_57

25. Jaworek-Korjakowska, J., Kłeczek, P., Grzegorzek, M., Shirahama, K.: Automatic detection of blue-whitish veil as the primary dermoscopic feature. In: Rutkowski, L., Korytkowski, M., Scherer, R., Tadeusiewicz, R., Zadeh, L.A., Zurada, J.M. (eds.) ICAISC 2017. LNCS (LNAI and LNB), vol. 10245, pp. 649–657. Springer, Cham (2017). https://doi.org/10.1007/978-3-319-59063-9_58

26. Celebi, M.E., et al.: Automatic detection of blue-white veil and related structures in dermoscopy images. CMIG **32**(8), 670–677 (2008)

27. Di Leo, G., Fabbrocini, G., Paolillo, A., Rescigno, O., Sommella, P.: Toward an automatic diagnosis system for skin lesions: estimation of blue-whitish veil and regression structures. In: International Multi-Conference on Systems, Signals & Devices, SSD 2009 (2009)

Ensembles of Randomized Neural Networks for Pattern-Based Time Series Forecasting

Grzegorz Dudek$^{(\boxtimes)}$ (ID) and Paweł Pełka (ID)

Electrical Engineering Faculty, Częstochowa University of Technology,
Częstochowa, Poland
{grzegorz.dudek,pawel.pelka}@pcz.pl

Abstract. In this work, we propose an ensemble forecasting approach based on randomized neural networks. Improved randomized learning streamlines the fitting abilities of individual learners by generating network parameters adjusted to the target function complexity. A pattern-based time series representation makes the ensemble model suitable for forecasting problems with multiple seasonal components. We develop and verify forecasting models with six strategies for controlling the diversity of ensemble members. Case studies performed on several real-world forecasting problems verified the superior performance of the proposed ensemble forecasting approach. It outperformed both statistical and machine learning models in terms of forecasting accuracy. The proposed approach has several advantages: fast and easy training, simple architecture, ease of implementation, high accuracy and the ability to deal with nonstationarity and multiple seasonality.

Keywords: Ensemble forecasting · Pattern representation of time series · Randomized neural networks · Short-term load forecasting · Time series forecasting

1 Introduction

Time series (TS) forecasting plays an important role in many fields, including commerce, industry, public administration, politics, health, medicine, etc. [1]. TS describing various processes and phenomena can express the nonlinear trend, multiple seasonal patterns of different lengths and random fluctuations. This makes the relationship between predictors and output variables very complex and places high demands on forecasting models. In recent years, many advanced models for this challenging problem have been developed. They include statistical models and more sophisticated machine learning (ML) and hybrid solutions.

Among ML models, neural networks (NNs) are the most commonly used [2]. In addition to classic NNs such as multilayer perceptron (MLP), generalized

---

Supported by Grant 2017/27/B/ST6/01804 from the National Science Centre, Poland.

T. Mantoro et al. (Eds.): ICONIP 2021, LNCS 13110, pp. 418–430, 2021.
https://doi.org/10.1007/978-3-030-92238-2_35

regression NN, radial basis function NN, and self-organizing Kohonen map [3], many models based on deep learning (DL) have been developed recently. These use basic building blocks, such as MLPs, convolutional NNs, and recurrent NNs (RNNs) [4,5]. Their success was achieved due to the increased complexity and ability to perform representation learning and cross-learning.

An effective way to increase the predictive power of the forecasting model is ensembling. It combines multiple learning algorithms in some way to improve both the accuracy and stability of the final response compared to a single algorithm. The challenge in ensemble learning, which to a large extent determines its success, is achieving a good tradeoff between the performance and the diversity of the ensemble members [6]. The more accurate and the more diverse the individual learners are, the better the ensemble performance. Depending on the base model type, the diversity of learners can be achieved using different strategies. For example, the winning submission to the M4 Makridakis forecasting competition, was a hybrid model combining exponential smoothing (ETS) and long-short term memory [7], which uses three sources of diversity. The first is a stochastic training process, the second is similar to bagging, and the third is using different initial parameters for individual learners. In another state-of-the-art DL forecasting model, N-Beats [8], in order to achieve diversity, each of the ensemble members is initialized randomly and is trained using a random batch sequence.

Randomization-based NNs are especially suitable for ensembling as they are highly unstable and extremely fast trained [9]. In randomized NNs, the hidden node parameters are selected randomly and the output weights are tuned using a simple and fast least-squares method. This avoids the difficulties associated with gradient-based learning such as slow convergence, local minima problems, and model uncertainties caused by the initial parameters. Ensemble methods based on randomized NNs were presented recently in several papers. In [10], an ensemble based on decorrelated random vector functional link (RVFL) networks was proposed using negative correlation learning to control the trade-off among the bias, variance and covariance in the ensemble learning. A selective ensemble of randomization-based NNs using successive projections algorithm (SPA) was put forward in [11]. SPA improves diversity by selecting uncorrelated members. In [12], to enhance the generalization capacities of the ensemble, a novel framework for constructing an ensemble model was proposed. It selected appropriate representatives from a set of randomization-based models such as RVFL or stochastic configuration networks.

In the forecasting domain, ensembles of randomization-based NNs are widely used. Examples include: [13], where ensemble members, which are extreme learning machines, learn on TS decomposed by wavelet transform; [14], where a hybrid incremental learning model is proposed for ensemble forecasting, which uses empirical mode decomposition, discrete wavelet transform, and RVFL networks as base learners; [15], where a new bagging ensemble model is proposed for online data stream regression based on randomised NNs; [16], where a data-driven evolutionary ensemble forecasting model is proposed using two optimization objectives (error and diversity) and RVFL as a base learner.

Motivated by the good performance of the randomized NN based ensemble forecasting models mentioned above, and new improvements in randomized learning [17] as well as pattern-based TS representation suitable for TS with multiple seasonality [3], this study contributes to the development of forecasting models in the following ways:

1. A new ensemble forecasting approach using randomization-based NNs is presented. Enhanced randomized learning improves the fitting abilities of individual learners. The model uses pattern representation of TS to deal with nonstationarity and multiple seasonality.
2. Six strategies for generating ensemble diversity are proposed. Each strategy governs diversity in a different way.
3. An experimental study using four real-world datasets demonstrates the superior performance of the proposed ensemble forecasting approach when compared to statistical and machine learning forecasting models.

The rest of the work is organized as follows. Section 2 presents the base forecasting model, RANDNN. Section 3 describes strategies for generating ensemble diversity. The performance of the proposed ensemble forecasting model is evaluated in Sect. 4. Finally, Sect. 5 concludes the work.

2   Forecasting Model

Figure 1 shows RANDNN, a base forecasting model, which was designed for forecasting TS with multiple seasonal patterns [18]. It is used as an ensemble member. RANDNN combines three components: an encoder, a randomized feedforward NN (FNN) and a decoder.

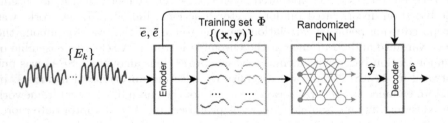

Fig. 1. Block diagram of the proposed forecasting model.

The encoder transforms TS $\{E_k\}_{k=1}^{K}$ into input and output patterns expressing seasonal sequences of the shortest length. These sequences are represented by vectors $\mathbf{e}_i = [E_{i,1}, E_{i,2}, \ldots, E_{i,n}]^T$, where n is a period of the seasonal cycle (e.g. 24 h for daily seasonality) and $i = 1, 2, \ldots, K/n$ is the sequence number. Input patterns $\mathbf{x}_i = [x_{i,1}, x_{i,2}, \ldots, x_{i,n}]^T$ are created from vectors $\mathbf{e}_i$ as follows:

$$\mathbf{x}_i = \frac{\mathbf{e}_i - \overline{\mathbf{e}}_i}{\widetilde{e}_i} \tag{1}$$

where $\bar{e}_i$ is the mean value of sequence $\mathbf{e}_i$, and $\widetilde{e}_i = \sqrt{\sum_{t=1}^{n}(E_{i,t} - \bar{e}_i)^2}$ is a measure of sequence $\mathbf{e}_i$ dispersion.

Input patterns $\mathbf{x}_i$ represent successive seasonal sequences which are centered and normalized. They have a zero mean, the same variance and unity length. Thus they are unified and differ only in shape (see Fig. 2 in [18]).

Forecasted seasonal sequences $\mathbf{e}_{i+\tau} = [E_{i+\tau,1}, E_{i+\tau,2}, \ldots, E_{i+\tau,n}]^T$, where $\tau \geq 1$ is a forecast horizon, are expressed by output patterns $\mathbf{y}_i = [y_{i,1}, y_{i,2}, \ldots, y_{i,n}]^T$ defined as follows:

$$\mathbf{y}_i = \frac{\mathbf{e}_{i+\tau} - \bar{e}_i}{\widetilde{e}_i} \tag{2}$$

Note that in (2), we use the same codding variables, $\bar{e}_i$ and $\widetilde{e}_i$, as for input patterns (1). This is because the coding variables for the forecasted sequence, $\bar{e}_{i+\tau}$ and $\widetilde{e}_{i+\tau}$, which should be used in (2), are unknown.

The randomized FNN learns to map the input patterns to output ones. The forecasted output pattern is converted into a TS seasonal sequence by the decoder. To do so, the decoder uses transformed Eq. (2) with the coding variables for the input query pattern:

$$\widehat{\mathbf{e}} = \widehat{\mathbf{y}}\widetilde{e} + \bar{e} \tag{3}$$

where $\widehat{\mathbf{e}}$ is the forecast of the seasonal sequence, $\widehat{\mathbf{y}}$ is the forecast of the output pattern, $\widetilde{e}$ and $\bar{e}$ are the coding variables determined from the TS sequence encoded in query pattern $\mathbf{x}$.

The randomized FNN is a single hidden layer network with m logistic sigmoid nodes. Output nodes are linear. The number of inputs and outputs correspond to the input and output pattern size (n in our case). The network learns on the training set composed of the paired x- and y-patterns: $\Phi = \{(\mathbf{x}_i, \mathbf{y}_i)|\mathbf{x}_i, \mathbf{y}_i \in \mathbb{R}^n, i = 1, 2, \ldots, N\}$. An improved randomized learning algorithm which we use in this study (see [17]) consists of the following three steps:

1. Generate hidden node parameters. The weights are selected randomly: $a_{t,j} = \sim U(-u, u)$ and biases are calculated from:

$$b_j = -\mathbf{a}_j^T \mathbf{x}_j^* \tag{4}$$

   where $j = 1, 2, \ldots, m$; $t = 1, 2, \ldots, n$; $\mathbf{x}_j^*$ is the x-pattern selected from the training set randomly.
2. Calculate the output matrix of the hidden layer $\mathbf{H}$.
3. Calculate the output weight matrix:

$$\beta = \mathbf{H}^+ \mathbf{Y} \tag{5}$$

   where $\beta \in \mathbb{R}^{m \times n}$, $\mathbf{Y} \in \mathbb{R}^{N \times n}$ is a matrix of target output patterns, and $\mathbf{H}^+ \in \mathbb{R}^{m \times N}$ is the Moore-Penrose generalized inverse of matrix $\mathbf{H}$.

In the first step, the weights are selected randomly from symmetrical interval $[-u, u]$. The interval bounds, u, decide about the steepness of the sigmoids. To make the bounds interpretable, let us express them by the sigmoid slope angle [17]: $u = 4 \tan \alpha_{max}$, where α_{max} is the upper bound for slope angles. Hyperparameter α_{max} as well as number of hidden nodes m control the bias-variance tradeoff of the model. Both these hyperparameters should be adjusted to the target function complexity.

The biases calculated according to (4) ensure the introduction of the steepest fragments of the sigmoid into the input hypercube [17]. These fragments are most useful for modeling target function fluctuations. Thus, there are no saturated sigmoids and wasted nodes in the network.

3   Ensembling

An ensemble is composed of M individual learners (RANDNNs). Each ensemble member learns from the training set Φ using one of the strategies for generating diversity, ENS1 − ENS6 described below. The ensemble prediction is an average of individual member predictions $\widehat{\mathbf{e}}_k$:

$$\widehat{\mathbf{e}}_{\mathrm{ens}} = \frac{1}{M} \sum_{k=1}^{M} \widehat{\mathbf{e}}_k \tag{6}$$

As an ensemble diversity measure, we define the average standard deviation of forecasts produced by the individual learners:

$$Diversity = \frac{1}{n|\Psi|} \sum_{i \in \Psi} \sum_{t=1}^{n} \sqrt{\frac{1}{M} \sum_{k=1}^{M} (\widehat{E}_{i,t}^k - \overline{\widehat{E}}_{i,t})^2} \tag{7}$$

where Ψ is a test set, $\widehat{E}_{i,t}^k$ is a forecast of the t-th element of the i-th seasonal sequence produced by the k-th learner, and $\overline{\widehat{E}}_{i,t}$ is an average of forecasts produced by M learners.

An ensemble diversity is generated using one of the following six strategies:

ENS1 generates diversity by using different parameters of hidden nodes. For each learner, new weights are randomly selected taking α_{max} as the upper bound for the sigmoid slope angles. Then, biases are calculated from (4) and output weights form (5). The diversity level is controlled by α_{max}. For larger α_{max} we get steeper sigmoids and higher diversity.

ENS2 controls diversity by training individual learners on different subsets of the training set. For each ensemble member, a random sample from the training set is selected without replacement. The sample size is $N' = \eta N$, where $\eta \in (0,1)$ is a diversity parameter. Each learner has the same hidden node parameters. Its output weights are tuned to the training subset.

ENS3 controls diversity by training individual learners on different subsets of features. For each ensemble member the features are randomly sampled without replacement. The sample size is $n' = \kappa n$, where $\kappa \in (0,1)$ is a diversity parameter. The ensemble members share the hidden node parameters. Their output weights are tuned to the training set.

ENS4 is based on hidden node pruning. The learners are created from the initial RANDNN architecture including m hidden nodes. For each learner, $m' = \rho m$ nodes are randomly selected and the remaining are pruned. Output weights β are determined anew for each learner. Parameter $\rho \in (0,1)$ controls the diversity level.

ENS5 is based on hidden weight pruning. The learners are created from the initial RANDNN architecture including m hidden nodes. For each learner, $p = \lambda mn$ hidden node weights are randomly selected and set to zero. Output weights β are determined anew for each learner. Parameter $\lambda \in (0,1)$ controls the diversity level.

ENS6 generates diversity by noising training data. For each learner the training patterns are perturbed by Gaussian noise as follows: $x_{i,t} = x_{i,t}(1 + \zeta_{i,t}), y_{i,t} = y_{i,t}(1 + \xi_{i,t})$, where $\zeta_{i,t}, \xi_{i,t} \sim N(0,\sigma)$. Standard deviation of the noise, σ, is a diversity parameter. Each learner has the same hidden node parameters. Its output weights are tuned to the noised training data.

4   Experiments and Results

In this section, the performance of the proposed ensembles of RANDNNs is evaluated on four real-world forecasting problems. These concern short-term load forecasting for Poland (PL), Great Britain (GB), France (FR) and Germany (DE) (data was collected from www.entsoe.eu). The hourly electrical load TS express yearly, weekly and daily seasonalities (see Fig. 2 in [18]). The data period is from 2012 to 2015 (4 years). We forecast the daily load sequence ($n = 24$ h) with horizon $\tau = 1$ for each day of 2015 excluding atypical days such as public holidays (about 10–20 days a year depending on the data set). For each forecasted day of 2015, a training set is built anew. It contains historical paired input and output patterns representing the same days of the week as the query pattern and forecasted pattern. The number of ensemble members was $M = 100$.

In the first experiment, to assess the impact of the RANDNNs hyperparameters on the forecasting accuracy of ENS1, we train the RANDNN members with $m = 10, 20, ..., 70$ and $\alpha_{\max} = 0°, 10°, ..., 90°$. Figure 2 shows the mean absolute percentage errors (MAPE) depending on the hyperparameters. The optimal values of $m/\alpha_{\max}$ were: $50/70°$ for PL, $30/60°$ for GB, $40/70°$ for FR, and $40/80°$ for DE. Based on these results, we select $m = 40$ and $\alpha_{\max} = 70°$ for each ensemble variant and dataset, except ENS1, where we control diversity by changing $\alpha_{\max}$, and ENS4, where we control diversity by changing the number of pruning nodes. In this latter case we use $m = 80$ nodes.

From Fig. 3, we can assess the impact of the hyperparameters on the diversity (7) of ENS1. As expected, greater numbers of hidden nodes and steeper sigmoids (higher $\alpha_{\max}$) cause an increase in diversity.

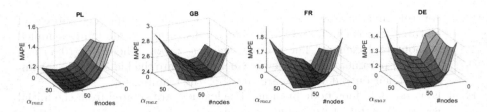

Fig. 2. MAPE depending on RANDNN hyperparameters for ENS1.

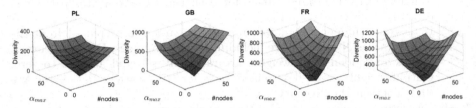

Fig. 3. Diversity of ENS1 depending on RANDNN hyperparameters.

Figure 4 shows MAPE and ensemble diversity depending on the diversity parameters for all ensemble variants and datasets. From this figure, we can conclude that:

- for ENS1 diversity increases with $\alpha_{\max}$. The optimal $\alpha_{\max}$ is around $70° - 80°$ for all datasets.
- for ENS2 diversity decreases with η. For lower η (small training sets), the diversity as well as MAPE are very high. The MAPE curves have an irregular character, so it is difficult to select the optimal η value. The error levels are much higher than for ENS1.
- for ENS3 diversity decreases with κ. MAPE reaches its minima for: $\kappa \in (8/24, 16/24)$ for PL, $\kappa = 20/24$ for GB, $\kappa \in (14/24, 18/24)$ for FR, and $\kappa \in (10/24, 18/24)$ for DE. Thus, the best ensemble solutions use from 8 to 20 features.
- for ENS4 diversity changes slightly with ρ. MAPE reaches its minima for: $\rho \in (40/80, 48/80)$ for PL, $\rho = 32/80$ for GB, $\rho \in (40/80, 48/80)$ for FR, and $\rho = 40/80)$ for DE. Thus, the best choice for the number of selected nodes is around $m' = 40$.
- for ENS5 diversity has its maximum for $\lambda = 0.1$ (10% of hidden weights are set to 0). MAPE increases with λ, having its lowest values for $\lambda = 0$ (no weight pruning). Thus, any weight pruning makes the error higher. Increasing the hidden node number to 80 did not improve the results.
- for ENS6 we observe low sensitivity of diversity as well as MAPE to diversity parameter σ. For smaller σ, the error only slightly decreases as σ increases. Then, for higher σ, the error starts to increase.

Table 1 shows the quality metrics of the forecasts for the optimal values of the diversity parameters: MAPE, median of APE, root mean square error

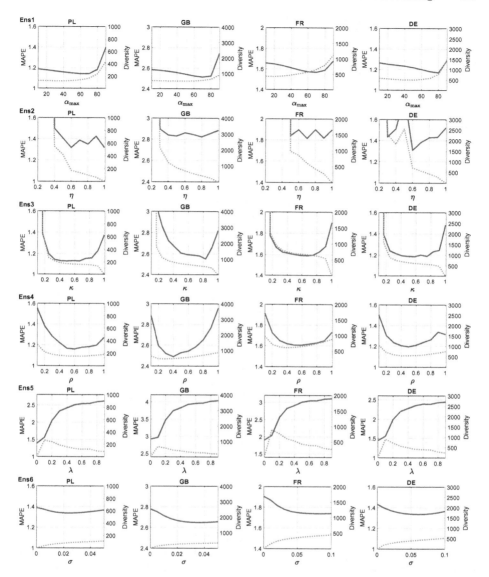

Fig. 4. MAPE (solid lines) and ensemble diversity (dashed lines) depending on the diversity parameters.

(RMSE), mean percentage error (MPE), and standard deviation of percentage error (Std(PE)) as a measure of the forecast dispersion. For comparison, the results for a single RANDNN are shown. In this case, for each forecasting task (each day of 2015) the hyperparameters of RANDNN, m and u, were optimized using grid search and 5-fold cross-validation [18]. The results for single

RANDNN shown in Table 1 are averaged over 100 independent training sessions. ENS7 shown in Table 1 is an ensemble of these 100 runs (calculated from (6)). ENS7 is similar to ENS1. The only difference is that for ENS7 the RANDNN hyperparameters were optimized for each forecasting task, and for ENS1 we set $m = 40$ and $\alpha_{max} = 70°$ for all forecasting tasks. Distributions of APE are shown in Fig. 5.

Table 1. Forecasting results.

		ENS1	ENS2	ENS3	ENS4	ENS5	ENS6	RANDNN	ENS7
PL	MAPE	1.14	1.32	**1.13**	1.16	1.39	1.34	1.32	1.24
	Median (APE)	0.79	0.94	**0.78**	0.81	0.98	0.94	0.93	0.89
	RMSE	304	342	**299**	313	380	351	358	333
	MPE	0.31	0.43	0.31	**0.25**	0.36	0.45	0.40	0.40
	Std (PE)	1.67	1.88	**1.64**	1.72	2.07	1.94	1.94	1.80
GB	MAPE	2.51	2.74	2.55	**2.49**	2.92	2.65	2.61	2.52
	Median (APE)	**1.76**	2.01	1.82	1.77	2.09	1.95	1.88	1.80
	RMSE	1151	1205	1160	**1131**	1325	1167	1187	1147
	MPE	**−0.53**	−0.83	**−0.53**	**−0.53**	−0.55	−0.65	−0.61	−0.61
	Std (PE)	3.48	3.64	3.51	3.42	4.02	3.56	3.57	3.44
FR	MAPE	**1.57**	1.80	1.59	1.61	1.90	1.74	1.67	1.60
	Median (APE)	**1.04**	1.24	1.07	1.06	1.33	1.24	1.15	1.07
	RMSE	**1378**	1505	1398	1402	1597	1459	1422	1385
	MPE	−0.28	−0.26	**−0.25**	−0.31	−0.33	−0.27	−0.42	−0.42
	Std (PE)	**2.53**	2.80	2.58	2.57	2.92	2.72	2.60	**2.53**
DE	MAPE	**1.18**	1.31	1.19	1.20	1.45	1.33	1.38	1.29
	Median (APE)	0.81	0.91	**0.80**	0.82	0.98	0.96	0.96	0.91
	RMSE	**1077**	1155	1111	1097	1423	1159	1281	1206
	MPE	0.11	0.03	0.10	0.09	**0.02**	0.17	0.14	0.14
	Std (PE)	**1.89**	2.01	1.96	1.92	2.44	2.04	2.22	2.11

To confirm that the differences in accuracy between ensemble variants are statistically significant, we performed a one-sided Giacomini-White test for conditional predictive ability [19]. This is a pairwise test that compares the forecasts produced by different models. We used a Python implementation of the Giacomini-White test in multivariate variant from [20,21]. Figure 6 shows results of the Giacomini-White test. The resulting plots are heat maps representing the obtained p-values. The closer they are to zero the significantly more accurate the forecasts produced by the model on the X-axis are than the forecasts produced by the model on the Y-axis. The black color indicates p-values larger than 0.10.

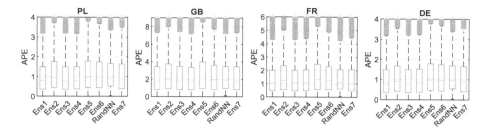

Fig. 5. Distribution of APE.

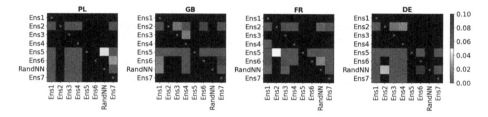

Fig. 6. Results of the Giacomini-White test for the proposed RANDNN ensembles.

From the results presented in Table 1 and Figs. 5 and 6 we can conclude that the most accurate ensembles are ENS1, ENS3, and ENS4. For PL and DE they significantly outperformed all other models including single RANDNN and ENS7, although, for GB and FR, ENS7 can compete with them in terms of accuracy. Note that the APE distributions for ENS1, ENS3, and ENS4 shown in a simplified form in Fig. 5 are very similar. The worst ensemble solution was ENS5 based on weight pruning. It was beaten by all other models for all datasets. ENS2 and ENS6 were only slightly better in terms of accuracy than ENS5.

MPE shown in Table 1 allows us to assess the bias of the forecasts. A positive value of MPE indicates underprediction, while a negative value indicates overprediction. As can be seen from Table 1, for PL and DE all the models underpredicted, whilst for GB and FR they overpredicted. The forecasts produced by the three best ensemble models were less biased than the forecast produced by a single RANDNN and ENS7.

In the next experiment, we compare ENS1 performance (one of the best ensemble solutions) with that of other models based on classical statistical methods and ML methods. The baseline models that we use in our comparative studies are outlined below (see [22] for details). Their hyperparameters were selected on the training set in grid search procedures.

- NAIVE – naive model: $\widehat{\mathbf{e}}_{i+\tau} = \mathbf{e}_{i+\tau-7}$,
- ARIMA – autoregressive integrated moving average model,
- ETS – exponential smoothing model,
- PROPHET – a modular additive regression model with nonlinear trend and seasonal components [23],

428   G. Dudek and P. Pełka

- MLP – perceptron with a single hidden layer and sigmoid nonlinearities,
- SVM – linear epsilon insensitive support vector machine (ϵ-SVM) [24],
- ANFIS – adaptive neuro-fuzzy inference system,
- LSTM – long short-term memory,
- FNM – fuzzy neighborhood model,
- N-WE – Nadaraya-Watson estimator,
- GRNN – general regression NN.

Table 2 shows MAPE for ENS1 and the baseline models. Note that for each dataset, ENS1 returned the lowest MAPE. To confirm the statistical significance of these findings, we performed a Giacomini-White test. The test results depicted in Fig. 7, clearly show that ENS1 outperforms all the other models in terms of accuracy. Only in two cases out of 44 were the baseline models close to ENS1 in terms of accuracy, i.e. N-WE for PL data and SVM for GB data.

Table 2. MAPE for ENS1 and baseline models.

	ENS1	NAIVE	ARIMA	ETS	PROPHET	MLP	SVM	ANFIS	LSTM	FNM	N-WE	GRNN
PL	**1.14**	2.96	2.31	2.14	2.63	1.39	1.32	1.64	1.57	1.21	1.19	1.22
GB	**2.51**	4.80	3.50	3.19	4.00	2.84	2.54	2.80	2.92	3.02	3.12	3.01
FR	**1.57**	5.53	3.00	2.79	4.71	1.93	1.63	2.12	1.81	1.84	1.86	1.81
DE	**1.18**	3.13	2.31	2.10	3.23	1.58	1.38	2.48	1.57	1.30	1.29	1.30

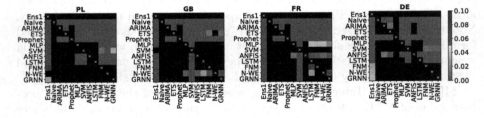

Fig. 7. Results of the Giacomini-White test for the proposed ENS1 and baseline models.

5 Conclusion

Challenging forecasting problems, such as forecasting nonstationary TS with multiple seasonality, need sophisticated models. In this work, to deal with multiple seasonal periods, we employ a pattern representation of the TS, in order to simplify the relationship between input and output data and make the problem easier to solve using simple regression models such as randomized NNs. RANDNNs have three distinct advantages: a simple single-hidden layer architecture, extremely fast learning (no gradient-based learning, no problems with vanishing and exploding gradients), and ease of implementation (no complex optimization algorithms, no additional mechanisms such as dilation, attention, and residual connections). RANDNN produces a vector output, which means the model is able to forecast the time series sequence at once.

We propose an ensemble of RANDNNs with six different methods of managing diversity. Among these, the most promising ones turned out to be: learning the ensemble members using different parameters of hidden nodes (ENS1), training individual learners on different subsets of features (ENS3), and hidden node pruning (ENS4). These three methods brought comparable results in an experimental study involving four real-world forecasting problems (short-term load forecasting). ENS1 has the additional advantage of having only two hyperparameters to tune, i.e. the number of hidden nodes and the interval bounds for random weights. In the comparative study, ENS1 significantly outperformed both classical statistical methods and machine learning methods.

In our further work, we plan to introduce an attention mechanism into our randomization-based forecasting models to select training data and develop probabilistic forecasting models based on RANDNNs.

References

1. Palit, A.K., Popovic, D.: Computational Intelligence in Time Series Forecasting: Theory and Engineering Applications. Springer, London (2005). https://doi.org/10.1007/1-84628-184-9
2. Benidis, K., et al.: Neural forecasting: introduction and literature overview. arXiv:2004.10240 (2020)
3. Dudek, G.: Neural networks for pattern-based short-term load forecasting: a comparative study. Neurocomputing **205**, 64–74 (2016)
4. Torres, J.F., Hadjout, D., Sebaa, A., Martínez-Álvarez, F., Troncoso, A.: Deep learning for time series forecasting: a survey. Big Data **9**(1), 3–21 (2021)
5. Hewamalage, H., Bergmeir, C., Bandara, K.: Recurrent neural networks for time series forecasting: current status and future directions. Int. J. Forecast. **37**(1), 388–427 (2021)
6. Reeve, H.W.J., Brown, G.: Diversity and degrees of freedom in regression ensembles. Neurocomputing **298**, 55–68 (2018)
7. Smyl, S.: A hybrid method of exponential smoothing and recurrent neural networks for time series forecasting. Int. J. Forecast. **36**(1), 75–85 (2020)
8. Oreshkin, B.N., Carpov, D., Chapados, N., Bengio, Y.: N-BEATS: neural basis expansion analysis for interpretable time series forecasting. In: 8th International Conference on Learning Representations, ICLR (2020)
9. Ren, Y., Zhang, L., Suganthan, P.N.: Ensemble classification and regression - recent developments, applications and future directions. IEEE Comput. Intell. Mag. **11**(1), 41–53 (2016)
10. Alhamdoosh, M., Wang, D.: Fast decorrelated neural network ensembles with random weights. Inf. Sci. **264**, 104–117 (2014)
11. Mesquita, D.P.P., Gomes, J.P.P., Rodrigues, L.R., Oliveira, S.A.F., Galvão, R.K.H.: Building selective ensembles of randomization based neural networks with the successive projections algorithm. Appl. Soft Comput. **70**, 1135–1145 (2018)
12. Huang, C., Li, M., Wang, D.: Stochastic configuration network ensembles with selective base models. Neural Netw. **264**, 106–118 (2021)

13. Li, S., Goel, L., Wang, P.: An ensemble approach for short-term load forecasting by extreme learning machine. Appl. Energy **170**, 22–29 (2016)
14. Qiu, X., Suganthan, P.N., Amaratunga, G.A.J.: Ensemble incremental learning random vector functional link network for short-term electric load forecasting. Knowl.-Based Syst. **145**, 182–196 (2018)
15. de Almeida, R., Goh, Y.M., Monfared, R., et al.: An ensemble based on neural networks with random weights for online data stream regression. Soft. Comput. **24**, 9835–9855 (2020)
16. Hu, Y., et al.: Short-term load forecasting using multimodal evolutionary algorithm and random vector functional link network based ensemble learning. Appl. Energy **285**, 116415 (2021)
17. Dudek, G.: Generating random parameters in feedforward neural networks with random hidden nodes: drawbacks of the standard method and how to improve it. In: Yang, H., Pasupa, K., Leung, A.C.-S., Kwok, J.T., Chan, J.H., King, I. (eds.) ICONIP 2020. CCIS, vol. 1333, pp. 598–606. Springer, Cham (2020). https://doi.org/10.1007/978-3-030-63823-8_68
18. Dudek, G.: Randomized neural networks for forecasting time series with multiple seasonality. In: Rojas, I., Joya, G., Català, A. (eds.) IWANN 2021. LNCS, vol. 12862, pp. 196–207. Springer, Cham (2021). https://doi.org/10.1007/978-3-030-85099-9_16
19. Giacomini, R., White, H.: Tests of conditional predictive ability. Econometrica **74**(6), 1545–1578 (2006)
20. Epftoolbox library. https://github.com/jeslago/epftoolbox. Epftoolbox documentation. https://epftoolbox.readthedocs.io
21. Lago, J., Marcjasz, G., De Schutter, B., Weron, R.: Forecasting day-ahead electricity prices: a review of state-of-the-art algorithms, best practices and an open-access benchmark. Appl. Energy **293**, 116983 (2021)
22. Dudek, G., Pełka, P.: Pattern similarity-based machine learning methods for mid-term load forecasting: a comparative study. Appl. Soft Comput. **104**, 107223 (2021)
23. Taylor, S.J., Letham, B.: Forecasting at scale. Am. Stat. **72**(1), 37–45 (2018)
24. Pełka, P.: Pattern-based forecasting of monthly electricity demand using support vector machine. In: International Joint Conference on Neural Networks, IJCNN 2021, pp. 1–8 (2021)

Grouped Echo State Network with Late Fusion for Speech Emotion Recognition

Hemin Ibrahim[1]📛, Chu Kiong Loo[1(✉)]📛, and Fady Alnajjar[2]📛

[1] Department of Artificial Intelligence, Faculty of Computer Science and Information Technology, Universiti Malaya, 50603 Kuala Lumpur, Malaysia
hemin.ibrahim@siswa.um.edu.my, ckloo.um@um.edu.my
[2] College of Information Technology, UAE University, Al Ain, United Arab Emirates

Abstract. Speech Emotion Recognition (SER) has become a popular research topic due to having a significant role in many practical applications and is considered a key effort in Human-Computer Interaction (HCI). Previous works in this field have mostly focused on global features or time series feature representation with deep learning models. However, the main focus of this work is to design a simple model for SER by adopting multivariate time series feature representation. This work also used the Echo State Network (ESN) including parallel reservoir layers as a special case of the Recurrent Neural Network (RNN) and applied Principal Component Analysis (PCA) to reduce the high dimension output from reservoir layers. The late grouped fusion has been applied to capture additional information independently of the two reservoirs. Additionally, hyperparameters have been optimized by using the Bayesian approach. The high performance of the proposed SER model is proved when adopting the speaker-independent experiments on the SAVEE dataset and FAU Aibo emotion Corpus. The experimental results show that the designed model is superior to the state-of-the-art results.

Keywords: Speech emotion recognition · Reservoir computing · Grouped echo state network · Time series classification · Recurrent neural network

1 Introduction

Emotion has a vital role in a human's life and can be recognized from different ways such as audio, face, gesture, and texts. Emotions are not easy to adapt, that is why detecting emotions from different channels are a challenging task. However, the extreme need of exploiting emotion in various applications like social robots and Human-Computer Interaction (HCI) motivate researchers to widely studying it [2]. Additionally, speech as a quick and effective way to communicate is considered as a valuable tool for HCI. Despite the high influence of emotion on speech, automatic emotion recognition from the speech is a challenging task in the artificial intelligence area [22].

© Springer Nature Switzerland AG 2021
T. Mantoro et al. (Eds.): ICONIP 2021, LNCS 13110, pp. 431–442, 2021.
https://doi.org/10.1007/978-3-030-92238-2_36

A major challenge in SER is to extract the highest related emotional features from the raw data. Many works aim to catch the most effective emotion features in speech [16]. There are two major approaches to extract features: handcrafted features and deep learned emotion features. The handcrafted features can be globally extracted to signify speech samples in one dimension representation. These features can also be locally extracted from the time series based data frames (frame-based features). The COVAREP [10] and openSMILE [11] toolkits are among these tools that are used to compute local and global features respectively. Many works [1] and [25] have used openSMILE toolkit to extract global features from speech. However, some recent researches are concentrating on deep learning based features that are straightly from the raw data [23]. Mustaqeem et al. [23] used convolutional neural network (CNN) for extracting high-level features from speech spectrograms and fed to 1D Convolutional Neural Network for SER model.

Besides having the right emotion features from speech signals, developing a strong model is an important step to get a high performance of SER systems [7]. The frame-based features can be utilized for models that is useful for multivariate sequence data such as Recurrent Neural Network (RNN). Authors in [17] and [23] adopted the use of high-level features fed to a Bidirectional Long Short-Term Memory (BiLSTM) model which is a special kind of RNN. However, multivariate time series features can be used with ESN as a part of the reservoir computing framework due to the sparse nature of emotion in speech. The ESN model has been used in a few works for SER systems, for instance, Scherer et al. [27] used ESN for real-time model for recognizing emotion from speech, and authors in [12,26] adopted ESN for emotion detection from speech signal, where [26] used only anger and neutral emotion classes.

The simplicity of the ESN architecture is reported as one of its main advantages. The ESN architecture started with an input layer followed by a reservoir layer with randomly connected neurons and non-trained weights, in addition to an output layer [33]. However, some works tackled the instability in ESN as the weights are assigned randomly only once and fixed in the reservoir layer [33]. Therefore, some researchers have proposed DeepESN to overcome this problem by having more than one reservoir layer that strikes a valuable effect arising from the combination of reservoirs [13].

Dimension reduction approaches convert the high dimensional features into lower-dimensional representation to reduce computation and help in decorrelating the transformed features. The PCA method is used in ESN [3] to improve the model performance by reducing the high dimensional sparse output from reservoirs.

The hyperparameters in the ESN have an important role on the model performance, some works assigned them manually with fixed values based on experience [5,19]. However, to obtain a better performance from ESN, tunning its parameters are necessary, therefore, various optimization methods like grid search, Bayesian optimization, and random search have been used. Bayesian optimization [24] method is utilized to optimize the ESN hyperparameters in [21] and [6].

In the current study, we proposed a parallel reservoir with grouped late fusion in ESN approach for SER using PCA for dimension reduction and optimizing hyperparameters by Bayesian optimization. Moreover, Mel-Frequency Cepstral Coefficients (MFCCs) and Gamma-Tone Cepstral Coefficients (GTCCs) as multivariate time series handcrafted features is used to feed the reservoirs.

The remainder of this paper is distributed as follows: In Sect. 2 the methodology of the proposed model is presented, while Sect. 3 shows the experiments and results. In Sect. 4 the discussion is presented, and followed by the conclusion which comes in Sect. 5.

2   Proposed SER Model

This section presents the architecture of the proposed model and explains each stage. It presents the major components and explains how the proposed model assists to obtain a better performance in ESN model for the SER system. Most researchers used non-sequence features on SER and few studies adopted time series local features. Furthermore, there are limited studies that used ESN model for the SER systems [12, 26, 27], but none of them exceeded an outstanding performance. This non-persuasive performance may be due to the untrained nature of ESN, the selection of the hyperparameters manually or based on experience, and the temporal data output from the reservoir layer which has a high dimensional representation that negatively affects the performance readout stage. To overcome these problems and inspired by the work of [5], we have adopted grouped late fusion ESN by having more than one reservoir computing layer. The contribution of this study leads to improve the speech emotion recognition model performance. The following subsections present the details of the designed model, as shown in Fig. 1.

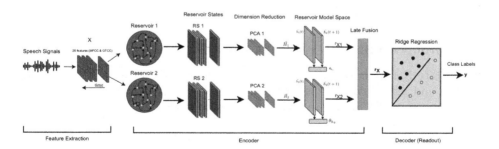

Fig. 1. Grouped ESN with late fusion, the proposed model.

2.1   Feature Extraction

Choosing the right features with discriminative emotion information in speech has a vital role in SER. In our study, handcrafted multivariate frame-based features have been used to feed the grouped ESN model. In the first step, 13 MFCC features are extracted. MFCC is one of the well-known features that is widely used for SER because of the suitable capability of extracting informative features. Moreover, another 13 GTCC features have been extracted which have a good performance under noisy conditions. A total of 26 handcrafted features are adopted as input to the proposed model. The MATLAB audioFeatureExtractor tool is applied to extract emotion features with windows of length 30 ms overlapped by 20 ms. The sample lengths in both involved datasets are varied as shown in Fig. 2. To overcome this issue, the padding with zeros is applied at the start and end for each short raw data and pruning at the start and end is used for the long samples. Therefore, the proposed model has used 600 frames for SAVEE dataset and 300 frames for FAU Aibo dataset. The number of frames in both datasets are selected based on the closely maximum length in each dataset.

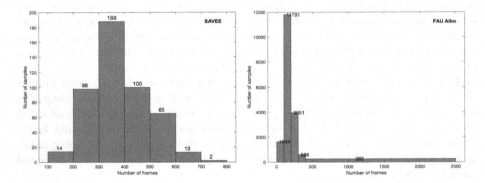

Fig. 2. Sample lengths (in frames) for SAVEE and FAU Aibo dataset

2.2   Grouped ESN

Echo State Networks (ESNs) are a special kind of RNN for learning nonlinear models and are a part of the Reservoir Computing (RC) framework [15]. In reservoir computing, weights are initially chosen randomly without any training, while the training stage falls only in the readout part [20].

The fixed (untrained) structure of ESN model helps it to reduce the available complexity in the trained structured models such as LSTM. Additionally, ESN consists of an input layer, a reservoir layer, and an output layer. Regarding the input layer, the input to the proposed model is multivariate time series data which will be feed later to two reservoirs. The advantage of having more than one reservoir is to capture more independent information from the input data with different initialize random weights from the reservoir layer. The reservoir layer initializes untrained sparsely connected neurons which are assigned randomly.

The input data is a multivariate time series which has a D-dimensional feature vector for each time step t, where $t = 1, 2, 3, ..., T$, and T is a time instant which represents the length of the sample in terms of time steps. The time t is denoted as $x(t) \in \mathbb{R}^D$ and $X = [x(1), x(2), ... x(T)]^T$. The multivariate time series data is suitable for the RNN based model and the state in the reservoir layer can be updated by the following equations:

$$h_1(t) = f(x(t), h_1(t-1); \theta_{enc1})$$
$$h_2(t) = f(x(t), h_2(t-1); \theta_{enc2})$$

(1)

The function f is a nonlinear activation function (normally hyperbolic tangent) that takes the values of $h_1(t-1)$ and the current input $x(t)$ in addition to θ_{enc1} to produce $h_1(.)$. This process will be repeated for the other reservoir represented by $h_2(.)$. $h_1(.)$ and $h_2(.)$ represent the RNN states at time t for both reservoirs and θ_{enc1} and θ_{enc2} are adaptable parameters for both reservoirs.

The Eq. (1) can be presented as follows:

$$h_1(t) = tanh(W_{in1} \, x(t) + W_{r1} h_1(t-1))$$
$$h_2(t) = tanh(W_{in2} \, x(t) + W_{r2} h_2(t-1))$$

(2)

where W_{in1} and W_{in2} are the input weights and (W_{r1} and W_{r2}) are the weights from reservoir1 and reservoir2 connections respectively, and the reservoir states ($RS1$ and $RS2$) are generated by both reservoirs over time, where $RS1 = [h_1(1), h_1(2), .., h_1(T)]^T$ and $RS2 = [h_2(1), h_2(2), .., h_2(T)]^T$. The θ_{enc1} and θ_{enc2} can be represented with $\theta_{enc1} = \{W_{in1}, W_{r1}\}$ and $\theta_{enc2} = \{W_{in2}, W_{r2}\}$ respectively.

The reservoirs are controlled by several hyperparameters that have an important influence on its performance such as (i) the size of internal units R, (ii) the spectral radius ρ which assists the model to be stable [4], (iii) the nonzero connections β, (iv) the scaling ω which is another parameter of the values in W_{in1} and W_{in2} which controls the total of nonlinearity in handling the internal units with ρ, (v) the leak is an amount of leakage in the reservoir states update, and (vi) applying dropout regularization particularly for recurrent architectures [5].

2.3   Dimension Reduction with PCA

The reservoir layers produce a high dimensional sparse output which is intractable and causes over-fitting and high computational cost. The well-known trainable dimension reduction PCA has been used to reduce the high dimensional sparse output into a non-correlated representation. PCA is a data-dependent projection that de-correlates the data representations from the original data by picking several eigenvectors of the covariance matrix with the greatest eigenvalues [2].

In this stage, the dimension of the data will be reduced to a pre-fixed number that can be fixed or tuned by one of the optimization approaches. Dimension deduction has an important effect in reducing the computation of the reservoir model space which will be applied in the next stage. The step of reducing dimension results in minimizing the volume of the data representation output features from both reservoirs and produces a new sequence $\breve{H}_1$ and $\breve{H}_2$ to feed the reservoir model space as an input.

2.4   Reservoir Model Space and Readout

The reservoir model space method introduced by [5], characterizes a generative process of the reservoir sequence and prompts a metric relationship between the samples. In this study, we used grouped ESN approach which has two reservoirs with a late fusion. Processing from different reservoirs in an independent way can supply different relational information about the sequences. The fusion after reservoir model space combines more varied representations of the data and makes the characteristics of each reservoir be more highlighted. Accordingly, the equation from [5] has been modified with two separate outputs from PCA as shown in the following equations:

$$\breve{h}_1(t+1) = U_{h_1}\ \breve{h}_1(t) + u_{h_1}$$
$$\breve{h}_2(t+1) = U_{h_2}\ \breve{h}_2(t) + u_{h_2} \tag{3}$$

where $\breve{h}_1(.)$ and $\breve{h}_2(.)$ are the columns of a frontal slice $\breve{H}_1$ and $\breve{H}_2$ respectively, $U_{h_1}, U_{h_2} \in \mathbb{R}^{D \times D}$ and $u_{h_1}, u_{h_2} \in \mathbb{R}^D$.

The late fusion of both reservoirs will be applied in this stage by combining both outputs from r_{X_1} and r_{X_2} where:

$$r_{X_1} = \theta_{h_1} = [vec(U_{h_1}); u_{h_1}]$$
$$r_{X_2} = \theta_{h_2} = [vec(U_{h_2}); u_{h_2}] \tag{4}$$

$$r_X = [r_{X_1}; r_{X_2}] \tag{5}$$

Equation 6 shows the learning part by minimizing a ridge regression loss function for both θ_{h_1} and θ_{h_2}:

$$\theta_{h_1}^* = \underset{\{U_{o_1}, u_{o_1}\}}{argmin}\ \frac{1}{2}\|\ \breve{h}_1(t)\ U_{o_1} + u_{o_1} - \breve{h}_1(t+1)\|^2 + \mu\|U_{o_1}\|^2$$
$$\theta_{h_2}^* = \underset{\{U_{o_2}, u_{o_2}\}}{argmin}\ \frac{1}{2}\|\ \breve{h}_2(t)\ U_{o_2} + u_{o_2} - \breve{h}_2(t+1)\|^2 + \mu\|U_{o_2}\|^2 \tag{6}$$

where the μ is the regularization parameter in the reservoir model space to set the number of the coefficient shrinkage.

In the classification level that is the decoding part, grouped ESN adopts a linear model which is usually represented as the following equation:

$$y = g(r_X) = V_o r_X + v_o \tag{7}$$

The decoder parameters $\theta_{dec} = \{V_o, v_o\}$ can be optimized by minimizing the loss function of the ridge regression which admits a closed form solution:

$$\theta_{dec}^* = \arg\min_{\{V_o, v_o\}} \frac{1}{2} \|r_X V_o + v_o - y\|^2 + \lambda \|V_o\|^2 \tag{8}$$

where λ is a ridge regression regularization parameter which has a role in reducing the overfitting during the training stage. The linear readout is implemented to perform the final stage of the classification by mapping the r_X representation into the emotion class labels y.

2.5 Bayesian Hyperparameter Optimization

Bayesian optimization is a gradient-free global optimization method to tune random functions [29] which has been adopted to optimize the size of internal unit and the number of dropouts in the reservoir layer, both regularization parameters μ in model space and λ in the ridge regression, in addition to dimensionality reduction size of PCA.

3 Experimental Setup and Results

In this section, the performance of the designed grouped ESN model to recognize emotion from the speech signals was evaluated on two publicly available datasets for SER system. In this work, 26 multivariate time series features divided equally among MFCCs and GTCCs feature types for each window of length 30 ms with an overlap of 20 ms have been used. The extracted features are fed to two reservoir layers parallelly, in which the size of hidden layers (internal units) have been tuned by adopting Bayesian optimization approach. The PCA method has been used to reduce the size of the high dimensional sparse data output from both reservoirs into a more compactive representation. The reservoir model space stage distinguishes a generative model of each reservoir and induces a metric relationship between the samples from the dimension reduction stage. Afterwards, the late fusion of both reservoir model spaces have been applied which came from both reservoirs. Hyperparameters of two reservoirs, PCA and ridge regression stages are tuned by the Bayesian optimization method.

For SAVEE dataset, our grouped ESN model applied Leave One Speaker Out (LOSO) speaker-independent method and we followed the adopted approach of the interspeech09 challenge [28] for FAU Aibo dataset. The results of each emotion class from both datasets are shown in Tables 1 and 2. The performance of the grouped ESN model is validated by the use of Surrey Audio-Visual Expressed Emotion (SAVEE) [14] as an acted dataset and FAU Aibo Emotion Corpus [30] as a non acted dataset.

3.1 SAVEE

SAVEE [14] is an SER dataset that includes audio and visual expression. The dataset includes seven emotion categories, consisting of anger, disgust, fear, happiness, sadness, surprise, and neutral. Four actors have participated when each recorded 120 utterances results in a total number of 480 files.

SAVEE is an acted dataset that is adopted in the validation of the proposed model in LOSO approach, where we select one speaker as a testing set and other speakers as a training set and repeat this process to assure the involvement of all the speakers in the testing set.

Table 1. The proposed model performance (%) for speaker-independent (LOSO approach) SER using SAVEE dataset.

Emotion	Precision	Recall	F1 score
Anger	71.83	85.00	77.86
Disgust	67.74	35.00	46.15
Fear	91.18	51.67	65.96
Happiness	80.85	63.33	71.03
Neutral	61.14	98.33	75.40
Sadness	80.00	40.00	53.33
Surprise	60.81	75.00	67.16
Unweighted	73.36	64.05	65.27
Weighted	71.84	68.33	66.54

Table 1 presents the results of precision, recall, F1 score, unweighted, and weighted percentage accuracy for each class in SAVEE dataset.

3.2 FAU Aibo Emotion Corpus

The FAU Aibo has been also adopted in this study since its an unprompted German dataset that includes 18216 chunk speech samples with five emotion classes including anger, emphatic, neutral, positive, and rest [30]. Following to the proposed approach of the interspeech09 challenge [28], we adopted the 'Ohm' part (9959 chunks) to train the proposed model and the 'Mont' part (8257 chunks) to test the model.

The FAU Aibo dataset classes are extremely unbalanced due to the lack of samples in some classes. To overcome this issue, we adopted the random under-sampling [18] approach on the classes with a higher number of samples by randomly removing a specific number of samples that help in balancing the classes.

Table 2 shows the precision, recall, F1 score, unweighted, and weighted percentage accuracy for all the emotion classes for FAU Aibo dataset. It is clearly observed that a large gap between the unweighted and weighted accuracy is obtained as a result of the imbalanced data.

Table 2. The performance (%) of the proposed model for FAU Aibo dataset.

Emotion	Precision	Recall	F1 Score
Anger	23.15	57.77	33.05
Emphatic	32.80	64.92	43.58
Neutral	84.47	25.18	38.80
Positive	09.98	59.07	17.07
Rest	13.09	20.88	16.09
Unweighted	32.70	45.56	29.72
Weighted	63.83	35.45	37.18

4   Discussion

This section presents the comparison of the grouped ESN performance with single ESN and other models. In this work, we are presenting the overall unweighted accuracy (UA), which is more representative of the real performance, especially when the dataset is imbalanced. Adopting the LOSO speaker-independent approach is utilized to compare the performance of the previous studies of both SAVEE and FAU Aibo dataset with the current proposed model. The classification UA of speaker-independent experiments for both SAVEE and FAU Aibo datasets are listed in Table 3.

Table 3. The comparison of unweighted accuracies (UA%) of the proposed model in SAVEE and FAU Aibo dataset with some recent studies using LOSO approach.

Dataset	Method	Year	UA%
SAVEE	Daneshfar et al. [8]	2020	55.00
	Wen et al. [32]	2017	53.60
	Single Reservoir	–	61.30
	Proposed model	–	**64.05**
FAU Aibo	Triantafyllopoulos et al. [31]	2021	41.30
	Deb & Dandapat [9]	2019	45.20
	Zhao et al. [34]	2019	45.40
	Single Reservoir	–	44.04
	Proposed model	–	**45.56**

To evaluate the effect of the adopted grouped ESN with two reservoirs model, a LOSO speaker-independent approach has been adopted for the single reservoir as well.

For the SAVEE dataset, the proposed model adopted LOSO approach and achieved 64.05% which is 9.05% higher than the second-best result from other studies and when we used only one reservoir the performance was only 61.3%. Table 3 shows the performance of our proposed model for FAU Aibo dataset based on the 2009 challenge protocol. It is noticeable that the grouped ESN model has outperformed normal ESN with single reservoir and other studies by 45.56% of UA accuracy.

5   Conclusion

This work proposed a recurrent based model for multivariate temporal SER classification using grouped late fusion ESN. Using more than one reservoir helps to strike a valuable effect arising from different reservoirs and the late fusion is applied to capture additional information independently from the reservoir output data.

The dimensionality reduction with PCA has been adopted to reduce the high dimension output from reservoirs. Therefore, with the limited number of feature representations and a nontrainable ESN method, the proposed model is computationally cheap and achieved a high performance. Another aspect that has a distinguished effect on rising the performance of the proposed model is optimizing the hyperparameters using the Bayesian optimization approach.

The disadvantage of one reservoir is producing a widespread representation and the randomly initiated weights assigned to it. Therefore, our grouped ESN, which has two reservoirs is able to extract more informative representation independently from the input data. The proposed model achieved the highest classification UA compared to the single reservoir and previous works on SER.

Acknowledgements. This work was supported by the Covid-19 Special Research Grant under Project CSRG008-2020ST, Impact Oriented Interdisciplinary Research Grant Programme (IIRG), IIRG002C-19HWB from University of Malaya, and the AUA-UAEU Joint Research Grant 31R188.

References

1. Al-Talabani, A., Sellahewa, H., Jassim, S.: Excitation source and low level descriptor features fusion for emotion recognition using SVM and ANN. In: 2013 5th Computer Science and Electronic Engineering Conference (CEEC), pp. 156–161 (2013). https://doi.org/10.1109/CEEC.2013.6659464
2. Al-Talabani, A., Sellahewa, H., Jassim, S.A.: Emotion recognition from speech: tools and challenges. In: Agaian, S.S., Jassim, S.A., Du, E.Y. (eds.) Mobile Multimedia/Image Processing, Security, and Applications 2015, vol. 9497, pp. 193–200. International Society for Optics and Photonics, SPIE (2015). https://doi.org/10.1117/12.2191623
3. Bianchi, F.M., Scardapane, S., Løkse, S., Jenssen, R.: Bidirectional deep-readout echo state networks. In: ESANN (2018)
4. Bianchi, F.M., Livi, L., Alippi, C.: Investigating echo-state networks dynamics by means of recurrence analysis. IEEE Trans. Neural Netw. Learn. Syst. **29**(2), 427–439 (2018). https://doi.org/10.1109/TNNLS.2016.2630802

5. Bianchi, F.M., Scardapane, S., Løkse, S., Jenssen, R.: Reservoir computing approaches for representation and classification of multivariate time series. IEEE Trans. Neural Netw. Learn. Syst. **32**(5), 2169–2179 (2021). https://doi.org/10.1109/TNNLS.2020.3001377

6. Cerina, L., Santambrogio, M.D., Franco, G., Gallicchio, C., Micheli, A.: EchoBay: design and optimization of echo state networks under memory and time constraints. ACM Trans. Archit. Code Optim. **17**(3), 1–24 (2020). https://doi.org/10.1145/3404993

7. Chen, L., Mao, X., Xue, Y., Cheng, L.L.: Speech emotion recognition: features and classification models. Digit. Signal Process. **22**(6), 1154–1160 (2012). https://doi.org/10.1016/j.dsp.2012.05.007. https://www.sciencedirect.com/science/article/pii/S1051200412001133

8. Daneshfar, F., Kabudian, S.J., Neekabadi, A.: Speech emotion recognition using hybrid spectral-prosodic features of speech signal/glottal waveform, metaheuristic-based dimensionality reduction, and Gaussian elliptical basis function network classifier. Appl. Acoust. **166**, 107360 (2020). https://doi.org/10.1016/j.apacoust.2020.107360. https://www.sciencedirect.com/science/article/pii/S0003682X1931117X

9. Deb, S., Dandapat, S.: Multiscale amplitude feature and significance of enhanced vocal tract information for emotion classification. IEEE Trans. Cybern. **49**(3), 802–815 (2019). https://doi.org/10.1109/TCYB.2017.2787717

10. Degottex, G., Kane, J., Drugman, T., Raitio, T., Scherer, S.: COVAREP - a collaborative voice analysis repository for speech technologies. In: 2014 IEEE International Conference on Acoustics, Speech and Signal Processing (ICASSP), pp. 960–964 (2014). https://doi.org/10.1109/ICASSP.2014.6853739

11. Eyben, F., Wöllmer, M., Schuller, B.: OpenSMILE: the munich versatile and fast open-source audio feature extractor. In: Proceedings of the 18th ACM International Conference on Multimedia, MM 2010, pp. 1459–1462. Association for Computing Machinery, New York (2010). https://doi.org/10.1145/1873951.1874246

12. Gallicchio, C., Micheli, A.: A preliminary application of echo state networks to emotion recognition (2014)

13. Gallicchio, C., Micheli, A.: Reservoir topology in deep echo state networks. In: Tetko, I.V., Kůrková, V., Karpov, P., Theis, F. (eds.) ICANN 2019. LNCS, vol. 11731, pp. 62–75. Springer, Cham (2019). https://doi.org/10.1007/978-3-030-30493-5_6

14. Haq, S., Jackson, P.: Multimodal emotion recognition. In: Machine Audition: Principles, Algorithms and Systems, pp. 398–423. IGI Global, Hershey, August 2010

15. Jaeger, H., Haas, H.: Harnessing nonlinearity: predicting chaotic systems and saving energy in wireless communication. Science **304**(5667), 78–80 (2004). https://doi.org/10.1126/science.1091277. https://science.sciencemag.org/content/304/5667/78

16. Kathiresan, T., Dellwo, V.: Cepstral derivatives in MFCCS for emotion recognition. In: 2019 IEEE 4th International Conference on Signal and Image Processing (ICSIP), pp. 56–60 (2019). https://doi.org/10.1109/SIPROCESS.2019.8868573

17. Lee, J., Tashev, I.: High-level feature representation using recurrent neural network for speech emotion recognition. In: INTERSPEECH (2015)

18. Lemaître, G., Nogueira, F., Aridas, C.K.: Imbalanced-learn: a python toolbox to tackle the curse of imbalanced datasets in machine learning. J. Mach. Learn. Res. **18**(17), 1–5 (2017). http://jmlr.org/papers/v18/16-365.html

19. Lukoševičius, M.: A practical guide to applying echo state networks. In: Montavon, G., Orr, G.B., Müller, K.-R. (eds.) Neural Networks: Tricks of the Trade. LNCS, vol. 7700, pp. 659–686. Springer, Heidelberg (2012). https://doi.org/10.1007/978-3-642-35289-8_36

20. Lukoševičius, M., Jaeger, H.: Reservoir computing approaches to recurrent neural network training. Comput. Sci. Rev. **3**(3), 127–149 (2009). https://doi.org/10.1016/j.cosrev.2009.03.005. https://www.sciencedirect.com/science/article/pii/S1574013709000173

21. Maat, J.R., Gianniotis, N., Protopapas, P.: Efficient optimization of echo state networks for time series datasets. In: 2018 International Joint Conference on Neural Networks (IJCNN), pp. 1–7 (2018)

22. Mao, Q., Dong, M., Huang, Z., Zhan, Y.: Learning salient features for speech emotion recognition using convolutional neural networks. IEEE Trans. Multimed. **16**(8), 2203–2213 (2014). https://doi.org/10.1109/TMM.2014.2360798

23. Mustaqeem, Sajjad, M., Kwon, S.: Clustering-based speech emotion recognition by incorporating learned features and deep BiLSTM. IEEE Access **8**, 79861–79875 (2020). https://doi.org/10.1109/ACCESS.2020.2990405

24. Nogueira, F.: Bayesian optimization: open source constrained global optimization tool for Python (2014). https://github.com/fmfn/BayesianOptimization

25. Özseven, T.: A novel feature selection method for speech emotion recognition. Appl. Acoust. **146**, 320–326 (2019)

26. Saleh, Q., Merkel, C., Kudithipudi, D., Wysocki, B.: Memristive computational architecture of an echo state network for real-time speech-emotion recognition. In: 2015 IEEE Symposium on Computational Intelligence for Security and Defense Applications (CISDA), pp. 1–5 (2015). https://doi.org/10.1109/CISDA.2015.7208624

27. Scherer, S., Oubbati, M., Schwenker, F., Palm, G.: Real-time emotion recognition using echo state networks. In: André, E., Dybkjær, L., Minker, W., Neumann, H., Pieraccini, R., Weber, M. (eds.) PIT 2008. LNCS (LNAI), vol. 5078, pp. 200–204. Springer, Heidelberg (2008). https://doi.org/10.1007/978-3-540-69369-7_22

28. Schuller, B., Steidl, S., Batliner, A.: The interspeech 2009 emotion challenge. In: Tenth Annual Conference of the International Speech Communication Association (2009)

29. Snoek, J., Larochelle, H., Adams, R.P.: Practical Bayesian optimization of machine learning algorithms. In: Proceedings of the 25th International Conference on Neural Information Processing Systems, vol. 2, NIPS 2012, pp. 2951–2959. Curran Associates Inc., Red Hook (2012)

30. Steidl, S.: Automatic Classification of Emotion Related User States in Spontaneous Children's Speech. Logos-Verlag (2009)

31. Triantafyllopoulos, A., Liu, S., Schuller, B.W.: Deep speaker conditioning for speech emotion recognition. In: 2021 IEEE International Conference on Multimedia and Expo (ICME), pp. 1–6 (2021). https://doi.org/10.1109/ICME51207.2021.9428217

32. Wen, G., Li, H., Huang, J., Li, D., Xun, E.: Random deep belief networks for recognizing emotions from speech signals. Comput. Intell. Neurosci. **2017** (2017)

33. Wu, Q., Fokoue, E., Kudithipudi, D.: On the statistical challenges of echo state networks and some potential remedies (2018)

34. Zhao, Z., et al.: Exploring deep spectrum representations via attention-based recurrent and convolutional neural networks for speech emotion recognition. IEEE Access **7**, 97515–97525 (2019). https://doi.org/10.1109/ACCESS.2019.2928625

Applications

Applications

MPANet: Multi-level Progressive Aggregation Network for Crowd Counting

Chen Meng[1], Run Han[1], Chen Pang[1], Chunmeng Kang[1,2(✉)], Chen Lyu[1,2], and Lei Lyu[1,2(✉)]

[1] School of Information Science and Engineering, Shandong Normal University, Jinan 250358, China
[2] Shandong Provincial Key Laboratory for Distributed Computer Software Novel Technology, Jinan 250358, China
lvlei@sdnu.edu.cn

Abstract. Crowd counting has important applications in many fileds, but it is still a challenging task due to background occlusion, scale variation and uneven distribution of crowd. This paper proposes the Multi-level Progressive Aggregation Network (MPANet) to enhance the channel and spatial dependencies of feature maps and effectively integrate the multi-level features. Besides, the Aggregation Refinement (AR) module is designed to integrate low-level spatial information and high-level semantic information. The proposed AR module can effectively utilize the complementary properties between multi-level features to generate high-quality density maps. Moreover, the Multi-scale Aware (MA) module is constructed to capture rich contextual information through convolutional kernels of different sizes. Furthermore, the Semantic Attention (SA) module is designed to enhance spatial and channel response on feature maps, which can reduce false recognition of the background region. Extensive experiments on four challenging datasets demonstrate that our approach outperforms most state-of-the-art methods.

Keywords: Crowd counting · Multi-scale feature extraction · Feature fusion · Computer vision · Deep learning

1 Introduction

In recent years, crowd counting has been widely applied in public safety, video surveillance, and traffic control. In particular, employing crowd counting to keep crowd densities within reasonable limits is effective to prevent the spread of some diseases. During COVID-19 in 2020, crowd counting and management play critical roles in interrupting transmission of the virus. A wide variety of approaches [3,27] have been proposed to pursue accurate prediction in the number of people from images or videos. Typically, the convolutional neural network (CNN) based methods [1,3] have become mainstream in crowd counting with the development of deep learning. However, there remain several challenges that can adversely

© Springer Nature Switzerland AG 2021
T. Mantoro et al. (Eds.): ICONIP 2021, LNCS 13110, pp. 445–457, 2021.
https://doi.org/10.1007/978-3-030-92238-2_37

(a)   (b)   (c)

Fig. 1. The challenges of crowd counting. (a) The uneven distribution. (b) The background occlusion. (c) The scale variation.

affect crowd counting accuracy, including scale variation, background occlusion, and uneven distribution of crowd. Especially the large scale variation as the result of the difference in the angle and position of cameras, manifested in that the heads in the images show different sizes, and the distribution of the crowd also shows diversity (as shown in Fig. 1).

To solve the problem of scale variation in the image, most previous methods employed multi-column scale-dependent features extraction architectures [12,25,27]. The MCNN [27] utilizes three columns of different-size convolutional kernels to obtain rich scale features. However, the deeper layers of the network can capture more semantic information but less original detail information. Hence, an effective approach is to connect features at different layers by skip connections. Inspired by U-Net [10], W-Net [17] adopts an encoder-decoder structure to aggregate features at different levels. Although these methods focus on extracting the multi-scale features, they ignore the differences between multi-level features and do not effectively utilize the rich scale information.

This paper proposes a novel network named Multi-level Progressive Aggregation Network (MPANet) consisting of four modules: the front-end network, the aggregation refinement module, the multi-scale aware module, and the semantic attention module. The front-end network is used to extract low-level features with rich original detail based on VGG-16 [13]. Based on the low-level features, the Multi-scale Aware (MA) module is constructed to capture the multi-scale high-level semantic information. Furthermore, considering the characteristics difference of features at different levels, we design the Aggregation Refinement (AR) module to hierarchically aggregate low-level original features and high-level semantic features. Finally, the Semantic Attention (SA) module is introduced to enhance the spatial and channel dependencies, which can suppress the influence of cluttered backgrounds and reduce false recognition.

To summarize, our key contributions are four-fold:

- We construct an innovative network for crowd counting, which can efficiently aggregate multi-level features to address scale variation and false recognition of the background region.

– We design the Aggregation Refinement (AR) module that utilizes complementarity between multi-level features to progressively generate density maps.
– We introduce the Multi-scale Aware (MA) module, which can extract multi-scale features to deal with scale variation in images effectively.
– We design the semantic attention (SA) module to enhance the spatial and channel dependencies on feature maps which can effectively suppress the background noises and highlight crowd areas.

2   Related Works

We divide CNN-based methods [1,3,8] into single-column based methods and multi-column based methods.

2.1   Single-Column Based Approaches

The single-column approach is a single deeper CNN structure, which usually uses convolution of different forms to capture scale information. SaCNN [26] gradually adds features of different scales via skip-connections from the front-end network to the back-end network to generate the final density map. SANet [1] employs an encoder to exploit multi-scale features, then uses up-sampling in the decoder to get density maps. CSRNet [7] uses dilated convolution to enlarge the receptive field without increasing the number of parameters. SPN [2] also utilizes dilated convolution with different dilation rates in parallel to exploit multi-scale features.

In general, single-column methods consider multi-scale features, but they usually ignore primitive detail information, including borders, spatial structure, which limits feature representation.

2.2   Multi-column Based Approaches

Multi-column networks usually employ multiple branch structure to extract multi-scale information. MCNN [27] and Switch-CNN [12] use multi-column network with convolutional kernels of different sizes to exploit multi-scale features. To improve the robustness of the model in dealing with scale variation, ANF [25] utilizes conditional random fields (CRFs) to aggregate multi-scale information. TEDNet [6] employs a novel trellis architecture to hierarchically aggregate multiple features, improving the representativeness of the features. SASNet [16] can automatically learn the relationship between scales and features and select the most appropriate feature level for different scales.

Multi-column networks can aggregate multi-level features that are effective for solving scale variation, but they ignore the fact that the gap between different features and their contributions is different.

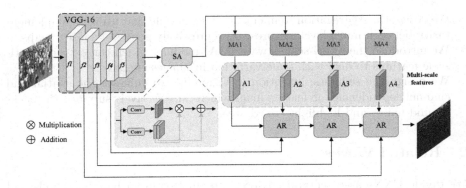

Fig. 2. The architecture of the proposed MPANet. The front-end network based on VGG-16 extracts low-level features. These feature maps are enhanced by the SA modules and then fed into the MA modules. The MA module extracts multi-scale features. Finally, the AR module is used to aggregate multi-level feature maps.

3 Approach

3.1 Overview

The framework of our proposed network is illustrated in Fig. 2, which contains four modules. Given the strong generalization ability of VGG-16, we employ its first 13 convolutional layers as the front-end network, and it includes 5 blocks $\{f_1, f_2, f_3, f_4, f_5\}$. We regard the outputs of f_2, f_3, and f_4 as low-level features from the front-end network. Specifically, we first employ a SA module to enhance the channel and spatial dependencies on feature maps, and four MA modules to generate multi-scale high-level features through different convolution kernels. Then we employ the AR module three times to progressively aggregate multi-level features and refine feature maps. In the AR module, a group of low-level spatial features and two groups of high-level semantic features are fused through multiplication operations.

3.2 Aggregation Refinement Module

Generally, the low-level features extracted from the backbone network contain more original information but also contain some noises caused by the background. Conversely, the high-level features contain more semantic information and fewer noises. Therefore, the combination of features at different levels is complementary. Hence, we propose the AR Module, which first integrates spatial information and semantic information from different levels, and then further refines the feature maps.

As Fig. 2 shows, each AR module has three inputs: the low-level features from the backbone network, the high-level features generated by the previous layer,

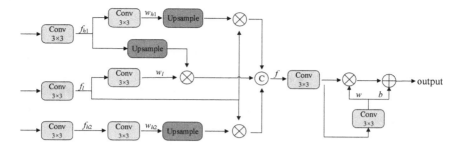

Fig. 3. Illustration of the AR module, where symbol "c" denotes the concatenation operation.

and the other high-level features generated by the MA module. In particular, both high-level inputs in the first AR module are generated by MA modules.

The detail of the AR module is shown in Fig. 3, which contains three interleaved branches. Specifically, the three inputs are firstly through three 3×3 convolutional layer, respectively, which make the number of channels of these feature maps consistent. Then the high-level feature maps f_{h1} are fed into a 3×3 convolution to obtain a semantic mask w_{h1}. Further, the w_{h1} after up-sampling and f_l are multiplied to obtain the output of the first branch. Similarly, the same operation is performed on another high-level feature maps f_{h2} to obtain the output of the second branch. The element-wise multiplication operation can more effectively focus the network on the foreground information, meanwhile suppress the influence of background noises. In addition, considering high-level features contain less detail information, we also use a 3×3 convolution on f_l to obtain a semantic mask w_l, then the w_l is multiplied with the f_{h1} after up-sampling to obtain the output which contains more complete detail information. Finally, the output features of the three branches are concatenated in channel to obtain feature maps f.

Further, we take the following operations to refine the generated density map. Firstly, a 3×3 convolution is employed to double the channel number of f, then one half of f acts as the mask w and the other half acts as bias b. Finally, we employ addition and multiplication operations to get the output.

3.3 Multi-scale Aware Module

Given that head scale vary largely in some scenes, a fixed-size convolution kernel cannot match all head sizes. To obtain rich multi-scale head features representation, we design the MA module which contains three sub-networks with different-size convolutional kernels, and the MA module can capture multi-scale information and enrich the diversity of features.

As illustrated in Fig. 4, we introduce four heterogeneous MA modules. Their outputs are regarded as high-level features and sent to the AR modules. The

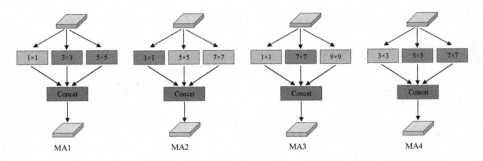

Fig. 4. Four different types of MA modules.

combination of multi-scale features can complement each other and learn more contextual information, and then we aggregate multi-scale features in AR modules.

3.4 Semantic Attention Module

To reduce false recognition of the background, we propose the SA module to enhance the channel and spatial dependencies of the feature maps and increase the attention to the crowd regions. The module is on the top layer of the backbone network.

In detail, we first feed input f into two 3×3 convolutional layers to get feature maps f_1 with 512 channels and feature maps f_2 with 256 channels, respectively. Then, we regard the first half of f_1 as the weights of f_2. After that, we add the other half of f_1 to the result from the previous multiplication operation to get the enhanced features f_{out}.

$$f_{out} = E_1(f_1) \otimes f_2 + E_2(f_1) \tag{1}$$

where $\otimes$ represents element-wise multiplication, E represents the operation of feature extraction.

3.5 Loss Function

To train the MPANet, we adopt the Euclidean loss as the loss function:

$$L(\Theta) = \frac{1}{M} \sum_{k=1}^{M} \left\| D\left(X^k; \Theta\right) - D^k \right\|_2^2 \tag{2}$$

where X^k is the k-th input image, D^k and $D\left(X^k; \Theta\right)$ denote the ground truth density map and predicted density map, respectively, Θ is the parameters of the network, and M is the size of the training set.

4   Experiments

In this section, we first describe the implementation details and evaluation metric of experiments, and then we evaluate the performance of MPANet by comparing it with the state-of-the-art methods on four challenging datasets. Finally, an ablation experiment is conducted to demonstrate the key components of the model.

4.1   Implementation Details

During the training process, we adopt the first 13 layers of the pre-trained VGG-16 as the front-end network and use Gaussian distribution with $\delta = 0.01$ to randomly initialize the other convolutional layers. In addition, we adopt the Adam optimizer with a learning rate of $1e-5$ in our network.

4.2   Evaluation Metric

Mean Absolute Error (MAE) and Mean Squared Error (MSE) are adopted to evaluate the performance of our method. They are defined as follows.

$$MAE = \frac{1}{N} \sum_{i=1}^{N} \left| C_i - \hat{C}_i \right| \tag{3}$$

$$MSE = \sqrt{\frac{1}{N} \sum_{i=1}^{N} \left| C_i - \hat{C}_i \right|^2} \tag{4}$$

where N represents the number of test images, in the i-th image, C_i represents the ground-truth count and $\hat{C}_i$ represents the estimated count. Generally, MAE reflects the accuracy of the model and MSE evaluates the robustness of the model.

4.3   Comparison with the State-of-the-art

ShanghaiTech Dataset [27]. The dataset contains 1,198 images with 330,165 annotated people heads, which is divided into Part A and Part B.

We compare the MPANet with fifteen state-of-the-art methods on Shang-haiTech dataset. As shown in Table 1, MPANet obtains advantageous performance on both Part A and Part B. Especially, compared with the SANet, our method makes significant improvement of 13.3% in MAE and 14.5% in MSE. Although other methods have a good effect on Part B, our method still obtains a promising result which decreases the MSE from 11.0 to 10.5.

UCF_CC_50 [4]. Only 50 images randomly obtained from the Internet are contained in this dataset. The annotated people count in the images varies from 94 to 4,543.

On the dataset, we employ a five-fold cross validation following the standard setting [4]. From Table 1, our proposed MPANet achieves the lowest MAE value, improving the performance by 2.1%, and it can also achieve superior performance in MSE. Although this dataset has only a few images, MPANet still obtains a better result.

UCF-QNRF [5]. This dataset is a large-scale crowd counting dataset, which contains more than 1.5 thousand images with more than 1.2 million annotated people heads.

Our method is compared to eight state-of-the-art methods and results are shown in Table 1. It can be viewed that MPANet achieves the best MSE, with a 1.6% reduction compared to the second-best, which indicates the excellent robustness of MPANet. Meantime MPANet delivers the second-best MAE, it is very close to the best AMRNet, which indicates that our MPANet performs excellently in occluded scenarios.

Table 1. Comparison with other state-of-the-art crowd counting methods on Shang-haiTech dataset, UCF_CC_50, and UCF-QNRF.

Dataset	Part A		Part B		UCF_CC_50		UCF-QNRF	
Method	MAE	MSE	MAE	MSE	MAE	MSE	MAE	MSE
MCNN [27]	110.2	173.2	26.4	41.3	377.6	509.1	277.0	426.0
CP-CNN [14]	73.6	106.4	20.1	30.1	295.8	320.9	–	–
Switch-CNN [12]	90.4	135.0	21.6	33.4	318.1	439.2	228.0	445.0
CSRNet [7]	68.2	115.0	10.6	16.0	266.1	397.5	135.4	207.4
SANet [1]	67.0	104.5	8.4	13.6	258.5	334.9	–	–
HA-CCN [15]	62.9	94.9	8.1	13.4	256.2	348.4	118.1	180.4
CAN [8]	61.3	100.0	7.8	12.2	212.2	**243.7**	107.0	183.0
SPN [2]	61.7	99.5	9.4	14.4	259.2	335.9	–	–
RANet [24]	59.4	102.0	7.9	12.9	239.8	319.4	111.0	190.0
PGCNet [22]	**57.0**	**86.0**	8.8	13.7	244.6	361.2	–	–
RPNet [23]	61.2	96.9	8.1	11.6	–	–	–	–
LSC-CNN [11]	66.4	117.0	8.1	15.7	–	–	–	–
MSCANet [20]	60.1	100.2	_6.8_	_11.0_	_181.3_	258.6	100.8	185.9
AMRNet [9]	61.59	98.36	7.0	_11.0_	184.0	265.8	**86.6**	_152.2_
PSODC [21]	65.1	104.4	7.8	12.6	–	–	–	–
MPANet	_58.1_	_89.3_	**6.7**	**10.5**	**177.5**	_245.3_	_88.3_	**149.8**

NWPU-Crowd [18]: The dataset is the largest and extremely challenging dataset at present. It is composed of 5,109 images, including 2,133,238 annotated instances. It also contains some negative samples which have no people.

We compare MPANet with five state-of-the-art methods on NWPU-Crowd. The comparison results are reported in Table 2, and it can be observed that

Table 2. Performance comparison on NWPU-Crowd dataset.

Method	MAE	MSE
MCNN [27]	224.5	708.6
CSRNet [7]	125.9	496.5
SANet [1]	195.8	512.7
CAN [8]	93.6	491.9
SFCN [19]	91.3	483.2
MPANet	**85.4**	**386.4**

MPANet obtains an MAE of 85.4 and an MSE of 386.4. Compared with the second-best performance, our method improves the results by 6.5% in MAE and 20% in MSE. The result demonstrates that our method is robust to scale variation.

We select some pictures from ShanghaiTech Part A and NWPU-Crowd, and compare the visualized results with those of the other three methods. The counts are recorded below each density map. As shown in Fig. 5, we can observe that the distribution predicted by our network is very close to the ground truth, and our method performs more robust in crowded areas.

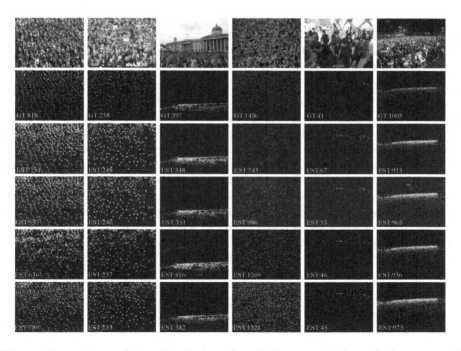

Fig. 5. Comparison of visualization results of three methods and the proposed MPANet. Line 1: input images; Line 2: the ground truth; Line 3: the result of CSRNet; Line 4: the result of CAN; Line 5: the result of SFCN; Line 6: the result of our network.

4.4 Ablation Study

To verify the effectiveness of the network components, we conduct an ablation study on ShanghaiTech Part B dataset. As shown in Table 3, the network containing all components obtains the highest accuracy, which shows that each component is necessary for the proposed model.

In the ablation study, our proposed MPANet adopts the first 13 layers of VGG-16 as the baseline to generate density maps. After introducing MA modules, the network achieves an MAE of 10.5 and an MSE of 15.3. It can be seen that the MAE is decreased by 41% compared to the baseline, which indicates that the MA module can effectively extract high-level features and improve the counting ability of the model. Further, we introduce the AR module, and the MAE is improved by 21%. This clearly shows that aggregating semantic information and spatial information can obtain more complete information. Finally, we add the SA module and achieve an MAE of 6.7 and an MSE of 10.5. Compared with the other three networks, MPANet achieves the best performance.

Table 3. Ablation study of the proposed network components.

Version	MAE	MSE
Baseline	17.8	20.5
Baseline+MA	10.5	15.3
Baseline+MA+AR	8.3	12.9
Baseline+MA+AR+SA (MPANet)	**6.7**	**10.5**

We display the visualized results of different architectures in Fig. 6. To display the effectiveness of each component more intuitively, we enlarge the background area in row 2. Obviously, the density map generated after the addition of MA modules effectively distinguishes the crowded and sparse areas in the image, which indicates that the model can solve the scale variation and uneven distribution of crowd in the image well. There is a problem with identifying background as crowd area in the density map generated by baseline, after adding SA module, the background region in the image can be accurately identified. It can be seen that our MPANet is more robust in identifying background areas and crowd areas and can effectively reduce counting errors.

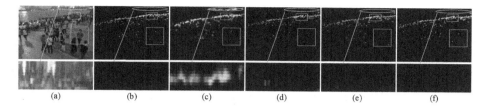

Fig. 6. Visualization results of various structures of our proposed method. (a) The original image. (b)The ground truth. (c) The result of baseline. (d) The result of baseline+MA. (e) The result of baseline+MA+AR. (f) The result of our network.

5   Conclusion

In this work, we introduce a novel architecture named MPANet which can effectively address scale variation and false recognition of the background. To effectively integrate features of different levels, we design the AR module to fuse multi-level features at different stages. Meanwhile, the MA module is constructed to capture the multi-scale information, which increases the richness of features and learns more representative features. Extensive experiments demonstrate that the MPANet achieves competitive results in comparison to the state-of-the-art methods. Especially, MPANet significantly improves the counting performance in extremely congested crowd scenes. In future work, we would like to explore crowd localization for extremely congested scenes.

Acknowledgments. This work is supported by the National Natural Science Foundation of China (61976127).

References

1. Cao, X., Wang, Z., Zhao, Y., Su, F.: Scale aggregation network for accurate and efficient crowd counting. In: Proceedings of the European Conference on Computer Vision, pp. 734–750 (2018)
2. Chen, X., Bin, Y., Sang, N., Gao, C.: Scale pyramid network for crowd counting. In: 2019 IEEE Winter Conference on Applications of Computer Vision, pp. 1941–1950 (2019)
3. Gao, G., Gao, J., Liu, Q., Wang, Q., Wang, Y.: CNN-based density estimation and crowd counting: a survey. arXiv preprint arXiv:2003.12783 (2020)
4. Idrees, H., Saleemi, I., Seibert, C., Shah, M.: Multi-source multi-scale counting in extremely dense crowd images. In: International Conference on Computer Vision and Pattern Recognition, pp. 2547–2554 (2013)
5. Idrees, H., Tayyab, M., Athrey, K., Zhang, D., Al-Maadeed, S., Rajpoot, N., Shah, M.: Composition loss for counting, density map estimation and localization in dense crowds. In: Proceedings of the European Conference on Computer Vision, pp. 532–546 (2018)

6. Jiang, X., Xiao, Z., Zhang, B., Zhen, X., Cao, X., Doermann, D., Shao, L.: Crowd counting and density estimation by trellis encoder-decoder networks. In: International Conference on Computer Vision and Pattern Recognition, pp. 6133–6142 (2019)

7. Li, Y., Zhang, X., Chen, D.: CSRNet: dilated convolutional neural networks for understanding the highly congested scenes. In: International Conference on Computer Vision and Pattern Recognition, pp. 1091–1100 (2018)

8. Liu, W., Salzmann, M., Fua, P.: Context-aware crowd counting. In: International Conference on Computer Vision and Pattern Recognition, pp. 5099–5108 (2019)

9. Liu, X., Yang, J., Ding, W.: Adaptive mixture regression network with local counting map for crowd counting. arXiv preprint arXiv:2005.05776 (2020)

10. Ronneberger, O., Fischer, P., Brox, T.: U-net: convolutional networks for biomedical image segmentation. In: Navab, N., Hornegger, J., Wells, W.M., Frangi, A.F. (eds.) MICCAI 2015. LNCS, vol. 9351, pp. 234–241. Springer, Cham (2015). https://doi.org/10.1007/978-3-319-24574-4_28

11. Sam, D.B., Peri, S.V., Sundararaman, M.N., Kamath, A., Radhakrishnan, V.B.: Locate, size and count: accurately resolving people in dense crowds via detection. IEEE Trans. Pattern Anal. Mach. Intell. (2020)

12. Sam, D.B., Surya, S., Babu, R.V.: Switching convolutional neural network for crowd counting. In: International Conference on Computer Vision and Pattern Recognition, pp. 4031–4039 (2017)

13. Simonyan, K., Zisserman, A.: Very deep convolutional networks for large-scale image recognition. arXiv preprint arXiv:1409.1556 (2014)

14. Sindagi, V.A., Patel, V.M.: Generating high-quality crowd density maps using contextual pyramid CNNs. In: Proceedings of the IEEE International Conference on Computer Vision, pp. 1861–1870 (2017)

15. Sindagi, V.A., Patel, V.M.: HA-CCN: hierarchical attention-based crowd counting network. IEEE Trans. Image Process. **29**, 323–335 (2019)

16. Song, Q., et al.: To choose or to fuse? Scale selection for crowd counting. In: Proceedings of the AAAI Conference on Artificial Intelligence, pp. 2576–2583 (2021)

17. Valloli, V.K., Mehta, K.: W-net: reinforced U-net for density map estimation. arXiv preprint arXiv:1903.11249 (2019)

18. Wang, Q., Gao, J., Lin, W., Li, X.: NWPU-crowd: a large-scale benchmark for crowd counting and localization. IEEE Trans. Pattern Anal. Mach. Intell. **43**(6), 2141–2149 (2020)

19. Wang, Q., Gao, J., Lin, W., Yuan, Y.: Learning from synthetic data for crowd counting in the wild. In: International Conference on Computer Vision and Pattern Recognition, pp. 8198–8207 (2019)

20. Wang, X., Lv, R., Zhao, Y., Yang, T., Ruan, Q.: Multi-scale context aggregation network with attention-guided for crowd counting. In: International Conference on Signal Processing, pp. 240–245 (2020)

21. Wang, Y., Hou, J., Hou, X., Chau, L.P.: A self-training approach for point-supervised object detection and counting in crowds. IEEE Trans. Image Process. **30**, 2876–2887 (2021)

22. Yan, Z., et al.: Perspective-guided convolution networks for crowd counting. In: Proceedings of the IEEE International Conference on Computer Vision (2019)

23. Yang, Y., Li, G., Wu, Z., Su, L., Huang, Q., Sebe, N.: Reverse perspective network for perspective-aware object counting. In International Conference on Computer Vision and Pattern Recognition, pp. pp. 4374–4383 (2020)

24. Zhang, A., et al.: Relational attention network for crowd counting. In: Proceedings of the IEEE International Conference on Computer Vision, pp. 6788–6797 (2019)

25. Zhang, A., et al.: Attentional neural fields for crowd counting. In: Proceedings of the IEEE International Conference on Computer Vision, pp. 5714–5723 (2019)
26. Zhang, L., Shi, M., Chen, Q.: Crowd counting via scale-adaptive convolutional neural network. In: 2018 IEEE Winter Conference on Applications of Computer Vision, pp. 1113–1121 (2018)
27. Zhang, Y., Zhou, D., Chen, S., Gao, S., Ma, Y.: Single-image crowd counting via multi-column convolutional neural network. In: International Conference on Computer Vision and Pattern Recognition, pp. 589–597 (2016)

AFLLC: A Novel Active Contour Model Based on Adaptive Fractional Order Differentiation and Local-Linearly Constrained Bias Field

Yingying Han[1,2], Jiwen Dong[1,2], Fan Li[1,2], Xiaohui Li[1,2], Xizhan Gao[1,2], and Sijie Niu[1,2(✉)]

[1] School of Information Science and Engineering, University of Jinan,
Jinan 250022, China
[2] Shandong Provincial Key Laboratory of Network-Based Intelligent Computing,
Jinan 250022, China

Abstract. In this work, we propose a novel active contour model based on adaptive fractional order differentiation and the local-linearly constrained bias field for coping with images caused by complex intensity inhomogeneity and noise. First, according to the differentiation properties of Fourier transform, we employ the Fourier transform and the Inverse Fourier transform to obtain a global nonlinear boundary enhancement image. In order to overcome the difficulty of setting the optimal order manually, an adaptive selection strategy of the order of fractional differentiation is presented by normalizing the average gradient amplitude. Then, according to the image model, the energy functional is constructed in terms of the level set by taking the fractional differentiation image as the guiding image. In energy functional, local linear functions are used to describe the bias field and construct the local region descriptor since they can flexibly deal with local intensity variety and ensure the overall data fitting. Finally, experimental results demonstrate our proposed model achieves encouraging performance compared with state-of-the-art methods on two datasets.

Keywords: Image segmentation · Level set · Fractional differentiation · Local linear model

1 Introduction

Image segmentation methods have been deeply developed and widely applied to various fields such as objection detection [13] and medical image analysis [17]. Among numerous segmentation methods, active contour models (ACMs) based on level set are widely used since they can deal with complex topological structures flexibly [15]. In general, the traditional ACMs are classified to edge-based [6,18] and region-based [1–3,7–9,11,12,14,16] according to the information concerned in the process of segmentation. Compared with the edge-based ACMs, the region-based ACMs evolve contour by minimizing the distance between the

T. Mantoro et al. (Eds.): ICONIP 2021, LNCS 13110, pp. 458–469, 2021.
https://doi.org/10.1007/978-3-030-92238-2_38

constructed region descriptor and the guide image, which are more robust to the weak boundary and initial condition than edge-based methods.

In real practice, the collected images are always destroyed by the complex intensity inhomogeneity, which leads to a large number of weak boundaries in images. It is a challenging task for the segmentation of weak boundaries. To cope with this problem, numerous region-based ACMs [1,3,7–9,11,12,14,16] are proposed by integrating image model to improve their segmentation performance. Among these methods, some are based on the assumption that the variation of the bias field is slow and the bias field can be depicted as a constant in the neighborhood, such as [8,12]. And others, like [7], adopt a linear combination of polynomials to describe the global bias field. However, for the image disturbed by severe intensity inhomogeneity, its intensity of bias field has a large variation even in one local window. Hence, the local region descriptor biased on the constant cannot characterize the local bias field correctly and accounts for the inaccurate segmentation result. And utilizing a set of polynomials to describe the global bias field often needs to lose some accuracy of the local intensity fitting for catering to the holistic intensity fitting. In general, when facing the image with complex intensity inhomogeneity, the above two types of methods are easily fallen into local optimum, and an appropriate approach for characterizing the bias field remain to be explored.

Compared with integer differentiation, fractional differentiation possesses with global property, i.e. the fractional differentiation in each pixel is defined by whole information. Fractional differentiation is capable of global-nonlinearly enhancing high-frequency boundary information as well as preserving low-frequency region information. Therefore, several ACMs [3,11] based on fractional differentiation are proposed to alleviate the issue of weak boundaries. In [3], the fractional differentiation is used to enhance high frequency boundary to improve image quality. However, the appropriate orders of fractional differentiation are hard to obtain. To avoid manual selection of the appropriate order, Li et al. [11] presented an ACM based on adaptive fractional order differentiation. And the fractional differentiation image is obtained by utilizing the first three neighborhoods on each pixel, which discards the global property of fractional differentiation. As a result, it strengthens local high-frequency information effectively but destroys the whole structure information.

In this work, to overcome the influence caused by complex intensity inhomogeneity and noise, we propose a novel ACM based on adaptive fractional order differentiation (AF) and the local-linearly constrained bias field (LLC). First, according to the differentiation properties of the Fourier transform, we employ the Fourier transform and the Inverse Fourier transform to obtain global non-linear boundary enhancement information. And an adaptive selection strategy of the order of fractional differentiation is presented by normalizing the average gradient amplitude to overcome the difficulty of setting the optimal order manually. Then, according to the image model, we construct an energy functional by taking the fractional differentiation image as the guiding image. Although fractional differentiation is able to enhance the information of weak boundaries, it is hard to eliminate the low-frequency intensity inhomogeneity. Therefore, we

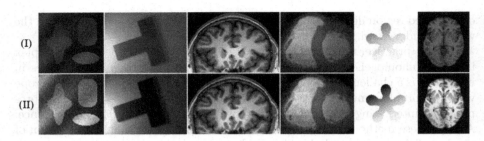

Fig. 1. Comparison of the fractional differentiation image (I) and the original image (II).

utilize the local linear function to describe the bias field and then construct a novel local region descriptor, which cannot only deal with local intensity change but accurately describe the whole image distribution. Finally, extensive experiments on two datasets demonstrate our proposed method can achieve superior performance compared with state-of-the-art methods. Especially, AFLLC possesses excellent robustness to initial active contour.

2 Methodology

2.1 Adaptive Fractional Order Differentiation

Compared with integer-based differentiation, fractional differentiation possesses the global property, which can global-nonlinearly enhance high-frequency boundary information as well as preserve low-frequency region information. As shown in Fig. 1, compared with the original image in Fig. 1(I), the boundary of object of fractional differentiation image (Fig. 1(II)) is clearer. To make full use of the global property of fractional differentiation, according to the differential properties of the Fourier transform and the equivalent differential property between the real number field and Fourier field, i.e. Eq. 1, we employ the Fourier transform and the Inverse Fourier transform to obtain global nonlinear boundary enhancement image.

$$D^n f = f^n(t) \leftrightarrow \mathcal{F}(f^n(t)) = (i\omega)^n \hat{f}(\omega) \tag{1}$$

where $n \in N$ represents the order of differentiation. According to the equivalent differential property described in Eq. 1, the fractional differentiation of the image $I(x, y) : \Omega \to \mathcal{R}$ in the x and y directions is calculated as follows [14].

$$\begin{cases} D_x^\alpha I(x, y) = F^{-1}\left(\left(1 - exp\left(\frac{-i2\pi\omega_1}{m}\right)\right)^\alpha exp\left(\frac{i\pi\alpha\omega_1}{m}\right)\hat{I}(\omega_1, \omega_2)\right) \\ D_y^\alpha I(x, y) = F^{-1}\left(\left(1 - exp\left(\frac{-i2\pi\omega_2}{n}\right)\right)^\alpha exp\left(\frac{i\pi\alpha\omega_2}{n}\right)\hat{I}(\omega_1, \omega_2)\right) \end{cases} \tag{2}$$

The fractional differential $D^\alpha I(x, y)$ of the image function $I(x, y)$ is defined by the gradient amplitude as follows.

$$D^\alpha I(x, y) = \sqrt{\left(D_x^\alpha I(x, y)\right)^2 + \left(D_y^\alpha I(x, y)\right)^2} \tag{3}$$

However, the reasonable order of fractional differentiation usually relies on consider-able repetitive experiments to manually select. Hence, in this work, based on the degree of average gradient information, a novel way to define the order adaptively is designed as Eq. 4.

$$\alpha = \frac{mean_{(x,y)\in\Omega}\left(|\nabla I\left(x,y\right)|\right)}{\max_{(x,y)\in\Omega}\{|\nabla I\left(x,y\right)|\}} \tag{4}$$

where $|\nabla I\left(x,y\right)|$ is the gradient image of the given image I. The α is a scalar in the range of $(0,1)$ and is used to characterize the normalizing average gradient amplitude of I. In the image domain, smaller and larger gradient amplitudes are corresponding to the image along with low-frequency and high-frequency, respectively. And images with different average frequencies are presumed to be treated unequally.

2.2   Local-Linearly Constrained Bias Field

Generally, a captured image can be modeled as a multiplicative model [8], i.e. $I\left(\boldsymbol{x}\right) = b\left(\boldsymbol{x}\right)J\left(\boldsymbol{x}\right) + n\left(\boldsymbol{x}\right), \boldsymbol{x}(x,y) \in \Omega$. The b is the bias field, which is composed of low-frequency information and assumed as slowly varying. The n is Gaussian noise with zero-mean. The J describes the intrinsic property of the image, which is usually represented by constants. Based on the image model, the fractional differentiation image $D^\alpha I\left(x,y\right)$ is formulated as $D^\alpha I\left(\boldsymbol{x}\right) = D^\alpha b\left(\boldsymbol{x}\right)D^\alpha J\left(\boldsymbol{x}\right) + D^\alpha n\left(\boldsymbol{x}\right), \boldsymbol{x} \in \Omega$. Fractional differentiation can nonlinearly enhance high-frequency boundary information and effectively preserve low-frequency information. Hence, the $D^\alpha I\left(x,y\right)$ has a similar distribution of the bias field with $I\left(x,y\right)$. For facing with complicatedly inhomogeneous image, in this work, the local linear functions are used to describe the bias field, since they combine the advantages of detail description of local operation and the flexibility of linear fitting.

$$b\left(x,y\right) = w_1 + w_2 x + w_3 y, (x,y) \in U\left(\boldsymbol{x},\delta\right) \tag{5}$$

where $U\left(\boldsymbol{x},\delta\right) = (x,y)\,|\,|x - x_0| < \delta, |y - y_0| < \delta$ represents the neighborhood centered as $\boldsymbol{x}\left(x_0,y_0\right) \in \Omega$. According to the multiplicative image model, in a neighborhood, we can construct the following local region descriptor $F(\boldsymbol{y})$.

$$F\left(\boldsymbol{y}\right) = \boldsymbol{w}^T\left(\boldsymbol{x}\right)\boldsymbol{g}\left(\boldsymbol{y}\right)c_i, \; \boldsymbol{y} \in U\left(\boldsymbol{x},\delta\right) \tag{6}$$

where $\boldsymbol{w} = \{w_1, w_2, w_3\}^T$, $\boldsymbol{g} = \{1, x, y\}^T$. Then, deriving by local intensity clustering property of the image intensities, a local clustering function based on the image model is constructed in a neighborhood. The corresponding global clustering function is given by integrating the local clustering function in image domain, i.e. Eq. 7.

$$E = \int_\Omega \sum_{i=1}^N \int_{\Omega_i} K_\sigma\left(\boldsymbol{x} - \boldsymbol{y}\right)|D^\alpha I\left(\boldsymbol{y}\right) - \boldsymbol{w}^T\left(\boldsymbol{x}\right)\boldsymbol{g}\left(\boldsymbol{y}\right)c_i|^2 d\boldsymbol{y}d\boldsymbol{x}. \; i = 1,...,N \tag{7}$$

where $K_\sigma(t) = \frac{1}{\sqrt{2\pi}\sigma} exp\left(-\frac{|t|^2}{2\sigma^2}\right)$ is used to obtain the local window. For the numerical solution, the level set functions are applied to represent the disjoint regions. Besides, two regularization terms [10], i.e. $\mathcal{L}(\phi) = \int |\nabla H(\phi(\boldsymbol{x}))|\, d\boldsymbol{x}$ and $\mathcal{R}(\phi) = \frac{1}{2}\int |\nabla\phi(\boldsymbol{x}) - 1|^2\, d\boldsymbol{x}$, are used to maintain the smoothness of curves and ensure stable evolution of contour, respectively. Hence, the energy functional with respect to the level set functions is rewritten as follows.

$$E = \int_\Omega \sum_{i=1}^{N} \int_\Omega K_\sigma(\boldsymbol{x}-\boldsymbol{y}) |D^\alpha I(\boldsymbol{y}) - \boldsymbol{w}^T(\boldsymbol{x})\boldsymbol{g}(\boldsymbol{y})c_i|^2 M_i(\phi(\boldsymbol{y}))\, d\boldsymbol{y}d\boldsymbol{x} + \nu\mathcal{L} + w\mathcal{R}$$

(8)

When $N = 2$, one level set function is used to express two-phase images by the Heaviside function $H(\phi) = \frac{1}{2}\left(1 + \frac{2}{\pi}arctan\left(\frac{\phi}{\varepsilon}\right)\right)$, i.e. $M_1(\phi) = H(\phi)$ and $M_2(\phi) = 1 - H(\phi)$ represent the disjoint region Ω_1 and Ω_2, respectively. When $N > 2$, two or more level set functions are used. Taking four-phase image as an example, three level set functions $\phi = \{\phi_1, \phi_2, \phi_3\}$ are needed and the four disjoint regions, i.e. $\Omega_1, \ldots, \Omega_4$, are expressed as $M_1(\phi) = (1 - H(\phi_1))H(\phi_2)H(\phi_3)$, $M_2(\phi) = (1 - H(\phi_1))(1 - H(\phi_2))H(\phi_3)$, $M_3(\phi) = (1 - H(\phi_1))(1 - H(\phi_2))(1 - H(\phi_3))$, $M_4(\phi) = H(\phi_1)$.

2.3   Model Optimization

In order to segment the region of interest of the image, we minimize the energy functional Eq. 8 with respect to weight vector $\boldsymbol{w}$, constant c_i, and the level set function ϕ with the iterative alternating way and the gradient descent method [9].

1) Optimize c_i: fixing the variables $(\boldsymbol{w}, \phi)$, the energy functional E is minimized with respect to c_i. The obtained optimized expression of $\hat{c}_i$ is formulated as

$$\hat{c}_i = \frac{\int_\Omega \int_\Omega K_\sigma(\boldsymbol{x}-\boldsymbol{y}) D^\alpha I(\boldsymbol{y})\boldsymbol{w}^T(\boldsymbol{x})\boldsymbol{g}(\boldsymbol{y}) M_i(\phi(\boldsymbol{y}))\, d\boldsymbol{x}d\boldsymbol{y}}{\int_\Omega \int_\Omega K_\sigma(\boldsymbol{x}-\boldsymbol{y})(\boldsymbol{w}^T(\boldsymbol{x})\boldsymbol{g}(\boldsymbol{y}))^T \boldsymbol{w}^T(\boldsymbol{x})\boldsymbol{g}(\boldsymbol{y}) M_i(\phi(\boldsymbol{y}))\, d\boldsymbol{x}d\boldsymbol{y}}$$

(9)

2) Optimize $\boldsymbol{w}$: fixing the variables (c_i, ϕ), the energy functional E is minimized with respect to $\boldsymbol{w}$. The obtained optimized expression of $\hat{\boldsymbol{w}}$ is formulated as

$$\hat{\boldsymbol{w}} = \frac{\sum_{k=1}^{N} \lambda_i \int_\Omega K_\sigma(\boldsymbol{x}-\boldsymbol{y}) D^\alpha I(\boldsymbol{y})\boldsymbol{g}(\boldsymbol{y}) c_i M_i(\phi(\boldsymbol{y}))\, d\boldsymbol{y}d\boldsymbol{x}}{\sum_{k=1}^{N} \lambda_i \int_\Omega K_\sigma(\boldsymbol{x}-\boldsymbol{y})\boldsymbol{g}^T(\boldsymbol{y})\boldsymbol{g}(\boldsymbol{y}) c_i^2 M_i(\phi(\boldsymbol{y}))\, d\boldsymbol{y}d\boldsymbol{x}}$$

(10)

3) Optimize ϕ: fixing the variables $(c_i, \boldsymbol{w})$, the energy functional E is minimized with respect to ϕ. The obtained optimized expression of $\hat{\phi}$ is formulated as

$$\frac{\partial\phi}{\partial t} = -\sum_{k=1}^{N} \lambda_i \frac{\partial M_i(\phi)}{\partial\phi} e_i + \nu\delta(\phi)\, div\left(\frac{\nabla\phi}{|\nabla\phi|}\right) + \mu\left(\triangle^2\phi - div\left(\frac{\nabla\phi}{|\nabla\phi|}\right)\right)$$

(11)

where $e_i(\boldsymbol{y}) = \int_\Omega K_\sigma(\boldsymbol{x}-\boldsymbol{y}) |D^\alpha I(\boldsymbol{y}) - \boldsymbol{w}^T(\boldsymbol{x})\boldsymbol{g}(\boldsymbol{y})c_i|^2 d\boldsymbol{x}$.

3   Experiment and Analysis

Dataset: We test the performance of the proposed method on two datasets. One synthetic image dataset is provided by [16], which contains 200 images with different level of intensity inhomogeneity. The SBD [5] is composed by 1830 multiphase MRI brain images, containing two kinds of images simulated by two anatomical models: normal and multiple sclerosis (MS), and each image has four disjoint regions, i.e. background, cerebrospinal fluid, gray matter and white matter. Each type of image corresponds to three levels of intensity non-uniformity ($RF = 0\%, 20\%, 40\%$) and five levels of noise ($n = 0\%$, 3%, 5%, 7%, 9%). Based on 5 levels of noise, we divide 1830 images into $5 + 5$ groups (5 groups of MS data, 5 groups of normal data), and each group contains 183 images. Due to the small structural difference between the two adjacent images, in each group, we select 27 images as validation, where one validation group contains 9 kinds of layer structure (70 to 110 brain images with an interval of 5) and each layer corresponds to three RF.

Experimental Setting: For synthetic inhomogeneous image dataset, our parameters are set as follows: $\nu = 0.055 \times 255^2$, $\sigma = 3$, $\triangle t = 0.001$, $\lambda_1 = 1.0045$, $\lambda_2 = 1$. For the SBD, under the same noise level, the parameter settings of MS data and normal data are the same. Parameters of AFLLC in SBD are set as Table 1. All the comparison ACMs use the optimal parameters. All the numerical experiments of ACMs are performed in MATLAB (R2020a) on a windows10 (64bit) desktop computer with an Intel Core i5 3.0 GHz processor and 8.0 GB of RAM.

Besides comparing with the traditional ACMs, we also compare AFLLC with the supervise-based U-Net with AC loss [3]. For the U-Net with AC loss experiments, a 5-fold cross validation procedure is performed on two data sets. And for the SBD, the data set were divided into 156 training images and 27 validation images. Pytorch framework with version 1.8.1 is adopted to implement the U-Net and a RTX 3090 GPU is utilized to train this network. We adopt AdamW algorithm with a learning rate of 5e−4 to optimize this network. The batch size is set to 32. The network weights are initialized by the Kaiming algorithm and weight decay is set to 1e−4.

Evaluation Metrics: Dice similarity coefficient (DSC), i.e. $Dice(S, G) = \frac{S \cap G}{S \cup G}$, is used to evaluate segmentation accuracy, in which DSC value is closer to 1,

Table 1. The parameters setting of the proposed method in SBD

Parameters	ν	μ	$\triangle t$	σ	λ_1	λ_2	λ_3
n = 0%	10	1	0.07	5	1	1	1
n = 3%	30	1	0.1	5	1	1	1
n = 5%	50	1	0.1	5	1	1	1
n = 7%	70	1	0.12	5	1	1	1
n = 9%	100	1	0.15	5	1	1	1

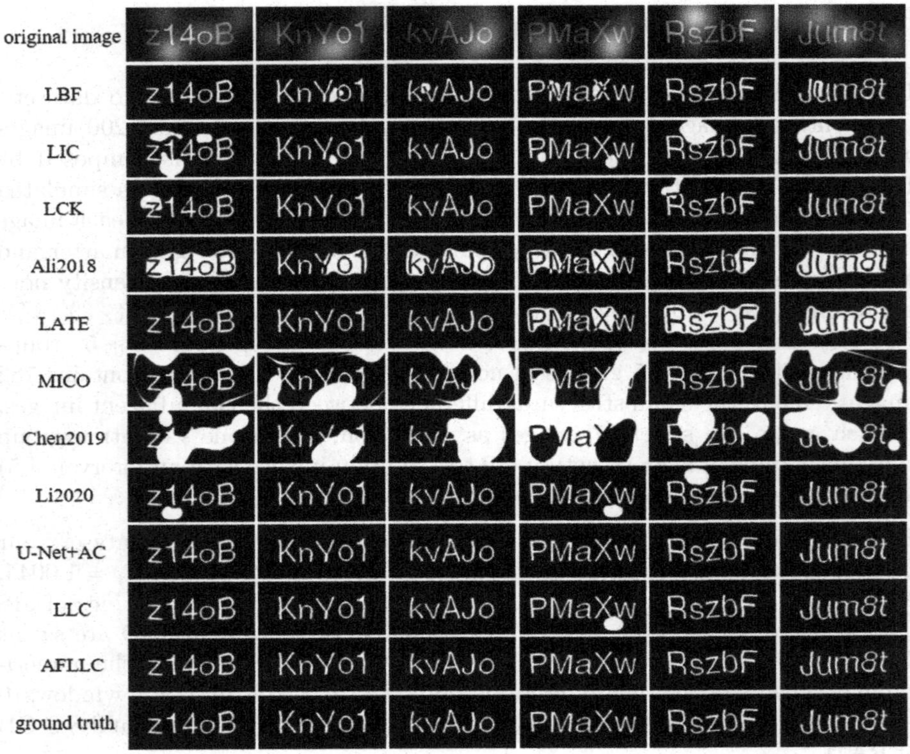

Fig. 2. From top to bottom, the example-original images, the segmentation results by LBF [9], LIC [8], LCK [16], Ali2018 [1], LATE [12], MICO [7], Chen2019 [3], Li2020 [11], U-Net with AC loss [4], LLC, AFLLC and ground truth respectively.

the result of segmentation is more outstanding. The codes of AFLLC and the trained U-Net with AC loss will be shared at https://github.com/23YingHan/AFLLC.

3.1   Experimental Analysis

In this subsection, we verify the performance of AFLLC on the synthetic dataset and SBD, and compare the segmentation results with the advanced methods. Besides, we add ablation experiments to demonstrate the effectiveness of the adaptive fractional order differentiation (AF) and the local-linearly constrained bias field (LLC).

Figure 2 presents visual comparisons of the segmentation results on synthetic dataset. It confirms that the segmentation results, given by the proposed methods, are cleaner than the compared ACMs. Figure 3(a) shows that the LLC achieves higher (larger means) and stabler (less variance) DSCs than the compared traditional ACMs. Especially, by adding the adaptive fractional differen-

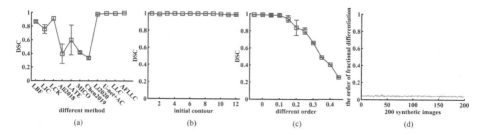

Fig. 3. (a)–(c) represent the mean and variance of DSC obtained by 11 compared methods, AFLLC with 12 different initial contours and AFFLC with different order of fractional differentiation, respectively; (d) shows the orders given by Eq. 4 on 200 synthetic images.

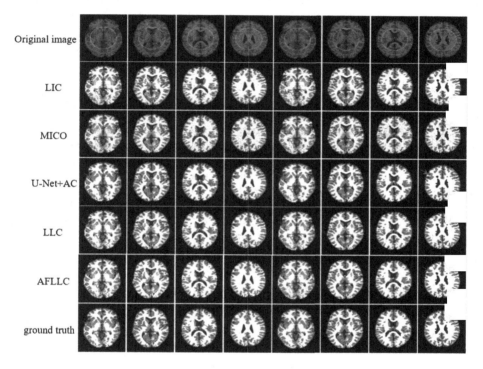

Fig. 4. From top to bottom, the original images with initial contour, the segmentation results obtained by LIC, MICO, U-Net with AC loss, LLC and AFLLC respectively, and the ground truth.

tiation to LLC, the DSCs obtained by AFLLC have further improvement and AFLLC outperforms the supervised-based U-Net with AC loss.

In Fig. 4, visual segmentation results of 8 example images on multiphase brain images with noise = 9% using LIC, MICO, U-Net with AC loss, LLC

Table 2. The DSCs of the comparison methods on SBD

		Multiphase normal brain images					Multiphase MS brain images				
Noise	Method RF	LIC	MICO	U-net +AC	LLC	AF +LLC	LIC	MICO	U-net +AC	LLC	AF +LLC
0%	0%	0.9567	0.9293	0.9585	0.9637	**0.9638**	0.9565	0.9630	0.9517	0.9637	**0.9638**
	20%	0.9560	0.9630	0.9563	0.9636	**0.9639**	0.9556	0.9630	0.9550	0.9636	**0.9637**
	40%	0.9525	0.9629	0.9588	0.9632	**0.9634**	0.9521	0.9629	0.9567	**0.9635**	0.9634
3%	0%	0.9443	0.9131	0.9480	0.9468	**0.9494**	0.9444	0.9470	0.9448	0.9465	**0.9498**
	20%	0.9450	0.9469	0.9487	0.9466	**0.9507**	0.9436	0.9475	0.9468	0.9476	**0.9506**
	40%	0.9446	0.9481	0.9484	0.9474	**0.9511**	0.9405	0.9485	0.9477	0.9482	**0.9514**
5%	0%	0.9268	0.9210	**0.9359**	0.9282	0.9317	0.9273	0.9229	**0.93**	0.9294	0.9298
	20%	0.9267	0.9242	**0.9382**	0.9316	0.9337	0.9271	0.9254	**0.9337**	0.9329	0.9329
	40%	0.9286	0.9263	**0.9376**	0.9338	0.9353	0.9288	0.9276	0.9351	0.9351	**0.9353**
7%	0%	0.9083	0.8978	**0.9253**	0.9126	0.9174	0.9089	0.9011	**0.9226**	0.9136	0.9177
	20%	0.9107	0.9037	**0.9223**	0.9151	0.9194	0.9106	0.9065	**0.9218**	0.9153	0.9207
	40%	0.9111	0.9093	**0.9275**	0.9169	0.9214	0.9120	0.9121	**0.9255**	0.9172	0.9223
9%	0%	0.8827	0.8480	**0.9157**	0.8941	0.9008	0.8849	0.8507	**0.9149**	0.8948	0.9023
	20%	0.8857	0.8547	**0.9145**	0.8963	0.9032	0.8865	0.8581	**0.9137**	0.8978	0.9047
	40%	0.8899	0.8626	**0.9157**	0.8981	0.9044	0.8902	0.8655	**0.9131**	0.8994	0.9050

and AFLLC are displayed respectively. As shown in Fig. 4, U-Net with AC loss generates smooth segmentation boundary but losses more detail, while LIC and MICO preserve detail information but the segmentation results carry with noise. The segmentation results obtained by LLC and AFLLC contain less noise and retain more details. Table 2 displays DSCs of 10 groups of experiments (5 groups of the MS dataset and 5 groups of the normal) obtained by five models. As shown in Table 2, LLC achieves more higher DSCs than LIC and MICO, and AFLLC has further improvement of the segmentation accuracy of LLC. As the noise increases, the segmentation accuracy of AFLLC remains above 0.9. In particular, when the noise = 0% and noise = 3%, the DSCs offered by the AFLLC outperform the supervised-based U-Net with AC loss.

3.2   Robustness Analysis for Initial Condition

We test AFLLC on synthetic image dataset with 12 various initial contours (as shown Fig. 5), including position difference and shape difference. The corresponding segmentation results are shown in Fig. 3(b). From Fig. 5 and Fig. 3(b), although the position and shape of the initial contours are various, it is seen that

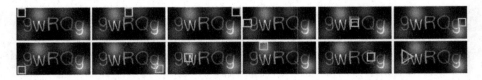

Fig. 5. 12 kinds of initial contours in the initial contour robustness experiment.

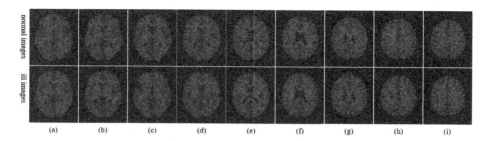

Fig. 6. Example images with random initial contours (9 ill images, 9 normal images) selected from the SBD dataset.

Table 3. The segmentation accuracy gained by AFLLC with random initialization on SBD.

Multiphase normal brain images					Multiphase ill brain images				
Noise RF	0%	3%	5%	7%	9%	0%	3%	5%	7%	9%
0%	0.9604	0.9406	0.9057	0.9111	0.8923	0.9472	0.9456	0.9252	0.8989	0.8936
20%	0.9639	0.9502	0.9274	0.9040	0.8873	0.9617	0.9396	0.9315	0.9084	0.8901
40%	0.9640	0.9482	0.9373	0.9150	0.8924	0.9537	0.9296	0.9296	0.9090	0.8912

the DSCs obtained by AFLLC are competitive and stable, and the maximum difference between minimum DSC and maximum is not more than 0.05.

We also test AFLLC on SBD with random initial contours (shown in Fig. 6) and the segmentation results are shown in Table 3. As shown in Table 3, despite the initial contours are random, AFLLC is still able to accurately capture the region of interests. According to above two experiments, it is easy to know that the proposed method has excellent robustness to the initial contour, first because the introduction of adaptive fractional differentiation nonlinearity enhances the high-frequency boundary information to improve the image quality, and second because the introduction of local linear functions flexibly deals with complex inhomogeneous images and accurately fit the image intensity.

3.3 The Validity Analysis of the Definition of Adaptive Order

We design a new way to define the order of fractional differentiation adaptively according to the normalized average gradient amplitude of the given image. For demonstrating its effectiveness, we quantitatively analyze the segmentation result of the AFLLC with different fractional orders on the synthetic image dataset. 11 orders are selected from 0 to 1 with 0.1 interval. The corresponding 11 kinds of segmentation results are shown in Fig. 3(c). According to Fig. 3(c), it can be seen that the segmentation accuracy is inversely proportional to the order, and the appropriate order is between 0 to 0.1. Figure 3(d) shows the corresponding orders of 200 synthetic images obtained by the adaptive fractional order definition, and the orders range vary from 0 to 0.1. Therefore, the experiments verify the effectiveness of the designed adaptive definition of fractional order.

4   Conclusion

In this work, a novel ACM is proposed to segment images disturbed by complex inhomogeneity and noise. Firstly, we use the fractional differential to global-nonlinearly enhance high frequency boundary, where the order of fractional differentiation is selected adaptively. Then, guiding by fractional differentiation image, we construct the energy functional in terms of level set function and the bias field, where the local linear function is used to describe the bias field since it can flexibly characterize local intensity changes and ensure the overall data fitting. In addition, the proposed model is capable of extending to the multi-level set model for dealing with multiphase images. Extensive experiments on two datasets with complex intensity and different level of noise prove the effectiveness and robustness of the proposed method.

Acknowledgments. This work is supported by the National Natural Science Foundation of China under Grant No. 61701192, No. 61872419, No.61873324, the Natural Science Foundation of Shandong Province, China, under Grant No. ZR2020QF107, No. ZR2020MF137, No. ZR2019MF040, ZR2019MH106, No. ZR2018BF023, the China Postdoctoral Science Foundation under Grants No. 2017M612178. University Innovation Team Project of Jinan (2019GXRC015), Key Science & Technology Innovation Project of Shandong Province (2019JZZY010324, 2019JZZY010448), and the Higher Educational Science and Technology Program of Jinan City under Grant with No. 2020GXRC057. The National Key Research and Development Program of China (No. 2016YFC13055004).

References

1. Ali, H., Rada, L., Badshah, N.: Image segmentation for intensity inhomogeneity in presence of high noise. IEEE Trans. Image Process. **27**(8), 3729–3738 (2018)
2. Chan, T.F., Vese, L.A.: Active contours without edges. IEEE Trans. Image Process. **10**(2), 266–277 (2001)
3. Chen, B., Huang, S., Liang, Z., Chen, W., Pan, B.: A fractional order derivative based active contour model for inhomogeneous image segmentation. Appl. Math. Model. **65**, 120–136 (2019)
4. Chen, X., Williams, B.M., Vallabhaneni, S.R., Czanner, G., Williams, R., Zheng, Y.: Learning active contour models for medical image segmentation. In: Proceedings of the IEEE/CVF Conference on Computer Vision and Pattern Recognition, pp. 11632–11640 (2019)
5. Cocosco, C.A., Kollokian, V., Kwan, R.K.S., Pike, G.B., Evans, A.C.: BrainWeb: olnline interface to a 3D MRI simulated brain database. In: NeuroImage. Citeseer (1997)
6. Gupta, D., Anand, R.: A hybrid edge-based segmentation approach for ultrasound medical images. Biomed. Signal Process. Control **31**, 116–126 (2017)

7. Li, C., Gore, J.C., Davatzikos, C.: Multiplicative intrinsic component optimization (MICO) for MRI bias field estimation and tissue segmentation. Magn. Reson. Imaging **32**(7), 913–923 (2014)

8. Li, C., Huang, R., Ding, Z., Gatenby, J.C., Metaxas, D.N., Gore, J.C.: A level set method for image segmentation in the presence of intensity inhomogeneities with application to MRI. IEEE Trans. Image Process. **20**(7), 2007–2016 (2011)

9. Li, C., Kao, C.Y., Gore, J.C., Ding, Z.: Implicit active contours driven by local binary fitting energy. In: 2007 IEEE Conference on Computer Vision and Pattern Recognition, pp. 1–7. IEEE (2007)

10. Li, C., Xu, C., Gui, C., Fox, M.D.: Distance regularized level set evolution and its application to image segmentation. IEEE Trans. Image Process. **19**(12), 3243–3254 (2010)

11. Li, M.M., Li, B.Z.: A novel active contour model for noisy image segmentation based on adaptive fractional order differentiation. IEEE Trans. Image Process. **29**, 9520–9531 (2020)

12. Min, H., Jia, W., Zhao, Y., Zuo, W., Ling, H., Luo, Y.: LATE: a level-set method based on local approximation of Taylor expansion for segmenting intensity inhomogeneous images. IEEE Trans. Image Process. **27**(10), 5016–5031 (2018)

13. Min, Y., Xiao, B., Dang, J., Yue, B., Cheng, T.: Real time detection system for rail surface defects based on machine vision. EURASIP J. Image Video Process. **2018**(1), 1–11 (2018). https://doi.org/10.1186/s13640-017-0241-y

14. Niu, S., Chen, Q., De Sisternes, L., Ji, Z., Zhou, Z., Rubin, D.L.: Robust noise region-based active contour model via local similarity factor for image segmentation. Pattern Recogn. **61**, 104–119 (2017)

15. Osher, S., Fedkiw, R.P.: Level set methods: an overview and some recent results. J. Comput. Phys. **169**(2), 463–502 (2001)

16. Wang, L., Pan, C.: Robust level set image segmentation via a local correntropy-based K-means clustering. Pattern Recogn. **47**(5), 1917–1925 (2014)

17. Xing, R., Niu, S., Gao, X., Liu, T., Fan, W., Chen, Y.: Weakly supervised serous retinal detachment segmentation in SD-OCT images by two-stage learning. Biomed. Opt. Express **12**(4), 2312–2327 (2021)

18. Zhu, G., Zhang, S., Zeng, Q., Wang, C.: Boundary-based image segmentation using binary level set method. Opt. Eng. **46**(5), 050501 (2007)

DA-GCN: A Dependency-Aware Graph Convolutional Network for Emotion Recognition in Conversations

Yunhe Xie, Chengjie Sun[(✉)], Bingquan Liu, and Zhenzhou Ji

School of Computer Science and Technology, Harbin Institute of Technology,
Harbin, China
{xieyh,sunchengjie,liubq,jizhenzhou}@hit.edu.cn

Abstract. Emotion Recognition in Conversations (ERC) has recently gained much attention from the NLP community. The contextual information and the dependency information are two key factors that contribute to the ERC task. Unfortunately, most of the existing approaches concentrate on mining contextual information while neglecting the dependency information. To address this problem, we propose a Dependency-Aware Graph Convolutional Network (DA-GCN) to jointly take advantage of these two kinds of information. The core module is a proposed dependency-aware graph interaction layer where a GCN is constructed and operates directly on the dependency tree of the utterance, achieving to consider the dependency information. In addition, the proposed layer can be stacked to further enhance the embeddings with multiple steps of propagation. Experimental results on three datasets show that our model achieves the state-of-the-art performance. Furthermore, comprehensive analysis empirically verifies the effectiveness of leveraging the dependency information and the multi-step propagation mechanism.

Keywords: Emotion recognition in conversations · Network representation learning · Natural language processing

1 Introduction

Emotion recognition has been very popular in Natural Language Processing (NLP) due to its wide application in opinion mining [18], recommendation systems [2], medical care [4], etc. Early research on emotion recognition focused on comprehending emotions in monologues [1]. However, the recent surge of open conversational data has caused serious attention towards emotion recognition in conversations (ERC) [6,25,28]. ERC is the task of detecting emotions from utterances in a conversation [9] and can be treated as a sequence classification task that maps the utterance sequence to the corresponding emotion label.

Intuitively, there are two key factors that contribute to the ERC task. One is *the contextual information* propagated according to the utterance chronological order and the other is *the dependency information* determined by the syntactic

© Springer Nature Switzerland AG 2021
T. Mantoro et al. (Eds.): ICONIP 2021, LNCS 13110, pp. 470–481, 2021.
https://doi.org/10.1007/978-3-030-92238-2_39

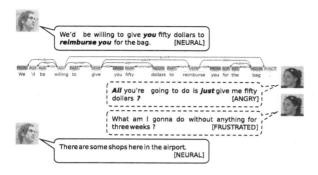

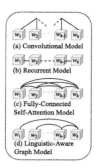

Fig. 1. A conversation clip from IEMOCAP. Below the first utterance is an example of a dependency tree where words are connected based on syntactic dependencies.

Fig. 2. Neural contextual encoders.

structure of the utterance itself. For the second utterance in Fig. 1, after capturing the contextual information "All", "just" and "?", models can infer the correct result (ANGRY). For the first utterance, tokens related to the key token "reimburse" are restricted and the distance between them is shortened by a single path based on their syntactic dependencies. Using dependency information alleviates the negativity of the key token and gets the correct judgment (NEURAL). Thus, it's critical to take the two sources of information into account.

To this end, [14,22] leverage convolutional models (Fig. 2(a)) to capture word's meaning by aggregating the local information but fail to capture the global contextual information. [10,13] utilize recurrent models (Fig. 2(b)) to learn contextual representations of words while affected by the long-term dependency problem. [16,26] exploit a fully-connected graph (Fig. 2(c)) to model the relation of every two words and let the model learn the structure by itself but easy to overfit on modestly-sized datasets. As analyzed in [21], linguistic-aware graph structure (Fig. 2(d)) can provide helpful inductive bias. Moreover, for the ERC task, as shown in Fig. 1, we find a dependency tree shortens the distance between the tokens, captures the syntactic relations between tokens, and offers syntactic paths of arbitrary utterances for information propagation.

Motivated by the above, we propose a **D**ependency-**A**ware **G**raph **C**onvolutional **N**etwork (DA-GCN) for ERC. The core module is a proposed dependency-aware graph interaction layer that considers the dependency information. In the dependency-aware graph, we perform information propagation by presenting a convolution over the dependency tree of the utterance. In summary, the main contributions of our work are concluded as follows: (i) We make the first attempt to simultaneously incorporate contextual- and dependency-information for ERC. (ii) Experiment results on three benchmarks show that our model achieves significant improvements compared to all baseline models. (iii) We thoroughly study different graph layers and present extensive experiments demonstrating the benefit of our proposed framework.

2   Related Work

Unlike the vanilla emotion recognition of sentences/paragraphs, ERC obviously needs to model the target utterance in each conversation to obtain the representation of the context. For utterances in a conversation, context refers to all the conversation records before the moment when the target utterance is uttered. According to the actual situation of the task, the context sometimes also includes all the conversation records after the moment when the target utterance is spoken. This context is determined by historical conversational information and depends on the temporal sequence of the utterances. Therefore, compared with the recently published work focusing on solving ERC tasks, neither lexicon-based [23] nor modern deep learning-based [12] emotion recognition methods can work well on ERC datasets.

Early work in the ERC field is dedicated to solving contextual perception and emotional dynamics to find contextualized conversational utterance representations. Most models adopt a hierarchical structure [7,11,19], combined with the powerful components of recurrent neural networks (RNNs), memory networks and attention mechanisms to achieve good results on the ERC task. Specifically, these models continuously input the context-independent initial word embeddings to the RNNs to obtain the context-independent and the context-aware utterance representations. Some works also use multiple memory networks and assist in a multi-hop mechanism to perform emotional reasoning [8]. Then the attention mechanism is leveraged to generate a summary of historical conversation information, which are integrated into the context-aware utterance representation to perform the final emotion judgment. The above works once again prove the importance of contextual modeling for conversation.

With the rise of the self-attention mechanism, some work based on the transformer structure [24] has emerged in recent years. Due to its rich representation and fast calculation, the transformer has been applied to many NLP tasks, such as document machine translation, response matching in dialogue systems and language modeling. Utilizing advanced pre-training language models (PLMs) to obtain word embeddings [27] or sentence embeddings [29] and exploiting generalized transformer structure for long-distance utterance feature capture has become a new paradigm for solving ERC task models, which further improves the lower limit of model capabilities.

Our proposed model differs from the existing models in the sense that we jointly take advantage of the contextual information and the dependency information rather than only mine the contextual information. Specifically, we firstly exploit a Bi-directional Long Short Term Memory (BiLSTM) to learn the contextual representation of the utterance. Then, we further enhance the embeddings with a GCN that operates directly on the dependency tree of the utterance, which is the core module named dependency-aware graph interaction layer. The proposed layer can be stacked to further enhance the embeddings with multiple steps of propagation. Such operations allow dependency information to be transferred from words to words by a single path based on their syntactic dependencies.

3   Task Definition

Assume that there are a set of conversations $\mathcal{D} = \{\mathbf{D}_j\}_{j=1}^{L}$, where L is the number of the conversation. In each conversation, $\mathbf{D}_j = \{(\mathbf{u}_i, s_i, e_i)\}_{i=1}^{N_i}$ is a sequence of N_i utterances, where the utterance $\mathbf{u}_i$ is spoken by the speaker $s_i \in \mathcal{S}$ with a predefined emotion $e_i \in \mathcal{E}$. All speakers compose the set $\mathcal{S}$ and the set $\mathcal{C}$ consists of all emotions, such as anger, happiness, sadness and neutral. Our goal is to train a model to detect each new utterance with an emotion label from $\mathcal{C}$ as accurately as possible.

4   Proposed Approach

In this section, we describe the architecture of our model, as illustrated in Fig. 3. It is mainly composed of four components: an utterance reader, an utterance encoder, a stack of dependency-aware graph layers to incorporate the contextual information and the dependency information through a residual connection, and a decoder for the final prediction. Briefly, our model takes as input a dependency tree of every utterance. Vertex embeddings of the dependency tree are initially modeled utilizing a BiLSTM, and the embeddings are further enhanced via a GCN. Finally, an aggregator is applied over the enhanced embeddings to distill a dense vector embedding for the classification task. In the following sections, the details of our framework are given.

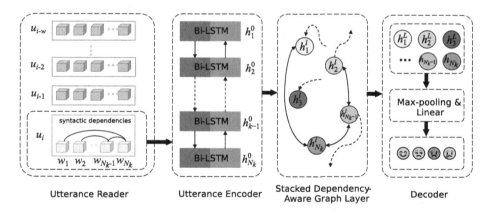

Fig. 3. The illustration of our proposed framework.

4.1   Utterance Reader

For the ith utterance in $\mathbf{D}_j$, $\mathbf{u}_i = \{w_k\}_{k=1}^{N_k}$, where w_k is the index of the word in the vocabulary and N_k is the number of tokens in the utterance $\mathbf{u}_i$, namely,

$$\mathbf{u}_i = \{w_1, w_2, \cdots, w_{N_k}\}. \tag{1}$$

$\mathbf{u}_i$ is then passed to the embedding layer to get a dense matrix $\mathbf{g}_i$ for each utterance:

$$\mathbf{g}_i = \{\mathbf{g}_1, \mathbf{g}_2, \cdots, \mathbf{g}_{N_k}\}, \tag{2}$$

where $\mathbf{g}_i \in \mathbb{R}^{D_g \times N_k}$ is treated as input hidden states to the followed utterance encoder and D_g denotes the output dimension of the word vector.

4.2 Utterance Encoder

Based on the hierarchical structure idea, we try to transfer the independent initial word embeddings into the context-sensitive word representations in this section. Though current works [8,17] utilize CNN to extract utterance features, we decide to adopt a BiLSTM. The BiLSTM can model the word sequence while gather the contextual information for each word in two directions, making it better for understanding an utterance sufficiently.

BiLSTM encodes the initial word embeddings forwardly from $\mathbf{g}_1$ to $\mathbf{g}_{N_k}$ and backwardly from $\mathbf{g}_{N_k}$ to $\mathbf{g}_1$ to produce a series of context-sensitive hidden states. Specifically, the LSTM learns hidden state representations in the forward direction on the word embeddings:

$$\overrightarrow{\mathbf{h}_k^0} = \overrightarrow{\mathrm{LSTM}}\left\{\mathbf{g}_k^0, \overrightarrow{\mathbf{h}_{k-1}^0}\right\}, k \in [1, N_k], \tag{3}$$

which allows contextual information to be captured in a forward direction. In a similar fashion, a backward LSTM will learn different representations:

$$\overleftarrow{\mathbf{h}_k^0} = \overleftarrow{\mathrm{LSTM}}\left\{\mathbf{g}_k^0, \overleftarrow{\mathbf{h}_{k+1}^0}\right\}, k \in [1, N_k]. \tag{4}$$

Finally, we can concatenate the corresponding parallel representations modeled by both forward and backward LSTMs into higher dimensional representations:

$$\mathbf{h}_k^0 = \left[\overrightarrow{\mathbf{h}_k^0}, \overleftarrow{\mathbf{h}_k^0}\right], \tag{5}$$

where $\mathbf{H}^0 = \left\{\mathbf{h}_1^0, \mathbf{h}_2^0, \cdots, \mathbf{h}_{N_k}^0\right\}$ and $\mathbf{H}^0 \in \mathbb{R}^{D_h \times N_k}$, D_h denotes the dimension of the BiLSTM hidden layer.

4.3 Dependency-Aware Graph Layer

We propose to use a dependency-aware graph convolutional network to leverage the dependency information, which enables the model to have a helpful inductive bias. A undirected graph is designed from the sequentially encoded words to explicitly incorporate the dependency information.

Graph Construction. The graph G is constructed from the utterance in the following way:

Vertices. Each word in the utterance is represented as a vertex v_k. Each vertex v_k is initialized with the corresponding sequentially encoded feature vector $\mathbf{h}_k^0$ as described above, for all $k \in [1, N_k]$. We denote this vector as the vertex feature. Vertex features are subject to change downstream, when the neighbourhood based transformation process is applied according to syntactic dependency paths between words in the graph.

Edges. Construction of the edges depends on the discriminative syntactic paths on arbitrary utterances from the dependency tree. The graph G for any arbitrary utterance can be represented as an $N_k \times N_k$ adjacency matrix $\mathbf{A}$, with entries A_{ij} signaling if vertex v_i is connected to vertex v_j by a single dependency path in G. Specifically, $A_{ij} = 1$ if vertex i is connected to vertex j, otherwise $A_{ij} = 0$.

Graph Feature Transformation. GCN effectively uses dependency paths to transform and propagate information on the path, and update the vertex embedding by gathering the propagated information. In such an operation, GCN only considers the first-order neighborhood of the vertex when modeling the embedding of that vertex. A single vertex embedding update takes the form:

$$\mathbf{h}_k^{l+1} = \text{ReLU}\left(\sum_{n=1}^{N_k} c^k A_{kn}\left(\mathbf{W}^l \mathbf{h}_n^l + \mathbf{b}^l\right)\right), \qquad (6)$$

where $\mathbf{h}_n^l$ is the hidden state representation for vertex n at the lth layer of the GCN, $\mathbf{W}^l$ is a parameter matrix, $\mathbf{b}^l$ is a bias term, c^k is a normalization constant, which we choose as $c^k = \frac{1}{d^k}$. d^k denotes the degree of vertex k in the graph calculated as:

$$d_k = \sum_{n=1}^{N_k} A_{kn}, \qquad (7)$$

and "ReLU" is a relu elementwise non-linear activation function. Note that $\mathbf{h}_k^{l+1}$ is the final output for vertex k at layer $l + 1$.

In order to learn deep features, we apply a stacked dependency-aware GCN with multiple layers. After stacking L layer, we obtain a final updated feature representation $\mathbf{h}_k^L$. Thus, the dependency-aware utterance representation is $\mathbf{H}^L = \{\mathbf{h}_1^L, \mathbf{h}_2^L, \cdots, \mathbf{h}_{N_k}^L\}$ and $\mathbf{H}^L \in \mathbb{R}^{D_h \times N_k}$.

4.4   Decoder

The context-aware encoded feature matrix $\mathbf{H}^0$ (from Sect. 4.2) and the dependency-aware encoded feature matrix $\mathbf{H}^L$ (from Sect. 4.3) are concatenated and a max-pooling is applied to obtain the final utterance representation for utterance $\mathbf{u}_i$:

$$\mathbf{h}_i = \text{maxpool}\left[\mathbf{H}^0; \mathbf{H}^L\right], \qquad (8)$$

where $\mathbf{h}_i \in \mathbb{R}^{2D_h}$ and then we perform LSTM upon the $\mathbf{h}_i$ to make the representation be conversation-dependent, where $\tilde{\mathbf{h}}_i = \text{LSTM}(\mathbf{h}_i)$ and $\tilde{\mathbf{h}}_i \in \mathbb{R}^{D_h}$.

We then adopt decoder to perform emotion prediction, which can be denoted as follows:

$$\hat{\mathbf{y}}^i = \text{softmax}\left(\mathbf{W}_e \tilde{\mathbf{h}}_i + \mathbf{b}_e\right), \tag{9}$$

where $\mathbf{W}_e \in \mathbb{R}^{h_e \times D_h}$, $\mathbf{b}_e \in \mathbb{R}^{h_e}$ are model parameters and h_e denotes the number of predefined emotions. We compute the loss of ERC task using standard cross-entropy loss:

$$loss_{erc} = -\sum_{j=1}^{L}\sum_{i=1}^{N_i}\sum_{e=1}^{h_e} y_e^i \log \hat{y}_e^i + \left(1 - y_e^i\right)\left(1 - \log \hat{y}_e^i\right), \tag{10}$$

where y_e^i denotes the ground truth value of $\mathbf{u}_i$ to emotion e.

5   Experimental Settings and Result Discussions

5.1   Datasets

We use three benchmark datasets to evaluate our DA-GCN, namely IEMOCAP [3], DailyDialog [15] and MELD [20].

1. **IEMOCAP** is a multimodal dataset with ten speakers involved in dyadic conversations. Each conversation video is segmented into utterances, with the following emotional tags: anger, happiness, sadness, neutral, excitement, and frustration.
2. **DailyDialog** is a daily multi-turn dialogue corpus and has a more extensive scale compare to IEMOCAP. The dataset contains 13118 multiple-turn of conversation, and each utterance in the conversation is manually labeled as an emotion, including neutral, happiness, surprise, sadness, anger, fear and disgust.
3. **MELD** is a large-scale multi-modal emotional dialogue database containing 1433 dialogues and more than 13708 utterances, and each dialogue involves more than two speakers. The emotional labels are the same as DailyDialog.

The details about the training/validation/testing split are provided in Table 1.

Table 1. Split of experimental datasets.

Dataset	Conv.(Train/Val/Test)	Utter.(Train/Val/Test)
IEMOCAP	100/20/31	4810/1000/1523
DailyDialog	11118/1000/1000	87170/8069/7740
MELD	1038/114/280	9989/1109/2610

5.2   Experimental Details

We conducted all experiments using Xeon(R) Silver 4110 CPU with 768 GB of memory and GeForce GTX 1080Ti GPU with 11 GB of memory. The input to this network is the 300 dimensional pretrained 840B GloVe vectors. All utterances are parsed by the Stanford parser. $D_h = 50$ and $L = 2$. The GCN model is trained for 100 epochs with batch size 32. We use the adam optimizer with learning rate 0.01 for all datasets. All the results are obtained using the text modality only. For IEMOCAP and MELD, we use weighted-F1 score as metric. For DailyDialog we use micro-F1 score as metric. The results reported in our experiments are all based on average of 5 random runs on the test set.

We also present restricted versions of our model denoted as DAB and DAG to perform the ablation study. Unlike our main model, DAB only exploits BiLSTM to model contextual information while DAG exploits a GCN to model dependencies between words.

5.3   Baselines

We compare our model with several of state-of-the-art baselines including:

1. **DialogueRNN** [17] employs three GRUs to model the speaker, the context and the emotion of the preceding utterances. The incoming utterance is fed into global GRU and party GRU and the updated speaker state is fed into the emotion GRU.
2. **DialogueGCN** [5] leverages speaker information and relative position by modeling conversation using a directed graph. The nodes represent individual utterances and the edges represent the dependency between the speakers and their relative positions.
3. **AGHMN** [11] proposes a hierarchical memory network with a BiGRU as the utterance reader and a BiGRU fusion layer for the interaction between historical utterances. An attention GRU is utilized to summarize and balance the contextual information from recent memories.
4. **KAITML** [26] attempts to incorporate commonsense knowledge from external knowledge bases and uses the related information in the Conceptnet by splitting the original graph attention mechanism into two steps according to the relation.
5. **QMNN** [14] provides a novel perspective on conversational emotion recognition by drawing an analogy between the task and a complete span of quantum measurement.
6. **CTNet** [16] models inter-modality and intra-modality interactions for multimodal features. In the meantime, CTNet also considers context-sensitive and speaker-sensitive dependencies in the conversation.

5.4   Overall Results

We evaluate our model against the state-of-the-art models, and the results are shown in the Table 2. From the results, we can observe that:

Table 2. Comparison of our model with baselines on three datasets. We highlight top-two values on each emotion in Bold. "Avg." means the weighted average of all individual emotion F1 scores on IEMOCAP and "-" means the original paper do not give the corresponding result.

Model	Happy	Sad	Neutral	Angry	Excited	Frustrated	Avg.	MELD	DailyDialog
DialogueRNN	33.18	78.8	59.21	**65.28**	**71.86**	58.91	62.75	57.03	50.65
DialogueGCN	42.75	**84.54**	**63.54**	64.19	63.08	**66.99**	64.18	58.1	–
AGHMN	**52.1**	73.3	58.4	61.9	69.7	**62.3**	63.5	58.1	-
KAITML	–	–	–	–	–	–	61.43	58.97	54.71
QMNN	39.71	68.3	55.29	62.58	66.71	62.19	59.88	–	–
CTNet	–	–	–	–	–	–	63.7	58.3	–
DAB	30.66	69.86	55.15	58.52	55.93	60.74	57.01	56.44	50.24
DAG	41.68	82.66	59.23	63.87	62.44	60.82	62.65	58.33	53.76
DA-GCN	**51.48**	**84.33**	60.21	**65.63**	**74.11**	61.22	**65.97**	**59.14**	**54.88**

1. DialogueRNN and AGHMN only use the recurrent models as the encoder at each stage, and capture the important contextual information in the conversation by carefully designing the structure of the model to obtain competitive results, even better than the model that uses external knowledge as auxiliary information. This shows that contextual information is very important for conversational tasks.
2. DialogueGCN takes into account the speaker dependency and temporal dependency, and obtains the best results by artificially defining the relationship between the utterances in the conversation and designing the graph structure. It is much stronger than the model effect based on quantum theory which also leverages a carefully designed module, demonstrating the powerful ability of graph neural network in conversation interaction.
3. Our DA-GCN model refreshes the current best results by considering both contextual information and dependency information. In particular, compared to DialogueGCN, our model does not need to predefine the complex relationship between utterances, which reduces the computational cost. At the same time, our model has achieved a very balanced performance in various emotions, and occupies the top-two positions in most emotions.

5.5 Effect of DA-graph Layer Number

In our experimentation, we find that as we increase the number of layers the performance increase to an extent. In particular, DA-GCN increase in model performance over one layer of the GCN. Since GCN passes information in the local neighborhood of any node, successive operations on the dependency tree allows DA-GCN to pass information to the furthest node. The problem of overfitting takes effect when the layers rises beyond a threshold, explaining the F1 score curve after the 2-th layer in the Fig. 4 and the horizontal axis represents the number of layers.

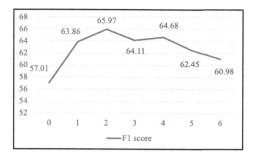

Fig. 4. F1 score curves for DA-GCN on IEMOCAP.

5.6   Ablation Study

In order to prove the effectiveness of the dependency-aware graph layer proposed in Sect. 4.3, we conduct an ablation study, as shown in the last three rows in Table 2. If the graph structure layer based on the utterance dependency tree is removed, it means that the continuous LSTM is used to model the utterance level and the conversational level context, and the result is not very satisfactory. This is mainly because there are many long utterances in the conversation, and only using the recurrent models will lose important information and affect the final emotional judgment. Compared with removing the graph structure layer, without considering the uterance context information has less impact on the model. This is mainly because for conversation tasks, the grasp of contextual information mainly depends on the capture of historical information in the conversation, and our model takes this into account.

6   Conclusion

In this paper, we propose a dependency-aware graph framework where a GCN is constructed and operates directly on the dependency tree of the utterance, achieving to jointly take advantage of the contextual information and the dependency information. Experiments on three datasets show the effectiveness of the proposed models and our model achieves state-of-the-art performance. In addition, we analyze the effect of the stacked DA-graph layer number and prove the capability of the multi-step propagation mechanism.

Acknowledgments. This work was supported by the National Key R&D Program of China via grant 2020YFB1406902.

References

1. Alswaidan, N., Menai, M.E.B.: A survey of state-of-the-art approaches for emotion recognition in text. Knowl. Inf. Syst. **62**(8), 2937–2987 (2020)
2. Ayata, D., Yaslan, Y., Kamasak, M.E.: Emotion based music recommendation system using wearable physiological sensors. IEEE Trans. Consum. Electron. **64**(2), 196–203 (2018)
3. Busso, C., et al.: IEMOCAP: interactive emotional dyadic motion capture database. Lang. Resour. Eval. **42**(4), 335–359 (2008). https://doi.org/10.1007/s10579-008-9076-6
4. Coleman, J.R., Lester, K.J., Keers, R., Munafò, M.R., Breen, G., Eley, T.C.: Genome-wide association study of facial emotion recognition in children and association with polygenic risk for mental health disorders. Am. J. Med. Genet. B Neuropsychiatr. Genet. **174**(7), 701–711 (2017)
5. Ghosal, D., Majumder, N., Poria, S., Chhaya, N., Gelbukh, A.: DialogueGCN: a graph convolutional neural network for emotion recognition in conversation. In: Proceedings of the 2019 Conference on Empirical Methods in Natural Language Processing and the 9th International Joint Conference on Natural Language Processing (EMNLP-IJCNLP), pp. 154–164 (2019)
6. Gu, Y., et al.: Human conversation analysis using attentive multimodal networks with hierarchical encoder-decoder. In: Proceedings of the 26th ACM International Conference on Multimedia, pp. 537–545 (2018)
7. Gu, Y., et al.: Mutual correlation attentive factors in dyadic fusion networks for speech emotion recognition. In: Proceedings of the 27th ACM International Conference on Multimedia, pp. 157–166 (2019)
8. Hazarika, D., Poria, S., Zadeh, A., Cambria, E., Morency, L.P., Zimmermann, R.: Conversational memory network for emotion recognition in dyadic dialogue videos. In: Proceedings of the 2018 Conference of the North American Chapter of the Association for Computational Linguistics: Human Language Technologies, Volume 1 (Long Papers), pp. 2122–2132 (2018)
9. Hazarika, D., Poria, S., Zimmermann, R., Mihalcea, R.: Conversational transfer learning for emotion recognition. Inf. Fusion **65**, 1–12 (2021)
10. Jiao, W., Lyu, M., King, I.: Exploiting unsupervised data for emotion recognition in conversations. In: Proceedings of the 2020 Conference on Empirical Methods in Natural Language Processing: Findings, pp. 4839–4846 (2020)
11. Jiao, W., Lyu, M., King, I.: Real-time emotion recognition via attention gated hierarchical memory network. In: Proceedings of the AAAI Conference on Artificial Intelligence, vol. 34, pp. 8002–8009 (2020)
12. Kratzwald, B., Ilić, S., Kraus, M., Feuerriegel, S., Prendinger, H.: Deep learning for affective computing: text-based emotion recognition in decision support. Decis. Support Syst. **115**, 24–35 (2018)
13. Li, J., Fei, H., Ji, D.: Modeling local contexts for joint dialogue act recognition and sentiment classification with Bi-channel dynamic convolutions. In: Proceedings of the 28th International Conference on Computational Linguistics, pp. 616–626 (2020)
14. Li, Q., Gkoumas, D., Sordoni, A., Nie, J.Y., Melucci, M.: Quantum-inspired neural network for conversational emotion recognition. In: Proceedings of the AAAI Conference on Artificial Intelligence, vol. 35, pp. 13270–13278 (2021)

15. Li, Y., Su, H., Shen, X., Li, W., Cao, Z., Niu, S.: DailyDialog: a manually labelled multi-turn dialogue dataset. In: Proceedings of the Eighth International Joint Conference on Natural Language Processing (Volume 1: Long Papers), pp. 986–995 (2017)
16. Lian, Z., Liu, B., Tao, J.: CTNet: conversational transformer network for emotion recognition. IEEE/ACM Trans. Audio Speech Lang. Process. **29**, 985–1000 (2021)
17. Majumder, N., Poria, S., Hazarika, D., Mihalcea, R., Gelbukh, A., Cambria, E.: DialogueRNN: an attentive RNN for emotion detection in conversations. In: Proceedings of the AAAI Conference on Artificial Intelligence, vol. 33, pp. 6818–6825 (2019)
18. Oramas Bustillos, R., Zatarain Cabada, R., Barrón Estrada, M.L., Hernández Pérez, Y.: Opinion mining and emotion recognition in an intelligent learning environment. Comput. Appl. Eng. Educ. **27**(1), 90–101 (2019)
19. Poria, S., Cambria, E., Hazarika, D., Majumder, N., Zadeh, A., Morency, L.P.: Context-dependent sentiment analysis in user-generated videos. In: Proceedings of the 55th Annual Meeting of the Association for Computational Linguistics (volume 1: Long papers), pp. 873–883 (2017)
20. Poria, S., Hazarika, D., Majumder, N., Naik, G., Cambria, E., Mihalcea, R.: MELD: a multimodal multi-party dataset for emotion recognition in conversations. In: Proceedings of the 57th Annual Meeting of the Association for Computational Linguistics, pp. 527–536 (2019)
21. Qiu, X.P., Sun, T.X., Xu, Y.G., Shao, Y.F., Dai, N., Huang, X.J.: Pre-trained models for natural language processing: a survey. Sci. China Technol. Sci. **63**(10), 1872–1897 (2020). https://doi.org/10.1007/s11431-020-1647-3
22. Ren, M., Huang, X., Shi, X., Nie, W.: Interactive multimodal attention network for emotion recognition in conversation. IEEE Signal Process. Lett. **28**, 1046–1050 (2021)
23. Shaheen, S., El-Hajj, W., Hajj, H., Elbassuoni, S.: Emotion recognition from text based on automatically generated rules. In: 2014 IEEE International Conference on Data Mining Workshop, pp. 383–392. IEEE (2014)
24. Vaswani, A., et al.: Attention is all you need. In: Proceedings of the 31st International Conference on Neural Information Processing Systems, pp. 6000–6010 (2017)
25. Wang, Z., Wan, Z., Wan, X.: BAB-QA: a new neural model for emotion detection in multi-party dialogue. In: Yang, Q., Zhou, Z.-H., Gong, Z., Zhang, M.-L., Huang, S.-J. (eds.) PAKDD 2019. LNCS (LNAI), vol. 11439, pp. 210–221. Springer, Cham (2019). https://doi.org/10.1007/978-3-030-16148-4_17
26. Zhang, D., Chen, X., Xu, S., Xu, B.: Knowledge aware emotion recognition in textual conversations via multi-task incremental transformer. In: Proceedings of the 28th International Conference on Computational Linguistics, pp. 4429–4440 (2020)
27. Zhang, R., Wang, Z., Huang, Z., Li, L., Zheng, M.: Predicting emotion reactions for human-computer conversation: a variational approach. IEEE Trans. Hum.-Mach. Syst. **62**(8), 2937–2987 (2021)
28. Zhang, Y., et al.: A quantum-like multimodal network framework for modeling interaction dynamics in multiparty conversational sentiment analysis. Inf. Fusion **62**, 14–31 (2020)
29. Zhang, Y., et al.: Learning interaction dynamics with an interactive LSTM for conversational sentiment analysis. Neural Netw. **133**, 40–56 (2021)

Semi-supervised Learning with Conditional GANs for Blind Generated Image Quality Assessment

Xuewen Zhang[1,4], Yunye Zhang[2,4], Wenxin Yu[1,4(✉)], Liang Nie[1,4],
Zhiqiang Zhang[3,4], Shiyu Chen[1,4], and Jun Gong[3,4]

[1] Southwest University of Science and Technology, Mianyang, Sichuan, China
yuwenxin@swust.edu.com
[2] University of Electronic Science and Technology of China, Chengdu, Sichuan, China
[3] Hosei University, Tokyo, Japan
[4] Beijing Institute of Technology, Beijing, China

Abstract. Evaluating the quality of images generated by generative adversarial networks (GANs) is still an open problem. Metrics such as Inception Score(IS) and Fréchet Inception Distance (FID) are limited in evaluating a single image, making trouble for researchers' results presentation and practical application. In this context, an end-to-end image quality assessment (IQA) neural network shows excellent promise for a single generated image quality evaluation. However, generated image datasets with quality labels are too rare to train an efficient model. To handle this problem, this paper proposes a semi-supervised learning strategy to evaluate the quality of a single generated image. Firstly, a conditional GAN (CGAN) is employed to produce large numbers of generated-image samples, while the input conditions are regarded as the quality label. Secondly, these samples are fed into an image quality regression neural network to train a raw quality assessment model. Finally, a small number of labeled samples are used to fine-tune the model. In the experiments, this paper utilizes FID to prove our method's efficiency indirectly. The value of FID decreased by 3.32 on average after we removed 40% of low-quality images. It shows that our method can not only reasonably evaluate the result of the overall generated image but also accurately evaluate the single generated image.

Keywords: Generative adversarial networks · Image quality assessment · Generated image

1 Introduction

Generative Adversarial Networks (GANs) [3] have made a dramatic leap in synthesizing images. However, how to evaluate the single generated image or how to provide a quality score for it is still an open problem. In this case, the absence of single-image evaluation methods brings the following challenges. Firstly, a large

ⓒ Springer Nature Switzerland AG 2021
T. Mantoro et al. (Eds.): ICONIP 2021, LNCS 13110, pp. 482–493, 2021.
https://doi.org/10.1007/978-3-030-92238-2_40

number of low-quality images in the results are hard to filter out, which hinders the practical application of image synthesis. Secondly, it makes trouble for researchers' results presentation. Since the inherent instability of GANs brings too much uncontrollable content to generated images, it is challenging to conduct supervised learning to capture the distortion. If someone wants to create a specific dataset for supervised learning, two questions are inevitable. On the one hand, the samples need to include enough generated images with various distortion types and various categories (Just like ImageNet [2]). It is difficult because so many GAN models need to be considered, which burdens the image collection progress. On the other hand, labeling these enormous numbers of images with a precise score is time-consuming. A bad generated image may due to the lack of authenticity instead of traditional degradation such as blur, low-resolution, or white noise. In contrast, authenticity distortion is challenging to define.

Recently, generated image quality assessment frequently focuses on the distribution of the features. Metrics such as Inception Score (IS) [14] and Fréchet Inception Distance (FID) [5] are employed to evaluate images base on the feature distribution. They may properly evaluate the overall quality, but neither can evaluate a single generated image. Besides, some full-reference metrics such as Peak Signal-to-Noise Ratio (PSNR) and Structural Similarity (SSIM) are limited because the reference images are frequently unobtainable in many image synthesis studies. For example, text-to-image synthesis [12] and image-to-image style transfer [23]. Although reference images exist in some special generation tasks, such as inpainting [11], one of our goals is to have a general quality assessment approach. Intuitively, using a deep neural network to learn the mapping from image features to the quality score is a simple but useful solution. However, the generated image dataset with the quality label is challenging to obtain, as mentioned above. Therefore, finding an effective training strategy to solve the problem caused by the scarcity of datasets is the key to using DNN to achieve GIQA.

In this context, we discover that conditional GANs [9] can constrain the generator's output with additional inputs. In the image-to-image transformation task [6], we observed that the output image is directly affected by the input image. Motivated by it, this paper hypothesizes that the more ground truth information the input image contains, the better the output image's quality, holding other network parameters constant. Based on this hypothesis, this paper proposes a semi-supervised method to train an image quality assessment (IQA) model for generated images. Our core idea is to train an image-to-image conditional GAN (CGAN) to produce images that include a quality label automatically. These samples can be used as the training data of the image quality prediction model. In detail, quality of the generated images by G can be controlled by the conditional input image. Therefore, we can obtain enough training samples with controllable quality while preserving the characteristics of the generated image (unstable and random), as shown in Fig. 1. Furthermore, this condition is utilized as the quality label of these generated images. It makes the

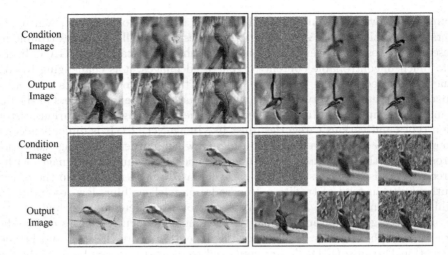

Fig. 1. Conditional GAN is utilized to generate images samples whose qualities are controllable. If other parameters are constant, the quality of the output image is directly influenced by the condition.

training of the IQA model for a single generated image at a low cost. For this paper, the main contributions are as follows.

1. A semi-supervised learning method for a single generated image quality assessment is proposed combining with the conditional GAN. We call it painter GAN which can produce a large number of quality-controllable generated images.
2. We propose a image filtering technique to remove low-quality generated images by evaluating the quality of a single image.
3. Extensive experiments are conducted to prove that our method can solve the evaluation problem of single-generated images to some extent.

2   Related Work

The evaluation of generated images is achieved by calculating the feature distribution and comparing it with a real image set. Inception Score (IS) [14] measures the image results from two aspects: the statistics level's recognizability and diversity. However, IS inherently has controversy [1], which leads to a lack of confidence in theory and practical applications. Fréchet Inception Distance (FID) [5] calculates the similarity of the feature distribution between the generated results and ground truth. These two methods measure the overall quality instead of the image itself. Therefore, methods that focus on a single image (such as PSNR, SSIM, etc.) are utilized to complete the evaluation system. These solutions directly compare the pixel-level difference with reference images. These kinds of methods are not flexible because reference images frequently unavailable in many studies. Besides, they difficultly match the human subjective perception in some cases. The existing no-reference image quality assessment methods (IQA) are mainly based on hand-craft features or learning-based features. Hand-crafted feature approaches utilize Natural Scene Statistic (NSS) models to capture distortion, which be used as the feature of the regressing model to achieve quality score [10,13]. Learning-based methods usually adopt well-designed network structures to map image features to scores [7,8,15,22]. [7] uses GANs to generate a illusory reference and then uses a phantom reference image to evaluate the target image. [8] utilize a Siamese network to learn the degradation distortion from rank order. In [22], they firstly concatenate the meta-learning method to blind image quality assessment to achieve small-sample learning. [15] employed a hyper-network to learn a specific quality representation for each input image and achieve a surprising result.

There are few works for generated images quality assessment. Gu et al. [4] first proposed GIQA to predict the scores of generated images based on CNNs. They saved the intermediate generated images before the GAN converged and used the number of iterations as the label of the quality to obtain a large number of training samples. However, due to the instability in the training process of GANs, the number of iterations is difficult to indicate the quality of the generated image accurately. In the experimental part, we will further compare our method with their work. Our previous work [20,21] evaluated the generated images using NSS and DNN-based methods combined with specific datasets, respectively. However, these methods rely heavily on the dataset itself. In this paper, a semi-supervised learning method is proposed to solve this problem by inputting various real images without labels and learning various quality representations using conditional GANs and CNNs.

3   Our Proposal

Annotating quality scores for large numbers of generated images is a strenuous task. To train a quality evaluation model without a large-scale labeled dataset, this paper proposed a semi-supervised strategy. As it's shown in Fig. 2, a conditional GAN (we metaphor it as a painter) is employed to generate images with different but quality-controllable image samples with unlabeled images. These images are fed into a convolution neural network (CNN) to know what a good or bad image is. Finally, a small number of images with quality labels are used to fine-tune the CNN.

3.1   The Painter GAN

As mentioned above, our purpose is to address the problem that there are not enough generated images with quality labels to train an image quality assessment (IQA) model. The solution is that we use a conditional GAN to get a large number of generated images as the training samples, and the input condition can play a role of score label. In order to better explain our proposal, we metaphor a generator network as a painter, and his ability is to paint images of what he

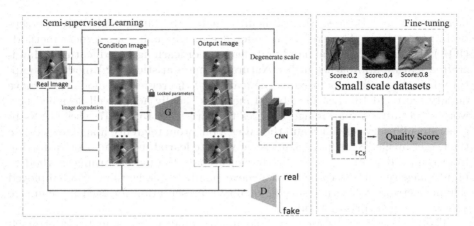

Fig. 2. The architecture of our method. The input of the generator is degraded images, while it outputs samples with stepped quality-level. As the condition image's degradation intensity increases, G will gradually generate low-quality generated images because the parameters of G are locked. These images with corresponding degradation intensity are sufficient training samples for CNN.

sees. If we assume that the painter's painting ability is constant, we can control the quality of his paintings by controlling what he can see. Therefore, we can get many quality-controllable paintings to teach a kid what painting is good or bad. Seriously, the painter in our paper is a conditional generator network for image-to-image tasks, and his paintings are the generated images. The kid is the generated IQA model we want to train.

Condition GANs [9] input conditions to both generator and discriminator to control the generated images. Its optimization goal display as follows.

$$\min_{G} \max_{D} V(G, D) = \mathbb{E}_{y \sim p_{data}(y)}(log D(y|c)) + \\ \mathbb{E}_{x \sim p_x(x)}(log(1 - D(G(x|c)))) \tag{1}$$

where x, y denote input and ground truth, respectively. The conditions c is seen by both generator and discriminator, which control results in customers' aspects.

In this paper, we use the down-sampled images as the conditions, and its original image is the ground truth. For each real image I_{real}, using average-pooling operations to reduce the information.

$$I_c = average\_pooling(I_{real}) \tag{2}$$

Subsequently, the down-sampled image and its original image are paired, and G aims to convert the down-sampled image into the original image. Following pixel-to-pixel [6], the objective includes adversarial loss and L1 loss.

$$L_{GAN}(G, D) = \mathbb{E}_{I_c, I_{real}}[log D(I_c, I_{real})] + \mathbb{E}_{I_c, z}[log(1 - D(I_c, G(I_c, z)))] \tag{3}$$

$$L_1(G) = \mathbb{E}_{I_c, I_{real}, z}(\|I_{real} - G(I_c, z)\|_1) \tag{4}$$

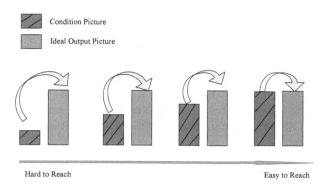

Fig. 3. An illustration of the condition image to the ideal output image in the image-to-image transformation task, the less information the input condition contains about the real image (such as contour or segmentation image), the more difficult it is for the generator to produce the ideal image. In the contrast, if the input condition image is the ideal output image itself, everything becomes easy. For the generator, the cases on the right produce better output than the left one when keeping other parameters unchanged (See Fig. 1). Thus, the condition image can be regarded as the quality label of input images.

$$Obj = \underset{G}{\overset{arg\,min}{}} \underset{D}{\overset{max}{}} L_{GAN}(G, D) + \lambda L_1(G) \tag{5}$$

where G is generator, D is discriminator, I_c is the down-sampled image, I_{real} is the real image, and z denotes random noises. Keeping the network training hyperparameters unchanged, we can determine the quality scores s of the generated image by the degree of image downsampling.

$$s = \frac{W_{I_c} \times H_{I_c}}{W_{I_{real}} \times H_{I_{real}}} \tag{6}$$

where W_{I_c} and H_{I_c} are the width and height of the down-sampled image. Similarly, $W_{I_{real}}$ and $H_{I_{real}}$ are the height and width of the real image.

In the design of this paper, the goal of conditional GAN is to generate the original image based on the down-sampled image (similar to super-resolution). If the condition contains less information, it will be more difficult to restore (See Fig. 3), and results tend to have more image artifacts. Therefore, the size of the down-sampled condition image implies the amount of information, which can influence the quality of the generated image.

3.2   Quality Evaluation Model

According to the strategy in Sect 3.1, a large number of samples with different qualities can be produced by the generator while adjusting the parameter of the down-sample operation. Therefore, an semi-supervised generated image quality assessment model (semiGIQA) is achieved through CNNs and FCs. CNNs aim to extract the features of images, while FCs map these features to the quality latent space.

$$s^* = f(I_g; \theta) \tag{7}$$

$$\mathbb{L} = \mathbb{E}(||s^* - s||_1) \tag{8}$$

where I_g denotes input image, f is the quality prediction model that includes the convolution layers and fully connected layers, θ is the model's parameters, s is defined in Eq. (6). According to the semi-supervised learning strategy, s also represent the quality label of sample in the small generated image quality assessment dataset. And L_1 loss is utilized to fine-tune the quality score regression model.

3.3 Evaluation and Optimization with Score

Our semiGIQA model can evaluate the quality of a single generated image without reference images, which is more flexible than methods such as IS, FID, or PSNR. The mean score of n generated images is defined as $mean = \frac{1}{n}\sum f(x_i; \theta)$.

Because every image's subjective score is obtained, it's possible to optimize the results by filtering out low-quality images. To not destroy the diversity, we filter out images with low scores in each category. Experiments show that FID decreased, and the image's subjective quality is effectively improved after the screening operation.

4   Experimental Results

To prove the method's effectiveness, we select a series of GANs-based generated images, used our method to predict quality scores, and display its superiority compare to FID, IS, or PSNR. Similar to our method, GIQA [4] is also a method for evaluating a single generated image. They saved the intermediate images before the GAN converged, and used the number of iterations as the label of the quality to obtain a large number of training samples. In the experiment, we will compare our method with the GIQA baseline.

4.1   Datasets

To simplify the process, this paper chooses GANs for the text-to-image synthesis tasks [12,17,18,24] and Caltech-UCSD Birds-200-2011(CUB) dataset [16] to verify the effectiveness of the proposed method. The reasons are as follows. Firstly, compared to the image-to-image conversion task, the text-to-image synthesis task is more complex, so the image quality in the results varies greatly, which is more conducive to verify our method. Secondly, our semi-supervised process requires a small-scale generated images dataset with quality labels. To the best of our knowledge, the only public generated IQA dataset MMQA [19] is based on the CUB dataset. CUB dataset contains 11788 images of birds in 100 different categories. All the images of this dataset are used for unsupervised learning without image quality labels. MMQA contains 5000 generated images by the GANs of text-to-image synthesis. 12 observers give each image quality scorers on 7 different aspects. This dataset is used to fine-tune our network, of which 80% is used for training and 20% is used for testing.

4.2   Implementation Details

All our experiments are implemented on PyTorch with two NVIDIA 1080Ti GPUs. First, each image in the CUB dataset is down-sampled 6 times with different sizes and then pairs them with the original image and marks according to formula (6). Therefore, there are 70,728 pairs used to train image-to-image GANs. Eventually, 70,728 generated images can produce with down-sampled images and automatically obtain their quality labels according to formula (6), and the structure and parameters of the GAN follow [6]. These samples are then used to train the image quality prediction network, of which 80% of images are used for training, and 20% of images are used for testing to prevent overfitting. Finally, the 5,000 images with artificial quality labels in MMQA were used to fine-tune the network. We randomly crop the images to 224×224, perform random flips, set the batch size to 64, and use the Adam optimizer with a 0.9 of momentum and a 0.0005 weight decay to train 10 epochs. Other parameters of IQA model follow [15].

Fig. 4. Example of quality score prediction for single generated images.

Table 1. Our method is used as a supplementary evaluation standard for IS and FID to make a comprehensive quantitative evaluation.

	GAN-CLS	StackGAN++	AttnGAN	DM-GAN
IS↑	2.93	4.08	4.32	4.70
FID↓	174.73	26.85	23.16	15.31
PSNR	–	–	–	–
Ours↑	0.27	0.47	0.50	0.51

4.3 Supplementary of Evaluation System

We choose four popular text-to-image synthesis methods, GAN-CLS [12], Stack-GAN++ [18], AttnGAN [17], DM-GAN [24]. As shown in Table 1, our method evaluates 4 methods (each method produces 30,000 generated images with the same condition). The evaluation trends of ours, GIQA [4], IS, and FID is identical, which matches the actual performance of these models. However, FID in low-quality images is a little biased so that it's not fair for low quality results (174.73 in GAN-CLS) and IS has some controversies. In contrast, our method is smoother, while evaluating images from human subjective perception is more intuitive. PSNR hardly works because it's difficult to find reference images in text-to-image synthesis. Besides, as shown in Fig. 4, our method is efficient in predicting scores for a single image, flexible and applicable. In contrast, IS and FID fail because they require a sufficient number of images to extract feature distributions. Therefore, our method is reasonably utilized as a supplement to the evaluation system, which shows that our method can not only reasonably evaluate the result of the overall generated image, but also accurately evaluate the single generated image (See Fig. 4).

Table 2. FID score of generated images after filter out low score images. The high-quality images from the top 100% to the top 60% are preserved.

GAN models	Top Percentage	1.0(vanilla)	0.9	0.8	0.7	0.6
GAN-CLS [12]	Random	174.73	174.61	174.30	173.70	173.05
	GIQA [4]	174.73	174.44	173.74	172.88	172.82
	Ours	174.73	**172.25**	**170.27**	**168.83**	**167.92**
StackGAN++ [18]	Random	26.85	26.51	26.68	26.90	27.24
	GIQA [4]	26.85	25.65	25.11	24.89	24.84
	Ours	26.85	**23.96**	**22.28**	**21.24**	**20.95**
AttnGAN [17]	Random	23.16	23.18	23.38	23.7	23.91
	GIQA [4]	23.16	22.86	22.57	22.3	22.13
	Ours	23.16	**21.02**	**19.49**	**18.63**	**18.38**
DM-GAN [24]	Random	15.34	15.48	15.61	15.86	16.22
	GIQA [4]	15.34	14.95	14.89	14.87	**15.06**
	Ours	15.34	**14.57**	**14.46**	**14.80**	15.46

4.4 Optimization of Generated Results

Compared with FID, our method is available for a single image, which makes results optimization possible. To verify the optimization strategy's rationality, we employ four popular text-to-image synthesis models to generate 30,000 images and filter out low-scoring images. A reasonable assumption is that after removing low-quality images, the overall quality of the rest images is better than before. Therefore, with this assumption, we can get images of high quality. As shown in

Table 2, we gradually remove 10% of low-quality images and calculate the FID score of the remaining images. The decrease in FID means an improvement in the overall quality. As a comparison, the same number of images are randomly removed at each step to get the FID score of the remaining images. It also shows that filtering images randomly rarely reduces FID stably and may even increase it. Besides, we use the same strategy in the results of GIQA [4], but its effect is not as evident as our method.

Our method has achieved promising results except for DM-GAN, we analyze the reasons as follows. We adopted a strategy of screening low-quality images proportionally. As shown in Table 1, the results in DM-GAN have the best quality, the high-quality images stay a large proportion of the results (maybe greater than 70%). Therefore, a lot of high-quality images will be removed, which will lead to an increase in FID. In contrast, the overall result becomes better in other GAN-based results because more low-quality images are removed.

4.5   Discussion

In this section, we will further discuss the motivation and justification of the proposed approach. GANs generally have two well-known problems; Instability and model collapse. The key of our method is to utilized CGANs to produce quality-controllable image samples. On the one hand, GAN's instability is reflected in the results and expressed in the training process. Therefore, it is not reasonable to directly use the number of training iterations as the standard for evaluating the generated images (GIQA adopts it). The experimental results also prove that, as shown in Table 2, GIQA cannot well screen out low-quality results. On the contrary, by observing the output of condition GAN (See Fig. 1), we find that sufficient conditions lead to stable results. In our task, GAN only needs to fill the gap between the given conditions and the ground truth, which allows us to control the generation of GANs with enough conditions. It is not a cheating method because our task is image quality evaluation rather than image generation. On the other hand, GAN often faces model collapse, which destroys the diversity of generated results. Fortunately, our approach avoids this problem; As mentioned earlier, we input sufficient conditions into GAN. Since the variety of the output is actually dependent on the diversity of the input conditions, the variety of generated images in our model can be guaranteed. Our semi-supervised learning strategy can learn quality representations from a high-quality unlabeled image, so the evaluation model's generalization can be improved by introducing various types of images.

5   Conclusion

The quality assessment algorithms for a single image usually rely heavily on the image dataset with quality labels. For generated images, it is tough to collect various generated images with quality labels. Therefore, this paper proposes a semi-supervised learning approach that allows models to learn quality representations from single unlabeled images. Firstly, we utilized conditional GAN to produce quality-controllable samples without any labels automatically. After these samples are fed into an IQA model, a small number of labeled images are used to fine-tune the model. The comprehensive experiment proves the effectiveness of the method. Subjectively, the prediction quality scores by our method are consistent with human perception. Objectively, our approach can improve the quality of overall results (according to FID) by accurately filtering low-quality images.

Acknowledgements. This research is supported by Sichuan Science and Technology Program (No. 2020YFS0307, No. 2020YFG0430, No. 2019YFS0146), Mianyang Science and Technology Program (2020YFZJ016).

References

1. Barratt, S.T., Sharma, R.: A note on the inception score. CoRR abs/1801.01973 (2018). http://arxiv.org/abs/1801.01973
2. Deng, J., Dong, W., Socher, R., Li, L.J., Li, K., Fei-Fei, L.: ImageNet: a large-scale hierarchical image database. In: 2009 IEEE Conference on Computer Vision and Pattern Recognition, pp. 248–255. IEEE (2009)
3. Goodfellow, I.J., et al.: Generative adversarial networks. arXiv preprint arXiv:1406.2661 (2014)
4. Gu, S., Bao, J., Chen, D., Wen, F.: GIQA: generated image quality assessment. In: Vedaldi, A., Bischof, H., Brox, T., Frahm, J.-M. (eds.) ECCV 2020. LNCS, vol. 12356, pp. 369–385. Springer, Cham (2020). https://doi.org/10.1007/978-3-030-58621-8_22
5. Heusel, M., Ramsauer, H., Unterthiner, T., Nessler, B., Hochreiter, S.: GANs trained by a two time-scale update rule converge to a local nash equilibrium. In: Advances in Neural Information Processing Systems 30: Annual Conference on Neural Information Processing Systems 2017, Long Beach, CA, USA, 4–9 December 2017, pp. 6626–6637 (2017)
6. Isola, P., Zhu, J.Y., Zhou, T., Efros, A.A.: Image-to-image translation with conditional adversarial networks. In: Proceedings of the IEEE Conference on Computer Vision and Pattern Recognition, pp. 1125–1134 (2017)
7. Lin, K., Wang, G.: Hallucinated-IQA: no-reference image quality assessment via adversarial learning. In: 2018 IEEE Conference on Computer Vision and Pattern Recognition, CVPR 2018, Salt Lake City, UT, USA, 18–22 June 2018, pp. 732–741. IEEE Computer Society (2018)
8. Liu, X., van de Weijer, J., Bagdanov, A.D.: RankIQA: learning from rankings for no-reference image quality assessment. In: IEEE International Conference on Computer Vision, ICCV 2017, Venice, Italy, 22–29 October 2017, pp. 1040–1049. IEEE Computer Society (2017)
9. Mirza, M., Osindero, S.: Conditional generative adversarial nets. CoRR 1411.1784 (2014). http://arxiv.org/abs/1411.1784
10. Mittal, A., Moorthy, A.K., Bovik, A.C.: No-reference image quality assessment in the spatial domain. IEEE Trans. Image Process. **21**(12), 4695–4708 (2012)

11. Pathak, D., Krahenbuhl, P., Donahue, J., Darrell, T., Efros, A.A.: Context encoders: feature learning by inpainting. In: Proceedings of the IEEE Conference on Computer Vision and Pattern Recognition, pp. 2536–2544 (2016)
12. Reed, S.E., Akata, Z., Yan, X., Logeswaran, L., Schiele, B., Lee, H.: Generative adversarial text to image synthesis. In: Balcan, M., Weinberger, K.Q. (eds.) Proceedings of the 33nd International Conference on Machine Learning, ICML 2016. JMLR Workshop and Conference Proceedings, vol. 48, pp. 1060–1069. JMLR.org (2016)
13. Saad, M.A., Bovik, A.C., Charrier, C.: Blind image quality assessment: a natural scene statistics approach in the DCT domain. IEEE Trans. Image Process. 21(8), 3339–3352 (2012)
14. Salimans, T., Goodfellow, I.J., Zaremba, W., Cheung, V., Radford, A., Chen, X.: Improved techniques for training GANs. In: Advances in Neural Information Processing Systems 29: Annual Conference on Neural Information Processing Systems 2016, Barcelona, Spain, 5–10 December 2016, pp. 2226–2234 (2016)
15. Su, S., et al.: Blindly assess image quality in the wild guided by a self-adaptive hyper network. In: 2020 IEEE/CVF Conference on Computer Vision and Pattern Recognition, CVPR 2020, Seattle, WA, USA, 13–19 June 2020, pp. 3664–3673. IEEE (2020)
16. Wah, C., Branson, S., Welinder, P., Perona, P., Belongie, S.: The Caltech-UCSD Birds-200-2011 dataset (2011)
17. Xu, T., et al.: AttnGAN: fine-grained text to image generation with attentional generative adversarial networks. In: 2018 IEEE Conference on Computer Vision and Pattern Recognition, CVPR 2018, Salt Lake City, UT, USA, 18–22 June 2018, pp. 1316–1324. IEEE Computer Society (2018)
18. Zhang, H., et al.: StackGAN++: realistic image synthesis with stacked generative adversarial networks. IEEE Trans. Pattern Anal. Mach. Intell. 41(8), 1947–1962 (2019)
19. Zhang, X., Yu, W., Jiang, N., Zhang, Y., Zhang, Z.: SPS: a subjective perception score for text-to-image synthesis. In: 2021 IEEE International Symposium on Circuits and Systems (ISCAS), pp. 1–5. IEEE (2021)
20. Zhang, X., Zhang, Y., Zhang, Z., Yu, W., Jiang, N., He, G.: Deep feature compatibility for generated images quality assessment. In: Yang, H., Pasupa, K., Leung, A.C.-S., Kwok, J.T., Chan, J.H., King, I. (eds.) ICONIP 2020. CCIS, vol. 1332, pp. 353–360. Springer, Cham (2020). https://doi.org/10.1007/978-3-030-63820-7_40
21. Zhang, Y., Zhang, X., Zhang, Z., Yu, W., Jiang, N., He, G.: No-reference quality assessment based on spatial statistic for generated images. In: Yang, H., Pasupa, K., Leung, A.C.-S., Kwok, J.T., Chan, J.H., King, I. (eds.) ICONIP 2020. CCIS, vol. 1332, pp. 497–506. Springer, Cham (2020). https://doi.org/10.1007/978-3-030-63820-7_57
22. Zhu, H., Li, L., Wu, J., Dong, W., Shi, G.: MetaIQA: deep meta-learning for no-reference image quality assessment. In: 2020 IEEE/CVF Conference on Computer Vision and Pattern Recognition, CVPR 2020, Seattle, WA, USA, 13–19 June 2020, pp. 14131–14140. IEEE (2020)
23. Zhu, J.Y., Park, T., Isola, P., Efros, A.A.: Unpaired image-to-image translation using cycle-consistent adversarial networks. In: Proceedings of the IEEE International Conference on Computer Vision, pp. 2223–2232 (2017)
24. Zhu, M., Pan, P., Chen, W., Yang, Y.: DM-GAN: dynamic memory generative adversarial networks for text-to-image synthesis. In: IEEE Conference on Computer Vision and Pattern Recognition, CVPR 2019, Long Beach, CA, USA, 16–20 June 2019, pp. 5802–5810. Computer Vision Foundation/IEEE (2019)

Uncertainty-Aware Domain Adaptation for Action Recognition

Xiaoguang Zhu, You Wu, Zhantao Yang, and Peilin Liu$^{(\boxtimes)}$

Shanghai Jiao Tong University, Shanghai 200240, China
{zhuxiaoguang178,chickandmushroom,y2242794082,liupeilin}@sjtu.edu.cn

Abstract. Domain Adaptation (DA) has been a crucial topic for action recognition, as the test set and training set are not always subject to the identical distribution, which will lead to significant performance degradation. Existing researches focus on DA methods based on the entire videos, ignoring the different contributions of different samples and regions. In this paper, we propose an uncertainty-aware domain adaptation method for action recognition from a new perspective. The aleatoric uncertainty is firstly used in the classifier to improve the performance by alleviating the impact of noisy labels. Then the aleatoric uncertainty calculated with Bayesian Neural Network is embedded in the discriminator to help the network focus on the spatial areas and temporal clips with lower uncertainty during training. The spatial-temporal attention map is generated to enhance the features with the guidance of backward passing. Extensive experiments are conducted on both small-scale and large-scale datasets, and the results indicate that the proposed method achieves competitive performance with fewer computational workloads. The code is available at: https://github.com/ChickAndMushroom/Uncertainty-aware-DA-for-AR.

Keywords: Uncertainty · Domain adaptation · Action recognition

1 Introduction

Action recognition (AR) has been studied widely in recent years, which aims to classify videos of human activity. It makes great sense in many practical fields including human-computer interaction, intelligent monitor systems, and intelligent transportation systems. However, the test scenario always suffers from the performance drop, as the test samples and training samples are not subject to the same distribution, i.e. there exists the domain gap between the training dataset and the test scenarios. Moreover, it is not practical to obtain manual labels for various test scenarios, thus it is extremely significant to study Domain Adaptation (DA) in AR, which focuses on solving the unsupervised domain shift problem.

Even though a large number of DA methods have been proposed as image-based approaches, the video-based DA remains to be challenging. The core idea of

© Springer Nature Switzerland AG 2021
T. Mantoro et al. (Eds.): ICONIP 2021, LNCS 13110, pp. 494–506, 2021.
https://doi.org/10.1007/978-3-030-92238-2_41

previous DA methods for AR comes from the feature adaptations with adversarial discriminator. Through the adversarial training, the feature representations of the source and target samples are pulled closer. Based on adversarial training, the previous works focus on exploring the relationship between the source and a target domain and using attention mechanisms to model the alignment of the videos [1–3,17]. However, existing researches focus on DA methods based on the entire videos, ignoring the different contributions of different samples, different regions, and different frames.

To address this issue, we propose an Uncertainty-Aware Domain adaptation (UADA) method for action recognition from a new perspective. Inspired by the study of uncertainty estimation [23], the aleatoric uncertainty is employed to both the classifier and discriminator. For the Bayesian classifier, the joint optimization of prediction and uncertainty is capable of alleviating the impact of noisy labels. For the Bayesian discriminator, it is used to help identify the regions and clips where the discriminator is with low uncertainty. We design a backward spatial-temporal attention mechanism to emphasize the critical part of videos with lower uncertainty, and in this way, we make the model pay more attention to parts tending to be better adapted. More specifically, we put forward the method with spatial attention, method with spatial-temporal attention, and that combined with LSTM serving as the frame-aggregation method. Then we conduct several experiments on both small-scale and large-scale datasets to verify the performance of our method. Our contributions are listed as follows:

- We build a framework applying uncertainty to domain adaptation in the field of action recognition, which is the first to the best of our knowledge.
- We design a backward uncertainty-aware spatial-temporal attention to focus on those certain areas and frames of videos which contributes to better adaptation.
- The experiments on popular benchmarks prove that the proposed method is competitive to the SOTA methods with fewer computational workloads and is suitable to plugin to other approaches.

2   Related Work

Domain Adaptation for Action Recognition. In recent years, an increasing number of DA approaches use deep learning architectures and most DA approaches follow the two-branch (source and target) architecture, and aim to find a common feature space between the source and target domains, which are therefore optimized with a combination of classification and domain losses. Although DA approaches for images are quite deeply explored, a small number of works study video DA problem with small-scale dataset [4–7]. Following methods conducted on large-scale dataset focus on exploring the relation of source and target videos from both clip-level and video-level [1–3,17]. However, these methods can hardly notice the areas and frames of a video that can be better adapted. Our method pays more attention to certain parts of videos according to spatial-temporal attention based on uncertainty.

Uncertainty-Aware Learning. Uncertainty measures the uncertainty of the data itself and the uncertainty of the prediction of the model, which is defined as aleatoric uncertainty and epistemic uncertainty. Recently, several works utilize uncertainty for DA. [9] calculates the uncertainty for each sample and progressively increases the number of target training samples based on the uncertainties to accurately characterize both cross-domain distribution discrepancy and other intra-domain relations. [10] uses a Bayesian neural network to teach a classifier to estimate the prediction uncertainty. By aligning data of two domains (via uncertainty), domain shift is greatly alleviated, leading the classifier to output consistent predictions between domains. [11] proposes to adapt from source and target domain through computing predictive uncertainties by applying variational Bayes learning. [12] proposes a very simple and efficient approach which only aligns predicted class probabilities across domains. [13] notices some certain areas of images can be well adapted more easily, such as foreground parts of an image. Inspired by [13], we consider applying spatial-temporal uncertainty-aware attention to video classification and therefore enable the model to concentrate on those certain areas and frames for videos.

3   Methodology

In this work, we only study the unsupervised domain adaptation (UDA) problem. In this way, the source dataset $\mathcal{D}_s = (X_i^s, y_i^s)$ includes video sample (X_i^s) and its label (y_i^s) and the target dataset $\mathcal{D}_t(X_i^t)$ includes video samples (X_i^t) without their labels. We also assume that $\mathcal{D}_s \in P_s$ while $\mathcal{D}_t \sim P_t$. P_s means the source distribution while P_t is target distribution. To address the problem, our work is based on the typical adversarial domain adaptation framework, and we train a discriminator to learn domain invariant features while training a classifier to learn class discriminative features. In our work, we propose a discriminator uncertainty based domain adaptation model shown in the Fig. 1, which is made up of four major modules: Domain-Invariant Feature Extractor, Bayesian Classifier, Uncertainty-Aware Attention Mechanism, and Bayesian Discriminator.

For a fair comparison, the same feature extraction strategy in TA^{3}N [17] is adopted as our baseline. The ResNet101 pre-trained on ImageNet is used to extract the frame-level features for both source and target data. In this way, our input for the model, the frame-level features for each video, can be referred as $X_i = \{x_i^1, x_i^2, ...x_i^K\}$, where x_i^j is the jth frame-level feature of the ith video. After that, here we elaborate on details for each module of our model.

3.1   Domain-Invariant Feature Extractor

Inputs are first fed into the feature extractor to extract the domain-invariant features. Here we just use the simple linear layers, because the inputs have been extracted by the pre-trained convolution network. The feature extractor G_f is fed with the input X_i, and outputs features f_i. This process can be defined as $f_i = G_f(X_i)$.

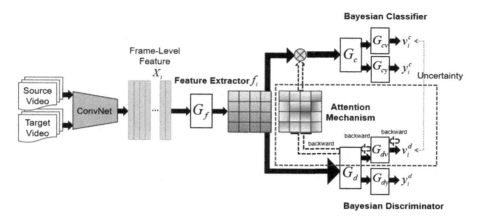

Fig. 1. The architecture of the framework of our method is made up of a shared feature extractor, uncertainty-aware attention module, Bayesian classifier, and Bayesian discriminator. Both the Bayesian classifier and discriminator need to output two items, that is, the uncertainty and prediction score. By calculating the aleatoric uncertainty of the discriminator, much more attention will be paid on parts where the discriminator is confident.

3.2 Bayesian Discriminator

The widely used approach to calculate uncertainty is Bayesian framework. [14] proves that adding dropout layer after each fully connected layer (fc) leads to probabilistic inference for networks. Therefore, we adopt a similar method to define the discriminator.

A typical discriminator only predicts the probability of each domain. However, in order to estimate uncertainty, we train the discriminator to output domain probabilities along with aleatoric uncertainty. More specifically, we predict a variance for each video sample representing the aleatoric uncertainty, where the regions or clips that contain large uncertainty cannot be easily adapted. If these uncertain areas are aligned, we will obtain a damaged feature representation, and result in terrible performance of the classifier. Therefore, according to the calculated aleatoric uncertainty, the discriminator can pay little attention to these uncertain parts, which reduces negative transfer [15] as well.

The discriminator generates logits and variance:

$$y_i^d = G_{dy}(G_d(f_i)), v_i^d = G_{dv}(G_d(f_i)), \tag{1}$$

where G_{dy} and G_{dv} output domain class logits y_i^d as well as domain aleatoric uncertainty v_i^d after utilizing features generated by G_d. Additionally, similar to [16], G_d contains a gradient-reversal layer, which takes the opposite number of the gradients during the backward propagation period. In this way, the optimization goal of the discriminator is to minimize the loss of itself, while that of the feature extractor is to maximize the loss of discriminator, forcing it to extract the domain-invariant features to confuse the discriminator.

We define the domain classification loss $\mathcal{L}_{dy}$ as:

$$\mathcal{L}_{dy} = \frac{1}{n_s + n_t} \sum_{X_i \in \mathcal{D}_s \cup \mathcal{D}_t} \mathcal{L}(y_i^d, d_i), \qquad (2)$$

where n_s is the number of source samples while n_t is that of target samples. d_i is the true label of domain.

Also, we need the aleatoric loss $\mathcal{L}_{dv}$ for the discriminator to restrict the model to predict correct aleatoric uncertainty:

$$\mathcal{L}_{dv} = -\frac{1}{n_s + n_t} \sum_{X_i \in \mathcal{D}_s \cup \mathcal{D}_t} log \frac{1}{T} \sum_t \mathcal{L}(\hat{y}_{i,t}^d, d_i), \qquad (3)$$

$$\hat{y}_{i,t}^d = y_i^d + \sigma_i^d * \epsilon_t, \epsilon_t \sim \mathcal{N}(0, I), \qquad (4)$$

where $v_i^d = (\sigma_i^d)^2$. During the optimization period, we need to combine these two kinds of loss.

3.3   Uncertainty-Aware Attention Mechanism

The discriminator can use uncertainty to tell the difference of areas and frames that can be adaptable, cannot be adapted, or have been adapted. The discriminator will output high uncertainty on those already aligned parts as well as parts that cannot be adapted. Therefore, we hope to focus more on those certain parts of videos fit for adaptation. In this way, we design a spatial-temporal attention mechanism to identify those certain parts.

Since the aleatoric uncertainty v_i^d is led by noise or damage in videos, we utilize the backward propagation period to identify those areas of feature contributing to the aleatoric uncertainty. Specifically, we calculate the gradients of aleatoric uncertainty v_i^d with respect to the features f_i when the gradients flow back through the discriminator. Because of the gradient reversal layer, we get the opposite number of the gradients. After that, we process the gradients $(-\frac{\partial v_i^d}{\partial f_i})$ as below:

$$a_i = ReLU(p_i) + c * ReLU(-p_i), \qquad (5)$$

$$p_i = f_i * (-\frac{\partial v_i^d}{\partial f_i}), \qquad (6)$$

where the positive parts of p_i represent certain regions requiring more attention, hence we use ReLU function to reserve them. For the negative parts, we multiply them with a large constant c, to magnify them. Also, we need softmax function to normalize a_i, making those magnified negative parts tend to zeros. Since we can determine the dimension where softmax is applied, we design both spatial attention and spatial-temporal attention according to where the dimensions are applied on. In the end, to get weighted features for better classification, we multiply the weights by the features f_i as below:

$$w_i = (1 - v_i^d) * Softmax(a_i), \tag{7}$$

$$h_i = (1 + f_i) * w_i, \tag{8}$$

where we also use $(1 - v_i^d)$ to pay different attention to different samples. In this way, we obtain the weighted features fit for DA.

3.4 Bayesian Classifier

The framework of the Bayesian classifier is quite similar to that of the Bayesian discriminator, using dropout after each linear layer. However, we should notice all the features above are in frame level, so we should aggregate the frames first. In other words, we need to turn the frame-level features into video-level features, of which the key is how to fuse the frames. We can choose **average pooling** or **LSTM** network, while the latter can aggregate the temporal information better.

Similarly, we enable the classifier to calculate aleatoric uncertainties to make it more robust. Class logits y_i^c and aleatoric uncertainty v_i^c are generated as:

$$y_i^c = G_{cy}(G_c(h_i)), v_i^c = G_{cv}(G_c(h_i)), \tag{9}$$

where we only use h_i as the input of the classifier during the training period and use f_i during the validation or test period. Losses of the classifier also contain two parts, $\mathcal{L}_{cy}$ and $\mathcal{L}_{cv}$:

$$\mathcal{L}_{cy} = \frac{1}{n_s} \sum_{X_i \in \mathcal{D}_s} \mathcal{L}(y_i^c, y_i), \tag{10}$$

$$\hat{y}_{i,t}^c = y_i^c + \sigma_i^c * \epsilon_t, \epsilon_t \sim \mathcal{N}(0, I), \tag{11}$$

$$\mathcal{L}_{cv} = -\frac{1}{n_s} \sum_{X_i \in \mathcal{D}_s} log \frac{1}{T} \sum_t \mathcal{L}(\hat{y}_{i,t}^c, y_i), \tag{12}$$

where we only compute losses for labeled data, that is, the source data, for we have no label for target data while training. A more detailed framework of the model is illustrated in Fig. 2.

3.5 Loss Function

We can define the total loss function J to optimize the model as:

$$J = \mathcal{L}_{cy} + \mathcal{L}_{cv} - \lambda * (\mathcal{L}_{dy} + \mathcal{L}_{dv}), \tag{13}$$

where we set λ as a trade-off parameter that can be adjusted.

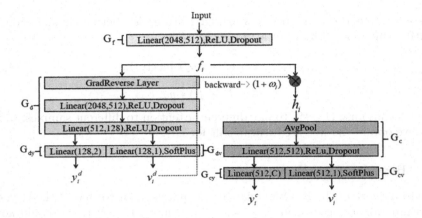

Fig. 2. The detailed implementation of our model.

Table 1. Comparison of the size of small-scale and large-scale datasets used in experiments.

	UCF-Olympic	UCF-HMDB$_{small}$	UCF-HMDB$_{full}$	Kinetics-Gameplay
Classes	5	6	12	30
Videos	1171	1145	3209	49998

4   Experiments

In this section, we elaborate on the experiments on both small-scale and large-scale datasets and prove the effectiveness of our method.

4.1   Dataset

There are very few benchmark DA datasets for AR, and they are also too small-scale to accurately measure the performance of models. These small-scale datasets include UCF-HMDB$_{small}$, and UCF-Olympic. In [17], Chen et al. propose two large-scale datasets, UCF-HMDB$_{full}$ and Kinetics-Gameplay. All these datasets study human actions, like climbing, walking, and golf. Table 1 gives the scale of these four datasets, and our experiments are conducted on them.

4.2   Implementation Details

For our experiments, we adopt the widely-used experimental protocol of UDA [18]: the training data includes the source samples with labels and the target samples with no label, while the validation data is the rest target samples. We test both the spatial and spatial-temporal attention mechanism, and then combine the latter with LTSM served as the frame-aggregation method. These are our three proposed models.

Table 2. The accuracy (%) for the SOTA work on UCF-Olympic and UCF-HMDB$_{small}$ dataset (U: UCF, O: Olympic, H: HMDB).

Method	Backbone	U→O	O→U	U→H	H→U
W. Sultani et al. [4]	–	33.3	47.9	68.7	68.7
T. Xu et al. [5]	–	87.0	75.0	82.0	82.0
AMLS (GFK) [6]	C3D	84.7	86.4	89.5	95.4
AMLS (SA) [6]	C3D	83.9	86.1	90.3	94.4
DAAA [6]	C3D	91.6	90.0	–	–
DANN [16]	C3D	**98.2**	90.0	**99.3**	98.4
TA^{3}N [17]	ResNet101	**98.2**	92.9	**99.3**	**99.5**
STCDA [1]	BNInception	94.4	93.3	97.4	99.3
TCoN [3]	B-TRN	96.8	**96.8**	–	–
Ours(UADA w/ Spatial)	ResNet101	**98.2**	87.5	**99.3**	96.8
Ours(UADA w/ Spatial-Temporal)	ResNet101	**98.2**	88.3	**99.3**	96.8
Ours(UADA w/ Spatial-Temporal+LSTM)	ResNet101	**98.2**	86.3	**99.3**	96.8

Our method is implemented on Pytorch framework and two Nvidia 2080Ti GPUs. The convolution network we use to extract frame-level features is ResNet-101 model, and we pick out five frames of each video to extract from. For each dataset, we compare the top-1 classification accuracy as the performance of each model. For optimization, we use the SGD optimizer and set the learning rate as 0.03 with the momentum and weight decay as 0.9 and 1×10^{-4}. The training epoch is set to 100. Then we set the trade-off parameter λ in loss function as 1. To obtain the same number of iterations for source and target domain in one epoch, the batch size of two domains is set as below:

$$b_t = b_s * n_t / n_s, \tag{14}$$

where b_s, b_t is the batch size of source and target domain data respectively.

During the training period, we train the classifier and the discriminator with data in both domains in one step. The input of the classifier is the weighted feature h_i determined by the attention mechanism on the discriminator. While during the testing period, we directly feed the classifier with original features f_i.

4.3   Comparison with State-of-the-Art Methods

We compare our method with other state-of-the-art methods on four datasets. Furthermore, we also do two extra experiments without DA, named as 'Target only' and 'Source only', which means the domain of data we use to train the model. The 'Target only' setting represents the upper bound of the performance without domain shift while the 'Source only' setting means that of the lower bound without domain adaptation.

We first analyze our models and other previous methods on small-scale datasets, as shown in Table 2. In these two small-scale datasets, UCF-HMDB$_{small}$, and UCF-Olympic, our methods achieve as good performance as

Table 3. The accuracy (%) for the SOTA work on UCF-HMDB$_{full}$ and Kinetics-Gameplay (U: UCF, H: HMDB, K: Kinetics, G: Gameplay). TCoN [3] method does not offer results of 'Source only' and 'Target only' settings.

Method	Backbone	Pre-train	U→H	H→U	K→G
Source only	ResNet101	ImageNet	70.3	75.0	17.2
DANN [16]	ResNet101	ImageNet	71.1	75.1	20.6
JAN [19]	ResNet101	ImageNet	71.4	80.0	18.2
AdaBN [20]	ResNet101	ImageNet	75.6	76.4	20.3
MCD [21]	ResNet101	ImageNet	71.7	76.2	19.8
TA^{3}N [17]	ResNet101	ImageNet	78.3	81.8	**27.5**
Target only	ResNet101	ImageNet	80.6	92.1	64.5
TCoN [3]	–	ImageNet	87.2	89.1	–
Source only	I3D	Kinetics400	80.3	88.8	-
SAVA [2]	I3D	Kinetics400	82.2	91.2	–
STCDA [1]	I3D	Kinetics400	81.9	91.9	–
Target only	I3D	Kinetics400	95.0	96.8	–
Source only	ResNet101	ImageNet	70.3	75.0	17.2
Ours(UADA w/ Spatial)	ResNet101	ImageNet	71.1	78.5	20.6
Ours(UADA w/ Spatial-Temporal)	ResNet101	ImageNet	71.4	79.3	21.2
Ours(UADA w/ Spatial-Temporal+LSTM)	ResNet101	ImageNet	72.2	81.1	23.1
Ours(UADA w/ Spatial-Temporal+TA^{3}N)	ResNet101	ImageNet	**78.6**	**83.2**	25.3
Target only	ResNet101	ImageNet	80.6	92.1	64.5

other SOTA methods. Although it indicates small-scale datasets can easily saturate and not fit for evaluate models, the results still prove the competitiveness of our methods.

Then we test our models on large-scale datasets, UCF-HMDB$_{full}$ and Kinetics-Gameplay, as shown in Table 3. As the SOTA methods are conducted on various baselines (e.g. 3D ResNet101 and C3D) and pretrained datasets (e.g. ImageNet and Kinetics400), we focus on comparisons with the methods with the same setting for fair comparison. In U→H, our methods behaves as well as DANN and JAN methods. We obtain gains of 0.83%, 1.11% and 1.94% compared with 'Source only' setting, which proves the effectiveness of our method. In H→U, our method outperforms any other SOTA method, with gains of 3.49%, 4.37% and 6.13%. In K→G, the domain shift is tremendous. We also obtain the gains of 3.34%, 4.01% and 5.87%, better than most SOTA methods, validating the competitiveness of our method. Besides, our model with spatial-temporal attention mechanism performs better than model with spatial attention, and LSTM network aggregates temporal information improving the performance of our models. Moreover, our methods can be embedded into the TA^{3}N to achieve better performance, which indicates the generalization ability to other methods.

Although our method does not achieve the best accuracy, we should notice that these SOTA methods including [1–3,17] focus on the video representations

Table 4. The accuracy (%) for models using different attention mechanism on UCF-HMDB$_{full}$ dataset (U: UCF, H: HMDB). Here we use average pooling as the frame-aggregation method.

Methods	H→U
UADA w/o Dicriminator Uncertainty w/o Classifier Uncertainty	76.73
UADA w/o Dicriminator Uncertainty	77.10
UADA w/Dicriminator Uncertainty w/Spatial Attention	78.45
UADA w/Dicriminator Uncertainty w/Spatial-Temporal Uncertainty	**79.33**

Fig. 3. The comparison of t-SNE visualization for our models using spatial-temporal attention. The right picture uses LSTM as the frame-aggregation method while the left not. The red dots and the blue dots mean the source and target data respectively. (Color figure online)

and try to utilize the correlations between video clips, while our method offers a new perspective to focus certain areas fit for DA. Besides, the computation costs of them can be tremendous, according to their backbones. From another aspect, our method just uses very simple backbone for feature extractor, and therefore these SOTA methods can be combined with our method to enhance our feature extractor. As the source code of TCoN, SAVA and STCDA are not public available, we only show the combined method with TA^{3}N.

Speed and Memory Costs. It should be noted that our model outperforms TA^{3}N [17] in terms of speed. During the test period, our model costs 0.0041 s to process a video while TA^{3}N [17] takes 0.0060 s for it. Hence, our method is much more suitable for large-scale datasets.

4.4 Ablation Study

To validate the effectiveness of our attention mechanism, we test our model without attention mechanism and the classifier uncertainty by comparing it with models above. The results are shown in Table 4. Obviously, model without attention mechanism and uncertainty estimation obtains lower accuracy than models

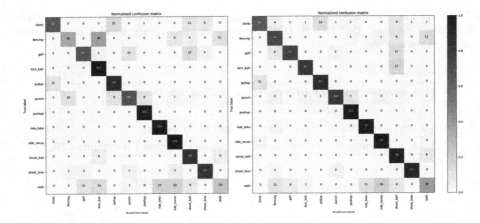

Fig. 4. The confusion matrix of the results of "Source only" method (left) and our model using spatial-temporal attention combined with LSTM (right). Both models are tested on UCF-HMDB$_{full}$ dataset. Vertical axis represents the true label while lateral axis represents the predicted one.

with spatial and spatial-temporal attention, which indicates the effectiveness of the uncertainty-based attention.

4.5   Visulization

Visualization of Distribution. In order to show how our approaches pull the distributions of source and target domain closer, we visualize the domain distribution using t-SNE [22] (Fig. 3). We can notice that our methods can make both source data and target data denser, which represents a good generalization.

Confusion Matrix. The confusion matrix of the results of our model is shown in Fig. 4. Most of the mistakes derive from the mistaken classification of walking. Compared with the result of 'Source only' method on the left, we can observe the gain our model obtains on the right.

5   Conclusion

In this paper, we apply the estimation of uncertainty to DA field for AR. We enable both the classifier and the discriminator to calculate aleatoric uncertainty and utilize the uncertainty of the discriminator to form an uncertainty-aware attention mechanism. By focusing on those regions and frames with lower uncertainty, we aid the classifier to achieve better adaptation. We test our models in both small-scale and large-scale datasets, and prove the competitiveness of our method together with the effectiveness of the proposed attention mechansim by ablation study. Further, the analysis consists of visualization, including t-SNE and confusion matrix. Our propose method offer a new aspect to address DA tasks for videos by using uncertainty. We hope to change the type of uncertainty and apply the model to practical scenes in the future.

References

1. Song, X., et al.: Spatio-temporal contrastive domain adaptation for action recognition. In: CVPR (2021)
2. Choi, J., Sharma, G., Schulter, S., Huang, J.-B.: Shuffle and attend: video domain adaptation. In: Vedaldi, A., Bischof, H., Brox, T., Frahm, J.-M. (eds.) ECCV 2020. LNCS, vol. 12357, pp. 678–695. Springer, Cham (2020). https://doi.org/10.1007/978-3-030-58610-2_40
3. Pan, B., et al.: Adversarial cross-domain action recognition with co-attention. In: AAAI (2020)
4. Sultani, I.W.: Human action recognition across datasets by foreground-weighted histogram decomposition. Phys. Lett. B **690**(2), 764–771 (2014)
5. Xu, T., et al.: Dual many-to-one-encoder-based transfer learning for cross-dataset human action recognition. Image Vis. Comput. **55**(PT.2), 127–137 (2016)
6. Jamal, A., Namboodiri, V.P., Deodhare, D., Venkatesh, K.S.: Deep domain adaptation in action space. In: BMVC (2018)
7. Zhang, X.Y., et al.: Learning transferable self-attentive representations for action recognition in untrimmed videos with weak supervision. In: AAAI (2019)
8. Girdhar, R., Ramanan, D., Gupta, A., et al.: ActionVLAD: learning spatio-temporal aggregation for action classification. In: CVPR (2017)
9. Liang, J., He, R., Sun, Z., et al.: Exploring uncertainty in pseudo-label guided unsupervised domain adaptation. Pattern Recognit. **96**, 1069(96) (2019)
10. Wen, J., Zheng, N., Yuan, J., et al.: Bayesian uncertainty matching for unsupervised domain adaptation. In: IJCAI (2019)
11. Han, L., Zou, Y., Gao, R., et al.: Unsupervised domain adaptation via calibrating uncertainties. In: CVPRW (2019)
12. Manders, J., Laarhoven, T.V., Marchiori, E.: Simple Domain Adaptation with Class Prediction Uncertainty Alignment. In: CORR, Arxiv/abs/1804.04448 (2018)
13. Kurmi, V.K., Kumar, S., Namboodiri, V.P.: Attending to discriminative certainty for domain adaptation. In: CVPR (2020)
14. Gal, Y.: Uncertainty in deep learning. PhD thesis, University of Cambridge (2016). 2, 3
15. Pei, Z., Cao, Z., Long, M., Wang, J.: Multi-adversarial domain adaptation. In: AAAI (2018)
16. Ganin, Y., et al.: Domain-adversarial training of neural networks. J. Mach. Learn. Res. **17**, 59:1–59:35 (2016)
17. Chen, M.H., et al.: Temporal attentive alignment for large-scale video domain adaptation. In: ICCV (2019)
18. Peng, X., Usman, B., Saito, K., Kaushik, N., Hoffman, J., Saenko, K.: Syn2real: A new benchmark for synthetic-to-real visual domain adaptation. arXiv preprint arXiv:1806.09755 (2018)
19. Long, M., et al.: Deep transfer learning with joint adaptation networks. In: ICML (2017)
20. Li, Y., et al.: Adaptive batch normalization for practical domain adaptation. Pattern Recognit. **80**, 109–117 (2018)

21. Saito, K., Watanabe, K., Ushiku, Y., et al.: Maximum classifier discrepancy for unsupervised domain adaptation. In: CVPR (2018)
22. Van der Maaten, L., Hinton, G.: Visualizing data using t-SNE. J. Mach. Learn. Res. **9**(Nov), 2579–2605 (2008)
23. Kendall, A., Gal, Y.: What uncertainties do we need in Bayesian deep learning for computer vision? In: NeuRIPS (2017)

Free-Form Image Inpainting with Separable Gate Encoder-Decoder Network

Liang Nie[1], Wenxin Yu[1(✉)], Xuewen Zhang[1], Siyuan Li[1], Shiyu Chen[1],
Zhiqiang Zhang[2], and Jun Gong[3]

[1] Southwest University of Science and Technology, Mianyang, Sichuan, China
`yuwenxin@swust.edu.cn`
[2] Hosei University, Tokyo, Japan
[3] Beijing Institute of Technology, Beijing, China

Abstract. Image inpainting refers to the process of reconstructing damaged areas of an image. For image inpainting, there are many means to generate not too bad inpainting results today. However, these methods either make the results look unrealistic or have complex structures and a large number of parameters. In order to solve the above problems, this paper designed a simple encoder-decoder network and introduced the region normalization technique. At the same time, a new separable gate convolution is proposed. The simple network architecture and separable gate convolution significantly reduce the number of network parameters. Moreover, the separable gate convolution can learn the mask (represents the missing area) from the feature map and update it automatically. After mask update, weights will be applied to each pixel of the feature map to alleviate the impact of invalid mask information on the completed result and improve the inpainting quality. Our method reduces 0.58M parameters. Moreover, our method improved the PSNR of Celeba and Paris Street View by 0.7–1.4 dB and 0.7–1.0 dB, respectively, in 10% to 60% damage cases. The corresponding SSIM has been increased 1.6 to 2.7 and 0.9 to 2.3%.

Keywords: Image inpainting · Separable gate convolution · Encoder-decoder network

1 Introduction

Image inpainting aims to reconstruct the missing contents of a damaged picture by a specific algorithm. This technique can find back the loss information, and it has a widespread application in image processing tasks (examples include image editing, old photo repair, 3D image synthesis, and more). However, an excellent restoration result requires that it look visually realistic. Hence, the inpainting algorithm must guarantee that it generates appropriate semantic contents for missing areas and maintain style consistency with the boundary.

© Springer Nature Switzerland AG 2021
T. Mantoro et al. (Eds.): ICONIP 2021, LNCS 13110, pp. 507–519, 2021.
https://doi.org/10.1007/978-3-030-92238-2_42

The existing approaches are mainly divided into traditional methods and deep learning-based methods. The traditional methods uses diffuse-based means (Such as Bertalmio et al. [1]) to gradually transfer the pixels around the missing region into its center region or patch-based methods (Darabi et al. [3]) that copy similar patches into them. These methods are acceptable for single-textured images, but they tend to be weak for complex scenes.

The learning-based methods mainly use a convolutional neural network to generate the semantic contents to fill the loss holes. Pathak et al. [9] were the first to apply convolutional neural network to image inpainting by designing a context encoder. At the same time, the researchers found that generative adversarial networks have excellent results in both image content generation and constrained image reconstruction. Therefore, many deep neural network architectures using adversarial learning strategy have emerged for image restoration, and this strategy is also used in this paper. Yang et al. [12], for example, designed a dual-branch codec network, where the two branches separately repair the texture and structure of an image and then fuse it. Yeh et al. [14] then used an adversarial learning network consisting of a pair of simple generator and discriminator. Finally, Iizuka et al. [4] proposed the concepts of the global discriminator and local discriminator, which further strengthened the constraints in image restoration.

With the rapid development of deep learning, more and more new technologies have emerged in image reconstruction, such as contextual attention mechanism [15], partial convolution [7], gate convolution [16], region normalization [17], etc. These new technologies provide the vigorous impetus for the development of image inpainting. Consequently, the development direction of image restoration also began to be diversified, including eye replacement repair with aesthetic indicators, multi-result generation network, super-resolution image inpainting, clothing change replacement, and so on.

However, due to the complexity of the image restoration task, the above methods are still short of the authenticity of the restoration results. Moreover, due to the characteristics of the multi-branch structure, they appear bloated and complex and are not suitable for portable equipment.

To solve these problems, we first simplified the network structure. The method presented in this paper has only one encoder-decoder as the generator and one discriminator. We named the network SGR-net. SGR-net is an end-to-end, single-branch network that significantly simplifies large-scale inpainting network architecture. Thus, the problem of a large number of network parameters can be effectively alleviated. Meanwhile, this paper proposed a novel separable-gate convolution to ensure the inpainting effect.

The main contributions in this paper are summarized as follows:

- We designed an end-to-end image inpainting network with the codec as the backbone, which reduced network parameters and ensured the improvement of restoration quality.
- At the same time, this paper proposes a separable-gate convolution, which can automatically learn the soft mask in the training process and assign weight to

the valid pixels. Moreover, compared with the original gate convolution [16], it has fewer parameters.

- Finally, to make full use of the learned mask and further reduce the impact of invalid information on the repair results, this paper introduces the region normalization technology. Because of the soft mask updating mechanism, the region normalization technology can play a better role.

2   Related Work

2.1   Image Inpainting

Previous image inpainting approaches can be generally divided into two categories: traditional and deep learning-based.

Traditional means mainly include diffusion-based and patch-based methods. The former is Bertalmio et al. [1] diffuse the pixels around the missing region like a stream of water toward its center. The latter, such as Darabi et al. [3], used to fill holes by speculating patches that the missing area is similar to the general area. Unfortunately, these approaches are computationally heavy and do not generate semantic content, limiting their effectiveness.

Deep learning-based methods have performed well and developed rapidly in recent years. They learn to extract semantic information through a large amount of data training to improve the repair effect significantly. For example, contextual attention (Yu et al.) [15] proposed the Coarse-to-Fine architecture, which first obtains the rough restoration result and then uses the result as a guide to obtaining the final inpainting result. This network architecture was subsequently used by various methods, such as Peng et al. [10] and Shin et al. [11]. Nazeri et al. [8] first predicts the edges of the damaged area and then uses the predicted edges to generate a complete image. Yang et al. [13] proposed a one-stage inpainting network to simplify the model structure while utilizing the information of image edge structure.

Although some existing heavy network architectures can achieve passable results, the complexity of their architecture and numerous parameters limit their deployment on lightweight devices.

2.2   Convolution Method

Convolution plays the most critical role in deep neural networks and is the cornerstone for accomplishing various tasks.

Since the emergence of convolutional neural networks, the convolutional mode has also experienced earth-shaking changes. When deep learning was first applied in image inpainting, people paid more attention to the network architecture. During this period, vanilla convolution was generally adopted for feature extraction, such as Pathakel. [9] and Yu et al. [15]. Subsequently, Liu et al. [7] found that for image restoration tasks, the vanilla convolution means would integrate invalid mask information into the reconstruction results, resulting in artifacts and reducing the quality of image generation. Therefore, Liu et al. [7] proposed partial convolution, in which the mask region was marked '0' and the mask was convoluted and updated to reduce the impact of invalid information on the repair results.

Inspired by Liu et al. [7], Yu et al. [16] improved the automatic mask update strategy and proposed gate convolution. Yu et al. [16] split the feature maps into two groups and activate one group with the sigmoid function to get the weight map(the smaller the value is, the higher the possibility that it is a mask). Then the other group is activated by the activation function and multiplied by the weight graph to get a new feature map. This soft mask can automatically learn and update, better represent the hidden mask, and reduce the negative impact of invalid values in holes on image content generation.

However, as the above convolution method gains power, the number of parameters also increases sharply. In addition, the gate convolution divides the feature maps into two groups equally, which diminishes the stability of the features. Therefore, under the inspiration of deepthwise separable convolution [2], this paper proposes a separable gate convolution. N feature maps (N represents the number of inchannels) were obtained through deepthwise convolution, and the feature graphs were divided into two groups according to $N - 1 : 1$ for activation. This operation decreases the number of parameters in the convolution process and ensures the utilization rate of feature information.

3   Approach

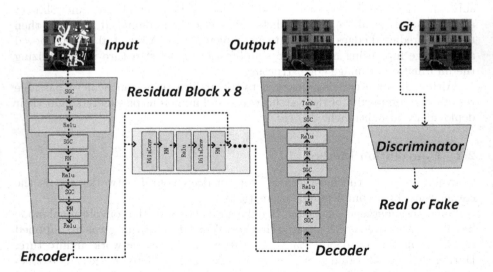

Fig. 1. The structure of SGR-net.

3.1   Network Structure

The backbone of SGR-net is a codec as the generator. As shown in Fig. 1, the net has a discriminator in addition to the generator. The encoder comprises separable gate convolution modules (SGM) and region normalization modules (RN). And then, the features go through eight residual blocks. A basic residual block consists of the convolution layers and region normalization modules. Then, the decoder is composed of separable gate convolution layers and RN modules. Finally, the output is activated by the tanh function to get the inpainting result. This result is then fed into the discriminator together with the ground truth to calculate the adversarial loss. At the same time, calculate the L1 distance between them. In this paper, L1 loss, adversarial loss, perceptual loss, and style loss are used to train and optimize the model (the loss function is consistent with [17]). The discriminator is PatchGAN, proposed by Isola et al. [5].

Separable Gate Convolution Modules. Image inpainting uses the information of the known area of the image to complete the filling of the missing holes. But the input is consists of both regions with valid pixels outside holes and invalid pixels (in shallow layers) or synthesized features (in deep layers) in masked regions. Hence, vanilla convolutions are ill-fitted for the task of free-form image inpainting. The vanilla convolutions process as shown in formula (1):

$$F_{x,y} = \sum_{i=-k'_w}^{k'_w} \sum_{j=-k'_h}^{k'_h} W_{k'_w+i,k'_h+j} \cdot I_{x+i,y+j} \tag{1}$$

(x, y) represent the position of pixels in the image. k_h and k_w means the kernel size of the filters W. $F_{x,y}$ is the output feature maps.

The equation shows that the same filters are applied for all spatial locations (x, y) to produce the output in vanilla convolutional layers. This operation is helpful in that all input image pixels are valid to extract local features in a sliding window, such as image super-resolution. However, for image inpainting, the input comprises both valid and invalid regions. That causes uncertainty and leads to visual artifacts such as color discrepancy, blurriness, and apparent edge responses during testing, as reported in [7].

It is crucial to reduce the interference caused by invalid pixels in the holes. Therefore, partial convolution was proposed by Liu et al. [7]. Partial convolution works by calculating the value of the current sliding window on M. If the $sum(M) > 0$, a scaling factor is applied to the input I of the current sliding window. Otherwise, the value of the current sliding window is set to zero (shown in formula (2).

$$F_{x,y} = \begin{cases} \sum\sum W \cdot \left(I \odot \frac{M}{\text{sum}(M)} \right), & \text{if sum(M)} > 0 \\ 0, & \text{otherwise} \end{cases} \tag{2}$$

M is the binary mask (1 for valid pixels, 0 for invalid). $\odot$ denotes element-wise multiplication. After each partial convolution operation, the mask is updated by the following rule: $m'_{y,x} = 1$, if $sum(M) > 0$.

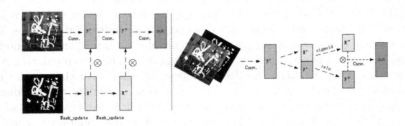

Fig. 2. Illustration of partial convolution (left) and gate convolution (right).

The complete process of partial convolution is shown in Fig. 2(*left*). Consistent with formula (2), the updated mask is multiplied with the features, and then convoluted the multiplied results, the final output is gradually obtained.

Nevertheless, there is still a problem with partial convolution. When in different layers, no matter how much the coverage of the filter of the upper layer is, the next layer will classify the mask obtained by the upper layer as 0 or 1. As shown in Fig. 3, the mask M convoluted by the filter W can get M'. The orange blocks represent the same area. When the M' through the convolution layer gets M'', it is easy to see that the orange block in M'' becomes 1. This means the 9 pixels orange area in M are credible. On the contrary, they are not. Cause the 9 pixels, just one is valid.

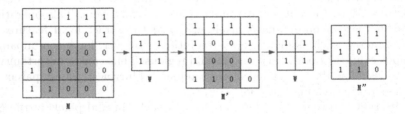

Fig. 3. Illustration of mask update mechanism of partial convolution. (Color figure online)

Inspired by this, Yu et al. [16] changed the mask update rules. They divided the feature maps into two equal groups according to the number of channels (shown in the right of Fig. 2). Sigmoid activation and ReLU activation were performed, respectively. The Sigmoid activation operation is named GATE, and a weight map (also known as 'soft mask', the smaller the weight value, the higher the possibility that the position is invalid information) is obtained through

GATE. Then the soft mask will multiply with the activated feature map to reduce the confidence of invalid pixels in the feature map.

Nevertheless, two problems remain with gate convolution. First, to split the feature maps into two groups on an average way, the convolution kernel is two times needed (cause M' * F', the channels are halved), and the number of parameters is directly doubled. Secondly, the equal divided method will reduce the features space used for learning. For mask updating, only one single channel feature map is sufficient. Therefore, separable gate convolution is proposed in this paper to solve the above problems.

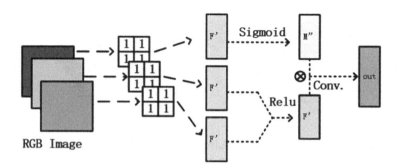

Fig. 4. Illustration of our separable gate convolution.

The operation principle of separable gate convolution is shown in Fig. 4. First, we set the group parameter of convolution kernels to the number of input channels, which makes the output feature maps equal to the input channels. Assuming the number of input channels is N, we get N feature maps. Next, divide them into two groups on an $N - 1 : 1$ proportion. The former is activated by relu function and the latter by sigmoid. After activation, multiplied them together to obtain the weighted feature maps. Finally, the weighted feature maps are sent into the convolutional layer, which consists of filters with kernel size equal to 1 for expanding the output channels.

Although separable gate convolution also requires double the convolution kernels. By separating the convolution operation, the number of its parameters does not increase.

The number of parameters required for gate convolution [16] is as follows:

$$
\begin{aligned}
N_{gc} &= K_s \times K_s \times O_c \times In_c \\
&= (K_s \times K_s \times In_c) \times O_c
\end{aligned}
\tag{3}
$$

The number of parameters required for separable gate convolution is as follows:

$$
\begin{aligned}
N_{sgc} &= K_s \times K_s \times In_c + O_c \times 1 \times 1 \times In_c \\
&= (K_s \times K_s + O_c) \times In_c \\
&= (K_s \times K_s \times In_c) \times (1 + \frac{O_c}{K_s \times K_s})
\end{aligned}
\tag{4}
$$

Among them, N_{gc} and N_{sgc} respectively represent the number of parameters required for the two kinds of convolution. K_s means kernel size. O_c and In_c represents the number of output and input channels respectively. As we can see, when $K_s = 1$, separable gate convolution (SGC) has In_c more parameters than gate convolution (GC). In other cases, the larger the size of the convolution kernel, the fewer parameters of SGC than GC (Fig. 5).

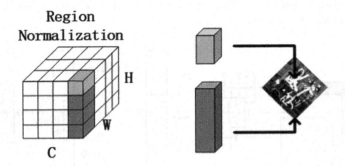

Fig. 5. Illustration of our Region Normalization (RN). Pixels in the same color (green or blue) are normalized by the same mean and variance. The corrupted and uncorrupted regions of the input image are normalized by different means and variances. (Color figure online)

Region Normalization. Region normalization (RN) [17] normalizes and transforms corrupted and uncorrupted regions separately. This can solve the mean and variance shift problem of normalization and avoid information mixing in affine transformation. RN is designed for use in the early layers of the inpainting network, as the input feature has large corrupted regions, which causes severe mean and variance shifts.

4   Experiment

4.1   Experiment Setup

We selected Paris Street View [9] and Celeba-HQ [6] datasets for training that are common in image inpainting. The former have 15000 building scenes pictures(100 for testing and the rest using for training). The latter contained 30,000 images of faces. In this paper, we divided it into 28000 images for training and 2000 for testing. The masks are from the irregular mask dataset [7]. Moreover, the masks are divided into 10–20%, 20–30%, 30–40%, 40–50%, and 50–60%, according to the proportion of the missing area. Each interval has 2000 masks. We trained with a 1080Ti and set the epoch to 10 until the model converged.

Table 1. The quantitative comparison of CelebA-HQ dataset.

	Mask	PC [7]*	EC [8]	RN [17]	Ours
PSNR↑	10–20%	29.339	29.842	29.868	**31.272**
	20–30%	26.344	26.550	27.154	**28.121**
	30–40%	24.060	24.652	24.993	**25.853**
	40–50%	22.072	23.122	23.185	**24.020**
	50–60%	20.274	20.459	20.455	**21.178**
SSIM↑	10–20%	0.919	0.935	0.933	**0.949**
	20–30%	0.866	0.878	0.889	**0.906**
	30–40%	0.811	0.832	0.838	**0.859**
	40–50%	0.749	0.778	0.780	**0.806**
	50–60%	0.667	0.686	0.687	**0.714**
Parameters↓		31.34M	10.26M(×2)	11.60M	**11.02M**

Table 2. The quantitative comparison of PSV dataset.

	Mask	PC [7]*	EC [8]	RN [17]	Ours
PSNR↑	10–20%	29.237	30.375	30.042	**31.032**
	20–30%	26.678	27.188	27.465	**28.280**
	30–40%	24.517	25.424	26.059	**26.134**
	40–50%	22.556	23.412	24.057	**24.192**
	50–60%	20.424	20.844	21.416	**21.643**
SSIM↑	10–20%	0.912	0.930	0.926	**0.935**
	20–30%	0.848	0.875	0.877	**0.888**
	30–40%	0.781	0.819	0.833	**0.833**
	40–50%	0.707	0.743	0.761	**0.762**
	50–60%	0.598	0.647	0.655	**0.659**
Parameters↓		31.34M	10.26M(×2)	11.60M	**11.02M**

4.2 Quantitative Comparison

In this paper, partial convolution (PC), edge-connect (EC), and region normalization (RN) are selected as comparison methods (∗ means the method comes from an unofficial version). These methods have performed well in the field of image restoration in recent years. The test results of these approaches on Celeba-HQ and Paris Street View datasets are shown in Table 1 and Table 2, respectively. Two commonly used metrics, PSNR and SSIM, were adopted. PSNR reflects the pixel similarity between the inpainting results and the original image, while SSIM shows the structural similarity. It can be seen that PSNR and SSIM have significant improvements in different damage scales of the two datasets with this

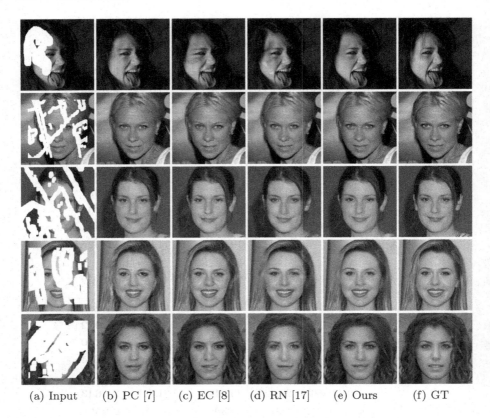

(a) Input   (b) PC [7]   (c) EC [8]   (d) RN [17]   (e) Ours   (f) GT

Fig. 6. The inpainting results in CelebA-HQ.

method. On Celeba-HQ, the PSNR improvement at 10%–60% scales was as high as 0.7–1.4 dB. On Paris Street View, the value was 0.7–1.0 dB.

In 10% to 60% of the damaged area, the SSIM lift is 0.016, 0.017, 0.020, 0.027, 0.027 and 0.009, 0.011, 0.019, 0.021, 0.023 on two datasets, respectively. The experimental data show that the larger the missing area of the image, the higher the lift of our method on SSIM. Therefore, in the face of repairing large missing areas, our method can recover more structural information. In addition, SGR-net has 0.58M fewer parameters than RN, which is a large network with the fewest parameters. Compared to edge-connect, our method has almost half the parameters. It is almost two-thirds less than the partial convolution.

4.3 Qualitative Comparison

We randomly selected reconstruction results of different damage degrees (10%–60%) from the test results of the two datasets to display (shown in Fig. 6 and Fig. 7). From left to right are input, partial convolution, edge-connect, domain normalization, our method, and the ground truth. As you can see, both the

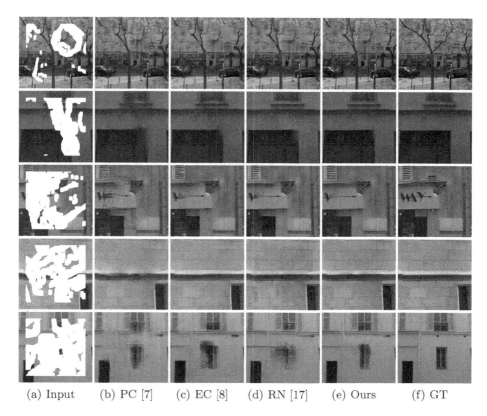

(a) Input   (b) PC [7]   (c) EC [8]   (d) RN [17]   (e) Ours   (f) GT

Fig. 7. The inpainting results in Paris Street View.

existing method and our method perform well at low levels of damage. The result looks authentic and very similar to the original image. On the face dataset Celeba-HQ, it can be seen that the pictures generated by SGR-net are more consistent in style, with softer facial contours. Subjectively, it looks more natural.

Our method can restore more texture details for objects such as branches and windows in the Paris Street View dataset. In the face of a large area of missing information, the best structural information can be recovered (the last line, where 60% of the area is missing, powerfully illustrates this problem). In addition, when we encounter objects such as challenging words to recover, our method also fails to repair them. However, compared to the other methods, SGR-net produces no artifacts, as shown in the third line.

5 Conclusion

This paper designed a simple encoder-decoder network and proposed a separable gate convolution to replace the vanilla convolution. The simple network architecture and separable convolution make the network parameters drop. Meanwhile, the gating mechanism gives weights in different positions of the feature maps to reduce the influence of invalid pixels, which improves the repair effect. Moreover, the region normalization technology is introduced to provide different mean values and variances for the masked area and the unmasked area to reduce the masked area's influence. Finally, compared with RN, our method reduces 0.58M parameters. Moreover, the PSNR of Celeba-HQ and Paris Street View improved by 0.7–1.4 dB and 0.7–1.0 dB, respectively, in 10% to 60% damage cases. The corresponding SSIM has been increased 1.6 to 2.7 and 0.9 to 2.3%.

Acknowledgement. This research is supported by Sichuan Science and Technology Program (No. 2020YFS0307, No. 2020YFG0430, No. 2019YFS0146), Mianyang Science and Technology Program (No. 2020YFZJ016).

References

1. Bertalmio, M., Vese, L., Sapiro, G., Osher, S.: Simultaneous structure and texture image inpainting. IEEE Trans. Image Process. **12**, 882–889 (2003)
2. Chollet, F.: Xception: deep learning with depthwise separable convolutions. In: Proceedings of the IEEE Conference on Computer Vision and Pattern Recognition, pp. 1251–1258 (2017)
3. Darabi, S., Shechtman, E., Barnes, C., Goldman, D.B., Sen, P.: Image melding: combining inconsistent images using patch-based synthesis. ACM Trans. Graph. **31**, 1–10 (2012)
4. Iizuka, S., Simo-Serra, E., Ishikawa, H.: Globally and locally consistent image completion. ACM Trans. Graph. **36**, 1–14 (2017)
5. Isola, P., Zhu, J.Y., Zhou, T., Efros, A.A.: Image-to-image translation with conditional adversarial networks. In: Proceedings of the IEEE Conference on Computer Vision and Pattern Recognition (CVPR), July 2017
6. Karras, T., Aila, T., Laine, S., Lehtinen, J.: Progressive growing of GANs for improved quality, stability, and variation. In: International Conference on Learning Representations (2018)
7. Liu, G., Reda, F.A., Shih, K.J., Wang, T.-C., Tao, A., Catanzaro, B.: Image inpainting for irregular holes using partial convolutions. In: Ferrari, V., Hebert, M., Sminchisescu, C., Weiss, Y. (eds.) ECCV 2018. LNCS, vol. 11215, pp. 89–105. Springer, Cham (2018). https://doi.org/10.1007/978-3-030-01252-6_6
8. Nazeri, K., Ng, E., Joseph, T., Qureshi, F.Z., Ebrahimi, M.: EdgeConnect: generative image inpainting with adversarial edge learning. arXiv preprint arXiv:1901.00212 (2019)
9. Pathak, D., Krahenbuhl, P., Donahue, J., Darrell, T., Efros, A.A.: Context encoders: feature learning by inpainting. In: Proceedings of the IEEE Conference on Computer Vision and Pattern Recognition (CVPR), June 2016
10. Peng, J., Liu, D., Xu, S., Li, H.: Generating diverse structure for image inpainting with hierarchical VQ-VAE. In: Proceedings of the IEEE/CVF Conference on Computer Vision and Pattern Recognition (CVPR), pp. 10775–10784, June 2021
11. Shin, Y.G., Sagong, M.C., Yeo, Y.J., Kim, S.W., Ko, S.J.: PEPSI++: fast and lightweight network for image inpainting. IEEE Trans. Neural Netw. Learn. Syst. **32**, 252–265 (2021)

12. Yang, C., Lu, X., Lin, Z., Shechtman, E., Wang, O., Li, H.: High-resolution image inpainting using multi-scale neural patch synthesis. In: Proceedings of the IEEE Conference on Computer Vision and Pattern Recognition (CVPR), July 2017
13. Yang, J., Qi, Z., Shi, Y.: Learning to incorporate structure knowledge for image inpainting. In: Proceedings of the AAAI Conference on Artificial Intelligence, pp. 12605–12612 (2020)
14. Yeh, R.A., Chen, C., Yian Lim, T., Schwing, A.G., Hasegawa-Johnson, M., Do, M.N.: Semantic image inpainting with deep generative models. In: Proceedings of the IEEE Conference on Computer Vision and Pattern Recognition (CVPR), July 2017
15. Yu, J., Lin, Z., Yang, J., Shen, X., Lu, X., Huang, T.S.: Generative image inpainting with contextual attention. In: Proceedings of the IEEE Conference on Computer Vision and Pattern Recognition (CVPR), June 2018
16. Yu, J., Lin, Z., Yang, J., Shen, X., Lu, X., Huang, T.S.: Free-form image inpainting with gated convolution. In: Proceedings of the IEEE/CVF International Conference on Computer Vision (ICCV), October 2019
17. Yu, T., et al.: Region normalization for image inpainting. In: Proceedings of the AAAI Conference on Artificial Intelligence, pp. 12733–12740, April 2020

BERTDAN: Question-Answer Dual Attention Fusion Networks with Pre-trained Models for Answer Selection

Haitian Yang[1,2(✉)], Chonghui Zheng[3], Xuan Zhao[4], Yan Wang[1(✉)],
Zheng Yang[1], Chao Ma[1], Qi Zhang[1], and Weiqing Huang[1,2]

[1] Institute of Information Engineering, Chinese Academy of Sciences, Beijing, China
{yanghaitian,wangyan}@iie.ac.cn
[2] School of Cyber Security, University of Chinese Academy of Sciences,
Beijing, China
[3] Hangzhou Institute for Advanced Study, University of Chinese Academy of
Sciences, Hangzhou, China
[4] York University, Ontario, Canada

Abstract. Community question answering (CQA) becomes more and more popular in both academy and industry recently. However, a large number of answers often amass in question-answering communities. Hence, it is almost impossible for users to view item by item and select the most relevant one. As a result, answer selection becomes a very significant subtask of CQA. Hence, we propose question-answer dual attention fusion networks with the pre-trained model (BRETDAN) for the task of answer selection. Specifically, we apply BERT model, which has achieved a better result in GLUE leaderboard with deep transformer architectures as the encoder layer to do fine-tuning for question subjects, question bodies and answers, respectively, then the cross attention mechanism selecting out the most relevant answer for different questions. Finally, we apply dual attention fusion networks to filter the noise caused by introducing question and answer pairs. Specifically, the cross attention mechanism aims to extract interactive information between question subject and answer. In a similar way, the interactive information between question body and answer is also captured. Dual attention fusion aims to address the noise problem in the question and answer pairs. Experiments show that the BERTDAN model achieves significant performance on two datasets: SemEval-2015 and SemEval-2017, outperforming all baseline models.

Keywords: Answer selection · Pre-trained models · Dual attention networks · Community Question Answering

1 Introduction

As a powerful product of the digital revolution, the Internet has revolutionized the way people obtain information and share knowledge. One of the most com-

T. Mantoro et al. (Eds.): ICONIP 2021, LNCS 13110, pp. 520–531, 2021.
https://doi.org/10.1007/978-3-030-92238-2_43

mon ways is to type keywords explicitly to express requirements, and then search engines will return a large amount of various web pages in relevance to different submitted keywords, which users can browse them one by one and select the best page catering for his expectation. However, conventional search engines can only provide general solutions to the domain-specific problems, while in some cases, there is no guarantee of an desired answer. Therefore, in order to overcome these shortcomings to a certain degree, Community Question Answering (CQA) has become popular in the past decade, which provides an interactive experience and a faster retrieve in many fields. Here we list some typical Question Answering platforms: as Yahoo! Answers, Baidu Knows, Quora, etc. We can note that these communities are fairly open that users can search information they are interested in, post questions of their own concerns, and provide answers to questions they know.

Users of CQA can be divided into two groups in general, the first one is the group who mainly ask questions in communities and rarely answer questions or just provide simple even unrelated answers to some questions while others devote themselves to giving informative answers which are known as experts generally in a certain field such as time plan or keeping fit. As a result, answers from different users show huge variance in quality that some are related to the topic or exactly desirable while others completely deviate from the actual intention of users. It is time-consuming for users to view all candidate answers and select the most relevant one. Furthermore, the number of repeated and unanswered questions increases significantly, but some of them in fact have been posted by other users and answered before just with different representation in words or syntax. Thus, answer selection becomes a significant task of CQA, which aims to find the most relevant answer in the repository, largely reducing users' time going through all candidate answers, so as to provide better experience for them and making it possible to solve some unsettled but similar questions with various representation which actually have been answered before.

In this paper, we propose question-answer dual qttention fusion networks With Pre-trained Models for Answer Selection in CQA. Our model uses the pre-trained model for words encoding, fully considering the contextual information of each word in question subject, question body, and answer. Then, we apply the cross attention between words in question subject and answer, as well as words in question body and answer, respectively, to capture interactive information between question and answer. Next, we apply dual attention fusion to address the noise issue. Finally, our proposed model integrates attention-question and attention-answer to compute the matching probability of question-answer pair to determine the best answer to a given question.

The main contributions of our work can be summarized as follows:

(1) We propose to treat question subjects and question bodies separately, integrating answer information into the neural attention model, hence the noise from answers is reduced and the performance is improved.

(2) We leverage the pre-trained model for question subjects, question bodies, and answers, respectively, comprehensively capturing the semantic information in each question and answer. In addition, we apply cross attention

mechanism to effectively obtain important interactive features between each question and answer pair. Further, we carry out an ablation study to verify the effectiveness of the pre-trained model and the cross-attention mechanism. The results demonstrate that they both effectively level the performance of our approach on the answer selection task.

(3) Our proposed model achieves state-of-the-art performance on both SemEval-2015 and SemEval-2017 datasets, outperforming all baseline models.

The rest of this paper is outlined as follows: Sect. 2 surveys some advanced techniques related to our work. Section 3 introduces our proposed answer selection model - BERTDAN, also describe the details of the structure; Sect. 4 discusses the implementation of BERTDAN, explaining various parameters and the training process, and further compare our model with other baselines; Sect. 5 summarizes the conclusion and suggests the further research potentials.

2   Related Work

In the early years, some researchers also studied various rules with naive Bayesian, perceptron, etc., as updating strategies and statistical features, then fed these features into machine learning models. Roth et al. [1] proposed a sparse network for multi-class classification, which uses linear functions to calculate the probability of questions in each category. Metzler et al. [2] applied SVM with radial basis kernel function for factoid questions and used multiple feature fusion methods to combine various syntactic and semantic features. However, even though these feature-guided techniques could achieve good performance, heavy dependence on feature engineering inevitably leads to the indispensability of domain knowledge and an enormous number of handcrafts.

In addition, some previous studies propose to use global ranking strategies. As far as we know, Barron-Cedeno et al. [3] is the first to use the structured prediction model for answer selection. Joty et al. [4] also use global inference for answers and represents them in fully connected graph. However, global ranking strategies may lose some unseen information that only can be captured by deep networks.

Recently, deep learning models [5–10] are proved to be a relatively more effective method. Deep learning methods can automatically capture various features through multi-layer networks instead, avoiding complex feature engineering. Wan et al. [11] proposed MV-LSTM based on bi-direction LSTM with representing questions and answers as tensors to capture positional information at each time step. Zhang et al. [12] introduced a novel cross attention mechanism to alleviate the redundancies and noises in answers to a certain degree which are usually prevalent in CQA.

Although these deep learning models have achieved better performance, none of these explored question subjects, question bodies, and answers respectively. More importantly, the BERT model is not applied by any of the above-listed methods. As an extremely powerful pre-trained model, BERT can make comprehensive semantic representations. Moreover, we don't find any paper using the attention mechanism after the pre-trained BERT model to further capture interactive features in questions and answers.

3   Proposed Model

3.1   Task Description

In this section, we first provide a description of the answer selection task. Specifically, we denote the the answer selection task in CQA as a tuple of four elements (S, B, A, y). In the tuple, $S = [s^1, s^2, \cdots, s^m]$ represents a question subject with m as its length; $B = [b^1, b^2, \cdots, b^g]$ represents the corresponding question body with g as its length; $A = [a^1, a^2, \cdots, a^n]$ represents the corresponding answer with n as its length; and $y \in Y$ represents the relevance degree, in other words, $Y = \{Good, Bad\}$ denotes whether a candidate can answer a question properly (Good) or not (Bad). Hence, our BERTDAN model on the answer selection task in CQA can be summarized as assigning a label to each answer based on the conditional probability $Pr(y \mid S, B, A)$ with the given set $\{S, B, A\}$.

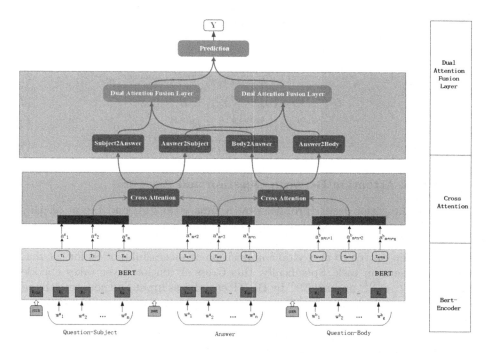

Fig. 1. The structure of our proposed BERTDAN model

3.2   Overview of Proposed Model

The structure of BERTDAN can be described as three layers, including Encoder layer which obtains contextual representation of question subjects, question bodies and answers with BERT model, Cross Attention layer which transforms the relationship between question subjects and answers as well as question bodies and answers into important interactive features between questions and answers, and dual attention fusion layer which outputs a label to each answer corresponding to the given question based on the conditional probability the model calculated. The pipeline of our proposed model is described in Fig. 1.

3.3   BERT Encoder

We use BERT model [13] as the encoder layer to convert question subjects, question bodies, and answers into H-dimensional encoded forms containing different contextual representation, respectively. We define $\{w_t^{subject}\}_{t=1}^m$, $\{w_t^{body}\}_{t=1}^g$ and $\{w_t^{answer}\}_{t=1}^n$ as word sets of all candidate question subjects, all candidate question bodies and all candidate answers, respectively. Noted, m, g, and n are the length of each question subject, each question body and each answer, respectively. We obtain $U^{subject}, U^{body}$ and U^{answer} by encoding question subjects $\{w_t^{subject}\}_{t=1}^m$, answers $\{w_t^{answer}\}_{t=1}^n$ and $\{w_t^{body}\}_{t=1}^g$ into encoded forms $u_t^{subject}$, u_t^{body} and u_t^{answer}, by BERT model, respectively.

The following equations demonstrate this layer:

$$u_t^{subject}, u_t^{body}, u_t^{answer} = BERT([w_t^{subject}; w_t^{answer}; w_t^{body}]) \tag{1}$$

$$U^{subject} = \{u_t^{subject}\}_{t=1}^m \in R^{m \times H} \tag{2}$$

$$U^{body} = \{u_t^{body}\}_{t=1}^g \in R^{g \times H} \tag{3}$$

$$U^{answer} = \{u_t^{answer}\}_{t=1}^n \in R^{n \times H} \tag{4}$$

3.4   Cross Attention Between Question and Answer

The Cross Attention layer (proposed by Adams et al. [14] and Lin et al. [15]), aims at fusing information of words between question subjects and answers as well as words between question bodies and answers. The cross attention mechanism has been proven to be a vital composition of successful models for the reading comprehension task. Specifically, this Cross Attention layer takes encoded forms $u_t^{subject}$, u_t^{body} and u_t^{answer} by BERT encoder as input and compute the relevance between each word in question subjects and answers as well as the counterpart in question bodies and answers, i.e., $Subject2Answer$, $Answer2Subject$, $Body2Answer$ and $Answer2Body$.

Taking the process of computing relevance between each word in question subjects and answers as an example, we get a matrix denoted as $SAS \in R^{m \times n}$. Then, two similarity matrices $\bar{S}_{Subject2Answer} \in R^{m \times n}$ and $\bar{S}_{Answer2Subject} \in R^{m \times n}$, are generated after normalization over each row and column by softmax function of the similarity matrix SAS. Let $s_{x,y} \in R$ be an element in similarity matrix $SAS \in R^{m \times n}$, where rows represent question subjects and columns represent answers. Also we get interactive features between question subjects and answers $A_{Answer2Subject}$ and $A_{Body2Subject}$ as the final output. The computation process can be described as the following equations:

$$s_{x,y} = w_{subject}^T \cdot [u_x^{subject}; u_y^{answer};$$
$$u_x^{subject} \odot u_y^{answer}] \tag{5}$$

$$\bar{S}_{Subject2Answer} = softmax_{row}(SAS) \tag{6}$$

$$\bar{S}_{Answer2Subject} = softmax_{col}(SAS) \tag{7}$$

$$A_{Subject2Answer} = \bar{S}_{Subject2Answer} \cdot U^{answer} \tag{8}$$

$$A_{Answer2Subject} = \bar{S}_{Subject2Answer}$$
$$\cdot \bar{S}_{Answer2Subject}^T \cdot U^{subject} \tag{9}$$

where $w_{subject}^T$ is a trainable parameter.

Similarly, we can obtain interactive features of the question bodies with answers as followings:

$$A_{Body2Answer} = \bar{S}_{Body2Answer} \cdot U^{answer} \tag{10}$$

$$A_{Answer2Body} = \bar{S}_{Body2Answer} \cdot \bar{S}_{Answer2Body}^T \cdot U^{body} \tag{11}$$

3.5   Dual Attention Fusion Mechanism

In question answering communities, the question and the corresponding answer information are given by users in the real life, both the question and the corresponding answer contain a lot of noise information, which reduces the performance of the answer selection task model. Therefore, We propose the use of a dual attention fusion module to fuse the features adaptively. Then, to reduce the possibility of noise introduced by paired answer information, we utilize the filtration gate to adaptively filter out some of the useless question and answer information. Inspired by previous works of Chen et al. [16] and Mou et al. [17], we obtain new representations of questions and answers within enhanced local semantic information from BERT encoder layer and cross attention layer. To be specific:

$$question_i^m = [U^{subject} \odot A_{answer2subject}; U^{body} \odot A_{answer2body}] \tag{12}$$

$$answer_i^n = [U^{answer} \odot A_{subject2answer}; U^{body} \odot A_{body2answer}] \tag{13}$$

where $[\cdot; \cdot; \cdot; \cdot]$ is the concatenation operation of vectors and $\odot$ indicates element-wise product.

Next, since the answer is a crucial signal, we propose a question-answer dual attention mechanism to emphasize those important answers and de-emphasize irrelevant answers. We utilize the above $question_i^m$, $answer_i^n$ and calculate the first attention weights as:

$$\alpha_i = \frac{exp(\delta(u_i^{answer}, question_i^m))}{\sum exp(\delta(u_i^{answer}, question_i^m))} \tag{14}$$

where δ is the attention function. In the same way We calculate the second attention weights as:

$$\beta_j = \frac{exp(\delta(u_i^{[subject;body]}, answer_i^n))}{\sum exp(\delta(u_i^{[subject;body]}, answer_i^n))} \tag{15}$$

where δ is the attention function. Specifically, $u_i^{[subject;body]} = [u_i^{subject}; u_i^{body}]$, where $[\cdot;\cdot;\cdot;\cdot]$ is the concatenation operation of vectors. We combine α_i and β_i. with the following formula to obtain the final attention weight of each state:

$$\psi_i = softmax(\gamma\alpha_i + (1-\gamma)\beta_i) \tag{16}$$

$$= \frac{exp(\gamma\alpha_i + (1-\gamma)\beta_i)}{\sum_{k\in[1,n]} exp(\gamma\alpha_k + (1-\gamma)\beta_k)} \tag{17}$$

In the above-displayed equation, ψ_i represents the final attention weight towards the question and answer i. Also, $\gamma \in [0,1]$ is a soft switch to adjust the importance of two attention weights, α_i and β_i. Specially, among the multiple ways to set the parameter γ, for example, treating γ as a hyper-parameter and manually adjusting it to obtain the best performance. Indeed, we choose to learn γ from a neural network automatically. The reason is that it ensures adaptively assigning of different values to γ on different scenarios, realizing better experimental results. The γ is calculated as follows:

$$\gamma = \sigma(w^T[\alpha;\beta] + b) \tag{18}$$

It is worth noting that both vectors ω and scalars b are learnable parameters while σ is the sigmoid function. Next, we implement the attention weights to calculate a weighted sum of the state vectors. A semantic vector that represents the context is required.

$$Q_i = \sum \psi_i[u_i^{subject}; u_i^{body}] \tag{19}$$

$$A_i = \sum \psi_i u_i^{answer} \tag{20}$$

Eventually, we use a single-layer softmax classifier with cross-entropy loss, where the input is the final context vector c_i:

$$y_i = softmax(W_o(tanh(W([Q_i; A_i]) + b))) \tag{21}$$

Here, W_0, b are trainable parameters, y_i is the probability of each candidate answers. The entire model is trained end-to-end.

4   Experimental Setup

4.1   DataSet

We use two corpora to train and evaluate our model. These datasets are SemEval-2015 and SemEval-2017 CQA datasets. We exclude SemEval-2016 dataset in this paper, because the SemEval-2017 dataset is an updated version of the SemEval-2016, sharing the same evaluation metrics. These datasets are specialized for the answer selection task in CQA. Our model aims to identify good answers from potentially useful answers and bad or useless answers. Table 1 is the statistics of the two corpora.

Table 1. Statistical Information of SemEval-2015 and SemEval-2017 Corpora.

Statistics	SemEval-2015			SemEval-2017		
	Train	Dev	Test	Train	Dev	Test
Number of questions	2376	266	300	5124	327	293
Number of answers	15013	1447	1793	38638	3270	2930
Average length of a question subject	6.36	6.08	6.24	6.38	6.16	5.76
Average length of a question body	39.26	39.47	39.53	43.01	47.98	54.06
Average length of an answer	35.82	33.90	37.33	37.67	37.30	39.50

4.2   Training and Hyperparameters

We choose Adam Optimizer proposed by Jimmy et al. [18] with the momentum coefficient β 0.001 for optimization. The per-minibatch L2 regularization parameter and batch size of the model are set to 1×10^{-5} and 100 respectively. The dropout value d is set to 0.3 to prevent overfitting. The maximum number of words in a question subject, a question body and an answer are set to 20, 110, 100, respectively. The proposed model is implemented in Pytorch and the cross attention layer has a dimension of 300. We use the best parameters on development sets, and evaluate the performance of our model on test sets.

4.3   Results and Analysis

Three evaluation metrics: F1, Acc (accuracy), and MAP (Mean Average of Precision) are adopted. The descriptions of baselines are demonstrated as follows.

JAIST [19] proposes an answer quality scoring approach with a feature-based SVM-regression model.

BGMN [20] uses an Bi-directional Gated Memory Network to model the interactions between question and answer by matching them in two directions.

ECUN [21] proposes traditional method of extracting features and deep learning models, such as training a classifier and learn the question comment representation based CNN.

CNN-LSTM-CRF [22] validates the effectiveness of label dependency with two neural network-based models, including a stacked ensemble of Convolutional Neural Networks as well as a simple attention-based model referred to as Long Short Term Memory and Conditional Random Fields.

QCN [23] investigates the subject-body relationship of community questions, taking the question subject as the primary part of the question representation, and the question body information being aggregated based on similarity and disparity with the question subject.

Table 2. Comparisons of different models on two corpora.

Dataset	Model	MAP	F1	Acc
SemEval-2015	(1) JAIST	NA	0.7896	0.7910
	(2) HITSZ-ICRC	NA	0.7652	0.7611
	(3) BGMN	NA	0.7723	0.7840
	(4) CNN-LSTM-CRF	NA	0.8222	0.8224
	(5) QCN	NA	0.8391	0.8565
	(6) **BERTDAN (ours)**	NA	**0.8496**	**0.8584**
SemEval-2017	(1) ECNU	0.8672	0.7767	0.7843
	(2) QCN	0.8851	0.7811	0.8071
	(4) **BERTDAN (ours)**	**0.9026**	**0.7939**	**0.8235**

The comparison results are shown in Table 2, the results demonstrate that:

(1) Both JAIST and HITTZ-ICRC perform well on the SemEval-2015 dataset. The experimental results show that JAIST outperforms deep learning model BGMN by 1.73% on F1 and 0.7% on Acc , which indicates that not all deep learning models can outperform conventional machine learning models.
(2) CNN-LSTM-CRF outperforms BGMN, HITSZ-ICRC, and JAIST on all evaluation metrics. The results demonstrate that a mixture of CNN, LSTM, and CRF can take advantage of each trait to fully integrate various features so that it achieves a better performance than conventional machine learning models on the answer selection task.
(3) QCN outperforms ECNU, CNN-LSTM-CRF, BGMN, HITZZ-ICRC, and JAIST on MAP, F1, and Acc. Due to QCN model investigates the subject-body relationship of community questions, taking the question subject as the primary part of the question representation, and the question body information being aggregated based on similarity and disparity with the question subject, hence achieves a better performance on the answer selection task.

(4) Our proposed model BERTDAN outperforms all baseline models on three evaluation metrics with advancement attributed to the pre-trained BERT model and attention mechanism. We use BERT model as pre-trained methods, fully fusing context information in question subjects, question bodies and answers, respectively. Then cross attention between question subjects and answers as well as question bodies and answers helps our model estimate the relevance of question-subject-answer pairs and question-body-answer pairs so as to effectively capture crucial interaction semantic features between questions and answers. BERTDAN studies the relationship of questions and answers, fully capturing semantic features at different angles, hence the performance of the answer selection task is greatly enhanced.

4.4   Ablation Study

To fully verify the improvement of our proposed model, we implement six variants of BERTDAN on the Yahoo! Answers dataset. The results are listed in Table 3.

Table 3. Ablation study of the seven models on the SemEval-2017 dataset.

Model	MAP	F1	Acc
(1) without BERT but with task-specific word embeddings	0.8069	0.7165	0.7632
(2) without BERT but with character embeddings	0.7986	0.6859	0.7482
(3) without cross attention	0.8875	0.7819	0.8065
(4) without the dual attention fusion layer but with simple combination	0.8752	0.7736	0.7948
(5) without cross attention and dual attention fusion layer but with simple combination	0.8679	0.7642	0.7859
(6) without treat question subjects and question bodies separately	0.8927	0.7865	0.8174
(7) **BERTDAN (ours)**	**0.9026**	**0.7939**	**0.8253**

As the results shown in Table 3, model (1) and (2) use traditional embeddings at word and character levels, respectively as pre-trained tools instead of BERT model, and the results decrease by 9.57% and 10.40% on MAP, 7.74% and 10.80% on F1, 6.21% and 7.71% on Acc compared with the best model BERTDAN (7)($p < 0.05$ based on student t-test), indicating that BERT model makes great contributions for the final BERTDAN model mainly attributed to the organic combination of masked language model, transformer mechanism and sentence-level representation in it. With the traits of transformer mechanism including self-attention, multi-head attention and position encoding, the pre-trained BERT model manages to capture semantic features at each dimension and takes full advantage of position information whose performance is far beyond traditional word embeddings and character embeddings.

When it comes to the effect of cross attention as well as the interaction and prediction layer, model (3) removes the cross attention layer while model (4) utilizes simple combination instead of the interaction and prediction layer. Model (5) only uses the BERT model to pre-train question subjects and answers as well as question bodies and answers, then combine the outputs as the final results. Compared with the best BERTDAN model (7), the above three models decrease by 1.51%, 2.74%, and 3.47% on MAP, 1.20%, 2.03%, and 2.97% on F1, 1.88%, 3.05%, and 3.94% on Acc ($p < 0.05$ based on student t-test). The results show that cross attention has abilities of capturing the relationship between questions and answers to obtain deeper semantic information as well as the interaction and prediction layer succeeds in integrating the interactive features output by cross attention to learn higher-level semantic representation.

Moreover, model (6) is slightly inferior to the BERTDAN model (7). The reason is that the mechanical connection of question subject and question body introduces too much noise, leading to performance degradation.

In general, our proposed model BERTDAN comprehensively learns the context features with the BERT model, fully utilizes the relationship of questions and answers with attention mechanism, and finally integrates semantic information of questions and answers to furthest enhance the performance of the answer selection task.

5   Conclusion

Answer selection is one of the most important parts of CQA. To automate the answer selection progress, we propose a method named BERTDAN. Specifically, we use the BERT model to capture context features of question subjects, question bodies, and answers, respectively. More importantly, we obtain comprehensive interactive information between question subjects and answers, question bodies and answers, through the cross attention mechanism. Through integrating attention questions and attention-answers, the BERTDAN model gets final results that achieve state-out-of-art performance.

In the future, our research group would like to test the BERTDAN model in other fields to improve its universality. Also, we would like to improve the computing speed of the BERTDAN model, leveling up the performance of our solution.

References

1. Roth, D.: Learning to resolve natural language ambiguities: a unified approach. In: AAAI/IAAI 1998, pp. 806–813 (1998)
2. Metzler, D., Croft, W.B.: Analysis of statistical question classification for fact-based questions. Inf. Retr. 8(3), 481–504 (2005)

3. Barrón-Cedeno, A., et al.: Thread-level information for comment classification in community question answering. In: ACL, pp. 687–693, Beijing, China (2015)

4. Joty, S., Màrquez, L., Nakov, P.: Joint learning with global inference for comment classification in community question answering. In: ACL, pp. 703–713, San Diego, California (2016)

5. Yang, M., et al.: Knowledge-enhanced hierarchical attention for community question answering with multi-task and adaptive learning, pp. 5349–5355. In: IJCAI (2019)

6. Deng, Y., et al.: Joint learning of answer selection and answer summary generation in community question answering, pp. 7651–7658. In: AAAI (2020)

7. Xie, Y., Shen, Y., et al.: Attentive user-engaged adversarial neural network for community question answering. In: AAAI, vol. 34, pp. 9322–9329 (2020)

8. Garg, S., Thuy, V., Moschitti, A.: Tanda: transfer and adapt pre-trained transformer models for answer sentence selection. In: AAAI, vol. 34, pp. 7780–7788 (2020)

9. Yang, M., Wenting, T., Qiang, Q., et al.: Advanced community question answering by leveraging external knowledge and multi-task learning. Knowl.-Based Syst. **171**, 106–119 (2019)

10. Yang, H., et al.: AMQAN: adaptive multi-attention question-answer networks for answer selection. In: Hutter, F., Kersting, K., Lijffijt, J., Valera, I. (eds.) ECML PKDD 2020. LNCS (LNAI), vol. 12459, pp. 584–599. Springer, Cham (2021). https://doi.org/10.1007/978-3-030-67664-3_35

11. Wan, S., Lan, Y., Guo, J., et al.: A deep architecture for semantic matching with multiple positional sentence representations. In: AAAI, pp. 2835–2841 (2016)

12. Zhang, X., Li, S., Sha, L., Wang, H.: Attentive interactive neural networks for answer selection in community question answering. In: AAAI, vol. 31 (2017)

13. Devlin, J., Chang, M.W., et al.: Bert: pre-training of deep bidirectional transformers for language understanding. In: NAACL (2019)

14. Yu, A.W., et al.: Fast and accurate reading comprehension by combining self-attention and convolution. In: ICLR (2018)

15. Lin, Z., et al.: A structured self-attentive sentence embedding. In: ICLR (2017)

16. Chen, Q., et al.: Enhanced lstm for natural language inference[c]. In: ACL, pp. 1657–1668 (2017)

17. Mou, L., et al.: Natural language inference by tree-based convolution and heuristic matching[c]. In: ACL, pp. 130–136 (2016)

18. Ba, J., Kingma, D.P.: Adam: a method for stochastic optimization. In: ICLR (2015)

19. Tran, Q.H., Tran, D.V., Vu, T., Le Nguyen, M., Pham, S.B.: Jaist: combining multiple features for answer selection in community question answering. In: SemEval-2015, pp. 215–219, Denver, Colorado (2015)

20. Wu, W., Wang, H., Li, S.: Bi-directional gated memory networks for answer selection. In: Chinese Computational Linguistics and Natural Language Processing Based on Naturally Annotated Big Data, pp. 251–262 (2017)

21. Wu, G., Sheng, Y., Lan, M., Wu, Y.: Ecnu at semeval2017 task 3: using traditional and deep learning methods to address community question answering task. In: SemEval-2017, pp. 365–369 (2017)

22. Xiang, Y., Zhou, X., et al.: Incorporating label dependency for answer quality tagging in community question answering via cnn-lstm-crf. In: COLING, pp. 1231–1241, Osaka, Japan (2016)

23. Wu, W., Sun, X., Wang, H., et al.: Question condensing networks for answer selection in community question answering. In: ACL, pp. 1746–1755 (2018)

Rethinking the Effectiveness of Selective Attention in Neural Networks

Yulong Wang[1], Xiaolu Zhang[1], Jun Zhou[1], and Hang Su[2(✉)]

[1] Ant Financial Group, Beijing, China
{frank.wyl,yueyin.zxl,jun.zhoujun}@antfin.com
[2] Department of Computer Science and Technology, Tsinghua University,
Beijing, China
suhangss@mail.tsinghua.edu.cn

Abstract. The introduction of the attention mechanism is considered as an important innovation in the recent development of neural networks. Specifically, in convolutional neural networks (CNN), selective attention is utilized in a channel-wise manner to recalibrate the output feature map based on different inputs dynamically. However, extra attentions and multi-branch calculations introduce additional computational cost and reduce model parallel efficiency. Therefore, we rethink the effectiveness of selective attention in the network and find that the bypass branch computation is redundant and unnecessary. Meanwhile, we establish an equivalent relationship between Squeeze-and-Excitation Networks (SENet) and Selective Kernel Networks (SKNet), which are two representative network architectures with the feature attention mechanism. In this paper, we develop a new network architecture variant called Elastic-SKNet by reducing the calculation in the bypass branch of SKNet. Furthermore, we utilize the differentiable Neural Architecture Search (NAS) method to quickly search for the reduction ratio in each layer to further improve model performance. In the extensive experiments on ImageNet and MS-COCO datasets, we empirically show that the proposed Elastic-SKNet outperforms existing state-of-the-art network architectures in image classification and object detection tasks with lower model complexity, which demonstrates its further application prospects in other fields.

Keywords: Computer vision · Neural networks · Efficient computing · Attention mechanism

1 Introduction

The design of deep neural network architecture has been proven to play an essential role in improving model performance in many machine learning fields. In recent works, the attention mechanism is considered to be an effective technique to improve the representation ability of deep learning models, which has been widely applied in many areas such as image recognition [3,6], machine translation [14], natural language generation [16], and graph neural network [17].

© Springer Nature Switzerland AG 2021
T. Mantoro et al. (Eds.): ICONIP 2021, LNCS 13110, pp. 532–543, 2021.
https://doi.org/10.1007/978-3-030-92238-2_44

Compared with fixed arithmetic operations such as convolution, pooling, and matrix multiplication, the attention mechanism can dynamically recalibrate feature outputs according to different inputs. Specifically, in a convolutional neural network, the attention unit transforms the input feature map into a set of weighting coefficients, which are then multiplied onto output feature maps along each channel [3,6]. More diverse adjustment methods for the output, such as weighting in spatial dimensions, have also been proposed in previous works [12].

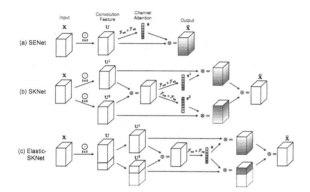

Model	FLOPs	Acc (%)
ResNeXt50	4.24G	77.77
SENet50	4.25G	78.88
SKNet50	4.47G	79.21
ESKNet50, 0.25×	4.29G	**80.14**
ESKNet50, 0.50×	4.35G	**80.01**
ESKNet50, 0.75×	4.41G	**79.96**
ResNeXt101	7.99G	78.89
SENet101	8.00G	79.42
SKNet101	8.46G	79.81
ESKNet101, 0.25×	8.10G	**80.87**
ESKNet101, 0.50×	8.21G	**80.68**
ESKNet101, 0.75×	8.33G	**80.84**

Fig. 1. Overview. **Left**: The model architecture differences between (a) SENet, (b) SKNet and (c) the proposed Elastic-SKNet, where $\odot$ denotes convolution operator with specific kernel size, $\oplus$ denotes element-wise summation, $\otimes$ denotes channel-wise multiplication, $\mathcal{F}_{ex} \circ \mathcal{F}_{sq}$ denotes the excitation and squeezing functions to generate attention weights. Our model design removes unnecessary bypass branch calculation to reduce model complexity. **Right**: The model performance comparisons between the proposed Elastic-SKNet and state-of-the-art models, where "ESKNet, $m\times$" denotes Elastic-SKNets with different elastic ratios, "FLOPs" measures the model complexity by the number of multiply-adds, "Acc" denotes the model top-1 accuracy on the ImageNet validation dataset.

Among the attention-based network architectures, Selective Kernel Networks (SKNet) use the selective attention unit to dynamically aggregate the output features from multiple computing branches with different kernel sizes (as shown in Fig. 1(b)). Such attention unit design has been verified to improve model performance. However, the multi-branch design introduced in SKNet causes much additional computational burden and reduces parallel computing efficiency. We want to explore whether there is redundancy in the bypass branch calculation and whether it can be further simplified while maintaining performance. Therefore, we rethink the necessity and effectiveness of selective attention in neural networks. Specifically, we empirically find that in a network structure that includes multi-branch calculations such as SKNet, the attention intensity of the bypass branch is relatively small, and part of its output results can be directly deleted without excessively affecting model performance. Meanwhile, we also establish an equivalent relationship between SKNet and Squeeze-and-Excitation Networks (SENet), that is, SENet can be regarded as a SKNet with all bypass branches deleted.

Based on the above results, in this paper, we develop a new network architecture variant called **Elastic-SKNet** (as shown in Fig. 1(c)), which reduces the calculation in the bypass branches by setting an "elastic ratio" to control the computational complexity. Since the proposed network structure reduces extra calculations and merges multiple branches into a single operation process, Elastic-SKNet can greatly improve model computational efficiency compared to SKNet. Moreover, the elastic ratio is quite flexible and can be set to the same value globally or adjusted according to different layers. In order to further optimize the ratio setting, we borrow the idea from the differentiable Neural Architecture Search (NAS) method [9], and formulate the layerwise ratio search as an optimization problem of learning the bypass branch channel importance. The resulting elastic ratios are obtained by calculating the layerwise proportions of important channels reserved to meet the computational budgets. In the extensive experiments on the ImageNet dataset, we empirically show that the proposed Elastic-SKNets achieve higher accuracy than the existing multiple network architectures with lower model complexity. We further validate that Elastic-SKNets have better feature transferability in the object detection task. In the experiments on the MS-COCO dataset, we improve the object detection performance by merely replacing the backbone network with the pre-trained Elastic-SKNet weights.

2   Related Work

Attention Networks. The attention mechanism is firstly applied in machine translation task [14], where the dynamic attention weights are utilized to guide the decoder to focus on different features from inputs. Afterward, more powerful language models [16] comprehensively utilize attention mechanisms as their basic computation units for feature extraction. The attention mechanism is also widely adopted beyond natural language processing fields. In the design of convolutional neural networks, attention units are applied to generate channel-wise weighting coefficients to recalibrate output feature maps [3,6]. Beyond the channel attention mechanism, other works design different attention units [12] to perform selective weighting operations on the spatial dimensions.

Multi-branch Computation. In the pioneering network design of GoogLeNet [15], multi-branch computation is utilized to extract information with different receptive fields. Afterward, introducing bypass branch computation and skip connection into the network design has become an effective method to improve model performance [2]. However, scattered multi-branch computation reduces the model's parallel efficiency and increases real-time inference latency. Grouped convolution, which was firstly introduced in AlexNet [4] as a memory reducing technique, has been found to effectively simulate multi-branch computation with higher computational efficiency [18].

Differentiable NAS. The neural architecture search aims to discover an optimal network structure design through algorithm learning. Early works utilize reinforcement learning [19] to conduct architecture search in a huge configuration space. Subsequent works [9] use differentiable optimization methods to trim out the optimized network structure from a redundant hyper-network. From this perspective of problem modeling, differentiable NAS methods are more similar to model pruning approaches, which aims to compress the model size while maintaining a similar model performance.

3   Methodology

Before introducing the proposed method, we recover the existing attention network methods, including SENet and SKNet. Then we report our observations in the section of rethinking the effectiveness of selective attention. Based on these analysis results, we elaborate on the design of the proposed architecture Elastic-SKNet. Finally, we present the solution to search for the optimized elastic ratios in Elastic-SKNet.

3.1   Preliminary

SENet. In the design of SENet, the authors propose a "Squeeze-and-Excitation" (SE) block to adaptively recalibrate the output channel-wise features by explicitly modeling the dependency between channels. Specifically, given input $\mathbf{X} \in \mathbb{R}^{H' \times W' \times C'}$ and the corresponding convolution feature maps $\mathbf{U} \in \mathbb{R}^{H \times W \times C}$, the SE block implements such a transformation $\mathcal{F}$, which maps $\mathbf{U}$ into a set of attention weights $\mathbf{s} \in \mathbb{R}^C$. The transformation mapping is usually composed of two consecutive operations, which are squeezing function $\mathcal{F}_{sq}$ and excitation function $\mathcal{F}_{ex}$. The $\mathcal{F}_{sq}$ function reduces the feature maps along spatial dimensions into channel-wise statistics vector $\mathbf{z} \in \mathbb{R}^C$ by global average pooling, which is

$$\mathbf{z}_c = \frac{1}{H \times W} \sum_{i=1}^{H} \sum_{j=1}^{W} \mathbf{U}_c(i, j), \tag{1}$$

where $\mathbf{z}_c$ and $\mathbf{U}_c$ denote the c-th element of $\mathbf{z}$ and $\mathbf{U}$ along channel dimension. The $\mathcal{F}_{ex}$ is instantiated as two fully-connected layers to generate channel-wise attention weights, which is

$$\mathbf{s} = \sigma(\mathbf{W}_2 \delta(\mathbf{W}_1 \mathbf{z})), \tag{2}$$

where $\sigma(\cdot)$ and $\delta(\cdot)$ are Sigmoid and ReLU functions respectively, $\mathbf{W}_1 \in \mathbb{R}^{\frac{C}{r} \times C}$ and $\mathbf{W}_2 \in \mathbb{R}^{C \times \frac{C}{r}}$ are weights for two fully-connected layers. r denotes a reduction ratio to control the SE block computational complexity. The final recalibrated output feature maps $\widetilde{\mathbf{X}}$ are generated by scaling feature maps $\mathbf{U}$ with channel-wise attentions $\mathbf{s}$

$$\widetilde{\mathbf{X}}_c = \mathbf{s}_c \mathbf{U}_c. \tag{3}$$

SKNet. SKNet is built upon by stacking multiple layers with "Selective Kernel" (SK) unit. In a SK unit, multi-branch output features resulting from convolutions with different kernel sizes are aggregated by a set of attention weights. What is different from the SE block is that the attention weights are not only used to scale output feature maps along the channel dimension, but also used to recalibrate the feature fusion from different branches. Similar to SE block, given convolution feature maps from two branches $\mathbf{U}^1$ and $\mathbf{U}^2$, SK unit first fuses these features into a channel-wise statistics vector $\mathbf{z}$ by global average pooling on the summation of feature maps

$$\mathbf{z}_c = \frac{1}{H \times W} \sum_{i=1}^{H} \sum_{j=1}^{W} (\mathbf{U}^1 + \mathbf{U}^2)_c(i,j). \tag{4}$$

Then, the selective attention weights $\mathbf{a}, \mathbf{b}$ are generated by two consecutive fully-connected layers for each branch

$$\mathbf{s}^1 = \mathbf{W}_2^1 \delta(\mathbf{W}_1^1 \mathbf{z}), \quad \mathbf{s}^2 = \mathbf{W}_2^2 \delta(\mathbf{W}_1^2 \mathbf{z})$$
$$[\mathbf{a}, \mathbf{b}] = \text{softmax}([\mathbf{s}^1, \mathbf{s}^2]). \tag{5}$$

The final recalibrated output feature maps $\widetilde{\mathbf{X}}$ is obtained by

$$\widetilde{\mathbf{X}}_c = \mathbf{a}_c \mathbf{U}_c^1 + \mathbf{b}_c \mathbf{U}_c^2, \tag{6}$$

where c denotes the c-th channel dimension.

3.2   Rethinking the Effectiveness of Selective Attention in SKNet

Relationship Between SENet and SKNet. Based on the above formulations, we can observe that SENet and SKNet share many similarities in the design of generating selective attention. Here we present a formal derivation to build an equivalent relationship between SENet and SKNet.

Suppose that in the SK unit, output feature maps from one bypass branch are all zeros (take $\mathbf{U}^2 = \mathbf{0}$ for example), then Eq. (4) in the fusion process of the SK unit reduces to Eq. (1) in the SE block

$$\mathbf{z}_c = \frac{1}{H \times W} \sum_{i=1}^{H} \sum_{j=1}^{W} \mathbf{U}_c^1(i,j). \tag{7}$$

Moreover, Eq. (6) for recalibrating output feature maps $\widetilde{\mathbf{X}}$ reduces to Eq. (3) in SE block

$$\widetilde{\mathbf{X}}_c = \mathbf{a}_c \mathbf{U}_c^1. \tag{8}$$

As for the selective attention weights $\mathbf{a}$ in Eq. (5), it can be further simplified as

$$\mathbf{a} = \frac{e^{\mathbf{s}^1}}{e^{\mathbf{s}^1} + e^{\mathbf{s}^2}} = \frac{1}{1 + e^{-(\mathbf{s}^1 - \mathbf{s}^2)}}$$
$$= \sigma\left(\mathbf{W}_2^1 \delta(\mathbf{W}_1^1 \mathbf{z}) - \mathbf{W}_2^2 \delta(\mathbf{W}_1^2 \mathbf{z})\right) \tag{9}$$

which can be considered as a more complex parameterization form to generate attention weights similar to Eq. (2) in the SE block. Therefore, we can conclude that ***SENet is a degenerated version of SKNet when bypass branch features are deleted.***

3.3 Elastic-SKNet

Based on the above analysis, we propose a new network architecture variant called Elastic-SKNet by reducing computation in the bypass branch of SK units. Specifically, we utilize the "elastic ratio" $r_e \in [0,1]$ to control the output feature maps in the bypass branch by setting output channel numbers as $C_e = \mathrm{round}(C \times r_e)$. In this situation, the fusion process in the SK unit is changed to

$$
\mathbf{z}_c = \frac{1}{H \times W} \sum_{i=1}^{H} \sum_{j=1}^{W} \begin{cases} (\mathbf{U}^1 + \mathbf{U}^2)_c(i,j), & 0 < C \le C_e \\ \mathbf{U}_c^1(i,j), & C_e < C \le C, \end{cases} \tag{10}
$$

and the recalibration process is changed to

$$
\widetilde{\mathbf{X}}_c = \begin{cases} \mathbf{a}_c \mathbf{U}_c^1 + (1 - \mathbf{a}_c) \mathbf{U}_c^2, & 0 < C \le C_e \\ \mathbf{a}_c \mathbf{U}_c^1, & C_e < C \le C, \end{cases} \tag{11}
$$

where we utilize the facts that the softmax outputs $\mathbf{a}$ and $\mathbf{b}$ satisfy $\mathbf{a}_c + \mathbf{b}_c = 1$.

To further reduce the computational redundancy in the original SKNet, we apply two more techniques. First, we replace the multi-branch computation with a single combined operation. In the original SKNet, the multi-branch output features with different kernel sizes are computed in separate processes, which hurts the model parallel efficiency. Instead, we follow the similar practice in [5] to use one 3×3 convolution with $(C + C_e)$ channels to replace two branches with 3×3 and 5×5 convolutions. Second, we simplify the generation process of selective attention as shown in Eq. (5) by only modeling $\mathbf{a}$ with one fully-connected layer. As demonstrated in Eq. (11), this suffices to conduct selective attention weighting process.

3.4 Architecture Search for Elastic-SKNet

The proposed Elastic-SKNet provides us a flexible hyperparameter "elastic ratio" to control the model complexity. The elastic ratio r_e can be not only set to the same value for all layers, but also adjusted for different layers. Inspired by the recent advances in Neural Architecture Search (NAS) method [9], we present a differentiable method to search for the optimized elastic ratios for different layers.

Following similar practices in [10], we associate channel gates on the bypass branches in the original SKNet. The channel gates are utilized to scale the corresponding output features by $\widehat{\mathbf{U}}_c^2 = \boldsymbol{\lambda}_c \mathbf{U}_c^2$, where $\boldsymbol{\lambda}_c$ is the associated channel gate variable on the c-th channel, and $\widehat{\mathbf{U}}^2$ is the scaled output feature maps used for the following computations in SK unit. If a channel gate λ_c is near to zero, it means that the corresponding output channel contributes little to the final prediction. Therefore, it can be safely deleted without affecting model performance. Then the elastic ratio is obtained by calculating the proportion of remaining important channels $r_e = \sum_c \mathbb{I}(\lambda_c \neq 0)/C$. To train the channel gate variables, we formulate the optimization objective as follows

$$\min_{\Lambda=\{\lambda^1,\cdots,\lambda^L\}} \sum_i \mathcal{L}(f(x_i;\Lambda),y_i) + \gamma \sum_l |\lambda^l|_1, \qquad (12)$$

where Λ denote the channel gate variables from all L layers, $\mathcal{L}$ denotes cross-entropy loss function, $f(x_i;\Lambda)$ denotes the network prediction for input x_i with channel gates associated, y_i is the sample label, γ is a balance factor to control the influence from ℓ_1-norm based sparsity regularization term.

After optimizing the channel gate variables, we choose the remaining important channels by setting a global threshold. We delete the output channels whose channel gate values are less than the threshold. As for how to determine the global threshold, we adopt a binary search algorithm to determine the best threshold τ given a predefined model computational budget. Specifically, given the model budget B (usually measured in FLOPs), the threshold τ is initialized to 0.5 and searched in $[0, 1]$. If the resulting architecture complexity B_t exceeds the budget B, it means current threshold τ_t should be adjusted to delete more unimportant channels. We set a maximum search iteration T and such a stopping criterion that when the searched architecture $|B_t - B|/B \leq \epsilon$, then the search process is stopped.

4   Experiments

In this section, we will elaborate on the experimental settings and results on the ImageNet classification task and MS-COCO object detection task.

4.1   Experimental Settings

Our experiments are all performed on the ImageNet dataset, including over 1.2M training images and 50K validation images. To train the proposed Elastic-SKNet with a fixed global elastic ratio, we conduct experiments on ResNeXt-50 and ResNeXt-101 models, which introduce grouped convolution operation in the standard ResNet model. We follow the original SKNet architecture settings with $M = 2$ branches and $G = 32$ convolution groups. We choose the elastic ratio from [0.25, 0.5, 0.75] to investigate the model performance under different complexity. We follow the standard data augmentation practices [6], which include random cropping to size 224×224, random horizontal flipping, and normalization across channels. Label-smoothing regularization is used in the total loss with mixing weight $\alpha = 0.1$. We follow the standard optimization settings, which include SGD optimizer with an initial learning rate of 0.1, momentum of 0.9, batch size of 256, and weight decay of 1e−4. To fully optimize the model weights, all the models are trained for 120 epochs with the cosine annealing learning rate scheduler [11]. To accelerate the overall training process, we use PyTorch framework to perform automatic mixed precision distributed training across 8 NVIDIA GeForce GTX 1080 Ti GPUs.

As for searching the optimized elastic ratios in Elastic-SKNet, we set the balance factor $\gamma = 0.01$ in Eq. (12). The associated channel gate variables Λ are optimized together with the model weights from scratch. The training epoch is reduced to 100 epochs, and other hyperparameter settings are the same as above. The binary search for the final elastic ratios is performed for a maximum iteration of $T = 10$ with a tolerance ratio $\epsilon = 0.01$.

4.2   ImageNet Classification Results

In this section, we compare the proposed Elastic-SKNet with state-of-the-art models on ImageNet classification performance. We report the model complexity measured by the number of parameters (Params) and the number of multiply-adds (FLOPs), model latency, and top-1 classification accuracy in crop size of 224×224. To calculate the model latency, we measured each model's inference time with an input batch size of 10 on a single NVIDIA GeForce GTX 1080 Ti GPU and repeated the same procedure for 50 times to report the average values. The results are shown in Table 1. Our models outperform all the other network architectures and achieve the highest accuracy. Moreover, the proposed Elastic-SKNets reduce much model computational cost and memory usage than SKNets, and achieve similar model inference speed with SENets, with more than 1.0% improvement on top-1 accuracy. These results empirically demonstrate that our models achieve a good trade-off between model complexity and performance, and are very promising to be deployed in the application scenarios.

4.3   Performance on Auto-ESKNets

We further report the classification performance of the networks searched by the method in Sect. 3.4. For a fair comparison, we set the FLOPs constraint of the searched network structure to the complexity of its corresponding Elastic-SKNet. We call the resulting network architecture as Auto-ESKNet. Table 2 summarizes the results. The reported metrics are the same as above. We empirically discover that the searched Elastic-SKNet architectures achieve faster inference speed and higher prediction accuracy than manually set networks with similar model complexity. To further visualize the architecture differences, we plot the channel numbers on Elastic-SKNet's bypass branches. Figure 2 shows the differences. We can discover that Auto-ESKNets tend to allocate more channels at the early stages and final stages, and use fewer channels during the middle stages. Therefore, we conclude that the differentiable architecture searching algorithm proposed in Sect. 3.4 can effectively optimize the Elastic-SKNet architecture configurations and further boost model performance.

Table 1. ImageNet classification performance for state-of-the-art models. "ESKNet" denotes the proposed Elastic-SKNet. "50" and "101" denote the depth of the networks. "r_e" denotes the elastic ratio. "Acc" denotes the top-1 accuracy when a center crop size of 224×224 is applied onto the validation image.

Model	FLOPs	Params	Latency	Acc (%)
ResNet50	4.14G	25.5M	19 ms	76.15
ResNeXt50	4.24G	25.0M	30 ms	77.77
ResNeXt50-BAM	4.31G	25.4M	35 ms	78.30
ResNeXt50-CBAM	4.25G	27.7M	56 ms	79.62
SENet50	4.25G	27.7M	34 ms	78.88
ResNetD-50	4.34G	25.6M	29 ms	79.15
SKNet50	4.47G	27.5M	47 ms	79.21
ESKNet50, $r_e = 0.25$	4.29G	26.1M	35 ms	**80.14**
ESKNet50, $r_e = 0.50$	4.35G	26.4M	37 ms	**80.01**
ESKNet50, $r_e = 0.75$	4.41G	26.7M	38 ms	**79.96**
ResNet101	7.87G	44.5M	32 ms	77.37
ResNeXt101	7.99G	44.3M	50 ms	78.89
ResNeXt101-BAM	8.05G	44.6M	57 ms	79.33
ResNeXt101-CBAM	8.00G	49.2M	93 ms	79.40
SENet101	8.00G	49.2M	59 ms	79.42
ResNetD-101	8.06G	44.6M	48 ms	80.54
SKNet101	8.46G	48.9M	83 ms	79.81
ESKNet101, $r_e = 0.25$	8.10G	46.1M	59 ms	**80.87**
ESKNet101, $r_e = 0.50$	8.21G	46.7M	63 ms	**80.68**
ESKNet101, $r_e = 0.75$	8.33G	47.4M	65 ms	**80.84**

Table 2. ImageNet classification performance for the searched architectures. "Auto-ESKNet" denotes the automatically searched network, where "r" denotes the manually set elastic ratio of its corresponding Elastic-SKNet with similar model complexity. All the other metrics are the same as the definition in Table 1.

Model		FLOPs	Params	Latency	Acc (%)
ESKNet50	$r_e = 0.25$	4.29G	26.1M	35.0 ms	80.14
	$r_e = 0.50$	4.35G	26.4M	37.3 ms	80.01
	$r_e = 0.75$	4.41G	26.7M	37.9 ms	79.96
Auto-ESKNet50	$r_e = 0.25$	4.29G	25.9M	36.2 ms	79.98
	$r_e = 0.50$	4.36G	26.4M	**36.9 ms**	80.04
	$r_e = 0.75$	4.43G	26.9M	**37.8 ms**	80.20
ESKNet101	$r_e = 0.25$	8.10G	46.1M	58.9 ms	80.87
	$r_e = 0.50$	8.21G	46.7M	63.2 ms	80.68
	$r_e = 0.75$	8.33G	47.4M	65.0 ms	80.84
Auto-ESKNet101	$r_e = 0.25$	8.10G	46.1M	61.0 ms	80.84
	$r_e = 0.50$	8.22G	46.8M	**62.2 ms**	80.76
	$r_e = 0.75$	8.34G	47.5M	**64.3 ms**	80.92

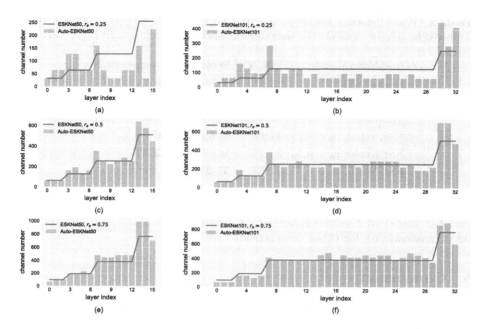

Fig. 2. The layerwise bypass branch output channel number comparison between ESKNet and Auto-ESKNet with optimized architecture design. **Left**: channel number comparison results on ResNeXt50 architecture. **Right**: channel number comparison results on ResNeXt101 architecture.

4.4 MS-COCO Object Detection Results

In order to further validate that the proposed Elastic-SKNets have better feature transferability, we migrate the pre-trained Elastic-SKNet weights on the ImageNet dataset to the object detection task. Following the standard practices, we adopt the widely used challenging MS-COCO dataset for all the experiments as the primary benchmark. All the models are trained under *minival* setting, i.e., on 115k MS-COCO 2017 training set and evaluated on 5k MS-COCO 2017 validation set. For a fair comparison, we simply replace the backbone networks with the proposed Elastic-SKNets, and all the other default settings remain unchanged, including feature pyramid network (FPN) technique [7], data augmentation, and "1×" training schedule. We conduct all the experiments based on MMDetection framework.

Table 3 summarizes the detection results based on Faster-RCNN [13], Cascade-RCNN [1] and RetinaNet [8] methods. We report the standard MS-COCO evaluation mean average precision (mAP) metric, where multiple IoU thresholds from 0.5 to 0.95 are applied. Compared to baseline methods with ResNet backbones, we improve the detection performance by simply using pre-trained Elastic-SKNets without extra techniques. These results demonstrate

Table 3. Object detection results on the MS-COCO dataset. All detection methods are improved by simply replacing the backbone network with the proposed Elastic-SKNets.

Backbone Model	FRCNN	RetinaNet	Cascade RCNN
ResNeXt50	33.8%	35.6%	40.4%
SENet50	34.2%	36.3%	40.5%
SKNet50	34.6%	34.5%	40.9%
ESKNet50, $r_e = 0.25$	34.6%	36.6%	40.3%
ESKNet50, $r_e = 0.50$	**34.7%**	36.5%	**42.3%**
ESKNet50, $r_e = 0.75$	34.2%	**37.3%**	41.6%

that the proposed Elastic-SKNets enjoy better representation ability and feature transferability for other downstream tasks.

5   Conclusion

In this paper, we rethink the effectiveness of selective attention in neural networks. We discover the redundant parts in the computation of bypass branches. We propose Elastic-SKNet with bypass calculation reduction. Moreover, we utilize a differentiable NAS method to optimize the layerwise elastic ratio. In the ImageNet classification experiments, Elastic-SKNets outperform state-of-the-art models and achieves faster inference speed with lower model complexity. In the MS-COCO object detection task, we improve the detection performance by simply replacing the backbone with pre-trained Elastic-SKNets. These results demonstrate that Elastic-SKNet has excellent feature representation ability, which is promising to be applied in other tasks.

References

1. Cai, Z., Vasconcelos, N.: Cascade R-CNN: delving into high quality object detection. In: Proceedings of the IEEE Conference on Computer Vision and Pattern Recognition, pp. 6154–6162 (2018)
2. He, K., Zhang, X., Ren, S., Sun, J.: Deep residual learning for image recognition. In: Proceedings of the IEEE Conference on Computer Vision and Pattern Recognition, pp. 770–778 (2016)
3. Hu, J., Shen, L., Sun, G.: Squeeze-and-excitation networks. In: Proceedings of the IEEE Conference on Computer Vision and Pattern Recognition, pp. 7132–7141 (2018)
4. Krizhevsky, A., Sutskever, I., Hinton, G.E.: Imagenet classification with deep convolutional neural networks. In: Advances in Neural Information Processing Systems, pp. 1097–1105 (2012)
5. Lee, J., Won, T., Hong, K.: Compounding the performance improvements of assembled techniques in a convolutional neural network. arXiv preprint arXiv:2001.06268 (2020)

6. Li, X., Wang, W., Hu, X., Yang, J.: Selective kernel networks. In: Proceedings of the IEEE Conference on Computer Vision and Pattern Recognition, pp. 510–519 (2019)
7. Lin, T.Y., Dollár, P., Girshick, R., He, K., Hariharan, B., Belongie, S.: Feature pyramid networks for object detection. In: Proceedings of the IEEE Conference on Computer Vision and Pattern Recognition, pp. 2117–2125 (2017)
8. Lin, T.Y., Goyal, P., Girshick, R., He, K., Dollár, P.: Focal loss for dense object detection. In: Proceedings of the IEEE International Conference on Computer Vision, pp. 2980–2988 (2017)
9. Liu, H., Simonyan, K., Yang, Y.: Darts: Differentiable architecture search. arXiv preprint arXiv:1806.09055 (2018)
10. Liu, Z., Li, J., Shen, Z., Huang, G., Yan, S., Zhang, C.: Learning efficient convolutional networks through network slimming. In: Proceedings of the IEEE International Conference on Computer Vision, pp. 2736–2744 (2017)
11. Loshchilov, I., Hutter, F.: Sgdr: Stochastic gradient descent with warm restarts. arXiv preprint arXiv:1608.03983 (2016)
12. Park, J., Woo, S., Lee, J.Y., Kweon, I.S.: Bam: Bottleneck attention module. arXiv preprint arXiv:1807.06514 (2018)
13. Ren, S., He, K., Girshick, R., Sun, J.: Faster R-CNN: towards real-time object detection with region proposal networks. In: Advances in Neural Information Processing Systems, pp. 91–99 (2015)
14. Sutskever, I., Vinyals, O., Le, Q.V.: Sequence to sequence learning with neural networks. In: Advances in Neural Information Processing Systems, pp. 3104–3112 (2014)
15. Szegedy, C., et al.: Going deeper with convolutions. In: Proceedings of the IEEE Conference on Computer Vision and Pattern Recognition, pp. 1–9 (2015)
16. Vaswani, A., et al.: Attention is all you need. In: Advances in Neural Information Processing Systems, pp. 5998–6008 (2017)
17. Veličković, P., Cucurull, G., Casanova, A., Romero, A., Lio, P., Bengio, Y.: Graph attention networks. arXiv preprint arXiv:1710.10903 (2017)
18. Xie, S., Girshick, R., Dollár, P., Tu, Z., He, K.: Aggregated residual transformations for deep neural networks. In: Proceedings of the IEEE Conference on Computer Vision and Pattern Recognition, pp. 1492–1500 (2017)
19. Zoph, B., Le, Q.V.: Neural architecture search with reinforcement learning. arXiv preprint arXiv:1611.01578 (2016)

An Attention Method to Introduce Prior Knowledge in Dialogue State Tracking

Zhonghao Chen and Cong Liu$^{(\boxtimes)}$

Sun Yet-Sen University, 135 Xin-gang Xi Lu, Haizhu, Guangzhou 510275, Guangdong, People's Republic of China
chenzhh66@mail2.sysu.edu.cn, liucong3@mail.sysu.edu.cn

Abstract. *Dialogue state tracking* (DST) is an important component in task-oriented dialogue systems. The task of DST is to identify or update the values of the given slots at every turn in the dialogue. Previous studies attempt to encode dialogue history into latent variables in the network. However, due to limited training data, it is valuable to encode prior knowledge that is available in different task-oriented dialogue scene. In this paper, we propose a neural network architecture to effectively incorporate prior knowledge into the encoding process. We performed experiment, in which entities belonging to the dialogue scene are extracted as the prior knowledge and are encoded along with the dialogue using the proposed architecture. Experiment results show significantly improvement in slot prediction accuracy, especially for slot types *date* and *time*, which are difficult to recognize by an encoder that is trained with limited data. Our results also achieve new state-of-the-art joint accuracy on the MultiWOZ 2.1 dataset.

Keywords: Natural language processing · Task-oriented dialogue system · Dialogue state tracking

1 Introduction

With the wide applications of conversation robots in business-to-customer scene, task-oriented dialogue system becomes a popular technology in natural language processing (NLP) in both academia and industry. Task-oriented dialogue systems aim to help users to fulfill their requests like booking hotel and navigation. *Dialogue state tracking* (DST) or belief tracking is an important component in task-oriented dialogue system, which identifies the user's intention and the related arguments. The role of the DST component is to identify, record and update the state at every turn of the dialogue, where the state of a turn is the set of slot-value pairs that indicates the current user intent [17]. A dialogue example from the MultiWOZ 2.1 dataset [4] is shown in Table 1.

As shown in Table 1, given a *dialogue* history and a *slot*, the goal of the DST model is to find a corresponding *value*. The output values of a DST task are words inside an ontology, which includes all possible candidates for each slot

© Springer Nature Switzerland AG 2021
T. Mantoro et al. (Eds.): ICONIP 2021, LNCS 13110, pp. 544–557, 2021.
https://doi.org/10.1007/978-3-030-92238-2_45

Table 1. An example dialogue on the *train* domain in the MultiWOZ 2.1 dataset. There are four turns in this dialogue. At each turn there is a user sentence and a system response. The DST model is required to predict the state of each turn. At the first turn, for instance, the DST predicts that the intent at the current turn is related to the slot *departure* and the value of this slot is *Birmingham*.

User : Hi, I'm looking for a train that is going to Cambridge and arriving there by 20:45, is there anything like that?
State : *destination = Cambridge; arriveBy = 20:45*
Sys : There are over 1,000 trains like that. Where will you be departing from?
User : I am departing from Birmingham new street.
State : *departure = Birmingham;* destination = Cambridge; arriveBy = 20:45
Sys : Can you confirm your desired travel day?
User : I would like to leave on Wednesday.
State : *departure = Birmingham;* destination = Cambridge; arriveBy = 20:45; departure day = Wednesday
Sys : I show a train leaving Birmingham new street at 17:40 and arriving at 20:23 on Wednesday. Will this work for you?
User : That will, yes. Please make a booking for 5 people please.
State : *departure = Birmingham*; *leaveAt = 17:40*; *day = Wednesday*; *arriveBy = 20:23*; *bookPeople = 5*; *reference = A9NHSO9Y*

defined by task. Some of these candidate values are extracted from the dialogue history with off-the-shelf named entity recognition (NER) tools, and therefore can be used as prior knowledge for the task. However, existing encoders, using pre-trained embeddings or not, are difficult to capture these kind of candidates from the input sentence due to limited in-domain training data. However, as shown in the example in Table 1, the output of a DST task usually includes special word types like day, number, time and location, which are simple to recognize by the NER tools. Therefore, the results extracted by the NER tools are expected to improve the performance of the DST model if they are properly encoded as prior knowledge. In this paper, we use the extracted named entities as an example to investigate how to incorporate prior knowledge into sentence encoding which can be extrapolated to other applications and other types of prior knowledge. Prior knowledge in NLP scenes may consist of entities, relations, constraints, etc. Existing work encodes database schema using graph neural network [22], where a single encoder is used to encode the sentence and the relation information. In this paper, we propose a neural network architecture, called *Prior Knowledge Attention* (PKA). Unlike the previous model that concatenates everything and encodes using a single encoder, our architecture adopts to DST task with prior knowledge with two tricks. First, we use different encoders for the inputs of different abstraction levels. Secondly, we disentangle the encoders between the implicit sub-tasks. Specifically, encoding a single long sentence and encoding multiple short phrases at the same time require the encoder to learn to perform two distinct functionalities. In our experiment, we use the named entities extracted from the dialogue turn as the prior knowledge. We use a fixed BERT pre-trained [3] model as the entity encoder, and use a trainable BERT pre-trained model as the dialogue turn encoder. The entity encoder is set to untrainable due to two reasons. First, further abstraction of the name entity is not necessary, since the same untrainable BERT model will encode the slot values to be selected from in the final stage of the model. Secondly, the training data is insufficient in term of per task domain knowledge and using additional trainable parameter to encode the entities will easily lead to over fitting.

To combine the encoded dialogue turn and the encoded prior knowledge, our model use a dynamic gating module, which assigns a larger weight to the encoded dialogue turn or the encoded entities, whichever will contribute more to the final result.

We run experiments on the MultiWOZ 2.1 dataset [4] and the MultiWOZ 2.0 dataset [1], and our model achieves state-of-the-art joint accuracies of 59.08% and 54.07%, respectively. We further compare the output of our model with that of the baseline model and discover significant improvement on certain slot types. Our major contributions are as follows:

- We purposed a novel method to introduce prior knowledge into the DST task.
- We perform experiment to compare and analyze the effectiveness of the proposed model.
- We achieve new state-of-the-art joint accuracy performance on the MultiWOZ 2.1 dataset.

2 Related Work

Early dialogue state tracking (DST) systems employ statical methods and rely on hand-crafted features. Some of them learn a DST model and a natural language understanding model (NLU) jointly from the training data [6,23], while others attempts to train a DST model as a down stream task of other NLU modules [14,16]. In recent years, both convolutional neural network (CNN) [13] and recurrent neural network (RNN) [12,23] are used as feature extractors. Further improvements are made thanks to the attention mechanism and the pre-trained language models (PLM) [2,8,9,18,19,21]. Recently, most work focus on the encoding contextual information from dialogue history [2,18]. Meanwhile, the transition of dialogue state is used as an inductive bias to improve model performance. [8] adopts four different state transition types as an auxiliary output and uses the predicted state transitions to determine whether or not to update the dialogue states. Database tools has also been introduced to DST task through generating query statement [7]. Graph models are also used in the DST tasks to encode schemas as the dialogue contexts [19]. While our model also encode named entities extracted from the dialogue, we mainly focus on the model architecture, by which dialogue history and other prior information can be properly encoded.

3   Introducing Named Entity Information

Named entity information consists of a named entity, which is usually a noun, and an attribute tag, which is the category of the entity. For example *(Birmingham, CITY)*, the first item is the named entity and the second item is the attribute tag of the former.

3.1   Extracting Named Entity Information

The named entity extractor in Stanford CoreNLP toolkit [10] is used to extract named entities. The input to the named entity extractor is a natural language sentence. The extractor output is a set of entity and entity-attribute-tag pairs, for example { *(Birmingham, CITY), (17:40 ,TIME)*}. Denote $X^{entity} = \{(n_1, t_1), (n_2, t_2), \ldots, (n_j, t_j)\}$ as the output of named entity extractor from a sentence in natural language.

The Stanford CoreNLP toolkit provides 25 different tags to represent the attribute of the entities, including one special tag "O", which represents the *"others"* attribute. All the "O" attribute entities are discarded and the other entities are used.

3.2   Formulating Named Entity Information

In order to encode the set of named entity information from the NER extractor using pre-trained language model, the named entity information is required to be packed into a sequence.

Pre-trained language model allows to introduce custom tokens into its vocabulary by replacing some [*unused*] place holder tokens with our custom tokens. 25 [*unused*] tokens are replaced by the 25 entity tags provided by the NER extractor. For instance, by replacing [*unused*1] with a new [*CITY*] token, the "CITY" entity tag is introduced into the vocabulary.

Denote $[TAG]_j$ the tag introduced for t_j in X^{entity}. The named entity information is formulated as the sequence: $[CLS] \oplus [TAG]_1 \oplus n_1 \oplus [TAG]_2 \oplus n_2 \oplus \cdots \oplus [SEP]$, where $\oplus$ denotes concatenation and $[CLS]$ and $[SEP]$ are the special tokens in the BERT vocabulary that represent the boundaries of a sequence. For example, in the 3rd turn of the dialogue example as shown in Table 1, the formulated sequence of the extracted named entity information $\{(Birmingham, CITY), (17 : 40, TIME), (20 : 23, TIME), (5, NUMBER)\}$, will be as follows:

$$X_3^{entity} =$$
$$[CLS][CITY]Birmingham[TIME]17 : 40[TIME]20 : 23[NUMBER]5[SEP] \tag{1}$$

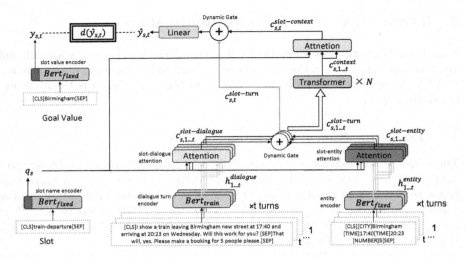

Fig. 1. The architecture of our model. The *gray* boxes are the BERT pre-trained modules, which are untrainable and are used to encode the prior entity information, the slot names, and the candidate slot values, respectively. The lower *blue* box is the dialogue turn encoder, which contains a unique trainable BERT module. Both BERT modules are initialized with the pre-trained parameters provided by Google.

4   Approach

The functionality of the DST model is to predict the value v of each slot s at each dialogue turn t. The input to the model includes the slot name s, the ontology of s, the dialogue history up to turn t, and the formulated named entities from every turn up to turn t. Our model, as shown in Fig. 1, consist of three major components. The first component uses $Bert_{train}$ and $Bert_{fixed}$ to encode the dialogue turns and the corresponding formulated named entities, respectively. The second component applies Slot-Dialogue and Slot-Entity Attention to capture the relevant information into the encoding of current turn. The third component uses the Historical Context Encoder and the Slot-Context Attention to encode the historical dialogue information and focus on that of the most relevant dialogue turn.

Finally the value selection sub-module predicts the *value* of given slot and computes the loss between the predicted value and the ground truth.

4.1   Problem Statement

The input of the DST task is a sequence of user sentence and system response pairs $\{(X_1^{sys}, X_1^{usr}), (X_1^{sys}, X_1^{usr})\dots,(X_t^{sys}, X_t^{usr})\}$ belonging to a certain dialogue, where the input at turn $t(1 \leq t \leq T)$ is defined as follows:

$$(X_t^{sys}; X_t^{usr}) = \{(w_{t,1}^{sys}, ..., w_{t,n}^{sys}); (w_{t,1}^{usr}, ..., w_{t,m}^{usr})\}. \tag{2}$$

In the above equation, $w_{t,i}^{usr}$ and $w_{t,i}^{sys}$ are the ith words in user sentence and system response at turn t, respectively, and n, m are the lengths of the user sentence and the system response.

Our task is to predict a value $v_{s,t}$ for every slot s at every turn t. A *slot*, as defined in the dataset MultiWOZ, is represented by a *domain* name and a *slot* name. Following [9,18], we use a special value *none* for the slots that do not concern the current turn, and the DST task becomes a classification task.

4.2 Encoding Dialogue Turn and Named Entity Information

Each dialogue turn consists of a system response and a user sentence pair. The pair will be represented by a concatenated sequence X_t:

$$X_t = [CLS] \oplus X_t^{sys} \oplus [SEP] \oplus X_t^{usr} \oplus [SEP] \tag{3}$$

The dialogue turn and the corresponding formulated named entity information will be encoded by two BERT modules. Both BERT modules are initialized with the pre-trained parameters. But the BERT module for named entity information encoding is not trainable due to two reasons. First, further abstraction of the name entity is not necessary since the same untrainable BERT model will encode the slot values to be selected from in the final stage of the model. Secondly, the training data is insufficient in term of per task domain knowledge and using additional parameter to encode the entities will easily lead to over fitting. Written as follows:

$$\begin{aligned} h_t^{dialogue} &= Bert_{train}(X_t). \\ h_t^{entity} &= Bert_{fixed}(X_t^{entity}). \end{aligned} \tag{4}$$

4.3 Slot-Dialogue and Slot-Entity Attention

Following [9] and [18], we use the current slot name as the query vector of the attention module to extract relevant words from the dialogue and named entity information. The information of the relevant words after extraction will eventually be combined by a dynamic gating module.

A untrainable BERT module is used to encode the current slot s, whose value is to be predicted by the model into an encoding q^s. This BERT module is also initialized with the pre-trained parameters.

Let $\mathcal{C}_s$ be the set of candidate values of the current slot s, as defined in the ontology of the current task, each candidate value $v' \in \mathcal{C}_s$ is encoded individually using the same untrainable BERT module as the slot as follows:

$$\begin{aligned} q^s &= Bert_{fixed}([CLS] \oplus s \oplus [SEP]). \\ y_{v'} &= Bert_{fixed}([CLS] \oplus v' \oplus [SEP]). \end{aligned} \tag{5}$$

The output vector at the $[CLS]$ token in the outputs of q^s and $y_{v'}$ will be used as the final representation of each *slot name* and candidate *value*. The BERT module $Bert_{fixed}$ is untrainable here so that the model can better handle unseen *slot* and *value* in any dataset in the future without needing to fine-tuning the model for each new task.

Slot-Dialogue and Slot-Entity Attention. The multi-head attention module is denoted as

$Multihead(Q, K, V)$. Please refer to [15] for more details about multi-head attention. The slot-dialogue attention module captures the information related to the encoded slot q_s from the encoded dialogue turn $h_t^{dialogue}$ as follows:

$$c_{s,t}^{slot\text{-}dialogue} = Multihead(q_s, h_t^{dialogue}, h_t^{dialogue}). \tag{6}$$

The slot-entity attention module captures the information related to the encoded slot q_s from the encoded entity information h_t^{entity} as follows:

$$c_{s,t}^{slot\text{-}entity} = Multihead(q_s, h_t^{entity}, h_t^{entity}). \tag{7}$$

Dynamic Gating Between Entity and Dialogue Turn. The dynamic gating sub-module learns to assign gating weights to the output of the slot-dialogue attention module and that of the slot-entity attention module according to their relative contribution, so the model is able to avoid the non-relative information in $c_{s,t}^{slot\text{-}entity}$. The encoded information $c_{s,t}^{turn}$ of current turn after gating is shown as follow:

$$\begin{aligned} \alpha_{de} &= \sigma(W_n \times [c_{s,t}^{slot\text{-}dialogue}; c_{s,t}^{slot\text{-}entity}]), \\ c_{s,t}^{turn} &= \alpha_{de} \times c_{s,t}^{slot-dialogue} + (1 - \alpha_{de}) \times c_{s,t}^{slot\text{-}entity}, \end{aligned} \tag{8}$$

where $W_n \in \mathbb{R}^{2d \times 1}$ are parameters, σ is the sigmoid function, and α_{de} is the gating weight that determines the relative importance between the dialogue and the entities.

4.4   Historical Context Encoder and Slot-Context Attention

In multi-turn dialogue, information to predict the value of a slot in the current turn may appear in the dialogue of any previous turn.

So the model need to be able to go back to the previous dialogue turn $t' < t$ for relevant context information.

The context information of a dialogue is defined as the combination of the encoded dialogue turns from turns 1 to t.

Historical Context Encoder. To encode the context information among different turns, we purpose the Historical Context Encoder. The dialogue turn encodings will be contextualized by an N-layer Transformer [15] encoder submodule. Each Transformer layer contains a multi-head self-attention layer, an add&normalization layer, a feed-forward layer and another add&normalization layer at the end. The output of the historical context encoder module is:

$$c_{s,1...t}^{context} = Transformer(\{c_{s,1}^{turn}, c_{s,2}^{turn} \dots, c_{s,t}^{turn}\}). \tag{9}$$

Slot-Context Attention. We use another multi-head attention to extract the relevance information between current input slot s and the context information we just computed, as follows:

$$c_{s,t}^{slot\text{-}context} = Multihead(q_s, c_{s,1...t}^{context}, c_{s,1...t}^{context}). \tag{10}$$

Dynamic Gating Between Historical Context and Present Dialogue. This dynamic gating module weights between the encoding of the current dialogue turn $c_{s,t}^{slot\text{-}dialogue}$ and the complete historical context encoding $c_{s,t}^{slot\text{-}context}$. The module computes a new context encoding:

$$
\begin{aligned}
\alpha_{ct} &= \sigma(W_m \times [c_{s,t}^{slot\text{-}context}; c_{s,t}^{slot\text{-}turn}]), \\
c_{s,t} &= \alpha_{ct} \times c_{s,t}^{slot\text{-}context} + (1 - \alpha_{ct}) \times c_{s,t}^{slot\text{-}turn},
\end{aligned} \tag{11}
$$

where $W_m \in \mathbb{R}^{2d \times 1}$ are parameters.

4.5 Outputs and Training Criteria

Following [9], we use Euclidean distance as the metric between the model output $\hat{y}_{s,t}$ and each of the candidate slot value encoding. The model output $\hat{y}_{s,t}$ is as follow:

$$\hat{y}_{s,t} = LayerNorm(Linear(Dropout(c_{s,t}))). \tag{12}$$

The probability $p(v'|X_{1...t}^{sys}, X_{1...t}^{usr}, s)$ of each candidate value v' of the current slot s at current turn t is calculated as the distances between model output $\hat{y}_{s,t}$ which is then normalized with a SoftMax, as:

$$p(v'|X_{1...t}^{sys}, X_{1...t}^{usr}, s) = \frac{\exp(-d(\hat{y}_{s,t}, y_{v'}))}{\sum_{v \in \mathcal{C}_s} \exp(-d(\hat{y}_{s,t}, y_v))}, \tag{13}$$

where d is the Euclidean distance and $\mathcal{C}_s$ is the set of candidate value of the current slot s.

With $v_{s,t}$ being the ground truth value of the current slot, we train the model to minimize the cross entropy loss of the probability distribution for all slot s and all turn t, shown as follow:

$$\mathcal{L} = -\sum_{s \in \mathcal{D}} \sum_{t=1}^{T} \log(p(v_{s,t}|X_{1...t}^{sys}, X_{1...t}^{usr}, s)), \tag{14}$$

where $\mathcal{D}$ is the set of slots and T is the total number of turns in the dialogue.

5 Experiment

5.1 Dataset

MultiWOZ 2.0 [1] is a multi-domain task-oriented dialogue dataset, which contains 10438 dialogues, 7 domains and over 30 different slots. In this paper, only 5 domains {*restaurant, hotel, train, attraction, taxi*} are used, and the dataset pre-processing code is borrowed from [9].

Table 2. Joint Accuracy (%) on the MultiWOZ 2.0 and MultiWOZ 2.1 test set. Results for the compared algorithms are borrowed from the origin papers. † indicates the best result of our reproduction of the current state-of-the-art model CHAN-DST using their source code in our experiment environment.

Model	Generative or Classifier	MultiWOZ 2.0	MultiWOZ 2.1
HyST [5]	Classifier	38.10	38.10
SUMBT [9]	Classifier	42.40	–
DS-DST [20]	Classifier	–	51.21
Som-DST [8]	Generative	52.32	53.68
Transformer-DST [19]	Generative	**54.64**	55.35
SimpleTOD [7]	Generative	–	57.47
CHAN-DST [18]	Classifier	52.68	58.55(† 58.34)
DialoGLUE [11]	Generative	–	58.70
PKA-DST (Ours)	Classifier	54.07	**59.08**

MultiWOZ 2.1 [4] shares the same dialogue data with the MultiWOZ 2.0 dataset, but it corrects a majority of annotation errors in the latter.

5.2 Metrics

Two evaluation metrics are used to draw comparison between different DST models:

– Joint Accuracy: the percentage of dialogues where all 35 slots are correctly predicted.
– Slot Accuracy: the percentage of correctly predicted slots.

5.3 Baselines

We compare our model with existing models that assume a predefined ontology as listed below:

SUMBT [9] uses a BERT module to encode sentences and an LSTM module to combine history information. Finally, it predicts the dialogue state using the distance between the model output and the encoded candidates.

HyST [5] uses a hierarchical RNN encoder and combines an ontology-based network and an open vocabulary-based network in a hybrid method.

DS-DST [20] uses two BERT modules to the encode context. This paper also designed a picklist-based mechanism, which is similar to SUMBT.

Som-DST [8] designed a model which also predict the value transition among different turns of dialogue.

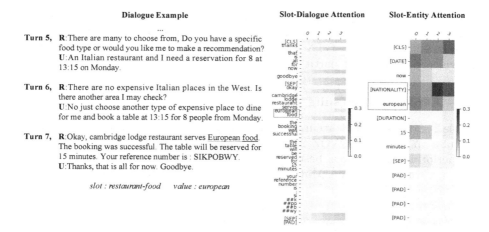

Fig. 2. Runtime value visualization of the slot-dialogue and the slot-entity attention module. The example is drawn from the MultiWOZ 2.1 test set. The ground truth *value* of the current slot, *restaurant food*, at the current turn ($t = 7$) is *european* . The color depths of four columns show the attention scores of all the four attention heads in the multi-head attention. The example show that the proposed slot-entity attention module is a good complementary to the slot-dialogue attention module. (Color figure online)

Transformer-DST [19] propose a purely Transformer-based framework and jointly optimize the model on value generation and state transition prediction.

SimpleTOD [7] builds a generative language model which sequentially output the dialogue state, query command to extra database and the system response.

CHAN-DST [18] uses a hierarchical attention-based contextual encoding structure to handle dialogue history information. This paper also implements a dynamic weighting loss function to solve the problem of sample size imbalance in the training data.

DialoGLUE [11] provides various latest baselines for many dialogue-related tasks. This paper adopts ConvBERT-DG pre-trained model on the DST task.

5.4   Experiment Settings

The BERT pre-trained module being used in our model has a hidden size of 768 in each of its 12 layers and 12 heads in its self-attention modules. We trained the model for 400 epochs with an early-stop patience of 20. The slot-dialogue, slot-entity and slot-context attention modules in our model have 4 heads and 768 hidden units. The Transformer in the historical context encoding has 6 layers. We use BertAdam as our optimizer. All learning rates are being set to 1e−4 and the learning rate warm up step is set to ten percent of the total training steps. Dropout rate is 0.1. Operation system is Ubuntu 18.04. We tried to train the model both on GTX 1070 and Titan RTX and the results end up better on the former. Batch size is 32. We report the average score over 5 randomly seeded experiments trained on GTX 1070. Average training time is 41 h. The dataset provided train-dev-test partition is used.

5.5   Experiment Results

Table 2 shows the joint accuracy of all compared DST models and our model. Results on both MultiWOZ 2.1 and MultiWOZ 2.0 are presented. Our model outperforms the previous best model by 0.38% on the MultiWOZ 2.1 dataset. Slot accuracy results on MultiWOZ 2.1 and MultiWOZ 2.0 are 98.13% and 97.78%, respectively.

5.6   Attention Visualization

Figure 2 visualize the slot-dialogue attention score and the slot-entity attention score in our model for an input example. In this example, although the slot-dialogue attention module failed to attend to the correct phrase *european food* in sentence, this phrase is successfully extracted by the entity toolkit and then captured by the slot-entity attention module.

5.7   Effectiveness of the Employed Named-Entity Recognition Toolkit

We present statistical data in Fig. 3 to analyze the hit rate of the named entity recognition separately for each slot in the ontology. As shown in Fig. 3, only half of the slots have a set of their candidate values recognized as named entities. For

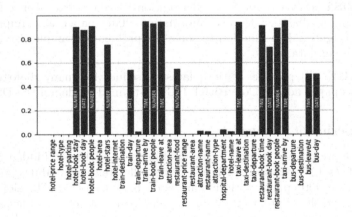

Fig. 3. The entity recognition hit rate of the NER toolkit. We show the hit rate by slot. The words on the blue bars are the most common tag of each slot. (Color figure online)

instance, the candidates of slots *departure* and *destination* are supposed to be recognized as *LOCATION* or *CITY* but they are not. We checked the dataset manually and find out the reason that the NER toolkit can not recognized them correctly is that a number of candidates of slots *departure* and *destination* are not capitalized. For instance, *cambridge* will be recognized as *other* instead of *CITY*. We believe that a better named entity recognition tool will bring further improvement to the performance of our model.

5.8 Effectiveness of Dynamic Gating Between Prior Entity Information and Dialogue

To verify the effectiveness of the dynamic gating module, we investigate the slot *"train-arrive by"* in test set of MultiWOZ 2.1. We find that the candidates of slot *"train-arrive by"* are recognized as *TIME* in about 90% of the cases. We compare the α_{de}'s in Eq. 8 in two groups of examples, the first of which have their *TIME* entity recognized in their sentences and the second of which do not. The average value of α_{de} is 0.448605 for the first group and it is 0.840443 for the second group. This verifies that the dynamic gating module can integrate prior entity information when needed.

5.9 Effectiveness of Prior Entity Information

As shown in Fig. 4, we pick the slots whose candidates can be recognized as named entities with high hit rate. We compared the slot accuracy of our model with that of the current SOTA model CHAN-DST. The figure shows that introducing prior entity information leads to significant improvement on slots related to number of people and days.

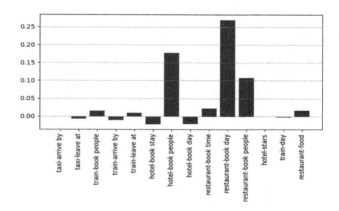

Fig. 4. The improvement (degeneracy) on slot accuracy(%) compared to the baseline model CHAN-DST. The number is collected on the test set of MultiWOZ 2.1 dataset.

6   Conclusion

We designed a method based on attention mechanism and dynamic gating to introduce prior knowledge to dialogue state tracking model. Experiment results showed that, by introducing prior named entity information to the model, the joint accuracy on dataset MultiWOZ 2.1 and MultiWOZ 2.0 reaches 59.09% and 54.07%, respectively. Our method outperforms previous state-of-the-art by 0.38% on MultiWOZ 2.1 dataset.

References

1. Budzianowski, P., et al.: Multiwoz-a large-scale multi-domain wizard-of-oz dataset for task-oriented dialogue modelling. In: Proceedings of the 2018 Conference on Empirical Methods in Natural Language Processing, pp. 5016–5026 (2018)
2. Chen, J., Zhang, R., Mao, Y., Xu, J.: Parallel interactive networks for multi-domain dialogue state generation. In: Proceedings of the 2020 Conference on Empirical Methods in Natural Language Processing (EMNLP), pp. 1921–1931. Association for Computational Linguistics, Online (November 2020). https://doi.org/10.18653/v1/2020.emnlp-main.151, https://www.aclweb.org/anthology/2020.emnlp-main.151
3. Devlin, J., Chang, M.W., Lee, K., Toutanova, K.: Bert: pre-training of deep bidirectional transformers for language understanding. In: Proceedings of the 2019 Conference of the North American Chapter of the Association for Computational Linguistics: Human Language Technologies, Volume 1 (Long and Short Papers), pp. 4171–4186 (2019)
4. Eric, M., et al.: Multiwoz 2.1: a consolidated multi-domain dialogue dataset with state corrections and state tracking baselines (2019)
5. Goel, R., Paul, S., Hakkani-Tür, D.: Hyst: a hybrid approach for flexible and accurate dialogue state tracking. In: Proceedings of the Interspeech 2019, pp. 1458–1462 (2019)
6. Henderson, M., Thomson, B., Young, S.: Word-based dialog state tracking with recurrent neural networks. In: Proceedings of the 15th Annual Meeting of the Special Interest Group on Discourse and Dialogue (SIGDIAL) (2014)
7. Hosseini-Asl, E., McCann, B., Wu, C.S., Yavuz, S., Socher, R.: A simple language model for task-oriented dialogue. arXiv preprint arXiv:2005.00796 (2020)
8. Kim, S., Yang, S., Kim, G., Lee, S.W.: Efficient dialogue state tracking by selectively overwriting memory. arXiv preprint arXiv:1911.03906 (2019)
9. Lee, H., Lee, J., Kim, T.Y.: Sumbt: Slot-utterance matching for universal and scalable belief tracking. arXiv preprint arXiv:1907.07421 (2019)
10. Manning, C.D., Surdeanu, M., Bauer, J., Finkel, J., Bethard, S.J., McClosky, D.: The Stanford CoreNLP natural language processing toolkit. In: Association for Computational Linguistics (ACL) System Demonstrations, pp. 55–60 (2014). http://www.aclweb.org/anthology/P/P14/P14-5010
11. Mehri, S., Eric, M., Hakkani-Tur, D.: Dialoglue: A natural language understanding benchmark for task-oriented dialogue. arXiv preprint arXiv:2009.13570 (2020)
12. Rastogi, A., Hakkani-Tur, D., Heck, L.: Scalable multi-domain dialogue state tracking. In: 2017 IEEE Automatic Speech Recognition and Understanding Workshop (ASRU) (2017)

13. Shi, H., Ushio, T., Endo, M., Yamagami, K., Horii, N.: Convolutional neural networks for multi-topic dialog state tracking. In: Jokinen, K., Wilcock, G. (eds.) Dialogues with Social Robots. LNEE, vol. 999, pp. 451–463. Springer, Singapore (2017). https://doi.org/10.1007/978-981-10-2585-3_37

14. Thomson, B., Young, S.: Bayesian update of dialogue state: a pomdp framework for spoken dialogue systems. Comput. Speech Lang. 562–588 (2010)

15. Vaswani, A., et al.: Attention is all you need. In: Advances in Neural Information Processing Systems, pp. 5998–6008 (2017)

16. Wang, Z., Lemon, O.: A simple and generic belief tracking mechanism for the dialog state tracking challenge: on the believability of observed information. In: Proceedings of the SIGDIAL 2013 Conference, pp. 423–432 (2013)

17. Williams, J.D., Henderson, M., Raux, A., Thomson, B., Black, A., Ramachandran, D.: The dialog state tracking challenge series. AI Mag. 35(4), 121–124 (2014)

18. Yong, S., Li Zekang, Z.J., Meng Fandong, F.Y., Niu Cheng, Z.J.: A contextual hierarchical attention network with adaptive objective for dialogue state tracking. In: Proceedings of the 58th Conference of the Association for Computational Linguistics (2020)

19. Zeng, Y., Nie, J.Y.: Multi-domain dialogue state tracking based on state graph. arXiv preprint arXiv:2010.11137 (2020)

20. Zhang, J.G., et al.: Find or classify? dual strategy for slot-value predictions on multi-domain dialog state tracking. arXiv preprint arXiv:1910.03544 (2019)

21. Zhong, V., Xiong, C., Socher, R.: Global-locally self-attentive encoder for dialogue state tracking. In: Proceedings of the 56th Annual Meeting of the Association for Computational Linguistics (Volume 1: Long Papers), pp. 1458–1467 (2018)

22. Zhu, S., Li, J., Chen, L., Yu, K.: Efficient context and schema fusion networks for multi-domain dialogue state tracking. arXiv preprint arXiv:2004.03386 (2020)

23. Zilka, L., Jurcicek, F.: Incremental lstm-based dialog state tracker. In: 2015 IEEE Workshop on Automatic Speech Recognition and Understanding (Asru), pp. 757–762. IEEE (2015)

Effect of Input Noise Dimension in GANs

Manisha Padala$^{(\boxtimes)}$, Debojit Das, and Sujit Gujar

Machine Learning Lab, International Institute of Information Technology (IIIT),
Hyderabad, Hyderabad, India
{manisha.padala,debojit.das}@research.iiit.ac.in, sujit.gujar@iiit.ac.in

Abstract. Generative Adversarial Networks (GANs) are by far the most successful generative models. Learning the transformation which maps a low dimensional input noise to the data distribution forms the foundation for GANs. Despite their application in various domains, they are prone to certain challenges like mode collapse and unstable training. To overcome the challenges, researchers have proposed novel loss functions, architectures, and optimization methods. Unlike the previous approaches, we focus on the input noise and its role in the generation in our work here.

We aim to quantitatively and qualitatively study the effect of the dimension of the input noise on the performance of GANs. For quantitative measures, typically *Fréchet Inception Distance (FID)* and *Inception Score (IS)* are used as performance measure on image data-sets. We compare the FID and IS values for DCGAN and WGAN-GP. We use three different image data-sets – each consisting of different levels of complexity. Our experiments show that the right dimension of input noise for optimal results depends on the data-set and architecture used. We also observe that the state of the art performance measures does not provide enough useful insights.

Keywords: Generative Adversarial Networks (GANs) · Neural networks · Generative models

1 Introduction

Generative Adversarial Networks (GANs) are the most popular generative models, which learn the data distribution. These models are useful for generating realistic data, thus helping in data augmentation. In computer vision, GANs are used for super-resolution of images [14,18], transferring domain knowledge from images of one domain to another [15,32,35], object detection [19], image editing [30], medical images [8]. Other applications include generation of music [7], paintings [20] and text [29,33]. These are but a few of the recently developed applications.

A vanilla GAN [10] primarily has two networks, the *generator* and the *discriminator*. As its name suggests, the generator generates samples resembling real data from a lower-dimensional input noise. The discriminator distinguishes the images generated by the generator from the actual images. While training,

© Springer Nature Switzerland AG 2021
T. Mantoro et al. (Eds.): ICONIP 2021, LNCS 13110, pp. 558–569, 2021.
https://doi.org/10.1007/978-3-030-92238-2_46

the discriminator is trained to improve its ability to discriminate real vs fake images while the generator is trained to fool the discriminator. The set-up is similar to a two-player zero-sum game, and at equilibrium, the generator samples realistic data. That is, it learns the data distribution.

GANs can be considered an experimental success. The novelty and simplicity in its set-up contribute to its popularity. According to the manifold hypothesis, the real-world high dimensional data lie on low dimensional manifolds embedded in high dimensional space [17]. Hence, we believe that the data can have an efficient lower-dimensional representation. Given access to high dimensional distribution of the data, we could efficiently perform dimensionality reduction using existing methods such as PCA [9], JL transform [13], or deep auto-encoders. However, the main challenge is that we do not know the higher dimensional distribution of the data, and hence mapping becomes difficult. In a GAN set-up, the training happens without any knowledge of the data distribution. The generator learns the mapping and corresponding lower-dimensional representation of the high dimensional data input.

Despite the success, GANs face major issues such as mode collapse. As a result of mode collapse, the images generated, although very sharp and realistic, lack diversity. The other issue is that the training is not smooth and may not converge sometimes. There has been quite a lot of research to overcome mode collapse and stabilize the training. Researchers have proposed different loss functions [3,23,26], architectures [25,34] and optimizers [5,24] for overcoming the above challenges. In this work, we take a different route and inspect the effect of input noise on the performance of GANs. The authors in [4] through unsupervised training make the latent variable (i.e., input noise) represent some visual concepts. Along similar lines the authors train the input noise using supervised learning in [6,31]. To the best of our knowledge, no paper rigorously explores the impact of input on the performance of GANs in general.

Contribution. In this work, we focus on studying the effect of noise on the performance of the model. We vary the dimension of the input noise and study its effect on the samples generated for two different GANs, namely (a) DCGAN [27] (b) WGAN-GP [11].). We consider three different data-sets: i) Gaussian Data ii) MNIST digits data [16] iii) CelebA face data [21]. We provide the following results,

1. Quantitative estimation by comparing the *Fréchet Inception Distance (FID)* [12] and *Inception Score (IS)* [28] for the generated samples.
2. Qualitative estimation: by comparing the samples of images generated after the training converges.

We believe that such an analysis would lead to the best set of parameters and provide valuable insights into the model's working. To best of our knowledge, this is the first kind of study that analyses the effect of input noise in GANs, in particular for image generation.

Organization. In Sect. 2, we discuss the GAN models and performance measures used in detail. Next, we describe the experimental set-up and compare the result in Sect. 3. Finally, we discuss the insights derived from the experiments in Sect. 4 before concluding in Sect. 5.

2 Preliminaries

Consider the data distribution is p_d, and the noise is sampled from another distribution $p_z(z)$. Let the dimension of z be denoted by d_z i.e., $z \sim \mathbb{R}^{d_z}$. The generated samples follow the distribution p_g, which is referred to as the model distribution. We denote the generator with G and discriminator with D. When dealing with images, G and D are convolutional multi-layered networks. G takes the $z \sim p_z$ and generates a vector $\hat{x}$, which has the same dimension as the data x. $\hat{x} \sim p_g$ is the distribution learnt by G. The weights/parameters of G are denoted by ϕ. The other network is the D, parameterized by θ. It takes either $\hat{x}$ or x as input and outputs a single value. The value is a score for the input; a higher score indicates that the input is likely to belong to real-data thus sampled from p_d and not p_g. The main challenge is to construct a suitable loss and optimize over the loss for achieving our objective of generating realistic images. In this paper, we consider two different kinds of loss i) DCGAN ii) WGAN-GP as further elaborated below,

2.1 DCGAN

DCGAN is a modification of vanilla GAN as introduced in [10]. The objective is a simple binary cross-entropy loss used for 2 class classification given below,

$$\min_{\phi} \max_{\theta} V_G(D_\theta, G_\phi) = \mathbb{E}_{x \sim p_d(x)}[log D_\theta(x)] + \mathbb{E}_{z \sim p_z(z)}[log(1 - D_\theta(G_\phi(z)))] \quad (1)$$

The above objective is equivalent to minimizing the *Jenson Shannon Divergence (JSD)* between p_d and p_g. Early in training, when the samples generated by G are very noisy, the discriminator can classify with high confidence, causing the gradients w.r.t. ϕ to be very small. Hence the authors propose to use the following loss for G,

$$\max_{\phi} \log(D_\theta(G_\phi(z))) \quad (2)$$

Using *simultaneous gradient descent*, the authors prove the convergence of the loss under specific assumptions. In this method, ϕ is fixed and one step gradient descent is performed over θ for Eq. 1. Then θ is fixed and one step gradient descent is performed over ϕ for Eq. 2.

2.2 WGAN-GP

In [1], the authors introduce the issue of *vanishing gradient*. If both p_g and p_d lie on different manifolds, the discriminator is easily able to achieve zero loss and the gradients w.r.t. G vanish, leading to the vanishing gradient problem.

The authors also prove that using Eq. 2 leads to mode collapse and unstable updates. Hence, in [2], the authors propose to use Wasserstein distance between the distributions. The loss is given as follows,

$$\max_{\theta} \min_{\phi} \; V_W(D_\theta, G_\phi) = \mathbb{E}_{x \sim p_d(x)}[D_\theta(x)] - \mathbb{E}_{z \sim p_z(z)}[D_\theta(G_\phi(z))] \tag{3}$$

The authors clamp θ to be within a specified range for stable training. Further in [11], the authors show that clamping weights leads to exploding and vanishing gradients. Hence they introduce a gradient penalty term (GP) in Eq. 3 to form the WGAN-GP loss as follows,

$$
\begin{aligned}
V_W(D_\theta, G_\phi) &= \mathbb{E}_{x \sim p_d(x)}[D_\theta(x)] - \mathbb{E}_{z \sim p_z(z)}[D_\theta(G_\phi(z))] \\
&\quad + \lambda \mathbb{E}_{\tilde{x} \sim p_{\tilde{x}}}[(\| \nabla_{\tilde{x}} D(\tilde{x}) \|_2 - 1)^2]
\end{aligned}
\tag{4}
$$

The $\tilde{x} \sim p_{\tilde{x}}$ is obtained by sampling uniformly along straight lines between pairs of points sampled from the data distribution p_d and the generator distribution p_g. The optimization method used is similar to GAN.

2.3   Measures to Evaluate GANs

In general, it is not possible to compute how close p_g is to p_d quantitatively, given that GANs do not provide the distribution explicitly. For the synthetic Gaussian data, we compute the empirical distance between p_g and p_d using, i) JSD ii) *Fréchet Distance* (FD). The definitions of these measures are as follows,

Definition 1 (Jenson Shannon Divergence (JSD)). *JSD is a symmetric distance metric between the two distribution $p_d(x), p_g(x)$ given by,*

$$JSD(p_d \parallel p_g) = \frac{1}{2}KL\left(p_d \parallel \frac{p_d + p_g}{2}\right) + \frac{1}{2}KL\left(p_g \parallel \frac{p_d + p_g}{2}\right) \tag{5}$$

where KL is the Kl-divergence.

Definition 2 (Fréchet Inception Distance (FD)). *Given p_g and p_d both multivariate continuous Gaussian distributions. The mean and variance of p_g is μ_g, Σ_g and p_d is μ_d, Σ_d respectively. The FD is then,*

$$\| \mu_g - \mu_d \|_2^2 + Tr(\Sigma_g + \Sigma_d - 2(\Sigma_g \Sigma_d)^{\frac{1}{2}}) \tag{6}$$

In our experiments on MNIST and CelebA datasets, it is not possible to use the above measures; hence, we use the following standard performance measures,

Inception Score (IS). This is the most widely used score and was proposed by [28]. The score function defined measures the average distance between the label distribution $p(y|x)$ and marginal distribution $p(y)$. Here the x is the image generated by G, and y is the label given by the pre-trained Inception Net. The distribution $p(y|x)$ needs to have less entropy, which indicates that the network can classify x with high confidence hence more likely to be a realistic image. At the same time, $p(y)$ needs to have high entropy to indicate diversity in the samples generated. Hence, the higher the inception score, the better is the performance.

Fréchect Inception Distance (FID). Proposed by [12] also uses a pre-trained Inception Net. The activations of the intermediate pooling layer serve as our feature embeddings. It is assumed that these embeddings follow a multivariate continuous Gaussian distribution. We pass multiple samples of $x \sim p_d$ and calculate the empirical mean and variance of their embeddings. Similarly, we sample p_g and calculate the empirical mean and variance. The FD, given by Eq. 6, is applied over the mean and variance of the two Gaussian embeddings. If the p_g is close to p_d, the FD will be low. Hence lower the FID, the better is the performance.

Apart from quantitative evaluation, we also provide the samples of images generated for each type of input noise, for visual comparison.

3   Experiments with Input Noise for GANs

In this section, we empirically study the effect of varying the dimension of input noise (d_z) on the performance. We used two different GANs in our experiments: (a) DCGAN and (b) WGAN-GP. We considered the following three data-sets for training and evaluation,

Synthetic Data-set. We show results on 1-dimensional synthetic Gaussian data. This enables us to compare the plots for the distribution of the real data samples and generated samples. We study the effect of varying the variance of the Gaussian. We also vary the modes in the Gaussian (i.e., have two peaks) and visualize the problem of mode collapse.

Architecture: We use a simple feed-forward multi-layered perceptron having two hidden layers each. Both the generator and discriminator have ReLU activation.

MNIST Data-set [16] the data-set consists of 28×28 dimensional black and white handwritten digit images.

Architecture: The discriminator has two convolution layers with 64 and 128 filters and stride 2 and leaky ReLU activation. There is a final dense layer to return a single value of probability. Generator has an architecture reverse to that of discriminator's but includes batch normalization layers after every transpose convolution layer. The activation used is ReLU.

CelebA Data-set [21] this data-set consists of more than 200K celebrity images, each of 128×128 resolution. We re-scale the images to 32×32 and 64×64, and train two different models for each of the resolutions.

Architecture: Here, the architecture is similar to that of MNIST, although the networks for 32×32 have 3 convolutional layers with 64, 128 and 256 filters and batch normalization layers even in the discriminator. For generating images of 64×64, we include one extra convolutional layer which has 512 filters.

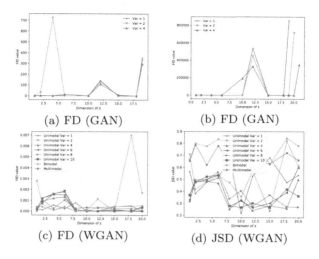

(a) FD (GAN)   (b) FD (GAN)

(c) FD (WGAN)   (d) JSD (WGAN)

Fig. 1. Performance measure plots for Gaussian distribution

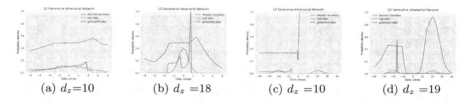

(a) $d_z=10$ (b) $d_z=18$ (c) $d_z=10$ (d) $d_z=19$

Fig. 2. Gaussian distributions generated using vanilla GAN

3.1 Results for Synthetic Data-Set

In Fig. 1 (a) and (b), we present the FD values (Eq. 6) between the real and generated Gaussian data, (a) the data has single mode and (b) bimodal data. It is observed that the variance in the data does not effect the trend in performance. The FD values stay low till the dimension of input noise is 10. Increasing the dimension only worsens the performance. And increasing it further from 20, for a fixed generator and discriminator architecture blows up the distance. The problem of mode collapse is also evident as we see the FD values for bimodal is greater than the values in the unimodal case. From Fig. 2, we can visually observe that the real data and generated data distributions are nearby with input noise dimension is 10. The distributions are far apart when the dimension is increased to around 20.

In Fig. 1 (c) and (d), we compare FD and JSD for WGAN-GP. We find that, there is no particular trend although there is small difference in the measures as the dimension increases and for all the variance and modes. We also observe that, the values are smaller than GAN. Hence, we conclude that WGAN performs better than the normal GAN on this data-set. This is also visible from the

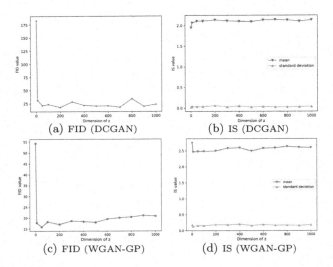

(a) FID (DCGAN)   (b) IS (DCGAN)

(c) FID (WGAN-GP)   (d) IS (WGAN-GP)

Fig. 3. Performance measure plots for MNIST

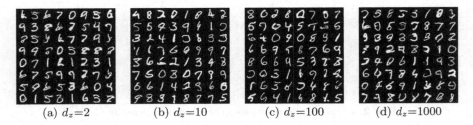

(a) $d_z=2$ (b) $d_z=10$ (c) $d_z=100$ (d) $d_z=1000$

Fig. 4. Images generated by DCGAN for MNIST

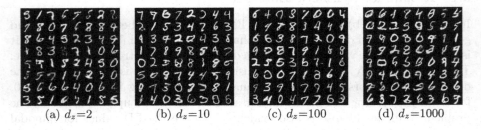

(a) $d_z=2$ (b) $d_z=10$ (c) $d_z=100$ (d) $d_z=1000$

Fig. 5. Images generated by WGAN-GP for MNIST

distribution plots in [22, Figure 4]. The problem of mode collapse is still not completely overcome even in WGAN-GP.

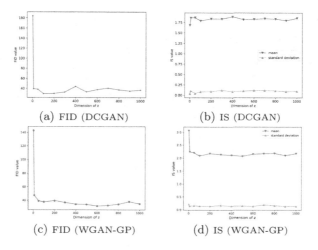

(a) FID (DCGAN)   (b) IS (DCGAN)

(c) FID (WGAN-GP)   (d) IS (WGAN-GP)

Fig. 6. Performance measure plots for 32×32 CelebA

(a) $d_z = 2$ (b) $d_z = 10$ (c) $d_z = 100$ (d) $d_z = 900$

Fig. 7. Images generated by DCGAN for 32×32 CelebA

3.2   Results for MNIST

In Fig. 3 (a) and (c), we compare the FID values for different dimensions of input noise for DCGAN and WGAN-GP. We fix the architecture of the generator and the discriminator and we find that the FID scores are very high for noise dimension 2 but do not change much for higher dimensions till 1000. At the same time, we find the FID values for WGAN-GP are much better than DCGAN. In Fig. 3 (b) and (d), we plot the IS values which are evaluated for batches of samples and the mean and the variance of the IS values across the batches is plotted for both DCGAN and WGAN-GP. We observe a similar trend as the FID.

We visually compare the results in Figs. 4, 5. We find that results are bad when noise dimension is 2 compared to the other dimensions. WGAN-GP performs worse compared to DCGAN at lower-dimensional input noise. Hence we conclude that having dimension of noise as 10 is sufficient for good performance.

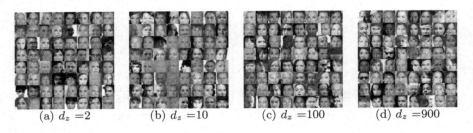

(a) $d_z = 2$ (b) $d_z = 10$ (c) $d_z = 100$ (d) $d_z = 900$

Fig. 8. Images generated by WGAN-GP for 32×32 CelebA

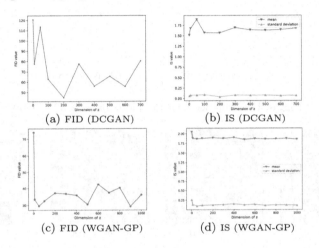

(a) FID (DCGAN)   (b) IS (DCGAN)

(c) FID (WGAN-GP)   (d) IS (WGAN-GP)

Fig. 9. Performance measure plots for 64×64 CelebA

3.3   Results for CelebA 32

In Fig. 6 (a) and (c), we compare the FID and values for generating CelebA images for different dimensions of input noise for DCGAN and WGAN-GP. We fix the architecture of the generator and the discriminator and we find that the FID scores are very high noise dimension 2 and then reduce drastically. Further increasing the dimension does not effect the FID values much. Both WGAN and DCGAN perform almost equally. Although WGAN is slightly better. In Fig. 6 (b) and (d), we plot the IS values which are evaluated for batches of samples and the mean and the variance of the IS values across the batches is plotted for both DCGAN and WGAN-GP. Figures 7, 8. We find that results when noise dimension is 2 is bad compared to the other dimensions. For dimension 2 and 10 the images generated using WGAN lack clarity while that of DCGAN lack variety. For dimension 100 and 900, visually their performance is similar.

3.4   Results for CelebA 64

In Fig. 9 (a) and (c), we compare the FID scores for generating CelebA images for different dimensions of input noise for DCGAN and WGAN-GP. We fix the architecture of the generator and the discriminator and we find that the FID scores are very high for very low noise dimension but decreases considerably after a threshold and then do not change much for higher dimensions. We find the FID values for WGAN-GP are way better as also indicated by the figure. In Fig. 9 (b) and (d), we plot the IS values which are evaluated for batches of samples and the mean and the variance of the IS values across the batches is plotted for both DCGAN and WGAN-GP. The IS values do not seem to be indicative of the results as much as the FID values. WGAN-GP performs better at celebaA 64 than celebA 32 while the opposite is true for DCGAN.

We visually compare the results in [22, Figures 12, 13]. We find that results when noise dimension is 2 is bad compared to the other dimensions. WGAN-GP performs better compared to DCGAN. The clarity as well as variety of images generated by WGAN-GP is better compared to DCGAN unlike the case for images of size 32.

4   Insights and Future Work

In the various recent analysis on GANs, there has been hardly any focus on the input noise. The transformation of input noise to the generated data is a crucial part of the generation process. We think it calls for more attention and further analysis. From the results above, we can see that, there is a significant effect on the results when the input noise dimension is changed. It is also observed that the optimal noise dimension depends on the data-set and loss function used.

Given that we intend to map the high dimensional data to a low dimensional distribution, we would like the dimension of noise to be as small as possible. Although very small values do not give good results, hence we find an optimal dimension after which the model does not perform better in case of CelebA and MNIST or performs worse in case of Gaussian data.

FID value is not so indicative for analyzing the effect of change in input noise for the CelebA and MNIST data-set. We believe a more theoretical study and further analysis will help in training hand faster and better results than starting with a random size of z. We may also need to come up with performance measures which are more indicative of the quality of generated images.

5   Conclusion

We studied the effect of changing input noise for GANs quantitatively and qualitatively. We conclude that the input noise dimension has a significant effect on the generation of images. To obtain useful and quality data generation, the input noise dimension needs to be set based on data-set, which GAN architecture used, and/or which loss function is used to train it. We leave the theoretical analysis of the relation between the low dimensional distribution and the high dimensional data for the future work.

References

1. Arjovsky, M., Bottou, L.: Towards principled methods for training generative adversarial networks. arXiv preprint arXiv:1701.04862 (2017)
2. Arjovsky, M., Chintala, S., Bottou, L.: Wasserstein generative adversarial networks. In: International Conference on Machine Learning, pp. 214–223. PMLR (2017)
3. Bińkowski, M., Sutherland, D.J., Arbel, M., Gretton, A.: Demystifying mmd GANs. In: International Conference on Learning Representations (2018)
4. Chen, X., Duan, Y., Houthooft, R., Schulman, J., Sutskever, I., Abbeel, P.: Infogan: interpretable representation learning by information maximizing generative adversarial nets. In: NIPS, pp. 2172–2180 (2016)
5. Daskalakis, C., Ilyas, A., Syrgkanis, V., Zeng, H.: Training GANs with optimism. In: International Conference on Learning Representations (2018)
6. Donahue, J., Simonyan, K.: Large scale adversarial representation learning. arXiv preprint arXiv:1907.02544 (2019)
7. Dong, H.W., Hsiao, W.Y., Yang, L.C., Yang, Y.H.: Musegan: multi-track sequential generative adversarial networks for symbolic music generation and accompaniment. In: Thirty-Second AAAI Conference on Artificial Intelligence (2018)
8. Frid-Adar, M., Diamant, I., Klang, E., Amitai, M., Goldberger, J., Greenspan, H.: Gan-based synthetic medical image augmentation for increased CNN performance in liver lesion classification. Neurocomputing **321**, 321–331 (2018)
9. F.R.S., K.P.: Liii. on lines and planes of closest fit to systems of points in space. Lond. Edinb. Dublin Philoso. Mag. J. Sci. **2**(11), 559–572 (1901)
10. Goodfellow, I., et al.: Generative adversarial nets. NIPS 27 (2014)
11. Gulrajani, I., Ahmed, F., Arjovsky, M., Dumoulin, V., Courville, A.C.: Improved training of wasserstein gans. In: NIPS (2017)
12. Heusel, M., Ramsauer, H., Unterthiner, T., Nessler, B., Hochreiter, S.: GANs trained by a two time-scale update rule converge to a local nash equilibrium. In: NIPS 30, pp. 6626–6637 (2017)
13. Johnson, W.B., Lindenstrauss, J.: Extensions of lipschitz mappings into a hilbert space. Contemp. Math. **26**(189–206), 1 (1984)
14. Karras, T., Aila, T., Laine, S., Lehtinen, J.: Progressive growing of gans for improved quality, stability, and variation. arXiv preprint arXiv:1710.10196 (2017)
15. Kim, T., Cha, M., Kim, H., Lee, J.K., Kim, J.: Learning to discover cross-domain relations with generative adversarial networks. In: Proceedings of the 34th International Conference on Machine Learning, vol. 70, pp. 1857–1865. PMLR, International Convention Centre, Sydney, Australia, 06–11 August 2017
16. Lecun, Y., Bottou, L., Bengio, Y., Haffner, P.: Gradient-based learning applied to document recognition. Proc. IEEE **86**(11), 2278–2324 (1998). https://doi.org/10.1109/5.726791
17. LeCun, Y., Chopra, S., Hadsell, R., Ranzato, M., Huang, F.: A tutorial on energy-based learning. Predict. Struct. Data **1** (2006)
18. Ledig, C., et al.: Photo-realistic single image super-resolution using a generative adversarial network. In: Proceedings of the IEEE Conference on Computer Vision and Pattern Recognition, pp. 4681–4690 (2017)
19. Li, J., Liang, X., Wei, Y., Xu, T., Feng, J., Yan, S.: Perceptual generative adversarial networks for small object detection. In: 2017 IEEE Conference on Computer Vision and Pattern Recognition (CVPR), pp. 1951–1959 (July 2017). https://doi.org/10.1109/CVPR.2017.211

20. Liu, Y., Qin, Z., Luo, Z., Wang, H.: Auto-painter: Cartoon image generation from sketch by using conditional generative adversarial networks. arXiv preprint arXiv:1705.01908 (2017)
21. Liu, Z., Luo, P., Wang, X., Tang, X.: Deep learning face attributes in the wild. In: Proceedings of International Conference on Computer Vision (ICCV) (December 2015)
22. Manisha, P., Das, D., Gujar, S.: Effect of input noise dimension in gans. CoRR abs/2004.06882 (2020)
23. Mao, X., Li, Q., Xie, H., Lau, R.Y.K., Wang, Z.: Multi-class generative adversarial networks with the L2 loss function. CoRR abs/1611.04076 (2016)
24. Mescheder, L., Nowozin, S., Geiger, A.: The numerics of gans. In: Proceedings of the 31st International Conference on Neural Information Processing Systems, pp. 1823–1833 (2017)
25. Nguyen, T.D., Le, T., Vu, H., Phung, D.: Dual discriminator generative adversarial nets. In: Proceedings of the 31st International Conference on Neural Information Processing Systems, pp. 2667–2677. NIPS 2017 (2017)
26. Qi, G.J.: Loss-sensitive generative adversarial networks on lipschitz densities. Int. J. Comput. Vis. **128**(5), 1118–1140 (2020)
27. Radford, A., Metz, L., Chintala, S.: Unsupervised representation learning with deep convolutional generative adversarial networks. arXiv preprint arXiv:1511.06434 (2015)
28. Salimans, T., Goodfellow, I., Zaremba, W., Cheung, V., Radford, A., Chen, X.: Improved techniques for training gans. NIPS **29**, 2234–2242 (2016)
29. Subramanian, S., Rajeswar, S., Dutil, F., Pal, C.J., Courville, A.C.: Adversarial generation of natural language. In: Rep4NLP@ACL (2017)
30. Wu, H., Zheng, S., Zhang, J., Huang, K.: Gp-gan: towards realistic high-resolution image blending. In: Proceedings of the 27th ACM International Conference on Multimedia, pp. 2487–2495 (2019)
31. Wu, J., Zhang, C., Xue, T., Freeman, B., Tenenbaum, J.: Learning a probabilistic latent space of object shapes via 3d generative-adversarial modeling. In: NIPS, pp. 82–90 (2016)
32. Yoo, D., Kim, N., Park, S., Paek, A.S., Kweon, I.S.: Pixel-level domain transfer. In: Leibe, B., Matas, J., Sebe, N., Welling, M. (eds.) ECCV 2016. LNCS, vol. 9912, pp. 517–532. Springer, Cham (2016). https://doi.org/10.1007/978-3-319-46484-8_31
33. Yu, L., Zhang, W., Wang, J., Yu, Y.: Seqgan: sequence generative adversarial nets with policy gradient. In: Proceedings of the AAAI Conference on Artificial Intelligence, vol. 31 (2017)
34. Zhao, J., Mathieu, M., LeCun, Y.: Energy-based generative adversarial networks. In: 5th International Conference on Learning Representations, ICLR 2017 (2017)
35. Zhu, J.Y., Park, T., Isola, P., Efros, A.A.: Unpaired image-to-image translation using cycle-consistent adversarial networks. In: Computer Vision (ICCV), 2017 IEEE International Conference on (2017)

Wiper Arm Recognition Using YOLOv4

Hua Jian Ling, Kam Meng Goh$^{(\boxtimes)}$, and Weng Kin Lai

Department of Electrical and Electronics Engineering, Faculty of Engineering and Technology, Tunku Abdul Rahman University College, Jalan Genting Kelang, 53300 Wilayah Persekutuan Kuala Lumpur, Malaysia

gohkm@tarc.edu.my

Abstract. Quantity control is as important as quality control for those products manufactured in the mass production phase. However, there is scarce work implementing deep learning methods in the manufacturing line, especially in wiper arm recognition. This paper proposed a deep learning-based wiper arm recognition for the windshield wiper manufacturer to reduce human error and workforce requirement in quantity control. The proposed method applied the state-of-the-art YOLOv4 object detection algorithm. Our proposed method able to achieve 100% in terms of precision, recall, F1-score, and mean average precision. Moreover, the proposed method can make correct predictions under several conditions: object occlusion, different scales of objects, and different light environments. In term of speed, the proposed method can be predicted up to 30.55 fps when using a moderate GPU.

Keywords: Deep learning · YOLOv4 · Recognition · Detection

1 Introduction

Windshield wipers are essential equipment in almost all motor vehicles, including cars, trucks, buses, trains, and even aircraft. It is used to remove rainwater and impurities from the windshield to provide a clearer view for the vehicle's operator. According to statistics [1], the total number of operational cars worldwide was estimated at 1.42 billion, including 1.06 billion passenger cars and 363 million commercial vehicles. In addition, the latest statistics on worldwide vehicle production released by the International Organization of Motor Vehicle Manufacturers (OICA) in 2020 show that the annual production of vehicles exceeds 77 million, including 55.8 million passenger cars and 21.7 million commercial vehicles [2]. Thus, the statistics above indicate that the windshield wiper had reached the mass production phase to sustain the enormous demand for wipers from the automotive industry.

Counting is one of the essential manufacturing processes in the production line in order to sustain the huge demands of wipers during the mass production phase. The wiper arm counting process is used to grasp the number of wipers produced and ensure the number of wipers in respective boxes is correct before delivering to customers. Besides, windshield wipers are not a universal product as they come in countless shapes, sizes, and variations to fit all types of vehicles [3]. Thus, the identification process of the wiper arm

T. Mantoro et al. (Eds.): ICONIP 2021, LNCS 13110, pp. 570–581, 2021.
https://doi.org/10.1007/978-3-030-92238-2_47

model needs to be carried out simultaneously with the counting process. However, the current model identification and counting processes are commonly performed manually by human operators using the naked eye which may result in wrong recognition or counting. To overcome this problem, we proposed a deep learning approach, *You Only Look Once* v4 (YoloV4), to recognize the wiper arm in this paper. Hence, the main contribution of this paper is

- to implement YOLOv4 for wiper arm recognition and counting that can be applied in industrial manufacturing. This is the first work to implement YOLOv4 in wiper arm recognition.

The paper is structured as follows: Sect. 2 discusses the prior work done in deep learning implementation in product/manufacturing. Then, we discuss our proposed method in Sect. 3, and we further elaborate our findings in Sect. 4. Lastly, the conclusion and recommendations are shown in Sect. 5.

2 Prior Work

In this section, we will discuss the previous work done for object recognition. To date, there is no work done on recognizing wiper arms. Hence, we will discuss some common methods used for object recognition in the manufacturing industries. As the essence of this research is on deep learning-based object detection and recognition, the related work on these two domains are discussed.

2.1 Object Recognition

A. O'Riordan et al. [4] introduced the opportunities and impacts of object recognition applications on smart manufacturing within industrial 4.0. Typically, robots are position-based, and the predetermined set of Cartesian coordinates limits all their motions. Therefore, object recognition could be applied to the robotic operations to gather information on the object's poses and distances from the cameras. On the other hand, I. Apostolopoulos et al. [5] proposed an object recognition method based on deep learning to evaluate industrial defect objects. In addition, data augmentation was applied in the datasets to provide an adequate amount of image data. As a result, the multipath VGG19 achieved 97.88%, while the original VGG19 achieved 70.7% for evaluating defect objects. A. Caggiano et at. [6] proposed a machine learning model based on bi-stream Deep Convolutional Neural Network (DCNN) to recognize the defects induced by improper Selective Laser Melting (SLM) process conditions achieving an accuracy of 99.4%. In recent years, DCNN has been demonstrated outstanding performance in various scenarios [7–9] such as image recognition, motion recognition, automatic speech recognition, natural language processing, and more. On the other hand, G. Fu et al. [10] proposed a deep learning-based approach that incorporates a pre-trained SqueezeNet CNN architecture with a multi-receptive field (MRF) module for steel surface defects classification. J. Wang et al. [11] proposed a product inspection method using deep learning that consists of three phases to identify and classify defective products in the production line. The proposed method achieves 99.60% accuracy in 35.80ms with a model size of 1176 kB. In the retail sector, deep learning has been used to recognize retail products [12]. The problem is now modeled as an object detection problem where one of the challenges is the large number of different categories. In their work, they used a Faster R-CNN object detector [13] to perform product recognition, and they achieved 75.9% mAP among 20 classes (VOC 2012) and 42.7% mAP among 80 classes (COCO). Another practical example named YOLO9000 [14] had been used to detect 9000 object classes by using revised Darknet, showing that CNN can be used to handle large-scale classification.

2.2 Object Detection

R.Cirshick et al. [15] proposed Region-based Convolutional Neural Networks (R-CNN) in early 2014 which works in four steps, namely region proposals extraction, size rescaling, and features extraction, and classification by regions. The R-CNN object detection algorithm was able to achieve 58.50% mean Average Precision (mAP) on the PASCAL VOC 2007 dataset and 53.70% mAP on the PASCAL VOC 2012. K. He et al. [16] proposed Spatial Pyramid Pooling Networks (SPP-Net) by introducing a spatial pyramid pooling (SPP) layer to supersede the rescaling process on R-CNN. This improvement made the detection speed ten times faster than R-CNN without sacrificing detection accuracy due to the SPP layer avoiding repeated computing of the convolution features. The SPP-Net achieved 59.2% mAP on PASCAL VOC 2007, with 60.0% mAP on PASCAL VOC 2012. Both R-CNN and SPP-Net are multi-stage pipelines that are not end-to-end training architecture. They use an SVM classifier to train the feature vectors and then classify the objects instead. On the other hand, R. Girshick et al. [17] proposed the Fast R-CNN to overcome the problem of multi-stage pipelines that occurred in R-CNN and SPP-Net. The formation of Fast R-CNN is designed as an end-to-end training architecture using multi-task loss function. The loss function improved the training time 84 h in R-CNN to 8.75 h in Fast R-CNN while achieving 70% mAP on the PASCAL VOC 2007, and 68.4% mAP for PASCAL VOC 2012. The improvements of Fast R-CNN enable all the convolutional layers to be updated during the training period. However, Fast R-CNN is still unable to identify objects in real-time due to the region proposals extracted from the selective search method, which is a time-consuming process. S. Ren et al. [13] proposed Faster R-CNN in late 2015. A Region Proposal Network (RPN) is introduced to replace the selective search method used in R-CNN and Fast R-CNN for object localization in the proposed work. Most of the individual component blocks unified into the RPN form an end-to-end learning framework to generate high-quality. The shared computation of Fast R-CNN and RFN achieved 73.2% mAP on the PASCAL VOC 2007 and 70.4% on the PASCAL VOC 2012 with 300 proposals per image. Instead of using the RCNN family for object detection and recognition. R. Joseph et al. [18] proposed *You Only Look Once* (YOLO). YOLO is an end-to-end training architecture and is the first one-stage detector in deep learning. The algorithm uses the features collected from the entire image to predict multiple bounding boxes and class probabilities for those boxes simultaneously through a single convolutional network. The detection pipeline of YOLO is designed as an end-to-end training architecture that enables the model to perform in real-time without complexity. YOLO achieved 63.4 mAP on the PASCAL

VOC 2007 and 2012. The detection speed taken is 45 fps which is extremely fast as compared to the R-CNN family. However, the YOLO makes 19.0% of localization errors due to its struggles to localize small objects that appear in the group precisely. Therefore, the YOLO's accuracy in terms of mAP is lower than the Fast R-CNN but higher than R-CNN. R. Joseph et al.'s [14] YOLOv2 improved YOLO by applying a new custom deep architecture to perform feature extraction. YOLOv2 with Darknet-19 improves the top 1 accuracy from 72.5 (YOLO) to 74.0 (YOLOv2) and the top 5 accuracies from 90.8 (YOLO) to 91.8 (YOLOv2) in ImageNet. Compared to the fully connected layer in YOLO, YOLOv2 replaces the fully connected layers with five anchor boxes to increase the total number of bounding boxes within an image from 98 (YOLO) to 845 boxes. R. Joseph et al. [19] subsequently introduced YOLOv3 to overcome the problem of small object detection on the previous versions. While its detection speed is slower, nevertheless YOLOv3 is able to make detection in three different scales. In addition, YOLOv3 uses an independent logistic classifier instead of a SoftMax classifier to perform multi-label classification for detecting the objects within the image. YOLOv3 with 416*416 input image resolution achieved 55.3 mAP with 35fps calculated at IOU threshold 0.5 on MS COCO 2017. In early 2020, B. Alexey et al. [20] introduced YOLOv4 which allows people to use conventional GPUs for training purposes with a small mini-batch size. YOLOv4 achieved 62.8 mAP with 54 fps calculated at IOU threshold 0.5 on MS COCO 2017 using a conventional GPU. Thus, the YOLOv4 achieved a 10% improvement in the accuracy and 12% in the prediction speed as compared to YOLOv3.

3   Proposed Method

We realized that the YOLOv4 [20] is the most suitable object detection algorithm that can be adopted to handle the manufacturing tasks in real-time. Most modern neural networks are still unable to perform in real-time and require Graphics Processing Units (GPU)s for training with a large mini-batch size. As YOLOv4 allows a single conventional GPU such as 1080Ti to be used, hence we proposed a deep learning-based wiper arm recognition using YOLOv4. After recognition, a simple object counting approach could be applied to calculate the number of wiper arms for their respective models. The proposed method consists of three phases: dataset construction, model training, and model testing.

3.1 Dataset Construction

In this work, we split the dataset into a training set and a testing set. The wiper arm datasets were captured using FLIR Blackfly S BFS-U3-13Y3C-C with a 1280 × 1024-pixel resolution. All the images were taken under two additional light sources to eliminate most of the shadows behind the actual object, where one image may consist of multiple wiper arms. In our work, we had collected a total of images, from four different classes and we divide all images into 80% for training and 20% for testing. As a result, the training set consisted of 714 images with 2869 wiper arms sample, while the testing set consisted of 143 images with 603 wiper arms. The number of wiper arms for each type is illustrated in Table 1. The data labelling process is crucial for the YOLOv4 algorithm to recognize those four different classes of wiper arms. All collected images were further annotated with precise labels for training and testing purposes.

Table 1. The detailed number of wiper arms for each class

Classes	Training set (714 images)	Testing set (143 images)
Wiper A	1000	207
Wiper B	815	176
Wiper C	711	166
Wiper D	343	54
Total	2869	603

3.2 Model Training - Data Augmentation and DropBlock Regularization

Data augmentation was used to increase the number of samples and increase the stability and diversity of the annotated dataset for training purposes. Photometric distortion and mosaic data augmentation were used in this project to produce special conditions for the algorithm to learn and identify. The photometric distortion methods are dealing with saturation, exposure, and hue. The parameters were set as: the saturation = 1.50; the exposure = 1.50; and the hue = 0.1. This enables the neural network to extract and classify the features from the input images under different light environments. The effects of the pixel-wise adjustments from the photometric distortion are shown in Fig. 1.

Fig. 1. Image after pixel-wise adjustment (saturation = 1.50; exposure = 1.50; hue = 0.1)

Mosaic data augmentation [20] provides the function of mixing four training images with different contexts to one image at specific proportions, as shown in Fig. 2. This enables the model to identify objects at a smaller/bigger scale than usual. Besides, it significantly reduces the need to use a large mini-batch size for training.

DropBlock regularization [21] is a technique to removes certain semantic information on the datasets. In this work, the removing parts will be the wiper head or wiper hood. The purpose of applying this technique is to enforce the YOLOv4 to learn spatially discriminating features to classify the input image. As a result, the YOLOv4 can still make an accurate prediction in the case of object occlusion. For example, this technique drops contiguous regions from the feature map, as shown in Fig. 3.

Fig. 2. Images after Mosaic data augmentation

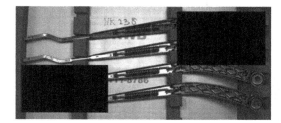

Fig. 3. Images after DropBlock regularization

3.3 YOLOv4 Object Detection

We proposed the use of YOLOv4 [20] to recognize the model of wiper arms. Figure 4 illustrates a YOLOv4-based wiper arm recognition structure consisting of 4 parts namely, input terminal, backbone, neck, and head. The images and annotations of the training set are prepared at the input terminal to adopt those mentioned data augmentations and regularization methods and then transmitted to the backbone. There are five essential components and two basic operations included in the YOLOv4 network structure shown here.

1. CBM is the smallest component consisted of convolutional layers, Cross Mini-Batch Normalization, and mish activation function.
2. CBL consists of convolutional layers, Cross Mini-Batch Normalization, and Leaky Rectified Linear Unit (ReLU) activation function.
3. The Res unit is represented as the residual structure in ResNet.
4. CSPX consists of three convolutional layers with several Res unit modules.
5. SPP is the neck of YOLOv4, which performed multi-scale fusion through maximum pooling of 1×1, 5×5, 9×9, and 13×13.

The basic operations are:

1. **Concatenation operation** (route operation) - tensor stitching is used to expand the dimension of convolutional layers.

2. **Add operation** (shortcut operation) - addition of tensors but does not expand the dimension of convolutional layers.

The adoption of the Cross Mini-Batch Normalization technique brings the advantage of allowing users to train models with small mini-batch sizes. In addition, the technique significantly improves the stability of the training process. The YOLOv4 uses the mish activation function combined with the CSPDarknet53, which substantially improves detection accuracy. The mish activation function is a non-monotonic function, preserving the small negative values to stabilize the network gradient flow. However, most commonly used activation functions like ReLU (used in YOLOv2) and Leaky ReLU (used in YOLOv3) failed to preserve a negative value, making the neurons unable to update. The mish function also provides a better generalization and effective optimization than the ReLU function used in YOLOv2.

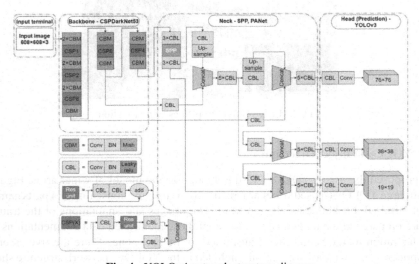

Fig. 4. YOLOv4 network structure diagram

The backbone used in YOLOv4 is CSPDarknet53 for extracting features from the input image, while the neck used in YOLOv4 is Spatial Pyramid Pooling Network (SPP-Net) and Path Aggregation Network (PANet). The SPP-Net is used to increase the receptive field and separate the most critical features from the backbone, while the PANet is to mix and combine the features. These features will be used for the prediction step. Lastly, complete Intersection over Union (CIou) loss was applied to the head of YOLOv4 to refine the object localization performance by minimizing the normalized distance of the central point. When all the potential bounding boxes are generated, the non-maximum suppression (NMS) technique filters those boxes with less confidence. Only the candidate bounding boxes with the highest response were retained to ensure that each object was detected only once.

4   Results and Discussion

All the wiper arms within the image, as shown in Fig. 5, were successfully detected and the correct models identified with high confidence scores. Each of the model number is shown in the upper left corner.

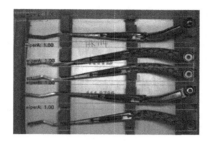

Fig. 5. Results from the YOLOv4 model

4.1   Experimental Setup

In this work, the parameters are configured as follows, the batch size and subdivision are set to 8 and 2 respectively, resulting in a mini-batch size of 4; the width and height are both set to 608; momentum of 0.9; weight decay at 0.0005; learning rate of 0.001; the warm-up steps (burn-in) at 1000; the maximum batches (iterations) at 8000; training steps at 6400 and 7200; the scales are .1,.1; 4 classes; 27 filters; and 90 epochs. Note that all these optimal parameters are obtained after several empirical testing.

4.2   Impact of the Proposed Method Under Several Conditions

The performance of predicting objects under several conditions such as different light conditions, different scales of objects, and object occlusion is shown in Fig. 6. To improve the effectiveness of YOLOv4 object detector under these conditions, data augmentation and regularization techniques were used to increase the diversity of input images before feeding into YOLOv4. One of the data augmentations named photometric distortion was applied to enable the YOLOv4 neural network to extract and classify the features from input images under different light environments as shown in the first row of Fig. 6. In addition, the mosaic data augmentation was used to combine four images with different scaling ratios into one, which effectively reduces the computational burden and enables the detector to predict the targets in different sizes as shown in second row of Fig. 6. DropBlock regularization technique was used to remove certain semantic informations in the input images to enforce the YOLOv4 to learn in this situation, therefore the YOLOv4 can still work well under the situation of object occlusion as shown in the last row of Fig. 6.

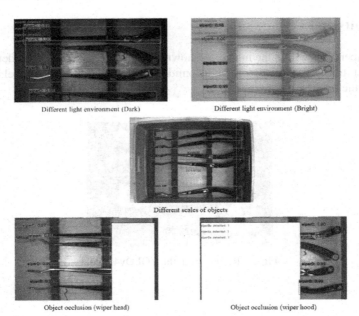

Different light environment (Dark)   Different light environment (Bright)

Different scales of objects

Object occlusion (wiper head)   Object occlusion (wiper hood)

Fig. 6. Predictions under challenging conditions

4.3 Comparison of Different Algorithms with Different Input Resolution at 8000ᵗʰ Iteration

Model training is essential for all deep learning models, and the value of the average loss is the only factor determining when to stop training. When the average loss no longer decreases—the acceptable range of the final average loss lies between 0.05 to 3.0. Table 2 shows that both YOLOv3 and YOLOv4 obtained an average loss that is close to 0 and a mAP of around 100. It can be concluded that the models are neither under-fitting nor over-fitting. Even though the average loss of YOLOv3 was lower than YOLOv4, but the mAP of YOLOv4 is still higher than YOLOv3.

A series of evaluation metrics were used to measure the performance of the proposed method. Precision is the probability of the predicted bounding boxes that matches the actual ground truth boxes, that is number of true positives (TP) divided by the sum of the number of true positives (TP) and false positives (FP). The other measure is recall which is the probability of the ground truth objects being detected correctly. It is the number of true positives (TP) divided by the sum of true positives and false negatives (FN). Figure 7 shows how TP, TF and FP are measured.

The *F1-score* is the harmonic mean of precision and recall, which is used to measure the proposed method's comprehensive performance. The equation is defined as in (1)

$$\mathbf{F1 - score} = \frac{2}{\frac{1}{\textbf{Recall}} + \frac{1}{\textit{Precision}}} \tag{1}$$

Table 2. Average loss and mAP of different algorithms

Algorithm	Average loss	Mean average precision (mAP, %)
YOLOv3 (416 × 416)	0.0991	99.9
YOLOv3 (608 × 608)	0.0717	99.5
YOLOv4 (416 × 416)	0.3443	100%
YOLOv4 (608 × 608)	0.3563	100%

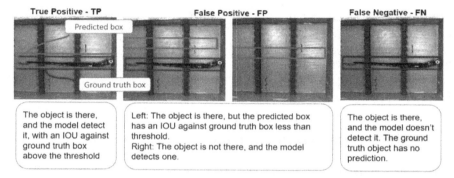

Fig. 7. True positive, false positive, and false negative

The *mean Average Precision* is used to measure the overall performance of the proposed method under 0.5 IOU confidence thresholds. The equation is defined in (2)

$$mAP = \frac{1}{n} \sum_{k=0}^{k=n} APk \qquad (2)$$

where n = number of classes, AP_k = average precision of class k. The algorithm uses the *Intersection over Union* to determine the similarity between the ground truth bounding box and the predicted bounding box, as stated in (3).

$$\textbf{Intersection over Union} = \frac{Area\ of\ overlap}{Area\ of\ Union} \qquad (3)$$

The updated version of YOLO such as YOLOv2, v3 and v4 can adjust the resolution of input image to any factor of 32. According to the results obtained as shown in Table 3, YOLOv4 with 608 × 608 input resolution was able to identify all the testing datasets without any false results while YOLOv3 has wrongly identified the defects on a few occasions. Moreover, YOLOv4 (608 × 608) is 60% slower than YOLO4 (416 × 416) but still 34% faster than YOLO3 (608 × 608). It can be concluded as the higher the resolution of input image, the slower the detection speed, therefore the adjustment of resolutions can achieve a good trade-off between accuracy and detection speed.

Table 3. Comprehensive performance analysis of different algorithms

Algorithm	TF	FP	FN	Precision (%)	Recall (%)	F1 score (%)	mAP (%)	Avg IOU (%)	Speed (fps)
YOLOv3 (416 × 416)	602	2	1	100	100	100	99.87	91.19	24.84
YOLOv3 (608 × 608)	599	0	4	100	100	100	99.52	92.33	14.16
YOLOv4 (416 × 416)	603	1	0	100	100	100	100	92.55	30.55
YOLOv4 (608 × 608)	603	0	0	100	100	100	100	93.07	18.99

5   Conclusions and Recommendations

In this paper, our proposed method could recognize up to 30.55 fps using a TESLA T4 GPU. The proposed method was able to increase the detection speed with this more powerful GPU. According to the results obtained, the overall performance of YOLOv4 is better than YOLOv3. Several attempts to optimize the parameter settings resulted in an increased input resolution network to 608 × 608, and the proposed method achieved 100% in different measurement. YOLOv4 was able to predict all testing datasets without any false detections. The proposed work was tested under several challenging conditions such as object occlusion, different scales of objects, and bad light conditions for robustness. In the future, more wiper arm models with different size, and shape but similar texture, are needed to further test the method's reliability.

Acknowledgments. The authors are grateful to DENSO WIPER SYSTEMS (M) SDN BHD for providing the wiper arms used in this research.

References

1. Okeafor JP: How many cars are there in the world. https://naijauto.com/market-news/how-many-cars-are-there-in-the-world-7100.
2. Organisation Internationale des Constructeurs d'Automobiles: 2020 Production Statistics. https://www.oica.net/category/production-statistics/2020-statistics/.
3. Parts, D.C.S.: Are Wiper Blades Universal. https://www.diycarserviceparts.co.uk/blog/2019/08/26/wiper-blade-types-are-wiper-blades-universal/.
4. O'Riordan, A.D., Toal, D., Newe, T., Dooly, G.: Object recognition within smart manufacturing. Procedia Manuf. Procedia Manuf. **38**, 408–414 (2019). https://doi.org/10.1016/j.promfg.2020.01.052

5. Apostolopoulos, I.D., Tzani, M.: Industrial object, machine part and defect recognition towards fully automated industrial monitoring employing deep learning. The case of multi-level VGG19. ArXiv preprint, pp. 1–17 (2020)
6. Caggiano, A., Zhang, J., Alfieri, V., Caiazzo, F., Gao, R., Teti, R.: Machine learning-based image processing for on-line defect recognition in additive manufacturing. Procedia CIRP **68**, 451–454 (2019). https://doi.org/10.1016/j.cirp.2019.03.021
7. Gu, J., et al.: Recent advances in convolutional neural networks. Patt. Recogn. **77**, 354–377 (2015)
8. Wang, P., Liu, H., Wang, L., Gao, R.X.: Deep learning-based human motion recognition for predictive context-aware human-robot collaboration. Procedia CIRP. **67**, 17–20 (2018). https://doi.org/10.1016/j.cirp.2018.04.066
9. Wang, J., Ma, Y., Zhang, L., Gao, R.X., Wu, D.: Deep learning for smart manufacturing: Methods and applications. J. Manuf. Syst. **48**, 144–156 (2018). https://doi.org/10.1016/j.jmsy.2018.01.003
10. Fu, G., et al.: A deep-learning-based approach for fast and robust steel surface defects classification. Opt. Lasers Eng. **121**, 397–405 (2019). https://doi.org/10.1016/j.optlaseng.2019.05.005
11. Wang, J., Fu, P., Gao, R.X.: Machine vision intelligence for product defect inspection based on deep learning and Hough transform. J. Manuf. Syst. **51**, 52–60 (2019). https://doi.org/10.1016/j.jmsy.2019.03.002
12. Wei, Y., Tran, S., Xu, S., Kang, B., Springer, M.: Deep learning for retail product recognition: challenges and techniques. Comput. Intell. Neurosci. **2020** (2020). https://doi.org/10.1155/2020/8875910.
13. Ren, S., He, K., Girshick, R., Sun, J.: Faster R-CNN: towards real-time object detection with region proposal networks. IEEE Trans. Pattern Anal. Mach. Intell. **39**, 1137–1149 (2017). https://doi.org/10.1109/TPAMI.2016.2577031
14. Redmon, J., Farhadi, A.: YOLO9000: better, faster, stronger. In: Proceedings - 30th IEEE Conference on Computer Vision and Pattern Recognition, CVPR 2017, pp. 6517–6525, January 2017. https://doi.org/10.1109/CVPR.2017.690.
15. Girshick, R., Donahue, J., Darrell, T., Malik, J.: Region-based convolutional networks for accurate object detection and segmentation. IEEE Trans. Pattern Anal. Mach. Intell. **38**, 142–158 (2016). https://doi.org/10.1109/TPAMI.2015.2437384
16. Msonda, P., Uymaz, S.A., Karaağaç, S.S.: Spatial pyramid pooling in deep convolutional networks for automatic tuberculosis diagnosis. Traitement du Signal. **37**, 1075–1084 (2020). https://doi.org/10.18280/TS.370620.
17. Girshick, R.: Fast R-CNN. Proceedings of the IEEE International Conference on Computer Vision, pp. 1440–1448 (2015). https://doi.org/10.1109/ICCV.2015.169
18. Redmon, J., Divvala, S., Girshick, R., Farhadi, A.: You only look once: unified, real-time object detection. In: Proceedings of the IEEE Computer Society Conference on Computer Vision and Pattern Recognition, pp. 779–788 (2016). https://doi.org/10.1109/CVPR.2016.91
19. Redmon, J., Farhadi, A.: YOLOv3: An Incremental Improvement. ArXiv preprint (2018)
20. Bochkovskiy, A., Wang, C.-Y., Liao, H.-Y.M.: YOLOv4: Optimal Speed and Accuracy of Object Detection. ArXiv preprint (2020)
21. Ghiasi, G., Lin, T.-Y., Le, Q.V.: DropBlock: A regularization method for convolutional networks. ArXiv preprint (2018)

Context Aware Joint Modeling of Domain Classification, Intent Detection and Slot Filling with Zero-Shot Intent Detection Approach

Neeti Priya[✉], Abhisek Tiwari, and Sriparna Saha

Department of Computer Science and Engineering, Indian Institute of Technology
Patna, Bihta 801103, Bihar, India

Abstract. Natural language understanding (NLU) aims to extract schematic information contained in user utterances, which allows down streaming module of dialogue system, i.e., Dialogue Manager (DM) to process user queries and serve users in accomplishing their goal. If NLU component detects information improperly, it will cause error propagation and failure of all subsequent modules. Although the development of an adequate conversation system is challenging because of its periodic and contextual nature, its efficacy, applicability, and positive impact continue to fuel its recent surge and attention in the research community. The proposed work is the first of its kind, which attempts to develop a unified, multitasking, and context-aware BERT-based model for all NLU tasks, i.e., Domain classification (DC), Intent detection (ID), Slot filling (SF). Additionally, we have also incorporated a zero-shot intent detection technique in our proposed model for dealing with new and emerging intents effectively. The experimental results, as well as comparisons to the present state-of-the-art model and other several baselines on a benchmark dataset, firmly establish the efficacy and necessity of the proposed model.

1 Introduction

Natural language understanding (NLU) is the foremost component of a pipelined goal-oriented dialogue system, having the responsibility of extracting semantic meaning from user messages. With the growing complexity of user messages/queries and the emergence of multi-purpose conversation systems, domain identification along with intent detection and slot tagging is imperative. Consider an example: " *"find a movie running in Francis nearby restaurant, which provides Chinese food""*. Here, the user query has two different domains as *"Movie"* and *"Restaurant"* with the user *"intents=FindMovie, FindRestaurant"* for domains "Movie" and "Restaurant", respectively, and the slots as *"MovieTheatre=Francis"* for "FindMovie" intent and *"Cuisine=Chinese"* corresponding to "FindRestaurant". The traditional way to deal with multi-domain scenarios is to first determine the domain [11] and then intents [6]

© Springer Nature Switzerland AG 2021
T. Mantoro et al. (Eds.): ICONIP 2021, LNCS 13110, pp. 582–595, 2021.
https://doi.org/10.1007/978-3-030-92238-2_48

Table 1. Sample conversation from Airline domain and its corresponding domain, intents and slot labels

Actor	Utterance	Domain	Intent	Slot Labels
Agent	Hello, how can i help you?	–	–	–
User	Hi, Please check upcoming flight seat assignment	Airline	ChangeAssignmentSeat	O O O O O O O O
Agent	Please share your booking id	–	–	–
User	123456	Airline	ContentOnly	booking_confirmation_number
Agent	Please confirm it	–	–	–
User	123456	Airline	ContentOnly	booking_confirmation_number
Agent	which seat do you want to get?	–	–	–
User	Please change it to window seat	Airline	ChangeSeatAssignment	O O O O seat_type seat_type
Agent	Are you sure ?	–	–	–
User	yes please change	Airline	ChangeSeatAssignment	O O O

and slots [2]. However, this pipelined approach suffers from a major disadvantage of error propagation from DC task to ID similarly from ID to SF. For an instance: *"USER - "show me Chinese language classes in the town""*. Here, the true user *domain = Places* with user *intent = FindPlaces* and slot *Chineselanguageclasses = ServiceProvider* but if it gets incorrectly classified as *"Fastfood""*, user intent would more likely be *"OrderFood""*, which further leads to error in slot tagging (*"cuisine"" - "Chinese"*). Table 1 illustrates an example in which the majority of user utterances are very relevant and founded on earlier utterances (dialogue context), as is typical behavior in a dialogue situation. All three, DC, ID, and SF are interrelated tasks. Henceforth, utilizing this relevant and dependent information may lead to a more intuitive semantic space. Motivated by the role of joint modeling of NLU tasks and importance of context information, we build a multitasking and context aware NLU framework, which utilizes dialogue context (previous utterance, domain, intents and slots) and leverages inter-relations among these tasks (DC, ID and SF) using multitasking framework. Such a multitask model may not only outperform individual model but also simplifies the NLU stage significantly by using only a unified model. The main contributions of this paper are as follows:

- We propose a context aware multitask NLU framework for joint domain classification, intent detection and slot labeling tasks in an end to end setting.
- We have also incorporated a context aware zero-shot intent detection module for detecting emerging/unseen intents with some precision, leading the model to deal with such unfavourable situation effectively.
- The obtained experimental results outperform the current state of the art model and several single task and non-contextual baselines by a significant margin.

2 Related Work

In [7], an attention based recurrent neural network (RNN) model has been proposed for joint modelling of ID and SF on single domain (Air Travel). In [1], authors have proposed a BERT [3] model for joint intent and slot detection, which outperforms other existing RNN and CNN-based baselines by a significant margin. The first-ever joint model for DC, ID, and SF was proposed by Kim et al. [5], which utilizes character and word embeddings provided by BiLSTMs for utterance representation as input to their multi-layer perceptron layer. In [4], authors proposed context aware joint modeling of two tasks, ID and SF. They had considered previous dialogue acts (DA), intents and slots as context signals using a self-attention approach based on recurrent neural networks (RNN). Zero shot intent learning is a challenging task in NLU, in which new intents are emerged. One of the most recent work in this area is [13], here authors utilized capsule networks to transfer knowledge of prediction vectors from seen to unseen classes. However, context information is missing in this approach.

From the above studies (we could highlight only the most relevant and recent works), it is evident that there is dearth of work in joint modeling of all the three tasks, DC, ID and SF taking into account context signals for goal-oriented dialogue systems. However, it is essential to have such a robust model which can perform with high accuracy and precision in multi-domain scenarios to identify the correct domain of any user utterance followed by its true intents and slots.

3   Proposed Methodology

The block diagram of the proposed context aware multitask model is shown in Fig. 1. It compromises of six sub modules as: semantic module, context module, domain module, intent module, slot module and zero shot intent module.

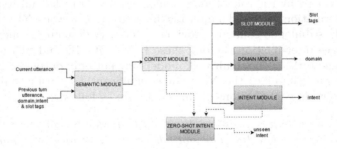

Fig. 1. The block diagram of the proposed context-aware multi-task model with zero-shot intent detection

The semantic module extracts features from current user utterance and dialogue context signals such as previous turn user utterance, domain, intent, and slot labels. This representation is then utilized by context module to encode all the context signals with the current utterance for providing a context aware representation. Different modules named as domain module, intent module and slot module learn to map this space to their correct classes for domain classification, intent detection and slot filling, respectively. Here, the zero shot intent module utilizes feature space produced by the context module, prediction done by intent module with its probability distribution over intents (softmax layer output) and similarity between seen and unseen intent labels to detect unseen intents. The detailed diagram of our proposed architecture is shown in Fig. 2.

3.1 Semantic Module

Utterance Encoding: To get the contextual representation of the current utterance and previous turn utterance, BERT model is used [3]. We have fine-tuned the pre trained BERT model for the task and have considered the outputs from entire sequence of utterances. The outputs of BERT layer for current utterance and previous utterance are U_t and U_{t-1}, respectively. Here $U_t, U_{t-1} \in \mathbb{R}^{l \times b_h}$ where l is the maximum sequence length of utterance and b_h is the hidden layer size of BERT.

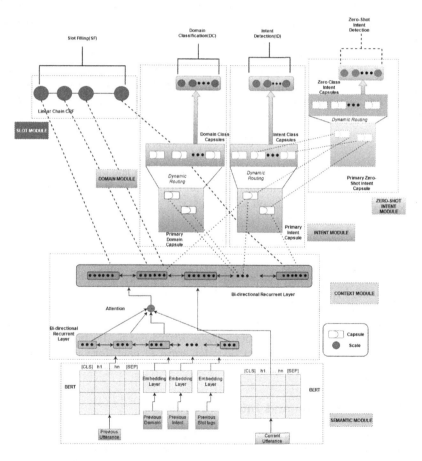

Fig. 2. The architectural diagram of the proposed context-aware multi-task model with zero-shot intent detection

Domain, Intent and Slots Encoding: One-hot vector encoding of previous turn's domain is being passed to embedding layer. The output of this layer is $D_{t-1} \in \mathbb{R}^{d \times d_h}$, where d is the number of domain classes. Similarly, one-hot encoding of intent label and slot tags are also passed to an embedding layer which outputs $I_{t-1} \in \mathbb{R}^{i \times i_h}$, where i is the number of intent classes and $S_{t-1} \in \mathbb{R}^{l \times s_h}$. Here, d_h, i_h and s_h are embedding dimensions for domain, intent and slot tags, respectively.

3.2   Context Module

All context signals, U_{t-1}, D_{t-1}, I_{t-1} and S_{t-1}, are concatenated together which gives a feature matrix, $F_t = (U_{t-1} \cdot D_{t-1} \cdot I_{t-1} \cdot S_{t-1})$. We have used an Encoder decoder with attention approach [8] to sequentially encode the dialogue context and the current user utterance. Firstly, **Bi-directional LSTM** is used as encoder to encode the obtained context signal features into hidden states where each forward and backward hidden states of the encoder are $\overrightarrow{h_e}$, $\overleftarrow{h_e}$, respectively. These two hidden states $\overrightarrow{h_e}$, $\overleftarrow{h_e}$ are concatenated to generate final encoder state as h_e. Now, the output of encoder is the set of hidden states $H_e = \{e_1, e_2, ..., e_n\}$, where $H_e \in \mathbb{R}^{l \times 2d_h}$. n is the number of encoder states, l is the maximum sequence length and d_h is the dimension of encoder state in each forward and backward LSTM. Then, attention weights are calculated by a softmax function as follows:

$$A_{xy} = \frac{\exp(e_{xy})}{\sum_{z=1}^{n} \exp(e_{xz})} \tag{1}$$

e_{xy} is given by a function $allign(m_x, e_y)$ that obtains the alignment between the input at position y of the encoder state and output at x of the decoder state. The context vector is calculated as follows:

$$c_x = \sum_{k=1}^{n} A_{xk} \cdot h_x. \tag{2}$$

The obtained c_x is now concatenated with the BERT representation of the current utterance, U_t, which is then passed to a decoder network. Decoder is another **Bi-directional LSTM** which sequentially encodes the context signal representation with the current user utterance. The output of decoder is the set of hidden states, $H_d = \{h_1, h_2, ..., h_n\}$, where $H_d \in \mathbb{R}^{l \times 2d_h}$. n is the number of decoder states and d_h is the dimension of decoder state in each forward and backward LSTM.

3.3   Domain Module and Intent Module

Capsule Network: Motivated by the recent success of capsule network for text classification [10], we have used capsule networks for domain classification and intent detection. The first layer in capsule network is primary capsule where each capsule is a vector, which contains semantic features as well as position of words. The higher layer contains capsules equal to number of classes in each capsule network. A **Dynamic Routing Mechanism** is followed [10] where, capsules from primary capsule layer send their outputs to higher layer capsules in an unsupervised way such that higher level capsules agree with lower level capsules. The steps for sending output from a capsule in lower level say m to capsule in higher level m + 1 are as follows:

- prediction vector is calculated from the capsule say g_a in layer m to every capsule in layer, $m + 1$ as

$$p_{z|a} = W_{za}g_a \tag{3}$$

W_{za} is the capsule weight matrix.
- next step is to obtain capsule say b_z at layer m + 1, i.e.,

$$b_z = \sum_a c_{az}p_{z|a} \tag{4}$$

where, c_{az} is the coupling coefficient calculated by dynamic routing algorithm. This c_{az} represents how the capsule a at layer m and capsule z at layer $m+1$ are related.
- The initial coupling coefficient is $b_{az} \leftarrow 0$.
- b_{az} is updated using routing by algorithm as described below:

$$b_{az} \leftarrow b_{az} + p_{z|a}.l_z \tag{5}$$

here, l_z is the activation vector which is being restricted between 0 and 1 and its length shows the probability and orientation of vector representing higher capsule features. l_z is obtained by applying a non-linear squash function as follows:

$$l_z = \frac{\|b_z\|^2}{1 + \|b_z\|} \frac{b_z}{\|b_z\|^2} \tag{6}$$

- Equations 4, 5, 6, are repeated to optimize the coupling coefficient value at m + 1th layer. Capsules use separate margin loss [14] for each class at final capsule layer.

3.4   Slot Module

Linear chain Conditional Random Field (CRF) [12] is utilized for slot labeling, which takes BiLSTM embedded current utterance representation along with attained context information. It utilizes conditional probability $P(Y/X)$ for assigning a sequence of slot labels (say $\{y_1, y_2, y_3, ..., y_n\}$, where $y_i \in S$ given a user utterance which is a sequence of words $\{x_1, x_2, x_3, ..., x_n\}$. Here, n is the sequence length.

3.5   Zero-Shot Intent Module

It utilizes the information obtained from Intent Module and similarity between the seen intent classes and unseen intent classes, for example: unseen intent as "enquireflight", which is somewhat similar to the seen intent "bookflight". Whenever a new utterance is encountered by the model, it passes through semantic module followed by context module. The obtained representations are utilized by the Intent capsule and prediction vector $p_{s/a}$ is calculated with respect to seen intent class, s, $s \in I$. Next step is to obtain the coupling coefficient, c_{as}, which is calculated as per Eqs. 4, 5, 6. Then, an activation vector, a_{sa}, is calculated for a seen intent in intent capsule layer. The process could be summarized as below:

$$a_{sa} = c_{as} \cdot p_{s/a} \tag{7}$$

This activation vector is now used to get prediction vector for unseen intent as described below:

$$p_{l/a} = \sum_{s=1}^{I} q_{ls} \cdot a_{sa} \tag{8}$$

where q_{ls} is the similarity between the seen intent, $s \in I$, and unseen intent, $l \in U$, computed as follows:

$$q_{ls} = \frac{\exp{-d(e_{i_s}, e_{u_l})}}{\sum_{s=1}^{I} \exp{-d(e_{i_s}, e_{u_l})}} \tag{9}$$

Here,

$$d(e_{u_l}, e_{i_s}) = (e_{u_l}, e_{i_s})\top \sum^{-1}(e_{u_l}, e_{i_s}) \tag{10}$$

where e_{u_l} and e_{i_s} are embeddings of seen and unseen intent labels, respectively, which are obtained by passing intent labels to BERT model. $\sum$ models the correlations among embedding dimensions of intent labels. Then dynamic routing algorithm is followed [10] to send features to higher zero-shot intent capsule in the similar manner as mentioned in Domain Module and Intent Module. Finally, the output of zero-shot intent module is used to classify unseen intent.

4   Implementation Details

4.1   Dataset

We have performed our experiments on the subset of 15000 conversations (annotated) of MultiDoGo [9] dataset. The dataset consists of six different domains, namely, Airline, Fastfood, Finance, Insurance, Media and Software. For zero-shot intent detection, we have prepared test samples consisting of 5 unseen intents similar to existing intents (enquireflight, changeallocatedseat, updatemobilenumber, reportfoundcard, pauseserviceintent) with a total of only 20 conversations. The detailed data statistics are provided in Table 2.

Table 2. Dataset statistics by domain

Domain	#conversations	#turns	#sentences	#intents	#slots
Airline	2500	39616	66368	11	15
Fastfood	2500	46246	73305	14	10
Finance	2500	46001	70828	18	15
Insurance	2500	41220	67657	10	9
Media	2500	35291	65029	16	16
Software	2500	40093	70268	16	15

4.2 Experimental Setup

Hyperparameters are as follows: X = 128 (no. of hidden units in BiLSTM network), dropout rate = 0.1, Optimizer = Adam, $\lambda = 0.5$, $q^+ = 0.9$, $q^- = 0.1$, epoch_size = 6, learning rate = 0.001. The dataset is split into 70% train set, 10% development set and 20% test set.

5 Results and Analysis

A rigorous set of experiments are performed to analyze the role of context for joint domain, intent and slot tagging task along with zero shot intent detection using the proposed multitask framework. We evaluated different baselines and the proposed model in terms of accuracy as well as F_1 score. We have also compared the proposed model with different SOTA (state of the art) models to comprehend the robustness and efficacy of the proposed model.

Baselines:

I. Non Contextual and Contextual Baselines: (a) *Joint model without context*: The model is jointly trained for all the three tasks and all domains without utilizing dialogue context. **(b)** *Context-joint model (only intent):* The model utilizes a limited dialogue context (only previous user turn's intent). **(c)** *Context-joint model (only user utterance):* The model is trained with only 1 context signal, i.e., previous user utterance for all six domains combined. **(d)** *Context-joint model (intent and slots):* It utilizes much broader context compared to previous models, i.e., previous user's intent and slot tag. **(e)** *context-joint model (window size 2):* This model considers context compromising of previous 2 turns' user utterances, domains, intents and slots. **(f)** *Context-joint model (window size 3):* It utilizes significantly depth context compared to all previous models, i.e., context of window size 3.

II. Non Multitasking and Different Combinations of Multitasking Baselines: Different combinations of multitasking are also evaluated individually, keeping context as previous turn user utterance, domain, intent and slot

Table 3. Performances of different baselines and the proposed model for all three tasks (Domain Classification, Intent detection and Slot filling)

Model	DC Results		ID Results		SF Results	
	Acc.	F1-score	Acc.	F1-score	Acc.	F1-score
Joint model without context	0.5962	0.5947	0.9257	0.9241	0.9462	0.8112
Context-joint (only intent)	0.6070	0.6031	0.930	0.9296	0.9486	0.7986
Context-joint (only utterance)	0.6217	0.6296	0.9256	0.9234	0.9512	0.8152
Context-joint (intent+ slots)	0.6850	0.6845	0.9268	0.9251	0.9542	0.8256
Context-joint (window size 2)	0.9235	0.9296	0.6758	0.6536	0.3254	0.3168
Context-joint (window size 3)	0.8547	0.8541	0.5790	0.5542	0.2045	0.1846
Proposed Model	**0.9237**	**0.9262**	**0.9345**	**0.9316**	**0.9561**	**0.8400**

tags, i.e., context of window size 1. **(g)** *Context-joint DC:* This model is trained for only domain classification tasks over all six domains combined. **(h)** *Context-joint ID:* It is trained for only intent detection task over all six domains combined. **(i)** *Context-joint SF:* Context-joint SF is trained for only slot filling task over all six domains combined. **(j)** *Context-joint (DC+ID):* This model is trained for domain classification and intent detection tasks together over all six domains combined. **(k)** *Context-joint (ID+SF):* Context-joint ID and SF model is trained for intent detection and slot filling tasks together over all six domains combined.

Table 4. Performances of different baselines and the proposed model without multi-tasking and multitasking with different combinations

Model	DC Results		ID Results		SF Results	
	Acc.	F1-score	Acc.	F1-score	Acc.	F1-score
Context-joint-DC	0.9164	0.9174	–	–	–	–
Context-joint-ID	–	–	0.9200	0.9190	–	–
Context-joint-SF	–	–	–	–	0.9543	0.8237
Context-joint-(DC+ID)	0.9175	0.9180	0.9269	0.9222	–	–
Context-joint-(ID+SF)	–	–	0.9282	0.9272	0.8237	0.7113
Proposed model	**0.9237**	**0.9262**	**0.9345**	**0.9316**	**0.9561**	**0.8400**

The reported results (accuracy and F1-Score) in Table 3, 4 and 5 are the average values over five iterations. From the obtained results, it firmly establishes the efficacy of our proposed context-aware joint model, which outperforms all baselines (including non-contextual, combinations of context signals, non multi tasking models) by a significant margin.

III. Ablation Study: (1) *GloVe (without BERT LM):* This model utilizes utterance representation (concatenation of words /tokens) using the well known GloVe

Table 5. Results of ablation study, which shows performances of the proposed model w and w/o different key components

Model	DC Results		ID Results		SF Results	
	Acc.	F1-score	Acc.	F1-score	Acc.	F1-score
GloVe (without BERT LM) (proposed)	0.8874	0.8856	0.9023	0.9012	0.7828	0.7693
Without capsule network) (proposed)	0.9156	0.9189	0.9054	0.8963	0.9512	0.8358
Without attention (proposed)	0.8963	0.8952	0.9254	0.9236	0.9384	0.8182
Without CRF) (proposed)	0.9184	0.9196	0.9145	0.9121	0.9274	0.7963
Proposed model	**0.9237**	**0.9262**	**0.9345**	**0.9316**	**0.9561**	**0.8400**

embedding technique. (**m**) *Without Capsule network:* For DC and ID tasks, the capsule layer is replaced with softmax function (Fig. 2). (**n**) *Without attention:* The model does not consider attention layer after getting the BERT representation of context from context module (Fig. 2, bottom left). (**o**) *Without CRF:* We have kept everything as it is, but instead of CRF in Slot Module, we have used softmax function (Fig. 2, top left).

Comparison with State-of-the-Art Models: We have performed different experiments for each domain individually for a fair comparison with the state-of-the-art models. **NC(non contextual)-(ID+SF)**: The NC-(ID+SF) model is trained without considering any context signal. **C(contextual)-ID**: The C-ID is trained only for ID task. **C(contextual)-SF**: The Context-SF is trained only for SF task. **C(contextual)-(ID+SF)(proposed)**: This model is trained for both the tasks of ID and SF together. The reported results in Table 6 clearly demonstrate that the proposed model outperforms all the single tasks and non-contextual models.

Table 6. Performance of non contextual and contextual models for different domains.–/– represents acc./F1 metrics

Model	Airline		Fastfood		Finance	
	ID	SF	ID	SF	ID	SF
NC(ID+SF)	0.905/0.900	0.970/0.865	0.890/0.889	0.899/0.782	0.934/0.932	0.956/0.804
C(ID)	0.904/0.903	–	0.885/0.883	–	0.945/0.944	–
C(SF)	-	0.964/0.849	–	0.902/0.789	–	0.966/0.817
C(ID+SF)	**0.909/0.907**	**0.979/0.890**	**0.896/ 0.895**	**0.905/0.792**	**0.949/0.949**	**0.967/0.819**
Model	Insurance		Media		Software	
	ID	SF	ID	SF	ID	SF
NC(ID+SF)	0.946/0.942	0.965/0.882	0.944/0.941	0.952/0.883	0.901/0.900	0.951/0.806
C(ID)	0.949/0.948	–	0.943/0.942	–	0.901/0.901	–
C(SF)	–	0.974/0.897	–	0.962/0.892	–	0.955/0.798
C(ID+SF)	**0.959/0.959**	**0.978/0.901**	**0.951/0.951**	**0.968/0.894**	**0.907/0.905**	**0.956/0.816**

Table 7 shows and compares the performances of different state-of-the-art works and the proposed model. The proposed model outperforms the state of the art model in all domains for both the tasks, i.e., intent detection and slot filling.

Table 7. Comparisons with SOTA at turn level. −/− represents acc./F1 metrics

Model	DC Results	ID Results	SF Results
Arline-MFC [9]	–	–/32.79	–/37.73
Airline-LSTM [9]	–	–/89.15	–/75.78
Airline-ELMO[9]	–	–/89.99	–/85.64
Airline-(proposed)	–	**90.85/90.73**	97.91/89.00
Fastfood-MFC [9]	–	–/25.33	–/61.84
Fastfood-LSTM [9]	–	–/87.35	–/73.57
Fastfood-ELMO [9]	–	–/88.96	–/79.63
Fastfood-(proposed)	–	**89.62/89.45**	90.54/79.20
Finance-MFC [9]	–	–/38.16	–/34.31
Finance-LSTM [9]	–	–/ 92.30	–/70.92
Finance-ELMO[9]	–	–/94.50	–/79.47
Finance-(proposed)	–	**94.91/94.89**	96.67/81.89
Insurance-MFC [9]	–	–/39.42	–/54.66
Insurance-LSTM [9]	–	–/94.75	–/76.84
Insurance-ELMO [9]	–	–/95.39	–/89.51
Insurance-(proposed)	–	**95.91/95.89**	97.77/90.06
Media-MFC [9]	–	–/31.82	–/78.83
Media-LSTM [9]	–	–/94.35	–/87.33
Media-ELMO [9]	–	–/94.76	–/91.48
Media-(proposed)	–	**95.08/95.04**	96.79/89.41
Software-MFC [9]	–	–/33.78	–/54.84
Software-LSTM [9]	–	–/89.78	–/72.34
Software-ELMO [9]	–	–/90.85	–/76.48
Software-(proposed)	–	**90.73/90.48**	95.63/81.59
Proposed-Airline	**92.96/96.30**	**90.85/90.73**	**97.91/89.00**
Proposed-Fastfood	**95.83/95.82**	**90.80/89.50**	**90.61/80.40**
Proposed-Finance	**94.15/96.38**	**95.24/94.65**	**96.67/81.89**
Proposed-Insurance	**91.35/95.40**	**96.12/96.10**	**97.78/90.06**
Proposed-Media	**91.75/93.37**	**95.17/95.25**	**96.80/86.45**
Proposed-Software	**88.89/94.12**	**91.70/91.46**	**95.12/81.85**

Zero-Shot Intent Detection Setting: To test the effectiveness of our model for unseen intents, we have experimented the model with conversations having some unseen intents, i.e., such user utterances are not being seen by model during training time. We have experimented with the following baselines: **1. Without zero-shot intent module:** The model is trained with seen intents and tested on conversations with unseen intent only. **2. Without context zero-shot intent detection setting:** The model does not utilize any of the context signals, considering all tasks together for all six domains.

3. Context zero-shot intent detection setting: The proposed model is tested on conversations with unseen intent only considering all tasks together.

Table 8 illustrates the performance of the proposed model for unseen intents. As expected if we remove zero-shot intent module and use the proposed intent module for prediction of unseen intents, its accuracy is 0.0 as the model is trained only for seen intents. Table 8 also shows that contextual model performs better than non-contextual model in zero-shot intent detection setting as well.

Error Analysis: We carried out an in-depth analysis to examine the probable reasons for failed situations. The key findings are as follows:

Table 8. Performances of the proposed model for unseen intents

Model	DC Results		ID Results		SF Results	
	Acc.	F1-score	Acc.	F1-score	Acc.	F1-score
Without zero-shot intent module	0.9013	0.9043	0.0	0.0	0.9167	0.7965
Without context zero-shot intent setting	0.5871	0.5865	0.6645	0.6523	0.9352	0.8038
Context zero-shot intent setting (proposed)	0.9216	0.9240	**0.6699**	**0.6689**	0.9486	0.8105

Table 9. Sample conversations with the predicted and true labels for the proposed model

S No.	Turn	Context	User-Utterance	Domain (True/Predict)	Intent (True/Predict)	Slot tags (True/Predict)
1	1	NULL	i need 1 veg pizza	fastfood/fastfood	orderpizzaintent/ orderpizzaintent	B-food_item I-food_item/O O
	2	Turn 1	i change my mind i like to order nonveg pizza	fastfood/fastfood	orderpizzaintent/ orderpizzaintent	✓
2	1	NULL	i want to order roasted red peppers	fastfood/fastfood	ordersideintent/ contentonly	✓
	2	Turn 1	i like to add arugula fresh fruit salad	fastfood/fastfood	ordersaladintent/ ordersaladintent	B-Ingredient I-ingredient I-ingredient B-food_item /B-food_item O I-food_item I-food_item

- **Similar utterances in different domains:** One of the possible reasons for wrong domain prediction is to have similar utterances in different domains such as "thank you", "Hello" etc. which are present in every domain.
- **Homogeneous intents:** In case of intent detection task, intents which are semantically similar contribute to wrong intent prediction. For example: "ExpenseReport" in "Software" domain is semantically similar to "ViewBillsIntent" in Media, both inquire similar query in different domain.
- **Homogeneous intents:** Similarly, in slot tagging, similar tags for example "Food_item" and "Ingredient" in "Fastfood" create confusion.
- **Imbalance distribution** of intents and slots at conversation level split as mentioned in [9]. Table 9 illustrates few sample conversations for error analysis.

6   Conclusion and Future Work

In this work, we have proposed a novel and unified context-aware multitask model for domain classification, intent detection, and slot filling tasks utilizing various context signals such as previous user utterance, domain, intent, and slot labels, which help to predict domain, intent, and slots more accurately. Additionally, we have incorporated a zero-shot intent detection module for dealing with new/unseen users' query/intent (emerging intent). The obtained experimental results on the bench-marked dataset, MultiDoGo, outperform several baselines and the current state-of-the-art models, which firmly establish the need and efficacy of the proposed system. In future, we would like to develop a context-aware multimodal-multilingual framework for joint domain classification, intent detection, and slot labeling.

Acknowledgement. Dr. Sriparna Saha gratefully acknowledges the Young Faculty Research Fellowship (YFRF) Award, supported by Visvesvaraya Ph.D. Scheme for Electronics and IT, Ministry of Electronics and Information Technology (MeitY), Government of India, being implemented by Digital India Corporation (formerly Media Lab Asia) for carrying out this research.

References

1. Chen, Q., Zhuo, Z., Wang, W.: Bert for joint intent classification and slot filling. arXiv preprint arXiv:1902.10909 (2019)
2. Chen, S., Wais, S.Y.: Word attention for joint intent detection and slot filling. In: Proceedings of the AAAI Conference on Artificial Intelligence, vol. 33, pp. 9927–9928 (2019)
3. Devlin, J., Chang, M., Lee, K., Toutanova, K.: BERT: pre-training of deep bidirectional transformers for language understanding. In: Proceedings of the 2019 Conference of the North American Chapter of the Association for Computational Linguistics: Human Language Technologies, NAACL-HLT 2019, Minneapolis, MN, USA, 2–7 June 2019, Volume 1 (Long and Short Papers), pp. 4171–4186 (2019)

4. Gupta, A., Zhang, P., Lalwani, G., Diab, M.: CASA-NLU: context-aware self-attentive natural language understanding for task-oriented chatbots. In: Proceedings of the 2019 Conference on Empirical Methods in Natural Language Processing and the 9th International Joint Conference on Natural Language Processing (EMNLP-IJCNLP), Hong Kong, China. Association for Computational Linguistics, November 2019
5. Kim, Y.-B., Lee, S., Stratos, K.: OneNet: joint domain, intent, slot prediction for spoken language understanding. In: 2017 IEEE Automatic Speech Recognition and Understanding Workshop (ASRU), pp. 547–553. IEEE (2017)
6. Kim, Y.-B., Stratos, K., Sarikaya, R.: Scalable semi-supervised query classification using matrix sketching. In: Proceedings of the 54th Annual Meeting of the Association for Computational Linguistics (Volume 2: Short Papers), pp. 8–13 (2016)
7. Liu, B., Lane, I.: Attention-based recurrent neural network models for joint intent detection and slot filling. In: Morgan, N. (eds.) Interspeech 2016, 17th Annual Conference of the International Speech Communication Association, San Francisco, CA, USA, 8–12 September 2016, pp. 685–689. ISCA (2016)
8. Luong, T., Pham, H., Manning, C.D.: Effective approaches to attention-based neural machine translation. In: Proceedings of the 2015 Conference on Empirical Methods in Natural Language Processing, Lisbon, Portugal, pp. 1412–1421. Association for Computational Linguistics, September 2015
9. Peskov, D., et al.: Multi-domain goal-oriented dialogues (MultiDoGO): strategies toward curating and annotating large scale dialogue data. In: Proceedings of the 2019 Conference on Empirical Methods in Natural Language Processing and the 9th International Joint Conference on Natural Language Processing (EMNLP-IJCNLP), Hong Kong, China. Association for Computational Linguistics, November 2019
10. Sabour, S., Frosst, N., Hinton, G.E.: Dynamic routing between capsules. In: Guyon, I., et al. (eds.) Advances in Neural Information Processing Systems, vol. 30. Curran Associates Inc. (2017)
11. Sarikaya, R., et al.: An overview of end-to-end language understanding and dialog management for personal digital assistants. In: 2016 IEEE Spoken Language Technology Workshop (SLT), pp. 391–397. IEEE (2016)
12. Sutton, C., McCallum, A.: An introduction to conditional random fields (2010)
13. Xia, C., Zhang, C., Yan, X., Chang, Y., Yu, P.: Zero-shot user intent detection via capsule neural networks. In: Proceedings of the 2018 Conference on Empirical Methods in Natural Language Processing, Brussels, Belgium. Association for Computational Linguistics, October–November 2018
14. Xiao, L., Zhang, H., Chen, W., Wang, Y., Jin, Y.: MCapsNet: capsule network for text with multi-task learning. In: Proceedings of the 2018 Conference on Empirical Methods in Natural Language Processing, pp. 4565–4574 (2018)

Constrained Generative Model
for EEG Signals Generation

Te Guo[1,2,3,4], Lufan Zhang[1,3], Rui Ding[1], Da Zhang[1,2,3,4], Jinhong Ding[5],
Ming Ma[6], and Likun Xia[1,2,3,4](✉)

[1] College of Information Engineering, Capital Normal University,
Beijing 100048, China
xlk@cnu.edu.cn
[2] International Science and Technology Cooperation Base of Electronic System
Reliability and Mathematical Interdisciplinary, Capital Normal University,
Beijing 100048, China
[3] Laboratory of Neural Computing and Intelligent Perception, Capital Normal
University, Beijing 100048, China
[4] Beijing Advanced Innovation Center for Imaging Theory and Technology,
Capital Normal University, Beijing 100048, China
[5] School of Psychology, Capital Normal University,
Beijing 100048, China
[6] Department of Computer Science, Winona State University,
Winona, MN 55987, USA

Abstract. Electroencephalogram (EEG) is one of the most promising
modalities in the field of Brain-Computer Interfaces (BCIs) due to its
high time-domain resolutions and abundant physiological information.
Quality of EEG signal analysis depends on the number of human subjects. However, due to lengthy preparation time and experiments, it is
difficult to obtain sufficient human subjects for experiments. One of possible approaches is to employ generative model for EEG signal generation. Unfortunately, existing generative frameworks face issues of insufficient diversity and poor similarity, which may result in low quality of
generative EEG. To address the issues above, we propose R^2WaveGAN,
a WaveGAN based model with constraints using two correlated regularizers. In details, inspired by WaveGAN that can process time-series
signals, we adopt it to fit EEG dataset and then integrate the spectral regularizer and anti-collapse regularizer to minimize the issues of
insufficient diversity and poor similarity, respectively, so as to improve
generalization of R^2WaveGAN. The proposed model is evaluated on one
publicly available dataset - Bi2015a. An ablation study is performed to
validate the effectiveness of both regularizers. Compared to the state-
of-the-art models, R^2WaveGAN can provide better results in terms of
evaluation metrics.

Keywords: EEG · Generative model · WaveGAN · Spectral
regularizer · Anti-collapse regularizer

T. Guo and L. Zhang—Contributed equally to the work.

T. Mantoro et al. (Eds.): ICONIP 2021, LNCS 13110, pp. 596–607, 2021.
https://doi.org/10.1007/978-3-030-92238-2_49

1   Introduction

Electroencephalogram (EEG) reflects the functional state of brain. Effectively extracting the information in EEG enables a deep understanding of its functional activities and provides the base for evaluating the psychological cognitive status of human beings. In recent years, deep neural network (DNN) has shown great potential in EEG analysis because of its remarkable feature extraction ability. However, such network requires a large number of samples at training stage; otherwise they are apt to overfit. On the other hand, since experiments for EEG acquisition are generally complicated and EEG is susceptible to noise, it is difficult to obtain a large number of available EEG.

One of the solutions is to perform EEG signal generation with machine learning based techniques such as autoregressive model [1], variational autoencoding Bayes [2], generative adversarial networks (GANs) [3], and flow-based model [4]. The former two techniques can relatively easily generate single sample during training process since they lack guidance of discriminant model; whereas the flow-based model has not been widely used due to the huge computational resources required for training. Comparably, GANs are favorable for EEG generation due to its relatively simple implementation and promising results in a range of domains. However, the original GAN built with fully connected (FC) layers has difficulty to capture the complex relationship between the characteristics and attributes of time-series signal [5], leading to the easy loss of the time correlation information when processing EEG signal. Such issues can be solved with the extensive techniques of GANs.

To preserve time correlation information in EEG signal, Abdelfattah et al. [6] proposed the Recurrent GAN (RGAN) to generate multi-channel EEG and utilized the mean square error (MSE) to classify EEG, showing good classification results. However, Recurrent Neural Network (RNN) in RGAN had problems such as gradient explosion, which resulted in instability to the generative model. To address the issue, Luo et al. [7] added auxiliary control conditions to the input of Conditional Wasserstein GAN (CWGAN) to improve the stability during training. But likes RGAN, because GAN was trained independently of the classification task, there was no guarantee that the generative EEG would always improve the classification. In contrast, one key advantage of Class-Conditioned WGAN-GP (CC-WGAN-GP) proposed by Panwar et al. [8] was that EEG generation and classification were trained together with a unified loss, so as to optimize generative EEG and improve classification performance. However, the evaluation of such EEG was only associated with classification results, rather than focused on their time-frequency characteristics. The problem was solved in motor imagery Electroencephalography-GAN (MIEEG-GAN) [9], which was composed of Bidirectional Long Short-Term Memory neurons (Bi-LSTM). It compared the time-frequency characteristics of real and generative EEG by using short-time Fourier transform (STFT) and Welch's power spectral density (PSD). However, the model did not capture variations of amplitude in the real EEG well. Furthermore, most analyses were based on visual inspection and lack quantitative evaluation.

The above methods have not completely solved the problems of lack of diversity and poor similarity of generative EEG. In this paper, we propose a generative model based on WaveGAN [10] with two correlated regularizers for EEG signal generation, namely R^2WaveGAN. The spectral regularizer (SR) [11] is used to improve the similarity of generative EEG, and the anti-collapse regularizer (AR) [12] is applied to ensure the diversity of generative EEG. It is worth mentioning that both regularizers are correlated using a weight in order to balance the similarity and diversity of the generative EEG.

The contributions of our work can be summarized as follows: (1) we present a WaveGAN based generative model for EEG signal generation; (2) two correlated regularizers acting as constraints are integrated into the generative model to improve its performance, focusing on diversity and similarity, respectively; (3) correlation between the regularizers is realized using a user defined weight to balance the similarity and diversity to fit various situations; (4) we have claimed that each of the regularizers contributes to the model, and also demonstrated that the proposed model outperforms the state-of-the-art models.

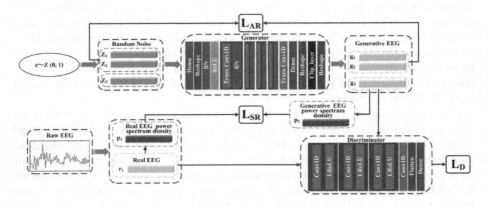

Fig. 1. The schematic diagram of R^2WaveGAN model

2   Methodology

In this section, we detail the proposed R^2WaveGAN as illustrated in Fig. 1. It includes two parts: generator and discriminator. The former takes the combination of AR and SR as the loss function, while the latter uses GAN loss for confrontation training.

Raw EEG is preprocessed to obtain the real EEG using phases of downsampling and min-max normalization, and then p_1 (the PSD of the real EEG) is obtained by the discrete Fourier transform (DFT). The discriminator discriminates between real EEG and generative EEG, so the GAN loss L_D is obtained.

The generator synthesizes three generative EEG g_1, g_2 and g_3 with three random noise z_1, z_2, and z_3. By analyzing the relationship between (z_1, z_2) and (g_1, g_2), we achieve the AR loss L_{AR}. And we obtain SR loss L_{SR} by calculating the relationship between p_1 and p_2 (the PSD of the generative EEG). We now elaborate our methodology in the rest of the subsections.

2.1   WaveGAN

WaveGAN is a popular generative model capable of synthesizing raw audio. It is successful in producing audio from different domains such as speech and musical instruments. Both the generator and the discriminator are similar to DCGAN [13], but differently, WaveGAN requires one-dimensional (1D) filters with length of 25 instead of two-dimensional (2D) filters with the size of 5×5 in DCGAN. WaveGAN uses the Wasserstein loss with gradient penalty (WGAN-GP) to optimize training. In general, WGAN-GP not only improves the training speed, but also ensures that the gradient update is within a controllable range, so as to reduce the possibility of gradient explosion in WGAN. The loss function of the WaveGAN generator is shown in Eq. (1):

$$L_G = -E_{x \sim P_g}[D(x)] \tag{1}$$

The loss function of the WaveGAN discriminator is given in Eq. (2):

$$L_D = E_{x \sim P_g}[D(x)] - E_{x \sim P_r}[D(x)] + \lambda E_{x \sim \mathbb{P}_x}\left[\left\|\nabla_x D(x)\right\|_p - 1\right]^2 \tag{2}$$

where $E_{x \sim P_r}$ and $E_{x \sim P_g}$ represent the probability of real sample and generative sample, respectively. λ is constant, and $P_{\hat{x}}$ is defined by sampling uniformly along straight lines, which between pairs of points sampled from the real sample distribution P_r and the generative sample distribution P_g.

Generator. It includes eighteen layers. We provide random noise with the same length as the real EEG, i.e., $l = 231$. A dense layer is then added, followed by a reshape layer to transform random noise dimensions 12×20. Subsequently, four trans Conv1D (converted by 2D convolution in DCGAN) layers are used to obtain the EEG with dimension 12×64, and the fifth layer Trans Conv1D increases the dimension to 240×1 in order to obtain the EEG features of 231 length completely. Finally, a clip layer and a reshape layer are introduced to ensure the dimension satisfies the input dimension of the discriminator. Table 1 details parameters of the generator.

Discriminator. It includes twelve layers. The first layer is the input layer with EEG of length 231, followed by four Conv1D with leaky ReLU (LReLU) activation function ($\alpha = 0.2$), which realizes the feature extraction operation of EEG. The fifth layer of Conv1D adjusts the EEG dimension to $1 \times 12 \times 1024$ ($s \times$

2^{n-1}, where s indicates inputsize, $s = 64$ in the case, n indicates number of layers and $n = 5$.), followed by a flatten layer that transforms multidimensional input into one dimensional array, obtains EEG with the size of 12288 ($1 \times 12 \times 1024$). Finally, a dense layer with one output neuron. Table 1 highlights the parameters corresponding to the output size of the discriminator.

Table 1. The architecture of generator and discriminator (B = Batch Size)

Generator		Discriminator	
Layer	Output Shape	Layer	Output Shape
Input layer	[B, 231]	Input layer	[B, 231, 1]
Dense	[B, 240]	Conv1D	[B, 231, 64]
Reshape	[B, 12, 20]	LReLU	[B, 231, 64]
Batch Normalization	[B, 12, 20]	Conv1D	[B, 231, 128]
ReLU	[B, 12, 20]	LReLU	[B, 231, 128]
Trans Conv1D	[B, 12, 512]	Conv1D	[B, 231, 256]
Batch Normalization	[B, 12, 512]	LReLU	[B, 231, 256]
Trans Conv1D	[B, 12, 256]	Conv1D	[B, 231, 512]
Batch Normalization	[B, 12, 256]	LReLU	[B, 231, 512]
Trans Conv1D	[B, 12, 128]	Conv1D	[B, 1, 12, 1024]
Batch Normalization	[B, 12, 128]	Flatten	[B, 12288]
Trans Conv1D	[B, 12, 64]	Dense	[B, 1]
Batch Normalization	[B, 12, 64]		
Trans Conv1D	[B, 240, 1]		
Dense	[B, 240, 1]		
Reshape	[B, 1, 240, 1]		
Clip layer	[B, 1, 231, 1]		
Reshape	[B, 231, 1]		

2.2   SR and AR

One of the major challenges of signal generation is model collapse, which forces the generator to generate identical samples repeatedly in some cases. Moreover, the convolutional neural network (CNN) in GAN can cause models incorrectly to reproduce the spectrum distribution of real samples. Therefore, we integrate SR and AR into WaveGAN to ensure the diversity and the similarity of generative EEG signal.

SR. It improves the similarity of generative EEG signal by calculating the ratio of PSD between the real and generative ones. In order to obtain the PSD of both EEG signals, we initially calculate theirs DFT, which discretizes the signals into the frequency distribution, as illustrated in Eq. (3). This is because computers can only store and process data with limited number of bits.

$$F[k] = \sum_{n=0}^{N-1} \mathrm{x}[n]e^{-\mathrm{j}2\pi\kappa n/\mathrm{N}} \tag{3}$$

Where $x[n]$ presents the $(n+1)^{th}$ discrete input signal, $n = 0, 1, ..., N-1$, and the number of sampling points is N; $F[k]$ denotes the $(k+1)^{th}$ DFT value, $k = 0, 1, ..., N-1$; j indicates the imaginary axis. Based on DFT, we then calculate PSD for the signals and apply them to spectral regularizer as shown in Eq. (4).

$$L_{SR} = -\frac{1}{(N/2-1)} \sum_{k=0}^{N/2-1} F_k^{real} \cdot \log\left(F_k^{out}\right) + \left(1 - F_k^{real}\right) \cdot \log\left(1 - F_k^{out}\right) \tag{4}$$

Where L_{SR} is the binary cross entropy between the output F_k^{out} of the generative EEG and the average F_k^{real} obtained from the real EEG. SR minimizes the discrepancies between two EEG signals by measuring their probability distribution in frequency domain, that is, to improve the similarity of the generative EEG.

AR. It improves the diversity of generative EEG signal by calculating the ratio of input noise and generative EEG, as illustrated in Eq. (5), where z_1 and z_2 are random noise, and g_1 and g_2 are corresponding generative EEG signals.

$$L_{AR} = E_{\mathrm{x}\sim p_g} \left[\frac{1 - \cos\left(g_1, g_2\right)}{1 - \cos\left(z_1, z_2\right)} \right] \tag{5}$$

It is seen that the numerator represents the similarity between two generative EEG g_1 and g_2, while the denominator represents the similarity between two corresponding noise vectors z_1 and z_2. Some studies have shown that as z_1 and z_2 are more similar, i.e., the denominator becomes smaller, g_1 and g_2 are more likely to fall into the same pattern [12], in other words, the above formula is equivalent to amplification of the dissimilarity between g_1 and g_2, thus enhancing the diversity of the generative EEG signals.

It is worth mentioning that the performance of the model is dependent on the correlation of two regularizers. Therefore, it is necessary to define a correlation parameter that can assign suitable weights to the regularizers, so as to maximize the performance of the model. This can be realized by observing the overall loss function of the generator in our proposed model, as described in Eq. (6), where is the correlation weight parameter, ranging between 0 and 1.

$$L_{G_{\mathrm{frat}}} = L_G + \lambda L_{SR} + (1 - \lambda)\frac{1}{L_{AR}} \tag{6}$$

3 Experiment and Results

R^2WaveGAN is implemented on Keras with single GPU (Tesla P100). Adaptive moment estimation (Adam) is employed for network optimization. The initial learning rate is set to $4e^{-4}$ with a weight decay of 10, and the maximum epoch is set to 3000. The batch size is set to 128 during training.

Dataset. A publicly available dataset - Bi2015a is used for visual P300 events. EEG is recorded by 32 channels at a sampling rate 512 Hz. It contains three types of EEG recordings by setting different flicker durations of 50 healthy subjects playing to a visual P300 Brain-Computer Interface (BCI) video game named "Brain Invaders" and 9 of whom are excluded from the study due to special circumstances [14].

Preprocessing. After preliminary screening, we firstly delete 12 subjects with corrupted channels; and then compare the P300 characteristics of 32 channels. It is found that the P300 characteristics of C3 channel are the most obvious. Therefore, this paper uses raw EEG signal from C3 channel with 29 subjects to conduct the experiment. In the preprocessing stage, the raw EEG signal is sampled down 256 Hz to increase computational speed. Finally, in order to retain the P300 characteristics (300 ms), we intercept the EEG signal from 100 ms before the occurrence of the event to 500 ms after the occurrence of the event, a total of 900 ms, and obtain a real EEG signal. Its length can be calculated using Eq. (7), where l is the length of the segmented signal, f presents the sampling rate of EEG, and t indicates the epoch time (time of interest), $t = 900$ ms in the case.

$$l = f \times t \tag{7}$$

The dimension of the signals is $l \times n$, where n indicates the number of channels, $n = 1$ in the case, and $l = 231$.

It is noted that the batch normalization (BN) layer causes the generator output mapped to the range $[0, 1]$, which leads to difference of range between generative sample and real sample [15]. Therefore, the range of EEG is mapped to $[-1v,1v]$ by using min-max normalization method to ensure that the amplitude of generative EEG is within the same range as that of real EEG; moreover, the convergence speed can be improved [16]. The min-max normalization method is shown in Eq. (8):

$$y = \frac{(x - x_{\text{mean}})}{x_{\text{max}} - x_{\text{min}}} \tag{8}$$

where, x_{mean} denotes the mean of the data, x_{min} is the minimum value in the data, x_{max} presents the maximum value in the data and y is the normalized data.

Evaluation Metrics. To measure the performance of different models, we observe two perspectives, including quantitative and qualitative analysis.

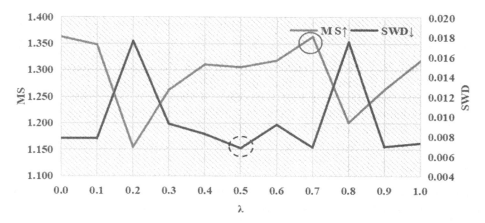

Fig. 2. Impact of weight parameter λ. X axis represents the value of λ, $\lambda \in [0,1]$; Y principal axis represents the evaluation result of MS; Y secondary axis represents the evaluation result of SWD.

Quantitative analysis we apply Mode Score (MS) [17] and Sliced Wasserstein distance (SWD) [18] methods to analyze generative samples. MS is a method to evaluate diversity by calculating the cross entropy between real and generative samples label distribution. Based on this, a higher MS score indicates better diversity. SWD measures the similarity by estimating Wasserstein-1 distance between real and generative samples [19]. As the result, a lower SWD score indicates greater similarity.

Qualitative analysis to be fidelity, we plot STFT to observe characteristics of generative samples inherited from real EEG, especially visual P300. And we utilize EEG waveform and Welch's PSD of the signals, aiming to visualize the degree to which the generative EEG distribution closely correlates with the real EEG distribution.

3.1   Selection of Weight Parameter

To select the most suitable value of λ, we accordingly calculate two evaluation matrices MS and SWD with respect to λ, and then plot their difference with respect to λ. It is discussed above (subsection for evaluation metrics) that the larger the value of MS is, the better the performance will be; the smaller the value of SWD is, the better the performance will be, so the selection of the weight is based on the maximum distance between two values, which is described in Eq. (9).

$$Dis_\lambda = MS_\lambda - SWD_\lambda, \quad \lambda \in [0, 1]$$
$$\lambda = \arg\max(Dis_\lambda)$$

$$(9)$$

After we plot the values as illustrated in Fig. 2, we find that MS at $\lambda = 0.7$ has the highest value highlighted in red circle, i.e., MS $= 1.3637$; for SWD, the lowest value that can be found is at $\lambda = 0.5$ highlighted in blue circle (broken line), i.e., SWD $= 0.0068$. Since we want the largest distance between them, Dis_λ, a number of calculations are performed according to Eq (9), so at $\lambda = 0.5$, $Dis_{\lambda,\lambda=0.5} = 1.3057 - 0.0068 = 1.2989$; at $\lambda = 0.7$, $Dis_{\lambda,\lambda=0.7} = 1.3637 - 0.0069 = 1.3568$, i.e., $1.3568 > 1.2989$. It is concluded that the distance at $\lambda = 0.7$ is the largest, so $\lambda = 0.7$ in the case.

3.2   Ablation Study

Quantitative Analysis. As shown in Table 2, we illustrate the effectiveness of each regularizer in R^2WaveGAN. On the one hand, as shown in the first and third rows of the table, compared with baseline, SR is useful to reduce SWD from 0.0085 to 0.0073. SR is effective in achieving progress of similarity by constraining spectral consistency to diminish difference between real EEG and generative EEG. When we use AR, SWD (0.0078) declines significantly less than with SR. On the other hand, as shown in the first and second rows of the table, AR increases MS from 1.3266 up to 1.3628, indicating that AR promotes diversity of generative EEG. This is because AR amplifying the gap between the generative EEG. Finally, as highlighted in the bottom row, by combining SR with AR, both similarity and diversity are much improved than single regularizer due to balancing interaction between them.

Table 2. Ablation study for two regularizers in WaveGAN.

SR	AR	MS↑	SWD↓
−	−	1.3266	0.0085
−	√	1.3628	0.0078
√	−	1.3180	0.0073
√	√	**1.3637**	**0.0069**

Qualitative Analysis. To further prove that the generative EEG of R^2WaveGAN is sufficient fidelity, we conduct STFT to investigate visual P300 features learned from real EEG as shown in Fig. 3 and 4. It is seen that, a brighter area means a more energetic signal interval, which suggests the presence of the visual P300 features. As highlighted by the red circles in Fig. 3, visual P300 features are observed in the time range of 250 ms–550 ms. In Fig. 4, they are in the range of 100 ms–400 ms, and their positions are approximately same as in Fig. 3. These results indicate that R^2WaveGAN can capture similar visual P300 features to real EEG, and signify that it has the ability to generate real-like EEG signals.

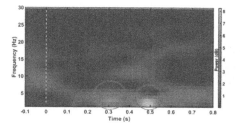 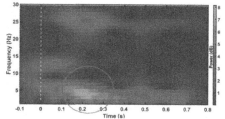

Fig. 3. Example of STFT generated image for real EEG.

Fig. 4. Example of STFT generated image for generative EEG.

3.3   Comparison with State-of-the-Art Model

Quantitative Analysis. In comparison with CC-WGAN-GP, we test the performance of both models by MS and SWD on Bi2015a dataset. As shown in Table 3, R^2WaveGAN increases the MS value from 1.0779 to 1.3637 and reduces the SWD to 0.0069. They indicate that R^2WaveGAN stands out by virtue of excellent similarity and plentiful diversity. This result is because WaveGAN captures time domain information, and SR improves spectrum characteristics from traditional upsampling methods to promote similarity. In addition, AR amplifies the distance of EEG to increase diversity.

Table 3. Performances with two models on Bi2015a dataset.

Model	Evaluation metrics	
	MS↑	SWD↓
CC-WGAN-GP	1.0779	0.2046
R^2WaveGAN	**1.3637**	**0.0069**

Qualitative Analysis. We compared the visual performance of R^2WaveGAN with CC-WGAN-GP to demonstrate the advantages of our model. As presented in Fig. 5, the waveform of the EEG signal visualized how similar the distribution of the generative EEG is to the real EEG. Generally, amplitude of EEG signal generated by R^2WaveGAN is approximate to real EEG and higher than CC-WGAN-GP. Particularly, as indicated by the yellow circles in the figure, in the range of where visual P300 features appear, CC-WGAN-GP waveform has a problem with amplitude distortion. Moreover, for Welch's PSD analysis as described in Fig. 6, all of three signals are close to each other. After 60 Hz, the curve of PSD by R^2WaveGAN is almost coincident to real EEG, whereas the power of CC-WGAN-GP shows obvious bias in the yellow circle. This implies

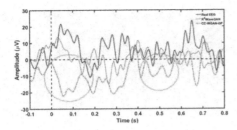
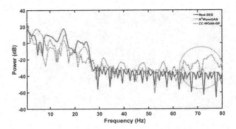

Fig. 5. EEG signal of real EEG and generative EEG.

Fig. 6. Welch's PSD of real EEG and generative EEG.

that R^2WaveGAN has the superiority in capturing underlying spectrum features. As described above, the proposed R^2WaveGAN performs better than CC-WGAN-GP on similarity.

4 Conclusion

In this paper, we present a WaveGAN based generative model with two correlated regularizers termed R^2WaveGAN for generating high quality EEG signal. The correlation is realized by a redistributed parameter to balance similarity and diversity. An ablation study is performed from both quantitative and qualitative perspectives, which demonstrates that each of the added components is effective to the proposed model. Compared with the state-of-the-art model CC-WGAN-GP on a publicly available dataset Bi2015a, R^2WaveGAN has achieved better performance in terms of similarity and diversity. It is believed that our proposed model enables to overcome small sample size issue and long EEG acquisition process in BCI applications.

References

1. Xu, F., Ren, H.: A linear and nonlinear auto-regressive model and its application in modeling and forecasting. In: Dongnan Daxue Xuebao (Ziran Kexue Ban)/J. Southeast Univ. (Nat. Sci. Ed.) **43**(3), 509–514 (2013)
2. Diederik, P.K., Welling, M.: Auto-encodingVariational Bayes. In: 2nd International Conference on Learning Representations, Banff, Canada, pp. 1–14 (2014)
3. Goodfellow, I.J., et al.: Generative adversarial nets. In: 28th Annual Conference on Neural Information Processing Systems, Montréal, Canada, pp. 2672–2680. MIT Press (2014)
4. Kingma, D.P., Dhariwal, P.: Glow: generative flow with invertible 1x1 convolutions. In: 24th Annual Conference on Neural Information Processing Systems, Vancouver, Canada, pp. 10236–10245. MIT Press (2010)
5. Yu, L.T., Zhang, W.N., Wang, J., Yong, Y.: SeqGAN: sequence generative adversarial nets with policy gradients. In: 31st Conference on Artificial Intelligence, San Francisco, USA, pp. 2852–2858. AAAI Press (2017)

6. Abdelfattah, S., Abdelrahman, M., Wang, M.: Augmenting the size of EEG datasets using generative adversarial networks. In: 31st International Joint Conference on Neural Networks, Rio de Janeiro, Brazil, pp. 1–6. IEEE Press (2018)
7. Luo, Y., Lu, B.: EEG data augmentation for emotion recognition using a conditional Wasserstein GAN. In: 40th Annual International Conference of the IEEE Engineering in Medicine and Biology Society, Honolulu, USA, pp. 2535–2538. IEEE Press (2018)
8. Panwar, S., Rad, P., Jung, T., Huang, Y.: Modeling EEG data distribution with a Wasserstein Generative Adversarial Network to predict RSVP Events. IEEE Trans. Neural Syst. Rehabil. Eng. **1**(1), 99 (2020)
9. Roy, S., Dora, D., Mccreadie, K.: MIEEG-GAN: generating artificial motor imagery electroencephalography signals. In: 33th International Joint Conference on Neural Networks (IJCNN), Glasgow, UK, pp. 1–8. IEEE Press (2020)
10. Donahue, C., Mcauley, J., Puckette, M.: Adversarial audio synthesis. In: 6th International Conference on Learning Representations, Vancouver, Canada, pp. 1–8. OpenReview.net Press (2018)
11. Durall, R., Keuper, M., Keuper, J.: Watch your up-convolution: CNN based generative deep neural networks are failing to reproduce spectral distributions. In: 38th Computer Vision and Pattern Recognition, pp. 1–10. IEEE Press (2020)
12. Li, K., Zhang, Y., Li, K., Fu, Y.: Adversarial feature hallucination networks for few-shot learning. In: 38th Computer Vision and Pattern Recognition, pp. 1–8. IEEE Press (2020)
13. Donahue, C., Mcauley, J., Puckette, M.: Adversarial audio synthesis. In: International Conference on Learning Representations, Vancouver, Canada, pp. 1–16. OpenReview.net Press (2018)
14. Korczowski, L., Cederhout, M., Andreev, A., Cattan, G., Congedo, M.: Brain Invaders calibration-less P300-based BCI with modulation of flash duration Dataset (bi2015a) (2019). https://hal.archives-ouvertes.fr/hal-02172347. Accessed 3 July 2019
15. Luo, P., Wang, X., Shao, W., Peng, Z.: Understanding regularization in batch normalization. In: 7th International Conference on Learning Representations, New Orleans, USA, pp. 1–8 (2019)
16. Ioffe, S., Szegedy, C.: Batch normalization: accelerating deep network training by reducing internal covariate shift. arXiv preprint arXiv:1502.03167 (2015)
17. Shmelkov, K., Schmid, C., Alahari, K.: How good is my GAN? In: Ferrari, V., Hebert, M., Sminchisescu, C., Weiss, Y. (eds.) ECCV 2018. LNCS, vol. 11206, pp. 218–234. Springer, Cham (2018). https://doi.org/10.1007/978-3-030-01216-8_14
18. Lee, C.Y., Batra, T., Baig, M.H., Ulbricht, D.: Sliced Wasserstein discrepancy for unsupervised domain adaptation. In: 38th Computer Vision and Pattern Recognition, Seattle, USA, pp. 1–8. IEEE Press (2020)
19. Oudre, L., Jakubowicz, J., Bianchi, P., Simon, C.: Classification of periodic activities using the Wasserstein distance. IEEE Trans. Biomed. Eng. 1610–1619 (2012)

Top-Rank Learning Robust to Outliers

Yan Zheng[1]($\boxtimes$), Daiki Suehiro[1,2]($\boxtimes$), and Seiichi Uchida[1]($\boxtimes$) (iD)

[1] Kyushu University, Fukuoka, Japan
yan.zheng@human.ait.kyushu-u.ac.jp, {suehiro,uchida}@ait.kyushu-u.ac.jp
[2] RIKEN, Tokyo, Japan

Abstract. Top-rank learning aims to maximize the number of absolute top samples, which are "doubtlessly positive" samples and very useful for the real applications that require reliable positive samples. However, top-rank learning is very sensitive to outliers of the negative class. This paper proposes a robust top-rank learning algorithm with an unsupervised outlier estimation technique called local outlier factor (LoF). Introduction of LoF can weaken the effect of the negative outliers and thus increase the stability of the learned ranking function. Moreover, we combine robust top-rank learning with representation learning by a deep neural network (DNN). Experiments on artificial datasets and a medical image dataset demonstrate the robustness of the proposed method to outliers.

Keywords: Top-rank learning · Learning to rank · Outlier

1 Introduction

The general (bipartite) ranking task aims to train the ranking function $r(\mathbf{x})$ that lets positive samples have a higher rank value than negative samples. Let $\Omega = \Omega^+ \cup \Omega^-$ denote the training set. The subsets $\Omega^+ = \{\mathbf{x}_1^+, \ldots, \mathbf{x}_m^+\}$ and $\Omega^- = \{\mathbf{x}_1^-, \ldots, \mathbf{x}_n^-\}$ denote the positive and the negative sample sets, respectively. Formally, the general ranking task aims to train the ranking function $r(\mathbf{x})$ to maximize the number of sample pairs $(\mathbf{x}_i^+, \mathbf{x}_j^-) \in \Omega_i^+ \times \Omega_j^-$ that satisfy $r(\mathbf{x}_i^+) > r(\mathbf{x}_i^-)$.

Top-rank learning has a different aim from the general ranking task. As shown in Fig. 1(a), it aims to maximize *Pos@Top* (positive-at-the-top), which is the percentage of positive samples that rank even higher than the top-ranked (i.e., the highest-ranked) negative sample. These positive samples are called *absolute tops*, and they are obviously different from the negative samples. Consequently, top-rank learning can find highly reliable positive samples as the absolute-top samples. In recent years, top-rank learning has attracted much attention in the machine learning field because of its theoretical interest [1,2,8,16]. Moreover, it is also promising and useful in many practical applications, such as medical image diagnosis, because it provides reliable decisions for positive samples.

Despite its usefulness, top-rank learning has a severe problem toward its practical application, that is, sensitivity to outliers. An outlier of the negative

© Springer Nature Switzerland AG 2021
T. Mantoro et al. (Eds.): ICONIP 2021, LNCS 13110, pp. 608–619, 2021.
https://doi.org/10.1007/978-3-030-92238-2_50

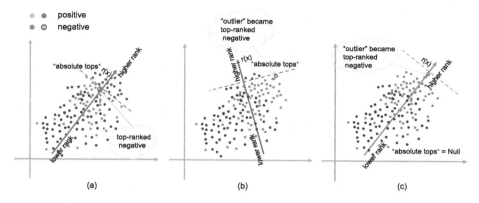

Fig. 1. (a) Top-rank learning, and (b) its weakness to the negative outlier in the training set. In (b), the negative outlier becomes the top-ranked negative and, consequently, Pos@Top becomes much less than (a). Moreover, the ranking function becomes very different from (a) just by the single outlier. (c) The proposed method can weaken the effect of the outlier and get a more suitable ranking function like (a). Note that this figure does not focus on representation learning (i.e., feature extraction), although the proposed method does it along with top-rank learning in an end-to-end manner.

class will drastically degrade Pos@Top. As shown in Fig. 1 (b), when the outlier of the negative class lies in the "very positive" area, it becomes difficult to maximize Pos@Top. In the worst case, the outlier might be selected as the top-ranked negative sample, and then the ranking function is totally affected by just the single outlier. Unfortunately, in some real applications, especially for medical image analysis tasks, the training set often contains such outliers due to measurement error or labeling error.

The purpose of this paper is to propose a new top-rank learning algorithm with sufficient robustness to outliers. More specifically, we propose a top-rank learning algorithm with an unsupervised outlier estimation technique. The unsupervised outlier estimation technique can estimate the degree of the anomaly of each sample based on the sample density around the sample. (Roughly speaking, if there is no other negative sample around a negative sample $\mathbf{x}_j^-$, its degree of anomaly becomes large.) By newly introducing the degree of anomaly into the objective function of the top-rank learning problem, we can weaken the unexpected effect of the outliers on the ranking function. In the following, we employ the standard *local outlier factor* (LoF) [3] as the outlier estimator, although we can use other (even more elaborated) outlier estimators that fit a specific application.

Figure 1 (c) illustrates the expected effect of the proposed method. Although there is the same negative outlier as (b), the proposed method will give a similar ranking function as (a); it is because the proposed method can weaken the effect of the negative outliers during training the ranking function.

We further introduce the representation learning mechanism into the above robust top-rank learning framework. It can be realized by training a deep neural network (DNN) with the objective function of the robust top-rank learning. In other words, we can train the ranking function and the feature extraction function in an end-to-end manner while weakening the effect of outliers. Very recently, Zheng et al. have proposed TopRank CNN (convolutional neural network) [25], which also combines representation learning with top-rank learning. Their algorithm, however, has no function to weaken outliers, and therefore their ranking function has a risk of degradation by the outliers.

We conduct experiments for observing the characteristics of the proposed method on artificial datasets and its performance on practical ranking tasks of medical images. As noted above, the medical image analysis task is a very important target of top-rank learning; however, it often suffers from outliers due to the nature of the labeling difficulty. Therefore, the application of the original top-rank learning to medical images encounters drastic performance degradation. We show that our method can weaken the effect of outliers in a medical image set and achieve more stable and superior performance than several comparative methods.

The main contributions of this paper are as follows:

- To the author's best knowledge, it is the first attempt to introduce unsupervised outlier estimation into the top-rank learning algorithm for increasing the robustness to outliers. (In fact, it is the first time to assert the importance of robustness against outliers in real-world applications of the top-rank learning algorithm.)
- We have realized an end-to-end learning framework by combining our robust top-rank learning and DNN-based representation learning, which will give a synergy to achieve more Pos@Top of the test set.
- We confirmed the usefulness of the proposed method via a medical image diagnosis task, after confirming its ability to avoid the severe degradation by outliers through experiments with artificial datasets.

2   Related Work

2.1   Top-Rank Learning with Deep Learning

Top-rank learning is, recently, combined with DNNs and used in metric-learning tasks, such as person re-identification [5,6,23] and visual search [12,19]. For example, Chen et al. [6] proposed a deep ranking framework to rank the candidate person images that match with a query person image. You et al. [23] proposed a top-push distance learning model and applied it to a video-based person re-identification task.

This paper is inspired by Top-rank Convolutional Neural Network (TopRank CNN) [25], which is proposed very recently. TopRank CNN is the first attempt to combine a loss function for maximizing pos@top and convolutional layers for representation learning. Although they reported good ranking performance at a certain dataset, its ranking function will become unstable due to outliers, as illustrated in Fig. 1 (b). In a later section, we also prove that our method has more robustness to outliers through a comparative experiment.

2.2 Outlier Estimation

To evaluate the degree of outliers, many unsupervised outlier estimation methods (, or unsupervised anomaly estimation methods,) are extensively studied [10,13, 15]. In general, these studies assume that inliers are located in dense areas while outliers are not. Therefore, the outlier estimation can be characterized by their density estimation method, such as statistical methods [10,13,15], neighbor-based methods [3,11], and reconstruction-based methods [21,22]. Robust model fitting algorithms are also prevalent in many computer vision applications where the dataset is contaminated with outliers [20] which aim to learn robust models with consensus maximization by maximizing the number of inliers and achieve efficient performance. Different from robust fitting methods, the purpose of this work is to weaken the effect of outliers in top-rank learning.

In this study, we will employ LoF [3], one of the most common unsupervised outlier estimation methods, for the proposed method. The strong characteristic of LoF is that it can evaluate the local anomaly [13], which is often overlooked by other common methods, such as the kNN-based outlier estimator. Note again that the proposed method can employ other outlier estimators by virtue of its flexibility.

Recently, outlier estimation methods have been employed in DNN learning tasks. Specifically, these methods are used for sample selection or sample reweighting for classification tasks [7,17,18]. It should be emphasized that the top-rank learning is further sensitive to outliers than classification tasks, as noted in Sect. 1, and this paper is the first attempt to introduce an outlier estimation method to the top-rank learning task.

3   Robust Top-Rank Deep Neural Network

3.1   Top-Rank Learning

As shown in Fig. 1 (a), top-rank learning aims to learn a real-valued ranking function r that gives the positive samples higher rank scores than top-ranked negative samples. Its objective function to be minimized is given as:

$$\mathcal{J}_{\text{TopRank}} = \frac{1}{m}\sum_{i=1}^{m}\ell\left(\max_{1\leq j\leq n} r(\mathbf{x}_j^-) - r(\mathbf{x}_i^+)\right) = \frac{1}{m}\sum_{i=1}^{m}\max_{1\leq j\leq n}\ell\big(r(\mathbf{x}_j^-) - r(\mathbf{x}_i^+)\big). \quad (1)$$

The negative sample $\mathbf{x}_j^-$ that gives the maximum in Eq. 1 is the top-ranked negative, as shown in Fig. 1. As the function ℓ, we choose $\ell(z) = \log\left(1 + e^{-z}\right)$, which is often assumed in the top-rank learning tasks[1].

---

[1] More precisely, the original motivation of top-rank learning is to maximize Pos@Top$= \frac{1}{m}\sum_{i=1}^{m}\mathbb{I}\left(r(\mathbf{x}_i^+) > \max_{1\leq j\leq n} r(\mathbf{x}_j^-)\right)$, where $\mathbb{I}$ is the indicator function. By replacing $\mathbb{I}$ by the surrogate loss $\ell(z)$, we have Eq. 1.

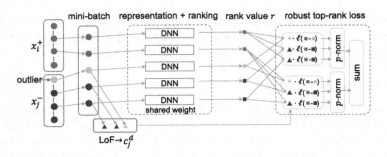

Fig. 2. The overview of the proposed method, robust top-rank learning.

3.2 Top-Rank Learning Robust to Outlier

As shown in Fig. 1 (b), top-rank learning is drastically degraded just by a single negative outlier in training data. The degradation appears as not only the decrease of absolute tops (i.e., "doubtlessly positive" samples) but also the instability of the ranking function r. The latter degradation is more serious, especially for using r for test samples.

We, therefore, newly introduce the LoF into the objective function for weakening the effect of outliers on the ranking function by top-rank learning. Specifically,

$$\mathcal{J}_{\text{RobustTopRank}} = \frac{1}{m} \sum_{i=1}^{m} \max_{1 \leq j \leq n} c_j^d \cdot \ell \left(r(\mathbf{x}_j^-) - r(\mathbf{x}_i^+) \right), \qquad (2)$$

where c_j^d is the weight given by

$$c_j^d = \left(\frac{1}{\max(\text{LoF}_k(\mathbf{x}_j^-), 1)} \right)^d. \qquad (3)$$

Here, $\text{LoF}_k(x_j^-)$ gives LoF value of $\mathbf{x}_j^-$. (The role of the hyperparameters $d \in \mathbb{N}^+$ and $k \in \mathbb{N}^+$ will be explained later.) A higher degree of anomaly gives a larger LoF value and thus a smaller c_j^d. Thus, the effect of outlier samples will be automatically weakened during the training of r.

There are two points to note. First, in Eq. 2, the weight c_j^d is imposed only on the negative samples. As discussed in Fig. 1, our purpose is to avoid that a negative outlier becomes the top-ranked negative. We, therefore, introduce the weight so that the effect of negative outliers is weakened. Second, $\text{LoF}_k(\mathbf{x}_j^-)$ (or, equivalently, the weight c_j^d) is pre-calculated in the original feature space of the training samples Ω and fixed during the training process. In other words, although we will introduce representation learning as discussed in Sect. 3.3, LoF is not evaluated in the learned feature space. This is because we need to avoid the unexpected representation learning result where all negative samples become outliers for minimizing the objective function with smaller c_j^d.

The hyperparameter d controls the degree of outlier weakening. A larger d will weaken outliers more strongly. Note that $c_j^d \in (0, 1]$ for any d. The other hyperparameter k defines the number of nearest neighbors to calculate LoF. In general, the hyperparameter k should be determined by watching the data distribution. In the later experiment with medical image data, we set $k = 30$ by observing the training samples.

3.3 Combination with Representation Learning

In this study, we combine the above robust top-rank loss (2) with representation learning, which is realized by a DNN, such as CNN. Equation 2 contains the max operation. It prevents us from the end-to-end collaborative training of the ranking function r and the feature representation. We, therefore, introduce the p-norm relaxation technique to replace the max operation with a differentiable operation:

$$\mathcal{J}_{\text{Relaxed}} = \frac{1}{m} \sum_{i=1}^{m} \left(\sum_{j=1}^{n} \left(c_j^d \cdot \ell(r(\tilde{\mathbf{x}}_i^+) - r(\tilde{\mathbf{x}}_j^-)) \right)^p \right)^{\frac{1}{p}}, \qquad (4)$$

where $\tilde{\mathbf{x}}$ denote the feature vector given by DNN. Although $p \to \infty$ will give a better approximation of Eq. 2 theoretically, it will cause the overflow problem in practice. Therefore, we set $p = 32$ at maximum in the later experiment.

As shown in Fig. 2, we can train the DNN, which gives the ranking function r and the representation $\tilde{\mathbf{x}}$ in an end-to-end manner. DNN in this figure plays both roles; its earlier layers extract appropriate features $\tilde{\mathbf{x}}$, and the final layers determine the rank value $r(\tilde{\mathbf{x}})$. This end-to-end training utilizes its representation ability for obtaining more absolute tops while increasing the stability of the ranking function by weakening the effect of negative outliers.

In practice, the loss of Eq. 4 is calculated for each mini-batch, and the mini-batch should consist of more negative samples than positive samples. It is because the purpose of top-rank learning is to give higher rank values to positive samples than the higher-ranked negative samples; therefore, it is better to increase the probability that the mini-batch contains higher-ranked negatives by increasing the ratio of negative samples. Note that, different from the classification task, class imbalance does not matter for the ranking task. This is because positive and negative samples are always paired in the objective function, and their effect is equivalent.

4 Experiment on Artificial Datasets

To show the robustness of the proposed method to outliers, we conduct experiments with two artificial datasets; one dataset contains 100 one-dimensional train samples, and the other has 100 two-dimensional train samples. The data distributions of their training sets are shown in Figs. 3 (a) and 4 (a), respectively.

In this experiment, we do not use the convolutional layers for representation learning; we directly use the original sample in Eq. 4 and use four fully connected layers to model the ranking function r. We set $d = 100$ and $k = 15$ (using the same-sized validation set, which is also used for determining the number of the

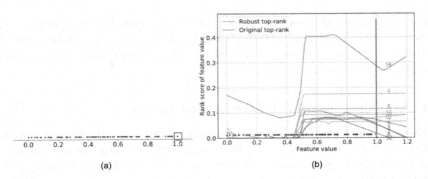

(a)   (b)

Fig. 3. (a) One-dimensional artificial dataset. The blue box indicates the outlier. (b) Results of the original top-rank learning and the robust top-rank learning. The red vertical red bar in (b) indicates the negative outlier location. A digit on a curve indicates the value of p. (Color figure online)

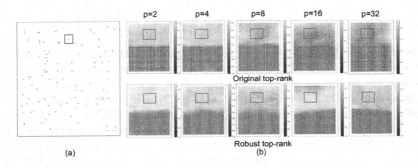

(a)   (b)

Fig. 4. (a) Two-dimensional artificial dataset. (b) Results of the original top-rank learning and the robust top-rank learning. In the heatmap representation, purple indicates a lower ranking score, and yellow indicates a higher ranking score. The blue box indicates the outlier location. (Color figure online)

training epochs). We examined different $p \in \{2, 4, 8, 16, 32\}$. For comparison, we realize the original top-rank learning by setting $c_j^d = 1, \forall j$.

Figure 3 (b) plots the learned ranking functions by the original top-rank learning and the proposed method, respectively. It proves that the ranking functions by the original top-rank learning show a big drop around the negative outlier (at $x^- = 1$). In fact, the leftmost positive (green) samples have a lower rank value than the positive samples in the middle. In contrast, the proposed method does not show such a drop successfully by weakening the effect of the negative outlier[2].

Figure 3 (b) shows the results of the two-dimensional artificial data. The rank function is represented as a heatmap between dark-blue (lower rank) and

---

[2] The ranking function r only specifies the relative relationship between two samples (such that $r(\mathbf{x}^+) > r(\mathbf{x}^-)$). Thus, its absolute value is not important. Therefore, the curve shape is more important when we observe Fig. 3 (b).

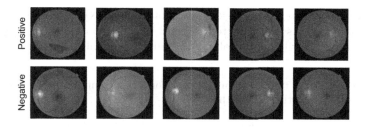

Fig. 5. Examples from the Messidor dataset. The first and the second row respectively show the disease images (positive) and the normal images (negative).

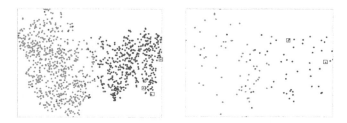

Fig. 6. t-SNE visualizations of initial feature vectors (by CABNET [14]) of the training (left) and test (right) samples of the Messidor dataset. The red dots denote negative samples, and the green dots denote positive samples. The red dots with the blue box indicates the negative outliers. (Color figure online)

yellow (higher rank). Like the one-dimensional case, the original top-rank learning suffers from a significant drop around the negative outlier; in other words, even a single outlier can degrade the ranking function seriously. However, the outlier effect is further weakened by the proposed method.

5   Experiment on Medical Image Diagnosis

5.1   Dataset

As a practical application of the proposed method, we conduct binary diabetes retinopathy (DR) ranking experiment on the Messidor dataset [9], which is public on website[3]. Figure 5 shows several samples from the Messidor dataset. As noted in Sect. 1, the datasets of medical image analysis tasks often contain outliers due to measurement error or labeling error, and the original top-rank learning will become unstable by the outliers.

Recently, He et al. [14] proposed a binary DR grade classification method, called CABNET, and achieved state-of-the-art performance. We, therefore, convert all image samples into the CABNET features in advance. In this experiment, we followed the same data division as [14]. Consequently, 501 positive samples and 699 negative samples are used in our experiments.

---

[3] https://www.adcis.net/en/third-party/messidor/.

Figure 6 shows the t-SNE visualization of the CABNET features of the train and test samples. As we expected, several negative outliers are found in the positive sample distribution, probably because of the labeling difficulty.

Table 1. Pos@top on the Messidor dataset. For a fair comparison, we have used the same network architecture for all the methods.

Method	Pos@Top↑	
	Train	Test
CABNET [14]	0.4286	0.5379
RankNet[4]	0.0803	0.2449
TopRank CNN[25]	0.3368	0.5896
Ours	0.5433	**0.6524**

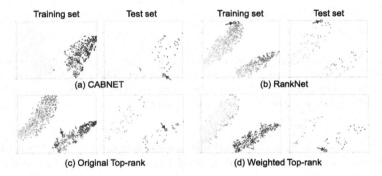

Fig. 7. t-SNE visualizations results of the comparative methods on the Messidor dataset. Red, green, and purple dots denote negative, positive, absolute top (i.e., doubtlessly positive) samples, respectively; yellow dots with a red circle (pointed by a red arrow) denote the top-ranked negative sample. (Color figure online)

5.2   Model Setup

We use a CNN as the DNN of Fig. 2. Each CNN has three convolutional layers and two fully connected layers. ReLU and max-pooling operation and batch-normalization followed each convolutional layer. Since we use the CABNET feature as the input feature, such a shallow CNN is sufficient. In training, we utilized the stochastic gradient descent (SGD) optimization algorithm. As noted in Sect. 3.3, we used a very unbalanced mini-batch with 5 positive samples and 45 negative samples for training.

5.3   Comparative Methods

To measure the performance of the proposed method, we conduct comparative experiments with CABNET, RankNet[4,24], and TopRank CNN[25]. RankNet is a very famous learning to rank algorithm that combines representation learning. For a fair comparison, RankNet and TopRank CNN used the same architecture and the same input (CABNET feature vectors) as the proposed method. In other words, they are different only at their loss function.

The ten-folds cross-validation method is used to tune the hyperparameters. The partitions of each fold data are followed the CABNET [14]. For the proposed method, we select hyperparameters p and d from $\{2, 4, 8, 16, 32\}$ and $\{1, 10, 100\}$, respectively. TopRank CNN also employs p-norm relaxation for its top-rank loss and thus has the same hyperparameter p; we also optimize its p in the same way.

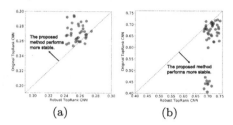

Fig. 8. (a) Kendall Tau distance and (b) Spearman correlation coefficient. Those values by TopRank CNN and the proposed method are depicted in vertical axes and horizontal axes, respectively.

5.4 Quantitative Evaluation by Pos@top

Table 1 presents the pos@top of the comparison method, which are average among ten-fold test data. (The soft-max values are used to calculate the pos@top for the CABNET.) It should be emphasized that the proposed method has *no guarantee* of achieving higher pos@top for *the training set*; Fig. 1 (c) depicts the extreme case where the weakened negative outlier is ranked at the top, and thus there is no absolute top (pos@top=0). However, it should be more emphasized that the ranking function of Fig. 1 (c) is more suitable than Fig. 1 (b) because it is similar to Fig. 1 (a), which shows the ranking function in the absence of the negative outliers. We, therefore, can expect that the proposed method will give a higher pos@top for *the test set*[4].

As shown in Table 1, the proposed method successfully outperforms the comparative methods for *the test set*. Especially, the comparison with TopRank CNN proves that the proposed top-rank learning method could weaken the effect of the outliers and get more stable results. Since the proposed method shows higher pos@top, it can derive more reliable positive (i.e., doubtlessly positive) samples—this is very beneficial for medical diagnosis tasks.

The t-SNE visualization of a one-fold dataset distribution on the learned feature space is shown in Fig. 7. We can confirm that the proposed method achieved better performance on the test set than the comparison methods. As noted above,

---

[4] If we can remove negative outliers in the evaluation, the ranking function of Fig. 1(c) will achieve more pos@top than Fig. 1(b), even for the training set.

the proposed method cannot have many absolute tops in *the training set* due to the outliers. However, the learned ranking function is still reasonable and thus could have more absolute tops in *the test set*.

5.5 Quantitative Evaluation of Ranking Stability

We also conduct another quantitative evaluation of the stability of the top-rank learning methods by using the Kendall tau distance and Spearman correlation coefficient. Among the ten-folds (for the cross-validation), some folds contain a certain negative outlier, and the others do not. If the ranking method is sensitive to the outlier, the ranking of the samples might change drastically. If not, the ranking will be almost the same. The Kendall tau distance and Spearman correlation coefficient can evaluate the dissimilarity and similarity of two ranking results of the same dataset, respectively.

Figure 8 shows (a) the Kendall tau distance and (b) Spearman correlation coefficient between every two folds for the proposed method and the original TopRank CNN on the ranking. Since we have 45 ($=_{10}C_2$) fold-pairs, both of (a) and (b) have 45 dots. In (a), the proposed method almost always shows a smaller distance. In (b), the proposed method almost always shows a larger correlation coefficient. We, therefore, can conclude that the proposed method is more stable by weakening the effect of the outliers.

6   Conclusion

This study proposed a novel top-rank learning method that can weaken the effect of outliers in the training process by using an unsupervised outlier detection technique. To the author's best knowledge, it is the first proposal to realize robust top-rank learning. Moreover, we combine the robust top-rank learning framework with representation learning in an end-to-end way. Experiments on artificial datasets and medical image datasets demonstrate the robustness of the proposed method for outliers. Benefits from the robustness to outliers, the proposed method obtains more reliable positive samples (higher pos@top) in the test set than the comparison methods on medical image diagnosis tasks. Also, the stability of the learned ranking function is confirmed experimentally. In the future, further analysis of the characteristic of robust top-rank learning will be given, and its applications on other datasets will be observed.

Acknowledgment. This work was supported by MEXT-Japan (Grant No. J17H06100), and JST, ACT-X, Japan (Grant No. JPMJAX200G), and China Scholarship Council (Grant No. 201806330079).

References

1. Agarwal, S.: The infinite push: a new support vector ranking algorithm that directly optimizes accuracy at the absolute top of the list. In: Proceedings of the ICDM, pp. 839–850 (2011)
2. Boyd, S.P., Cortes, C., Mohri, M., Radovanovic, A.: Accuracy at the top. In: Proceedings of the NIPS, pp. 962–970 (2012)

3. Breunig, M.M., Kriegel, H.P., Ng, R.T., Sander, J.: LOF: identifying density-based local outliers. In: Proceedings of the ICMD, pp. 93–104 (2000)
4. Burges, C.J.C., et al.: Learning to rank using gradient descent. In: Proceedings of the ICML, pp. 89–96 (2005)
5. Chen, C., Dou, H., Hu, X., Peng, S.: Deep top-rank counter metric for person re-identification. In: Proceedings of the ICPR, pp. 2732–2739 (2020)
6. Chen, S., Guo, C., Lai, J.: Deep ranking for person re-identification via joint representation learning. IEEE Trans. Image Process. **25**(5), 2353–2367 (2016)
7. Chen, X., et al.: Sample balancing for deep learning-based visual recognition. IEEE Trans. Neural Netw. Learn. Syst. **31**(10), 3962–3976 (2020)
8. Clémençon, S., Vayatis, N.: Ranking the best instances. J. Mach. Learn. Res. **8**, 2671–2699 (2007)
9. Decenciére, E., et al.: Feedback on a publicly distributed database: the Messidor database. Image Anal. Stereol. **33**(3), 231–234 (2014)
10. Eskin, E.: Anomaly detection over noisy data using learned probability distributions. In: Langley, P. (ed.) Proceedings of the ICML, pp. 255–262 (2000)
11. Eskin, E., Arnold, A., Prerau, M., Portnoy, L., Stolfo, S.: A geometric framework for unsupervised anomaly detection. In: Barbará, D., Jajodia, S. (eds.) Applications of Data Mining in Computer Security. ADIS, vol. 6, pp. 77–101. Springer, Boston (2002). https://doi.org/10.1007/978-1-4615-0953-0_4
12. Geng, Y., et al.: Learning convolutional neural network to maximize pos@top performance measure. arXiv preprint arXiv:1609.08417 (2016)
13. Goldstein, M., Uchida, S.: A comparative evaluation of unsupervised anomaly detection algorithms for multivariate data. PLoS ONE **11**(4), e0152173 (2016)
14. He, A., Li, T., Li, N., Wang, K., Fu, H.: CABNet: category attention block for imbalanced diabetic retinopathy grading. IEEE Trans. Med. Imaging **40**(1), 143–153 (2021)
15. Kim, J., Scott, C.D.: Robust kernel density estimation. J. Mach. Learn. Res. **13**, 2529–2565 (2012)
16. Li, N., Jin, R., Zhou, Z.: Top rank optimization in linear time. In: Proceedings of the NIPS, pp. 1502–1510 (2014)
17. Lin, T., Goyal, P., Girshick, R.B., He, K., Dollár, P.: Focal loss for dense object detection. IEEE Trans. Pattern Anal. Mach. Intell. **42**(2), 318–327 (2020)
18. Oquab, M., Bottou, L., Laptev, I., Sivic, J.: Learning and transferring mid-level image representations using convolutional neural networks. In: Proceedings of the CVPR, pp. 1717–1724 (2014)
19. Song, D., Liu, W., Ji, R., Meyer, D.A., Smith, J.R.: Top rank supervised binary coding for visual search. In: Proceedings of the ICCV, pp. 1922–1930 (2015)
20. Truong, G., Le, H., Suter, D., Zhang, E., Gilani, S.Z.: Unsupervised learning for robust fitting: a reinforcement learning approach. In: Proceedings of CVPR, pp. 10348–10357 (2021)
21. Xia, Y., Cao, X., Wen, F., Hua, G., Sun, J.: Learning discriminative reconstructions for unsupervised outlier removal. In: Proceedings of the ICCV, pp. 1511–1519 (2015)
22. Xu, H., Caramanis, C., Mannor, S.: Outlier-robust PCA: the high-dimensional case. IEEE Trans. Inf. Theory **59**(1), 546–572 (2013)
23. You, J., Wu, A., Li, X., Zheng, W.: Top-push video-based person re-identification. In: Proceedings of the CVPR, pp. 1345–1353 (2016)
24. Zheng, Y., Zheng, Y., Ohyama, W., Suehiro, D., Uchida, S.: RankSVM for offline signature verification. In: Proceedings of the ICDAR, pp. 928–933 (2019)
25. Zheng, Y., Zheng, Y., Suehiro, D., Uchida, S.: Top-rank convolutional neural network and its application to medical image-based diagnosis. Pattern Recogn. **120**, 108138 (2021)

Novel GAN Inversion Model with Latent Space Constraints for Face Reconstruction

Jinglong Yang⬤, Xiongwen Quan$^{(\boxtimes)}$⬤, and Han Zhang⬤

College of Artificial Intelligence, Nankai University, Tianjin, China
quanxw@nankai.edu.cn

Abstract. This paper considers how to encode a target face image into its StyleGAN latent space accurately and efficiently, with applications to allow the various image editing method being used on the real images. Compared with optimization-based methods using gradient descent on latent code iteratively, the learning-based method we adopt can encode target images with one forward propagation, which is better suited for real-world application. The key advances in this paper are: adopting the face recognition model as a constraint to keep the identity information intact and adding a classifier to encourage latent code to retain more attributes possessed in the original image. Experiments show our method can achieve an excellent reconstruction effect. The ablation study indicates the proposed design advances the GAN Inversion task qualitatively and quantitatively. However, the method may fail when there are other objects around the target face and generate a blurry patch around that object.

Keywords: StyleGAN · GAN inversion · Encoder · Face reconstruction

1 Introduction

In recent years, face reconstruction related applications are gaining increasing attention. It attracts people with the impressive special effects in TikTok and FaceApp. It can also act as an important role in many other places, such as social security area where face reconstruction model can generate time-corrected criminal images given images recorded years ago. With the development of deep learning, the neural network has been proved to be better than the traditional manual-designed feature extraction methods in many computer vision tasks, especially in face reconstruction. Among the well-known deep learning models, Generative Adversarial Network (GAN) [6] is an unsupervised learning model that can produce diversified and high-quality pictures of large resolution. It is pivotal in the field of image generation and face reconstruction.

This work was supported by the National Natural Science Foundation of China through Grants (No.61973174).

T. Mantoro et al. (Eds.): ICONIP 2021, LNCS 13110, pp. 620–631, 2021.
https://doi.org/10.1007/978-3-030-92238-2_51

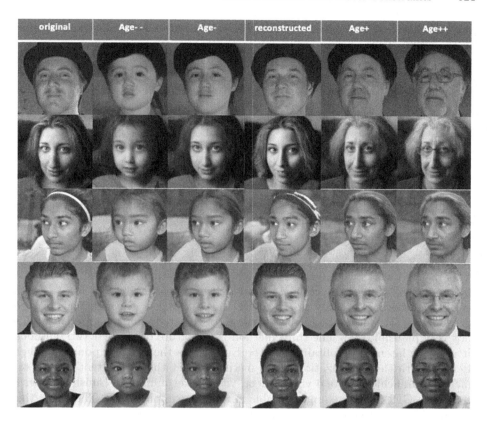

Fig. 1. This figure shows the effect of combining our GAN inversion method with the image editing method [16]. Note the image editing method here is a linear modifier, so the edited attribute is inevitably coupled with other attributes.

The task we try to tackle is face reconstruction based on GAN inversion. GAN Inversion, also known as GAN encoder, is the prerequisite of any application involving the power of GAN. Since GAN itself just starts with a random vector Z, and passes through the Generator to produce an exquisite image yet of little use. For example, in DCGAN [13], given two images and their latent codes, we can interpolate between two latent codes, and get a series of smooth, incremental transitions between the two images. With GAN inversion, we are able to get the interpolated images of our interest. Otherwise, we are limited to the images that the GAN model randomly generates.

There are also plenty of image editing methods based on GAN [1,2,14]. Since the image we want to edit in the real-world application is not generated by a latent code randomly sampled in Gaussian distribution, instead a real image taken by users. So to edit an image in its latent space, we need to obtain the latent code of our target image in the first place. That is why GAN inversion is an inevitable task to take on. In Fig. 1, we show the result of combining our GAN inversion method with InterfaceGAN [16], which can linearly edit the attributes in the latent space.

The model that can map the input image to the corresponding latent code is called Encoder $E(x; \theta_E)$. The Encoder is essentially the inverse mapping of GAN Generator G, and the training of Encoder is as follows:

$$\theta_E^* = argmin_{\theta_E} \sum_n loss(G(E(x_n; \theta_E)), x_n) \qquad (1)$$

where x_n denotes the n-th image of our dataset, θ_E denotes the parameters of Encoder.

There are three ways to do GAN inversion. The most popular one is the optimization-based method [1, 21], by which gradient descent method is used to optimize the latent code iteratively:

$$z^* = argmin_z loss(x, G(z; \theta)) \qquad (2)$$

This method works well, but it needs many computation resources. The second method is to train an encoding neural network as described in Eq. 1, which is efficient because it only needs to let the image pass through one forward propagation. However, the quality of the reconstructed image via an Encoder is not very plausible [11]. The third method is a mixture of the first two methods. It uses the latent code obtained by the Encoder as the starting point and then iterates with the optimization-based method to get the final latent code. This method can achieve a good visual effect, yet it takes minutes to finish even with a modern GPU, which is unacceptable in real-time applications. Therefore, this paper makes progress based on the learning-based method.

This paper has two main contributions. Firstly, the face recognition model is used as the constraint to train the Encoder. We proposed identity loss (id loss) to capture the distance between the original image and the reconstructed image in the identity space. Using the pre-trained neural network to extract features already exists in some literature. For example, VGG [17], which is trained in the image classification task, can effectively extract high-level semantic features. The distance between the extracted features of the original image and the reconstructed image is used in our loss functions as the perceptual loss. However, a network that focuses on extracting features from faces of different identities is more suitable for the GAN inversion task. The second contribution is to add a classifier structure to encourage that the latent code retains all the attributes information in the original image, so that Generator can output more accurate images. On top of that, inspired by InfoGAN [4], the classifier can also help the encoding of each attribute to be more decoupled.

Experiments show that the method proposed in this paper can achieve a good reconstruction effect. Our method takes less than 100ms to encode a target image with a 2080Ti GPU, which can meet the requirements of many real-time applications.

2   Related Work

In recent years, the most crucial development of GAN lies in the field of face generation. StyleGAN [9] borrows some ideas from the field of neural style transfer. It utilizes a disentangling multi-layer perceptron (MLP) structure to map latent vectors from a Gaussian space (Z space) to a disentangled W space. This MLP serves as a disentangling module that makes W space more linearly separated, so manipulating one attribute will have less influence over other attributes. StyleGAN also utilizes adaptive instance normalization (AdaIN), which is a normalization method that aligns the mean and variance of the style features y to those of the content features x. Combined with other techniques, the StyleGAN series models have achieved an incredible effect that most recent GAN inversion works are based on these models.

InterfaceGAN [16] uses the attributes of the image generated by GAN as labels, such as gender and smile. Then it trains an SVM [3] linear classifier with paired data composed of the latent codes and these labels. Thus the normal vector of the SVM's classification plane is the editing direction of this attribute, and images can be edited with respect to that attribute by changing the latent code of the image along that direction. We use this work as the following procedure to examine the effectiveness of our GAN inversion model as in Fig. 1.

2.1   Establishment of GAN Inversion

Zhu et al. first introduced the GAN Inversion task into the field of image editing [22]. Before that, image editing was a task in the field of computer graphics and image processing. It is difficult for users to control the high-level semantics directly, like gender, smile. That work establishes the paradigm of three kinds of GAN inversion methods. Since then, we can apply those manipulations on the latent space of the GAN model onto real-world images.

Learning-based GAN Inversion methods [11,22] trains an encoding model $E(x; \theta_E)$, which maps the target image to the latent space of GAN as in Eq. 1. The common loss functions used here are MSE loss, perceptual loss [8] and adversarial loss.

$$MSE(x,y) = \frac{1}{mn} \sum_{i=0}^{m-1} \sum_{j=0}^{n-1} [x(i,j) - y(i,j)]^2 \tag{3}$$

$$Perceptual\ loss(x,y) = \sum_i MSE(VGG_i(x), VGG_i(y)) \tag{4}$$

$$Adversarial\ loss(y) = D(y) \tag{5}$$

where x denotes the target image, y denotes the reconstructed image, i in perceptual loss denotes the selected layers of the pre-trained VGG network, D denotes the Discriminator of GAN. MSE loss penalizes the deviation of the reconstructed image from the target image at the pixel level. Perceptual loss is constructed by sending image pairs into the VGG network and using certain layers to represent the semantic feature of the images, then calculate the L2 norm of the distance as the perceptual loss. This way of representing the difference of pictures is first introduced in [8], and complements the MSE loss with the high-level semantics representation. The Discriminator is used here to provide adversarial loss. Normally, the Discriminator and Generator are pre-trained models, and their parameters are frozen in the training of Encoder. This adversarial loss ensures the reconstructed image are of high quality.

While the image generated from the former one is always inferior to the latter by a distinct margin [1,21]. Thus, most works use the optimization-based method in the face editing field as their choice for GAN Inversion. They start with an initial latent code, and iteratively optimize it to make the generated image closer to the target image, as in Eq. 2. In [1], the author discussed two ways of finding the initial latent code. One is random sampling, and the other is using the mean vector of the latent space. This mean vector initialization tends to have better image quality, while it is easier to fall into the local optima. Apart from these two methods, training an Encoder to get the initial latent code is adopted by many other works.

2.2 Face Recognition Network

Among the various attributes describing the image, what we are most concerned about are not those common attributes, such as smile, gender, or sunglasses. Because it is more often those subtle variations that are difficult to quantify that change the identity of the face. So, in addition to MSE loss, perceptual loss, and adversarial loss, the retention of human identity information is also an effective constraint. Fortunately, due to the huge commercial value of face recognition, the research on face recognition has been very mature. We can use the identity features extracted by these pre-trained face recognition networks as constraints to make the reconstructed image maintain the identity of the original image.

Modern face recognition network functions like metric learning. The vector obtained from the image through the metric network contains the unique identity of the face. To determine whether two faces are of the same identity, we need to calculate the distance between two vectors. When the distance is less than the threshold, we think they are of the same person. A good face recognition backbone should suffice that the distance within the class is small and the distance between different classes is large. In order to make the network of such a property, FaceNet [15] uses triplet loss. Each data group has three images. The first one and the second one should be the same person, and the third one should be a different person. This triplet loss tries to minimize the distance of the first two images and maximize the distance between the first and the third images. Another kind of improvement on face recognition is the variant of softmax. Sphereface [12] puts forward the concept of angular softmax loss. Cosface [19] promotes a additional cosine margin loss and Arcface [5] comes up with a additional angular margin loss. They make the boundary between different classes wider and each class more compact. The proposed identity loss in our paper uses the ability of Arcface to extract facial features.

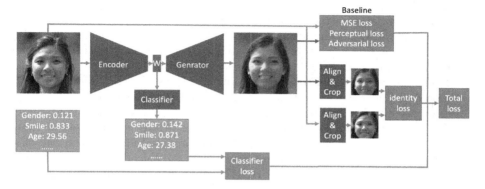

Fig. 2. Network architecture. The reconstructed image in this figure is selected during the training process, so as to distinguish it from the original image.

2.3   Interpretability and Decoupling Effect of Latent Space

An unsolved problem of GAN is the interpretability of its latent space. Although we can obtain continuous image transformation through interpolation between two latent codes in latent space and find the direction vector of the attributes of interest in latent space, these attributes are coupled in each dimension of latent space. The coupling problem causes that when one property is changed, other related or unrelated properties are also changed, which is something we do not want. We hope to decouple each attribute in the latent space or go further to allow a single dimension to control a single attribute separately. For example, in the task of generating handwritten numbers, the first dimension controls which number is 0–9. The second dimension controls the thickness of the strokes, and the third dimension controls the inclining angle of the numbers. This problem has been improved in InfoGAN [4]. In InfoGAN, by adding a classifier network to the generated image, the value of certain dimensions in the input latent code can be reconstructed. Such constraints are added to the GAN training procedure, and the final model can match the most significant attributes to the reconstructed dimension. The reason for this phenomenon is that if we want the classifier to reconstruct specific dimensions accurately, then these dimensions are best to be the most significant attributes in the image. In this way, without any attribute supervision information, the latent space can still be decoupled to the specific dimensions.

3   Approach

The dataset used in our work is FFHQ, and FFHQ paired attributes. FFHQ is a dataset proposed in StyleGAN [9], which includes 70,000 high-definition face images. There are multiple resolution versions of this dataset. The highest version is of 1024×1024. In the training process, we divide the dataset into 65,000 training sets and 5,000 validation sets. In addition to pictures, our Encoder needs to use face attributes as supervision information. Since FFHQ does not have attribute information attached to it, we choose to use Face++'s face attribute recognition tool to get the attributes corresponding to each picture of FFHQ. Since the label information here is annotated with other models instead of human experts, it can be considered as semi-supervised learning.

Model architecture and training pipeline are shown in Fig. 2. The network is composed of Encoder, Generator, Classifier, identity loss module, and the baseline loss module. The input of our model are the target image and its attribute label. We get the total loss through the network, which is composed of three kinds of losses: losses of the baseline model, our proposed identity loss, and classifier loss. Firstly, the original image is sent into the Encoder to get the latent code W. On the one hand, the W code is input into the classifier module to get the prediction attribute, then the classifier loss is calculated with the input attribute label. On the other hand, the W code is sent into the Generator to get the reconstructed image. Secondly, with each pair of the target image and reconstructed image, we can get the losses of the baseline method by sending the image pairs into its loss functions. Thirdly, we conduct the "align and crop" operations to make the image agree with the requirements of the face recognition network, and then send images to the face recognition network to get the identity feature vector. Then calculate the identity loss. Finally, the total loss is obtained by summing up baseline losses, classifier loss, and identity loss.

The implementation of our Encoder borrows the code of [11], while we have made many improvements. These improvements include building blocks of the backbone network, the architecture of the classifier, and the identity loss function to retain the identity of the original face. We freeze the parameters of the Generator of StyleGAN, since the pre-trained StyleGAN has a stable latent space.

The classifier placed after the W code constrains the Encoder to preserve more attributes in the W code. Accurate image reconstruction needs the latent code to contain the information of the original image. This kind of coding of image attributes is imperceptible to our human senses. We cannot judge whether the attribute information is preserved by looking at the value in the latent code. However, extracting and understanding useful information from complex data is just the strength of the neural network. Therefore, after we get the latent code, we do not just send it to the Generator. We also input it into a classifier to extract the information in the latent code.

Apart from the attributes that can be quantified and classified by the classifier, there is a lot of nuanced information and subtle variation that are difficult to describe in numbers. They determine the unique identity of a face. To capture these, the identity loss module is devised to complement the classifier module.

We input the original image and reconstruction image into the backbone network of Arcface [5] respectively to get the feature vector. The distance between the two vectors can be used as our identity loss. However, the pipeline of the face recognition network does not directly input the image into CNN. Instead, it first detects and crops the face, then conducts an affine transform according to the five key points in the standard template. After all these preprocessing procedures, it is then sent into the Arcface network. Therefore, we need to walk through the whole pipeline when utilizing the face recognition network as a constraint. However, there is a difficulty in the implementation, which is how to build a computational graph to make identity loss be backpropagated smoothly. Before the whole training, we need to find the affine transformation matrix of the 65,000 training images to the standard template, as well as the specific coordinates of the crop box. Then during training, the original image is used as the computing node without gradient track, and the reconstructed image is used as the computing node that needs gradient track. Therefore, we need to implement this transformation to be capable of tracking gradient in the computation graph ourselves in PyTorch. After the transformation, the target face box is cropped, and then sent to the face recognition network. In this way, the input image of the face recognition network is of the same distribution as its training data. Thus, the identity loss can be backpropagated to constrain the Encoder to retain the unique face identity information in the process of generating the latent code.

Another improvement is to remove the truncation module in StyleGAN. The truncation module can improve the quality of GAN-generated images by reducing the variability of the randomly sampled latent code. However, our task starts with a real image. Thus, it is unnecessary to use truncation to "pull" the W code to mean vector. If the truncation module exists, the Encoder will not map the image to the corresponding W code, rather $W\ code + \frac{1}{truncation\ ratio}(mean\ vector - w\ code)$ It increases the complexity of the learning task since the target is not the difference between W code and mean vector.

During implementation, We also improve the building block of the CNN structure by adding a Batch Normalization (BN) layer. BN layer can change feature map into a regular distribution with mean value of 0 and variance of 1, and it almost always can make the gradient backpropagation more efficient and easier to learn. However, there is no BN layer in the implementation of [11]. So we also fixed that in the building blocks, and it leads to a noticeable better effect.

4   Experiments

In this section, we conducted various experiments on different model architecture and constraints. The evaluation of these methods includes the traditional metrics in the image reconstruction field, like Peak Signal-to-Noise Ratio (PSNR), and the more advanced learning-based Learned Perceptual Image Patch Similarity (LPIPS) [20]. On top of that, we also compare these methods using those distribution distance metrics, which are popular in GAN evaluation, like Frechet Inception Distance (FID) [7]. The performance is summarized in Table 1.

Fig. 3. This figure shows the effect of face reconstruction by each method. These images are mapped to the latent space by different encoders, and then generated by the same Generator. These images are randomly sampled from the validation set. We can see a clear trend, that with the introduction of new model structure and constraint function, the quality of reconstructed images is getting better and better.

Table 1. Quantitative evaluation

Methods	FID↓	PSNR↑	LPIPS↓
Baseline	42.207	18.164	0.288
+noTrunc	40.785	18.379	0.279
+BN	38.525	18.763	0.249
+classifier loss	34.093	17.763	0.264
+identity loss	24.134	17.395	0.256

PSNR is one of the most widely used metrics in the field of image reconstruction. It is the ratio of the peak value of the signal to the noise between the two images. For the original image x and the reconstructed image y with the size of m × n:

$$MSE = \frac{1}{mn} \sum_{i=0}^{m-1} \sum_{j=0}^{n-1} [x(i,j) - y(i,j)]^2 \qquad (6)$$

$$PSNR = 10 \cdot log_{10}(\frac{MAX^2}{MSE}) \qquad (7)$$

where MAX here is 255.

LPIPS measures image similarity beyond the traditional methods using the power of deep learning networks. This evaluation has been proved to be more in line with human perception. Images pass through a CNN network, such as

Fig. 4. Bad cases. The first row contains the original images. The second row are the reconstructed images.

the AlexNet [10] used in this calculation, and then the distance of the weighted feature layer is calculated.

$$LPIPS(x,y) = \sum_l \frac{1}{H_l W_l} \sum_{h,w} \|\omega_l \odot (x_{hw}^l - x_{hw}^l)\|_2^2 \qquad (8)$$

where L represents the extracted feature layer, ω is used to adjust weight.

FID calculates whether the distribution of the two groups of images is similar, which is the most commonly used method to evaluate the image quality generated by GAN. This method uses the inceptionV3 [18] as the model of feature extraction.

$$FID = \|\mu_r - \mu_g\|^2 + Tr(\Sigma_r + \Sigma_g - 2\sqrt{\Sigma_r \Sigma_g}) \qquad (9)$$

where Tr is the trace of the matrix.

In Table 1, we show the evaluation score of each module added incrementally. Since the baseline model uses truncation, we first remove the truncation module. Then we add the BN layer in the building blocks. Then we add a classifier. Then we add identity loss. Note that the scores of FID and LPIPS agree with our subjective evaluation shown in Fig. 3. The score of PSNR does not change much, and it does not agree with the actual visual effect.

Admittedly, when there are other objects around the face, such as hats, headscarves, hands, or other faces, the performance of our algorithm will degrade as in Fig. 4. Such problems exist in all GAN Inversion methods. Since the training data FFHQ is very clean around the face, when other things are showing near the face, this part of the coding information is not in the decoding ability of the Generator, so the final decoded image is very fuzzy in the corresponding position.

5   Conclusion

Our article provides a novel method to do GAN Inversion based face reconstruction, which significantly improves the image quality compared with the baseline model. This work shows that the face recognition model can be used as a good constraint of our GAN Encoder, and also shows that adding a classifier constraint to the latent space can make the encoding process retain more image information. We show that learning-based methods can also achieve relatively high reconstruction results. The successful implementation of this method makes the high-level semantic editing be used on real images. Compared with the optimization-based method, the computation efficiency of this method provides the possibility of encoding an image and editing it in the real world application. Although our experimental results show that this is an improvement qualitatively and quantitatively, it can degrade when there are objects around the face. We speculate that this is because the weights of the Generator are frozen during the training of our Encoder, so it can only generate a quality image with no other objects around the face. Therefore in future work, we may train the GAN model from scratch together with Encoder, rather than using the pre-trained GAN.

References

1. Abdal, R., Qin, Y., Wonka, P.: Image2StyleGAN: how to embed images into the StyleGAN latent space? In: Proceedings of the IEEE/CVF International Conference on Computer Vision, pp. 4432–4441 (2019)
2. Abdal, R., Zhu, P., Mitra, N.J., Wonka, P.: StyleFlow: attribute-conditioned exploration of StyleGAN-generated images using conditional continuous normalizing flows. ACM Trans. Graph. (TOG) **40**(3), 1–21 (2021)
3. Boser, B.E., Guyon, I.M., Vapnik, V.N.: A training algorithm for optimal margin classifiers. In: Proceedings of the Fifth Annual Workshop on Computational Learning Theory, pp. 144–152 (1992)
4. Chen, X., Duan, Y., Houthooft, R., Schulman, J., Sutskever, I., Abbeel, P.: InfoGAN: interpretable representation learning by information maximizing generative adversarial nets. arXiv preprint arXiv:1606.03657 (2016)
5. Deng, J., Guo, J., Xue, N., Zafeiriou, S.: ArcFace: additive angular margin loss for deep face recognition. In: Proceedings of the IEEE/CVF Conference on Computer Vision and Pattern Recognition, pp. 4690–4699 (2019)
6. Goodfellow, I.J., et al.: Generative adversarial networks. arXiv preprint arXiv:1406.2661 (2014)
7. Heusel, M., Ramsauer, H., Unterthiner, T., Nessler, B., Hochreiter, S.: GANs trained by a two time-scale update rule converge to a local Nash equilibrium (2018)
8. Johnson, J., Alahi, A., Fei-Fei, L.: Perceptual losses for real-time style transfer and super-resolution. In: Leibe, B., Matas, J., Sebe, N., Welling, M. (eds.) ECCV 2016. LNCS, vol. 9906, pp. 694–711. Springer, Cham (2016). https://doi.org/10.1007/978-3-319-46475-6_43
9. Karras, T., Laine, S., Aila, T.: A style-based generator architecture for generative adversarial networks. In: Proceedings of the IEEE/CVF Conference on Computer Vision and Pattern Recognition, pp. 4401–4410 (2019)
10. Krizhevsky, A., Sutskever, I., Hinton, G.E.: ImageNet classification with deep convolutional neural networks. In: Pereira, F., Burges, C.J.C., Bottou, L., Weinberger, K.Q. (eds.) Advances in Neural Information Processing Systems, vol. 25. Curran Associates, Inc. (2012). https://proceedings.neurips.cc/paper/2012/file/c399862d3b9d6b76c8436e924a68c45b-Paper.pdf

11. Lee, B.: StyleGAN2-encoder-pytorch (2020). https://github.com/bryandlee/stylegan2-encoder-pytorch

12. Liu, W., Wen, Y., Yu, Z., Li, M., Raj, B., Song, L.: SphereFace: deep hypersphere embedding for face recognition. In: Proceedings of the IEEE Conference on Computer Vision and Pattern Recognition, pp. 212–220 (2017)

13. Radford, A., Metz, L., Chintala, S.: Unsupervised representation learning with deep convolutional generative adversarial networks. arXiv preprint arXiv:1511.06434 (2015)

14. Richardson, E., et al.: Encoding in style: a StyleGAN encoder for image-to-image translation. arXiv preprint arXiv:2008.00951 (2020)

15. Schroff, F., Kalenichenko, D., Philbin, J.: FaceNet: a unified embedding for face recognition and clustering. In: Proceedings of the IEEE Conference on Computer Vision and Pattern Recognition, pp. 815–823 (2015)

16. Shen, Y., Yang, C., Tang, X., Zhou, B.: InterfaceGAN: interpreting the disentangled face representation learned by GANs. IEEE Trans. Pattern Anal. Mach. Intell. (2020)

17. Simonyan, K., Zisserman, A.: Very deep convolutional networks for large-scale image recognition. arXiv preprint arXiv:1409.1556 (2014)

18. Szegedy, C., Vanhoucke, V., Ioffe, S., Shlens, J., Wojna, Z.: Rethinking the inception architecture for computer vision (2015)

19. Wang, H., et al.: CosFace: large margin cosine loss for deep face recognition. In: Proceedings of the IEEE Conference on Computer Vision and Pattern Recognition, pp. 5265–5274 (2018)

20. Zhang, R., Isola, P., Efros, A.A., Shechtman, E., Wang, O.: The unreasonable effectiveness of deep features as a perceptual metric. In: CVPR (2018)

21. Zhu, J., Shen, Y., Zhao, D., Zhou, B.: In-domain GAN inversion for real image editing. In: Vedaldi, A., Bischof, H., Brox, T., Frahm, J.-M. (eds.) ECCV 2020. LNCS, vol. 12362, pp. 592–608. Springer, Cham (2020). https://doi.org/10.1007/978-3-030-58520-4_35

22. Zhu, J.-Y., Krähenbühl, P., Shechtman, E., Efros, A.A.: Generative visual manipulation on the natural image manifold. In: Leibe, B., Matas, J., Sebe, N., Welling, M. (eds.) ECCV 2016. LNCS, vol. 9909, pp. 597–613. Springer, Cham (2016). https://doi.org/10.1007/978-3-319-46454-1_36

Edge Guided Attention Based Densely Connected Network for Single Image Super-Resolution

Zijian Wang[1], Yao Lu[1(✉)], and Qingxuan Shi[2]

[1] Beijing Laboratory of Intelligent Information Technology, School of Computer Science and Technology, Beijing Institute of Technology, 100081 Beijing, China
vis_yl@bit.edu.cn
[2] School of Cyber Security and Computer, Hebei University, 071000 Baoding, China

Abstract. Densely connected neural networks have achieved superior results in the single image super-resolution (SISR) task benefited from the rich feature representations provided by the dense connection structure. However, the rich feature representations also include redundant information, which is useless for final image reconstruction. Besides, during the image reconstruction process, the low-resolution images cannot directly learn the high-frequency information from their corresponding high-resolution images, which can't constrain the high-frequency part of super-resolved images and hinders reconstruct quality. To solve these problems, we apply the neural attention mechanism to extract effective features from the densely connected network and utilize the image edge-prior to guide the super-resolution network to learn high-frequency information effectively. Therefore, we propose an edge-guided attention-based densely connected network (EGADNet) for SISR. EGADNet mainly includes three parts: content branch (CT branch), edge reconstruction branch (ER branch), and information fusion branch (IF branch). Each low-resolution image is first processed by the CT branch and the ER branch to extract the content information and the high-frequency information corresponding to the high-resolution image. After that, the IF branch processes fuse the above information and regress the final super-resolution result. We perform experiments on commonly-used image super-resolution datasets. Both qualitative and quantitative results have proved the effectiveness of our method.

Keywords: Super-resolution · Densely connected network · Edge prior · Neural attention

---

This work is supported by the National Natural Science Foundation of China (No. 61273273), by the National Key Research and Development Plan (No. 2017YFC0112001), by China Central Television (JG2018-0247), by Science and Technology Project of Hebei Education Department (QN2018214), and by the Natural Science Foundation of Hebei Province (F2019201451).

T. Mantoro et al. (Eds.): ICONIP 2021, LNCS 13110, pp. 632–643, 2021.
https://doi.org/10.1007/978-3-030-92238-2_52

1   Introduction

Single image super-resolution (SISR) is a classical computer vision task which reconstructs the high-resolution (HR) image from the low-resolution (LR) image input. With the renaissance of deep learning, many super-resolution networks are proposed in recent years and have achieved great progress.

Most current SISR works aim to increase the non-linear mapping abilities of super-resolution networks to improve reconstruction accuracy. They are inspired by the image classification task and borrow the model design philosophy to construct super-resolution networks. Representative works include residual-based super-resolution networks [2,22,23] and dense connection based super-resolution networks [10,26,31,39], which are motivated by ResNet [11] and DenseNet [14], respectively.

Although the above methods obtain excellent performance, they usually have complex model structures and redundant features, which need a long time to be optimized and bring an obstacle to get satisfying image reconstruction results. Besides, directly learning the high-frequency information of the HR image from LR-HR image pairs is not the optimal choice because the high-frequency information of SR is prone to be affected without explicit supervised information.

To solve these problems, we introduce the neural attention mechanism and image edge-prior into a super-resolution network. The neural attention mechanism, which has been widely used in the computer vision field [32,41–44], helps the super-resolution network focus on representative features, and image edge-prior gives a supervision signal on high-frequency information of SR image. Based on these motivations, we propose an edge-guided attention-based densely connected network (EGADNet) for SISR. EGADNet mainly consists of three sub-networks: content branch (CT branch), edge-reconstruction branch (ER branch), and information fusion branch (IF branch). CT branch and ER branch first generate the content information and the high-frequency information of SR image, respectively. Then, the extracted information is fed to the IF branch for effective fusion and final image reconstruction. We apply neural attention module [12] in the CT branch and ER branch to extract the representative features and reduce the effect of redundant features. We conduct experiments on several commonly-used super-resolution benchmarks and achieve superior results, which proves our method's effectiveness.

To summarize, our contributions are three folds:

- We proposed an effective way to extract representative features of the content information and high-frequency information from LR-HR pairs, which are useful for final image reconstruction.
- We proposed an effective method to introduce the image edge-prior to the densely connected super-resolution network for providing the high-frequency supervision of SR.
- We evaluate our method on Set5 [3], Set14 [37], BSD100 [24], Urban100 [15] and Manga109 [25], and our method achieves promising results compared with other state-of-the-art methods.

The rest of this paper is as follows. We review some related works of our method in Sect. 2. In Sect. 3, we demonstrate each component of EGADNet in detail and present experiment set up and result analysis in Sect. 4. We conclude this paper in Sect. 5.

2 Related Work

Many super-resolution neural networks are proposed in recent years. SRCNN [6] is the first deep learning network for super-resolution. It first up-sampled LR input into the desired HR output size, and trained the network to learn the feature representations. However, early up-sampling operation increased the memory and computation cost of the neural network, which was not satisfied with the real-application need. Therefore, most of the super-resolution networks [7,29] utilized the late up-sampling operation to process the extracted features. Besides, for generating more detailed texture content, some super-resolution networks [4,21] aim to improve the high-frequency information of learning process via the low-frequency guidance and the octave convolution [5]. Although the above methods achieve satisfied reconstruction accuracy, they are prone to generate over-smoothed results. Different from them, some Generative Adversarial Networks (GAN) [19,27,33] for SISR are proposed, which can generate more photo-realistic images. In addition, there are some open problems of SISR should be solved, such as arbitrary up-sampling scale super-resolution [13], perceptual extreme super-resolution [28], and real-world super-resolution [9,20]. In this paper, we add edge-prior to the super-resolution network for better learning the high-frequency information of the HR image.

3 Proposed Method

The overview of our proposed EGADNet is shown in Fig. 1. It mainly consists of three subnets: CT branch, ER branch, and IF branch. CT branch and ER branch generate content information and high-frequency information of SR image, and the IF branch fuse the above information for final image reconstruction. In this section, we first introduce the details of the neural attention for content information and high-frequency information. Then, we utilize the edge from the HR image to provide the high-frequency information supervision and demonstrate the detail generation of edge. After that, the details of each subnet are presented. Finally, the loss function of EGADNet is introduced.

3.1 Neural Attention

Squeeze and Excitation block (SE block) [12] is a widely used neural attention module, which applies squeeze $f_{sq}(\cdot)$ and excitation $f_{ex}(\cdot)$ operations to exploit the independence relationships between feature channels and suppress the redundant information. The process of SE block is as follows:

$$M = f_{ex}\left(\mathbf{S}_{avg}\right) = fc_2\left(fc_1\left(\mathbf{S}_{avg}\right)\right) = W_2\delta\left(W_1\mathbf{S}_{avg}\right), \tag{1}$$

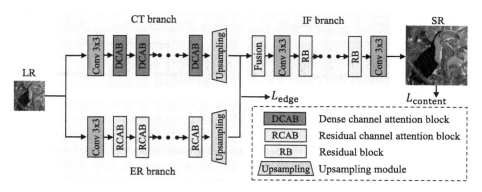

Fig. 1. Overview of our proposed EGADNet. It consists of three subnets: content branch (CT branch), edge-reconstruction branch (ER branch), and information fusion branch (IF branch). Each low-resolution input (LR) is simultaneously fed to CT branch and ER branch for extracting the content information and high-frequency information of super-resolution image (SR). IF branch receives the extracted information and process them for final image reconstruction.

where $\mathbf{S}_{avg} = f_{sq}(\mathbf{X}) \in \mathbb{R}^{1 \times 1 \times C}$. $\mathbf{X} \in \mathbb{R}^{H \times W \times C}$ is the input features and $f_{sq}(\cdot)$ is the global average pooling layer. fc_1 and fc_2 denotes two fully connect layers, and $W_1 \in \mathbb{R}^{\frac{C}{r} \times C}$ and $W_2 \in \mathbb{R}^{C \times \frac{C}{r}}$ are their corresponding weight parameters. C is the feature channel numbers, r is the channel reduction ratio, and $\delta(\cdot)$ is the ReLU activation function. In this paper, we replace two fully connected layers of SE block as two 1×1 convolutional layers, add the modified SE block into densely connected block [14] and residual block [11] for effectively extracting content information and high-frequency information.

Content Attention. We combine SE block and densely connected block into a new block, dense channel attention block (DCAB), which is applied into CT branch for extracting effective features and suppressing redundant information. The architecture of DCAB is shown in Fig. 2, and the process of DCAB block is as follows:

$$F_{g,0} = M_{g,b}(F_{g-1,b})F_{g-1,b}, \tag{2}$$

$$F_{g,b} = H_\ell\left([F_{g,0}, F_{g,1}, \ldots, F_{g,b-1}]\right), \tag{3}$$

where g and b denotes the group and block of the network, and $F_{g,b}$ represents the g group and b block features. H_ℓ represents a single 3×3 convolutional layer, and $M_{g,b}$ denotes the SE block.

High-Frequency Attention. The redundant information of densely connected block will overwhelm the high-frequency information, we choose a simple block, residual block (RB), to extract high-frequency information. Based on RB, we combine SE block and RB to form a new block, residual channel attention block

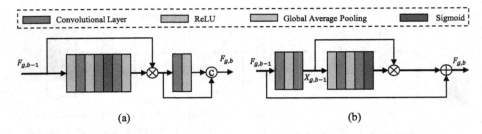

Fig. 2. The architecture of (a) the dense channel attention block (DCAB) and (b) the residual channel attention block (RCAB).

(RCAB), which is the same with [38]. The structure of RCAB is shown in Fig. 2, and the operation process is as follows:

$$F_{g,b} = F_{g,b-1} + M_{g,b}(X_{g,b}) \cdot X_{g,b}, \tag{4}$$

$$X_{g,b} = W_{g,b}^2 \delta \left(W_{g,b}^1 F_{g,b-1} \right), \tag{5}$$

where $X_{g,b}$ represents the learned residual features. $W_{g,b}^1$ and $W_{g,b}^2$ are the weight parameters of two stack convolutional layers from RCAB.

3.2 Edge Generation

Edge is an important image feature, which contain abundant image information and has been widely used in image restoration task [35,36,40]. In this paper, we apply image edge as the prior knowledge to guide the super-resolution network to learn high-frequency information. Following [8], we apply the soft-edge extracted from HR image to provide the high-frequency supervision information. The extraction process of soft-edge is as follows:

$$I_{Edge} = \text{div}(u_x, u_y), \tag{6}$$

where $u_i = \frac{\nabla_i I_{HR}}{\sqrt{1+|\nabla I_{HR}|^2}}$, $i \in \{x, y\}$. x, y, ∇ and $\text{div}(\cdot)$ denote the horizontal direction, vertical direction, the gradient operation and the divergence operation, respectively.

3.3 Network Structure

Content Branch. The CT branch mainly consists of three parts: model head, model body, and model tail. The model head is a single 3×3 convolutional layer, which converts the LR input into feature space. The model body is constructed with dense channel attention blocks (DCAB), which can reduce the redundant features generated by the densely connected structure. We apply 8 DCABs in the CT branch. The model tail is the upsampling module and is implemented with convolutional layers and the subpixel convolution [29].

Edge Reconstruction Branch. The ER branch has similar structures with the CT branch. It also includes the model head, model body, and model tail. Unlike the CT branch, the model body is mainly built on residual channel attention blocks (RCAB) [38]. The generated high-frequency information of the ER branch is optimized by the edge loss and fed to the IF branch for final image reconstruction. There are 5 RCABs in the ER branch.

Information Fusion Branch. The IF branch mainly concludes two parts: fusion block (FB) and residual block (RB). FB is implemented with a single 1×1 convolutional layer. RB is implemented with two 3×3 convolutional layers with residual structure [11], which can be avoided gradient vanish problem. We implement 40 RBs in the IF branch.

3.4   Loss Function

The total loss L_{total} of EGADNet includes two parts: content loss L_{content} for low-frequency information learning, and edge loss L_{edge} for high-frequency learning. We apply $L1$ loss to optimize the CT branch. The L_{content} is as follows:

$$L_{\text{content}} = \|I_{SR} - I_{HR}\|_1 , \qquad (7)$$

where I_{SR} represents the output of EGADNet, and I_{HR} is the counterpart high-resolution image. $L1$ loss is also chosen as the L_{edge}:

$$L_{edge} = \|E\left(I_{LR}\right) - I_{Edge}\|_1 , \qquad (8)$$

where $E(\cdot)$ is the function of the ER branch, I_{LR} denotes the LR input, and I_{Edge} represents the edge image generated from Eq. (6). The total loss of EGADNet is as follows:

$$L_{\text{total}} = L_{\text{content}} + \lambda L_{\text{edge}}, \qquad (9)$$

where λ is the coefficient parameter. During the training stage, we set $\lambda = 0.1$.

4   Experiments

In this section, we present the experiment setup and results analysis in detail. Dataset and evaluation metrics are first introduced. Then , we introduce the implementation details of EGADNet. After that, we compare our method with other state-of-the-art methods on several popular super-resolution datasets. Finally, we perform ablation studies to analyze the effectiveness of each component of EGADNet.

4.1   Datasets and Evaluation Metrics

We choose 800 images from the train set of DIV2K [1] to train our model and evaluate the performance of our method on Set5 [3], Set14 [37], BSD100 [24], Urban100 [15] and Manga109 [25]. We convert the output result into YCbCr space and evaluate PSNR and SSIM [34] on Y channel only.

4.2   Implementation Details

The I_{LR} is generated from the corresponding I_{HR} with the Bicubic interpolation. We augment the train data with horizontal and vertical flips. The LR patches are randomly cropped with the size of 64×64 for $\times 2$ and $\times 4$ scale super-resolution, and the size of 96×96 for $\times 3$ scale super-resolution. The optimizer is chosen as Adam. The initial learning rate is set as 1×10^{-4} and halved every 200 epochs. We train our model with a total of 600 epochs, and the batch size is set to 16. Our method is implemented with the Pytorch framework and is trained on a single Nvidia Tesla V100 GPU card.

4.3   Comparisons with State-of-The-Arts

We compare our EGADNet with other state-of-the-art methods, such as SRCNN [6], VDSR [16], LapSRN [18], DRCN [17], DRRN [30], SRDenseNet [31], and SeaNet [8]. The quantitative results are displayed in Table 1. We can see that EGADNet almost outperforms the other state-of-the-art methods. Specifically, compared with the other densely connected super-resolution networks, EGADNet achieves the better results than SRdensenet [31] for $\times 4$ scale super-resolution, which benefit from the high-frequency information learned by ER branch. Compared with SeaNet [8], EGADNet achieves the better results on all test sets, which benefits from the neural attention providing strong feature representative ability of CT branch and ER branch. Figure 3 displays the visualization comparisons of different methods on Urban100 for $\times 4$ scale super-resolution. We can see that EGADNet can reconstruct more texture details and achieve more photo-realistic results than other methods.

4.4   Ablation Studies

The Effectiveness of Neural Attention Mechanism. To evaluate the effectiveness of neural attention module, we modify EGADNet into three different variants: EGADNet without neural attention (EGADNet *w/o* NA), EGADNet without high-frequency attention (EGADNet *w/o* HFA), and EGADNet without content attention (EGADNet *w/o* CTA). We remove all the channel attention layers (CA) of EGADNet, denoted as EGADNet *w/o* NA, and remove only CA layers in the ER branch or CT branch, denoted as EGADNet *w/o* HFA and EGADNet *w/o* CTA, respectively. The experiments of $\times 4$ SR are performed on commonly-used testing datasets, and the comparison results are displayed in

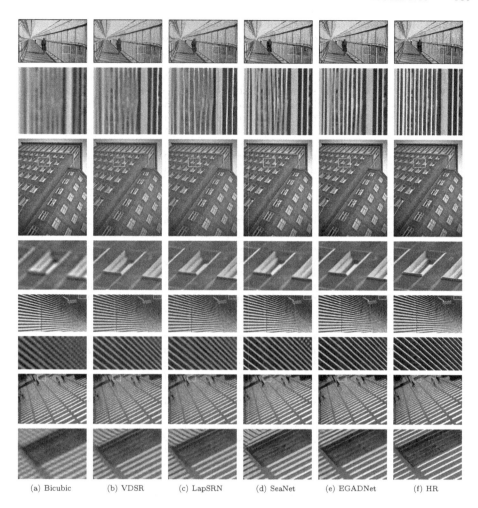

(a) Bicubic   (b) VDSR   (c) LapSRN   (d) SeaNet   (e) EGADNet   (f) HR

Fig. 3. Visual comparisons of different methods on Urban100 for ×4 SR. From left to right: the results of Bicubic, VDSR [16], LapSRN [18], SeaNet [8], EGADNet and HR. Our proposed EGADNet can reconstruct more realistic texture details than other methods.

Table 2. We see that EGADNet w/o neural attention achieves the lowest reconstruction accuracy. With the introduction of neural attention added to the CT branch and ER branch, the performance is better than EGADNet w/o neural attention, especially on the Manga109 dataset. EGADNet achieves the best results.

The Effectiveness of Each Component of EGADNet. We perform experiments to evaluate the effectiveness of each component of EGADNet. The quantitative results are shown in Table 3. The first row presents the results of the EGADNet $w/$ CT branch. We only use the $L_{content}$ to optimize the network.

Table 1. Performance comparisons of the state-of-the art methods for ×2, ×3 and ×4 image super-resolution. '–' denotes the results are not provided. Red and blue indicate the best results and the second best results, respectively.

Scale	Method	Set5		Set14		BSD100		Urban100		Manga109	
		PSNR	SSIM	PSNR	SSIM	PSNR	SSIM	PSNR	SSIM	PSNR	SSIM
×2	Bicubic	33.68	0.9304	30.24	0.8691	29.56	0.8435	27.39	0.8410	30.30	0.9339
	SRCNN [6]	36.66	0.9542	32.45	0.9067	31.36	0.8879	29.52	0.8950	35.60	0.9663
	VDSR [16]	37.53	0.9597	33.05	0.9127	31.90	0.8960	30.77	0.9140	37.22	0.9750
	LapSRN [18]	37.52	0.9591	32.99	0.9124	31.80	0.8949	31.05	0.9100	37.27	0.9855
	DRCN [17]	37.63	0.9588	33.04	0.9118	31.85	0.8942	30.76	0.9130	37.63	0.9723
	DRRN [30]	37.74	0.9591	33.23	0.9136	32.05	0.8973	31.23	0.9190	37.60	0.9736
	SeaNet [8]	38.15	0.9611	33.86	0.9198	32.31	0.9013	32.68	0.9332	38.76	0.9774
	SRDenseNet [31]	–	–	–	–	–	–	–	–	–	–
	EGADNet	38.17	0.9611	33.77	0.9184	32.28	0.9009	32.71	0.9337	39.07	0.9778
×3	Bicubic	30.40	0.8686	27.54	0.7747	27.21	0.7389	24.46	0.7349	26.95	0.8556
	SRCNN [6]	32.75	0.9090	29.29	0.8215	28.41	0.7863	26.24	0.7989	30.48	0.9117
	VDSR [16]	33.66	0.9213	29.78	0.8318	28.83	0.7976	25.18	0.7530	32.01	0.9340
	LapSRN [18]	33.82	0.9227	29.79	0.8320	28.82	0.7973	27.07	0.8272	32.21	0.9318
	DRCN [17]	33.82	0.9226	29.76	0.8311	28.80	0.7963	27.16	0.8311	32.31	0.9328
	DRRN [30]	34.03	0.9244	29.96	0.8349	28.95	0.8004	27.53	0.8378	32.42	0.9359
	SeaNet [8]	34.55	0.9282	30.42	0.8444	29.17	0.8017	28.50	0.8594	33.73	0.9463
	SRDenseNet [31]	–	–	–	–	–	–	–	–	–	–
	EGADNet	34.65	0.9290	30.52	0.8459	29.21	]0.8078	28.67	0.8621	33.93	0.9471
×4	Bicubic	28.43	0.8109	26.00	0.7023	25.96	0.6678	23.14	0.6577	24.89	0.7866
	SRCNN [6]	30.48	0.8628	27.50	0.7513	26.90	0.7103	24.52	0.7221	27.58	0.8555
	VDSR [16]	31.35	0.8838	28.02	0.7678	27.29	0.7252	25.18	0.7530	28.83	0.8870
	LapSRN [18]	31.54	0.8866	28.09	0.7694	27.32	0.7264	25.21	0.7553	29.09	0.8845
	DRCN [17]	31.53	0.8854	28.02	0.7670	27.23	0.7233	25.14	0.7520	28.98	0.8816
	DRRN [30]	31.68	0.8888	28.21	0.7720	27.38	0.7284	25.44	0.7638	29.18	0.8914
	SeaNet [8]	32.33	0.8970	28.72	0.7855	27.65	0.7388	26.32	0.7942	30.74	0.9129
	SRDenseNet [31]	32.02	0.8934	28.50	0.7782	27.53	0.7337	26.05	0.7819	–	–
	EGADNet	32.36	0.8964	28.66	0.7827	27.61	0.7368	26.32	0.7927	30.69	0.9113

Table 2. Ablation study of the effectiveness of neural attention mechanism on commonly-used SR benchmarks for ×4 scale.

Method	Set5	Set14	BSD100	Urban100	Manga109
EGADNet *w/o* NA	32.28/0.8964	28.64/0.7833	27.61/0.7374	26.21/0.7899	30.55/0.9101
EGADNet *w/o* HFA	32.34/0.8963	28.63/0.7829	27.62/0.7372	26.34/0.7938	30.65/0.9116
EGADNet *w/o* CTA	32.36/0.8964	28.65/0.7832	27.61/0.7371	26.19/0.7891	30.61/0.9105
EGADNet	32.36/0.8964	28.66/0.7827	27.61/0.7368	26.32/0.7927	30.69/0.9113

The second row displays the results of the EGADNet *w/* CT branch and ER branch. We only apply a single 3 × 3 convolutional layer to replace the IF branch. The third row is the performance of EGADNet. We can see that the ER branch improves the EGADNet *w/* CT branch on Urban100 and Manga109, which benefits from the high-frequency information ER branch provided. We can also see that the IF branch can effectively fuse the content information and the high-frequency information from comparisons between the second and the results of the third row.

Table 3. Ablation study of the effectiveness of each component from EGADNet on commonly-used SR benchmarks for ×4 scale.

Method	CT branch	ER branch	IF branch	Urban100	Manga109
	✓			26.24/0.7909	30.65/0.9100
EGADNet	✓	✓		26.24/0.7904	30.61/0.9104
	✓	✓	✓	26.32/0.7927	30.69/0.9113

5   Conclusion

In this paper, we propose an EGADNet for SISR. EGADNet consists of the CT branch, the ER branch, and the IF branch. The content information of SR is generated from the CT branch, and the high-frequency information of SR is obtained from the ER branch. IF branch fuse content information and high-frequency information for reconstructing the final output. We perform our method on several super-resolution benchmarks, and the experiment results demonstrate our method's effectiveness. Although our method performs well on synthesis data, we don't evaluate the performance on real-world super-resolution tasks whose down degradation model is unknown. In future work, we will extend our proposed EGADNet to handle the real-world super-resolution task with various degradation models.

References

1. Agustsson, E., Timofte, R.: Ntire 2017 challenge on single image super-resolution: Dataset and study. In: CVPRW, pp. 126–135 (2017)
2. Ahn, N., Kang, B., Sohn, K.-A.: Fast, accurate, and lightweight super-resolution with cascading residual network. In: Ferrari, V., Hebert, M., Sminchisescu, C., Weiss, Y. (eds.) ECCV 2018. LNCS, vol. 11214, pp. 256–272. Springer, Cham (2018). https://doi.org/10.1007/978-3-030-01249-6_16
3. Bevilacqua, M., Roumy, A., Guillemot, C., Alberi-Morel, M.L.: Low-complexity single-image super-resolution based on nonnegative neighbor embedding. In: BMVC, pp. 135.1–135.10 (2012)
4. Chen, W., Liu, C., Yan, Y., Jin, L., Sun, X., Peng, X.: Guided dual networks for single image super-resolution. IEEE Access **8**, 93608–93620 (2020)
5. Chen, Y., et al.: Drop an octave: reducing spatial redundancy in convolutional neural networks with octave convolution. In: ICCV, pp. 3435–3444 (2019)
6. Dong, C., Loy, C.C., He, K., Tang, X.: Image super-resolution using deep convolutional networks. IEEE TPAMI **38**(2), 295–307 (2015)
7. Dong, C., Loy, C.C., Tang, X.: Accelerating the super-resolution convolutional neural network. In: Leibe, B., Matas, J., Sebe, N., Welling, M. (eds.) ECCV 2016. LNCS, vol. 9906, pp. 391–407. Springer, Cham (2016). https://doi.org/10.1007/978-3-319-46475-6_25
8. Fang, F., Li, J., Zeng, T.: Soft-edge assisted network for single image super-resolution. IEEE TIP **29**, 4656–4668 (2020)

9. Guo, Y., et al.: Closed-loop matters: dual regression networks for single image super-resolution. In: CVPR, pp. 5407–5416 (2020)

10. Haris, M., Shakhnarovich, G., Ukita, N.: Deep back-projection networks for super-resolution. In: CVPR, pp. 1664–1673 (2018)

11. He, K., Zhang, X., Ren, S., Sun, J.: Deep residual learning for image recognition. In: CVPR, pp. 770–778 (2016)

12. Hu, J., Shen, L., Sun, G.: Squeeze-and-excitation networks. In: CVPR, pp. 7132–7141 (2018)

13. Hu, X., Mu, H., Zhang, X., Wang, Z., Tan, T., Sun, J.: Meta-SR: a magnification-arbitrary network for super-resolution. In: CVPR, pp. 1575–1584 (2019)

14. Huang, G., Liu, Z., Van Der Maaten, L., Weinberger, K.Q.: Densely connected convolutional networks. In: CVPR, pp. 4700–4708 (2017)

15. Huang, J.B., Singh, A., Ahuja, N.: Single image super-resolution from transformed self-exemplars. In: CVPR, pp. 5197–5206 (2015)

16. Kim, J., Kwon Lee, J., Mu Lee, K.: Accurate image super-resolution using very deep convolutional networks. In: CVPR, pp. 1646–1654 (2016)

17. Kim, J., Kwon Lee, J., Mu Lee, K.: Deeply-recursive convolutional network for image super-resolution. In: CVPR, pp. 1637–1645 (2016)

18. Lai, W.S., Huang, J.B., Ahuja, N., Yang, M.H.: Deep laplacian pyramid networks for fast and accurate super-resolution. In: CVPR, pp. 624–632 (2017)

19. Ledig, C., et al.: Photo-realistic single image super-resolution using a generative adversarial network. In: CVPR, pp. 4681–4690 (2017)

20. Li, G., Lu, Y., Lu, L., Wu, Z., Wang, X., Wang, S.: Semi-blind super-resolution with kernel-guided feature modification, pp. 1–6 (2020)

21. Li, S., Cai, Q., Li, H., Cao, J., Wang, L., Li, Z.: Frequency separation network for image super-resolution. IEEE Access **8**, 33768–33777 (2020)

22. Lim, B., Son, S., Kim, H., Nah, S., Mu Lee, K.: Enhanced deep residual networks for single image super-resolution. In: CVPRW, pp. 136–144 (2017)

23. Liu, J., Zhang, W., Tang, Y., Tang, J., Wu, G.: Residual feature aggregation network for image super-resolution. In: CVPR, pp. 2359–2368 (2020)

24. Martin, D., Fowlkes, C., Tal, D., Malik, J.: A database of human segmented natural images and its application to evaluating segmentation algorithms and measuring ecological statistics. In: ICCV, pp. 416–423 (2001)

25. Matsui, Y., et al.: Sketch-based manga retrieval using manga109 dataset. Multimed. Tools Appl. **76**(20), 21811–21838 (2017). https://doi.org/10.1007/s11042-016-4020-z

26. Qin, J., Sun, X., Yan, Y., Jin, L., Peng, X.: Multi-resolution space-attended residual dense network for single image super-resolution. IEEE Access **8**, 40499–40511 (2020)

27. Sajjadi, M.S., Scholkopf, B., Hirsch, M.: EnhanceNet: single image super-resolution through automated texture synthesis. In: ICCV, pp. 4491–4500 (2017)

28. Shang, T., Dai, Q., Zhu, S., Yang, T., Guo, Y.: Perceptual extreme super-resolution network with receptive field block. In: CVPRW, pp. 440–441 (2020)

29. Shi, W., et al.: Real-time single image and video super-resolution using an efficient sub-pixel convolutional neural network. In: CVPR, pp. 1874–1883 (2016)

30. Tai, Y., Yang, J., Liu, X.: Image super-resolution via deep recursive residual network. In: CVPR, pp. 3147–3155 (2017)

31. Tong, T., Li, G., Liu, X., Gao, Q.: Image super-resolution using dense skip connections. In: ICCV, pp. 4799–4807 (2017)

32. Wang, W., Zhou, T., Qi, S., Shen, J., Zhu, S.C.: Hierarchical human semantic parsing with comprehensive part-relation modeling. IEEE TPAMI (2021)

33. Wang, X., et al.: ESRGAN: enhanced super-resolution generative adversarial networks. In: Leal-Taixé, L., Roth, S. (eds.) ECCV 2018. LNCS, vol. 11133, pp. 63–79. Springer, Cham (2019). https://doi.org/10.1007/978-3-030-11021-5_5

34. Wang, Z., Bovik, A.C., Sheikh, H.R., Simoncelli, E.P.: Image quality assessment: from error visibility to structural similarity. IEEE TIP **13**(4), 600–612 (2004)

35. Xie, J., Feris, R.S., Sun, M.T.: Edge-guided single depth image super resolution. IEEE TIP **25**(1), 428–438 (2015)

36. Yang, W., et al.: Deep edge guided recurrent residual learning for image super-resolution. IEEE Trans. Image Process. **26**(12), 5895–5907 (2017)

37. Zeyde, R., Elad, M., Protter, M.: On single image scale-up using sparse-representations. In: Boissonnat, J.-D., et al. (eds.) Curves and Surfaces 2010. LNCS, vol. 6920, pp. 711–730. Springer, Heidelberg (2012). https://doi.org/10.1007/978-3-642-27413-8_47

38. Zhang, Y., Li, K., Li, K., Wang, L., Zhong, B., Fu, Y.: Image super-resolution using very deep residual channel attention networks. In: Ferrari, V., Hebert, M., Sminchisescu, C., Weiss, Y. (eds.) ECCV 2018. LNCS, vol. 11211, pp. 294–310. Springer, Cham (2018). https://doi.org/10.1007/978-3-030-01234-2_18

39. Zhang, Y., Tian, Y., Kong, Y., Zhong, B., Fu, Y.: Residual dense network for image super-resolution. In: CVPR, pp. 2472–2481 (2018)

40. Zhou, Q., Chen, S., Liu, J., Tang, X.: Edge-preserving single image super-resolution. In: Proceedings of the 19th ACM International Conference on Multimedia, pp. 1037–1040 (2011)

41. Zhou, T., Li, J., Wang, S., Tao, R., Shen, J.: MATNet: motion-attentive transition network for zero-shot video object segmentation. IEEE TIP **29**, 8326–8338 (2020)

42. Zhou, T., Qi, S., Wang, W., Shen, J., Zhu, S.C.: Cascaded parsing of human-object interaction recognition. IEEE TPAMI (2021)

43. Zhou, T., Wang, S., Zhou, Y., Yao, Y., Li, J., Shao, L.: Motion-attentive transition for zero-shot video object segmentation. In: AAAI, pp. 476–483 (2020)

44. Zhou, T., Wang, W., Qi, S., Ling, H., Shen, J.: Cascaded human-object interaction recognition. In: CVPR, pp. 4263–4272 (2020)

An Agent-Based Market Simulator for Back-Testing Deep Reinforcement Learning Based Trade Execution Strategies

Siyu Lin[✉] and Peter A. Beling

University of Virginia, Charlottesville, VA 22903, USA
{sl5tb,pb3a}@virginia.edu

Abstract. Researchers have reported success in developing autonomous trade execution systems based on Deep Reinforcement Learning (DRL) techniques aiming to minimize the execution costs. However, they all back-test the trade execution policies on historical datasets. One of the biggest drawbacks of back-testing on historical datasets is that it cannot account for the permanent market impacts caused by interactions among various trading agents and real-world factors such as network latency and computational delays.

In this article, we investigate an agent-based market simulator as a back-testing tool. More specifically, we design agents which use the trade execution policies learned by two previously proposed Deep Reinforcement Learning algorithms, a modified Deep-Q Network (DQN) and Proximal Policy Optimization with Long-Short Term Memory networks (PPO LSTM), to execute trades and interact with each other in the market simulator.

Keywords: Agent-based market simulator · Deep Reinforcement Learning · Autonomous trade execution systems · Permanent market impacts

1 Introduction

Algorithmic trading becomes more and more efficient with the advancement in technology and it is prevalent in the modern financial market. The optimal trade execution problem is one of the most important problems in algorithmic trading. Given a certain amount of shares within a specified deadline, it studies how to minimize trade execution costs.

Nevmyvaka, Feng, and Kearns are the pioneers to apply RL to optimize trade execution costs [11]. Rather than adopting a pure RL method, Hendricks and Wilcox propose to create a hybrid framework, which combines the Almgren and Chriss model (AC) and RL algorithm, to map the states to the proportion of the AC-projected trading volumes [4]. Ning et al. [12] apply Deep Q-Network (DQN) [10] for optimal trade execution problem with high dimensional space. Lin and Beling [5,7] analyze and point out the flaws that could happen when applying a generic Q-learning algorithm to the optimal trade execution problem. To enforce the zero-ending inventory constraint, they slightly modify the DQN algorithm by combining the Q function estimation at the last two steps [8].

© Springer Nature Switzerland AG 2021
T. Mantoro et al. (Eds.): ICONIP 2021, LNCS 13110, pp. 644–653, 2021.
https://doi.org/10.1007/978-3-030-92238-2_53

All the researchers mentioned above back-test the learned policies on historical datasets, which ignores the permanent market impacts and interactions among various trading agents. In this article, we investigate an agent-based market simulator (ABIDES), which was developed by Byrd et al. [2].

2   DRL for Optimal Trade Execution

In this section, we review two DRL based trade execution systems proposed in the literature and illustrate how the researchers formulate optimal trade execution as a DRL problem [8]: 1) DQN: a value-based Q learning algorithm [10]; and 2) PPO: a policy gradient algorithm proposed by OpenAI [13].

2.1   Preliminaries

Deep Q-Network. The DQN agent aims to maximize the cumulative discounted rewards obtained by making sequential decisions. It uses a deep neural network to approximate the Q function. The optimal Q function at each step obeys the *Bellman equation*. The rationale behind *Bellman equation* is that the optimal policy at current step would be to maximize $E[r + \gamma Q^*(s', a')]$ given the optimal action-value function $Q^*(s', a')$ was completely known at next step [8], where γ is the discounted factor [9]

$$Q^*(s, a) = E_{s' \sim \varepsilon}[r + \gamma \max_{a'} Q^*(s', a') | s, a] \tag{1}$$

The DQN algorithm trains a neural network model to approximate the optimal Q function by minimizing a sequence of loss function $L_i(\theta_i) = E_{s, a \sim \rho(\cdot)}[(y_i - Q(s, a; \theta_i))^2]$ iteratively, where $y_i = E_{s' \sim \varepsilon}[r + \gamma \max_{a'} Q^*(s', a'; \theta_{i-1}) | s, a]$ is the target function and $\rho(s, a)$ refers to the probability distribution of states s and actions a. The model weights could be estimated by optimizing the loss function through stochastic gradient descent algorithms. In addition to the capability of handling high-dimensional problems, the DQN algorithm is also *model-free* and has no assumption about the dynamics of the environment. It learns about the optimal policy by exploring the state-action space [8].

Proximal Policy Optimization. The PPO algorithm, proposed by OpenAI [13], is sample-efficient, easy to implement, and has demonstrated outstanding performance on some difficult RL problems. It alternates between data sampling and optimizing a "surrogate" objective function to obtain optimal policies [8].

PPO is an on-policy algorithm and is well known for its robust performance. It clips updates policies via the equation below.

$$\theta_{k+1} = \arg \max_{\theta} E_{s, a \sim \pi_{\theta_k}}[L(s, a, \theta_k, \theta)] \tag{2}$$

where π is the policy, θ is the policy parameter, k is the kth step, a and s are action and state respectively. We can use SGD algorithm to optimize the objective function [13].

2.2   Problem Formulation

In this session [8], we describe the problem formulation for the two DRL based systems both proposed by Lin and Beling: 1) A modified DQN algorithm [5,7]; 2) PPO with Long short-term memory (LSTM) algorithm [6].

States. The state is a vector to describe the current status of the environment. In the optimal trade execution problem, the states consist of 1) Public state: market microstructure variables including top 5 bid/ask prices and associated quantities, bid/ask spread, 2) Private state: remaining inventory and elapsed time, 3) Derived state: features derived from historical LOB states to characterize the temporal component in the environment[1]. The derived features could be grouped into three categories: 1) Volatility in VWAP price; 2) Percentage of positive change in VWAP price; 3) Trends in VWAP price. More specifically, we derive the features based on the past 6, 12, and 24 steps (each step is a 5-s interval), respectively. Additionally, we use several trading volumes: 0.5*TWAP, 1*TWAP, and 1.5*TWAP[2] to compute the VWAP price. It is straightforward to derive features in 1). For 2), we record the steps that the current VWAP price increases compared with the previous step and compute the percentage of the positive changes. For 3), we calculate the difference between the current VWAP prices and the average VWAP prices in the past 6, 12, and 24 steps.

The researchers leverage the LSTM networks to process level 2 microstructure data to characterize the temporal components in the data [8].

Actions. We define the number of shares to trade at each step as the action. In the optimal trade execution framework, we limit the lower and upper bounds of actions from 0 and 2TWAP, respectively [8]. The total shares to trade for each stock are the same as in [5] and [6].

Rewards. As the reward structure is crucial to the DRL algorithm, we should carefully design it to facilitate the agent in learning the optimal policies. Previously [4,11,12], researchers use the IS[3], a commonly used metric to measure execution gain/loss, as the immediate reward.
Shaped Rewards. In [5,7], Lin and Beling adopt a shaped reward structure. They state that IS reward is noisy and nonstationary, which makes it an inappropriate choice as the reward function [8].
Sparse Rewards. In [6], Lin and Beling use a sparse reward function, which only provides a reward signal at the end of the trading period.

---

[1] For example, the differences of immediate rewards between different time steps, indicators to represent various market scenarios (i.e., regime shift, trends in price changes, etc.).

[2] TWAP represents the volume projections of the TWAP strategy in one step.

[3] Implementation Shortfall = (arrival price − executed price) × traded volume.

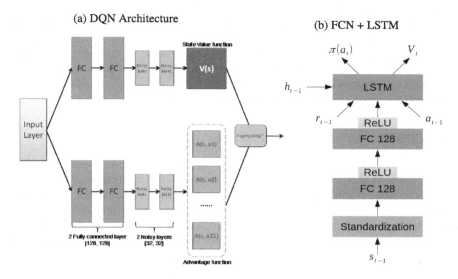

Fig. 1. DQN and PPO architectures.

Zero Ending Inventory Constraint. In this article, we adopt Lin and Beling's modified DQN algorithm [5,7], which combines the last two steps for Q-function estimation, to address the zero-ending inventory constraint [8].

2.3 Architecture

The same architectures and hyperparameters as in Lin and Beling's articles are used in this article [5–7]. The architectures are illustrated in Fig. 1 and Fig. 2, respectively.

3 A Real-Time Simulation Back-Testing Platform

One of the most significant drawbacks of back-testing on historical datasets is that it completely ignores the interactions among the market participants, which relies on the assumption that the market is resilient and will always bounce back to its previous equilibrium price levels next step. However, in the real world, the market resilient assumption does not always hold, since the liquidity is not infinite. Therefore, the models' performances on historical data may not reflect the actual performances.

We desire a high-fidelity market simulation environment for back-testing. Byrd et al. have developed a simulation environment named ABIDES, an Agent-Based Interactive Discrete Event Simulation environment, in 2019 [2]. ABIDES is modeled after NASDAQ's published equity trading protocols ITCH and OUCH and supports continuous double-auction trading at the same nanosecond time resolution as real markets [2]. Furthermore, it accounts for real-world factors such as network latency, agent computation delays, and allows interactions among agents, making it a desirable simulation environment for back-testing.

In this dissertation, we develop a real-time simulation back-testing platform by integrating the ABIDES simulation environment with Rllib. It allows the trained agents in Rllib to participate in the ABIDES simulation environment so that we could evaluate the agents' performances while they interact with other market participants. Such an evaluation could provide us insights on the robustness of the learned policy and how it would behave when interacting with other market players.

3.1   ABIDES

As mentioned earlier, ABIDES was developed by researchers from Gatech and JPMorgan AI Research in 2019 to support AI agent research in financial market applications. It includes a customizable configuration system, a simulation kernel, and a rich hierarchy of agent functionality. The simulation kernel is discrete event based, and it requires all agent messages to pass through its event queue. Also, it maintains a global current simulation time, while tracking a "current time" for each agent to ensure no inappropriate time travel exists. The simulation kernel follows the process in Fig. 2.

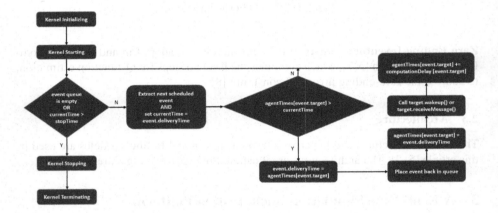

Fig. 2. ABIDES simulation Kernel event queue process

Fundamental Value and Its Properties.   In an agent-based financial market simulation, an exogenous price time series is needed to represent the stock's fundamental value. The fundamental value is derived from external factors and somehow reflects the true value of the stock. It changes over time in response to events or news about the company. In the simulation, there should be exactly one fundamental value time series for each stock. The fundamental value time series is predetermined, and agents do not affect it [1].

Megashock Ornstein-Uhlenbeck Process.   In our experiment, we use a continuous mean reverting series based on a modified Ornstein-Uhlenbeck (OU) process with "megashock events". The OU process, which we denote by Q_t, follows the differential equation below:

$$dQ_t = -\gamma(Q_t - \mu)dt + \sigma dW_t \qquad (3)$$

where W_t is a standard Brownian motion, γ and μ are positive constants, $\gamma < 1$. γ is the rate of mean reversion, and μ is the long term mean of the OU process around which the price fluctuates. The coefficient of dt, $-\gamma(Q_t - \mu)$, is called drift, and σ is called volatility. From Eq. 3, we observe that the drift term is negative when $Q_t > \mu$ and it is positive when $Q_t < \mu$. This explains why the OU process tends to revert towards μ when it is away from it. Additionally, the OU process is memoryless and has the Markov property. Given an initial value Q_0, Q_t is known to be normally distributed, and its expected value and variance are demonstrated as below [3].

$$E[Q_t] = \mu + (Q_0 - \mu)e^{-\gamma t} \qquad (4)$$

$$Var[Q_t] = \frac{\sigma^2}{2\gamma}(1 - e^{-\gamma t}) \qquad (5)$$

The ABIDES simulator developers have introduced the concept of "megashock events" to produce a price series that looks more like a real market price time series. Megashock events represent extrinsic news that occasionally occurs to change the valuation of stock significantly. They are drawn from a bimodal distribution with mean zero and arrive via a Poisson process [1].

4   Methodology

In this section, we compare the performances of the DQN and PPO LSTM models and observe that both models converge fast and have significantly outperformed the TWAP strategy on most stocks during the backtesting [8].

4.1   Data Sources

To create the training data, we use the NYSE daily millisecond TAQ data from January 1st, 2018 to September 30th, 2018, downloaded from WRDS. The TAQ data is used to reconstruct the LOB. Only the top 5 price levels from both seller and buyer sides are kept and aggregated at 5 s, which is the minimum trading interval. We trained the policies on the reconstructed LOB data. After training, we construct DRL agents based on the learned policies and have them execute trades while interacting with other agents such as zero-intelligence agents, TWAP agents, and other DRL agents in the ABIDES simulation environment for 6, 149 episodes, which is equivalent to 102.5 trading hours.

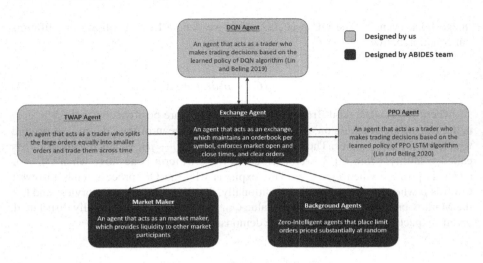

Fig. 3. Agents interactions in the ABIDES simulation Kernel

4.2   Experimental Methodology and Settings

In our experiments, we apply the two aforementioned algorithms on 14 US stocks. The experiment follows the steps below [8].

1. We obtain 2018 January–2018 September millisecond Trade and Quote (TAQ) data of 14 stocks above from WRDS and reconstruct it into the LOB. We set the execution horizon to be 1 min, and the minimum trading interval to be 5 s.
2. We apply the DQN/PPO architecture and hyperparameters to the 14 US stocks for training.
3. Upon the completion of training, we check the average episode rewards' progression. We construct the DRL agents based on learned policies and apply them to the ABIDES simulation environment. We let the two DRL agents and the TWAP agents execute 600 shares per minute for $6,149$ episodes, which is equivalent to 102.5 trading hours. Then, we compare their average episode rewards and the distribution of rewards against TWAP [9].

4.3   Algorithms

In this section, we compare the performances of the three algorithms below.

TWAP: The shares are equally divided across time.
DQN (Lin2019): The modified DQN algorithm tailored to the optimal trade execution problem [5].
PPO LSTM: The PPO algorithm uses LSTM to extract temporal patterns within market data.

4.4   Configuration for the Simulation Back-Testing Platform

The simulation back-testing platform is based on Rllib and ABIDES, a market simulator developed around a discrete event-based kernel. All agents' interactions are through the kernel's message system. In our experiments, we have included the following agents.

- 1 Exchange Agent. The exchange agent is the core of the market simulator. It plays a role as a stock exchange such as NASDAQ or NYSE in the simulator. It maintains the limit order book and allows trading agents to trade under the continuous double auction mechanism.
- 1 Market Maker Agent. The market maker agent is designed to provide liquidity to the simulated market.
- 1200 Zero Intelligence (ZI) agents. The ZI agents place limit orders with prices randomly determined. They do not possess much memory, intelligence, or visibility into the order stream. In our experiments, we use a particular type of ZI agents proposed by [14]. The ZI agents have a "strategic threshold" parameter $\eta \in [0, 1]$, which determines the surplus that the ZI agents demand to fill an order.
- 1 TWAP execution agent. An execution agent that splits and trades the shares equally across time (every 5 s for 1 min).
- 1 DQN execution agent. A trained execution agent that is based on the modified DQN algorithm proposed in Sect. 3.
- 1 PPO LSTM execution agent. A trained execution agent that is based on the PPO LSTM algorithm proposed in Sect. 4.

In the simulation, we use the default configuration for the network latency to be 20 ms. For computation delay, we set it to 50 ms. We use a generative sparse discrete mean-reverting process to simulate the stock prices. However, we set the stock prices within the minimum and maximum range of the historical data by letting the ZI agents to randomly choose the prices within the range [2]. In Fig. 3, we illustrate the agents interactions in our experiments. All the agents such as market maker, TWAP, DQN, PPO LSTM, and ZI agents interact with each other through the exchange agent. Market maker, DQN, PPO LSTM, and ZI both send and receive messages from the exchange agent, while the TWAP agent only sends messages to the exchange agent. The reason is that the TWAP agent does not need to query the order book information from the exchange to make trading decisions, while other agents do.

4.5   Main Evaluation and Back-Testing

We apply the trained DRL algorithms to the ABIDES simulation environment for 6, 149 episodes, which is equivalent to 102.5 trading hours, and compare their performances against TWAP. We report the mean and the median of $\Delta \text{IS} = \text{IS}_{\text{Model}} - \text{IS}_{\text{TWAP}}$ (in US dollars).

The statistical results for all the stocks are summarized in Table 1. We observe that both the proposed PPO LSTM and the DQN algorithms outperform the baseline model, TWAP, in most cases, while maintaining smaller standard deviations. The results suggest that the trained policies are robust, which provides us more confidence in their performances when they go live and operate in the real world financial markets [8].

Table 1. Model performances comparison. Left: mean is based on ΔIS (in US dollars); Right: median is based on ΔIS (in US dollars).

Ticker	DRL_LSTM	DRL_DENSE	TWAP	Ticker	DRL_LSTM	DRL_DENSE	TWAP
FB	**−8.33**	−9.35	−9.63	FB	**−7.25**	−7.81	−8.35
GOOG	−12.95	**−12.14**	−12.95	GOOG	−10.58	**−9.86**	−11.10
NVDA	**−11.66**	−11.95	−11.94	NVDA	**−9.34**	−9.54	−10.35
MSCI	3.92	**8.05**	4.41	MSCI	−0.79	**0.48**	−1.05
TSLA	**−10.62**	−11.29	−11.42	TSLA	**−8.78**	−9.46	−9.95
PYPL	**−9.11**	−10.80	−10.80	PYPL	**−7.28**	−8.90	−9.25
QCOM	**−8.39**	−9.48	−9.70	QCOM	**−7.06**	−7.81	−8.10
GS	**−12.81**	−14.12	−13.37	GS	**−10.40**	−11.60	−11.50
CRM	**−9.23**	−11.19	−11.15	CRM	**−6.91**	−9.12	−9.20
BA	**−12.58**	−13.60	−12.84	BA	**−10.21**	−11.04	−11.25
MCD	**−9.43**	−11.65	−11.62	MCD	**−7.42**	−9.61	−9.75
PEP	**−8.65**	−9.95	−10.18	PEP	**−7.25**	−8.54	−8.90
TWLO	**−7.64**	−8.95	−9.09	TWLO	**−6.11**	−7.82	−7.80
WMT	−13.99	**−12.20**	−13.14	WMT	−11.30	**−9.72**	−11.30

5   Conclusion and Future Work

In this section, we investigate the robustness of two DRL based trade execution models in a simulation environment. In our experiments, we have demonstrated that both DQN and PPO LSTM models are robust and can perform well in the simulation environment while trained on historical LOB data only. The experimental results are significant and indicate that the learned policies are robust and can adapt to a simulation environment in which market participants interact with each other. The development of such a simulation-based market environment is essential since we would like to evaluate the model performances before they go live in the trading systems.

References

1. Byrd, D.: Explaining agent-based financial market simulation. arXiv preprint arXiv:1909.11650 (2019)
2. Byrd, D., et al.: ABIDES: towards high-fidelity market simulation for AI research. arXiv preprint arXiv:1904.12066 (2019)
3. Chakraborty, T., Kearns, M.: Market making and mean reversion. In: Proceedings of the 12th ACM Conference on Electronic Commerce, San Jose, CA, USA (2011)
4. Hendricks, D., Wilcox, D.: A reinforcement learning extension to the Almgren-Chriss framework for optimal trade execution. In: Proceedings of the IEEE Conference on Computational Intelligence for Financial Economics and Engineering, London, UK, pp. 457–464 (2014)
5. Lin, S., Beling, P.-A.: Optimal liquidation with deep reinforcement learning. In: Proceedings of the 33rd Conference on Neural Information Processing Systems, Deep Reinforcement Learning Workshop, Vancouver, Canada (2019)

6. Lin, S., Beling, P.-A.: An end-to-end optimal trade execution framework based on proximal policy optimization. In: Proceedings of the 29th International Joint Conference on Artificial Intelligence (2020)
7. Lin, S., Beling, P.A.: A deep reinforcement learning framework for optimal trade execution. In: Dong, Y., Ifrim, G., Mladenić, D., Saunders, C., Van Hoecke, S. (eds.) ECML PKDD 2020. LNCS (LNAI), vol. 12461, pp. 223–240. Springer, Cham (2021). https://doi.org/10.1007/978-3-030-67670-4_14
8. Lin, S., Beling, P.A.: Investigating the robustness and generalizability of deep reinforcement learning based optimal trade execution systems. In: Arai, K. (ed.) Intelligent Computing. LNNS, vol. 284, pp. 912–926. Springer, Cham (2021). https://doi.org/10.1007/978-3-030-80126-7_64
9. Mnih, V., et al.: Playing Atari with deep reinforcement learning. arXiv preprint arXiv:1312.5602 (2013)
10. Mnih, V., et al.: Human-level control through deep reinforcement learning. Nature **518**, 529–533 (2015)
11. Nevmyvaka, Y., et al.: Reinforcement learning for optimal trade execution. In: Proceedings of the 23rd International Conference on Machine Learning, Pittsburgh, PA, pp. 673–68. Association for Computing Machinery (2006)
12. Ning, B., Ling, F.-H.-T., Jaimungal, S.: Double deep Q-learning for optimal execution. arXiv arXiv:1812.06600 (2018)
13. Schulman, J., et al.: Proximal policy optimization algorithms. arXiv:1811.08540v2 (2017)
14. Wah, E., et al.: Welfare effects of market making in continuous double auctions. J. Artif. Intell. Res. **59**, 613–650 (2017)

Looking Beyond the Haze: A Pyramid Fusion Approach

Harsh Bhandari and Sarbani Palit[✉][ID]

Indian Statistical Institute, 203, B.T. Road, Kolkata 700108, India
sarbanip@isical.ac.in
http://www.isical.ac.in

Abstract. Haze is a natural phenomenon caused mainly due to wide spread of dust particles from construction activities, smoke, water vapour causing poor visibility in the surrounding. Traditional methods normally involve enhancing the contrast of the image which fails in most of the cases except in light and uniform haze. Many neural network based algorithms have been developed which use hazy images and their corresponding non-hazy images i.e. ground truth, to create their model. However, such models are generally data specific and do not yield good results in every case as any specific benchmark data set with wide level of non-uniform haze is not available in public domain. In this paper, a simple and effective dehazing technique to increase the visual perception of a hazy image without any prior knowledge of the outdoor scene, has been proposed. Its most significant contribution is the formulation of an unique approach for extracting haze relevant features followed by an image fusion technique to finally obtain the resultant unhazed image with minimizing resource requirement.

Keywords: De-hazing · Pyramid fusion · Colorfulness · Saliency detection · Visibility · Entropy · Chromaticity

1 Introduction

The picture captured in the image in the presence of haze is often degraded in quality and visibility. Scattering of sunlight from particulate matter, water vapour present in air causes haze in the atmosphere. Many computer vision algorithms are based on assumptions that the images to be processed are without any sort of degradation including haze. It naturally follows that robust dehazing techniques are required for haze removal. The method of eliminating the impact of haze from hazy or foggy images is known as Image Dehazing, an example of which, using the proposed approach is presented in (1). As a consequence, different dehazing techniques have been established to restore the original haze-free image. Due to non-uniformity of the haze in the image and varying distance of the lens from the object, hence it cannot be addressed as a classical degradation problem and proper depth or density estimation of the haze is required for

© Springer Nature Switzerland AG 2021
T. Mantoro et al. (Eds.): ICONIP 2021, LNCS 13110, pp. 654–665, 2021.
https://doi.org/10.1007/978-3-030-92238-2_54

haze removal. Many Machine Learning based algorithms have also been proposed that uses prior information such as non-hazy ground-truth images to enhance the hazy image [3]. Different filtering methods have been designed to increase the contrast of an image [10]. Many algorithms have been proposed that estimate different atmospheric models and have used reverse technique to get unhazed image from a hazy image [12,13].

In this paper, an algorithm for dehazing which derives its inspiration from various characteristics of a hazy image, has been proposed. The major aim of the research work is to develop a novel approach which neither requires any prior information of hazy image nor depends on knowledge regarding any atmospheric parameter.

The proposed dehazing algorithm can be divided into three phases:

- Preprocessing of the color hazy image using white balancing and contrast enhancement.
- Extraction of features.
- Image Fusion

Section 2 presents a brief review of related works in this area. Section 4 describes the algorithm proposed by us. In Sect. 5, we report the results obtained using the proposed algorithm and also the comparison with the existing methods. Concluding remarks have been made in Sect. 6.

2   Related Work and Background

Several dehazing algorithms have been proposed till now. Dehazing requires accurately estimating the depth of the haze since degradation caused because of presence of haze increases with increase in distance from camera lens to object. Prior works in this field used multiple images pertaining to same background. Fang in [1] proposed a dehazing technique with the help of polarization effects due to both airlight and radiance of the scene. Ancuti in [20] proposed a multiscale fusion method for dehazing the hazy image. Tan in [6] proposed a technique for enhancing the quality of a degraded image caused because of severe weather. The technique makes two assumptions, first, the non-degraded images have better contrast in nature and secondly, the changes in the airlight tends to be smooth. Fattal in [2] developed a single image dehazing technique by considering separation of object shading and medium transmission. He in [4] developed a dehaze algorithm based on dark channel prior. A more better and efficient method is to use only single hazy image for dehazing, but estimating the fog density from the single image is an arduous task. Thus majority of the dehazing algorithms use estimated transmission map to enhance the visibility of the hazy image using the Koschmieder's atmospheric scatttering model [8]. Bovik in [15] have developed a new technique named Fog Aware Density Evaluator(FADE) on the basis of certain fog related statistical features. It estimates the visibility of hazy scene without any reference ground-truth. Though the algorithm substantially removes effects of haze from the image, but it also incorporates color distortions in the unhazed image.

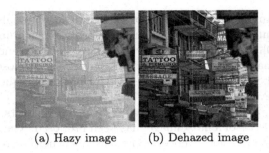

(a) Hazy image   (b) Dehazed image

Fig. 1. Image dehazing using the proposed algorithm

3   Haze Imaging Model

The atmosphere is filled with particles that affect the passage of light, leading to degradation in visibility and consequently, detection of objects more difficult. Efficiently modeling the scattering of light due to particles is a complex problem due to the fact that it depends on multiple factors such as varying sizes, type and distribution of the particles of the medium, wavelength of the incident light and many more. In such case, the visibility of such object is controlled by two concurrent events: first, light from the object gets attenuated due to scattering in different directions and absorption by dust, smoke and other particles. Second, the airlight coming from the Sun, acting as light source is scattered by particulate matters towards the lens, along the line of sight. Additionally, both transmission and airlight not only depends on the weather and angle of sunlight, but also on the concentration of air particles. The most widely used haze creation model was given by McCartney in [7]. According to the model, the hazy image is a linear combination of Direct Attenuation and Airlight.

$$I(x) = J(x)t(x) + A(1 - t(x)). \tag{1}$$

where $I(x)$ is the image and A is the Airlight. The transmission $t(x)$ depends on the distance of the object from the camera and the scattering coefficient β of the particles [8] such that

$$t(x) = e^{-\beta.d(x)}. \tag{2}$$

where $t \in \varepsilon[0, 1]$. When $t = 1$, light reaches the camera without any attenuation while $t = 0$, indicates extinction of object by the fog.

4   Proposed Algorithm

To begin with, features are selected such that they can be computed from intermediary or pre-processed images with the primary objective of using them to distinguish between natural images and images affected by the presence of haze. A fusion strategy for combining these is implemented to increase the visual perception of the hazy image I.

Two images are obtained from the original hazy image. The first image I_{WB} is derived by performing white balancing on the original image. The second input image I_{CE} is obtained by enhancing the contrast of the original image as outlined in Algorithm 1. It may be noted that the direct use of CLAHE algorithm or any other histogram equalization algorithm adds artificial colours to the output image and is hence not resorted to here.

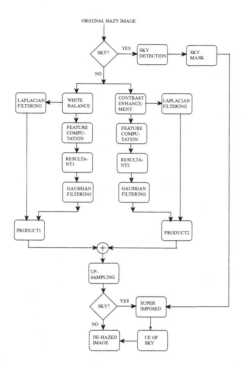

Fig. 2. The block diagram of the proposed approach

As already mentioned, features which shows significant correlation with the level of haze present in the image are extracted from each the derived input images. The obtained images and the extracted features are combined by performing fusion operation to get the final dehazed image.

The density of haze is non-uniform in nature. Thus, applying dehazing on the entire image does not yield a good result. To overcome this problem, patch based dehazing is proposed. This is because the depth of the scene and the transmittance are assumed to be constant in the patch provided the size of the patch is reasonable. However, the effect of dehazing on patches having smooth areas or poor contrast *e.g.* a typical sky area, is neither apparent nor measurable and does not make a significant difference in the quality of the image. Therefore, while processing the images, regions having smooth areas are not considered for dehazing.

Input: Hazy Image I
Output: Contrast Enhanced I_{CE}
1: $I_{gray} \leftarrow rgb2gray(I)$
2: $MeanIm_c \leftarrow Mean(I_c)$ where c ϵ [r,g,b]
3: $K_c \leftarrow (I_c * I_{gray})/MeanIm_c$
4: $I_1 \leftarrow concat(K_c)$
5: $I_2 \leftarrow imadjust(K_c)$
6: $G \leftarrow imgaussfilt(I_1)$
7: $J \leftarrow I_1 + (I_1 - G)$
8: $X \leftarrow WaveletFusion(I_2, J)$
9: $I_{CE} \leftarrow histeq(X)$

Algorithm 1: Contrast Enhancement

Outdoor hazy images usually contain both sky and non-sky regions. It is observed that sky regions are highly smooth areas which predominantly do not provide any vital output after applying dehazing algorithms. So, we focus on non-sky regions for application of the dehazing algorithm. To segment the sky region from the non-sky region, a mask needs to be constructed for which a sky detection algorithm such as [19] is used. For images with negligible sky region, the sky-non sky segmentation is bypassed by directly extracting features from the derived hazy images.

As shown in Fig. 2, the input hazy image is preprocessed to derive a white balanced image as well as an contrast enhanced image. The white balanced image eliminates the color constancy present in the image such that the objects that appear white in person are rendered white in image. The contrast enhanced image is obtained by employing Algorithm 1 rather than using histogram equalization or CLAHE which work only for light and uniform hazy images. Thus contrast enhancement is used as a preprocessing tool in order to further improve the quality of the dehazed image. The computation of features for the hazy image has been done from non-overlapping patches. It is assumed that the transmission value t of a hazy image is not constant throughout the image due to the density of haze being non-uniform and also varying with the depth of the objects in the image.

4.1 Feature Extraction

As shown in Fig. 3, for an image without significant sky region, two images are computed referred to as the White Balanced and Contrast Enhanced Image. Through out this paper, these will be referred to as input derived images with k = 1 denoting the White Balanced derived image and k = 2 indicating the other. Five window based attributes namely Colorfulness, Visual saliency, Entropy, Visibility and Chromaticity are computed for each of the input derived images. The spatial distribution of a particular attribute for each of the input derived images forms the featuremap of that attribute. Normalized weightmaps for each attribute are obtained by considering weightmaps of both input derived images

Fig. 3. Overall description of the proposed algorithm using an example of hazy image. The first column indicates the input hazy image. The second column provides the pre-processed images with the upper one being White-Balanced and lower one being Contrast Enhanced. The third till the seventh column being the featuremaps representing Chromaticity, Colorfulness, Entropy, Saliency and Visibility respectively on the derived images respectively. The eighth column is the normalized resultant featuremap of each input derived image. Finally, the dehazed output generated after applying fusion process on the two resultant images shown in Fig. 1(b).

in order to have the same scale as that of the input. The normalized value of the jth feature($j \in (1,5)$) for each patch i is computed as in (3).

$$W_{k,j}^i = \frac{feature_k^i}{feature_1^i + feature_2^i} \tag{3}$$

4.1.1 Colorfulness

The presence of haze degrades the color of the image and causes alteration of scene color which leads to bad visual perception of it. Colorfulness is defined as the degree of difference between the gray and the color. Thus Colorfulness is expressed as a linear combination of the local mean and local standard deviation. These parameters are computed as stated in [4] by splitting the image into small patches. Figure 4 shows the results of such an analysis by taking into consideration non-overlapping 5×5 patch for both haze free and the corresponding hazy image. Figure 4(e) illustrates the correlation between the sample standard deviation and sample mean for image in Fig. 4(a). Presence of fog causes shift in the color of the scene as evident from fall in the local mean and local standard deviation compared to the ground truth leading to decrease in colorfulness.

$$ColorFulness^k = w_1 \cdot \sqrt{\sigma_a^2 + \sigma_b^2} + w_2 \cdot \sqrt{\mu_a^2 + \mu_b^2} \tag{4}$$

where σ_a and σ_b are standard deviation and μ_a and μ_b are sample mean of the channel a and b of CIEL*a*b colorspace, w_1 and w_2 are weights assigned to the sample standard deviation and sample mean.

4.1.2 Visual Saliency

The human visual system pays varied attention to different objects and regions of the scene. Regions with more unique and well defined objects grab more attention. Haze in the scene causes the closer objects to be more prominent since saliency falls drastically with even small increase in the depth. Saliency of an image highlights the object in an image with respect to its surrounding. The Visual Saliency is obtained by fusing three priors obtained from the color image: color, frequency and location [16].

$$Saliency^k = SF.SC.SD$$
$$SF = (I_L * g)^2 + (I_a * g)^2 + (I_b * g)^2$$
$$SC = 1 - \exp(-\frac{(I_a)^2 + (I_b)^2}{\sigma_c^2})$$
$$SD = \exp\left(-\frac{(k - c)^2}{\sigma_d^2}\right)$$

where I_L, I_a and I_b are the three channels obtained from color image I to CIE L*a*b* colorspace. σ_c and σ_d are parameters, c is the center of the input image window W_k and normalized values of I_a and I_b are used in the equations. g is 2-D circular symmetric log-Gabor filter. Further objects closer to the central area of the image are more apparent to human perception, hence they are more likely to be salient than ones far away from the center. For the construction of the Visual Saliency Map, computation of the Visual Saliency is performed as given by (5).

$$VisualSaliency^k = Saliency^k * G_{r,\sigma} \tag{5}$$

where $G_{r,\sigma}$ is a Gaussian Filter with mean r and sigma σ.

As can be analysed from Fig. 4(d), histogram of the saliency map of the images in Fig. 4(b) containing natural scenes with different levels of haze, as the level of haze reduces the saliency distribution generally follow a definitive regular pattern with the reduction in skewness. On the other hand histogram of the saliency map of the images containing haze shows irregular pattern with increase in distortions as the density of haze increases. Thus the saliency very well tries to capture the distortion in the image caused by the presence of haze.

4.1.3 Entropy

Entropy of an image provides the measure of amount of information contained in an image. With increase in the density of fog, the entropy of the image decreases. The entropy feature map captures the gray level distribution variation of the image . The image is split into multiple patches, with each patch providing separate information. With $M \times N$ being the window size, the entropy of a window can be computed in the spatial domain. This window based entropy computation for the entire image gives its weightmap.

$$Entropy = \sum_{i=0}^{L-1} p_i \log \frac{1}{p_i} \tag{6}$$

where p_j = probabiity distribution of intensity level j in the window having L gray levels.

4.1.4   Visibility

Visibility of an image indicate the clarity of it. A color image can be transparent or non-transparent based on the medium. The visibility of an image is linked to the maximum distance at which an object of suitable size from the ground can be seen and recognized by the observer against the sky horizon. Visibility feature map is defined using a blur estimation technique. The color image is initially blurred using a Gaussian function as in (7). Then root squared blurred difference between the color image and its blurred version is used to measure the visibility as described in (8).

$$I_{blur} = I(x,y) \otimes G_M(x,y,\sigma_1)$$
$$V_1 = \sqrt{(I(x,y) - I_{blur}(x,y))^2 \otimes G_M(x,y,\sigma_2)}$$
$$V_2 = rgb2gray(V_1) \tag{7}$$

$$Visibility^k = \sum_{i=0}^{M-1} \sum_{j=0}^{N-1} \frac{V_2(i,j) - \mu}{\mu^{\alpha+1}} \tag{8}$$

Here $G_p(x,y,\sigma)$ is a 2D circular symmteric weighting function of size $p \times p$ patch and standard deviation σ. The operator $\otimes$ indicates a 2D convolution operation. $V_2(i,j)$ denotes gray intensity at position (i, j). α is visual coefficient and μ is the average gray intensity in the patch.

4.1.5   Chromaticity

Chromaticity is a specification of the quality of color consisting of hue and saturation. The presence of fog in an image does not have any impact on the hue of the color but severely affects the saturation. With an increase in the density of haze, the saturation of the color decreases leading to a drop in the chromaticity. Thus, to adjust the saturation gain, chromaticity feature map computed according to 9 has been adopted using a window W of size M×N.

$$Chromatic^k = \exp\left(-\frac{(S_j - S_{max})^2}{2\sigma_e^2}\right) \tag{9}$$

S_j is the saturation of the jth patch, S_{max} is the maximum saturation of the corresponding patch and σ_e is the standard deviation.

$$F_l = \sum_k \{G_l\{W^k\}\}L_l\{I_k\} \tag{10}$$

where l indicates the pyramid levels. Value of, $l = 7$ have been set experimentally to create the Gaussian and Laplacian levels represented by G_l and L_l respectively with the operations being performed successively on each level in a bottom-up manner. Finally, a dehazed image is obtained by performing Laplacian pyramid fusion using an upsampling operator with factor of 2^{l-1}.

4.2   Image Fusion

In order to generate a consistent result, each feature map is obtained by normalizing the features computed for each derived input. Thus two resultant feature

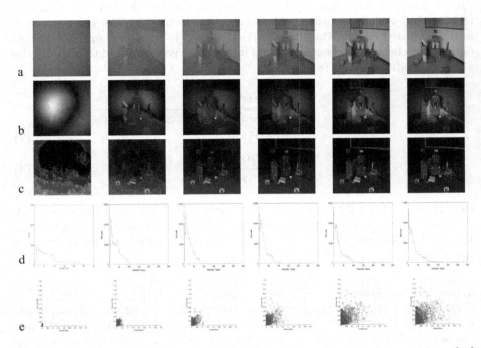

Fig. 4. Variation in the Saliency and Colorfulness Map using the CHIC Database [17] with the haze level increasing right to left with the last column representing the ground truth

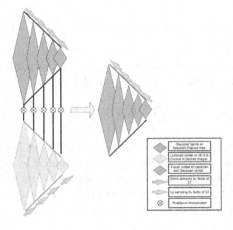

Fig. 5. The block diagram of the Pyramid Fusion approach

maps are obtained one for each of the derived input image by combining the five features as proposed above and the resultant feature map is indicated below:

$$Resultant feature Map_k = \prod_{i=1}^{5} \lambda^i W_k^i \qquad (11)$$

where λ^i is a multiplicative factor assigned to each weight map and assumed to be equal for all weightmaps.

Image Fusion is a technique where information extracted from multiple images are fused to derive a single image. Direct application of the normalized weightmaps introduces halo artifacts. To overcome the problem, multi-scale Gaussian band pass filter with down sampling is used on each of the derived input images and a band pass down sampling laplacian filter is applied on the resultant Weight Map. Finally, fusion is performed in a pyramid hierarchical manner. The sizes of frequency bands will decrease with the decomposition level. The Laplacian weights and the Gaussian derived inputs are fused at each level followed by up-sampling separately to get an inverse fused laplacian pyramid as shown in Fig. 5.

5   Experimental Results

In this section, the performance of the proposed method is evaluated by making a qualitative as well as quantitative study. For this purpose, certain benchmark datasets are used and results are compared with those of state-of-the-art methods whose codes are available in public domain. Quantitative evaluation of a dehazing technique can be made using three metrics proposed in [5]. Metric e computes ratio of visible edges present in a dehazed image against the edges in a hazy image. Metric $\sum$ indicates the percentage of pixels which become black or white upon performing dehazing. Metric $\bar{r}$ indicates the mean ratio of the gradient obtained before and after dehazing. Further patches of various sizes were used to analyze the performance of the algorithm.

5.1   Parameter Settings and Simulation Environment

Based on the feature weightmaps, the parameters used in our approach were empirically tuned as: $M = 11$, $N = 11$, $\omega_1 = 0.3$, $\omega_2 = 1$, $\sigma_c = 0.25$, $\sigma_d = 114$, $\alpha = 0.65$, $\sigma_e = 0.7$. The matlab program is simulated under the stated computer configuration: Intel(R) Core(TM) $i7 - 9700KU$ CPU @ 3.6 GHz, 8 Core(s) with 32 GB RAM having version MATLAB 2019b.

5.2   Performance Analysis

As can be observed from Table 1, taking into consideration severely dense, hazy images, the proposed algorithm restores the image in a much better way than many of the state-of-the-art algorithms. Results obtained from [18] though showing good results faces a major drawback i.e. the selection of the most haziest region in the image has to be done by the user thus making it an interactive algorithm.

Hazy Image DehazeNet[14] DEFADE[15] BCCR[18]   Proposed   Ground Truth

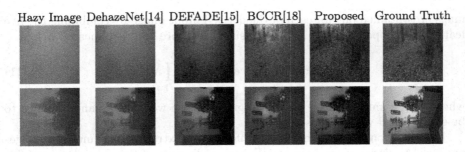

Fig. 6. Comparison of single image dehazing algorithms for densely hazed images using the O-haze Dataset [11] and the I-haze Dataset [9]

Table 1. Comparison of dehazed images shown in Fig. 6

Foggy Image [11]	DehazeNet [14]			Defade [15]			BCCR [18]			Proposed		
	e	$\sum$	$\bar{r}$	e	$\sum$	$\bar{r}$	e	$\sum$	$\bar{r}$	e	$\sum$	$\bar{r}$
a	6.2432	0%	1.4112	0.6705	0%	2.2222	37.8797	0%	**7.1776**	**43.1058**	0%	4.5276
g	2.8093	0%	1.6081	9.1045	0%	1.8996	11.7851	0%	3.8604	**11.9848**	0%	**4.512**

6   Conclusion and Future Work

A new framework for dehazing problem has been presented. The proposed algorithm consists of two derived input images from which five featuremaps are generated depicting the features extracted from the images differentiating between the haze-free and the hazy images. The experimental results indicate that the developed algorithm indicate good results as compared to many state-of-the-art algorithm even though no complex networks requiring training images in large numbers have been deployed. In future the algorithm can be extended to perform dehazing on non-uniformed as well as night time hazy images.

References

1. Fang, S., Xia, X.S., Huo, X., Chen, C.W.: Image dehazing using polarization effects of objects and airlight. Opt. Exp. **22**, 19523–19537 (2014)
2. Fattal, R.: Single Image dehazing. ACM Trans. Graph. SIGGRAPH **27**(3), 72 (2008)
3. Fang, S., Zhan, J., Cao, Y., Rao, R.: Improved single image dehazing using segmentation. In: 2010 17th IEEE International Conference on Image Processing (ICIP), pp. 3589–3592. IEEE (2010)
4. He, K., Sun, J., Tang, X.: Single image haze removal using dark channel prior. In: Proceedings IEEE Computer Society Conference on Computer Vision and Pattern Recognition, pp. 1956–1963 (2009)
5. Hautière, N., Tarel, J.-P., Aubert, D., Dumont, E.: Blind contrast enhancement assessment by gradient ratioing at visible edges. Image Anal. Stereol. **27**, 87–95 (2008). https://doi.org/10.5566/ias.v27.p87-95

6. Tan, R.T.: Visibility in bad weather from a single image. In: IEEE Conference on Computer Vision and Pattern Recognition, CVPR 2008, pp. 1–8. IEEE (2008)
7. McCartney, E.J.: Optics of the Atmosphere Scattering by Molecules and Particles, p. 421. Wiley, New York (1976)
8. Koschmieder, H.: Theorie der horizontalen Sichtweite: Kontrast und Sichtweite. Keim and Nemnich, Munich (1925)
9. Ancuti, C., Ancuti, C.O., Timofte, R., De Vleeschouwer, C.: I-HAZE: a dehazing benchmark with real hazy and haze-free indoor images. In: Blanc-Talon, J., Helbert, D., Philips, W., Popescu, D., Scheunders, P. (eds.) ACIVS 2018. LNCS, vol. 11182, pp. 620–631. Springer, Cham (2018). https://doi.org/10.1007/978-3-030-01449-0_52 arXiv:1804.05091
10. He, K., Sun, J., Tang, X.: Guided image filtering. IEEE Trans. Pattern Anal. Mach. Intell. **35**(6), 1397–1409 (2013)
11. Ancuti, C.O., Ancuti, C., Timofte, R., De Vleeschouwer, C.: O-HAZE: a dehazing benchmark with real hazy and haze-free outdoor images. In: IEEE/CVF Conference on Computer Vision and Pattern Recognition Workshops (CVPRW) (2018)
12. Park, H., Park, D., Han, D.K., Ko, H.: Single image haze removal using novel estimation of atmospheric light and transmission. In: 2014 IEEE International Conference on Image Processing (ICIP), pp. 4502–4506. IEEE (2014)
13. He, K., Sun, J., Tang, X.: Single image haze removal using dark channel prior. IEEE Trans. Pattern Anal. Mach. Intell. **33**(12), 2341–2353 (2011)
14. Berman, D., Treibitz, T., Avidan, S.: Non-local image dehazing. In: 2016 IEEE Conference on Computer Vision and Pattern Recognition (CVPR), Las Vegas, NV, pp. 1674–1682 (2016). https://doi.org/10.1109/CVPR.2016.185
15. Choi, L., You, J., Bovik, A.: Referenceless prediction of perceptual fog density and perceptual image defogging. IEEE Trans. Image Process. **24**, 3888–3901 (2015). https://doi.org/10.1109/TIP.2015.2456502
16. Zhang, L., Gu, Z., Li, H.: SDSP: a novel saliency detection method by combining simple priors. In: 2013 IEEE International Conference on Image Processing, Melbourne, VIC, pp. 171–175 (2013). https://doi.org/10.1109/ICIP.2013.6738036
17. El Khoury, J., Thomas, J.B., Mansouri, A.: A database with reference for image dehazing evaluation. J. Imaging Sci. Technol. **62**, 10503–1 (2017). https://doi.org/10.2352/J.ImagingSci.Technol.2018.62.1.010503
18. Meng, G., Wang, Y., Duan, J., Xiang, S., Pan, C.: Efficient image dehazing with boundary constraint and contextual regularization. In: 2013 IEEE International Conference on Computer Vision, Sydney, NSW, pp. 617–624 (2013). https://doi.org/10.1109/ICCV.2013.82
19. Song, Y., Luo, H., Ma, J., Hui, B., Chang, Z.: Sky detection in hazy image. Sensors (Basel) **18**(4), 1060 (2018). https://doi.org/10.3390/s18041060
20. Ancuti, C.O., Ancuti, C., Bekaert, P.: Effective single image dehazing by fusion. In: 2010 17th IEEE International Conference on Image Processing (ICIP), pp. 3541–3544. IEEE (2010)

DGCN-rs: A Dilated Graph Convolutional Networks Jointly Modelling Relation and Semantic for Multi-event Forecasting

Xin Song[1], Haiyang Wang[1], and Bin Zhou[2(✉)]

[1] College of Computer, National University of Defense Technology, Changsha, China
{songxin,wanghaiyang19}@nudt.edu.cn
[2] Key Lab. of Software Engineering for Complex Systems, School of Computer, National University of Defense Technology, Changsha, China
binzhou@nudt.edu.cn

Abstract. Forecasting *multiple co-occurring events of different types* (a.k.a. *multi-event*) from open-source social media is extremely beneficial for decision makers seeking to avoid, control related social unrest and risks. Most existing work either fails to jointly model the entity-relation and semantic dependence among multiple events, or has limited long-term or inconsecutive forecasting performances. In order to address the above limitations, we design a Dilated Graph Convolutional Networks (DGCN-rs) jointly modelling relation and semantic information for multi-event forecasting. We construct a temporal event graph (TEG) for entity-relation dependence and a semantic context graph (SCG) for semantic dependence to capture useful historical clues. To obtain better graph embedding, we utilize GCN to aggregate the neighborhoods of TEG and SCG. Considering the long-term and inconsecutive dependence of social events over time, we apply dilated casual convolutional network to automatically capture such temporal dependence by stacked the layers with increasing dilated factors. We compare the proposed model DGCN-rs with state-of-the-art methods on five-country datasets. The results exhibit better performance than other models.

Keywords: Multi-event forecasting · Relation and semantic · GCN · Dilated casual convolutional network

1 Introduction

Social events initiated by different organizations and groups are common in reality, and they are usually *co-occurring and of different types*, a.k.a. *multi-event*, such as *rally broke out at different cities simultaneously*, or *protests happened periodically within several weeks or months*, etc. The ability to successfully forecast multi-event thus be extremely beneficial for decision makers seeking to avoid, control, or alleviate the associated social upheaval and risks. Traditional methods tend to focus on the prediction of some single type of event or event scale

© Springer Nature Switzerland AG 2021
T. Mantoro et al. (Eds.): ICONIP 2021, LNCS 13110, pp. 666–676, 2021.
https://doi.org/10.1007/978-3-030-92238-2_55

[2,4]. The most intuitive way for *multi-event* forecasting is that a single event type forecasting method is repeated for different types. Although this approach simplifies the multi-event forecasting model, it ignores the potential relation and semantic dependence among multi-event.

Recently, researchers [3] attempt to model *multi-event* as *temporal event graph* (TEG), which can be viewed as a sequence of event graphs splited in time ascending order. Figure 1 shows an example subgraph of a TEG, the event (*Protester, Demonstrate or rally, Isfahan Gov, 2018-04-23*) mentioned *Escalation of Protests in Iran's Drought-Ridden Isfahan Province*. The previous events contain some useful information to provide historical clues for future prediction: (1) potential relation and semantic information in periodically and correlated events, such as rallies on 2018-04-11 and 2018-04-23; (2)temporal dependence, including the long temporal range (from *2018-04-05* to *2018-04-23*) and inconsecutive temporal dependence across different time intervals. How to identify such information to help predict multi-event in the future is still challenging.

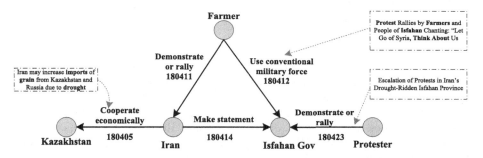

Fig. 1. An example subgraph of temporal event graph, where entities denote event actors and relations (edges) with a timestamp denote event types between two entities.

Obviously, the relation information between entities and the semantic information of events are inseparable, and jointly modeling them is vital for multi-event forecasting. Existing methods [7] consider relation information but ignore semantic information. Besides, some methods [18] using the pre-defined feature of events without dynamic development of events over time. Recently, graph features [2] constructed from text data have proven to be advantageous. In the paper, we extract event keywords under each timestamp to jointly construct a semantic context graph (SCG). Then, we introduce knowledge-aware attention based GCN [15] to capture the relation dependence of TEG at each timestamp for better relation embedding. Meanwhile, we utilize GCN to capture the semantic relevance of SCG at each timestamp for better semantic embedding.

In addition to preserve relation and semantic dependence, temporal dependence is also important, such as long-term or inconsecutive dependence with different temporal intervals. However, traditional methods [9,12] based on RNN, usually learn all features of historical timestamps equally, which is limited to capture the inconsecutive dependence across different time intervals and long-term dependence. On the contrary, CNN-based methods, such as dilated convolution, enjoy the advantages of parallel computing and can capture very long sequences by stacking layers. To ensure the sequence of data, some methods [14] combine dilated convolution with causal convolution. Hence, we introduce a dilated casual convolution(DCC) to capture the inconsecutive dependency by setting different dilated factors and capture the long-term dependencies by stacked convolution layers. Our contributions are summarized as follows:

1 We construct the TEG and SCG to jointly model the relation and semantic dependence of social events, which is helpful to capture the dynamic changes of events over time and provide historical clues for future forecasting.
2 We design a dilated casual convolution network to automatically capture temporal dependence among events, including long-term and inconsecutive dependence by stacked the layers with increasing dilated factors.
3 We evaluate the proposed model on five-country datasets with state-of-the-art models and demonstrate the effectiveness of our model by ablation test and sensitivity analysis.

2   Related Work

There has been extensive research to predict social events. Linear regression models use the frequency of tweets to predict when future events will occur [1]. More sophisticated features (e.g., topic-related keywords [13]) and multi-task multi-class deep learning models (for example, SIMDA [5], MITOR [4]) to predict sub-types and the scale of spatial events. Recently, GCN-based methods have also been studied. For example, DynamicGCN [2], REGNN [10], etc. Recently, the temporal knowledge graph modeling methods have applied in event forecasting. Existing work mainly follows two directions: The first is a combination of GNN and RNN, for example, glean [3], Renet [7], etc. Then, some more complex methods are studied, such as DCRNN [9], GaAn [17]. However, the RNN-based method is inefficient for long sequences and its gradient may explore when they are combined with GCNs. Therefore, CNN-based methods are proposed to encode temporal information, which expand the fields of neural network by stacking many layers or using global pooling. For example, STGCN [16], Graph WaveNet [14], etc.

3   Methodology

Figure 2 provides the system framework of DGCN-rs. Firstly, we construct TEG and SCG to jointly model social events. Then we utilize the knowledge-aware attention based GCN and GCN to aggregating neighborhood information of TEG and SCG at each timestamp, respectively, for better relation and semantic embedding. To capture the temporal dependence in TEG and SCG, such as the long-term or inconsecutive dependence, we design a dilated casual convolution for temporal embedding and obtain the global historical embedding X_t. We finally add a (MLP) to predict the probability of co-occurring events at $t + 1$.

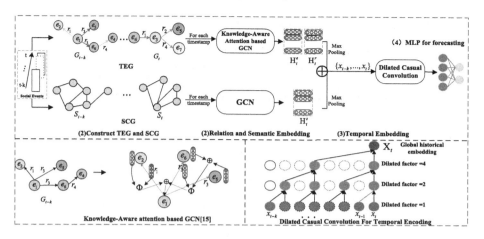

Fig. 2. System framework of our proposed model DGCN-rs.

3.1 Preliminaries

Temporal Event Graph (TEG). We construct a TEG based on event entities, event types and time to model the relation information between events. A TEG is defined as $\mathcal{G} = \{\mathcal{G}_{t-k}, \ldots, \mathcal{G}_{t-1}, \mathcal{G}_t\}$, where $\mathcal{G}_t$ is a multi-relation, directed graph with a time-stamped edges (event types) between nodes (entities). Let $\mathcal{E}$ and $\mathcal{R}$ be the finite set of nodes and edges. Then, a social event is defined as a quadruple (s, r, o, t) or $(s, r, o)_t$. where $s, o \in \mathcal{E}$ and $r \in \mathcal{R}$ and the direction of edge is pointing from s to o. We denoted a set of events at time t as $\mathcal{G}_t = \{(s, r, o)_t\}$.

Semantic Context Graph (SCG). We construct a SCG based on the event content text to model the context semantic information. A SCG is defined as $\mathcal{S} = \{\mathcal{S}_{t-k}, \ldots, \mathcal{S}_{t-1}, \mathcal{S}_t\}$, where each semantic graph $\mathcal{S}_t$ is a simple undirected graph, where nodes denote keywords extracted by removing very common words and rare words from the content text, the edges are based on word co-occurrence in the collection of content texts at timestamp t. The edges weights between two nodes are calculated by PMI [6] to measure the semantic relevance between words. We define there is no edge between word i and j when the $PMI(i, j) < 0$.

Problem Formulated. Given an observed TEG and SCG, we aim to encode the historical social events and learn a function f which can forecast the co-occurring probability of different event types at the future timestamp $t + 1$, denoted as:

$$\{G_{t-k:t}, S_{t-k:t}\} \overset{encode}{\rightarrow} X_t \overset{model}{\rightarrow} P\left(Y_{t+1} \mid G_{t-k:t}, S_{t-k:t}\right) \qquad (1)$$

where $Y_{t+1} \in R^{|\mathcal{R}|}$ is a vector of event types. X_t is a global historical embedding based on all social events of k historical consecutive time steps.

3.2   GCN for Relation and Semantic Embedding

After constructing the TEG and SCG, we utilize the knowledge-aware attention based GCN [15] and GCN [8] to aggregate the local neighborhood information of TEG and SCG, respectively, so that we can capture the relation dependence between entities and the semantic relevance between words.

Relation Embedding. Each event graph $\mathcal{G}_t$ expresses relation information among *multi-event*. Based on our previous work [15], we utilize knowledge-aware attention based GCN to effectively aggregate neighborhood information of an entity s, which can distribute different importance scores to neighboring relations and neighboring entities:

$$\mathrm{h}_s^{(l+1)} = f\left(\sum_{(r,o)\in N(s)} W_q^{(l)} \Phi\left(\alpha_{s,r}\mathrm{h}_r^{(l)}, \beta_{o|s,r}\mathrm{h}_o^{(l)}\right)\right) \tag{2}$$

Here, $\Phi : R^d \times R^d \to R^d$ is a composition operator. We choose multiplication as Φ. $\mathrm{h}_r^{(l)}$ and $\mathrm{h}_o^{(l)}$ denote feature embedding in l-th layer for relation r and entity o, respectively. W_q is a relation-specific parameter. $f(\bullet)$ is the ReLU activation function. The $\alpha_{s,r}$ represents the weight of relation r and $\beta_{o|s,r}$ represents the weight of object entity o under the same s and r, which are calculated as follows:

$$\alpha_{s,r} = \frac{\exp\left(\sigma\left(m^{\mathrm{T}} \cdot \mathrm{W}_1\left(h_s, h_r\right)\right)\right)}{\sum_{r_j \in N_s} \exp\left(\sigma\left(m^{\mathrm{T}} \cdot \mathrm{W}_1\left(h_s, h_{r_j}\right)\right)\right)} \tag{3}$$

$$\beta_{ols,r} = \frac{\exp\left(\sigma\left(n^T \cdot \mathrm{W}_2\left(\left(h_s, h_r\right), h_o\right)\right)\right)}{\sum_{o_j \in N_{s,r}} \exp\left(\sigma\left(n^T \cdot \mathrm{W}_2\left(\left(h_s, h_r\right), h_{o_j}\right)\right)\right)} \tag{4}$$

where σ is LeakyRelu and W_1 and W_2 are the trainable parameter. Next, we update the embedding vector of relation r:

$$\mathrm{h}_r^{(l+1)} = W_{rel}^{(l)}\mathrm{h}_r^{(l)} \tag{5}$$

where, $W_{rel}^{(l)}$ is a learnable transformation matrix in the l-th layer, which can project all the relations to the same embedding space as entities. By calculating the embedding of all entities and relations at each timestamp in TEG, we can get a matrix sequence $\left\{\left[H_{t-k}^e, H_{t-k}^r\right], \ldots, \left[H_t^e, H_t^r\right]\right\}$.

Semantic Embedding. In addition to relation embedding, the content text of social events are also indispensable when describing social events. Generally, each event is attached with a paragraph of text describing more semantic information. For example, in the introduction example Fig. 1, the event content of the quadruple (*Iran, Cooperate economically, Kazakhstan, 2018-04-05*) mentioned *Iran may increase imports of grain from Kazakhstan and Russia due to drought*, which contains more semantic information from keywords, such as *drought, imports, grain* and the event semantic are changing with time. Thus, we construct a SCG and utilize traditional GCN to capture the semantic relevance between words more deeply and obtain a better semantic context of events, as follows:

$$H^{(l+1)} = g\left(\hat{A}H^{(l)}W^{(l)} + b^{(l)}\right) \tag{6}$$

where l denotes the layer number. $\hat{A} = \tilde{D}^{-\frac{1}{2}}\tilde{A}\tilde{D}^{-\frac{1}{2}}$ is the normalized symmetric adjacency matrix and $\tilde{A} = A + I_N$. $\tilde{D}$ is the degree matrix. $H_t^{(l)}$ is the node embedding matrix passing by GCN layers. Intuitively, nodes aggregate semantic information from their local neighbors. As the number of layers increases, nodes can receive the semantics of further words from SCG. Finally, we can get a matrix sequence $\{H_{t-k}^s, \ldots, H_{t-1}^s, H_t^s\}$ to describe the event semantic context.

Relation and Semantic Fusion. We have obtained a sequence of embeddings of entities, event types, as well as the semantic embedding, represented as $\left\{\left[H_{t-k}^e, H_{t-k}^r, H_{t-k}^s\right], \ldots, \left[H_t^e, H_t^r, H_t^s\right]\right\}$. We first integrate above information at each timestamp t, as follows:

$$x_t = [p(H_t^e) : p(H_t^e) : p(H_t^e)] \tag{7}$$

where: denotes the concatenate operation. p represents the max pooling operation to reduce the dimension of the feature embedding matrix.

3.3   Dilated Casual Convolutions for Temporal Embedding

In addition to the relation and semantic information of social events, events also preserve temporal dependence, such as long-term or inconsecutive dependence with different tintervals. For the example of introduction in Fig. 1, a series of social event caused by *drought and water shortage* have two characteristics in temporal dependence: (1) The involved temporal range is relatively long, from *2018-04-05* to *2018-04-23*, and may continue until the end of the drought. (2) The time of social events is not consecutive and there are different time intervals.

Inspired by WaveNet [14], we apply a CNN-based method to capture such temporal dependence, named dilated casual convolution (DCC). The receptive field of DCC increases exponentially with the increase of the hidden layer, so we can capture the long-term dependence in social events by increasing the number of hidden layers. Based on x, we utilize the dilated causal convolution with filter f to encode temporal dependence:

$$X_t = \sum_{k=0}^{K-1} x(t - d \times k)f(k) \tag{8}$$

where d is the dilated factor, which means the input is selected every d time steps (or $d-1$ time intervals). s is the kernel size of the filter. To capture the long-term dependence and the periodic dependence across different time intervals, we stack the dilated causal convolutional layers in the order of increasing the dilation factor, which enable our model to capture longer sequences with less layers and save computation resources. Finally, we get the global historical embedding X_t, which includes three aspects of information: relation information of multi-event, semantic context information, long-term and periodic temporal dependence across different intervals

3.4 Multi-event Forecasting

Forecasting. We have obtained the historical embedding X_t up to time t. Then, we transform the task of multi-event forecasting into a multi-label classification problem to model the occurrence probability of different events at $t + 1$:

$$P\left(Y_{t+1} \mid G_{t-k:t}\right) = \sigma\left(W_\mu X_t\right) \qquad (9)$$

where $Y_{t+1} \in R^{|\mathcal{R}|}$ is a vector of event types. $X_{t-k:t}$ is the global historical embedding up to t. We further feed $X_{t-k:t}$ into a multi-layer perceptron (MLP) to calculate the probability of different event types. σ is a nonlinear function.

Optimization. We optimize the categorical cross-entropy loss [11]:

$$\mathcal{L} = -\frac{1}{L} \sum_{i \in L} y_i \ln\left(\frac{\exp\left(\hat{y}_i\right)}{\sum_{j \in L} \exp\left(\hat{y}_j\right)}\right) \qquad (10)$$

where $\hat{y}_i$ is the model prediction for event type i before the nonlinear function σ. $L \in N_+$ represents the total number of Event types.

4   Experiments and Results

We evaluate the performance of DGCN-rs for multi-event forecasting. Our results indicate that our model achieves significant performance gains.

4.1 Datasets and Settings

The experimental evaluation was conducted on the Global Database of Events, Language, and Tone event data (GDELT). These events are coded using 20 main types and 220 subtypes such as Appeal, Yield, Protest, etc. Each event is coded into 58 fields including date, actor attributes (actor1, actor2), event type, source (event URL), etc. In this paper, we focus on all types of events and select country-level datasets from five countries (Iran, Iraq, Saudi Arabia, Syria, and Turkey). We choose a wide range time period social events from January 1, 2018, to June 20, 2020. In our experiments, The time granularity is one day.

In our experiments, we pre-train a 100-dimensional sent2vec [6] for entities, relations and keywords in the vocabulary using the event content from each country. Then, we split the dataset of each country into three subsets, i.e., train(80%), valid(10%), test(10%). For hyper-parameter setting, the historical time step (day) k is set to 14 but for Glean and RE-NET,the k is 7. The layers of GCN are set to 2. All the parameters are trained using the Adam optimizer. For dilated casual convolution, dilated factor d is set 1,2,4.

4.2 Baselines

We compare our methods with several state-of-the-art baselines as follows:

- **DNN:** It consist of three dense layers. We use non-temporal TF-IDF text features extracted from all the event contents in historical time steps.
- **RE-NET** [7]: It contains a recurrent event encoder and a neighborhood aggregator to infer future facts.
- **Glean** [3]: A temporal graph learning method with heterogeneous data fusion for predicting co-occurring events of multiple types.

We also conduct some ablation tests:

- **−attention**: we remove the knowledge-aware attention, and only use the basic CompGCN to get the TEG embedding matrix at each timestamp.
- **−semantic**: we remove the SCG, only utilizing the relation embedding based on TEG as the global historical information.
- **+GRU**: we utilize GRU instead of DCC to capture temporal dependence.
- **+RNN**: similar to +GRU, we change the GRU module to a RNN module.
- **+LSTM**: we replace the GRU module with a simpler LSTM module.

Table 1. Forecasting results of our proposed model DGCN-rs

Method	Iran		Iraq		Turkey		Syria		Saudi Arabia	
	F1	Recall	F1	Recall	F1	Recall	F1	Recall	F1	Recall
DNN	49.08	59.62	53.07	65.57	57.71	65.67	54.86	60.59	47.21	55.81
RE-NET	56.20	62.99	55.46	68.82	60.04	70.52	56.08	68.97	54.82	66.25
Glean	57.20	73.06	59.05	74.60	61.55	73.47	58.65	73.21	56.18	70.16
−attention	70.31	82.38	62.45	80.65	**69.96**	**87.72**	62.76	85.46	**69.62**	**87.85**
−semantic	69.21	80.64	63.24	79.56	67.98	82.08	63.33	83.95	67.71	83.53
+GRU	65.53	73.07	60.04	74.41	68.04	81.72	61.73	80.07	67.59	84.58
+RNN	65.58	75.31	60.69	75.52	68.64	83.98	61.29	78.00	67.07	82.77
+LSTM	65.50	73.25	60.02	73.94	67.03	79.21	61.24	77.64	66.58	81.65
DGCN-rs	**71.60**	**84.95**	**64.87**	**83.07**	69.41	85.26	**64.56**	**85.42**	69.00	86.93

4.3 Experiments Results

We evaluate forecasting performance of our proposed model across datasets of five countries. Table 1 presents comparison and ablation results.

Forecasting Performance. From the comparison results, we observe:

- Overall, our model outperforms other baselines across all datasets in terms of F1-score and recall. The result difference of different datasets may be due to the different distribution of event types in each country, leading to strength differences in relation and semantic dependence between multi-event.

- The classic deep learning model DNN perform poorly, which shows ignoring graph structure and time dependence, and only considering summation of static historical text features is less effective for multi-event forecasting.
- The RE-NET model utilizes the temporal event graph and produces better performance than DNN, which indicates the relation dependence and temporal dependence of TEG can be helpful for multi-event forecasting.
- The Glean model achieves better results than RE-NET, because glean adds event semantic, which indicates that both relation and semantic information are indispensable for multi-event forecasting.
- Our model achieves the best performance. There may be three reasons: 1) we introduce the knowledge attention when capturing the relation dependence of TEG; 2) We built SCG, which can capture the semantic context of social event for better forecasting; 3) we utilize a DCC instead of RNN to capture the temporal dependence. Next, we will use the results of ablation tests to verify the above reasons.

Ablation Tests. From above ablation results, we have following observations:

- The difference in F-score and Recall of different datasets may be due to the different distribution of event types in each country, leading to strength differences in the relation and semantic dependence of multi-event.
- The results of -attention drop slightly, especially in the Turkey and Saudi Arabie dataset. The reason is that capturing unique temporal dependence is more important than the aggregation ways of local neighbors for relation embedding when forecasting long-term sequence data.
- The semantic is also essential, which can improve the performance of multi-event forecasting by enriching the context semantic of social events.
- The results of +GRU, +RNN, or +LSTM drops significantly and the performance of GRU and LSTM is not better than a simple RNN. The reason may be they increase the number of parameters leading to more complexity. Besides, we find the performance of our method is better than traditional temporal encoding methods. The likely reason is that it can capture long-term and inconsecutive temporal dependence with different intervals.

4.4   Sensitivity Analysis

We mainly conduct sensitivity analysis on the important parameter k (Historical Time Step) in temporal embedding module to prove that our model performs the advantages of our model for long-term forecasting. We evaluated the difference in forecasting performance of historical time steps from 7 to 28 days. The results of F1-score and Recall are shown in Fig. 3. We observe DCC is more suitable for capturing long-term dependence among events. The performance improves with k becoming larger. However, when we use RNN, the performance drop slightly as the k becomes larger. This result indicates that social events hide potential long-term temporal dependence and utilizing DCN is more conducive to improving the forecasting performance of such events.

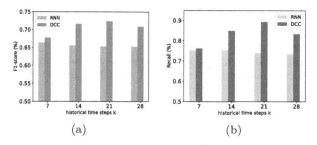

Fig. 3. Sensitivity analysis.

5   Conclusion

In the paper, we propose a novel dilated graph convolutional networks jointly modelling relation and semantic for multi-event forecasting. We construct a temporal event graph (TEG) and semantic context graph (SCG) based on social events to model the entity-relation and semantic information among multi-event. We employ our previous work of knowledge-aware attention based GCN to capture the relation dependence in TEG for better relation embedding. We use GCN to capture the semantic dependence in SCG for better semantic embedding. Considering the limitations of traditional temporal (such as RNN) encoding methods in capturing long-term dependence, we use dilated casual convolutional network to automatically capture the long-term and inconsecutive temporal dependence with different intervals between social events by stacking layers with increased dilated factors. Finally, we conduct extensive experiments to prove that our model outperforms other baselines. We also verify the effectiveness of our model through ablation experiments and sensitivity analysis. In the future, we will study more scalable methods to apply our proposed model on larger datasets in other domains.

Acknowledgement. This work was supported by the National Key Research and Development Program of China No. 2018YFC0831703.

References

1. Bollen, J., Mao, H., Zeng, X.: Twitter mood predicts the stock market. J. Comput. Sci. **2**(1), 1–8 (2011)
2. Deng, S., Rangwala, H., Ning, Y.: Learning dynamic context graphs for predicting social events. In: Proceedings of the 25th ACM SIGKDD International Conference on Knowledge Discovery & Data Mining, pp. 1007–1016 (2019)
3. Deng, S., Rangwala, H., Ning, Y.: Dynamic knowledge graph based multi-event forecasting. In: Proceedings of the 26th ACM SIGKDD International Conference on Knowledge Discovery & Data Mining, pp. 1585–1595 (2020)
4. Gao, Y., Zhao, L.: Incomplete label multi-task ordinal regression for spatial event scale forecasting. In: Proceedings of the AAAI Conference on Artificial Intelligence, vol. 32 (2018)

5. Gao, Y., Zhao, L., Wu, L., Ye, Y., Xiong, H., Yang, C.: Incomplete label multi-task deep learning for spatio-temporal event subtype forecasting. In: Proceedings of the AAAI Conference on Artificial Intelligence, vol. 33, pp. 3638–3646 (2019)

6. Gupta, P., Pagliardini, M., Jaggi, M.: Better word embeddings by disentangling contextual n-gram information. In: NAACL-HLT (1), pp. 933–939. Association for Computational Linguistics (2019)

7. Jin, W., Qu, M., Jin, X., Ren, X.: Recurrent event network: autoregressive structure inference over temporal knowledge graphs (2019)

8. Kipf, T.N., Welling, M.: Semi-supervised classification with graph convolutional networks (2016)

9. Li, Y., Yu, R., Shahabi, C., Liu, Y.: Diffusion convolutional recurrent neural network: data-driven traffic forecasting (2017)

10. Luo, W., et al.: Dynamic heterogeneous graph neural network for real-time event prediction. In: Proceedings of the 26th ACM SIGKDD International Conference on Knowledge Discovery & Data Mining, pp. 3213–3223 (2020)

11. Menon, A.K., Rawat, A.S., Reddi, S., Kumar, S.: Multilabel reductions: what is my loss optimising? (2019)

12. Pareja, A., et al.: EvolveGCN: evolving graph convolutional networks for dynamic graphs. In: Proceedings of the AAAI Conference on Artificial Intelligence, vol. 34, pp. 5363–5370 (2020)

13. Wang, X., Gerber, M.S., Brown, D.E.: Automatic crime prediction using events extracted from twitter posts. In: Yang, S.J., Greenberg, A.M., Endsley, M. (eds.) SBP 2012. LNCS, vol. 7227, pp. 231–238. Springer, Heidelberg (2012). https://doi.org/10.1007/978-3-642-29047-3_28

14. Wu, Z., Pan, S., Long, G., Jiang, J., Zhang, C.: Graph wavenet for deep spatial-temporal graph modeling (2019)

15. Song, X., Wang, H., Zeng, K., Liu, Y., Zhou, B.: KatGCN: knowledge-aware attention based temporal graph convolutional network for multi-event prediction. In: SEKE, pp. 417–422 (2021)

16. Yu, B., Yin, H., Zhu, Z.: Spatio-temporal graph convolutional networks: a deep learning framework for traffic forecasting (2017)

17. Zhang, J., Shi, X., Xie, J., Ma, H., King, I., Yeung, D.Y.: Gaan: gated attention networks for learning on large and spatiotemporal graphs (2018)

18. Zhao, L., Sun, Q., Ye, J., Chen, F., Lu, C.T., Ramakrishnan, N.: Multi-task learning for spatio-temporal event forecasting. In: Proceedings of the 21th ACM SIGKDD International Conference on Knowledge Discovery and Data Mining, pp. 1503–1512 (2015)

Training Graph Convolutional Neural Network Against Label Noise

Yuxin Zhuo, Xuesi Zhou, and Ji Wu$^{(\boxtimes)}$

Department of Electronic Engineering, Tsinghua University, Beijing, China
{zhuoyx19,zhouxs16}@mails.tsinghua.edu.cn, wuji_ee@mail.tsinghua.edu.cn

Abstract. For node classification task, graph convolutional neural network (GCN) has achieved competitive performance on graph-structured data. Under semi-supervised setting, only a small portion of nodes are labeled for training. Many existing works have a perfect assumption that all the class labels used for training are completely accurate. However, noises are inevitably involved in the process of labeling, which can cause a degraded model performance. Yet few works focus on how to deal with noisy labels on graph data. Techniques against label noise on image domain can't be applied to graph data directly. In this paper, we propose a framework, called super-nodes assisted label correction and dynamic graph adjustment based GCN (SuLD-GCN), which aims to reduce the negative impact of noise via label correction to obtain a higher-quality labels. We introduce the super-node to construct a new graph, which contributes to connecting nodes with the same class label more strongly. During iterations, we select nodes with high predicted confidence to correct their labels. Simultaneously, we adjust the graph structure dynamically. Experiments on public datasets demonstrate the effectiveness of our proposed method, yielding a significant improvement over state-of-art baselines.

Keywords: Graph Convolutional Neural Network · Label noise · Label correction

1 Introduction

Graph-structured data is broadly existing in real world, such as social networks, chemical compounds, proteins and knowledge graphs. Data represented in the forms of graph can be irregular, because the number of nodes in the graph is variable and the neighbor structure around each node is also different from each other [1]. Recently, there has been an increasing interest in dealing with graph-structured data. To learn a better graph representation, it's important to encode both node features and graph topology. With the remarkable progresses achieved by Graph Neural Networks (GNNs), such as Graph Convolutional Network (GCN) [2], Graph Attention Network (GAT) [3] and GraphSAGE [4], people can tackle various real-world graph datasets more effectively.

© Springer Nature Switzerland AG 2021
T. Mantoro et al. (Eds.): ICONIP 2021, LNCS 13110, pp. 677–689, 2021.
https://doi.org/10.1007/978-3-030-92238-2_56

Node classification is one of the main graph related tasks. The success of models like GCN on this task heavily relies on large and high-quality datasets with human-annotated labels [5]. However, it's time-consuming and really expensive to accurately annotate millions of nodes. Therefore, graph-based semi-supervised node-level classification has been proposed. Under this setting, GCN leverages only a small portion of nodes along with other unlabeled nodes in large graphs for training. GCN has the abilities to propagate, transform and aggregate node features from local neighborhoods. Hence, node representations are similar among nearby nodes [1]. The intuition behind GCN is smoothing node features across the edges of graphs. However, noises are inevitably involved during the labeling process. Then this smoothing operation will be corrupted by label noises on graphs. As the training progresses, GCN will fit all the labels including noisy labels, which results in poor classification performance and model bias. Therefore, the key challenge is how to train GCN against label noise on graphs.

There have been many attempts in image classification domain to deal with noisy labels. Label noise mainly comes from the variability among the observers, the errors of manual annotations and the errors of crowd-sourcing methods [6]. Researchers attempt to model the noise transition matrix by adding a noise adaptation layer [7], or to correct the forward or backward loss by estimating the noise transition matrix [8]. However, it's difficult to estimate the noise transition matrix especially in complex noise situation. Re-weighting methods are also widely studied to assign smaller weights to the training samples with false labels and greater weights to those with true labels [9]. Co-teaching framework [10] is used to select samples whose labels are likely to be correct to update the model, which helps to reduce the confirmation bias. However, due to label sparsity and label dependency [11], many existing works on image classification to deal with label noise can't be adapted to the graph data. Therefore, training a model against label noise on graph-structured data is a key challenge.

In this paper, we propose a novel framework SuLD-GCN, which introduces **Su**per-nodes to construct a graph and operate **L**abel correction and **D**ynamic graph adjustment to train a **GCN** model against label noise on graphs. In node classification task, the best node representation is that the embeddings of nodes with the same class are close to each other and the embeddings of nodes with different classes are separated largely [12]. Intuitively, connecting nodes with the same class more strongly benefits for classification. Based on this theory, we propose a kind of node called super-node to strengthen the relationships among nodes with the same label when constructing a graph. The super-node forms the dense connection of the intra-class nodes, which helps to propagate information across the edges linked by the same class nodes. And we select a set of nodes with high predicted confidence to correct their labels during training. Meanwhile, we adjust the graph structure dynamically according to the corrected labels. Experimental results on two real-world networks illustrate the effectiveness of our proposed approach. Our main contributions are summarized as follows:

- We design a novel framework to train GCN against label noise on graph data, which makes the applications of GCN on noisy real-world datasets more reliable.
- We are the first to introduce the super-node for strengthening the relationships among nodes with the same class label. We select nodes with high predicted confidence to correct their labels and simultaneously adjust the graph structure dynamically to avoid overfitting noisy labels.
- Experiments on two benchmark datasets illustrate that our model can be trained on GCN against label noise with different types and ratios. It also demonstrates our model's superiority over competitive baselines.

2 Related Work

2.1 Graph Convolutional Neural Network

In recent years, many graph neural networks [2–4] have been widely applied to graph-structured data. Graph Convolutional Neural Network (GCN) [2] is a generalization of the traditional convolutional neural network (CNN), which is applied to learn on graph-structured data. GCN was first proposed for node classification task under semi-supervised setting. It can propagate, transform and aggregate information from node features and graph structure, which achieves a competitive performance. However, it relies on high-quality labels of nodes. Label noise on nodes may hurt the performance of GCN model and result in poor performance. [13] empirically analyzed the accuracy of different graph-based semi-supervised algorithms in the presence of label noise, but they didn't propose a new approach to address this problem. [5] studied the robustness of training graph neural network in the presence of symmetric label noise. And they presented loss correction methods which are tolerant to noise for graph classification task. In this paper, we design a novel framework to tackle the label noise on graph-structured data for node classification task.

2.2 Learning with Noisy Labels

Learning with noisy labels has been widely studied on image domain. It can be categorized into five groups, which are robust architecture, robust regularization, robust loss function, loss adjustment and sample selection [14–16]. Based on robust architecture, changes have been made to model the noise transition matrix between clean and noisy labels [8], such as adding a noise adaptation layer [7]. However, this family can't identity false-labeled examples, which may result in estimation error for noise transition matrix. Based on robust regularization, methods have been widely studied to improve the generalizability of a model by avoiding overfitting, such as weight decay, dropout [7], batch normalization [17], data argumentation, etc. Based on robust loss function, various loss functions [18–21] have been designed to minimize the risk even if there are noisy labels in the training set. Generalized cross entropy (GCE) [22] is a generation of mean absolute error (MAE) loss [23] and categorical cross entropy (CCE) loss. The robustness of these methods is theoretically supported. However it can't perform well when the number of classes is large. Based on loss adjustment, four categories are proposed to reduce the negative impact of noisy labels, including loss correction, loss re-weighting, label refurbishment and meta learning [11,24,25]. Loss re-weighting [9] attempts to assign smaller weights to data with false labels and greater weights to those with true labels. However, it's hard to apply this method to practice due to the complexity of noise type. Based on sample selection, many works aim to select true-labeled samples from noisy training data. Co-teaching [10] trains two deep neural networks. Each network selects some small-loss samples and feed them into the other peer network for training. This method can help reduce confirmation bias, but it doesn't work well when true-labeled and false-labeled samples are overlapped largely. In this paper, we propose SuLD-GCN to improve the model performance in the presence of noise with different types and ratios, by introducing the super-nodes, correcting labels and adjusting graph dynamically.

3   Preliminary

Let $G = (V, E, X)$ be a graph, where V denotes a set of n nodes, E denotes a set of edges linking nodes and X denotes the node features. GCN propagates and transforms information to learn the representation vectors of nodes. We first add an identity matrix I to the adjacency matrix A to get $\tilde{A}$. Here $\tilde{D}$ is the diagonal node degree matrix of $\tilde{A}$. Then $\hat{A}$ is a symmetric normalization of the self-connections added adjacency matrix. W is the parameter to be updated by the network.

$$\tilde{A} = A + I, \tag{1}$$

$$\tilde{D}(i, i) = \sum_j \tilde{A}(i, j), \tag{2}$$

$$\hat{A} = \tilde{D}^{-\frac{1}{2}} \tilde{A} \tilde{D}^{-\frac{1}{2}}, \tag{3}$$

$$Z = f(X, A) = \text{softmax}\left(\hat{A} \, \text{ReLU} \left(\hat{A} X W^{(0)} \right) W^{(1)} \right). \tag{4}$$

We consider node classification under semi-supervised setting, where only partial nodes are labeled and the others remain unlabeled. The standard cross entropy loss is used as the objective function for node classification task, where $\mathcal{Y}_L$ is labels of labeled nodes, Z_{lf} is predicted results of models, Y_{lf} is the one-hot encoding of a node's label.

$$\mathcal{L} = -\sum_{l \in \mathcal{Y}_L} \sum_{f=1}^{F} Y_{lf} \ln Z_{lf}. \tag{5}$$

However, noise in the set of labeled nodes would severely degrade the model performance. Therefore, it's significant for the model to avoid overfitting the incorrect labels. Our work proposes a new method to construct graph and provides label correction along with dynamic graph adjustment to obtain a better classification performance, which improves the performance of GCN.

4   Methodology

In this section, we propose a novel framework SuLD-GCN, as shown in Fig. 1. Firstly, we propose a new method to construct a graph including both common-nodes and super-nodes. Super-nodes are introduced to strengthen the intra-class node connections with the same class label. Secondly, the GCN model are applied to learn a node representation. We select nodes with high predicted confidence to correct their labels. Meanwhile, the graph structure is adjusted dynamically. Then the corrected labels are fed into the neural network for training in the next iteration. As iterations stop, we obtain the latest corrected labels. Finally, we use the latest labels to train GCN model on the graph which only includes common nodes as usual to achieve better classification performance.

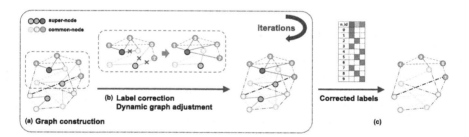

Fig. 1. Overview of the SuLD-GCN Framework. (a) shows that we first introduce super-nodes to construct a graph as input. (b) shows that during training, we select nodes with high predicted confidence to correct their labels and adjust the graph simultaneously. Finally, the corrected labels are used to train a GCN model which only includes common nodes as usual.

4.1   Graph Construction with Super-Nodes

In node classification, we aim to divide nodes into different groups. The best node representation is that the embeddings of nodes with the same class are close to each other and the embeddings of nodes with different classes are separated largely [12]. Intuitively, to achieve this goal, the model should connect nodes with the same class more strongly so that they can be pushed together by GCN models. Therefore, in this section, we propose a new method to construct a graph by introducing super-nodes.

As shown in Fig. 1(a), there are two kinds of nodes, common-nodes and super-nodes respectively. Firstly, common-nodes represent specific entities on graph-structured data. For example, in a citation network, each common-node represents a paper and papers are linked to each other via citationships. This is how the edges are connected among common-nodes. Secondly, each super-node can be considered as a collection of the same class. That is to say, if the nodes on the graph are divided into a certain number of classes, there will be the same number of super-nodes. Then edges between common-nodes and super-nodes are connected depending on labels. Each common-node is linked to the corresponding super-node according to its label.

Label noise on graph data would cause a degraded model performance, since information propagation is corrupted with noise and it makes the inter-class boundary less clear. [12] proved that GCN is to smooth the features of nodes through transforming information across the edges. And the performance of node classification task mainly benefits from edges among nodes with the same class in a graph. The key to improve the performance of GCN is to make nodes with the same class connect with each other more strongly. Thus, super-nodes are introduced to strengthen the intra-class node connections explicitly. It forms dense connection of nodes with the same class and contributes to learning a better intra-class node representation.

4.2   Label Correction and Dynamic Graph Adjustment

Under semi-supervised setting, a set of labeled nodes and graph structure are used to learn representation of all nodes and to predict the classes of unlabeled nodes. However, training the GCN model to overfit incorrect labels results in poor performance and hurts the robustness of models. Researchers [26] proved that at the beginning of training, models tend to fit the easy samples which are more likely to be clean. Inspired by methods of tackling noisy labels in image domain, we propose a novel framework to deal with label noise on graph data. The standard cross entropy loss is used as objective function. Our model stacks graph convolution layers followed by a softmax layer to output the predicted probability. It is denoted as

$$y_k = \frac{\exp(a_k)}{\sum_{i=1}^{n} \exp(a_i)}, \quad y = [y_1, \cdots, y_k, \cdots, y_n]. \tag{6}$$

We consider it as the distribution over n classes. The k-th position of the distribution represents the probability of a node belonging to the k-th class. First, the model tends to learn the distribution of cleaner data at the beginning of training. In the process of label correction, we set a threshold α, if the max value of the predicted probability distribution is larger than α, we correct the node's label to the corresponding class since these nodes are likely to be mislabeled and their labels can be corrected by information propagation of neighborhoods. We consider that the max value of the predicted probability distribution is larger, the node belongs to the class more likely. Note that, we only correct labels of the labeled nodes in the training set.

With the correction of labels, our model also adjusts the graph structure dynamically. In Sect. 4.1, we have introduced super-nodes to construct a graph, where the edges between super-nodes and common-nodes are linked according to common-nodes' original labels, where include noisy labels. Once a common-node's label is corrected, the original edge will be disconnected and the new edge between it and the new super-node is connected. Label correction and dynamic graph adjustment are matched with each other, carried out simultaneously, as shown in Fig. 1(b).

4.3   Training Iterations

At first iteration, we select a certain number of nodes which are most likely the false-labeled ones to correct their labels based on the predicted confidence. Meanwhile, the graph structure is adjusted dynamically along with the edges between super-nodes and common-nodes disconnected or connected. Note that, we didn't train GCN model to fit all labels, where noises still remain. Therefore, We adopt early stopping technique. Then the corrected labels and the adjusted graph are fed into GCN for the next training iteration. The training epochs increase with the iterative process. In other words, the model can fit the distribution of data deeper due to the corrected cleaner data. Therefore, the negative impact of label noise on graph data can be reduced by iterations. After the final iteration stop, the latest corrected labels of partial labeled nodes are obtained, whose quality is better than the original. Then we take the cleaner labels of partial labeled nodes and the graph only including common-node as input for training GCN as regular to predict classes for the unlabeled nodes, which achieves a better classification performance.

5   Experiments

5.1   Datasets

We use the following two real-world datasets in our experiments.

Cora: It is a small-scale citation network which contains a certain number of machine-learning papers divided into seven classes [27]. In this graph-structured data, nodes represent papers and edges represent citation links. Each node has a feature vector of sparse bag-of-words and a corresponding class label. We use the same data split in [1], with 20 nodes per class for training, 500 nodes for validation, 1000 nodes for testing.

Coauthor-Phy: It is a large-scale co-authorship network based on the Microsoft Academic Graph from the KDD Cup 2016 challenge. Here, nodes represent authors and if two authors co-authored a paper, the edge between the two nodes is connected. The keywords of author's papers are used as node features and the most active study fields of author are used as class labels. And we use the same train/validation/test split with [28].

Statistics of the two datasets are shown in Table 1. The intra-class edge rate is the proportion of edges connecting two nodes within the same class. Note that, in these two datasets, the intra-class edge rate is much higher than the inter-class edge rate.

Table 1. Datasets statistics.

	Nodes	Edges	Features	Classes	Intra-class edge rate
Cora	2708	5278	1433	7	81.0%
Coauthor-Phy	34493	247962	8415	5	93.1%

5.2 Baselines

We compare against three strong competing baselines for dealing with noisy labels on image classification. For a fair comparison, all these methods are adapted to work with the same GCN architecture.

Decoupling. [29] aims to figure out the problem of "when to update" from "how to update". Two base classifiers are trained and update steps are performed only on examples that lie in the disagreement area.

Co-teaching. [10] trains two deep networks simultaneously. Each network selects some small-loss samples and feed them into the other peer network for training. Two networks are taught by each other through selected useful knowledge.

Generalized Cross Entropy (GCE). [22] proposes a noise-robust loss function to combat the problem of errors in training labels. It can be seen as a generalization of mean absolute error (MAE) and categorical cross entropy (CCE).

5.3   Experimental Setup

Since the public graph datasets are clean without label noise, we manually generate noisy labels by noise transition matrix on these two public datasets following [10, 30]. Here, we mainly generate two types of noise. 1). **Symmetric noise:** Given a transition matrix, ε is the noise ratio, each label is flipped to other labels with a random uniform probability. 2). **Asymmetric noise:** labels are flipped between two classes. Humans tend to make mistake only within similar classes. It's worth noting that, the asymmetric noise is harder to deal with than the symmetric noise.

$$
Q^s = \begin{bmatrix} 1-\varepsilon & \frac{\varepsilon}{n-1} & \cdots & \frac{\varepsilon}{n-1} & \frac{\varepsilon}{n-1} \\ \frac{\varepsilon}{n-1} & 1-\varepsilon & \frac{\varepsilon}{n-1} & \cdots & \frac{\varepsilon}{n-1} \\ \vdots & & \ddots & & \vdots \\ \frac{\varepsilon}{n-1} & \cdots & \frac{\varepsilon}{n-1} & 1-\varepsilon & \frac{\varepsilon}{n-1} \\ \frac{\varepsilon}{n-1} & \frac{\varepsilon}{n-1} & \cdots & \frac{\varepsilon}{n-1} & 1-\varepsilon \end{bmatrix}, Q^a = \begin{bmatrix} 1-\varepsilon & \varepsilon & 0 & \cdots & 0 \\ 0 & 1-\varepsilon & \varepsilon & & 0 \\ \vdots & & \ddots & \ddots & \vdots \\ 0 & & & 1-\varepsilon & \varepsilon \\ \varepsilon & 0 & \cdots & 0 & 1-\varepsilon \end{bmatrix}. \quad (7)
$$

Our experiments are conducted under symmetric and asymmetric noise setting with four different noise ratios. The noise transition matrix is only applied to the labeled nodes in the training set, while the validation and test sets remain clean. We use a two-layer GCN which has a hidden dimension of 16. The learning rate for Adam optimizer is set to 0.01. We apply $L2$ regularization to GCN with a weight decay of 5e−4 and we use dropout technique with a dropout rate of 0.5. The threshold α depends on the performance of the validation set. We train our models for 200 epochs with 10 iterations.

5.4   Results

Table 2 shows the performance (test accuracy) of node classification for all models and datasets with different noise types and noise ratios. The results show that our proposed SuLD-GCN model outperforms the baselines by a large margin under different noise setting. Compared with GCN on Coauthor-Phy in the case of symmetric noise, SuLD-GCN improves the classification performance by 6.2%, 2.7%, 9.2%, 5.8% under the noise ratio of 20%, 30%, 40%, 60%, respectively. And it shows similar superiority on the Cora dataset.

In Sect. 4.2, we assume that models learn a cleaner distribution of data at the beginning of training. Under the extremely noisy settings, most of nodes are mislabeled and the labels are corrupted completely. This explains why the performance improvement is limited in the case of the asymmetric noise under 60% ratio setting.

It is worth mentioning that in the Coauthor-Phy dataset, with the noise ratio of 20%, our model achieves the classification accuracy to 91.1%, 90.4% under the symmetric and asymmetric noise setting, which is comparable to test accuracy (91.7%) reported by models trained on clean labels. All the represented results demonstrate the competitive performance of our proposed method.

Table 2. Performance comparison (test accuracy) on node classification.

Datasets	Models	Noise type							
		Symmetric noise				Asymmetric noise			
		Noise ratio							
		0.2	0.3	0.4	0.6	0.2	0.3	0.4	0.6
Cora	Decoupling	0.713	0.634	0.512	0.368	0.627	0.522	0.396	0.133
	Co-teaching	0.704	0.617	0.498	0.217	0.598	0.411	0.372	0.127
	GCE	0.737	0.676	0.539	0.353	0.657	0.591	0.422	0.158
	GCN	0.718	0.645	0.628	0.364	0.641	0.578	0.416	0.169
	SuLD-GCN	0.745	0.697	0.667	0.376	0.690	0.620	0.462	0.172
Coauthor-Phy	Decoupling	0.851	0.805	0.647	0.452	0.831	0.653	0.251	0.136
	Co-teaching	0.816	0.745	0.604	0.392	0.796	0.609	0.139	0.091
	GCE	0.843	0.821	0.675	0.433	0.853	0.647	0.373	0.127
	GCN	0.849	0.809	0.632	0.408	0.821	0.627	0.389	0.160
	SuLD-GCN	0.911	0.836	0.724	0.466	0.904	0.731	0.389	0.158

5.5   Ablation Studies

To further illustrate the effectiveness of our model, we conduct ablation studies on the two datasets. Our ablation studies are represented with three versions, as shown in Table 3. **S-GCN** corresponds to graph construction with super-nodes. **SL-GCN** corresponds to graph construction with super nodes and label correction. **SuLD-GCN** corresponds to graph construction with super nodes, label correction and dynamic graph adjustment. The results show that super-nodes, label correction and dynamic graph adjustment work well to reduce the negative impact of noise. The accuracy improvement on Coauthor-Phy is larger than that on Cora, which illustrates that the higher intra-class edge rate contributes to node classification against label noise. It also demonstrates that super-nodes introduced by our model to connect nodes with the same label more strongly benefit for this task. The results also illustrate the effectiveness of our proposed three components.

Table 3. Classification performance of ablation studies.

Datasets	Models	Noise type							
		Symmetric noise				Asymmetric noise			
		Noise ratio							
		0.2	0.3	0.4	0.6	0.2	0.3	0.4	0.6
Cora	GCN	0.718	0.645	0.628	0.364	0.641	0.578	0.416	0.169
	S-GCN	0.727	0.648	0.614	0.357	0.635	0.607	0.419	0.146
	SL-GCN	0.734	0.683	0.649	0.353	0.684	0.624	0.438	0.143
	SuLD-GCN	0.745	0.697	0.667	0.376	0.690	0.620	0.462	0.172
Coauthor-Phy	GCN	0.849	0.809	0.632	0.408	0.821	0.627	0.389	0.160
	S-GCN	0.861	0.803	0.656	0.417	0.844	0.658	0.362	0.143
	SL-GCN	0.904	0.825	0.692	0.459	0.877	0.713	0.357	0.145
	SuLD-GCN	0.911	0.836	0.724	0.466	0.904	0.731	0.389	0.158

6 Conclusions

In this study, we propose a method SuLD-GCN for learning with noisy labels on graph-structured data under semi-supervised setting. We are the first to introduce super-nodes for constructing a graph to connect nodes with the same class label more strongly. Furthermore, we progressively reduce the negative impact of label noise via label correction and dynamic graph adjustment after a few iterations. Then the corrected labels are used to train a regular GCN for node classification. Experiments on public datasets of different scales demonstrate the effectiveness of our proposed model. It can deal with noise with different types and ratios and outperform the competitive baselines. For future work, we can generalize our method to other GNN architectures and investigate how to improve the classification performance under extremely noise settings.

Acknowledgements. This work is sponsored by the National Natural Science Foundation of China (Grant No. 61571266), Beijing Municipal Natural Science Foundation (No. L192026), and Tsinghua-Foshan Innovation Special Fund (TFISF) (No. 2020THFS0111).

References

1. Wu, Z., Pan, S., Chen, F., et al.: A comprehensive survey on graph neural networks. IEEE Trans. Neural Netw. Learn. Syst. **32**(1), 4–24 (2020)
2. Kipf, T.N., Welling, M.: Semi-supervised classification with graph convolutional networks. arXiv preprint arXiv:1609.02907 (2016)
3. Veličković, P., Cucurull, G., Casanova, A., et al.: Graph attention networks. arXiv preprint arXiv:1710.10903 (2017)
4. Hamilton, W.L., Ying, R., Leskovec, J.: Inductive representation learning on large graphs. In: Proceedings of the 31st International Conference on Neural Information Processing Systems, pp. 1025–1035 (2017)

5. Hoang, N.T., Choong, J.J., Murata, T.: Learning graph neural networks with noisy labels (2019)
6. Karimi, D., Dou, H., Warfield, S.K., et al.: Deep learning with noisy labels: exploring techniques and remedies in medical image analysis. Med. Image Anal. **65**, 101759 (2020)
7. Srivastava, N., Hinton, G., Krizhevsky, A., et al.: Dropout: a simple way to prevent neural networks from overfitting. J. Mach. Learn. Res. **15**(1), 1929–1958 (2014)
8. Xiao, T., Xia, T., Yang, Y., et al.: Learning from massive noisy labeled data for image classification. In: Proceedings of the IEEE Conference on Computer Vision and Pattern Recognition, pp. 2691–2699 (2015)
9. Liu, T., Tao, D.: Classification with noisy labels by importance reweighting. IEEE Trans. Pattern Anal. Mach. Intell. **38**(3), 447–461 (2015)
10. Han, B., Yao, Q., Yu, X., et al.: Co-teaching: robust training of deep neural networks with extremely noisy labels. arXiv preprint arXiv:1804.06872 (2018)
11. Li, Y., Yin, J., Chen, L.: Unified robust training for graph neural networks against label noise. In: Karlapalem, K., et al. (eds.) PAKDD 2021. LNCS (LNAI), vol. 12712, pp. 528–540. Springer, Cham (2021). https://doi.org/10.1007/978-3-030-75762-5_42
12. Wang, H., Leskovec, J.: Unifying graph convolutional neural networks and label propagation. arXiv preprint arXiv:2002.06755 (2020)
13. de Aquino Afonso, B.K., Berton, L.: Analysis of label noise in graph-based semi-supervised learning. In: Proceedings of the 35th Annual ACM Symposium on Applied Computing, pp. 1127–1134 (2020)
14. Yu, X., Han, B., Yao, J., et al.: How does disagreement help generalization against label corruption? In: International Conference on Machine Learning, pp. 7164–7173. PMLR (2019)
15. Wang, Y., Liu, W., Ma, X., et al.: Iterative learning with open-set noisy labels. In: Proceedings of the IEEE Conference on Computer Vision and Pattern Recognition, pp. 8688–8696 (2018)
16. Nguyen, D.T., Mummadi, C.K., Ngo, T.P.N., et al.: Self: learning to filter noisy labels with self-ensembling. arXiv preprint arXiv:1910.01842 (2019)
17. Ioffe, S., Szegedy, C.: Batch normalization: accelerating deep network training by reducing internal covariate shift. In: International Conference on Machine Learning, pp. 448–456. PMLR (2015)
18. Krogh, A., Hertz, J.A.: A simple weight decay can improve generalization. In: Advances in Neural Information Processing Systems, pp. 950–957 (1992)
19. Mnih, V., Hinton, G.E.: Learning to label aerial images from noisy data. In: Proceedings of the 29th International Conference on Machine Learning (ICML 2012), pp. 567–574 (2012)
20. Manwani, N., Sastry, P.S.: Noise tolerance under risk minimization. IEEE Trans. Cybern. **43**(3), 1146–1151 (2013)
21. Van Rooyen, B., Menon, A.K., Williamson, R.C.: Learning with symmetric label noise: the importance of being unhinged. arXiv preprint arXiv:1505.07634 (2015)
22. Zhang, Z., Sabuncu, M.R.: Generalized cross entropy loss for training deep neural networks with noisy labels. In: 32nd Conference on Neural Information Processing Systems (NeurIPS) (2018)
23. Ghosh, A., Kumar, H., Sastry, P.S.: Robust loss functions under label noise for deep neural networks. In: Proceedings of the AAAI Conference on Artificial Intelligence, vol. 31, no. 1 (2017)

24. Patrini, G., Rozza, A., Krishna Menon, A., et al.: Making deep neural networks robust to label noise: a loss correction approach. In: Proceedings of the IEEE Conference on Computer Vision and Pattern Recognition, pp. 1944–1952 (2017)
25. Finn, C., Abbeel, P., Levine, S.: Model-agnostic meta-learning for fast adaptation of deep networks. In: International Conference on Machine Learning, pp. 1126–1135. PMLR (2017)
26. Huang, J., Qu, L., Jia, R., et al.: O2U-Net: a simple noisy label detection approach for deep neural networks. In: Proceedings of the IEEE/CVF International Conference on Computer Vision, pp. 3326–3334 (2019)
27. Sen, P., Namata, G., Bilgic, M., et al.: Collective classification in network data. AI Mag. 29(3), 93–93 (2008)
28. Shchur, O., Mumme, M., Bojchevski, A., et al.: Pitfalls of graph neural network evaluation. arXiv preprint arXiv:1811.05868 (2018)
29. Malach, E., Shalev-Shwartz, S.: Decoupling "when to update" from "how to update". arXiv preprint arXiv:1706.02613 (2017)
30. Jiang, L., Zhou, Z., Leung, T., et al.: MentorNet: learning data-driven curriculum for very deep neural networks on corrupted labels. In: International Conference on Machine Learning, pp. 2304–2313. PMLR (2018)

An LSTM-Based Plagiarism Detection via Attention Mechanism and a Population-Based Approach for Pre-training Parameters with Imbalanced Classes

Seyed Vahid Moravvej[1]([✉]), Seyed Jalaleddin Mousavirad[2],
Mahshid Helali Moghadam[3,4], and Mehrdad Saadatmand[3]

[1] Department of Computer Engineering, Isfahan University of Technology, Isfahan, Iran
sa.moravvej@ec.iut.ac.ir
[2] Department of Computer Engineering, Hakim Sabzevari Univesity, Sabzevar, Iran
[3] RISE Research Institutes of Sweden, Västerås, Sweden
[4] Mälardalen University, Västerås, Sweden

Abstract. Plagiarism is one of the leading problems in academic and industrial environments, which its goal is to find the similar items in a typical document or source code. This paper proposes an architecture based on a Long Short-Term Memory (LSTM) and attention mechanism called LSTM-AM-ABC boosted by a population-based approach for parameter initialization. Gradient-based optimization algorithms such as back-propagation (BP) are widely used in the literature for learning process in LSTM, attention mechanism, and feed-forward neural network, while they suffer from some problems such as getting stuck in local optima. To tackle this problem, population-based metaheuristic (PBMH) algorithms can be used. To this end, this paper employs a PBMH algorithm, artificial bee colony (ABC), to moderate the problem. Our proposed algorithm can find the initial values for model learning in all LSTM, attention mechanism, and feed-forward neural network, simultaneously. In other words, ABC algorithm finds a promising point for starting BP algorithm. For evaluation, we compare our proposed algorithm with both conventional and population-based methods. The results clearly show that the proposed method can provide competitive performance.

Keywords: Plagiarism · Back-propagation · LSTM · Attention mechanism · Artificial bee colony

1 Introduction

Plagiarism is one of the most important problems in educational institutions such as universities and scientific centers. The purpose of an automated plagiarism detection system is to find similar items at the level of a word, sentence, or document. There are different goals for plagiarism detection. For example, some of these studies only identify

T. Mantoro et al. (Eds.): ICONIP 2021, LNCS 13110, pp. 690–701, 2021.
https://doi.org/10.1007/978-3-030-92238-2_57

duplicate documents [1]. However, low accuracy is a main problem because they do not recognize copied sentences. Several other detectors are designed to find similar source codes in programming environments. It is worthwhile to mention that most research takes into account plagiarism detection at the sentence level [2].

Generally speaking, the methods presented for plagiarism detection are based on statistical methods or in-depth learning. Statistical methods usually utilize Euclidean distance or cosine similarity to calculate the similarity between two items [3, 4]. Convolutional Neural Network (CNN) and Recurrent Neural Network (RNN), as two leading deep learning models, have attracted much attention of researchers for plagiarism detection [5]. Uses a Siamese CNN to analyze the content of words and selects a representation of a word relevance with its neighbors. In [6], the representation of each word is made using Glove (a word embedding method) [7], and then the representation of the sentences is obtained using a recursive neural network. Finally, similar sentences are identified using cosine similarity [8]. Employed two attention-based LSTM networks to extract the representation of two sentences. In [9], the similarity between the sentences is considered for the answer selection task. Two approaches are proposed for this purpose. The first method uses two methods of embedding Language Models (ELMo) [10] and the Bidirectional Encoder Representations from transformers (BERT) and combines them with a transformer encoder. In the second approach, the model is tuned using two pre-trained transformer encoder models. In [11], the authors presented a method based on the context-aligned RNN called CARNN. This paper suggests embedding context information of the aligned words in hidden state generation. They showed that this technique could play an effective role in measuring similarity. In addition, from the literature, there are some few papers focused on the attention mechanism for plagiarism detection [12, 13].

One of the most important reasons for the convergence of neural networks is the initial value of the parameters. The gradient-based algorithms such as back-propagation are extensively used for deep learning models. However, these algorithms have some problems such as sensitivity to initial parameters and getting stuck in local optima [14, 15]. Population-based metaheuristic (PBMH) algorithms can be considered as an alternative to these problems. Artificial Bee Colony (ABC) is an efficient PBMH which has achieved many successes in optimizing a diverse range of applications [16].

In this study, a new architecture, LSTM-AM-ABC, based on the attention mechanism for plagiarism detection at the sentence level is proposed. The proposed algorithm benefits from three main steps including pre-processing, word embedding, and model construction. LSTM-AM-ABC employs LSTM and feed-forward networks as the core model, attention-based mechanism for changing the importance of all inputs, and an ABC algorithm for parameter initialization. Here, the main responsibility of the ABC algorithm is to find a promising point to commence the BP algorithm in LSTM, attention mechanism, and feed-forward network. The proposed model learns two pairs of positive and negative inputs. Negative pairs are dissimilar sentences, while positive pairs are similar sentences. In addition, we use several methods to overcome the data imbalance problem. We evaluate our results on three benchmark datasets based on different criteria. The evaluation results show that the proposed model can be superior to the compared models that use the random value for the parameters.

2 Long Short-Term Memory (LSTM)

In 1997, LSTM networks [17, 18] were first proposed. An LSTM unit includes an input gate, a memory gate, and an output gate that make it easy to learn long dependencies [12]. The update relations of an LSTM unit in step t are as follows [19]:

$$i_t = \sigma(W_i x_t + U_i h_{t-1} + b_i) \tag{1}$$

$$f_t = \sigma(W_f x_t + U_f h_{t-1} + b_f) \tag{2}$$

$$c_t = f_t c_{t-1} + i_t tanh(W_j x_t + U_j h_{t-1} + b_j) \tag{3}$$

$$o_t = \sigma(W_o x_t + U_o h_{t-1} + b_o) \tag{4}$$

$$h_t = o_t tanh(c_t) \tag{5}$$

Where i, f, o, and c are the input gate, forget gate, output gate and, cell input, respectively. $W \in \mathbb{R}^{h \times d}$. $U \in \mathbb{R}^{h \times h}$, $b \in \mathbb{R}^h$ are network parameters that should be learned during the learning process. Note that the input size x and hidden size h are d and h, respectively.

LSTM networks process input from start to finish or vice versa. It has been proven that it can be more effective if the processing is done from both sides simultaneously [20]. Bidirectional Long Short-Term Memory (BLSTM) networks are a type of LSTM networks that process input from both sides and produce two hidden vectors $\overrightarrow{h}_t$ and $\overleftarrow{h}_t$. In BLSTM, the combination of two hidden vectors, $h_t = [\overrightarrow{h}_t . \overleftarrow{h}_t]$, is considered as the final hidden vector.

Although LSTM networks consider long sequences, they give the same importance to all inputs. It can confuse the network in making decisions. Consider the following sentence: "Despite being from Uttar Pradesh, as she was brought up in Bengal, she is convenient in Bengali". Some words such as "Bengali", "brought up" and "Bengal" should have more weight because it has more related to the word "Bengali". The attention mechanism for this problem was later introduced [21]. In the attention mechanism, for each hidden vector, a coefficient is considered that the final hidden vector is calculated as:

$$h = \sum_{t=1}^{T} \alpha_t h_t \tag{6}$$

where α_t, h_t is the coefficient of significance and the hidden vector extracted in step t. T is the number of inputs.

3 LSTM-AM-ABC Approach

The proposed method, LSTM-AM-ABC, consists of three main steps, including pre-processing, word embedding, and model construction (According to Fig. 1). The details of each step are described below.

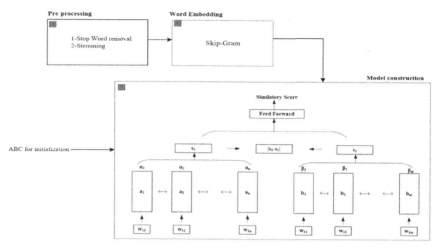

Fig. 1. Steps of the proposed model.

3.1 Pre-processing

Pre-processing means removing unnecessary and unimportant words, which reduces the computational load and increases the speed. Two techniques are used for this purpose.

Stop-Word Removal. Words such as 'or' and 'with' lack semantic information due to their repetition and presence in most documents. Eliminating these words plays a crucial role in the performance of plagiarism detection.

Stemming. The process of returning words to their root form is called the stemming operation (for example, the root form of *looking* is the word of *look*).

3.2 Word Embedding

One of the most important steps in natural language processing is word embedding because it is used as input and the embedding of sentences is made based on it. For this purpose, we use the well-known algorithm Skip-Gram [22]. It applies a simple neural network model to learn word vectors. Vectors are carefully generated so that the similarity of the two words can be estimated using a similarity function.

3.3 Model Construction

This paper proposes a plagiarism detection method, LSTM-AM-ABC, based on LSTM and feed-forward neural network, as the core models, the attention mechanism for altering the importance of the inputs, and ABC for parameter initialization. An LSTM is provided for each sentence S. In this research, two pairs of data have been used to learn the model. In positive pair (S_1, S_2), S_1 and S_2 are two copy sentences. The degree of similarity of the sentences depends on the dataset. In Negative pairs (S_1, S_2'), S_1 and

S_2' are not similar. For a positive pair, class label is one, while for a negative pair, the class label is zero. Let $S_1 = \{w_{11}.w_{12}.....w_{1n}\}$ and $S_2 = \{w_{21}.w_{22}.....w_{2m}\}$ be two sentences, where w_{ij} is the $j - th$ word in $i - th$ sentence. The two sentences S_1 and S_2 are limited to n and m words, respectively. The embedding of sentences s_1 and s_2 is formulated based on the attention mechanism as:

$$s_1 = \sum_{i=1}^{n} \alpha_i h_{a_i} \tag{7}$$

$$s_2 = \sum_{i=1}^{m} \beta_i h_{b_i} \tag{8}$$

where $h_{a_i} = [\overrightarrow{h}_{a_i}.\overleftarrow{h}_{a_i}] \in \mathbb{R}^{2d_1}$, $h_{b_i} = [\overrightarrow{h}_{b_i}.\overleftarrow{h}_{b_i}] \in \mathbb{R}^{2d_2}$ are the $i - th$ output in BLSTM. Each BLSTM output plays a role in the output with a coefficient in the range [0,1]. These coefficients are calculated for both networks as:

$$\alpha_i = \frac{e^{u_i}}{\sum_{i=1}^{n} e^{u_i}} \tag{9}$$

$$\beta_i = \frac{e^{v_i}}{\sum_{i=1}^{m} e^{v_i}} \tag{10}$$

$$u_i = tanh(W_u h_{a_i} + b_u) \tag{11}$$

$$v_i = tanh(W_v h_{b_i} + b_u) \tag{12}$$

where $W_u \in \mathbb{R}^{2d_1}.b_v \in \mathbb{R}$, $W_v \in \mathbb{R}^{2d_2}$. and $b_v \in \mathbb{R}$ are the parameters of the attention mechanism for two sentences. After calculating the embedding of sentences, they, along with their differences $|s_2 - s_1|$, enter a feed-forward network, and their similarity is calculated.

Parameter Optimization. There is a plethora of parameters in the proposed model including parameters in LSTM, feed-forward networks, and attention mechanism. This paper proposes a novel approach for parameter initialization using ABC algorithm. To this end, two main issues should be considered including encoding strategy and fitness function. Encoding strategy represents the structure of each candidate solution, while fitness function is responsible to calculate the quality of each candidate solution.

Encoding Strategy. The proposed model consists of three main parts including two LSTM networks, Two attention mechanism systems, and one feed-forward network. Figure 2 shows a typical encoding strategy for a two- layer feed-forward network, two single-layer LSTM networks and two attention mechanisms.

Fitness Function. Fitness function calculates the quality of each candidate solution. In this paper, we propose an objective function based on similarity as:

$$Fitness = \frac{1}{1 + \sum_{i=0}^{N} (y_i - \tilde{y}_i)^2} \tag{13}$$

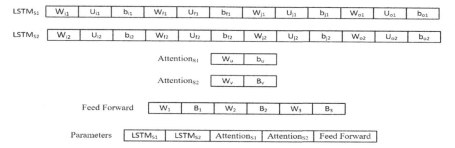

Fig. 2. Illustration of encoding strategy.

where N is training data, y_i is the $i-th$ target, and $\tilde{y}_i$ is the predicted similarity value for the $i-th$ data. The goal of optimization here is to find the optimal initial seeds for the BP algorithm.

Imbalance Classification. One of the most important challenges of machine learning is the problem of data imbalance. Data imbalance means that the number of data in the classes is not the same and, one of the classes (or even more) has more data. It reduces system performance because it causes the classifier to bias the output to one side. In our case, the problem of imbalance is due to a large number of negative pairs. This paper proposes a combination of imbalance methods to tackle the problem: Augmentation and Penalty.

The augmentation goal is to increase the positive pair. For this purpose, we combine the embedding of each word in the sentence with Gaussian noise and produce new sentences that are similar to positive sentences.

In the penalty technique, the minority and majority class error values are applied with different coefficients in the Loss function. Equation 14 illustrates the concept of this technique.

$$Loss = \alpha Loss_{C_1} + \beta Loss_{C_2} \tag{14}$$

where C_1 and C_2 are the minority and majority classes, respectively, and the coefficients α and β are the importance of the Loss function. If α and β are equal, we have a common classification problem.

4 Experiment and Analysis

In this section, we evaluate LSTM-AM-ABC algorithm compared to other competitors and different criteria.

4.1 Corpus

Plagiarism detection is a classification problem

$$sim(S_1.S_2) = \begin{cases} \geq \varepsilon & S_1 \text{ is a copy of } S_2 \\ < \varepsilon & S_1 \text{ is not a copy of } S_2 \end{cases} \tag{15}$$

According to Eq. 15, when the proposed model detects the degree of similarity of S_1 and S_2 above 0.5, a copy is detected. In this research, three common plagiarism datasets are utilized for this purpose.

SemEval2014. This dataset is taken from the Sentences Involving Com-positional Knowledge (SICK) dataset [23] for semantic evaluation of English sentences. It has 5000 pairs of sentences for training, and 5,000 sentences for testing. Each pair of sentences has a similarity label $\in \{1.2.3.4.5\}$, which 1 means that the sentences are irrelevant and 5 meaning the most similar sentences. We consider the label $= 0$ as class 0 and the label $\in \{2.3.4.5\}$ as class 1.

STS Semantic. Microsoft Research Paraphrase corpus (MSRP) [12] is a paraphrasing corpus containing 4076 pairs of sentences for training and 1725 pairs of sentences for testing. The sentences in this dataset are tagged by humans.

MSRP. Text similarity (STS) [8] is based on image captions, news headlines, and user forums. This database has 6928 sentences for training, and 1500 sentences for testing. To make unrelated pairs, we put unrelated sentences together.

4.2 Metrics

Recall. For plagiarism systems, it is vital not to recognize sentences that are copies so that copies can easily pass through the filter without being detected. The recall criterion is one of the valuable criteria for this problem because this criterion considers the number of copied sentences is not recognized.

Pearson's Correlation. This criterion distinguishes negative correlation and negative correlation numerically in the interval $[-1,1]$ as:

$$r = \frac{Cov\left(sim_{y'}.sim_y\right)}{\sqrt{var\left(sim_{y'}\right)var\left(sim_y\right)}} \qquad (16)$$

Mean Square Error (MSE). The MSE criterion indicates the difference between the degree of actual and predicted similarity defined as:

$$\text{MSE} = sum((sim_{y'} - sim_y)^2) \qquad (17)$$

4.3 Result

We compare the proposed method with a series of previous researches. To this end, we use k-fold cross-validation (k = 10 or 10CV) in all experiments, in which the dataset is divided into k subsets. One of the subsets is used for test data, while the remaining is employed for training. This procedure is repeated k times, and all data is used exactly once for testing. We report statistical results including mean, standard deviation, and median for each criterion and each dataset. Table 1 shows the parameter settings of the proposed model. In addition, ε was set to 0.515, 0.525, and 0.52 for SemEval2014, STS dataset, and MSRP datasets, respectively.

We compare our method with seven methods, including Siamese [5, 6, 8, 9, 11, 12], and [13]. Also, we compare our algorithm with the LSTM-AM that employ random number as the initial point of parameters to show that our proposed ABC approach can effectively improve the results. The results of the evaluation are shown in Tables 1, 2, and 3. For the SemEval2014 dataset, LSTM-AM-ABC has been able to overcome other methods in the recall and r criteria. Comparing LSTM-AM-ABC with LSTM-AM clearly indicates the effectiveness of our proposed initialization approach. For the STS dataset, LSTM-AM-ABC again presented the best results compared to other algorithms. By comparing LSTM-AM-ABC, we can observe that LSTM-AM-ABC could decrease the error more than 40%, indicating the effectiveness of initialization approach. The results of Table 3 are consistent with other tables. For MSRP dataset, our proposed algorithm obtained the highest mean recall followed by CETE algorithm.

The proposed algorithm employs ABC in conjunction with BP algorithm for training. In the following, we indicate the proposed training algorithm is effective compared to others. To have a fair comparison, we fix all remaining parts of our proposed algorithm including LSTM, feedforward network, attention-based mechanism and only the trainer is changed. To this end, we compare our proposed trainer with five conventional algorithms, including Gradient Descent with simple Momentum (GDM) [24], Gradient Descent with Adaptive learning rate backpropagation (GDA) [25], Gradient Descent with Momentum and Adaptive learning rate backpropagation (GDMA) [26], One-Step Secant backpropagation (OSS) [27], and Bayesian Regularization backpropagation (BR) [28], And four metaheuristic algorithms, including Grey Wolf Optimization (GWO) [29], Bat Algorithm (BA) [30], Cuckoo Optimization Algorithm (COA) [31], and Whale Optimization Algorithm (WOA) [32]. The results of the proposed algorithms compared to other trainers are shown Tables 4, 5, and 6. For the SemEval2014 dataset, as expected, metaheuristic algorithms generally work better than conventional algorithms. BR algorithm has been able to overcome metaheuristic algorithms including GWO, BAT, COA, and WOA. It can be seen that LSTM-AM-ABC outperformed all metaheuristic and conventional algorithms. In SemEval2014 datasets, the proposed trainer can reduce error more than 34% compared to the second best algorithm, LSTM-AM-BR. Such a difference exists in two other datasets, so that LSTM-AM-ABC improved error more than 25% and 32% for STS and MSRP datasets, respectively.

Table 1. 10CV classification results on SemEval2014 dataset.

Method	Recall			MSE			r		
	Mean	Std.dev.	Median	Mean	Std.dev.	Median	Mean	Std.dev.	Median
[5]	82.136	5.360	86.403	0.286	0.096	0.292	0.506	0.314	0.741
[6]	82.563	2.068	83.128	0.164	0.092	0.183	0.721	0.217	0.863
[9]	91.153	1.143	92.119	0.055	0.063	0.057	0.709	0.206	0.817
[11]	85.436	1.890	85.763	0.034	0.067	0.062	0.754	0.145	0.786
[8]	88.477	1.683	89.809	0.059	0.041	0.060	0.791	0.262	0.826
[12]	84.016	0.935	84.639	0.036	0.053	0.039	0.759	0.249	0.790
[13]	86.103	2.360	88.509	0.062	0.092	0.099	0.776	0.174	0.820
LSTM-AM	93.129	3.390	94.208	0.076	0.068	0.089	0.812	0.170	0.912
LSTM-AM-ABC	95.268	1.791	97.018	0.053	0.047	0.062	0.804	0.183	0.963

Table 2. 10CV classification results on STS dataset.

Method	Recall			MSE			r		
	Mean	Std.dev.	Median	Mean	Std.dev.	Median	Mean	Std.dev.	Median
[5]	85.153	4.712	88.106	0.125	0.088	0.147	0.549	0.229	0.570
[6]	86.100	3.190	88.247	0.099	0.081	0.121	0.769	0.187	0.775
[9]	95.014	1.371	96.053	0.043	0.082	0.056	0.784	0.225	0.809
[11]	89.056	0.441	89.610	0.029	0.054	0.035	0.771	0.215	0.797
[8]	92.101	3.943	94.105	0.048	0.072	0.059	0.820	0.042	0.826
[12]	86.283	1.800	87.120	0.031	0.062	0.039	0.782	0.162	0.798
[13]	89.156	1.089	89.664	0.049	0.070	0.054	0.801	0.140	0.819
LSTM-AM	96.206	3.610	97.421	0.061	0.052	0.072	0.832	0.129	0.840
LSTM-AM-ABC	97.410	3.811	98.163	0.041	0.051	0.054	0.840	0.091	0.849

Table 3. 10CV classification results on MSRP dataset.

Method	Recall			MSE			r		
	Mean	Std.dev.	Median	Mean	Std.dev.	Median	Mean	Std.dev.	Median
[5]	87.089	2.790	88.119	0.096	0.093	0.106	0.713	0.181	0.723
[6]	89.207	1.341	89.690	0.043	0.080	0.051	0.787	0.210	0.817
[9]	97.296	0.910	97.429	0.016	0.073	0.024	0.819	0.189	0.820
[11]	90.179	1.207	91.092	0.018	0.091	0.043	0.787	0.163	0.799
[8]	91.396	1.179	91.647	0.057	0.087	0.069	0.809	0.187	0.816
[12]	87.493	3.410	89.190	0.025	0.067	0.049	0.801	0.018	0.809
[13]	89.269	2.107	91.018	0.029	0.058	0.057	0.818	0.196	0.839
LSTM-AM	95.208	1.874	96.410	0.069	0.059	0.072	0.829	0.100	0.841
LSTM-AM-ABC	97.379	1.270	97.941	0.035	0.064	0.057	0.857	0.073	0.869

Table 4. Results of 10CV classification of metaheuristic algorithms on SemEval2014 dataset.

Method	Recall			MSE			r		
	Mean	Std.dev.	Median	Mean	Std.dev.	Median	Mean	Std.dev.	Median
LSTM-AM-GDM	89.126	1.142	90.250	0.072	0.096	0.084	0.774	0.125	0.850
LSTM-AM-GDA	88.473	1.480	88.892	0.076	0.024	0.081	0.759	0.107	0.800
LSTM-AM-GDMA	88.421	3.189	90.547	0.082	0.035	0.086	0.752	0.114	0.792
LSTM-AM-OSS	87.634	5.103	89.420	0.085	0.042	0.089	0.743	0.120	0.810
LSTM-AM-BR	92.169	1.300	92.962	0.070	0.029	0.075	0.788	0.103	0.824
LSTM-AM-GWO	90.145	1.250	91.123	0.072	0.025	0.077	0.775	0.102	0.836
LSTM-AM-BAT	91.160	0.146	91.532	0.068	0.012	0.072	0.780	0.094	0.792
LSTM-AM-COA	92.790	1.365	93.475	0.057	0.062	0.061	0.792	0.106	0.821
LSTM-AM-WOA	91.756	1.250	92.750	0.061	0.020	0.065	0.782	0.112	0.835

Table 5. Results of 10CV classification of metaheuristic algorithms on STS dataset.

Method	Recall			MSE			r		
	Mean	Std.dev.	Median	Mean	Std.dev.	Median	Mean	Std.dev.	Median
LSTM-AM-GDM	90.100	1.967	91.250	1.020	0.047	1.046	0.775	0.092	0.781
LSTM-AM-GDA	92.580	1.500	93.473	0.075	0.042	0.082	0.791	0.096	0.802
LSTM-AM-GDMA	91.140	2.450	93.485	0.062	0.051	0.067	0.782	0.112	0.791
LSTM-AM-OSS	89.263	3.593	91.530	0.066	0.068	0.071	0.760	0.132	0.773
LSTM-AM-BR	96.256	2.850	97.253	0.064	0.062	0.070	0.836	0.115	0.842
LSTM-AM-GWO	93.020	1.740	93.863	0.025	0.052	0.034	0.818	0.082	0.826
LSTM-AM-BAT	95.418	1.425	95.920	0.072	0.059	0.079	0.831	0.079	0.838
LSTM-AM-COA	96.520	3.475	97.835	0.053	0.072	0.062	0.838	0.121	0.846
LSTM-AM-WOA	93.120	2.148	95.128	0.061	0.068	0.069	0.824	0.118	0.836

Table 6. Results of 10CV classification of metaheuristic algorithms on MSRP dataset.

Method	Recall			MSE			r		
	Mean	Std.dev.	Median	Mean	Std.dev.	Median	Mean	Std.dev.	Median
LSTM-AM-GDM	89.180	2.893	90.658	0.082	0.090	0.089	0.782	0.126	0.819
LSTM-AM-GDA	93.185	2.459	94.150	0.072	0.089	0.081	0.810	0.093	0.827
LSTM-AM-GDMA	90.163	3.485	92.635	0.079	0.093	0.086	0.786	0.150	0.824
LSTM-AM-OSS	91.280	2.963	92.895	0.064	0.082	0.072	0.792	0.092	0.080
LSTM-AM-BR	96.000	1.285	96.142	0.042	0.060	0.050	0.819	0.035	0.082
LSTM-AM-GWO	95.183	1.590	95.740	0.053	0.072	0.062	0.804	0.020	0.816
LSTM-AM-BAT	96.180	2.010	97.005	0.038	0.094	0.049	0.832	0.081	0.843
LSTM-AM-COA	96.138	2.583	96.935	0.045	0.085	0.052	0.826	0.070	0.831
LSTM-AM-WOA	92.052	1.390	93.128	0.068	0.072	0.076	0.796	0.068	0.802

5   Conclusions

The goal of plagiarism is to find the similar items in a typical document or source code. This paper proposes a novel model for plagiarism detection based on LSTM-based architecture, feedforward neural networks, attention mechanism incorporating a population-based approach for parameter initialization (LSTM-AM-ABC). Gradient-based optimization algorithms such as back-propagation (BP) is so popular for learning process in LSTM, attention mechanism, and feed-forward neural network, whereas have some problems such as being sensitive in the initial conditions. Therefore, this paper proposed an artificial bee colony (ABC) mechanism to find initial seed for BP algorithm. ABC algorithm is employed on LSTM, attention mechanism, and feed-forward neural network, simultaneously. For evaluation, we compared LSTM-AM-ABC with both conventional and population-based methods. The experimental results on three datasets demonstrate that LSTM-AM-ABC is superior to previous systems. As future work, we intend to provide a suitable solution for imbalanced classification. A simple solution could be to use reinforcement learning. Reinforcement learning can be effective because of the rewards and punishments involved.

References

1. El Moatez Billah Nagoudi, A.K., Cherroun, H., Schwab, D.: 2L-APD: a two-level plagiarism detection system for Arabic documents. Cybern. Inf. Technol. **18**(1), 124–138 (2018)
2. He, H., Gimpel, K., Lin, J.: Multi-perspective sentence similarity modeling with convolutional neural networks. In: Proceedings of the 2015 Conference on Empirical Methods in Natural Language Processing (2015)
3. Joodaki, M., Dowlatshahi, M.B., Joodaki, N.Z.: An ensemble feature selection algorithm based on PageRank centrality and fuzzy logic. Knowl. Based Syst. **223**, 107538 (2021)
4. Joodaki, M., Ghadiri, N., Maleki, Z., Shahreza, M.L.: A scalable random walk with restart on heterogeneous networks with Apache Spark for ranking disease-related genes through type-II fuzzy data fusion. J. Biomed. Inf. **115**, 103688 (2021)
5. Pontes, E.L., Huet, S., Linhares, A.C., Torres-Moreno, J.-M.: Predicting the semantic textual similarity with siamese CNN and LSTM. arXiv preprint arXiv:1810.10641 (2018)
6. Sanborn, A., Skryzalin, J.: Deep learning for semantic similarity. In: CS224d: Deep Learning for Natural Language Processing. Stanford University, Stanford (2015)
7. Pennington, J., Socher, R., Manning, C.D.: Glove: Global vectors for word representation. In: Proceedings of the 2014 Conference on Empirical Methods in Natural Language Processing (EMNLP) (2014)
8. Moravvej, S.V., Joodaki, M., Kahaki, M.J.M., Sartakhti, M.S.: A method based on an attention mechanism to measure the similarity of two sentences. In: 2021 7th International Conference on Web Research (ICWR). IEEE (2021)
9. Laskar, M.T.R., Huang, X., Hoque, E.: Contextualized embeddings based transformer encoder for sentence similarity modeling in answer selection task. In: Proceedings of The 12th Language Resources and Evaluation Conference (2020)
10. Peters, M.E., Neumann, M., Iyyer, M., Gardner, M., Clark, C., Lee, K., Zettlemoyer, L.: Deep contextualized word representations. arXiv preprint arXiv:1802.05365 (2018)
11. Chen, Q., Hu, Q., Huang, J.X., He, L.: CA-RNN: using context-aligned recurrent neural networks for modeling sentence similarity. In: Proceedings of the AAAI Conference on Artificial Intelligence (2018)

12. Bao, W., Bao, W., Du, J., Yang, Y., Zhao, X.: Attentive Siamese LSTM network for semantic textual similarity measure. In: 2018 International Conference on Asian Language Processing (IALP). IEEE (2018)
13. Chi, Z., Zhang, B.: A sentence similarity estimation method based on improved siamese network. J. Intell. Learn. Syst. Appl. **10**(4), 121–134 (2018)
14. Ashkoofaraz, S.Y., Izadi, S.N.H., Tajmirriahi, M., Roshanzamir, M., Soureshjani, M.A., Moravvej, S.V., Palhang, M.: AIUT3D 2018 Soccer Simulation 3D League Team Description Paper
15. Vakilian, S., Moravvej, S.V., Fanian, A.: Using the cuckoo algorithm to optimizing the response time and energy consumption cost of fog nodes by considering collaboration in the fog layer. In: 2021 5th International Conference on Internet of Things and Applications (IoT). IEEE (2021)
16. Vakilian, S., Moravvej, S.V., Fanian, A.: Using the artificial bee colony (ABC) algorithm in collaboration with the fog nodes in the Internet of Things three-layer architecture. In: 2021 29th Iranian Conference on Electrical Engineering (ICEE) (2021)
17. Hochreiter, S., Schmidhuber, J.: Long short-term memory. Neural Comput. **9**(8), 1735–1780 (1997)
18. Sartakhti, M.S., Kahaki, M.J.M., Moravvej, S.V., Javadi Joortani, M., Bagheri, A.: Persian language model based on BiLSTM model on COVID-19 Corpus. In: 2021 5th International Conference on Pattern Recognition and Image Analysis (IPRIA). IEEE (2021)
19. Graves, A.: Generating sequences with recurrent neural networks. arXiv preprint arXiv:1308.0850 (2013)
20. Graves, A., Schmidhuber, J.: Framewise phoneme classification with bidirectional LSTM and other neural network architectures. Neural Netw. **18**(5–6), 602–610 (2005)
21. Bahdanau, D., Cho, K., Bengio, Y.: Neural machine translation by jointly learning to align and translate. arXiv preprint arXiv:1409.0473 (2014)
22. Moravvej, S.V., Kahaki, M.J.M., Sartakhti, M.S., Mirzaei, A.: A method based on attention mechanism using bidirectional long-short term memory (BLSTM) for question answering. In: 2021 29th Iranian Conference on Electrical Engineering (ICEE) (2021)
23. Marelli, M., Menini, S., Baroni, M., Bentivogli, L., Bernardi, R., Zamparelli, R.: A SICK cure for the evaluation of compositional distributional semantic models. In: Lrec. Reykjavik (2014)
24. Phansalkar, V., Sastry, P.: Analysis of the back-propagation algorithm with momentum. IEEE Trans. Neural Netw. **5**(3), 505–506 (1994)
25. Hagan, M., Demuth, H., Beale, M.: Neural Network Design (PWS, Boston, MA). Google Scholar Google Scholar Digital Library Digital Library (1996)
26. Yu, C.-C., Liu, B.-D.: A backpropagation algorithm with adaptive learning rate and momentum coefficient. In: Proceedings of the 2002 International Joint Conference on Neural Networks. IJCNN'02 (Cat. No. 02CH37290). IEEE (2002)
27. Battiti, R.: First-and second-order methods for learning: between steepest descent and Newton's method. Neural Comput. **4**(2), 141–166 (1992)
28. Foresee, F.D., Hagan, M.T.: Gauss-Newton approximation to Bayesian learning. In: Proceedings of international conference on neural networks (ICNN'97). IEEE (1997)
29. Mirjalili, S., Mirjalili, S.M., Lewis, A.: Grey wolf optimizer. Adv. Eng. Softw. **69**, 46–61 (2014)
30. Yang, X.-S.: A new metaheuristic bat-inspired algorithm. In: Nature Inspired Cooperative Strategies for Optimization (NICSO 2010), pp. 65–74. Springer (2010)
31. Yang, X.-S., Deb, S.: Cuckoo search via Lévy flights. In: 2009 World Congress on Nature & Biologically Inspired Computing (NaBIC). IEEE (2009)
32. Mirjalili, S., Lewis, A.: The whale optimization algorithm. Adv. Eng. Softw. **95**, 51–67 (2016)

Author Index

Printed in the United States
by Baker & Taylor Publisher Services